71 Springer Series in Solid-State Sciences

Edited by Klaus von Klitzing

Springer Series in Solid-State Sciences

Editors: M. Cardona P. Fulde K. von Klitzing H.-J. Queisser

40 **Semiconductor Physics** An Introduction
3rd Edition By K. Seeger

41 **The LMTO Method**
Muffin-Tin Orbitals and Electronic Structure
By H. L. Skriver

42 **Crystal Optics with Spatial Dispersion, and Excitons**
By V. M. Agranovich and V. L. Ginzburg

43 **Resonant Nonlinear Interactions of Light with Matter**
By V. S. Butylkin, A. E. Kaplan,
Yu. G. Khronopulo, and E. I. Yakubovich

44 **Elastic Media with Microstructure II**
Three-Dimensional Models
By I. A. Kunin

45 **Electronic Properties of Doped Semiconductors**
By B. I. Shklovskii and A. L. Efros

46 **Topological Disorder in Condensed Matter**
Editors: F. Yonezawa and T. Ninomiya

47 **Statics and Dynamics of Nonlinear Systems**
Editors: G. Benedek, H. Bilz, and R. Zeyher

48 **Magnetic Phase Transitions**
Editors: M. Ausloos and R. J. Elliott

49 **Organic Molecular Aggregates,** Electronic Excitation and Interaction Processes
Editors: P. Reineker, H. Haken, and H. C. Wolf

50 **Multiple Diffraction of X-Rays in Crystals**
By Shih-Lin Chang

51 **Phonon Scattering in Condensed Matter**
Editors: W. Eisenmenger, K. Laßmann,
and S. Döttinger

52 **Superconductivity in Magnetic and Exotic Materials**
Editors: T. Matsubara and A. Kotani

53 **Two-Dimensional Systems, Heterostructures, and Superlattices**
Editors: G. Bauer, F. Kuchar, and H. Heinrich

54 **Magnetic Excitations and Fluctuations**
Editors: S. Lovesey, U. Balucani, F. Borsa,
and V. Tognetti

55 **The Theory of Magnetism II**
Thermodynamics and Statistical Mechanics
By D. C. Mattis

56 **Spin Fluctuations in Itinerant Electron Magnetism** By T. Moriya

57 **Polycrystalline Semiconductors,**
Physical Properties and Applications
Editor: G. Harbeke

58 **The Recursion Method and Its Applications**
Editors: D. Pettifor and D. Weaire

59 **Dynamical Processes and Ordering on Solid Surfaces**
Editors: A. Yoshimori and M. Tsukada

60 **Excitonic Processes in Solids**
By M. Ueta, H. Kanzaki, K. Kobayashi,
Y. Toyozawa, and E. Hanamura

61 **Localization, Interaction, and Transport Phenomena**
Editors: B. Kramer, G. Bergmann,
and Y. Bruynseraede

62 **Theory of Heavy Fermions and Valence Fluctuations**
Editors: T. Kasuya and T. Saso

63 **Electronic Properties of Polymers and Related Compounds**
Editors: H. Kuzmany, M. Mehring, and S. Roth

64 **Symmetries in Physics:** Group Theory Applied to Physical Problems
By W. Ludwig and C. Falter

65 **Phonons: Theory and Experiments II**
Experiments and Interpretation of
Experimental Results By P. Brüesch

66 **Phonons: Theory and Experiments III**
Phenomena Related to Phonons
By P. Brüesch

67 **Two-Dimensional Systems: Physics and New Devices**
Editors: G. Bauer, F. Kuchar, and H. Heinrich

68 **Phonon Scattering in Condensed Matter V**
Editors: A. C. Anderson and J. P. Wolfe

69 **Nonlinearity in Condensed Matter**
Editors: A. R. Bishop, D. K. Campbell,
P. Kumar and S. E. Trullinger

70 **From Hamiltonians to Phase Diagrams**
The Electronic and Statistical-Mechanical
Theory of sp-Bonded Metals and Alloys
By J. Hafner

71 **High Magnetic Fields in Semiconductor Physics**
Editor: G. Landwehr

Volumes 1–39 are listed on the back inside cover

High Magnetic Fields in Semiconductor Physics

Proceedings of the International Conference,
Würzburg, Fed. Rep. of Germany,
August 18–22, 1986

Editor: G. Landwehr

With 378 Figures

Springer-Verlag Berlin Heidelberg New York
London Paris Tokyo

Professor Dr. Gottfried Landwehr

Physikalisches Institut, Universität Würzburg, Röntgenring 8
D-8700 Würzburg, Fed. Rep. of Germany

Series Editors:

Professor Dr., Dres. h. c. Manuel Cardona
Professor Dr., Dr. h. c. Peter Fulde
Professor Dr. Klaus von Klitzing
Professor Dr. Hans-Joachim Queisser

Max-Planck-Institut für Festkörperforschung, Heisenbergstrasse 1
D-7000 Stuttgart 80, Fed. Rep. of Germany

ISBN-13: 978-3-642-83116-4 **e-ISBN-13: 978-3-642-83114-0**
DOI: 10.1007/978-3-642-83114-0

Library of Congress Cataloging-in-Publication Data. High magnetic fields in semiconductor physics. (Springer series in solid-state sciences ; 71) „Contributiuons presented at the International Conference on the Application of High Magnetic Fields in Semiconductor Physics ... held at the University of Würzburg from August 18 to 22, 1986"–Pref. Includes index. 1. Semiconductors–Congresses. 2. Magnetic fields–Congresses. I. Landwehr, G. (Gottfried), 1929-. II. International Conference on the Application of High Magnetic Fields in Semiconductor Physics (1986: University of Würzburg) III. Series. QC610.9.H54 1987 537.6'22 87-9812

Offset printing: Druckhaus Beltz, 6944 Hemsbach/Bergstr.
Bookbinding: J. Schäffer GmbH & Co. KG., 6718 Grünstadt
2153/3150-543210

Preface

This volume contains the contributions presented at the International Conference on the Application of High Magnetic Fields in Semiconductor Physics which was held at the University of Würzburg from August 18 to 22, 1986.

The first conference on this subject was held in Würzburg in 1972. The purpose of the meeting was to bring together researchers who were active in that area in order to discuss at some length new results and developments and, of course, existing problems. A conference seemed timely because superconducting magnets had become widely available and high magnetic fields were being used more and more as a tool for the investigation of the electronic band structure of semiconductors. The concept of a meeting with emphasis on invited papers turned out to be useful so that subsequent conferences were held in Würzburg in 1974 and 1976, Oxford in 1978, Hakone in 1980 and Grenoble in 1982. The alternate years pattern was broken in 1984, but the tradition was resumed in 1986 with another Würzburg Conference on the Application of High Magnetic Fields in Semiconductor Physics.

An innovation of the 1972 meeting was that the full manuscripts of the invited talks were put together in a conference book, which was given to all participants at the beginning of the conference. This approach was very successful because it led to very intense discussions during and after the lectures. Consequently, lecture notes of this kind were produced at all subsequent high magnetic field meetings of the series. It was regretted by many participants and other scientists active in the field that no official proceedings were published after the first four "Würzburg Conferences". Because of the high quality of the lecture notes, the issues became rather like collectors items. In order to remedy this situation, conference proceedings were published after the meetings in Hakone and in Grenoble. Because these proceedings have been widely welcomed, this practice has been continued.

Owing to the rapid development of semiconductor physics in conjunction with high magnetic fields, it was necessary even at the 1976 conference to restrict the scope of the meeting. Three main subjects were chosen: semiconductor inversion layers, selected topics of magneto-optics, and transport properties in high magnetic fields. It is interesting to note that these are also the main topics of the present meeting. It is well known that the study of semiconductor inversion layers in high magnetic fields has been of special importance and that it culminated in the discovery of the quantum Hall effect by Klaus von Klitzing in 1980. Looking through the 1976 lecture notes, one can trace the presence of the quantum Hall effect, although unrecognized at that time. At the same meeting, localization in 2D systems in high magnetic fields was already being discussed by Prof. Uemura (Tokyo).

Going through the contents of these proceedings one finds that a substantial part is devoted to the quantum Hall effect. The experimental and theoretical work on this subject has increased enormously in the last few years. Despite all efforts, a detailed microscopic theory of the effect still seems to be lacking.

This volume contains 41 invited papers, many of them not only presenting new results but also giving a review of the respective areas. The 40 contributed papers printed in this issue were presented during a lively poster session. I am convinced that this book will be of wide interest, because it gives an up-to-date overview of a rapidly developing section of semiconductor physics.

The organizing committee consisted of G. Landwehr (Chairman), J. Hajdu, K. von Klitzing, and M. von Ortenberg.

Financial support of the conference by the following sponsors is gratefully acknowleged:
Deutsche Forschungsgemeinschaft,
Bayerisches Staatsministerium für Unterricht and Kultus,
Regionalverband Bayern der Deutschen Physikalischen Gesellschaft,
European Research Office, US Army,
Bruker Analytische Meßtechnik, Karlsruhe,
IBM, Stuttgart,
Siemens AG, München,
Philips Research Laboratories, Eindhoven,
Oxford Instruments, Wiesbaden.

Würzburg, January 1987 *G. Landwehr*

Contents

Part I — Integral Quantum Hall Effect, Localization in High Magnetic Fields, Density of States

Part II **Fractional Quantum Hall Effect**

Part III Heterostructures and Superlattices: Optics

Part V Metal-Insulator Transition

Part VI Semimagnetic Semiconductors, IV-VI Materials

Part VII Magneto-Optics in 3D Systems

Part VIII **Magneto-Transport in 3D Systems, Bandstructure**

Part IX **Very High Field Work**

Part I

Integral Quantum Hall Effect, Localization in High Magnetic Fields, Density of States

Localization in Landau Levels of 2D Systems and the Quantum Hall Effect

T. Ando

Institute for Solid State Physics, University of Tokyo, 7-22-1 Roppongi, Minato-ku, Tokyo 106, Japan

A recent development in understanding the problem of electron localization in two-dimensional systems in strong magnetic fields is reviewed mainly from a theoretical point of view with emphasis on the relation to the quantum Hall effect. It is shown that all the states are localized exponentially except those just at the center of the Landau level in sufficiently strong magnetic fields. The energy of extended states is shifted to the higher energy side with decreasing magnetic field. There is a scaling relation between the diagonal and off-diagonal components of the conductivity tensor.

1. INTRODUCTION

In a two-dimensional system in strong magnetic fields, the orbital motion of electrons is completely quantized and the energy spectrum comprises discrete Landau levels /1/. The transport properties of this system with such a singular density of states is of fundamental importance. One of the exciting phenomena is the quantum Hall effect /2,3/. This effect has been suggested to provide a method of high precision measurement of the universal quantity e^2/h and can also be used as a supreme resistance standard /2/.

The quantized Hall effect requires the presence of both localized and extended states in strong magnetic fields and therefore is closely related to the localization. The localization problem itself is quite interesting in its own light. In this paper, a brief review is given on the theory of the quantized Hall effect and the localization effect in a two-dimensional system in strong magnetic fields.

There have been several attempts to relate the quantization of the Hall conductivity to a topological invariant called the winding number. In Sec. 2 we shall review one such attempt briefly. Sec. 3 deals with results of extensive numerical study on the Anderson localization and its relation to the quantization of the Hall conductivity. It is demonstrated that states are always localized exponentially except those just at the center of the Landau level in sufficiently strong magnetic fields and that the energy of such extended states is shifted to the higher energy side with the decrease of magnetic fields. The system-size dependence of the diagonal and off-diagonal components of the conductivity tensor is also studied and shows that there exists a scaling relation between σ_{xx} and σ_{xy}.

2. QUANTIZATION OF HALL CONDUCTIVITY AND TOPOLOGICAL INVARIANT

The quantization of the Hall conductivity was first theoretically suggested by ANDO, MATSUMOTO and UEMURA in 1975 /4/. Later in 1982, LAUGHLIN has provided a convincing argument, which shows that the Hall conductivity is quantized into an integer multiple of $-e^2/h$ whenever the Fermi level lies in localized states /5/. There appeared various modified versions of Laughlin's argument. A rigorous mathematical proof based on the Kubo formula appeared recently.

We consider a two-dimensional system with a finite size LxL in a magnetic field H. The Hall conductivity σ_{xy} is written in terms of the velocity operators v_x and v_y as

$$\sigma_{xy} = \frac{\hbar e^2}{i\pi L^2} \int f(E)dE\,\mathrm{Tr}[v_x(\frac{\partial}{\partial E}\mathrm{Re}G(E+i0))v_y\mathrm{Im}G(E+i0)-(x\leftrightarrow y)], \qquad (2.1)$$

where $G(E)=(E-H)^{-1}$ is Green's function or a resolvent, E is energy, and H is the Hamiltonian. Now, we introduce a vector potential $\mathbf{A}^0=(A_x{}^0,A_y{}^0)$ independent of positions and time. In the Laughlin geometry with the system wound into a cylinder, a magnetic flux Φ_x penetrates the opening of the cylinder. The full vector potential $\mathbf{A}^0$ introduced here may be thought of as two magnetic fluxes $(\Phi_x,\Phi_y)=(A_x{}^0L,A_y{}^0L)$ which penetrate, respectively, inside and the opening of a torus when we impose periodic boundary conditions in both x and y directions. The system assumes its original state every time A_x or A_y increase by Φ_0/L with Φ_0 being the magnetic flux quantum ch/e. We can easily show that when states in the vicinity of the Fermi level are localized the above can be rewritten as

$$\sigma_{xy} = -\frac{\hbar e^2}{4\pi L^2}(\frac{e}{c})^2 \int_C dz\,\mathrm{Tr}[G\frac{\partial G^{-1}}{\partial A_x{}^0}G\frac{\partial G^{-1}}{\partial A_y{}^0}G\frac{\partial G^{-1}}{\partial z} - (x\leftrightarrow y)]\,. \qquad (2.2)$$

The path C for the energy in the complex plane may be taken as an infinite line parallel to the imaginary axis, since when the states at the Fermi level are localized, the matrix elements of the current vanish. The path C may be regarded as a closed contour in the z plane with $G(\mathrm{Im}(z)=\infty)=G(\mathrm{Im}(z)=-\infty)$. When the system size is sufficiently large, the physical quantities do not depend on the vector potential $\mathbf{A}^0$ and the observed conductivity is equal to the average over the vector potential $\mathbf{A}^0$. We have

$$\sigma_{xy} = -\frac{\hbar e^2}{4\pi L^2}(\frac{e}{c})^2(\frac{eL}{ch})^2 \int_C dz\int dA_x{}^0\int d_y{}^0\,\mathrm{Tr}[G\frac{\partial G^{-1}}{\partial A_x{}^0}G\frac{\partial G^{-1}}{\partial A_y{}^0}G\frac{\partial G^{-1}}{\partial z} - (x\leftrightarrow y)]\,. \quad (2.3)$$

Now, the quantity

$$\frac{1}{8\pi^2}\int_C dz\int dA_x{}^0\int dA_y{}^0\,\mathrm{Tr}[G\frac{\partial G^{-1}}{\partial A_x{}^0}G\frac{\partial G^{-1}}{\partial A_y{}^0}G\frac{\partial G^{-1}}{\partial z} - (x\leftrightarrow y)] \qquad (2.4)$$

is known to be a topological invariant called "the winding number" of the mapping from the space of $(\mathbf{A}^0,z)$ to the space of G. Similar expression for the winding number appears in the gauge-field theory in which the mapping is from a sphere in the 4D space-time to matrices of the gauge group. The quantization of the Hall conductivity into the integer multiple of $-e^2/h$ is equivalent to the mathematical theorem that the winding number is an integer.

Equation (2.3) can be cast into another form. If we integrate the formula by z and make use of

$$\frac{\partial u^p}{\partial A^0} = \sum_{q\neq p} \frac{u^q}{E_p-E_q} \left(q\left|\frac{\partial H}{\partial A}\right|p\right) \tag{2.5}$$

by the first-order perturbation, where u^p is the pth orthonormal eigenstate of the Hamiltonian, we can rewrite σ_{xy} for a fixed number of electrons as

$$\sigma_{xy} = -\frac{e^2}{h}\frac{1}{2\pi i} \sum_p^{occ} \int dA_x^0 \int dA_y^0 \left[\left(\frac{\partial u^p}{\partial A_x^0} \,\middle|\, \frac{\partial u^p}{\partial A_y^0}\right) - (x\leftrightarrow y)\right], \tag{2.6}$$

where $(u|v)$ stands for an inner product of two vectors. This shows that nonzero Hall conductivity arises because each wavefunction when the gauge is changed as $A_x \to A_x+\Delta A_x$ followed by $A_y \to A_y+\Delta A_y$ can be essentially different from the wavefunction when we first let $A_y \to A_y+\Delta A_y$ followed by $A_x \to A_x+\Delta A_x$. The integral in (2.6) is again a winding number corresponding to the mapping from the space $\mathbf{A}^0$ to that of the wavefunction. Equation (2.6) is exactly of the same form as discussed by NIU et al. /6/.

3. LOCALIZATION IN LANDAU LEVELS

3.1 Symmetry and Dimensionality

The dimension and the symmetry of the system play important roles in the localization problem. This can most easily be understood if we consider the scaling argument. Let us consider a d-dimensional cube with a size L. By combining different cubes of the same size, we can construct a cube with size 2L. By repeating such a procedure starting from a small system, we can obtain energy levels and wave functions of a system having arbitrary size. Combination of two cubes causes mixings of energy levels. In this case couplings of levels closest to each other in energy are most important. Let V(L) be the resonance energy of such energy levels and W(L) their energy difference. We can safely assume W(L) to be the average energy separation, i.e., $W(L) \sim 1/L^d D(E)$ with D(E) the density of states.

The ratio $g(L)=V(L)/W(L)$ is called the Thouless number. The Thouless number can easily be shown to be related to the conductance of the system /7/. Let τ be the time for an electron to diffuse away to the boundary. Then, there is an uncertainty relation $V(L)\tau \sim h$ and the diffusion constant is given by $D^* \sim L^2/\tau$. The Einstein relation tells us that the conductivity is given as

$$\sigma \sim e^2 \frac{L^2}{\tau}\frac{1}{L^d W} \sim L^{2-d}\frac{e^2}{\hbar}\frac{V}{W} \sim \frac{e^2}{\hbar} g(L) L^{2-d} . \tag{3.1}$$

That is, $(e^2/\hbar)g(L)$ is the conductance of the system and becomes the conductivity σ in two dimensions. If states are localized, g(L) decreases with the increase of the system size L reflecting how the wave function decays with the distance from the localization center. For extended states, g(L) approaches $L^{d-2}\sigma$ in the limit of infinitely large L.

It can readily be understood that the localization depends strongly on the dimension and the symmetry of the system. When the wave function can be chosen as real, i.e., the Hamiltonian is a real symmetric matrix, the resonance integral describing couplings between cubes is real, i.e., the "dimension" of the interaction is $\beta=1$. This corresponds to the system in the absence of a magnetic field. When a magnetic field is applied, the time reversal symmetry is broken and the wave function becomes complex or the

Hamiltonian a complex Hermitian matrix. Then we have $\beta=2$. If we consider electron spins, we have another symmetry. In the presence of spin-orbit interactions, interactions are now allowed between degenerate levels and $\beta=4$. We call $\beta=1$ orthogonal ensemble, $\beta=2$ unitary, and $\beta=4$ symplectic. It is clear that states are less easily localized for larger β.

ABRAHAMS et al. introduced the single-parameter scaling theory in 1979 /8/. In their theory with g(L) being the scaling variable states are always localized in two dimensions. Fortunately, the theory of AOKI and ANDO shows that the single-parameter scaling theory is invalid and there should exist some delocalized states in each Landau level at least in the strong-field limit, where mixings among Landau levels can be neglected /9/.

3.2 Numerical Study

One of the most direct approaches to study the localization problem is numerical computor simulations. Although the system size considered is limited, an appropriate criterion of localization enables the extrapolation to infinitely large systems. One of the most powerful methods is to use the Thouless number g(L) and it has been employed in extensive investigations for the determination of the energy dependence of the inverse localization length $\alpha(E)$ /10/.

We consider a system with a finite size LxL. Scatterers are distributed at random. Periodic boundary conditions are used for both x and y directions. The coupling V(L) can be estimated by the energy shift E due to the change in boundary conditions, i.e., by replacing periodic boundary conditions in the y direction by antiperiodic conditions. The coupling V(L) is the geometric mean of ΔE of a given energy interval and over different samples.

Another powerful approach is the finite-size scaling method. Consider a two-dimensional system having a form of long strips, $L_x x L_y$ ($L_x \gg L_y$). The wave function is always localized exponentially in the x direction, since the system becomes essentially one-dimensional as long as $L_x \gg L_y$. Let $\alpha(L_y)$ be the inverse localization length in the y direction for a given L_y. We can determine $\alpha(E)$ by detailed study of L_y dependence. This method has also been applied to the present case /11/.

It has been demonstrated that there is no single-parameter scaling relation

$$\alpha(E,L_y)L_y = F(\alpha(E)L_y), \tag{3.2}$$

where $\alpha(E)=\alpha(E,L_y\to\infty)$ and F is a hypothetical function independent of E. This is in contradiction to the conclusion reached by SCHWEITZER et al. /12/, who employed the same finite-size scaling method for a square lattice system and concluded that there is a mobility edge near $E=0.2\Gamma$ for the lowest Landau level where Γ is the broadening calculated in the self-consistent Born approximation.

Figure 1 shows the resulting inverse localization length as a function of energy for systems containing short-range scatterers of high concentration. The results obtained by the two independent methods, i.e., the Thouless number and the finite-size scaling, agree quite well. The inverse localization length exhibits a power-law dependence on the energy and the critical exponent s, defined by $\alpha(E)\propto|E|^s$, is slightly less than 2 for the lowest Landau level. The figure contains also the results for the first excited

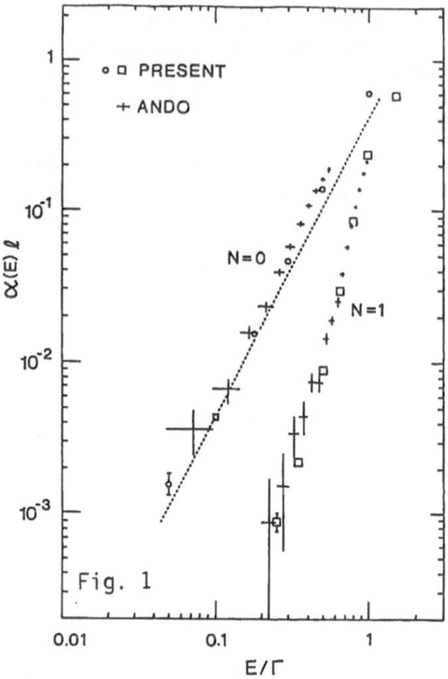

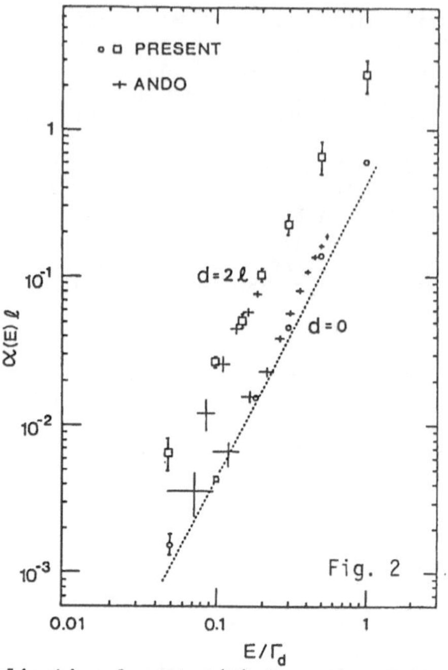

Fig. 1 Log-log plot of the inverse localization length $\alpha(E)$ in units of the inverse of the cyclotron radius of the lowest Landau level for N=0 and 1 versus energy measured from the center of each Landau level and normalized by the broadening Γ calculated in the self-consistent Born approximation. Scatterers with short-range potentials are distributed densely in the system. The crossed error bars are the result of the Thouless number study /10/ and the squares that of the finite-size scaling study /11/

Fig. 2 Log-log plot of the inverse localization length $\alpha(E)$ in units of the inverse of the cyclotron radius of the lowest Landau level for scatterers with long-range Gaussian potential (d/l=2) versus energy measured from the center of each Landau level and normalized by the broadening Γ calculated in the self-consistent Born approximation. The crossed error bars are the result of the Thouless number study /10/ and the squares that of the finite-size scaling study /11/

Landau level N=1, which have been obtained in similar procedures.

The same method can easily be applied to the case of scatterers with long-range potentials. Figure 2 shows similar results for scatterers with a long-range Gaussian potential ($v(r)\ \exp(-r^2/d^2)$ with $d/l=2$ where $l^2=c\hbar/eH$). The inverse localization length is much larger than that for short-range scatterers, indicating strong localization effects, but has a similar energy dependence.

In the above example, equal amounts of attractive and repulsive scatterers are assumed. In such a case, the density of states and the inverse localization length are symmetric about the center of each Landau level. Consequently, extended states, if they exist, should lie at the center. This is no longer true if only attractive scatterers are present. In this case the density of states and other quantities can depend on impurity concentration strongly. The Thouless number study has been performed to determine the energy of extended states in such a case /10/. It has been shown that the energy of extended states E_c is given by the condition $\mathrm{Re}(G(E_c))=0$ or $E_c=\mathrm{Re}(\Sigma(E_c))$, where $G(E)$ and $\Sigma(E)$ are Green's function and its self-energy, respectively, of the lowest Landau level. This means that E_c is determined by the maximum of the conductance obtained by neglecting localization effects.

It is widely accepted that states are localized in two dimensions in the absence of a magnetic field. When a weak magnetic field is applied to the system, the magnetic field tends to increase the extent of localized wave functions. This effect gives rise to the negative magnetoresistance widely observed in various quasi-two-dimensional systems. It is believed, however, that the magnetic field does not destroy the localization as long as it is sufficiently weak. Therefore, an important question arises, what happens when mixings between different Landau levels are appreciable with decreasing magnetic fields. The Thouless-number method can be extended to study such effects of level mixings /10/.

Figure 3 shows the histogram of the density of states for the lowest Landau level /10/. The three lowest Landau levels N=0,1,2 have been included in the calculation. For $\hbar\omega_c/\Gamma=1.5$ (ω_c is the cyclotron frequency), the density of states of each Landau level are separated from each other in the single-site approximation. On the contrary, the numerical density of states has a small tail and does not vanish in the gap region. With decreasing magnetic field, the energy corresponding to the peak of the density of states is shifted to the lower energy side due to the quantum-mechanical

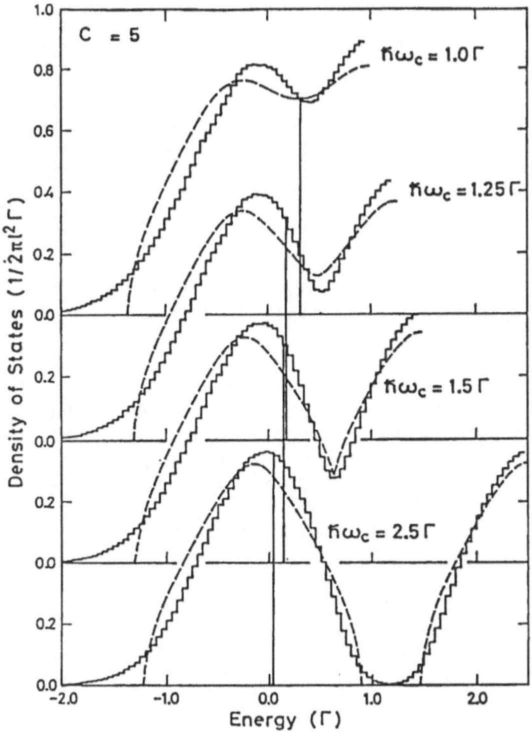

Fig. 3 Histograms of the density of states when mixings among different Landau levels are taken into account /10/. The dashed lines represent the results calculated in the single-site approximation. The energy is normalized by the broadening calculated in the single-site approximation and its origin is chosen at the position of the lowest Landau level. The vertical straight lines represent the energy of delocalized states, i.e., the energy where $\alpha(E)$ vanishes

level-repulsion effect. On the other hand, the energy of extended states, denoted by vertical straight lines, is shifted to the higher energy side.

This result suggests the interesting possibility: There always exist some extended states in the two-dimensional system in magnetic fields. Those extended states move up in energy with decreasing magnetic fields and disappear only in the limit of vanishing magnetic field. Whether this is correct or not remains to be seen by future study.

The exponent s plays a crucial role in determining the temperature dependence of the dc conductivity σ_{xx}. At nonzero temperatures, the presence of inelastic scattering processes can destroy the localization effect and the system has a cutoff length L_ε determined by electron-electron scatterings and electron-phonon scatterings. Let τ_ε inelastic scattering time ($\tau_\varepsilon \propto T^{-p}$). Then, L_ε is estimated to be $(\tau_\varepsilon/\tau)^{1/2} l$, where $\tau \sim \hbar/T$. An effective "mobility edge" E_ε can be obtained by the condition $\alpha(E_\varepsilon)L_\varepsilon \sim 1$ or $E_\varepsilon \propto T^{p/2s}$. It is shown that the peak value of the diagonal conductivity either vanishes like $T^{p/2s-1}$ or converges to a finite constant as $T \to 0$ according as $p/2s>1$ or <1. At sufficiently low temperatures inelastic processes are dominated by electron-electron scatterings and p is expected to be close to 1. Since s is close to 2 for the lowest Landau level, we can predict that the peak value of the conductivity approaches a constant as we approach T=0. This is in agreement with experimental results observed in Si inversion layers /13,14/.

3.3 Scaling Relation between Conductivities

Recently unitary nonlinear model with σ_{xy} as a θ term in the Yang-Mills theory has been proposed to explain the localization and quantum Hall effect /15/. According to this theory σ_{xx} and σ_{xy} are independent scaling variables and the flow-diagram in the σ_{xy}-σ_{xx} space determines the localization in Landau levels. The system-size dependence of the Hall conductivity can also be studied numerically /16/.

Figure 4 shows an example of calculated σ_{xy} for different system sizes L. With increasing L, the Hall conductivity approaches a step function as is expected. Figure 5 shows $\sigma_{xx}(E,L)$, proportional to the Thouless number g(E,L), versus $\sigma_{xy}(E,L)$ for a system with short-range scatterers. The diagonal and off-diagonal conductivities for different E's and system-size L's are plotted in the same diagram. The figure clearly demonstrates that σ_{xx} and σ_{xy} are mutually correlated and cannot be considered independent in the present system. This scaling relation between σ_{xx} and σ_{xy}, although dependent on N, clearly contradicts the conclusion obtained in the nonlinear model.

Quite recently, the study of system size dependence has been extended to the case in which only scatterers with attractive potentials are distributed and the density of states is no longer symmetric around the center, and to the case that magnetic fields become small and mixings between different Landau levels are appreciable /17/. It has been demonstrated that the scaling function is independent of signature of scatterers as long as the magnetic field is sufficiently strong, but is drastically modified by mixings among different Landau levels.

In principle this scaling relation between conductivities can be observed experimentally. One way is to study the temperature "flow diagram". This is possible since inelastic scatterings present at nonzero temperatures introduce an effective system size and by varying temperature we can study

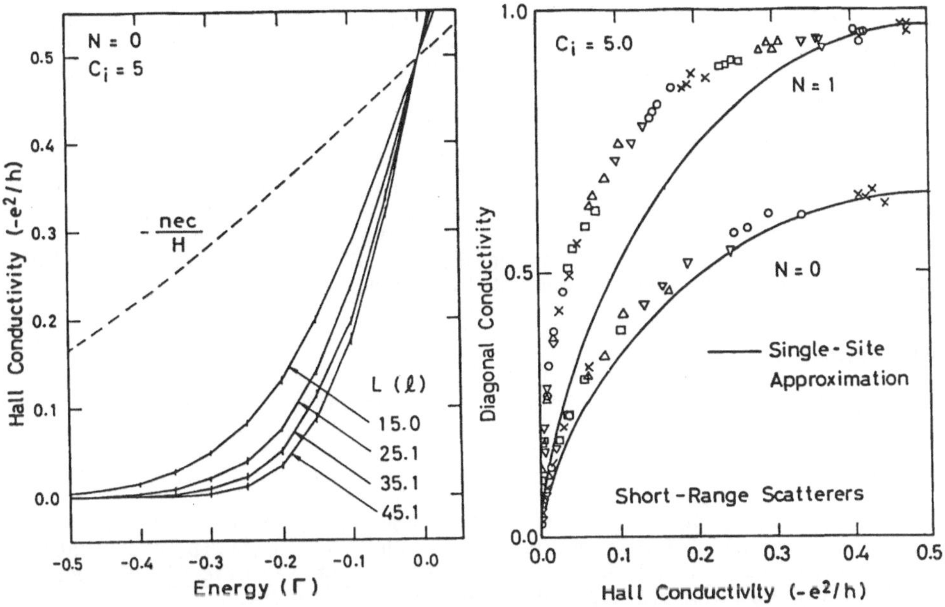

Fig. 4 Calculated Hall conductivity versus energy for different system sizes /16/. Scatterers with short-range potentials are distributed densely in the system and the Fermi level lies in the lowest Landau level. Only the low energy side is shown because of the symmetry. The dashed lines represent -nec/H, where n is the electron concentration

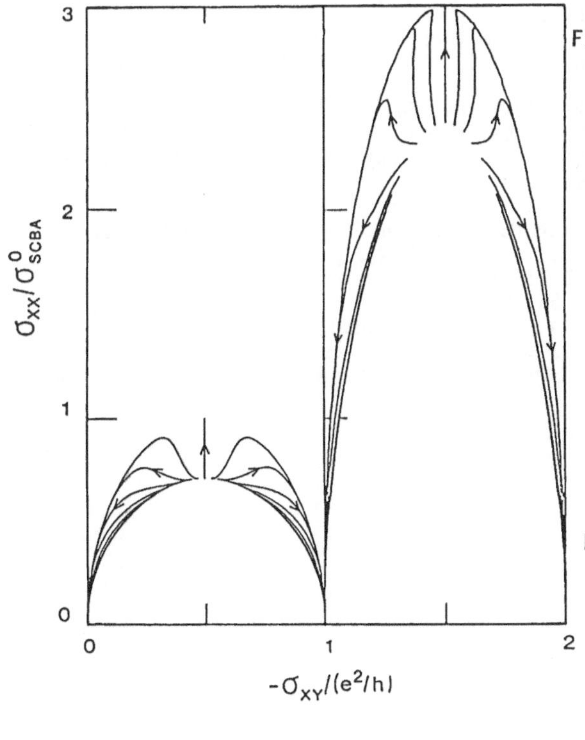

Fig. 5 Flow diagrams for σ_{xy} and σ_{xx} for the lowest two Landau levels /16/. The Thouless number, proportional to σ_{xx}, and the Hall conductivity of different energies and four different system sizes L's are plotted (15 < L/1 < 50)

Fig. 6 Calculated flow diagrams for $\sigma_{xy}(T)$ and $\sigma_{xx}(T)$ for varying temperature in a model system /18/

the system-size dependence of the conductivities. Actually, however, we have to take into account the temperature broadening of the Fermi distribution function, and the temperature flow diagram becomes dependent on the Fermi level. Figure 6 gives an example of calculated flow lines /18/. The envelope of flow lines for different Fermi levels becomes the scaling function of the conductivities.

There have been reported several experiments which support the presence of the scaling relation. KAWAJI has obtained the temperature flow lines in inversion layers on the Si surface and shown that they have an envelope line qualitatively in agreement with Fig. 6 /19/. WEI et al. reported similar results in two-dimensional systems at InGaAs/InP heterostructure /20/.

Acknowledgments

The author would like to thank H. Aoki for valuable discussions.

References

1 T. Ando, A.B. Fowler and F. Stern: Rev. Mod. Phys. **54**, 437 (1982).
2 K. von Klitzing, G. Dorda and M. Pepper: Phys. Rev. Lett. **45**, 494 (1980).
3 S. Kawaji and J. Wakabayashi: In Physics in High Magnetic Fields, ed. S. Chikazumi and N. Miura (Springer, Berlin, 1981), p.284.
4 T. Ando, Y. Matsumoto and Y. Uemura: J. Phys. Soc. Jpn. **32**, 859 (1972).
5 R.B. Laughlin: Phys. Rev. B **23**, 5632 (1981).
6 Q. Niu, D.J. Thouless and Y.-S. Wu: Phys. Rev. B **28**, 3372 (1985).
7 D.C. Licciardello and D.J. Thouless: J. Phys. C **8**, 4157 (1975); **11**, 925 (1978).
8 E. Abrahams, P.W. Anderson, D.C. Licciardello and T.V. Ramakrishnan: Phys. Rev. Lett. **42**, 673 (1979).
9 H. Aoki and T. Ando: Solid State Commun. **38**, 1079 (1981).
10 T. Ando: J. Phys. Soc. Jpn. **52**, 1893 (1983); **53**, 3101 (1983); **53**, 3126 (1983).
11 T. Ando and H. Aoki: J. Phys. Soc. Jpn. **54**, 2238 (1985).
12 L. Schweitzer, B. Kramer and A. MacKinnon: J. Phys. C **17**, 4111 (1984).
13 T. Ando, Y. Matsumoto, Y. Uemura, M. Kobayashi and K.F. Komatsubara: J. Phys. Soc. Jpn. **32**, 859 (1972).
14 S. Kawaji and J. Wakabayashi: Surf. Sci. **58**, 238 (1976).
15 H. Levine, S.B. Libby and A.M.M. Pruisken, Phys. Rev. Lett. **51**, 1915 (1983); Nucl. Phys. B **240**, 30; 49; 71 (1984).
16 T. Ando: Surf. Sci. **170**, 243 (1986).
17 T. Ando: J. Phys. Soc. Jpn. **55**, No.9 (1986).
18 H. Aoki and T. Ando, Surf. Sci. **170**, 249 (1986).
19 S. Kawaji: Prog. Theor. Phys. Suppl. **94**, 178 (1986).
20 H.P. Wei, A.M. Chang, D.C. Tsui, A.M.M. Pruisken and M. Razeghi: Surf. Sci. **170**, 238 (1986).

Scaling of the Integral Quantum Hall Effect

H.P. Wei[1], *D.C. Tsui*[1], *M.A. Paalanen*[2], *and A.M.M. Pruisken*[3]

[1]Department of Electrical Engineering, Princeton University, Princeton, NJ 08544, USA
[2]AT&T Bell Laboratories, Murray Hill, NJ 07974, USA
[3]Pupin Physics Laboratory, Columbia University, New York, NY 10027, USA

Abstract

A review is given of recent experiments testing the two-parameter scaling theory of the integral quantum Hall effect. The temperature driven flow diagram in the $\sigma_{xx}(T)$ vs. $\sigma_{xy}(T)$ of the two-dimensional electrons in InGaAs-InP, measured from 10K to 50mk, shows scaling behavior and is consistent with the theory.

The concept of scaling has provided an important framework in recent years to understand transport in disordered electronic systems. One most outstanding consequence of the one-parameter scaling theory of Abrahams et al. [1] is that all electronic states in two-dimensional (2D) systems are localized. This concept of weak localization in 2D has been essential to our current understanding of transport of the 2D electron gas in semiconductor heterojunctions in the low temperature (T) and weak magnetic field (B) limit. The integral quantum Hall effect (IQHE), on the other hand, is observed in these same physical systems in the presence of a strong B, when Landau quantization dominates. It follows from the fact, that the Hall conductance is quantized to integral multiples of e^2/h, that extended states must now exist. Since the introduction of B breaks the time-reversal symmetry, it is not surprising that the one-parameter scaling no longer holds in this IQHE limit.

Very recently, a two-parameter scaling theory of the IQHE has evolved from the field-theoretical work of Pruisken and collaborators [2,3]. This theory assumes a collection of non-interacting electrons in a random potential and the two independent physical parameters are the diagonal and the off-diagonal conductivity tensors, σ_{xx} and σ_{xy}, in units of e^2/h. The effect of the scale transformation on these magneto-transport coefficients is illustrated by the flow diagram shown in Fig. 1. The prediction that any initial set of system parameters, $(\sigma^o_{xx}, \sigma^o_{xy})$, will renormalize to the quantized values, (0,n), after successive length-scale transformations, is represented by the lines that flow towards the fixed points $(\sigma_{xx}, \sigma_{xy}) = (0,n)$. In addition to these "localization fixed points", which describe localized wavefunctions of the electrons near the Fermi energy E_f, there are intermediate coupling fixed points denoted by $\otimes$ on $\sigma_{xy} = n + ½$. They describe the singular behavior in the renormalized transport coefficients, corresponding microscopically to the occurrence of a diverging localization length. These "delocalization" fixed points are associated with the extended states at E_f. The localization to delocalization transition is characterized by a universal critical exponent. The theory is developed for T = 0 and the length scale transformations are, in principle, accomplished by varying the sample size. In practice, the experiment [4] is carried out at finite T and the effective sample size is related to an inelastic

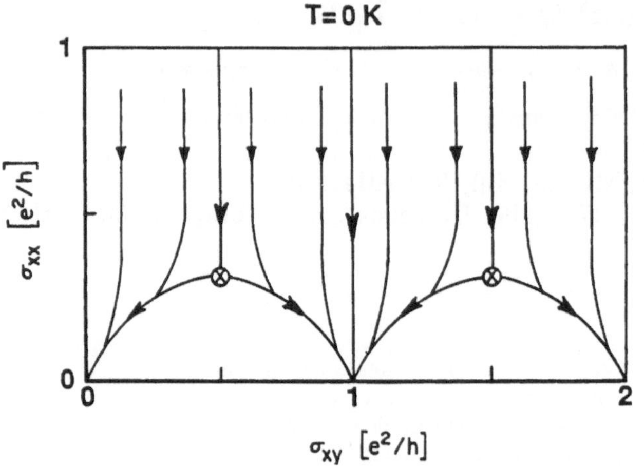

Fig. 1: A renormalization group flow diagram showing two parameter scaling of the integral quantum Hall effect for the N = 0 spin-split Landau levels. Both σ_{xy} and σ_{xx} are in units of e^2/h.

scattering length, which can be varied by varying T. At the present, there is no microscopic theory on the relation of the length scale to T.

Fig. 2 is a plot, in units of e^2/h, of the experimental data of $\sigma_{xx}(T)$ vs. $\sigma_{xy}(T)$, taken on an $In_{.53}Ga_{.47}As$-InP heterojunction, from T = 10K to 0.5K at various fixed values of B. The 2D electrons have a density $n = 3.4\times10^{11}/cm^2$, a mobility $\mu = 35{,}000 cm^2/Vsec$, and an effective mass m^*=0.047m. Each flow line corresponds

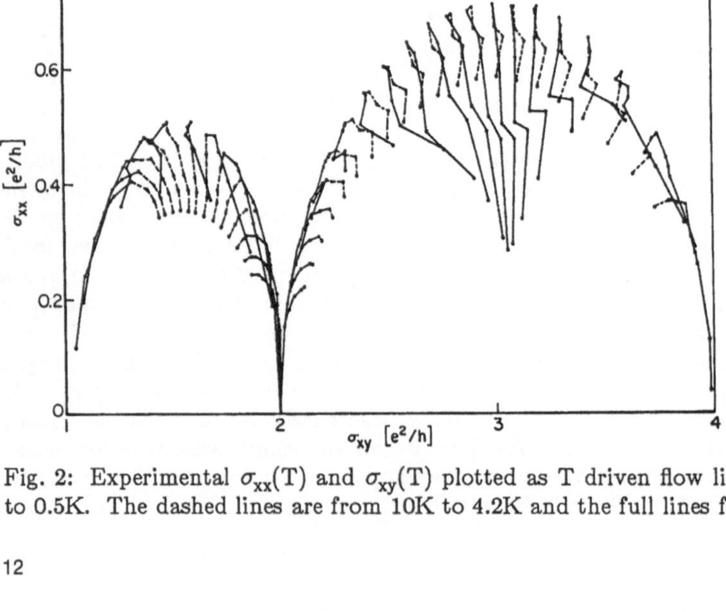

Fig. 2: Experimental $\sigma_{xx}(T)$ and $\sigma_{xy}(T)$ plotted as T driven flow lines from T = 10K to 0.5K. The dashed lines are from 10K to 4.2K and the full lines from 4.2K to 0.5K.

to a fixed B. The dashed portion of each line is from 10K to 4K and the solid portion from 4K to 0.5K.

The initial rise in σ_{xx} with decreasing T from 10K to 4K, as indicated by the broken lines in Fig. 2, is not due to scaling. It results from the T dependence of the Fermi-Dirac distribution function, f, according to

$$\sigma^{o}_{\mu v}(T) = \int dE \, \frac{\partial f(T)}{\partial E} \, \sigma^{o}_{\mu v}(T=0) \ , \qquad (1)$$

where $\sigma^{0}_{\mu v}(T = 0)$ is the magneto-conductivity tensor given by Ando's self-consistent Born approximation (SCBA)[5]. Figure 3 shows the T driven flow lines from 50K to 0K calculated from Eq. (1). It is clear that the calculation reproduces the flow pattern of the data at high T. In other words, in the absence of the expected scaling behavior, the two classical conductivity tensor components, due to the T dependence of the Fermi-Dirac distribution function alone, will follow the flow observed from 10K to 4K, but, as T $\rightarrow$ O, will approach the SCBA trajectory, indicated by the dash-dot curve in Fig. 3.

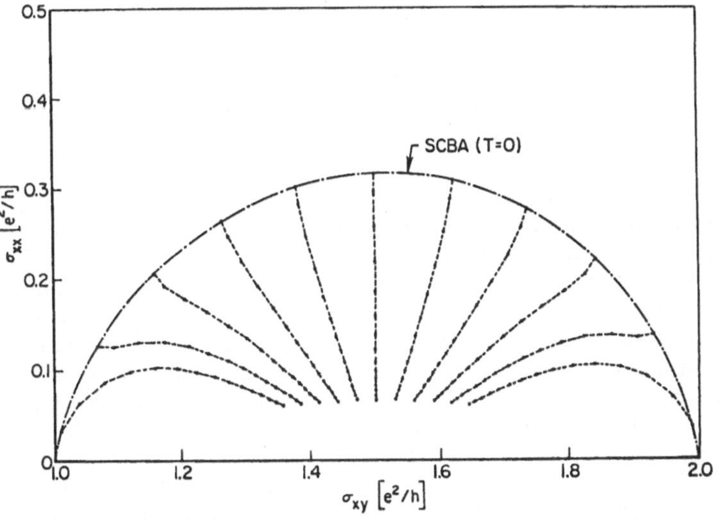

Fig. 3: T driven flow lines of $\sigma^{o}_{xx}(T)$ vs. $\sigma^{o}_{xy}(T)$ calculated from Eq. (1) for the spin resolved N = 0, $\downarrow$ Landau level. (Ref. 4).

The flow due to the T dependence of f is dominant only at high T when kT is comparable to the Landau level broadening. In the lower T regime (T < 4K) the data, as indicated by the full lines in Fig. 2, show a tendency to flow away from the SCBA trajectory and towards the fixed points (0,1), (0,2), (0,3) and (0,4). This deviation from the "classical" behavior is attributed to genuine scaling of the conductances. It is more clearly seen in the data from 770mK to 50mK shown in Fig. 4. Here, the dotted curve and the solid curve are obtained by sweeping B at T = 770mK and T = 50mK, respectively. The dashed flow lines in between are obtained by fixing B and sweeping T from 770mK to 50mK. The flow is down and out towards the fixed points as T decreases, consistent with the theory.

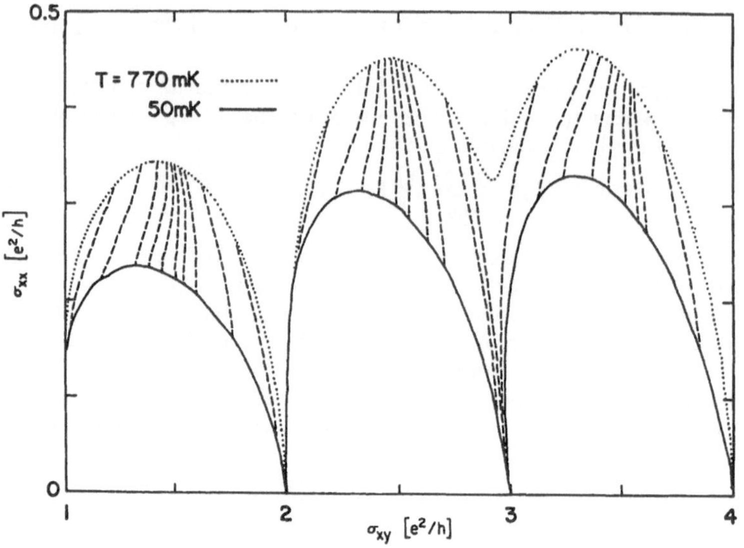

Fig. 4: Experimental $\sigma_{xx}(T)$ and $\sigma_{xy}(T)$ plotted as T driven flow lines (dashed lines) from T = 770mK to 50 mK. The upper dotted line and the lower solid line are taken by sweeping B, at T = 770mK and 50mK respectively.

Several comments are in order. First, the flow lines in Fig. 4 are somewhat twisted and the σ_{xx} maxima at the lowest T is shifted from $\sigma_{xy} = n + ½$. This behavior is sample dependent and is attributed to inhomogeneities in the sample. Second, the data for $2 \leq \sigma_{xy} \leq 3$ and the data for $3 \leq \sigma_{xy} \leq 4$ are from electrons in the two different spin states of the same Landau level (with Landau quantum number N = 1). The fact that there is no appreciable difference in the data and in their T dependences is consistent with the assumption of noninteracting electrons and that the spin quantum number is not relevant. Third, similar measurements are made on the 2D electrons in GaAs/$Al_xGa_{1-x}As$. The data, however, show no indication of the scaling behavior observed in the InGaAs-InP. This lack of scaling is consistent with the explanation that in this system the transport in the regime of the IQHE is dominated by classical percolation [6] and quantum mechanical localization plays only a rather insignificant role. Moreover, in this system, striking differences are observed in the data from the two different spin states of the same Landau levels [7]. The microscopic mechanism for this spin-dependent transport is unknown at the present. Fourth, Kawaji and Wakabayashi [8] recently studied the T driven flow of $\sigma_{xx}(T)$ vs. $\sigma_{xy}(T)$ of the 2D electrons in Si-MOSFETs from 4.2K to 1.49K in the range $\sigma_{xy} = 2$ to 2.7. They observed only the flow due to the T dependences of the Fermi-Dirac distribution function in this range of T. Lower T is apparently needed to probe into the scaling regime in this system. Finally, Ando [9] recently performed numerical calculations and his results contradict two-parameter scaling. The experiments, as discussed above, are consistent with the scaling theory, but they are not sufficient to be a definitive test of the theory. Clearly a more definitive test will require experiments at lower T and on more homogeneous samples to allow unambiguous probing into the critical region in the $\sigma_{xx}(T)$ vs. $\sigma_{xy}(T)$ flow diagram.

We acknowledge the contributions from Drs. M. Razeghi and A.M. Chang. The work at Princeton University is supported by the National Science Foundation through grant No. DMR-8212167.

References

1. E. Abrahams, P.W. Anderson, D.C. Licciardello, and T.V. Ramakrishnan, Phys. Rev. Lett. **42**, 673 (1979).
2. H. Levine, S.B. Libby, and A.M.M. Pruisken, Phys. Rev. Lett. **51**, 1915 (1983).
3. A.M.M. Pruisken, Phys. Rev. B**32**, 2636 (1985) and the references therein.
4. H.P. Wei, D.C. Tsui, and A.M.M. Pruisken, Phys. Rev. **B33**, 1488 (1986).
5. T. Ando and Y. Uemura, J. Phys. Soc. Jpn. **36** 959 (1974).
6. M.A. Paalanen, D.C. Tsui, and A.C. Gossard, Phys. Rev. **B25**, 5566 (1982).
7. H.P. Wei, D.C. Tsui, M.A. Paalanen, and G. Weimann, Bull. Amer. Phys. Soc. **31**, 606 (1986).
8. See S. Kawaji, Prog. Theor. Phys. Suppl. No. 84, 178 (1985).
9. T. Ando, Surf. Sci. **170**, 243 (1986).

Percolation Approach to the Quantum Hall Effect

S. *Luryi*

[2]AT&T Bell Laboratories, Murray Hill, NJ 07974, USA

1. Introduction

The percolation description of electron states in a 2D electron gas in strong magnetic fields was developed in refs. [1-5]. Being essentially semi-classical, this picture alone is not sufficient to give a rigorous account of the most fundamental property of the integral QHE: the extraordinary precision of quantization of the Hall conductance. Nevertheless, the percolation description appears to be quite useful in that it allows to interpret in a simple manner a number of other experimental features of the QHE and relate them to a specific model of electronic states in the presence of disorder in the sample. In this work I shall review the basic picture, following refs. [3,4] as well as some unpublished work, mostly done in collaboration with R. F. Kazarinov.

1.1 *Ideal System*

Consider the quantum-mechanical problem of the electronic motion in an (x,y) plane transverse to a uniform magnetic field **B**. Operators $\hat{X}$ and $\hat{Y}$ of the cyclotron-orbit-center coordinates are defined by $\hat{X} = \hat{x} + \hat{v}_y/\omega$ and $\hat{Y} = \hat{y} - \hat{v}_x/\omega$, where $\omega = eB/mc$ is the cyclotron frequency and $\hat{v}_x$, $\hat{v}_y$ are the components of the canonical velocity operator. The following commutation relations hold:

$$[\hat{v}_x, \hat{v}_y] = i\hbar\omega/m \tag{1}$$

$$[\hat{X}, \hat{v}_x] = [\hat{Y}, \hat{v}_y] = [\hat{X}, \hat{v}_y] = [\hat{Y}, \hat{v}_x] = 0 \tag{2}$$

$$[\hat{Y}, \hat{X}] = i\ell^2 \tag{3}$$

where $\ell \equiv (\hbar c/eB)^{1/2}$ is the magnetic length. In virtue of (1) the spectrum of the electron kinetic energy operator $\hat{H}_0 = (m/2)[\hat{v}_x^2 + \hat{v}_y^2]$ is given by $E_n = \hbar\omega(n + ½)$ and because of (2) the Landau levels n are degenerate with respect to the position of the center of the cyclotron orbit. The commutation relation (3) implies that the density of states N_1 in each Landau level is given by $N_1 = (2\pi\ell^2)^{-1} = eB/hc$ states per unit area of the sample.

If there is an electric field **F** in the plane of the inversion layer, then this degeneracy is lifted. The resultant electronic states are localized in the direction of **F** by the magnetic length ℓ. In the direction of **B** the states are localized by the quantum well, while in the third direction the states are delocalized. The electronic waves propagate along the equipotential lines like light in an optical fiber. The propagation velocity turns out to be the same as the average velocity in the classical problem — the Hall velocity $v_H = c(F/B)$. It is quite easy to write down an explicit expression for these eigenstates, which I shall refer to as *fibers*. All that is important for us, however, is that they represent narrow tubes extended along equipotentials. This property remains true even in the presence of disorder, when the equipotentials are wiggly curves – contours on a topographic map – and no exact solution is available. Fortunately, we shall not need it anyway.

The QHE is an interplay between the electron density N and N_1. The *plateaus* are thought to occur when $N = N_1 \times$ *integer*, i.e., when the Fermi level E_F is positioned between two successive Landau levels, n th and $(n+1)$st. Leaving aside the question about what holds it there (the most important question!), consider what happens. First of all, there is no dissipation — large energy separates the filled and empty states. This implies $\sigma_{xx} = 0$. Electrons under the Fermi level contribute to a nondissipative Hall current. Linear density of this current J is given by $J = eN\,v_H = n\,(e^2/h)F$ and the total Hall current in the area between two electrodes biased by V is $I = n\,(e^2/h)V$. It is carried by electrons which are in stationary states.

1.2 *Disorder Required*

Disorder is of fundamental importance for the QHE. At $T=0$ the Fermi level E_F by definition coincides with the highest occupied state. In an ideal case with a uniform electron density N, the Fermi level is pinned to a Landau level at all B except for discrete values, $B_n = N\,hc/en$, at which E_F jumps between the n-th and the $(n+1)$-st level. This means that in an ideal system with no disorder, the plateaus are reduced to discrete points. The finite width of the plateaus must be attributed to pinning of the Fermi level by localized states. The nature of the localization in inversion layers in strong B is one of the most interesting consequences of studying the QHE. In what follows I shall describe a microscopic model of both localized and delocalized states, which I believe accounts for most aspects of the *integral* effect.

2. Model of Electronic States in a Disordered Sample

In our model *all electron states are fibers,* i.e., are confined within narrow tubes extended along equipotential lines. Strictly speaking, the fibers are extended along the lines of constant classical energy. The latter includes also the kinetic energy, $mv_H^2/2$, which depends on the local electric field F. Throughout this work, speaking of the equipotentials, this comment is left understood. The potential is assumed to vary smoothly on the scale of ℓ, so that at every point the sample has a well-defined Landau-level system. The existence of truly localized states with energies continuously distributed in the gaps between the local Landau levels is not required.

2.1 *Global and Local Fibers*

All equipotentials (and associated fiber states) distinctly fall into two classes: *global* and *local*. This distinction is especially clear in the Corbino ring geometry, where the global fibers are those which encircle the central electrode while the local can be contracted to a point by a continuous deformation. The distinction between the global and local fibers is purely *topological* and has little to do with the fiber length. A local fiber can, in fact, be quite extended — even longer than a global fiber — but it does not contribute to the Hall current.

Even though the number of current-carrying states is reduced by disorder, the current remains the same as in the ideal situation, see Fig. 1. *The entire applied voltage drops on the global states. Similarly, when the current is given, the entire Hall e.m.f. develops across the global subsystem.* Regions bounded by a local equipotential are in thermodynamic equilibrium. They may contain nonfiber states as well as local fibers revolving around potential hills and inside potential "volcanos". These regions are macroscopic and have a well-defined chemical potential E_F and a fluctuating electron

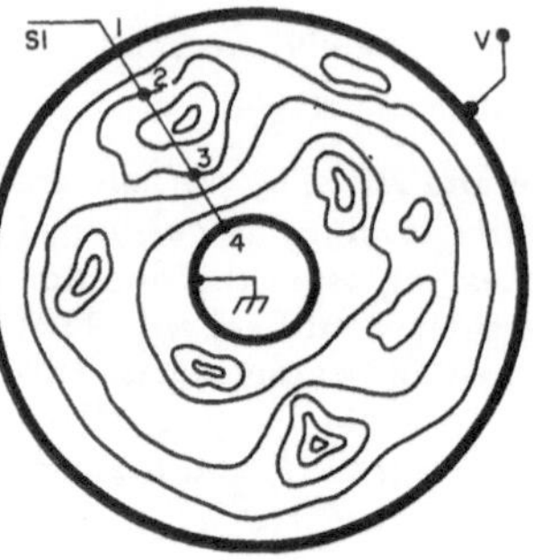

FIGURE 1: Illustration of a Corbino sample and a radial section S1, which crosses one or more isolated closed loops. Because points 2 and 3 lie on an equipotential the sum of voltages dropping in regions 1→2 and 3→4 equals the applied voltage V. The effective width of the Corbino ring in section S1 is reduced by the distance 2→3 but for a given V the Hall current does not depend on the width of the ring.

concentration. Even at $T = 0$ the Fermi level can be pinned to some of the local loops inside a macroscopic region, while the boundary of this region and an adjacent global fiber (if it exists), corresponding to a particular Landau level, can be either above or below E_F.

2.2 *Charge Fluctuations and Perfect Screening*

Let us discuss what gives rise to potential fluctuations in the inversion layer and why they can be expected to be smooth on the scale of ℓ. To be specific, we consider the GaAs/AlGaAs QHE samples. The spatial inhomogeneity of the self-consistent potential is brought about by fluctuations in the fixed positive charge responsible for the creation of the inversion layer in a heterojunction system. The fact that the Landau-level energy does indeed fluctuate, thus giving rise to patches with occupation numbers 0 or 1, is rather subtle.

Consider first an ideal situation with uniform N_+ giving rise to a partially filled Landau level and then impose a fluctuation δN_+. So long as $\delta N_+ << N_1$, this charge fluctuation is perfectly screened by the inversion layer, producing absolutely no spatial variation in the single-electron energy level which remains tied to E_F. This *perfect screening* is a consequence of the multiple *degeneracy* of Landau levels. A partially filled level can accommodate extra charge without changing its energy.

When $\delta N_+ > N_1$, then perfect screening does not occur and the self-consistent potential in the inversion layer fluctuates. Assume that fluctuations in the number of fixed charges N_+ per unit area are uncorrelated, $(\delta N_+/N_+) \sim (\lambda^2 N_+)^{-1/2}$, where $\lambda = \lambda(\delta N_+)$ is the spatial scale of the fluctuation. For $\delta N_+ \approx N_1$ one then has $\lambda(N_1) = \sqrt{N_+}/N_1 = \ell\sqrt{2\pi N_+/N_1}$. We see that $\lambda(N_1) >> \ell$ provided $N_+ >> N_1$. The crux of the matter is that the surface density N_+ of the fixed positive charge much exceeds the electron density in the inversion layer, i.e., $N_+ >> N$ (in GaAs QHE samples N_+ is mainly compensated by a negative surface charge). Typically, N_+ results from doping a layer of thickness ~ 500 Å with donors of volume density $\sim 2 \cdot 10^{18}\,\text{cm}^{-3}$, i.e., $N_+ \sim 10^{13}\,\text{cm}^{-2}$, while N_1 even at $B = 10\,\text{T}$ is only $\sim 2.4 \cdot 10^{11}\,\text{cm}^{-2}$. Thus the potential indeed varies smoothly on the scale of the magnetic length ℓ, which makes our model self-consistent. If N_+ were not so large, the breakdown of perfect screening would occur only at short fluctuation wavelengths and the fiber description would not be adequate.

The characteristic local fields due to the fluctuating charge by the order of magnitude are given by

$$\delta F = \frac{e\,\delta N_+}{\epsilon} = \frac{e}{\lambda\epsilon}\sqrt{N_+} \tag{4}$$

For $\lambda \sim \lambda(N_1)$ the fluctuating field $\delta F \sim eN_1/\epsilon \gtrsim 10^4 \mathrm{V/cm}$. Smaller fluctuating fields – those arising from large-area charge fluctutations – are typically screened by the electron gas. On the other hand, $\delta F \lesssim e\sqrt{N_+}/\epsilon d \sim 10^5 \mathrm{V/cm}$, where d is the average distance from the inversion layer to the fluctuating fixed charges – including the thickness of an undoped AlGaAs *spacer* layer. The field due to charge fluctuations of wavelength less than d averages out without reaching the inversion layer.

2.3 *Existence of Global Fibers*

It is not at all obvious that global states do exist in a large macroscopic sample. In fact, I shall argue that in equilibrium they do not! Global states only appear in the presence of an applied Hall voltage. To see the problem, consider a random potential surface $\psi(x,y)$. Equipotentials are contour lines on the topographic map. You can probe the topology of equipotentials at any energy by filling the terrain with water and looking at the *shoreline*. At low water levels there will be *lakes in a continent*. At high levels there will be *islands in the sea*. In both limits the shoreline represents local loops. This means that neither low-energy nor high-energy equipotentials are global. There is one definite energy at which the islands-in-the-sea topology goes over into that corresponding to lakes in the continent. This energy equals $<\psi>$ — assuming symmetric statistical properties of the random function $\psi(x,y)$ — and is called the *percolation threshold*. As we approach the transition point from the continent (sea) side the area of certain lakes (islands) diverges. It is clear that the length of the largest shoreline must also diverge at the threshold. In a finite-size sample this implies the existence of equipotentials connecting opposite edges of the sample. However, the coastline percolation in one direction (north-south) excludes the possibility of percolation in the other (east-west) direction. It is easy to see that for an uneven sample the percolation will be necessarily established in the shortest direction — radial direction in a Corbino ring.

The above argument shows that for a randomly disordered sample in equilibrium there are no global equipotentials. They begin to appear when an external voltage is applied. This is easy to visualize thinking about a *funnel* with crimped surface [4]. It has been shown [5] that at low and uniform external fields F the *fraction* of global states goes as $\sim F^p$ with $p = 41/84$. The energy range of global states may be referred to as the percolation band. Since in the presence of global equipotentials, the entire Hall voltage V develops across the global states, it follows that the percolation band emerges with a finite width equal to eV. It is the existence of global fibers that embodies the long-range order in QHE samples. Some disorder is required to establish the no-scattering situation (recall that an ideal 2DEG would give no plateaus), but too much anarchy is no good either: not every inversion layer exhibits the QHE! If one can imagine turning on the disorder at a given applied Hall voltage, then at some point one would eliminate all global fibers and the Hall current would cease. This phenomenon can be interpreted as a phase transition, in which the Hall current plays the role of an order parameter.

2.4 *Equal Occupation of Global States*

We have shown that global states constitute a small fraction of the total number of states in the inversion layer. On the other hand, the variation of the chemical potential E_F (the quasi-Fermi level) on the global states follows exactly the variation of the self-consistent potential ψ, in other words, all global states corresponding to the same Landau level are equally populated.

This can be seen as follows, Fig. 2a. Streams of the Hall current break the inversion layer into disjoint regions. Each of these regions (labelled i) is surrounded by a local

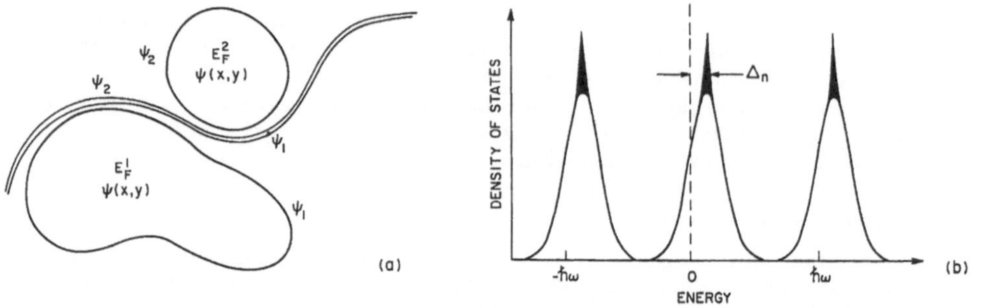

FIGURE 2: Equal occupation of global states: (a) illustration; (b) implication for the DOS

(though quite extended) equipotential ψ_i and is, therefore, in equilibrium. These regions are macroscopic, and have a well-defined chemical potential E_F^i and a fluctuating electron concentration. The average electron concentration is given by the surface density N and is the same in each macroscopic region, and hence the quantity $E_F^i - e <\psi>_i$ has the same value throughout the sample. The gist of this argument is to note that $\psi_i = <\psi>_i$. Indeed, the boundary of a macroscopic region is a very extended local equipotential. As such it must be close in energy to the percolation threshold $<\psi>_i$ of region i. Since the edges of each Hall stream have the same energy ψ and chemical potential E_F as the adjacent local equipotentials, we come to the conclusion that the difference $E_F - e\psi$ is the same for all global states.

Let us plot (Fig. 2b) the density of states $D(E)$ per unit area of the sample in the presence of an applied Hall voltage V. We shall be *counting the energy of fibers from the local value of the Fermi level,* which is probably the only meaningful way to describe a system consisting of nearly independent subsystems — each of which is in a thermodynamic equilibrium by itself but not with respect to the other subsystems. If $D(E)$ is plotted in this way, it is clear that the *global* states contribute a δ-function to each $\hbar\omega$ period. The shape of $D(E)$ in the local-state region is determined by the statistical properties of the random surface $\psi(x,y)$ corresponding to the self-consistent potential. All equilibrium regions i represent statistical realizations of the same system, and hence give rise to a density of states of the same form as that in the absence of an applied voltage. In a crude approximation (neglecting correlations introduced by the screening) we can expect it to be Gaussian. With increasing V, the total area under the local-state curve changes to account for an increasing fraction of the *global* states.

2.5 *Edge States*

It has been suggested that global states may be associated with the sample edges. Indeed, the potential surface in any sample is not entirely random. In order to confine the inversion layer laterally it has to look like a trough with steep walls at the edges. There is always a global equipotential on the wall, at any energy. The associated states can give a contribution to Hall currents — revolving clockwise on one edge and counterclockwise on the other. In a particular experiment, certain fraction of the total Hall current may flow near the edges, but description of the QHE in terms of the edge currents alone, in my view, physically amounts to an untenable assumption that the quasi-Fermi level E_F is flat in the interior of the sample. In this sense, the situation is different from the description of electron diamagnetism in terms of edge currents, which is a possible, though inconvenient, description of the Landau diamagnetism [6].

In my opinion, edge states play little role in the QHE. Their number varies from one sample to another depending on the boundary conditions, but in all macroscopic samples it remains statistically insignificant. You can think of boundary conditions for which all states near the edges are in an immediate contact with a three dimensional electron gas (e.g., a Corbino disk with source/drain implantation around the edges) — so that edge electrons can scatter into the third dimension. This will not change any experimentally observable aspect of the QHE, which is essentially a *surface* effect.

3. Consequences of the Model for the QHE

With respect to the filling of a Landau level n at a particular magnetic field and zero temperature, the sample area contains *three,* in general multiply connected, regions, Fig. 3. There are 2D patches where the n th level is filled (the n-phase), patches where this level is empty and complementary *metallic* regions – corresponding to intersecting Landau and Fermi levels.

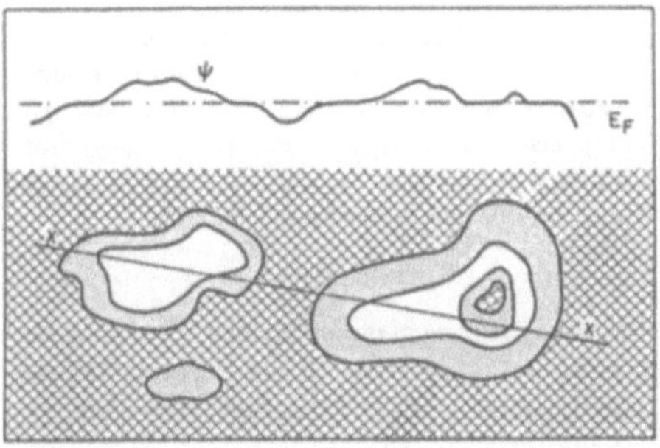

FIGURE 3: (a) Variation of the local single-electron energy $\psi(\mathbf{r})$ along a line x-x in a sample at equilibrium; (b) Three regions defined with respect to the filling of a Landau level n at zero temperature: blank area indicates the absence of n th-level electrons, cross-hatched area corresponds to completely filled regions (the n-phase), and dotted area to partially filled regions (metallic phase).

Within the n-phase the macroscopic current density J is given in terms of the gradient of the chemical potential E_F as follows:

$$J = \sigma_{xy} \nabla E_F \times \mathbf{B}/eB , \tag{5}$$

where $\sigma_{xy} = n e^2/h$. This can be proven as follows. Within the n-phase there are both global and local fibers. Consider a thin strip s along a global fiber. If it can be regarded as a linear conductor, its current I_s and the associated flux Φ_s of magnetic field through the contour I_s are complementary thermodynamic variables, $I_s = c\, \partial G_s/\partial \Phi_s$, where G_s is the free energy of electrons in the given strip. The single-electron contribution to the total current is given by $\delta I_s = \partial I_s/\partial N_s$, with N_s being the number of electrons in the strip. On the other hand, $\partial G_s/\partial N_s \equiv E_F^s$. Differentiating, we have $\delta I_s = c\, \partial E_F^s/\partial \Phi_s$. The minimum flux variation equals $\delta\Phi_s = hc/e$. It should be emphasized that the flux of the magnetic field through any fixed area of the ring is *not* quantized and in contrast to the situation familiar in superconductivity it can vary continuously. What is quantized in the present case is the magnetic flux through a variable area bounded by two global orbits on the chosen strip. The magnitude of the flux "quantum" follows from the periodic boundary conditions on the wave-functions of current-carrying states (which implies that the flux increment must be hc/e times an integral number $\delta\ell$) and the Principle of Least Action (whence $\delta\ell = 1$). The corresponding quantum δE_F of the chemical potential at zero temperature represents the variation of the Fermi energy across one global fiber. Therefore, $\delta I_s = (e/h)\delta E_F$ for each filled Landau level, whence we obtain (5). Note that all local fibers belong to *equilibrium* regions (surrounded by an equipotential), and hence across any local fiber $\delta E_F \equiv 0$.

Defining complex quantities $J \equiv J_x + iJ_y$ and $\tilde{F} = \nabla_x E_F + i\nabla_y E_F$, we can write (5) in the form $J = \sigma \tilde{F}$, where $\sigma \equiv \sigma_{xx} - i\sigma_{xy}$. Inasmuch as $\nabla \cdot \mathbf{J} = 0$, $\nabla \times \tilde{\mathbf{F}} = 0$, and σ is constant in the n-phase, it follows that both J and $\tilde{F}$ are analytic functions of the complex coordinate $z^* = x - iy$. Consider then a disordered sample schematically shown in Fig. 4. The total current I between the source and drain contacts equals the flux of J through any contour connecting, say, points 1 and 2. This contour can be chosen entirely within the n-phase, if the latter *percolates*, as shown in the figure. Because of the analyticity, the current is independent of the choice of a contour and equals $\sigma_{xy}\,[E_F(2) - E_F(1)]/e \equiv n\,(e^2/h)\,V$.

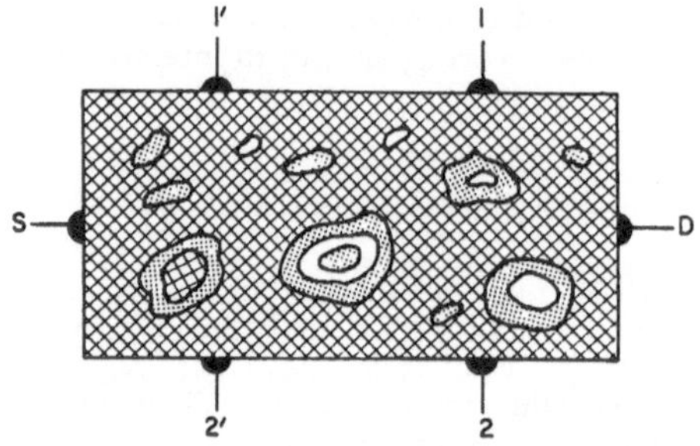

FIGURE 4: Illustration of the *on-plateau* situation. The n-phase (cross-hatched) percolates between the Hall probes. Within that phase the complex current density and chemical potential gradient are *analytic* functions of the complex coordinate.

When the magnetic field B is increased, the area of the n-phase shrinks. This occurs because $N_1 \propto B$, so that when B grows electrons fall onto the lower levels. When $N \approx (n - \frac{1}{2})N_1$, the n-phase disconnects and the metallic phase becomes percolating. The range ΔB of this inter-plateau region is proportional to the density N_G of global states in the sample, $\Delta B = (hc/e)N_G$, and therefore depends on the applied voltage (Sect. 3.2). Even at $T = 0$ the inter-plateau range can be finite. When B is increased still further, the metallic phase disconnects, the $(n-1)$-phase becomes percolating, and we have reached the next plateau.

3.1 *Plateau Width at Finite Temperatures*

Percolation model gives a simple explanation to the observed temperature dependence of the QHE plateaus. At finite T, instead of summing over *filled* global states, as we did in the preceding Section, we include the contribution of *all* global states – but weighted by their occupation probability given by the Fermi-Dirac distribution function $f = f\,(E_F - \psi)$ [this means we take $\delta I_s = (e/h)\,f\,\delta E_F$ for each global state]. In doing so, we can bring the function f out of the sum since its argument is constant on global states (Sect. 2.4). The result is an expression of the form $I = \bar{n}\,(e^2/h)\,V$, with $\bar{n}$ given by

$$\bar{n} = \sum_{n=1}^{\infty} f\,(\Delta_n)\,, \tag{6}$$

where $\Delta_n = E_F(x) - \psi(x) - n\,\hbar\omega$ (cf. Fig. 2) is constant for each Landau level n. Transitions between the Hall plateaus are due to the variation of Δ_n with B. For an ideal situation with no disorder, global states would constitute 100% of all states and the Hall plateaus would reduce to a set of discrete points. In a real system the opposite limit occurs. The Fermi level is pinned to the global states only in a small interval ΔB.

At a nonzero T the plateaus shrink due to the washing-out of the Fermi step-function. Varying the temperature at a fixed B outside ΔB we are simply tracing the tail of Fermi's distribution. Therefore, the temperature dependence of the deviation of R_{xy} from a quantized value at a fixed value of B must have an exponential form $\sim\exp(-E_a/kT)$, characterized by an activation energy $E_a(B) = \Delta_n$. Such behavior is indeed observed in the low-temperature experiments. The activation energy is vanishing for B inside ΔB and it grows as B is moved away from the inter-plateau range. When E_a is large, any deviation of $\bar{n}$ from an integer becomes intangible. This explains the high precision of the Hall resistance quantization.

3.2 *Voltage Dependence of the Plateau Width; Breakdown*

For typical QHE experiments the average applied field ($\sim 10^{-2}$ V/cm) is several orders of magnitude lower than the fluctuating field δF. This means that the local topography of the potential surface is only slightly modified by the applied field so that the area occupied by the local states changes very little compared to 100% at equilibrium. The *energy* range of global states was referred to as the quantum percolation band. We already know that the width of this band equals the applied Hall voltage V. Hence we can deduce qualitatively that the density N_G of global states increases with V and so does, therefore, ΔB. At the same time the plateaus *shrink*.

The exact dependence of the plateau width on V is not known — the function $N_G(V)$ may actually vary from one sample to another. Moreover, the Hall current density may be strongly nonuniform over the sample. Consider what happens if we begin to increase the voltage V applied to a Corbino ring at a fixed value of B. At a sufficiently large V the fraction of global states will approach unity $(N_G/N) \to 1$ in some section of the sample, and the Fermi level will be pinned to global fibers. This results in an *insulator-metal transition* induced by the applied field, which is observed as a breakdown of the nondissipative current flow. According to our model, it occurs when the applied field exceeds the characteristic δF of a particular sample, given by (4). Typically, $10^5 \gtrsim \delta F \gtrsim 10^4$ V/cm.

3.3 *Dissipative Current*

At $T=0$ the n-phases can support only a non-dissipative (Hall) current, since scattering is suppressed in the completely filled Landau levels. *Metallic* regions — those corresponding to intersecting Landau and Fermi levels — are, generally, disconnected, except for discrete ranges of the magnetic field. Thus, even at a value of B for which the average N in the inversion layer corresponds to a partially filled Landau level, the potential fluctuations induce a peculiar *metal-insulator transition*. At $T=0$ the longitudinal conductivity vanishes completely. No current can flow across filled global states due to the absence of scattering. Although, as discussed above, the area occupied by global states is small, streams of the quantum Hall current *slice* the sample into disjoint regions each of which remains at equilibrium.

At a finite temperature T in addition to the Hall current there is a longitudinal current due to generation of mobile carriers, i.e., thermal excitation of "electrons and holes" across the Landau gap $\hbar\omega$. With decreasing temperature this current goes to zero as $\exp(-\hbar\omega/kT)$. *Hopping* between local fibers across global streams also contributes to a dissipative current along the electric field. Indeed, the macroscopic equilibrium regions i are not at equilibrium *with respect to each other*. Some of the local fibers states associated with a potential hill or a potential "volcano" with its top above the Fermi level, will not be occupied. These states with energies close to the Fermi level may take part in a variable-range hopping current [7]. Temperature dependence of such conduction would be described by Mott's law $\sigma_{xx} \propto \exp[-(T_0/T)^{1/3}]$ for a two-dimensional system. At a sufficiently low temperature this path of current dominates over the generation current. A possible indication of the hopping nature of the dissipative current is provided by experiments in which the longitudinal conductance is measured in a Corbino geometry, at a given T and a fixed applied voltage. When we measured [8] the total current between two contact islands within a high-mobility ($3 \cdot 10^5 \mathrm{cm/V \cdot s}$) GaAs/AlGaAs QHE sample, separated by 1 mm and biased by 0.2 V, as a function of the magnetic field at $T = 2.1\,\mathrm{K}$, we found that in the range $1 \lesssim B \lesssim 4\mathrm{T}$, the B dependence of the current at the *minima* of SdH oscillations was strictly exponential, viz. $\sigma_{xx}^{\min} \propto \exp(-B/B_0)$, with $B_0 = 0.328\mathrm{T}$. (The conductivity was suppressed at still faster rate at higher B.) The exponential magnetoresistance is usually an indication of the hopping nature of the conduction [7].

4. Other Consequences

4.1 The Density of States

As discussed in Sect. 2.4, in the presence of a macroscopic number of global states the function $D(E)$ looks like a herd of unicorns — with one δ-function in each $\hbar\omega$ range. Now I shall argue that the animal may have another horn — at the position of the Fermi level, Fig. 5 — which moves with respect to the first when N or N_1 or even N_G are varied by external fields.

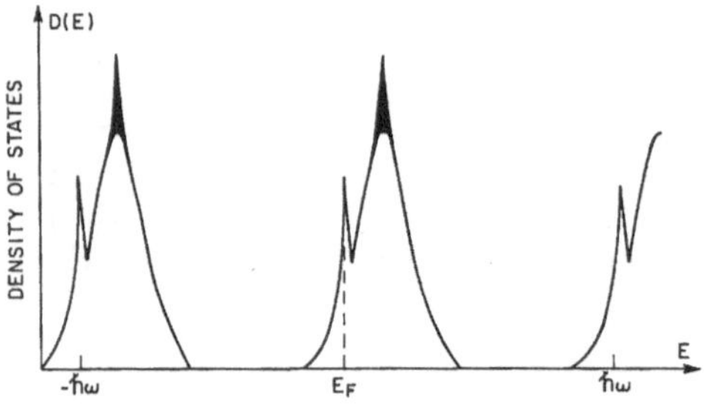

FIGURE 5: Qualitative picture of the density of states. In the presence of an applied bias V, the electronic-state energies are counted from the local value of the Fermi level. The overall shape of the peak is determined by statistical properties of the self-consistent potential at given values of B and V. The top δ-peak is proportional to the fraction of global states in the sample and hence strongly depends on V. The peak at E_F, due to the perfect screening effect, is broadened by collisions.

The appearance of a second peak in each period of the $D(E)$ is owing to the *perfect screening mechanism*, discussed in Sect. 2.2. Consider variation of the self-consistent potential $\psi(\mathbf{r})$ along some line within one of the macroscopic regions i, bounded by a local equipotential. Because region i is in equilibrium, it has a perfectly well defined E_F which is, of course, constant (dashed line in the figure). Note that a particular Landau level cannot simply cross E_F — because of the perfect screening it will be pinned to E_F until either filled or emptied. This gives a finite *measure* to the metallic regions in the sample. Their manifold still represents disconnected loops but it has acquired a finite area and contains a macroscopic fraction of electrons.

It should be realized that in this manifold *scattering is not suppressed.* Elsewhere in the sample there is no collision broadening of the Landau levels and the overall shape of $D(E)$ in our model is determined only by the inhomogeneous broadening, i.e., by the statistics of the random surface ψ. Not so in the metallic manifold — where there is a genuine lifetime broadening due to collisions. Broadening of the second peak may be describable in the self-consistent Born approximation for scattering [9] which predicts a semi-elliptic shape and a width Γ_2, which can be estimated from the relaxation time τ (related to the low-field mobility μ by $\omega\tau = \mu B/c$):

$$\Gamma_2^2 \sim \frac{2}{\pi}\hbar\omega\frac{\hbar}{\tau} = \frac{2}{\pi}(\hbar\omega)^2\frac{1}{\omega\tau} \,. \tag{7}$$

In the highest-mobility samples used in QHE experiments $\mu \sim 10^6$ cm ^{2}/Vsec $= 3 \cdot 10^8$ in gaussian units. For $B \sim 10$ T $= 10^5$ gauss, $\omega\tau \sim 10^3$ and hence $\Gamma_2 \sim 0.02\hbar\omega$.

Within the metallic manifold, the *electric* field F need not vanish – even if we neglect completely the width Γ_2. Indeed, it is the *total* single-electron energy ψ which is tied to the flat Fermi level. As mentioned at the beginning of Sect. 2, the local value of ψ includes besides the *electrostatic* potential ϕ also the kinetic energy $mv_H^2/2 \equiv (mc^2/2B^2)(\nabla\phi)^2$. Thus, in the absence of broadening ϕ must satisfy within the metallic region a nonlinear equation of the form $e\phi + (mc^2/2B^2)(\nabla\phi)^2 = \psi = E_F$. Equivalently, there is a relation for the electric field:

$$e\,\mathbf{F} = \frac{mc^2}{2B^2}\,\nabla(F^2)\,. \tag{8}$$

It appears that for a finite (though small) broadening, the presence of an electric field in the metallic phase is *necessary* in order to balance the diffusion flux of electrons between regions of different partial Landau-level filling.

4.2 *Fiber Inductance and Finite-Frequency Effects*

On a QHE plateau the entire externally applied voltage drops on the global fibers and drives their non-dissipative Hall current. There is a finite kinetic energy associated with the Hall current, $E = mv_H^2/2$ per electron. Because of the electron inertia it takes a finite time to charge and discharge the global fibers. The simplest way to describe these processes is to regard the fiber as an *inductive* impedance which is being charged through the Hall resistance R_{xy}.

Consider a strip of thickness ℓ along a global equipotential of length Λ. The number of filled global fibers in this strip equals $\bar{n}N_1 \cdot \ell\Lambda = \bar{n}\Lambda/2\pi\ell$. The strip contributes a Hall current

$$I_s = \frac{e\,v_H}{\Lambda} \times \frac{\bar{n}\Lambda}{2\pi\ell} \tag{9}$$

The associated kinetic energy of electrons in the strip is given by

$$E_s = \frac{m}{2}v_H^2 \times \frac{\bar{n}\Lambda}{2\pi\ell} \equiv \frac{1}{2}L\,I_s^2\,, \tag{10}$$

whence the "fiber inductance" L is of the form

$$L = \frac{2\pi\ell\,m}{\bar{n}\,e^2}\,\Lambda \equiv \tau R_{xy}\,; \qquad \tau = \frac{m\,\ell\,\Lambda}{\hbar}\,. \tag{11}$$

For $\ell \sim 100$ Å, a delay $\tau \sim 1\,\mu$sec corresponds to $\Lambda \sim 10$ cm. Note that a global fiber because of its wiggly nature can be longer than the circumference of the sample. The reactive delay (11) can be expected to be dominant in high-frequency measurements of the conductivities from I-V characteristics.

4.3 *Magnetic Field Induced Threshold Shift in Silicon MOSFETs*

Let us first briefly discuss the MOSFET capacitance in the absence of magnetic field. The typical dependence of the differential capacitance on the gate voltage V_G is shown in the top insert to Fig. 6. Below threshold the minimum capacitance is determined by the combined thicknesses of the oxide and the semiconductor depletion region weighted by the respective permittivities. At high V_G an inversion layer appears at the

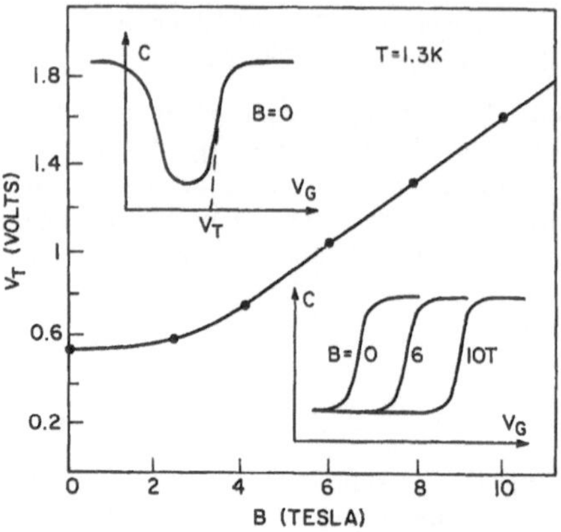

FIGURE 6: Magnetic-field induced shift of the capacitance threshold in Si MOSFET (Kazarinov and Luryi, 1982, unpublished). Experimentally [10], for $B \gtrsim 3$T the threshold is shifted linearly with B at the rate 143 mV/T. Taking the quoted value $d = 2900$ Å, formula (12) gives a slope 158 mV/T.

silicon/oxide interface, which screens further penetration of the electric field into the semiconductor. In this case the differential capacitance is determined by the oxide thickness only. The gate to channel capacitance per unit area in strong inversion is given by $\mathrm{d}(eN)/\mathrm{d}V_G = \epsilon/4\pi d$.

The nature of capacitance threshold in strong magnetic fields is quite different. Firstly, we note that it occurs when the MOSFET is already in the strong inversion. Therefore, the threshold is associated not with a screening effect but rather with the *conductivity* of the inversion layer. In a strong magnetic field the threshold voltage V_T was found [10] to increase linearly with B, see Fig. 6. This phenomenon is another aspect of the metal-insulator transition which occurs when the *sea* corresponding to a filled Landau level breaks into disjoint *lakes*. We know that at low temperatures the partially filled Landau levels experience a percolation transition at half filling. At the lowest Landau level this transition involves the vanishing of both the longitudinal and Hall conductivities, i.e., vanishing of the total current. In this situation the inversion layer does not respond to an AC signal on the gate (of course, it does respond to DC variations in V_G). You simply cannot quickly charge the disconnected lakes by a generation or a hopping current.

The transition occurs at a critical value $N_{cr} \approx N_1/2$, The threshold voltage $V_T = V_G(N_{cr})$ is, therefore, shifted by the magnetic field as follows:

$$\frac{\mathrm{d}V_T}{\mathrm{d}B} = \frac{\mathrm{d}V_G}{\mathrm{d}N}\frac{\mathrm{d}(N_{cr})}{\mathrm{d}B} = \frac{e^2}{\hbar c}\frac{d}{\epsilon} = \frac{d}{137\epsilon} . \tag{12}$$

This agrees with the experimentally measured slope [10] to better than 10%.

Acknowledgement

The percolation picture of the quantum Hall effect in the form presented here has been developed in 1981-1982 in collaboration with R. F. Kazarinov, who had originated many of the ideas described above. I am grateful to him for numerous invaluable discussions.

References

1. D.C. Tsui and S.J. Allen: Phys. Rev. B24, 4082 (1981)
1a. Y. Ono in: Anderson Localization, ed. by Y. Nagaoka and H. Fukuyama, Springer, Berlin, Heidelberg, New York (1982)

2. S. V. Iordansky: Solid State Comm. *43*, 1 (1982)
3. R. F. Kazarinov and S. Luryi: Phys. Rev. B*25*, 7626 (1982)
4. S. Luryi and R. F. Kazarinov: Phys. Rev. B*27*, 1386 (1983)
5. S. A. Trugman: Phys. Rev. B*27*, 7539 (1983).
6. R. Peierls: *Surprises in Theoretical Physics*, Chap. 4 (Princeton University Press, Princeton, New Jersey 1979)
7. B. I. Shklovskii and A. L. Efros: *Electronic Properties of Doped Semiconductors*, Springer Ser. Solid-State Sci., Vol. 45 (Springer, Berlin, Heidelberg 1984)
8. S. Luryi and K. K. Ng: (1982), unpublished
9. T. Ando, A. B. Fowler, and F. Stern: Rev. Mod. Phys. *54*, 437 (1982)
10. M. Kaplit and J. N. Zemel: Phys. Rev. Lett.*21*, 212 (1968)

Electrons in a Random Potential and Strong Magnetic Field: Lowest Landau Level

F. Wegner

Institut für Theoretische Physik, Ruprecht-Karls-Universität, D-6900 Heidelberg 1, Fed. Rep. of Germany

In this review the properties of two-dimensional independent electrons in a strong perpendicular magnetic field and a random potential are considered in the lowest Landau level. For the density of states of point scatterers an exact expression exists. For the d.c. conductivity and the inverse participation ratio series expansions in the Green's functions are available. They yield an estimate for the d.c. conductivity in the band centre and for the exponent which describes the vanishing of the inverse participation ratio in the band centre.

INTRODUCTION

The quantized Hall effect has attracted much theoretical interest in the properties of two-dimensional disordered materials subject to a strong perpendicular magnetic field. Calculations of the properties of electrons in an impurity potential are particularly easy for the lowest Landau level, since the propagator of free electrons in this level is a Gaussian. An expansion of the ensemble-averaged Green's functions in powers of the strength of the impurity potential yields a diagrammatic expansion which for uncorrelated point scatterers and for the more general case of Gaussian correlated random potentials reduces to integrals over Gaussians and thus to the evaluation of determinants. Moreover, in the special case of the one-particle Green's function in uncorrelated potentials the whole series can be completely summed, so that an exact expression for the density of states can be given. This will be discussed in the first section.

Quantities like the d.c. conductivity and the inverse participation ratio can be deduced from the two-particle Green's function. Although no closed expressions are available for these quantities, they can also be expressed in terms of diagrams. It is particularly useful to express them as power series of the Green's functions. By means of Borel-Padé techniques one is able to obtain estimates for the d.c. conductivity and the behaviour of the participation ratio in the centre of the band (second section).

This is an updated version of a contribution given in /1/. The reader is also referred to the review by Hikami /2/.

1. DENSITY OF STATES

1.1 Free Electron in a Magnetic Field

A free two-dimensional electron in a magnetic field is governed by the Hamiltonian

$$H_0 = \frac{1}{2m} (p - \frac{e}{c} A)^2. \tag{1}$$

In the symmetric gauge the vector potential reads $\vec{A} = B(-y,x)/2$. The cyclotron frequency is given by $\omega = eB/mc$, the cyclotron length by $\ell = (\hbar c/eB)^{1/2}$, and the degeneracy (density of states) by $\rho_0 = 1/(2\pi\ell^2)$. The energies of the Landau levels are $\hbar\omega(n+1/2)$, n=0,1,2,... . From now on units $\hbar\omega$=1 and ℓ=1 are used.

Introducing the complex coordinate $\zeta = x + iy$ the eigenstates of the lowest Landau level can be written

$$\varphi_{0,m}(r) = (2^{m+1}\pi m!)^{-1/2}\zeta^m \exp(-\zeta\zeta^*/4), \quad m=0,1,2,... \tag{2}$$

Denoting the projector on the lowest Landau level by P_0 we introduce the unperturbed Green's function of this level

$$G_0(r,r',z) := \langle r|P_0\,(z-H_0)^{-1}\,P_0|r'\rangle = (z-1/2)^{-1}\,C_0(r,r') \quad \text{with} \tag{3}$$

$$C_0(r,r') = \langle r|P_0|r'\rangle = \sum_m \varphi_{0,m}(r)\,\varphi^*_{0,m}(r')$$

$$= (2\pi)^{-1}\exp(-(\zeta\zeta^* + \zeta'\zeta'^* - 2\zeta\zeta'^*)/4). \tag{4}$$

1.2 Electron in a White Noise Potential /3/

Next we add a white noise potential to the Hamiltonian

$$H = H_0 + V(r)$$

$$\overline{V(r)} = 0; \quad \overline{V(r)V(r')} = W\delta(r-r'). \tag{5}$$

This potential is spatially uncorrelated and Gaussian distributed. It describes a situation where there are many uncorrelated scatterers within the distance of the cyclotron length. Provided the broadening of the Landau levels is small in comparison to $\hbar\omega$, the averaged Green's function

$$G(r,r',z) = \langle r|P_0\overline{(z-H_0-P_0VP_0)^{-1}}P_0|r'\rangle \tag{6}$$

is a good approximation of the full Green's function (without the projectors P_0) for $z\approx 1/2$. Expansion of G in powers of V yields

$$G(r,r',z) = \sum_{m=0}^{\infty} \langle r|P_0(z-H_0)^{-1}\overline{(P_0VP_0(z-H_0)^{-1})^m}P_0|r'\rangle. \tag{7}$$

From this expression diagrams are obtained by connecting the factors V pairwise in all possible ways. Then

$$G(r,r',z) = \sum_{\text{diagrams}} \frac{(W/2\pi)^n}{(z-1/2)^{2n+1}\lambda}\,C_0(r,r'), \tag{8}$$

where C_0/λ is the integral of 2n+1 factors C_0 over n intermediate points of scattering. λ is given by the number of Euler trails through the graph which is obtained when the pairwise connections are contracted to points /3/. To my present knowledge this was first shown in /4/. One observes /3/ that the graphs are those of a zero-dimensional ϕ^4 theory, since λ/m diagrams correspond to a graph with symmetry factor m yielding

$$G(r,r',z) = C_0(r,r')\sum_{\text{graphs}} \frac{(W/2\pi)^n}{(z-1/2)^{2n+1}m} = C_0(r,r') \text{ is } \langle\phi^*\phi\rangle. \tag{9}$$

Here s = - sign Im z and the expectation value has to be evaluated for the Hamiltonian $\mathcal{H} = is(z-1/2)\phi^*\phi + W(\phi^*\phi)^2/8\pi$. The density of states is obtained from the imaginary part of the Green's function, which yields

$$\rho(E) = (2\pi^2)^{-1}\, d\chi/dE \tag{10}$$

$$\tan\chi = \frac{2}{\sqrt{\pi}} \int_0^{(E-1/2)(2\pi/W)^{1/2}} \exp(\eta^2)d\eta . \tag{11}$$

This technique can also be used for three-dimensional systems. Then the evaluation of the averaged Green's function reduces to a one-dimensional problem which can be handled by transfer operator methods /5/.

1.3 General Point Scatterers /6/

Brézin, Gross, and Itzykson /6/ have generalized this solution to the case of general spatially uncorrelated scatterers. This is a good approximation for potentials whose range is small in comparison to the cyclotron length. The ensemble of potentials is described by a characteristic function $g(\alpha)$

$$\overline{\exp(-i\int d^2r\, \alpha(r)\, V(r))} = \exp(\int d^2r\, g(\alpha(r))). \tag{12}$$

Then (9) applies with

$$\mathcal{H} = is(z-1/2)\phi^*\phi - 2\pi h(\phi^*\phi/2\pi) \tag{13}$$

where

$$h(\alpha) = \int_0^{\alpha} \frac{d\beta}{\beta}\, g(\beta). \tag{14}$$

For the white noise potential one has $g(\alpha) = -W\alpha^2/2$, $h(\alpha) = -W\alpha^2/4$.

For scatterers $V(r) = \lambda \sum_i \delta(r-r_i)$ of density $\rho_0 f$, $f<1$, a portion $(1-f)$ of the eigenstates is unshifted. See figures in /6/. The authors have also pointed out that the problem can be represented by a supersymmetric theory which yields the reduction from two to zero dimensions.

One can also perform expansions for Gaussian correlated random potentials /7,8/. In this case no closed form expression is available. We will return to these potentials in section 2.

2. D.C. CONDUCTIVITY AND PARTICIPATION RATIO

2.1 Diagrammatic Expansion

While the density of states is obtained from the one-particle Green's function one has to consider the averaged two-particle Green's functions in order to obtain the diffusion constant and via the Einstein relation the d.c. conductivity, as well as the inverse participation ratio. For these quantities no closed form results are available. Since, however, the two-particle Green's function is obtained as integral over Gaussians as long as one restricts oneself to the lowest Landau level, perturbation expansions can be given.

In the case of an electron in a white noise random potential one obtains an expression similar to (8) for the two-particle Green's function /9/

$$\overline{\langle r_1|(z_+-H)^{-1}|r_1'\rangle\langle r_2|(z_--H)^{-1}|r_2'\rangle}$$

$$= \sum_{\text{diagrams}} (W/2\pi)^n(2\pi)^{-2}(z_+-1/2)^{-n_1}(z_--1/2)^{-n_2}(\lambda+\lambda')^{-1} \quad (15)$$

$$\cdot\exp[-\frac{1}{4}\sum\zeta_i\zeta_i^* + \frac{\lambda}{2(\lambda+\lambda')}(\zeta_1\zeta_1'^*+\zeta_2\zeta_2'^*) + \frac{\lambda'}{2(\lambda+\lambda')}(\zeta_1\zeta_2'^*+\zeta_2\zeta_1'^*)],$$

where $\lambda(\lambda')$ is the number of Euler trails if r_1' is connected with r_2 (r_1). Thus for single diagrams it is sufficient to determine λ and λ'.

2.2 Skeleton Diagram Expansion for the D.C. Conductivity

The diagrammatic sum (15) is an expansion in powers of $W^{1/2}(z_+-1/2)^{-1}$ and $W^{1/2}(z_--1/2)^{-1}$. Instead one can perform an expansion in powers of the one-particle Green's functions /10/

$$G(r,r',z_\pm) = \Gamma_\pm\, C_0(r,r'), \quad (16)$$

that is in powers of $W^{1/2}\Gamma_\pm$. This has two advantages: (i) the number of contributing diagrams, called skeleton diagrams, is smaller, (ii) Γ is a better expansion parameter, since it stays finite in the band centre, whereas $(z-1/2)^{-1}$ diverges. With

$$z_\pm = E \pm i\eta/2 \quad (17)$$

one may write

$$\Gamma_+ = \Gamma_-^* = Ce^{i\theta} = 2\pi/(A_1 \pm iA_2). \quad (18)$$

In the hydrodynamic limit $\eta \to 0$, $q \to 0$ the two-particle Green's function

$$K(q,E,\eta) := \int d^2r' \overline{\langle r|(z_+-H)^{-1}|r'\rangle\langle r'|(z_--H)^{-1}|r\rangle} e^{iq(r-r')} \quad (19)$$

behaves like

$$K(q,E,\eta) = \frac{i(\Gamma_+-\Gamma_-)}{\pi(\eta+Dq^2)} = \frac{\Gamma_+\Gamma_-A_2}{\pi^2(\eta+Dq^2)} \quad (20)$$

where D is the diffusion coefficient. From this one obtains

$$\frac{\pi}{A_2}(\eta+Dq^2+\dots) = \frac{\Gamma_+\Gamma_-}{\pi K} = 1 - \frac{q^2}{2} \pm \dots \quad (21)$$

which allows an expansion in powers of $x = 2\pi Wc^2$,

$$\frac{\pi\eta}{A_2} = 1 - x - \frac{x^2}{2}\frac{\sin 3\theta}{\sin\theta} + O(x^3) \quad (22)$$

$$\frac{2\pi D}{A_2} = 1 - \frac{x^2}{2}\cos 2\theta + O(x^3). \quad (23)$$

Hikami /10/ has performed this expansion to order x^6, more recently Singh and Chakravarty /8/ continued to order x^9. The series (22) is Borel summable, probably the same is true for (23). The series are evaluated by means of a Borel-Padé transformation. They are written

$$f(x) = \Sigma\, a_n x^n = \int_0^\infty e^{-z/x} g(z)/x dz \tag{24}$$

$$g(z) = \Sigma\, a_n z^n/n! = \sum_{i=0}^{k} b_i z^i / \sum_{j=0}^{\ell} c_j z^j + O(z^{k+\ell+1}). \tag{25}$$

The Borel-Padé result from (22) can be checked against the exactly known result, in particular for energies at the real axis, $\eta=0$. Equation (23) then determines the diffusion constant. Reference /10/ reports via the Einstein relation $\sigma_{xx} = e^2\rho D$ a value of $\sigma_{xx} = 1.4e^2/2\pi^2\hbar$, ref. /8/ $\sigma_{xx} = 1.43e^2/2\pi^2\hbar$ at the band centre.

A similar calculation can be performed for spatially correlated potentials /7,8/. In this case (22) is used to determine x for the band centre. Reference /8/ reports $\sigma_{xx} = 1.03e^2/2\pi^2\hbar$ if the correlation length for the potential equals the cyclotron length.

2.3 Participation Ratio

The inverse participation ratio

$$P = \sum_x \overline{|\psi_i(x)|^4} \tag{26}$$

can be expressed as

$$P = \lim_{\eta \to 0} \overline{(\eta \langle r|(z_+ - H)^{-1}|r\rangle\langle r|(z_- - H)^{-1}|r\rangle)} \tag{27}$$

with $z_\pm = E \pm i\eta/2$. Using the expansion for K and for η Hikami has given an expansion for P

$$P = 1 - \frac{1}{2}x - (\frac{5}{12} + \frac{1}{3}\cos 2\theta)x^2 + O(x^3) \tag{28}$$

up to order x^9. Assuming that P vanishes at the band centre with a power law like $|E-1/2|^{\pi_2}$ he estimates $\pi_2=3.8$.

In summary, these expansions are useful to obtain information on the Anderson localization in the lowest Landau level of systems in strong magnetic fields.

1. F. Wegner: In 2. DFG-Rundgespräch über den Quanten-Hall-Effekt (Schleching), PTB- und Universität Köln Preprint, p.67 (1985)
2. S. Hikami: Prog. Theor. Phys. Suppl. 84, 120 (1985)
3. F. Wegner: Z. Phys. B51, 279 (1983)
4. G.P. McCauley, D.J. Thouless: J. Phys. F6, 109 (1976)
5. H. Homeier: Effektive Feldtheorie für das unterste Landau-Niveau in einem dreidimensionalen Zufallspotential, Diplomarbeit Heidelberg (1986)
6. E. Brézin, D. Gross, C. Itzykson: Nucl. Phys. B235 [FS11], 24 (1984)
7. S. Hikami, E. Brézin: J. Physique 46, 2021 (1985)
8. R.P.R. Singh, S. Chakravarty: Nucl. Phys. B265 [FS15], 265 (1986)
9. T. Streit: J. Physique Lett. 45, 713 (1984)
10. S. Hikami: Phys. Rev. B29, 3726 (1984)
11. S. Hikami: Prog. Theor. Phys. and preprint (1986)

Calculation of the Conductivity of Two-Dimensional Disordered Systems in the Presence of a Strong Magnetic Field

B. Kramer

Physikalisch-Technische Bundesanstalt, Bundesallee 100,
D-3300 Braunschweig, Fed. Rep. of Germany

1. INTRODUCTION

Due to the discovery of the quantized Hall effect (QHE) [1] the interest in the theoretical calculation of the transport properties of inversion layers in strong magnetic fields has increased considerably during recent years [2]. Of particular interest is the inclusion of a random potential.

The starting point is linear response theory in which the conductivity tensor σ is determined from a linear relationship between the current density $\vec{j}$ and the electric field $\vec{E}$

$$\vec{j} = \sigma\vec{E}\,. \tag{1}$$

For an isotropic 2D system in a magnetic field σ is a 2x2 tensor of the form

$$\sigma = \begin{pmatrix} \sigma_{xx} & \sigma_{xy} \\ -\sigma_{xy} & \sigma_{xx} \end{pmatrix}; \tag{2}$$

σ_{xx} is the magneto-conductivity which describes the response parallel to the field. The Hall conductivity σ_{xy} describes the response perpendicular to the field. σ_{xx} is related to dissipative transport, whereas σ_{xy} is associated with dissipationless transport.

σ_{xx} and σ_{xy} are material parameters. They are defined in the thermodynamic limit. Boundary or surface effects are excluded by this condition, as well as contact effects. Since linear response theory describes a stationary state the non-stationary fluctuations of the current related to switching on the electric field are also excluded. Microscopically, σ may be calculated by using Kubo's theory [3]. For an ideal 2D system of noninteracting electrons, a Landau system without disorder, one obtains

$$\begin{aligned} \sigma_{xx} &= 0 \\ \sigma_{xy} &= -\frac{ne}{B}\,. \end{aligned} \tag{3}$$

2. SURVEY OF THE STATUS OF THE FIELD

A number of attempts have been made in order to include the effect of disorder:

(i) Ando and Uemura have developed the selfconsistent Born approximation (SCBA) for the calculation of σ in the presence of short-range scatterers [4]. The most important result of these calculations was that the peak values of σ_{xx} (near the centres of the Landau bands) are given by

$$\sigma_{xx} = \frac{e^2}{\pi^2 h}(N+½), \tag{4}$$

whereas the Hall conductivity is

$$\sigma_{xy} = -\frac{\mathrm{Im}\Sigma(E_F)}{h\omega_c}\sigma_{xx} . \tag{5}$$

Here N is the number of the Landau band, $h\omega_c$ the cyclotron energy, E_F the Fermi Energy, and $\Sigma(E_F)$ the self-energy due to the random scatterers. These calculations do not yield the plateaus in σ_{xy} which are observed in the quantized Hall effect. However, if E_F lies in the gap between two Landau bands, $\sigma_{xy}=-Ne^2/h$. These values are not affected by the presence of the impurities.

(ii) Aoki and Ando have argued that σ_{xy} is quantized in units of e^2/h in the presence of localized states in the tails of the Landau bands provided the afore-mentioned "special values" of σ_{xy} exist [5]. Their argument is based on the center-of-motion approximation developed by Kubo et al [3] which was also used in the SCBA, and which should be a good approximation in a strong magnetic field. An important conclusion of this work was that <u>not</u> all of the states can be localized if the QHE exists.

(iii) Ono generalized the self-consistent perturbation theory for the density response function of Vollhardt and Wölfle to the case of strong B [6]. The main result of this work was that the localization length diverges near the centres of the Landau bands, E_N, with an essential singularity

$$\sim \exp\left(\frac{\text{const}}{(E-E_N)^2}\right) . \tag{6}$$

Unfortunately, it was not possible to calculate σ_{xy} within this approach.

(iv) For finite systems with disorder, i.e. containing short range or long-range scatterers σ_{xy} and σ_{xx} were determined numerically by Ando using again the center-of-motion approximation. In this work periodic boundary conditions were applied to the system [4]. These calculations establish one of the first explicit indications for the existence of Hall plateaus in the regions of localized states. In these calculations the plateaus in σ_{xy} must be interpreted as a volume effect.

(v) Halperin pointed out that edge states might play an important role in elucidating certain aspects of the transport properties in the quantum Hall regime [7]. Aoki demonstrated numerically the close relationship of the quantized values of σ_{xy} and a gauge transformation [8] for a system with edges. The importance of the edge states was further evaluated and related to topological properties by Hajdu in the limit of strong B [9].

(vi) The Hall conductivity for strip-like disordered systems subject to Dirichlet boundary conditions was calculated recursively starting from the Kubo formula [10]. An attempt to perform the thermodynamic limit was made. It was shown that in this case only the states near the Fermi energy contribute to σ_{xy} for $T \to 0$. Thus, the existence of the plateaus in σ_{xy} is again related to the presence of boundary-induced states.

All of the above approaches yield results which are consistent with the essential fact of the QHE, namely that σ_{xx} vanishes in the plateau regions of σ_{xy}. Each of the methods is approximate. In particular, most of them are applied consistently only to one of the components of the conductivity (except (i) and (iv)). The most severe restriction in the numerical calculations is the fact that it is extremely difficult to perform the thermodynamic limit. In order to achieve this it is necessary to establish a scaling theory, as was successfully done for disordered systems without magnetic field [11]. So far the only attempt which treats both components of the conductivity on an equal footing is by A. M. M. Pruisken starting from the non-linear σ-model [12]. This theory is still to be verified numerically. In order to perform such an explicit verification one has to define suitable scaling parameters.

3. THE KUBO-FORMULA

Starting from linear response theory the components of σ may be written as [13]

$$\sigma_{\alpha\beta}(T) = \int_{-\infty}^{+\infty} dE\, f_T(E)\, \sigma_{\alpha\beta}(E) \qquad (\alpha, \beta = x, y) \tag{7}$$

$$\sigma_{\alpha\beta}(E) = \lim_{\eta \to 0} \lim_{F \to \infty} \sigma_{\alpha\beta}(E; \eta, F) \tag{8}$$

$$\sigma_{\alpha\beta}(E; \eta, F) = \frac{e^2\hbar}{2\pi F} \mathrm{Tr} \left\{ j_\alpha \frac{dG^-}{dE} j_\beta \,\mathrm{Im}G^+ - j_\alpha \,\mathrm{Im}G^+ j_\beta \frac{dG^+}{dE} \right\} . \tag{9}$$

Here $j_{\alpha,\beta}$ are the components of the current operator, F the area of the system, and $f_T(E)$ the Fermi distribution. The resolvents $G^\pm$ are defined as usual

$$G^\pm = (E \pm i\eta - H)^{-1} , \tag{10}$$

where H is the Hamiltonian of the system, for instance

$$H = \frac{1}{2m} (\vec{p} - e\vec{A})^2 + V(\vec{r}) \tag{11}$$

in the case of a free electron subject to a vector potential A and a random potential V(r).

The occurrence of the random potential in eq. (11) implies that in general $\sigma_{\alpha\beta}$ have to be configurationally averaged, i.e. products of one-electron propagators have to be averaged. This introduces a complication into the problem which is comparable to the one which originates in interaction effects. Assuming that the conductivity is self-averaging the configurational average can be omitted if the thermodynamic limit is considered.

A form of the Kubo formula which is especially useful in numerical calculations for a disordered system has been obtained by MacKinnon [14] and MacKinnon et al [15]. The generalization of this idea to the present problem was done by MacKinnon [16] and by Schweitzer et al [10]. The essential idea is to replace the current operators in eq. (9) by the position operator via

$$i\hbar\vec{p} = \left[G^{\pm}, \vec{r}\right] , \tag{12}$$

which is justified for any finite system without periodic boundary conditions. The result is

$$\sigma_{\alpha\beta}\ (T=0) = \frac{e^2}{h} \lim_{\eta\to 0} \lim_{F\to\infty} S(E;\eta,F) \tag{13a}$$

$$S_{\alpha\beta}(E;\eta,F) = \frac{1}{F} \mathrm{Tr} \left\{-4\eta^2 \beta G^+ \alpha G^- + 2i\eta(G^+ - G^-)\alpha\beta\right\} . \tag{13b}$$

It is important to note that in this formula only the states at the Fermi energy contribute to the conductivity at T=0. Thus, one cannot use it for the Landau system. However, if we consider a system with boundaries, it is possible to obtain non-vanishing components of the conductivity, due to the presence of boundary-induced states. The advantage of eqs. (13) is that it may be used recursively, i.e. that for a strip-like system

$$S_{\alpha\beta}(E;\eta,M(L+\Delta L)) = f(S_{\alpha\beta}(E;\eta,ML) , \tag{14}$$

where L and M denote the length and the width of the strip respectively. ΔL is the increase in the length of the system within one step of the recursion. Equation (14) means that the conductivity of the system of length L+ΔL may be calculated directly as a function of the conductivity of the system of length L. Thus, the thermodynamic limit may be performed in a systematic manner for any given width M.

Equations (13) have been applied to disordered systems without magnetic field [14,15]. In this case there are only diagonal elements. For a strip-like system in a perpendicular magnetic field described by a tight-binding model the Hall conductivity σ_{xy} has been calculated without disorder [10]. In this case the analytical results obtained earlier [17] for a 2D electron subject to a periodic potential could be reproduced. First results for a system containing in addition to the periodic a random potential have also been reported. These indicate that, consistent with the experimental observations and Ando's numerical results [4], the Hall conductivity above the centres of the magnetic subbands is higher than the classical value in the presence of disorder [10] (Fig. 1).

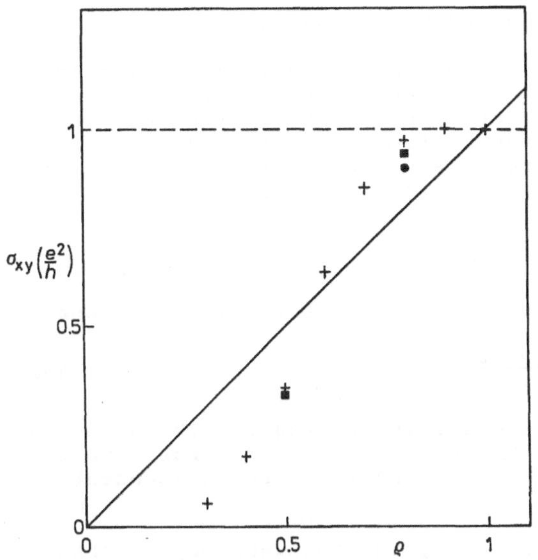

Fig. 1: σ_{xy} in units of e^2/h as a function of the filling factor ρ (in units of eB/ha^2) for a strip-like system described by a disordered tight-binding model (on a square lattice with lattice constant a) generalized to include a perpendicular magnetic field (Peierls substitution). The random potential (site energies in the tight-binding model) is taken according to a box distribution of the width W=0.5V where V is the hopping matrix element between nearest neighbors. The data points are (+) M=64, η=0.001; ■ M=48, η=0.001; ● M=48, η=0.005

4. AN EXAMPLE: THE HALL CONDUCTIVITY FOR BOUNDARY-INDUCED STATES

As mentioned above, the use of eqs. (13) suggests the interpretation of the plateaus in the QHE as a boundary effect. On the other hand, the afore-mentioned calculations by Ando (cf. section 2) show that the plateaus are related to bulk properties requiring in particular the existence of current carrying bulk states. For a better understanding of the quantum mechanical properties of the system it would certainly be useful to have a relation between the two descriptions. Pruisken [12] has obtained such a relation in a rather obscured manner by applying a generalization of Stoke's theorem. A more direct proof should be highly desirable. Can we learn something about this question by using the Kubo formula of eqs. (7) to (9) and/or eqs. (13) together with Landau states and boundary-induced states, respectively?

Ono has shown that eqs. (7) to (9) yield the classical result for σ_{xy} even in the case with a finite width in the x-direction and periodic boundary condition in the y-direction (see below) [22]. In this calculation the current flows within the bulk. Edge effects are negligible.

In this section I will concentrate on the edge aspect.

The application of eqs. (13) to the calculation of the conductivity is not straightforward because after having performed the limit $F\to\infty$ via $L\to\infty$ (M=const) one has to perform the limit $\eta\to 0$, and, in order to reach the 2D limit, to take $M\to\infty$. The calculations in the following are also intended to illustrate the problems which can arise in a numerical calculation.

4.1 The model

We consider a system described by a single particle resolvent

$$G^{\pm}(E) = \sum_{nk} \frac{|nk\rangle\langle nk|}{E\pm i\eta - E_{nk}} + \sum_{\nu} \frac{|\nu\rangle\langle\nu|}{E\pm i\eta - E_{\nu}} \quad ; \tag{15}$$

E_{nk}, $|nk\rangle$ are the energies and states related to the boundary of the system whereas E_{ν}, $|\nu\rangle$ correspond to disorder-induced states. We assume further, that E_{nk} and $|nk\rangle$ are independent of the disorder.

Examples of the states induced by the boundary may be obtained for a system which is finite in the x-direction, extending from $-L_x/2$ to $+L_x/2$, and subject to periodic boundary conditions in the y-direction, i.e.

$$\langle x = \frac{L_x}{2}, y|nk\rangle = \langle x = -\frac{L_x}{2}, y|nk\rangle = 0$$

$$\langle x, y+L_y|nk\rangle = \langle x, y|nk\rangle \ . \tag{16}$$

The state $|nk\rangle$ is taken to be of the form (in the Landau gauge such that $A=(0,Bx,0)$)

$$\psi_{nk}(x,y) = \langle xy|nk\rangle = \frac{1}{\sqrt{L_y}}\, e^{iky}\, \Phi_{nk}(x) \ , \tag{17}$$

where $\Phi_{nk}(x)$ is given by the solution of the differential equation

$$\frac{d^2\Phi_{nk}}{dx^2} - \left[\frac{1}{4}(x - x_k)^2 - E_{nk}\right] \Phi_{nk} = 0 \tag{18}$$

(Weber's differential equation [18]) and

$$x_k = k\ell \qquad ; \quad \ell^2 = \frac{2\hbar}{eB} \ . \tag{19}$$

Here, x and E_{nk} are in units of ℓ and $\hbar\omega_c = \hbar eB/m$, respectively. The properties of the spectrum and the corresponding wave functions are discussed explicitly in [18]. Qualitatively they are shown in Fig. 2 and Fig. 3.

For the discussion in the following we only need that

$$E_{nk} = E_{n-k} \tag{20}$$

$$\frac{dE_{nk}}{dk} = -\frac{dE_{n-k}}{dk} \tag{21}$$

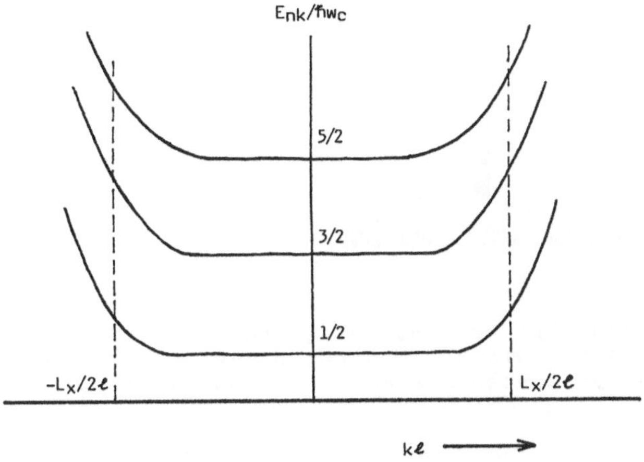

Fig. 2: Qualitative picture of the energy spectrum E_{nk} (in units of $\hbar w_c=\hbar eB/m$) of a system confined to $-L_x/2<x<L_x/2$

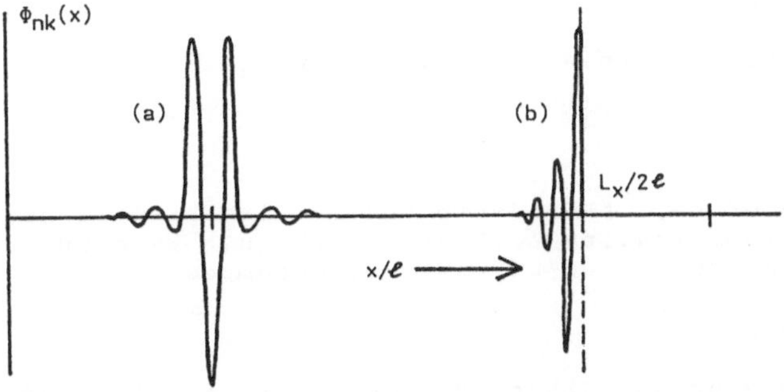

Fig. 3: Qualitative shape of the wave function corresponding to E_{nk} for (a) $|k\ell|<<L_x/2\ell$ (approximately Landau states), and (b) $|k\ell|>L_x/2\ell$ (approximately Weber functions localized at $L_x/2\ell$)

$$\int_{-\infty}^{+\infty} \Phi_{nk}^{*}(x)\ \Phi_{n'k}(x)\ dx = \delta_{nn'} \tag{22}$$

$$\int_{-\infty}^{+\infty} |\Phi_{n\pm k}^{*}(x)|^2\ x\ dx \approx \pm \frac{L_x}{2\ell} \text{ for } k > \frac{L_x}{2\ell} . \tag{23}$$

The last equation can certainly be justified for large B and energies well between the "bulk" energy levels 1/2, 3/2, 5/2, ... (in units of $\hbar w_c=\hbar eB/m$).

Furthermore, we take the plane wave part of $\psi_{nk}(x,y)$ to be normalized within the interval $[-L_y/2, L_y/2]$ such that for continuous k orthogonality is only guaranteed in the limit $L_y\to\infty$.

For the disorder-induced states we assume

$$\langle\nu|\nu'\rangle = \delta_{\nu\nu'} , \tag{24}$$

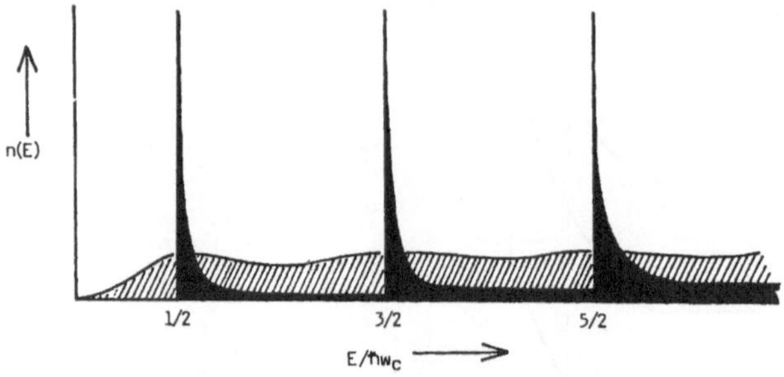

Fig. 4: Qualitative shape of the density of states of a system confined within a strip of width L_x in the presence of disorder. Full line: boundary-induced states, dotted area: disorder-induced states.

independent of the spatial extent of the system (localization in the sense of normalizability) and some (random) distribution of the energy levels E_ν.

The density of states has then the shape shown in Fig. 4.

4.2 The Hall conductivity

For the calculation of σ_{xy} the thermodynamic limit shall be performed by taking $L_y \to \infty$, leaving L_x finite, although we assume $L_x \gg \ell$. According to eqs. (13) and (15) we may decompose

$$S_{xy} = S^b_{xy} + S^d_{xy} + S^{db}_{xy} , \tag{25}$$

where $S^d{}_{xy}$ depends only on the disorder-induced states, $S^b{}_{xy}$ on the boundary-induced states, and $S^{db}{}_{xy}$ contains both, disorder induced and boundary-induced states.

Because of the condition eq. (24) the matrix elements in $S^d{}_{xy}$ and $S^{db}{}_{xy}$ are finite and independent of the size of the system in the limit $L_y \to \infty$, $L_x \to \infty$. Therefore, using the same arguments as previously for the case B=0 we obtain [14,15]

$$S^d_{xy} + S^{db}_{xy} = \mathrm{const}\ \eta . \tag{26}$$

The boundary-induced part consists of two terms

$$S^b_{xy} = S^{(1)}_{xy} + S^{(2)}_{xy} , \tag{27}$$

where

$$S^{(1)}_{xy} = \frac{1}{L_x L_y} \mathrm{Tr} \left\{ -4\eta^2 y G^+ x G^- \right\} \tag{28}$$

$$S^{(2)}_{xy} = \frac{1}{L_x L_y} \mathrm{Tr} \left\{ 2i\eta (G^+ - G^-) xy \right\} . \tag{29}$$

Using eq. (15) we obtain for $S_{xy}{}^{(2)}$

$$S_{xy}^{(2)} = \frac{4\eta}{L_x L_y} \sum_{nk} \frac{\eta}{(E-E_{nk})^2+\eta^2} \times$$

$$\times \frac{1}{L_y} \int_{-c}^{+c} dy\; y \int_{-\infty}^{-\infty} dx |\Phi_{nk}(x)|^2 x \tag{30}$$

where $c=L_y/2$. Since the integral over y vanishes $S_{xy}{}^{(2)}$ does not contribute to the Hall conductivity.

For the first term one obtains

$$S_{xy}^{(1)} = -\frac{4\eta^2}{L_x L_y} \sum_{nkk'} \frac{2i}{L_y^2} \frac{d}{d(k-k')} \left[\frac{\sin(k-k')\frac{L_y}{2}}{(k-k')}\right]^2$$

$$\times \frac{\int_{-\infty}^{+\infty} dx\; \Phi_{nk}^*(x)\; \Phi_{nk'}(x) \int_{-\infty}^{+\infty} \Phi_{nk}^*(x)\; x\; \Phi_{nk'}(x)\; dx}{(E-E_{nk}+i\eta)\;(E-E_{nk'}-i\eta)} . \tag{31}$$

In the limit $L_y\to\infty$ only the terms $k\approx k'$ contribute. Thus

$$\int_{-\infty}^{+\infty} dx\; \Phi_{nk}^*(x)\; \Phi_{nk}(x) = 1 \tag{32}$$

and using eq. (23)

$$\int_{-\infty}^{+\infty} dx\; \Phi_{nk}^*(x)\; x\; \Phi_{nk}(x) \approx \frac{L_x}{2} \operatorname{sgn}(k) . \tag{33}$$

Writing $q=k-k'$ and expanding the denominator for $|q|<<k$

$$\frac{1}{E-E_{nk-q}-i\eta} \approx \frac{1}{E-E_{nk}-i\eta+q\,a_{nk}} \approx \frac{1}{E-E_{nk}-i\eta}\left(1-\frac{a_{nk}\,q}{E-E_{nk}-i\eta}\right) \tag{34}$$

the k-, k'- sums may be evaluated.

One obtains for $L_y\to\infty$ (using eq. (33))

$$\lim_{L_y\to\infty} S_{xy}^{(1)} = -\frac{1}{2} \sum_n \frac{1}{\pi} \int_{\infty}^{\infty} dk \frac{\eta \operatorname{sgn}(k)}{(E-E_{nk})^2+\eta^2} \frac{i\eta\, a_{nk}}{E-E_{nk}-i\eta} . \tag{35}$$

In the limit $\eta\to 0$ we may write further

$$\lim_{\eta\to 0} \lim_{L_y\to\infty} S_{xy}^{(1)} = \sum_{nk_0} \frac{1}{2} \frac{a_{nk_0}}{|a_{nk_0}|} \operatorname{sgn}(k_0) . \tag{36}$$

Because of eq. (21) this yields

$$\lim_{\eta\to 0} \lim_{L_y\to\infty} S_{xy}^{(1)} = \sum_n (\text{occupied bands}) . \tag{37}$$

Thus, for energies well between the "bulk" bands (flat regions of E_{nk}) the result for σ_{xy} is

$$\sigma_{xy} = i\,\frac{e^2}{h} + \mathrm{const}\;\eta \quad ; \quad i = 0,\ 1,\ 2,\ 3,\ \ldots\ . \tag{38}$$

The second term which stems from the disorder-induced states vanishes in the limit of $\eta \to 0$. The result eq. (38) means that in a numerical calculation of σ_{xy} the deviations from the integer values due to finite η must vanish linearly if the width of the system L_x is sufficiently large. This is consistent with the afore-mentioned numerical calculations.

The above arguments do not necessarily apply for energies near $E_{nk=0}$ since here we have a degeneracy in the limit $L_x \to \infty$. To what extend boundary-induced states of the form eq. (17) depend on a random potential has still to be investigated.

5. CONCLUSION

Although the various attempts to obtain quantitative results for the Hall conductivity of 2D disordered systems in a perpendicular magnetic field have been quite successful, the physical understanding of the conduction mechanism is still missing.

Localization theory which is per definition only valid in an <u>infinite</u> system requires at least one (extraordinarily) good conducting extended state within each of the magnetic subbands, in order to be able to account for the occurrence, and the height of the plateaus observed in the QHE. There are theoretical models which can account for such states [12] although their physical origin is somewhat obscure. It is hard to imagine an extended one-particle state in a disordered system which can carry a macroscopically large current. However, there are predictions from scattering theory which are consistent with such an idea [19].

In infinite systems which are confined in one direction, as the example discussed in this paper, the states which carry the current are associated with the boundary. These states must then be required to be extremely insensitive to the disorder. Although there is numerical evidence for this [20], a theoretical argument is still missing.

Considering a system which is confined in one direction causes yet another difficulty. As the boundary may be considered as an inhomogeneity, it is a priori not obvious whether or not the Kubo formula may be applied. Strictly speaking, the distribution of the electric field in the stationary state has to be calculated. This problem is the subject of the paper by Johnston and Schweitzer at this conference [21].

Summarizing one may state that there are essentially two questions which have to be answered before any progress can be made in the understanding of the QHE. They are possibly connected to each other. Firstly, what means localization in the presence of a magnetic field? There are good reasons to doubt the applicability of the theories of localization which have been developed for disordered systems without magnetic field [12]. Secondly, what is the nature of the current-carrying states? As mentioned above, in a system without boundaries these must be able to carry an increasing amount of current if the disorder increases. Such a behaviour is opposite to what one expects for "normal" extended states in disordered systems without magnetic field. Are these states related to those which we have introduced in this paper somewhat artificially via the boundary ? What is the physical mechanism in a system without boundaries which can produce such states? It is possible that their existence in a 2D system subject to a perpendicular magnetic field can be explained by topological arguments in the classical limit [9]. However, a physical argument for their existence in a quantum mechanical system is not yet at hand.

6. ACKNOWLEDGEMENTS

Useful discussions with Wolfgang Wöger, Ludwig Schweitzer and Robert Johnston are gratefully acknowledged. It is a particular pleasure for me to thank Yoshiyuki Ono for numerous critical remarks and constructive ideas.

7. REFERENCES

1. K. v. Klitzing, G. Dorda, M. Pepper: Phys. Rev. Letters 45, 437 (1982)
2. J. Hajdu, G. Landwehr: In Strong and Ultrastrong Magnetic Fields and Their Applications, ed. by F. Herlach, Topics in Appl. Phys., Vol. 57 (Springer, Berlin, Heidelberg 1985)
3. R. Kubo, S. J. Miyake, N. Hashitsume: In Solid State Physics, ed. by F. Seitz and D. Turnbell, Vol. 17, 269 (Academic Press, New York 1965)
4. T. Ando, Y. Uemura: J. Phys. Soc. Japan 36, 959 (1974); for a recent review see T. Ando: In Anderson Localization, ed. by Y. Nagaoka, Progr. Theor. Physics Suppl. 84, 69 (1985)
5. H. Aoki, T. Ando: Sol. State Commun. 38, 1079 (1981)
6. See the review article by Y. Ono: In Anderson Localization, ed. by Y. Nagaoka, Progr. Theor. Physics 84, 138 (1985) and J. Phys. Soc. Japan 51, 2342 (1984)
7. B.I. Halperin: Phys. Rev. B25, 5849 (1982)
8. H. Aoki: J. Phys. C15, L1227; C16, 1893 (1983)
9. J. Hajdu: In Two-Dimensional Systems: Physics and New Devices, ed. by G. Bauer, F. Kuchar, H. Heinrich, Springer Series in Sol. State Sciences 67, 228 (1986)
10. L. Schweitzer, B. Kramer, A. MacKinnon: Z. Phys. B59, 379 (1985)
11. A. MacKinnon, B. Kramer: Phys. Rev. Letters 47, 1546 (1981); Z. Phys. B53, 1 (1983); see also A. MacKinnon: In Localization, Interaction and Transport Phenomena, ed. by B. Kramer, G. Bergmann and Y. Bruynseraede, Springer Series in Solid State Sciences 61, 90 (Springer, Berlin, Heidelberg 1985)
12. A. M. M. Pruisken: In Localization, Interaction and Transport Phenomena, ed. by B. Kramer, G. Bergmann and Y. Bruynseraede, Springer Series in Solid State Sciences 61, 188 (Springer, Berlin, Heidelberg 1985); Nucl. Phys. B235 FS [11], 277 (1984)
13. A. Bastin, C. Lewiner, O. Betbeder-Matibet, P. Nozieres: J. Phys. Chem. Solids 32, 1811 (1971)

14. A. MacKinnon: J. Phys. C13, L1031 (1980); Z. Phys. B59, 385 (1985)
15. A. MacKinnon, B. Kramer, W. Graudenz: J. Physique, Colloque C4, Suppl. n°10, 42, C4-63 (1981)
16. A. MacKinnon: Z. Phys. B59, 385 (1985)
17. D.J. Thouless, M. Kohmoto, M.P. Nightingale, M. den Nijs: Phys. Rev. Letters 49, 485 (1982)
18. M. Heuser: Diplomarbeit (Köln 1972)
19. R.E. Prange: Phys. Rev. B23, 4802 (1981);
J. Chalker: J. Phys. C16, 4297 (1983);
W. Brenig: Z. Phys. B50, 305 (1983)
20. L. Schweitzer, B. Kramer, A. MacKinnon, J. Phys. C17, 4111 (1984)
21. R. Johnston, L. Schweitzer: This Volume
22. Y. Ono: Private communication

Quantum Hall Effect: From the Winding Number to the Flow Diagram

H. Aoki[1] *and T. Ando*[2]

[1]Institute of Materials Science, University of Tsukuba, Sakura, Ibaraki 305, Japan
[2]Institute for Solid State Physics, University of Tokyo, 7-22-1 Roppongi, Minato-ku, Tokyo 106, Japan

To answer a query as to what extent the quantised Hall effect is universal, we show that (i) an exact proof for the quantisation of Hall conductivity is given by the topological invariant (winding number), (ii) there exists a definite distribution of winding numbers as a function of energy for given system size and degree of disorder giving rise to a well-behaved σ_{xy} for lattice systems with Landau-level mixing and (iii) the flow diagram for σ_{xy} vs σ_{xx} obtained by the Thouless number exhibits characteristic behaviours.

1. INTRODUCTION

The quantised Hall effect is a remarkable probe of unusual intrinsic properties of the electronic structure in magnetic fields. The most major problem here is: (i) Is the quantised Hall effect exact, (ii) If so, what is the relation to the observable σ_{xy}, which is a continuous function of energy rather than a series of exact step functions. The purpose of the present article is to clarify this query in three steps: First, an exact proof of the quantised Hall effect in terms of a topological invariant in the wavefunction, second, the relation of the exact quantisation and the observable σ_{xy} over the whole energy spectrum, third, the σ_{xx}-σ_{xy} flow diagram for lattice systems with non-parabolic band effects and Landau-level mixing.

2. WINDING NUMBER

Making use of the Kubo formula/1,2/, we can express the Hall conductivity in a form

$$\frac{\langle \sigma_{xy} \rangle}{(e^2/h)} = \frac{1}{8\pi^2} \int_C dz \iint_0^{\phi_0/L} dA_x dA_y \, \mathrm{Tr} \left[G\frac{\partial G^{-1}}{\partial A_x} G\frac{\partial G^{-1}}{\partial A_y} G\frac{\partial G^{-1}}{\partial z} - (x \leftrightarrow y) \right],$$

Here G is the Green's function, L the system size, ϕ_0=hc/e the magnetic flux quantum and we assume two Aharonov-Bohm magnetic fluxes, Φ_x and Φ_y, which gives rise to an extra vector potential, $A = (\Phi_x/L, \Phi_y/L)$. The right-hand side of the above expression is just a topological invariant called Pontrjagin number/3/, which is the winding number of a mapping from (A,z) to the complex wavefunction. From this formula, we can conclude that the Hall conductivity is exactly quantised in units of e^2/h for the Fermi energy, E_F, in a localised regime in infinite systems or for any E_F in finite systems with a fixed number of electrons as is illustrated in Fig. 1.

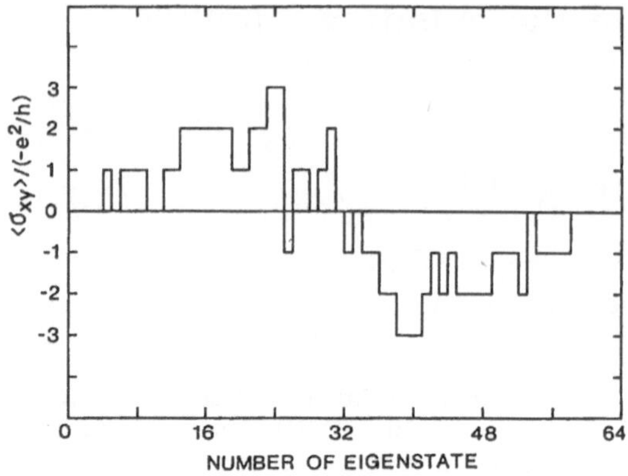

Fig. 1.
The Hall conductivity, σ_{xy}, averaged over A_x and A_y, is plotted for all the eigenstates in an 8×8 lattice with $Ha^2/\phi_0=1/8$ and W = 2.

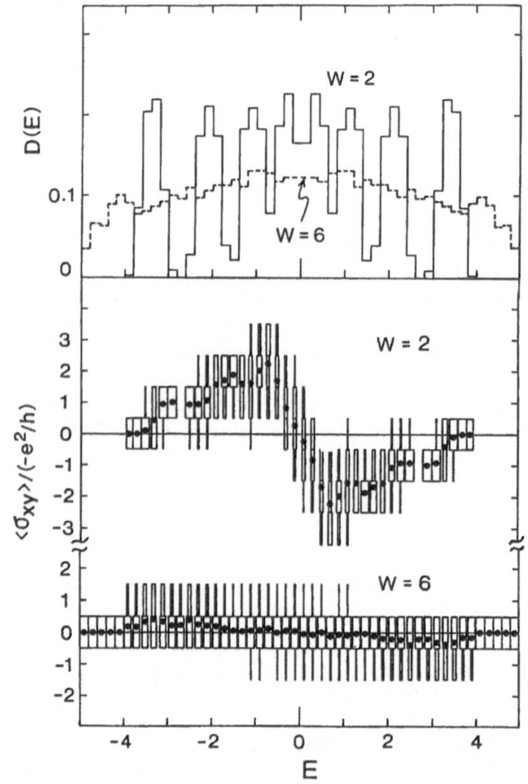

Fig. 2. The distribution (histogram) and the average (●) of the winding number are plotted for each energy bin together with the density of states for W=2 and W=6 with ensemble average over 36 samples, respectively. The magnetic field is specified by Ha^2/ϕ_0 (magnetic flux within a unit cell normalised by flux quantum) = 1/8 here with eight Landau levels. The site energy fluctuates with width W relative to transfer energy. The ratio of broadening and spacing of Landau levels, $\Gamma/\hbar\omega_c$, amounts to 0.18(0.55) for W=2(6) at the spectrum edges.

Although the winding number can vary from one eigenstate to another for individual samples, we can show that the winding number has a definite distribution function versus energy for given sample size and degree of randomness. Figure 2 shows the numerical result for finite lattice systems for varied degree of randomness. We have deliberately chosen the lattice system/4/, since we can study wide range of (a)magnitude of magnetic field, H, (b)degree of randomness, W and (c)effect of non-parabolic band. The distribution of the winding number is the key quantity, since, when σ_{xy} is averaged in each energy bin according to the distribution, a well-behaved σ_{xy} results. The averaged σ_{xy} exhibits a number of plateaux for small W with dips in between due to the Landau-level mixing, and suppressed values for larger W with small $\omega_c\tau$. It can be shown that the ensemble-averaged winding number has virtually the same energy dependence as the ensemble-averaged σ_{xy} for a fixed flux (A=0) with the difference between the two quantities vanishing with increasing system size.

The diagonal conductivity, σ_{xx}, is also closely related to the dependence of the system on the AB flux, and σ_{xx} is obtained from the Thouless number via the A-dependence of energy levels/5,6/. The behaviour of both σ_{xx} and σ_{xy} depends strongly on the localisation length determined by W and the sample size.

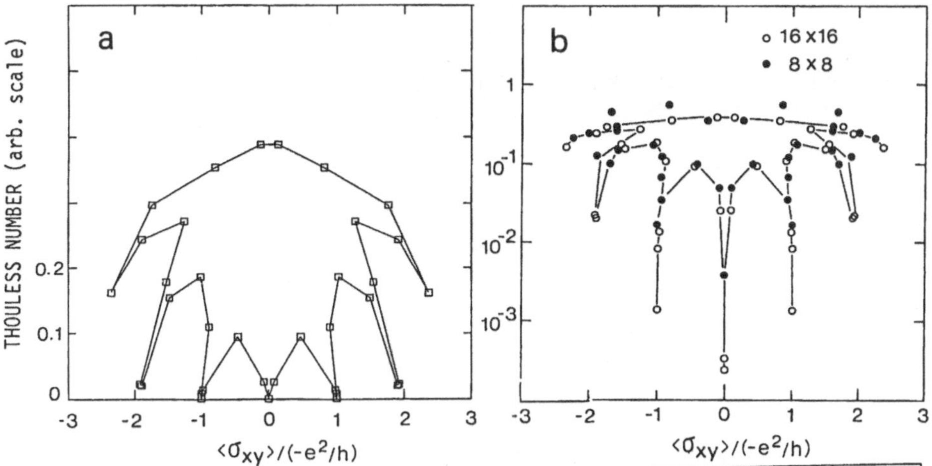

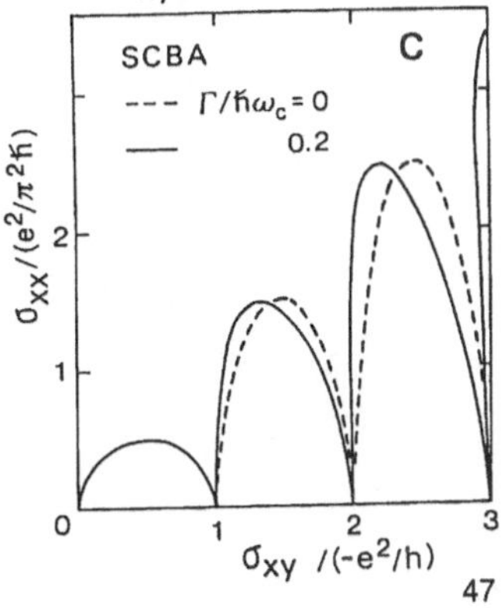

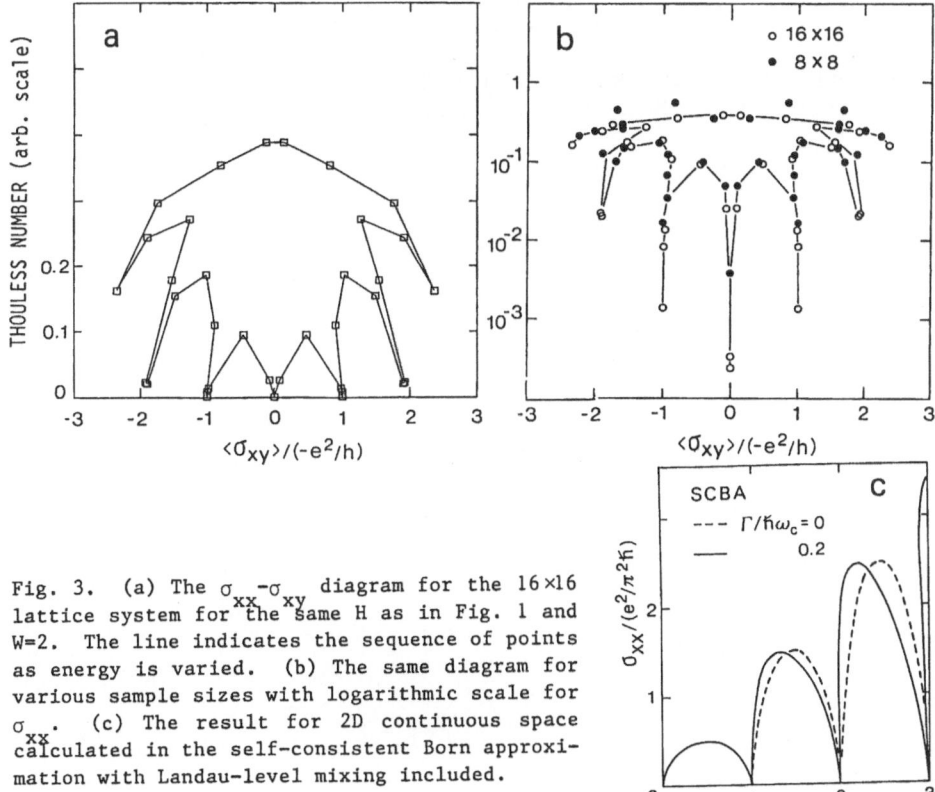

Fig. 3. (a) The σ_{xx}-σ_{xy} diagram for the 16×16 lattice system for the same H as in Fig. 1 and W=2. The line indicates the sequence of points as energy is varied. (b) The same diagram for various sample sizes with logarithmic scale for σ_{xx}. (c) The result for 2D continuous space calculated in the self-consistent Born approximation with Landau-level mixing included.

3. FLOW DIAGRAM FOR σ_{xx} vs σ_{xy}

To explore the behaviour and scaling properties of σ_{xx} and σ_{xy} over the whole energy spectrum, it is convenient to plot the σ_{xx}-σ_{xy} diagram/7,8/. Figure 3 shows the numerical result for 8×8 and 16×16 lattice systems. We have calculated σ_{xy} from the ensemble-averaged winding number and σ_{xx} from the Thouless number. Since there is an electron-hole symmetry in a square lattice, the diagram becomes a symmetric, closed curve. We have a very characteristic result, in which the peak in σ_{xx} between $|\sigma_{xy}|/(e^2/h)$ = N and N+1 increases with N and the oscillation curve is deformed towards the σ_{xy}=0 axis. This agrees with Ando's study/9/ of the diagram for the parabolic band in continuous 2D space with multiple Landau levels considered. There exist differences between the two results, however, arising from non-parabolicity effects in the lattice, which become stronger as E approaches zero. The result indicates that the only universal feature in the diagram for various systems is its topology.

1. T. Ando, Y. Matsumoto, Y. Uemura: J. Phys. Soc. Jpn 39, 279 (1975)
2. H. Aoki and T. Ando: Solid State Commun. 38, 1079 (1981)
3. A.A. Belavin, A.M. Polyakov, A.S. Schwartz, Y.S. Tyupkin: Phys. Lett. 59B, 85 (1975)
4. H. Aoki: Phys. Rev. Lett. 55, 1136 (1985)
5. H. Aoki: J. Phys. C 16, 1893 (1983); 18, L67 (1985)
6. T. Ando: J. Phys. Soc. Jpn 52, 1740 (1983); 53, 3101, 3126 (1984)
7. T. Ando: Surf. Sci. 170, 243 (1986)
8. H. Aoki and T. Ando: Surf. Sci. 170, 249 (1986)
9. T. Ando: J. Phys. Soc. Jpn, to be published

Scattering Approach to the Quantum Hall Effect

W. Brenig

Physik-Department der Technischen Universität München,
D-8046 Garching, Fed. Rep. of Germany

In the scattering approach, electronic transport properties of a sample are expressed in terms of scattering phases or scattering matrices governing the amplitude and phase relations between incoming and outgoing electronic wave functions outside the sample [1-4]. We use the approach to derive expressions for both the diagonal and off-diagonal conductivity of a two dimensional electron gas in a strong perpendicular magnetic field.

1. Introduction

Transport processes usually consist of a succession of scattering events. Transport coefficients, therefore, are usually expressed in terms of scattering cross sections. In the low density limit of scattering centres, in particular, only cross sections of a single scattering centre occur. In this paper we want to apply a more general point of view to the quantum Hall effect in which the whole sample is considered as one large scattering centre [2-4]. No low density (or weak coupling) assumption for the scattering impurities will be made [5].

2. Hall Conductivity

The scattering approach becomes particularly efficient, when combined with gauge transformation arguments [4,6,7]. One considers a 2D sample in the (x,y) plane (length L_x,L_z) placed in an electric field in the x-direction, a magnetic field B in the z-direction and a vector potential A=B(x+a) in the y-direction. If periodic boundary conditions in the y-direction are assumed the energy levels ϵ_n become functions of the flux $\Phi = BaL_z = \nu\Phi_0$ ($\Phi_0 = hc/e$ the flux quantum): $\epsilon_n = \epsilon_n(\nu)$. The Hall current then can be written as

$$I = (e/h) \sum_n (\partial\epsilon_n/\partial\nu) f_n \quad (f_n \text{ the occupation of level n}) . \tag{1}$$

If the potential outside the sample is independent on y the simplest solutions of the scattering problem have only one (incoming) momentum p for y going to minus infinity. The random potential then induces transitions to other momenta so that in the outgoing region (for y going to plus infinity) in general several momenta will appear, namely all those which are energetically allowed. In the simplest case all transitions except to the incoming momentum are neglected. This occurs for instance if inter-Landau-band transitions are neglected, and the energy $\epsilon(p)$ of the single band under consideration is monotonous in p. Then the outgoing wave can differ from the incoming one by at most a phase shift $\eta(p)$. The periodic boundary condition then requires

$$pL_z/\hbar + \eta(p) = 2\pi (n + \nu) . \tag{2}$$

This fixes the allowed p-values as functions of n and ν: $p=p_n(\nu)$. The allowed energy eigenvalues then are given by $\epsilon_n(\nu)=\epsilon(p_n(\nu))$. Since n and ν according to (2) occur only in the combination $n+\nu$ the energies $\epsilon_n(\nu)$ can be considered as sections $i<\nu<i+1$ of a single continuous function of the variable $n+\nu$ and the ν-derivatives in (1) can be replaced by n-derivatives. The result of such a replacement can be read off from fig.1: Whenever the distance between adjacent levels in the region of extended states is large the ν-derivative of ϵ is large, too. Or, in more quantitative terms: The contribution of an extended state to the Hall current for a large system is inversely proportional to the density of states. Fig.1 is only a schematic drawing. But detailed numerical studies [7] exhibit precisely the same behavior.

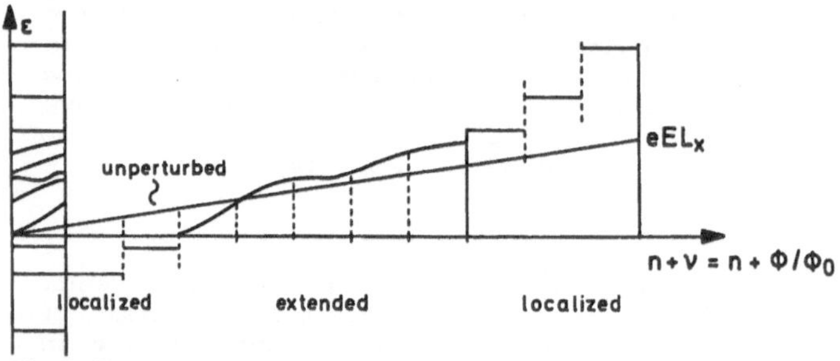

Fig.1 Energy ϵ as a function of flux Φ for a system with six localized and five extended states (schematically). The localized states have been incorporated stepwise, ordered according to their energy. The whole curve $\epsilon(\Phi)$ then is folded back into the first section $0<\Phi<\Phi_0$ in order to allow a direct comparison with [7].

For large systems the n-sum in (1) can be approximated by an integral, the corrections being of order $1/N^2$ [4]. For filling factors $f_n=1$ one finds

$$I = e^2U/h + O(1/N^2). \quad (3)$$

Here N is the number of occupied extended states. In (2) we have tacitly assumed that the scattering phase $\eta(p)$ in (2) is independent on the flux number ν. This assumption is indeed correct and a direct consequence of gauge invariance. In this sense (3) is a consequence of gauge invariance as well. The advantages of the scattering approach over straight forward gauge arguments [6] are: (i) finite size corrections to the quantised value of the Hall plateaus can be calculated. (ii) No assumptions concerning energy gaps between Landau levels or absence of edge states have to be made. (iii) The transition region between plateaus can be treated, for instance with filling factors $f_n=f$ one finds f times the quantised value (3).

Local filling factors smaller than one will occur for global filling factors smaller than one because of Coulomb interactions between electrons. Even if the correlation effects leading to fractional quantisation are neglected one expects a tendency for *spatially homogeneous* filling due to Coulomb interactions in order to neutralise a spatially homogeneous background charge. Thus in a gradual increase of the global filling of a Landau level one expects a gradual increase of an approximately spatially homogeneous local filling factor f from zero to one.

We mention without proof that (2) can be generalised to the multiband situation when interband transitions are taken into account [4]. The phase factor $\exp(i\eta(p))$ then has to be replaced by an S-matrix s_{pq} governing the amplitude and phase relations between ingoing and outgoing momenta allowed by energy conservation and the left-hand side of (2) has to be replaced by the eigenphases of the matrix $\exp(ipL_z/\hbar)s_{pq}$ [4].

3. Diagonal Conductivity

For weak random potentials the momentum p and the corresponding centre of gravity $x_p=cp/(eB)$ in x-direction of the electron scatter only weakly around their incoming values with no average drift velocity in the direction of the electric field. For strong randomness the wave function of the electron drifts away from its incoming centre of gravity at the "lower" end of the sample and spreads with increasing y over the hole sample. If the potential above the "upper" end would depend on x only because of the electric field (and perhaps a "confining potential") the wave function in the single band limit would have to contract to its incoming value again for y above the upper end because of energy conservation.

In order to avoid such unphysical edge effects at the upper and lower end of the sample we continue the random potential from its value $V(x,0)$ to negative y- values and from its value $V(x,L_z)$ to y-values larger than L_z. Then the average drift velocity in the direction of the electric field can be read off from the behavior of the wave function in the asymptotic regime.

A possible starting point is the Lorentz force balance equation

$$u_p = \langle u_x\rangle_p = -c\langle \partial V/\partial y\rangle_p/(eB), \tag{4}$$

where the brackets ⟨,⟩ indicate averages, i.e. integrals over the sample. Using the Schrödinger equation and Gauss' theorem these 2D integrals can be transformed into 1D integrals over the upper and lower end of the sample where the wave functions can be represented by their asymptotic behavior

$$\psi_p \to \exp(ipy/\hbar)\chi_p(x) \quad \text{and} \quad \psi_p \to \sum s_{pq} \exp(iqy/\hbar)\chi_q(x) \,. \tag{5}$$

The final result is [4]

$$u_p = \sum |s_{pq}|^2 v_q(x_p-x_q)/L_z \,, \tag{6}$$

where v_q is the average velocity in y-direction for the channel q, the q-sum runs only over those q's which conserve energy (i.e. $\epsilon_p=\epsilon_q$) and the unitarity relation

$$v_p = \sum |s_{pq}|^2 v_q \tag{7}$$

has been used [8].

(6) may now be compared with a similar formula which had been derived in slightly different context using lowest order perturbation theory for the scattering [9]:

$$u_p = \int w_{pq}(x_p-x_q)/L_z dq \,, \tag{8}$$

where the transition probability w_{pq} between the unperturbed states $|p\rangle$ and $|q\rangle$ in lowest order is given by

$$w_{pq} = 2\pi|\langle p|V|q\rangle|^2 \, \delta(\epsilon_p-\epsilon_q)/\hbar \,. \tag{9}$$

If one inserts this expression into (8) and uses $v_p = \partial\epsilon_p/\partial p$ one ends up with (6) (with the q-sum running over those q's which conserve energy). s_{pq} in this case, of course, is replaced by its first order approximation. Note that it is essential that the x-dependence of the random potential in the asymptotic regime is included (as explained in connection with eq. (4)). Otherwise energy would only be conserved for p=q leading to u_p=0.

(9) is valid for arbitrary electric fields. Usually one is interested only in linear response. This can be obtained by expanding the delta function in (9) to first order in the electric field E using

$$\epsilon_p-\epsilon_q = \epsilon_p^{\,o}-\epsilon_q^{\,o} - eE(x_p-x_q) \tag{10}$$

leading to

$$w_{pq} = r_{pq}\delta(\epsilon_p-\epsilon_q) = r_{pq}\{\delta(\epsilon_p^{\,o}-\epsilon_q^{\,o}) - eE(x_p-x_q)\delta'(\epsilon_p^{\,o}-\epsilon_q^{\,o})\} \tag{11}$$

up to first order in E. Here $r_{pq}=2\pi|\langle p|V|q\rangle|^2/\hbar$ and the prime (') indicates differentiation. The diagonal conductivity then can be written as

$$\sigma_{xx} = -e^2 \,\Sigma\, r_{pq}(x_p-x_q)^2\, f'(\epsilon_p)/(2L_z), \tag{12}$$

where $f'(\epsilon)$ is the derivative of the Fermi function. (12) is the analogue of Titeica's formula [9] for the case of elastic scattering.

Acknowledgement

The author is endebted to J.Hajdu and K.Wysokinski for stimulating discussions.

References

1. Fisher, D.S., Lee, P.A.: Phys.Rev. B 23,6851 (1981) (and related earlier work quoted in this paper)
2. Brenig, W.: Z.Phys. B 50,305 (1983)
3. Chalker, J.T.: J.Phys. C 16,4297 (1983)
4. Brenig, W.: Z.Phys. B 63,149 (1986)
5. Joynt, R.,Prange, R.E.: Phys. Rev. B 29,3303 (1984)
6. Laughlin, R.B.: Phys. Rev. B 23,5632 (1981)
7. Aoki, H.: J.Phys. C 15,L1227 (1982) and J.Phys. C 16,1893 (1983)
8. In [4] the factors v_p and v_q had been ommitted erroneously and a result slightly different from (6) had been obtained
9. Titeica, S.: Ann. d. Phys. 22, 129 (1935)

The Quantum Mechanical Hall Effect

R. Johnston and L. Schweitzer

Physikalisch-Technische Bundesanstalt, Bundesallee 100,
D-3300 Braunschweig, Fed. Rep. of Germany

The quantum dynamics of a two-dimensional electron gas at very low temperatures in applied electric and magnetic fields is investigated using a tightbinding model. The ratio of the current to the Hall potential is calculated and it is found that the Hall conductance is not proportional to the particle density. This contradicts results obtained using the Kubo theory for the same model.

1. INTRODUCTION

The Hall conductivity (σ_{xy}) is usually obtained using Kubo's linear response theory where the current response perpendicular to an applied electric field is calculated. In the experiment, however, both the Hall voltage and the current are measured responses to the applied electric field (battery). The Hall potential influences the current in the system's development towards a stationary state. A complete theory of the Hall effect must be capable of obtaining both the current and the Hall potential. In applications of the Kubo theory it is assumed that the Hall potential is a linear function of position across the sample. This corresponds to a position-independent Hall field.

In this paper a model is considered where electrons are confined to move on the surface of a finite cylinder (Teller-system; IxT). An applied magnetic field is everywhere normal to the surface. HAJDU et al. [1] have calculated σ_{xy} for the Teller system without disorder using the Kubo formula. They obtained $\sigma_{xy}=en/B$, which has no quantized plateaus.

As noted above it is closer to the experimental situation to investigate the time development of the Hall potential and current. Due to the topology of the system the electric field, which represents the battery, must be derived from a time dependent vector potential. The electric and magnetic fields are expected to cause a drift of electrons towards one of the edges of the cylinder. This should give rise to a current parallel to the edges and a potential difference between the edges, the ratio of which is the Hall conductance S_H. It is found, that σ_{xy} and S_H do not coincide and S_H exhibits plateaus although there is no disorder in the model system.

2. MODEL AND CALCULATION

The model one particle Hamiltonian is of tightbinding form

$$(H\Phi)(x,y) = e^{iA_x(t)}\Phi(x+1,y) + e^{-iA_x(t)}\Phi(x-1,y)$$
$$+ \Phi(x,y-1) + \Phi(x,y+1) + V(x,y,t). \quad (1)$$

$A_x(t)$ is the x-component of the vector potential,

$$A_x(t) = -By + F(t), \quad (2)$$

where B is the strength of the magnetic field and the electric field in the x-direction is

$$E_x(t) = -\frac{\partial}{\partial t} A_x(t). \quad (3)$$

The cylinder topology is equivalent to imposing on the wave function Φ periodic boundary conditions in the x-direction and Dirichlet boundary conditions in the y-direction.

The time-dependent Schrödinger equation can be integrated numerically using the "time slicer" technique [2]. The initial density matrix describes the equilibrium state of the electrons confined to the surface of a finite cylinder in a normal magnetic field. The density matrix at a given time $\Omega(t)$ can thus be calculated. From $\Omega(t)$ the total current and the electron density are obtained

$$I_x(t) = Sp\{\Omega(t)\; j(t)\} \quad (4)$$

and

$$n(t,y) = \langle y|\Omega(t)|y\rangle. \quad (5)$$

The current operator $j(t)$ is obtained through functional differentiation of the Hamilton operator with respect to the x component of the vector potential. From the electron density the electric potential is derived by solving the Poisson equation. The electric potential $V(y,t)$ is fed back into the Schrödinger equation and so influences the further development of the system.

There is no dissipation mechanism in the model, so the stationary state is achieved by varying the applied electric field $E_x(t)$. It is switched on and off in the following fashion

$$0\leq t\leq 2\pi\ ;\ E_x(t) = -\frac{\partial}{\partial t}F(t) = E(1-\cos(t)) \quad (6a)$$

$$t>2\pi\ ;\ E_x(t) = 0. \quad (6b)$$

The smooth turning on and off of the applied field is important for the stability of the method, because there are no damping mechanisms to dissipate the effects of sudden changes in the control parameters.

3. RESULTS

The charge density, current density and the Hall potential have been calculated in the stationary state at very low temperature. These provide information on the position dependence of the current flow as well as the local electric fields in the system. The Hall conductance S_H is the ratio of the total current to the potential difference as it is in the experiment.

In Fig. 1 the charge density $\rho(y)$, the electric potential $V(y)$, and the induced current density $j_E(y)$ ($j_E(y) = j_x(y) - j_0(y)$, where j_x and j_0 are the total and the diamagnetic currents, respectively) are shown as a function of position across the sample. Notice that the current flows at the edges and especially important is that the electrical potential is not of the form $-Ey$ as is assumed in calculations from the Kubo formula.

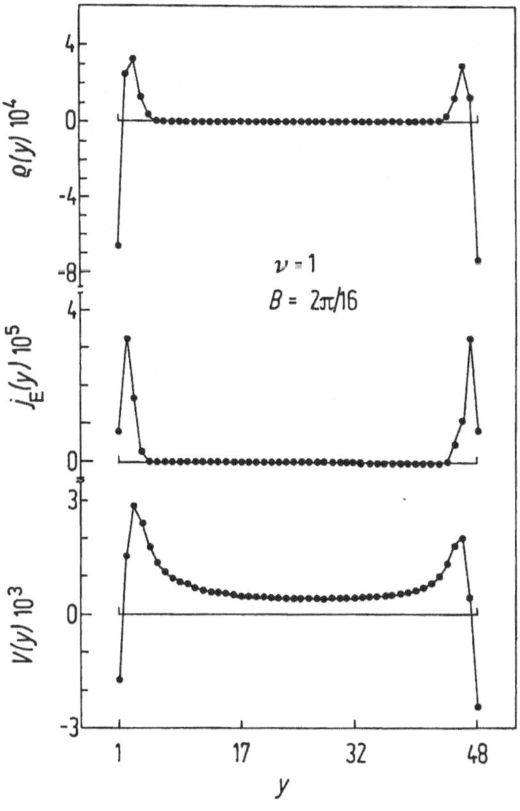

Fig. 1: The position dependence of the charge density ρ, the current density j_E and the electric potential V calculated for filling factor $\nu=1$ using the stationary state density matrix.

The conductance S_H is shown in Fig. 2 as a function of the filling factor ν. The result is not consistent with the "classical" result obtained previously [1] for the same system. S_H shows plateaus as a function of ν, although there is no disorder in the model. Notice that the jump takes place at $\nu=1$ and not at $\nu=½$ as seen in the experiments. This can be understood by realizing that first the bulk states, which contribute no net current, are occupied and only when ν is almost one the current carrying edge states are occupied. The deviation from the ideal plateaus at $S_H = e^2/2\pi$ is due to the lattice model considered here and is expected to be absent for free electrons.

4. DISCUSSION

It has been shown that S_H calculated above and σ_{xy} as calculated by HAJDU et al. [1] are completely different. S_H is derived in a way that is closer to the experimental situation than the derivation of σ_{xy}. It is possible that the position dependence of the electric field across the sample is responsible for the difference, however, this position dependence is not given and has to be calculated. To carry out the Kubo calculation with a position-dependent electric field would be very much more difficult and certainly the method of HAJDU et al. [1] does not carry over immediately. To explain the Hall conductance as given by I_x/U_H it is essential to consider a system with edges, so that the Hall potential can develop and be measured. Therefore U_H is not simply an input quantity but has to be calculated within the theory, because the Hall-effect is the result of an interplay between I_x and U_H.

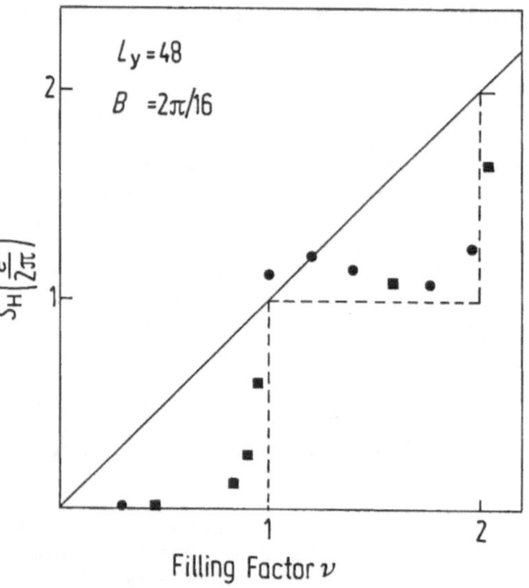

Fig. 2: The stationary state Hall conductance $S_H=I_x/U_H$ versus filling factor ν. The dashed curve is the expected T=0 free electron result.

It remains to be investigated why the plateaus start at $\nu=1$ and not at $\nu=\frac{1}{2}$ as observed in experiments.

5. REFERENCES

1. J. Hajdu and U. Gummich, Sol. State Comm. 52, 985 (1984)
2. J.O. Hirschfelder and R.W. Pyzalski, Phys.Rev.Lett. 55, 1244 (1985)

Density of States of Landau Levels in Two-Dimensional Systems from Activated Transport, Magnetocapacitance and Gate Current Experiments

D. Weiss and K. v. Klitzing*

Max-Planck-Institut für Festkörperforschung,
D-7000 Stuttgart 80, Fed. Rep. of Germany

In this publication we demonstrate that a combination of capacitance and gate current experiments together with an analysis of thermally activated conductivity seems to be useful for the determination of the density of states (DOS) of Landau levels in two-dimensional systems. The experimental results suggest a Landau level width not far away from the predictions of the self-consistent Born approximation (SCBA) if the Fermi level is close to the center of a Landau level. The DOS between Landau levels however cannot be explained with such a narrow linewidth and the experiments suggest the existence of a background DOS or an increased linewidth broadening for integer filling factors.

1. Introduction

A microscopic theory of the quantum Hall effect should give a correct description not only of the quantized resistivity values $\rho_{xy}=h/ie^2$ but also of the transitions between the plateaus and the values of the finite resistivity ρ_{xx}. Such transport calculations are extremely complicated since the theory itself is complicated and in addition not enough information is available about the scattering centers. The published theories are based on certain approximations and assumptions about the distribution, the strength and the range of the scattering potential. A first test whether such assumptions are realistic should be available from a comparison between the calculated and the measured density of states D(E) since calculations of D(E) are much easier than a transport theory for $\rho_{xx}(B)$ which includes complicated phenomena like localization and correlation. One of the first theories of the density of states (DOS) assumed short-range scatterers which leads within the self-consistent Born approximation (SCBA) to a broadening of the discrete energy spectrum (expected for an ideal two-dimensional electron gas without scattering) into an elliptic lineshape for the DOS [1]. Higher order approximations show that an exponentially decaying DOS is expected for energies $E-E_n$ larger than the linewidth of the Landau levels E_n [2], so that a real energy gap with vanishing DOS may not be present but the DOS at midpoint between two Landau levels should decrease drastically if the magnetic field (energy separation between adjacent Landau levels) is increased. Experimental information about the DOS can be obtained from measurements of the specific heat [3], from magnetization measurements [4], from temperature-dependent resistivity measurements in the regime of the Hall plateaus [5], from magnetocapacitance measurements [6,7] or from gate current measurements [8]. In this article we compare the results we have obtained from an analysis of the thermally activated resistivity, magnetocapacitance and gate current measurements carried out on one and the same sample. The following discussion is based on a picture which does not include many-body effects. The notation "density of states (DOS)" in this paper is used to characterize the electronic properties within a single particle picture. All experiments described in the following have been carried out on AlGaAs-GaAs heterostructures.

2. Activated resistivity

The temperature dependence of ρ_{xx}^{min} (where ρ_{xx}^{min} means the minimum in the resistivity which corresponds to a Fermi level position very close to the midpoint between two Landau levels) in the temperature range 2K<T<20K is usually dominated by an exponential term corresponding to

$$\rho_{xx}^{min} \sim \exp\{-\frac{E_{a,max}}{kT}\}, \quad (1)$$

where $E_{a,max}$ denotes the measured activation energy. Measured activation energies $E_{a,max}$ for different samples at different magnetic field values are shown in Fig.1. The filling factor i, defined as $i=n_s \cdot \frac{h}{eB}$ corresponds always to a fully occupied lowest Landau level (i=4 for (100) silicon MOSFETs and i=2 for GaAs-AlGaAs heterostructures). Since the measured activation energy $E_{a,max}$ agrees fairly well with half of the cyclotron energy $\hbar\omega_c$, this activation energy is interpreted as the energy difference between the Fermi energy E_F and the center of the Landau level E_n. For the sake of simplicity we assume that the mobility edge of the Landau level is located at the center of the Landau level, in agreement with calculations of the localization length [9] and percolation theories [10]. Furthermore the mobility edge is assumed to remain fixed, independent of the temperature and the carrier density. Changing the position of a Landau level E_n relative to the Fermi energy E_F(by changing the magnetic field) results in a reduced activation energy $E_a=|E_n-E_F|$. This motion of the Landau levels relative to the Fermi level if the filling factor of the Landau levels is varied is clearly visible in Fig.2. A change of the filling factor corresponds to a shift of the Fermi level, equivalent to a change Δn in the carrier density at fixed magnetic field. Measuring now the activation energy as a function of the magnetic field allows us to deduce a mean value for the DOS:

$$D(E) \approx \frac{\Delta n}{\Delta E}, \quad (2)$$

where ΔE is the energy difference between activation energies determined at consecutive magnetic field values. This analysing technique is restricted to the tails of Landau levels and has been described in more detail in a previous publication [8]. Figure 3 shows the reconstructed DOS obtained from sample 1 (n_s= 2.60·$10^{11}cm^{-2}$, μ= 158,000 cm^2/Vs).

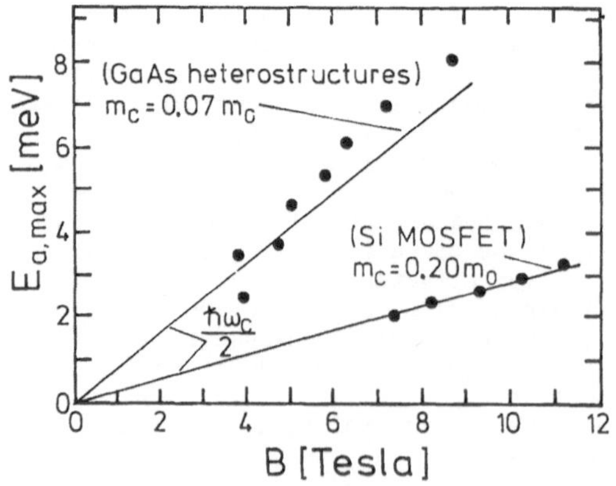

Fig. 1:

Measured activation energies $E_{a,max}$ in the resistivity at a filling factor corresponding to a fully occupied lowest Landau level as a function of the magnetic field B. The solid lines correspond to half of the cyclotron energy.

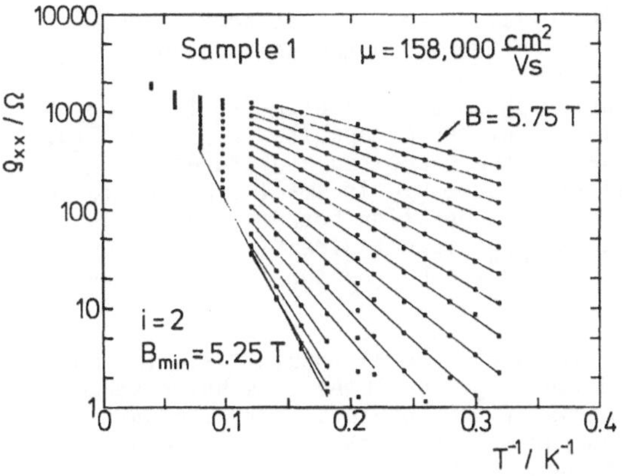

Fig. 2:

Temperature dependence of the resistivity ρ_{xx} at different magnetic fields close to a filling factor i=2.

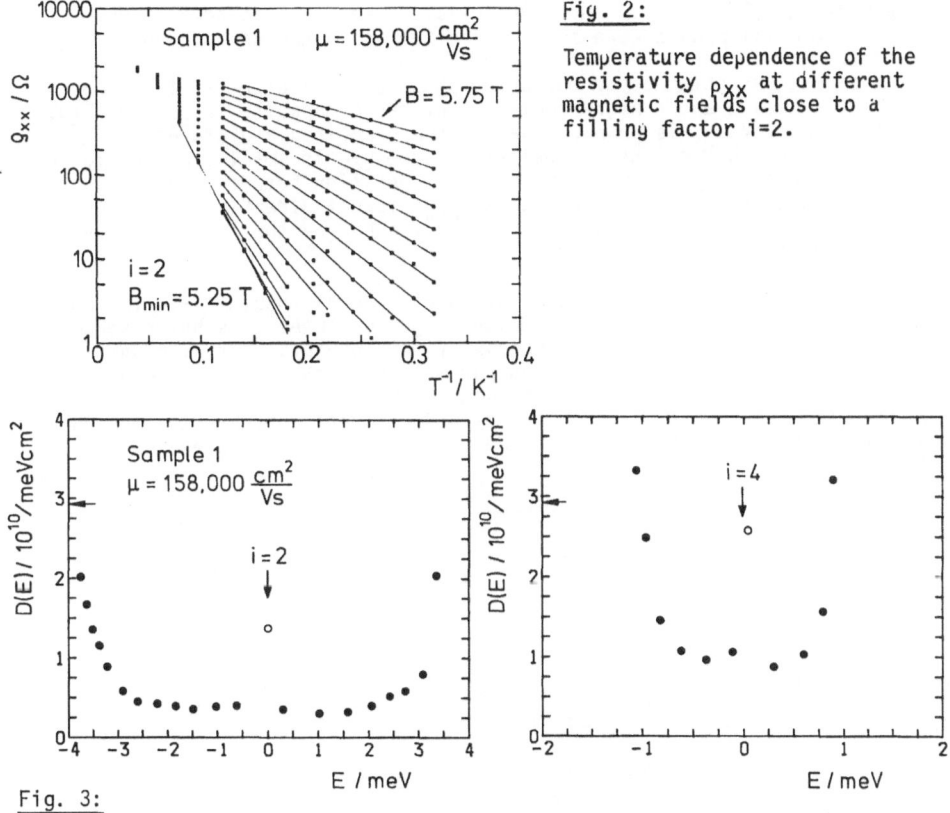

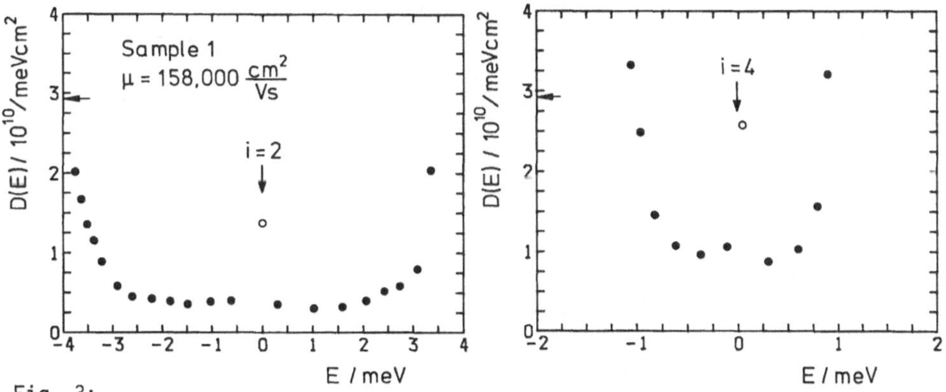

Fig. 3:

Reconstructed DOS for filling factors close to i=2 and i=4. The arrow marks the zero magnetic field DOS. The energy scale is taken relative to the mid-point between two Landau levels

The DOS between Landau levels does not vanish but shows a value depending on the mobility of the sample and on the magnetic field. Decreasing mobility and decreasing magnetic field results in an increased DOS between Landau levels. The magnetic field dependence of the DOS between Landau levels - obtained from an analysis at filling factors close to i=2 and i=4 - is contrary to our previous statements. The high DOS close to E=0 (Fig. 3) is an artefact of the analysis since for a Fermi energy at E=0 two Landau levels contribute to ρ_{xx}. The reconstruction of the DOS from activated resistivity measurements is restricted to the tails of the Landau levels as mentioned above. Information about the DOS for a Fermi level position close to the center of a Landau level can be obtained from magnetocapacitance measurements described in the next chapter.

3. Magnetocapacitance

The capacitance experiments were carried out on gated GaAs-AlGaAs heterostructures with a Hall geometry. The mobilities of the samples described here are between 83,000 and 480,000 cm²/Vs for carrier densities in the range between $2.27 \cdot 10^{11} cm^{-2}$ and $2.90 \cdot 10^{11} cm^{-2}$. For capacitance measurements all the potential probes were short-circuited and acted as a channel contact.

The signal was obtained by measuring phase sensitive the voltage drop between the sample and a high-precision reference capacitor C_{ref} (see Fig. 4b). The signal V_{meas} is proportional to the capacitance difference $C_{sample}-C_{ref}$ as long as the channel resistance R_{ch} is small compared to the AC resistance of C_{sample}. For a Fermi level position between two Landau levels σ_{xx} goes to zero and the signal V_{meas} is not directly proportional to the capacitance of the system but is influenced by the low conductivity state of the channel.

The capacitance of a system consisting of a metal-insulator-(with ionized impurities) semiconductor-sandwich (e.g. Au-AlGaAs-GaAs-heterostructure) depends not only on the thickness of the insulator but also on the DOS at the semiconductor side and on parameters of the material. Fig.4a shows the band diagram of a heterostructure including a Schottky gate in contact with the AlGaAs. If the two depletion layers interpenetrate each other the total capacitance at a given magnetic field can be expressed as [8,11]:

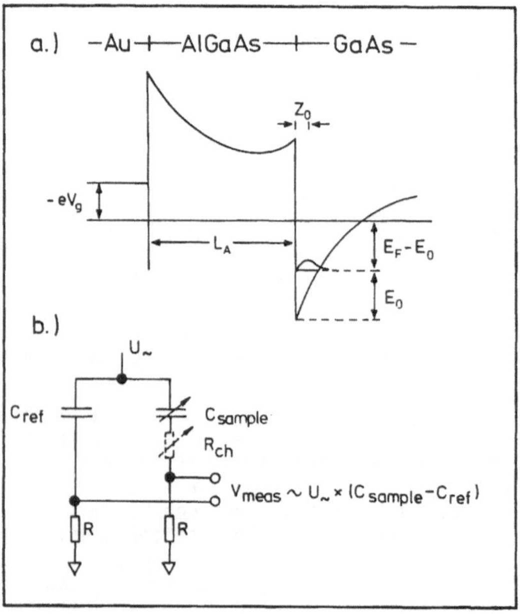

Fig. 4: Schematic diagrams of the conduction band edge for a gated GaAs-AlGaAs heterostructure showing the quantities used in the derivation (a) and the experimental set up (b). $U_\sim$ is the AC component of the applied voltage with an amplitude of about 7 mV and a frequency of 223 Hz

$$\frac{1}{C} = \frac{1}{C_A} + \frac{\gamma z_0}{\varepsilon_s} + \frac{1}{e^2 \left.\dfrac{dn_s}{d(E_F-E_0)}\right|_{E_F}} , \qquad (3)$$

where C_A is the capacitance of the insulating AlGaAs layer, ε_s is the dielectric constant of GaAs, z_0 is the average position of the electrons in the channel, γ is a constant numerical factor between 0.5 and 0.7, and $dn_s/d(E_F-E_0)$ is the thermodynamic DOS at the Fermi level, in the following denoted as dn_s/dE_F. The first two terms on the right-hand side of (3) are assumed to be constant in a magnetic field, and thus changes of the capacitance are directly related to changes in the thermodynamic DOS of the 2DEG. At T=0 the total inverse capacitance in a magnetic field can be expressed as

$$\frac{1}{C} = \frac{1}{C_0} - \frac{1}{e^2 D_0} + \frac{1}{e^2 D} \quad , \tag{4}$$

where C_0 denotes the value of the total capacitance at B=0, D is the DOS at the Fermi level in the presence of a magnetic field and D_0 is the DOS within the lowest subband, equal to $2.9 \times 10^{10} cm^{-2} meV^{-1}$ in the absence of a magnetic field. At finite temperatures D has to be replaced by dn_s/dE_F.

The experimental results were compared with calculations of C(B) assuming a Gaussian-like DOS of the form:

$$D(E) = \frac{e}{\pi\hbar} \cdot \frac{1}{\sqrt{2\pi}} \cdot \frac{B}{\Gamma} \cdot \sum_n \exp\{-\frac{(E-(n+\frac{1}{2})\hbar\omega_c)^2}{2\Gamma^2}\} \quad , \tag{5}$$

where Γ is the broadening parameter of the Gaussian distribution. First the position of the Fermi level E_F is determined by solving numerically the equation:

$$n_s = \int_{-\infty}^{\infty} D(E)\, f(E-E_F)\, dE \quad , \tag{6}$$

where $f(E-E_F)$ is the Fermi distribution function. The carrier density n_s is assumed to be independent of magnetic field and temperature.

In the next step the thermodynamic DOS

$$\left.\frac{dn_s}{dE_F}\right|_{E_F} = \int_{-\infty}^{\infty} D(E) \cdot \left.\frac{df(E-E_F)}{dE_F}\right|_{E_F} dE \tag{7}$$

is calculated numerically. With the temperature-dependent form of (4) and (7) one obtains C(B). Spin splitting which is small compared to the cyclotron energy for GaAs is neglected in the calculations.

Fig. 5 shows the capacitance data (for the same sample as discussed in Fig. 3) at different temperatures together with a theoretical curve calculated on the basis of a magnetic field independent linewidth Γ=0.48 meV. We have adjusted the fit to the magnetocapacitance maxima since the observed minima may be falsified - at least at high magnetic fields and low temperatures - by the small channel conductance. This argument cannot be used to explain the reduced depth of the capacitance minima at magnetic fields below 2 Tesla since the phase shift due to the channel resistance is negligibly small. However, inhomogeneities may explain the experimental data as shown in Fig. 6 where the change in the capacitance due to a Gaussian distribution of the carrier density n_s with a broadening parameter Δn_s=0.015 n_s is shown. A remarkable reduction of the depth of the capacitance minima is visible whereas the maxima remain unchanged.

The influence of inhomogeneities has been considered in a more sophisticated way by Gerhardts and Gudmundsson [12] in their statistical model for inhomogeneities. Their model is based on the assumption of a Gaussian-shaped DOS and the result of the calculations can be described using an effective linewidth Γ shown in Fig. 7. This effective linewidth oscillates and a maximum is always obtained for a Fermi level position between two

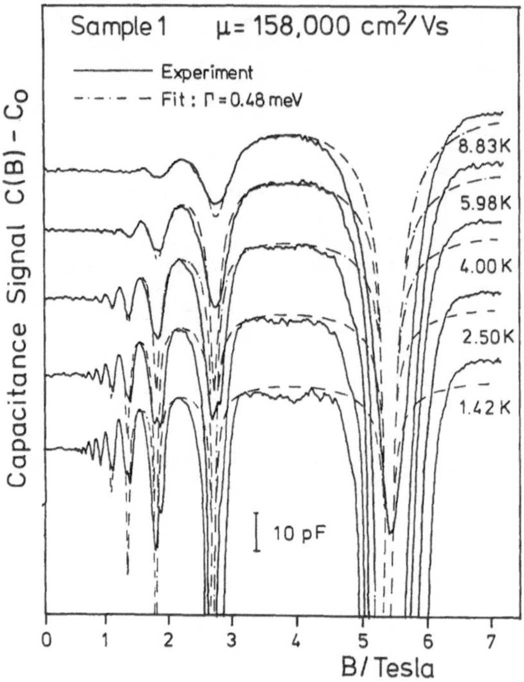

Fig. 5: Measured magnetocapacitance and corresponding fit using a broadening parameter $\Gamma = 0.48$ meV in the model DOS. For the sake of clarity the curves are shifted vertically

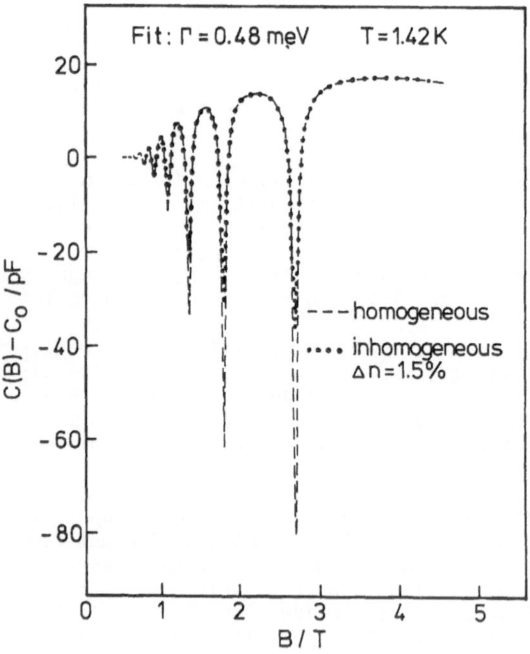

Fig. 6: Calculated magnetocapacitance showing the influence of inhomogeneities assuming a Gaussian distribution of the carrier density n_S

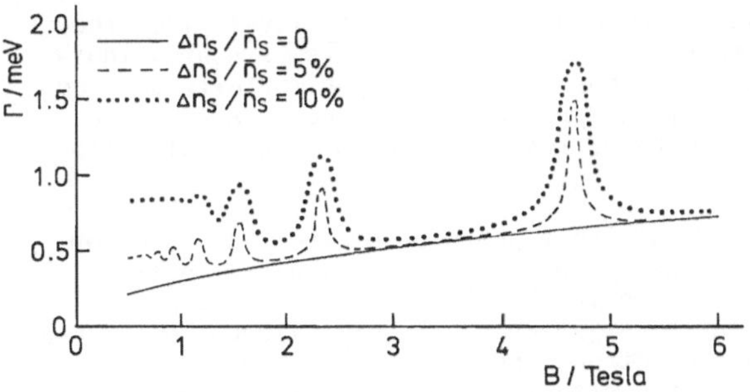

Fig. 7:
Effective Landau level broadening as a function of B. After [13]

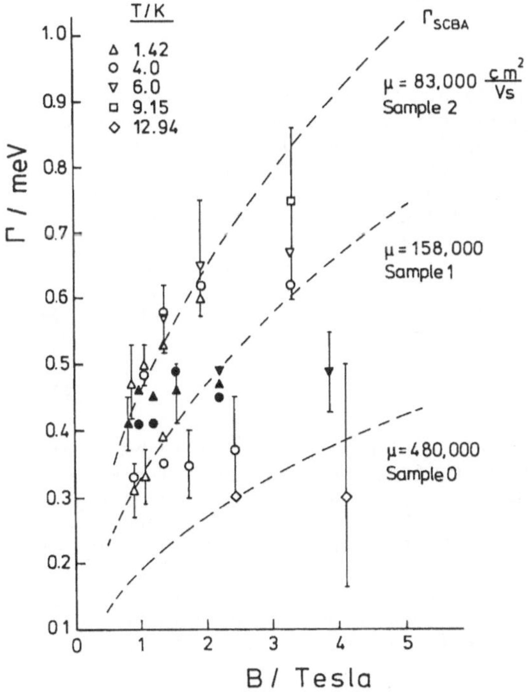

Fig. 8:
Landau level width Γ vs. B. Points are obtained by comparing magnetocapacitance maxima with model calculations. The dashed line corresponds to the SCBA- linewidth (8) where μ is the mobility of the samples. Full symbols correspond to sample 1 and lie in the range between open symbols corresponding to sample 0 and sample 2

Landau levels. This aspect is similar to calculations of the oscillating level broadening due to screening effects [14-16].

Figure 8 summarizes the results obtained from an analysis of the magnetocapacitance maxima of three samples. In this figure the broadening parameter Γ (see (5)) is plotted as a function of the magnetic field and compared with the linewidth Γ_{SCBA}, obtained from the selfconsistent Born approximation (SCBA) [1]:

$$\Gamma_{SCBA} = \frac{e\hbar}{m^*} \sqrt{\frac{2}{\pi} \cdot \frac{B}{\mu}} \tag{8}$$

where m* is the effective mass and μ the mobility of the sample. Fitting the magnetocapacitance maxima means that each point in Fig. 8 corresponds to a Fermi level position close to the center of a Landau level. The linewidth has been extracted from measurements at different temperatures indicated by different symbols. Figure 8 shows that the experimentally deduced linewidths are not so far away from the predictions of the SCBA - the $\sqrt{B}$ dependence of the linewidth Γ is more or less visible for the sample with the lowest mobility.

Up to now the model calculation of the magnetocapacitance was based on the assumption that the carrier density in the channel remains constant. This is incorrect, since the difference in the electrochemical potential across the capacitor is fixed and a variation in the capacitance leads to a charge transfer between the gate electrode and the channel. This small change in the carrier density is unimportant in an analysis of capacitance measurements but is a first order contribution in gate current experiments which will be discussed in the following chapter.

4. Gate Current

The assumption that the carrier density n_s remains constant changing the magnetic field is not correct. Actually not the carrier density n_s but the Fermi level is kept constant during capacitance experiments. Using the notation of Fig.4a this means that the gate voltage V_g is kept constant. Varying the magnetic field B then leads to oscillations of the surface potential (bottom of the potential well) and to a charge transfer between gate and channel of the heterostructure. Since the amount of transferred charge is small compared to the two-dimensional carrier density n_s the subband edge (taken relative to the bottom of the potential well) is assumed to be constant. The starting point for the model calculations is now no longer (5) but the following equation [8]:

$$\int_{-\infty}^{\infty} D(E)f(E-E_F)\, dE + \frac{C_A}{e^2} E_F = \text{const} \;, \tag{9}$$

where the constant can be determined at B=0. Equation (9) has to be solved numerically to give the correct position of the subband edge relative to the Fermi level and then the magnetocapacitance can be calculated using (7) and the temperature dependent form of (4). Calculating the magnetocapacitance in the way described above results in a broadening of the width of the capacitance minima compared to calculations assuming a constant carrier density n_s. The difference however is small and cannot be resolved in Fig. 5. The charge flow mentioned above can be determined by measuring the current between gate and channel as a function of the magnetic field B. The current flow is given by

$$I(B) = A \cdot e \cdot \frac{dn_s}{dt} = A \cdot e \frac{dn_s}{dB} \cdot \frac{dB}{dt} \;, \tag{10}$$

where A is the area of the two-dimensional electron gas and dB/dt the sweep rate of the magnetic field. dn_s/dB can be determined by solving (9) at different magnetic fields since the first term on the left-hand side is equal to the carrier density n_s. The current flow versus magnetic field is shown in Fig. 9. The upper curve shows the experiments (carried out again on sample 1) where a current minimum corresponds to a DOS maximum and a maximum in the current flow corresponds to a Fermi level position in a minimum of the DOS. The origin of the reverse current peak at about 5.2 Tesla is not clear yet. The experiment is compared with model calculations using a Landau level linewidth (Gaussian) of Γ= 0.48 meV (see Fig. 5 and Fig. 8) and 1.35 meV. The smaller linewidth Γ= 0.48 meV obtained from an

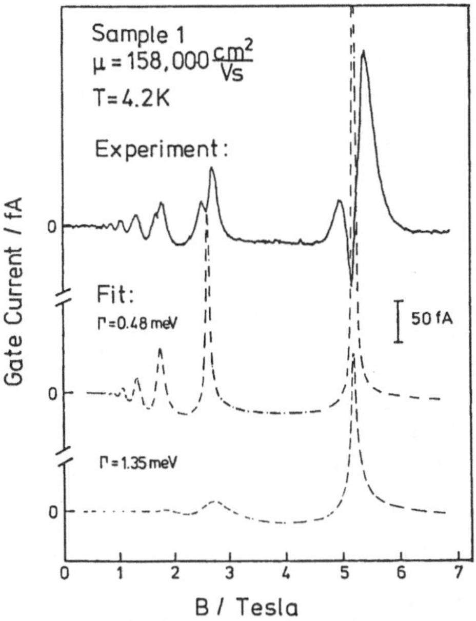

Fig. 9:
Measured and calculated current flow between gate and channel of sample 1 vs. magnetic field

analysis of the magnetocapacitance maxima of the same sample - again describes correctly the experiment for a Fermi level position close to the center of a Landau level. A larger linewidth Γ= 1.35 meV cannot fit the oscillations at lower magnetic field but fits approximately the height of the measured current maximum at about 5.4 Tesla (filling factor $i \approx 2$).

A more quantitative analysis however is hindered by the reverse current peak. It should be noted that the statistical model [12,13] or a background DOS at i=2 produces more broadened current maxima, comparable to the experimental ones.

5. Summary

Three different experimental methods have been carried out on one and the same sample. These three methods are sensitive to different energetical regions of the DOS. An analysis of the thermally activated resistivity - restricted to the tails of the Landau levels - shows a nonvanishing DOS between Landau levels depending on mobility and magnetic field which cannot be explained within the SCBA or higher order approximations. Magnetocapacitance measurements however are mainly restricted to the maxima of the DOS.

If the Fermi level position is close to the center of a Landau level the magnetocapacitance data can be explained with a Gaussian-shaped DOS where the linewidth Γ follows roughly the SCBA predictions. Gate current experiments in principle are sensitive to maxima as well as to minima in the DOS. For a Fermi level position in a maximum of the DOS the gate current measurements show the same result as magnetocapacitance measurements. The explanation of the gate current measurements for a Fermi level position between two Landau levels requires a background DOS or an increased linewidth broadening.

Acknowledgement

We would like like to thank R. Gerhardts and V. Gudmundsson for stimulating discussions and their interest in this work. We are grateful to K. Ploog and G. Weimann for providing the samples and we appreciate the cooperation with E. Stahl and V. Mosser at earlier stages of this work.

References

* Present address: Physik-Department E16, Technische Universität München, D-8046 Garching

1. T. Ando, Y. Uemura: J.Phys.Soc.Jap. 36, 959 (1974)
2. R.R. Gerhardts: Surf.Sci. 58, 227 (1976)
3. E. Gornik, R. Lassnig, G. Strasser, H.L. Störmer, A.C. Gossard, W. Wiegmann: Phys.Rev.Lett. 54, 1820 (1985)
4. J.P. Eisenstein, H.L. Störmer, V. Narayanamurti, A.Y. Cho, A.C. Gossard: Phys.Rev.Lett. 55, 875 (1985)
5. E. Stahl, D. Weiss, G. Weimann, K. v.Klitzing, K. Ploog: J.Phys. C18, L783 (1985)
6. T.P. Smith, B.B. Goldberg, P.J. Stiles, M. Heiblum: Phys.Rev. B32, 2696 (1985)
7. V. Mosser, D. Weiss, K. v.Klitzing, K. Ploog, G. Weimann: Solid State Commun. 58, 5 (1986)
8. D. Weiss, K. v. Klitzing, V. Mosser: in Two-Dimensional Systems: Physics and New Devices, ed. by G. Bauer, F. Kuchar, H. Heinrich, Springer Ser. Solid State Sci., Vol. 67 (Springer, Berlin, Heidelberg 1986) p. 204
9. T. Ando: J.Phys.Soc.Jap. 53, 3101 (1984)
10. S.A. Trugman: Phys.Rev. B27, 7539 (1983)
11. F. Stern: Phys.Rev. B5, 4891 (1972)
12. R.R. Gerhardts, V. Gudmundsson: Phys.Rev. B34, 2999 (1986)
13. V. Gudmundsson, R.R. Gerhardts: to be published
14. R. Lassnig, E. Gornik: Solid State Commun. 47, 959 (1983)
15. T. Ando, Y. Murayama: J.Phys.Soc. Japan. 54, 1519 (1985)
16. W. Cai, T.S. Ting: Phys.Rev. B33, 3967 (1986)

A Statistical Model for Inhomogeneities Explaining the Apparent Density of States Between Landau Levels of a Two-Dimensional Electron Gas

V. Gudmundsson and R.R. Gerhardts

Max-Planck-Institut für Festkörperforschung,
D-7000 Stuttgart 80, Fed. Rep. of Germany

A statistical model is presented for a spatially inhomogenous two-dimensional electron gas (2DEG) in a quantizing magnetic field, which simulates the effects of Poisson's equation and some essential properties of self-consistent screening. The model yields an effective background density of states (DOS) between Landau levels (L.L.'s) and is used to explain a number of recent experimental observations. A three-dimensional model of a heterojunction is introduced, which can account for the possible charge transfer between the gate and the 2DEG.

Recently, measurements of the capacitance /1,2/, the activation energy /3-5/, the gate current /1/, the specific heat /6/ and the magnetization /7/ of a 2DEG, in a strong quantizing magnetic field, have produced evidence for an unexpectedly large DOS between the L.L.'s. Theories of short range impurity scattering /8,9/ predict that the DOS between L.L.'s should be at least exponentially small for sufficiently strong magnetic fields. Self-consistent screening theories /10-12/ for longrange impurities, on the other hand, indicate that if the Fermi energy is located between two L.L.'s, then the level broadening may become so large that adjacent L.L.'s overlap even in the large magnetic field regime. In this situation the predictions of these theories are, however, not reliable since they are based on assumptions, which lead to a semi-elliptical shape for the DOS of an individual L.L. and breakdown for the overlapping L.L.'s. In this paper we will present a statistical model for inhomogeneities in the electron density n_s, caused, e.g., by inhomogeneities in the donor distribution, which can produce an effective DOS in between L.L.'s. We show that the statistical model is at least as effective in explaining experimental results as the ad hoc model of a constant background DOS superimposed on Gaussian-shaped L.L.'s, which has been used to interpret the data /1,4,6/.

In order to evaluate the measurable quantities, a simple model of the heterostructure in the effective mass approximation is considered /13-15/. The Schrödinger and the Poisson equations, in the direction perpendicular to the interface, are solved self-consistently using the variational method of FANG and HOWARD /16/ within the Hartree approximation /17/. Only one electrical subband is assumed occupied, and since the extent of the wavefunction into the bulk GaAs is much less than the depletion length, we consider a linear confining potential, which is determined by the electric field at the interface. The resulting set of non-linear equations for the gate voltage, the bottom of the confining potential, the energy of the lowest subband, the depletion and the inversion charges are then used to derive the chemical potential μ and the measurable quantities. In order to interpret capacitance and gate current measurements, charge transfer between gate and 2DEG via an external circuit can be considered, while the depletion charge can be assumed constant to an excellent approximation /17/.

So far we have assumed the heterostructure to be homogeneous parallel to the interface. Now introducing inhomogeneities in the 2DEG does require that the Poisson equation would have to be solved self-consistently with the Schrödinger equation in the 2D-layer to derive the statistical distributions of the two spatially fluctuating variables n_s and μ. Instead of solving this problem, we consider n_s and μ as random variables and select a simple form of the probability distribution which simulates some requirements of Poisson's equation. The statistical model introduced here considers therefore an ensemble of independent homogeneous subregions, rather than spatial fluctuations within one sample.

In the model with variable n_s, the n_{so}-model, we choose, for convenience $n_{so}= n_s(B=0)$ to be Gaussian distributed and calculate the distributions for $n_s(B)$ and $\mu(B)$ which then in general are not Gaussian. The standard deviation of n_{so}, Δn_{so}, is set constant , $\Delta n_s(B)$ is then close to Δn_{so} except at integer filling factors of the L.L.'s where it decreases, while $\Delta\mu$ develops peaks at the same locations. This is in accordance with Poisson's equation, since μ is the energy difference between the Fermi level and the electrical subband energy which is closely related to the electrostatic potential at the interface. The Poisson equation requires large but smooth variations of μ if two L.L.'s contribute to the n_s-distribution.

When n_s is kept constant, a Gaussian distribution can be selected either for $n_s(B)$ or $\mu(B)$ /17/. Due to the dependence of n_s on μ, selecting $\Delta\mu$ constant would lead to vanishing fluctuations of n_s at integer filling factors and thereby conflict with the Poisson equation for an inhomogeneous sample. To avoid this in a simple manner, we require for the μ-Gaussian model that Δn_s is fixed and independent of B, not $\Delta\mu$. When Δn_s is fixed in the n_s-Gaussian model, μ will be strongly pinned to the L.L.'s, which might require too large fluctuations of μ in a real sample. But due to the pinning, the effective thermodynamic DOS (TDOS) is not small between L.L.'s. The μ-Gaussian model, which seems to be more in accordance with the requirements at the Poisson equation, with respect to the fluctuations of μ, can be solved analytically yielding an effective DOS which is of the same form as the Gaussian input DOS, except the broadening has been changed to /18/:

$$\Gamma_{eff} = (\Gamma^2 + (\Delta\mu)^2)^{\frac{1}{2}} . \tag{1}$$

The deviation of μ peaks at integer filling factors due to the fact that Δn_s is constant, showing clearly the origin of the effective background, seen in Fig. 1. This dependence of Γ_{eff} on the location of the Fermi level is similar to the properties of the self-consistent screening theories.

In Fig. 2 the differential capacitance and the current between the gate and the 2DEG are shown as functions of the magnetic field B for the n_{so}-Gaussian model. It is evident that the statistical model can broaden the minima in the capacitance as effectively as the ad hoc background model /1,6/, with only modest amounts of inhomogeneities. The gate current of the statistical model is much closer to experimental results /1/ than that of the simple background model, since it predicts uniformly increasing height of the current peaks with increasing B, in contrast to the background model.

The specific heat C_v of the n_s-Gaussian model is seen in Fig. 3. Again it is clear that the statistical model is at least as good as the simple background model. The experimental results of Gornik et al. /6/ can be easily fitted with the n_s-Gaussian model for $\Gamma= 0.6\sqrt{B(T)}$ meV and $\Delta n_s/n_s= 0.01$. In addition to the ability to interpret experiments, the importance of the statistical model lies in the simple underlying physical ideas and

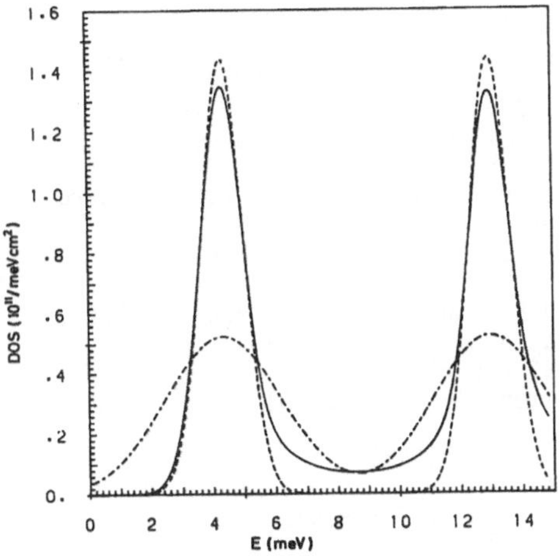

Fig. 1:
Average TDOS $\langle dn/d\bar{\mu}\rangle$ vs $\bar{\mu}$ (solid line) and effective DOS vs. E for μ in the center of the lowest L.L. (dashed line), and for μ in a Landau gap (dash-dotted line), calculated for the parameters: m=0.067 m_0, B=5T, T=1.64 K, $\Gamma=0.3\cdot\sqrt{B}$ meV, $\Delta n_s/n_s=0.05$ and μ= 10.7meV

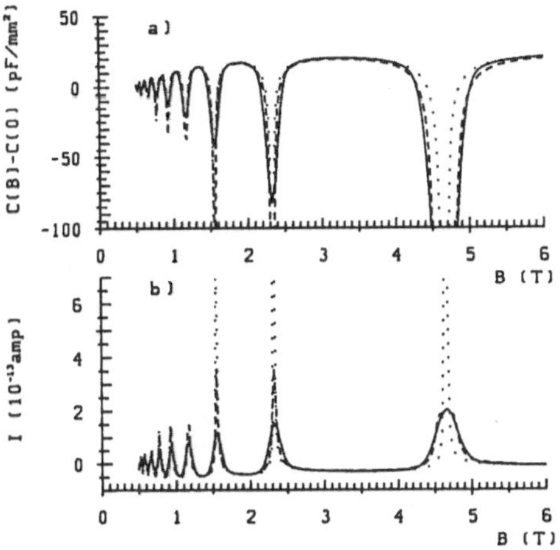

Fig. 2:
a) Capacitance, b) gate current according to the n_{so}-model without fluctuations, $\Delta n_{so}=0$ (dotted line), with 10% constant background DOS (dashed line), with $\Delta n_{so}/n_{so}$= 0.03 (solid line) vs. magnetic field. The parameters m, T and Γ are as in Fig. 1 with $n_{so}=2.25\cdot10^{11}cm^{-2}$ and the depletion charges: $N_A d$= $1.44\cdot10^{11}cm^{-2}$

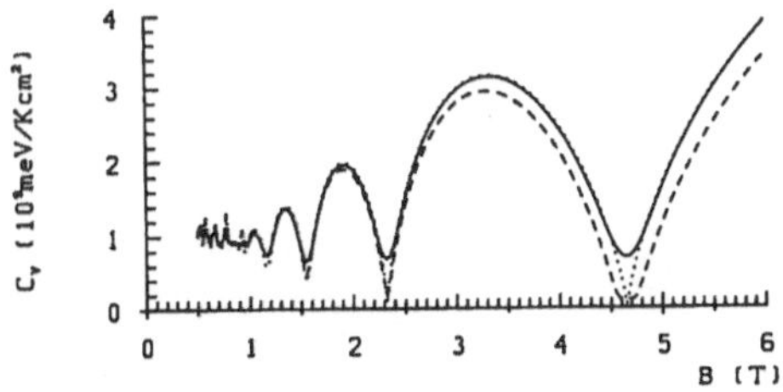

Fig. 3:
Specific heat according to the n_s-model without fluctuations (dotted line), with 10% constant background DOS (dashed line), with $\Delta n_s/n_s$= 0.03 (solid line) vs. magnetic field. The parameters m, T, and Γ are as in Fig. 1 with n_s= $2.25\cdot10^{11}cm^{-2}$

in its ability to simulate the behavior of the more complex self-consistent screening theories without the inherent inconsistencies concerning the shape of the DOS.

1 D. Weiss, K. v. Klitzing, V. Mosser, Springer Series in Solid State Sciences 67, Springer Berlin, Heiderberg (1986)
2 T.P. Smith, B.B. Goldberg, P.J. Stiles, M. Heiblum, Phys. Rev. B32, 2696 (1985)

3 B. Tausendfreund, K. v. Klitzing, Surf. Sci. 142, 220 (1984)
4 D. Weiss, E. Stahl, G. Weimann, K. Ploog, K. v. Klitzing, Proc. Int. Conf. EP2DS VI, Japan 1985, p. 307, J. Phys. C18, L783 (1985)
5 M.G. Gavrilov, I.V. Kukushkin, Pis'ma, Zh. Eksp. Fiz. 43, 79 (1986)
6. E. Gornik, R. Lassnig, G. Strasser, H.L Störmer, A.C. Gossard, W. Wiegmann, Phys. Rev. Lett. 54, 1820 (1985)
7 J.P. Eisenstein, H.L. Störmer, V. Narayanamurti, A.Y. Cho, A.C. Gossard, W.C. Tu, Phys. Rev. Lett. 55, 875 (1985)
8 T. Ando, Y. Uemura, J. Phys. Soc. Japan 36, 959 (1974); T. Ando, J. Phys. Soc. Japan 37, 622 (1974)
9 R.R. Gerhardts, Z. Physik B21, 275 (1975); Surf. Sci. 58, 234 (1976)
10 T. Ando, Y. Murayama, J. Phys. Soc. Japan 54, 1519 (1985)
11 W. Cai, T.S. Ting, Phys. Rev. B33, 3967 (1986)
12 R. Lassnig, E. Gornik, Solid State Commun. 47, 959 (1983)
13 F. Stern, Phys. Rev. B5, 4891 (1972)
14 F. Stern, S. Das Sarma, Phys. Rev. B30, 840 (1984)
15 G. Bastard, Surf. Sci. 142, 284 (1984)
16 F.F. Fang, W.E. Howard, Phys. Rev. Lett. 16, 797 (1966)
17 V. Gudmundsson, R.R. Gerhardts, to be published
18 R.R. Gerhardts, V. Gudmundsson, Phys. Rev. B34, in press

The Density of States of a Two-Dimensional Electron Gas in a Perpendicular Magnetic Field Under the Influence of a Correlated Random Potential

R. Johnston and L. Schweitzer

Physikalisch-Technische Bundesanstalt, Bundesallee 100,
D-3300 Braunschweig, Fed. Rep. of Germany

The influence of statistical correlations in the random potential on the density of states of the two-dimensional electron gas is investigated. The results are compared with the recently derived experimental density of states.

1. INTRODUCTION

In previous studies devoted to the calculation of the density of states D(E) of a two-dimensional electron gas in a normal magnetic field the random potential was modelled using independent random variables [1], [2] and [3]. In producing the samples for the experiments, it is sought to achieve the minimum disorder. It is to be expected that the potential in the sample will show short-range order and long-range fluctuations due to inhomogeneities and doping atoms. Recent experiments [4], [5] where the density of states is derived from capacitance measurements seemed to find considerable deviations from the calculated density of states using independent random variables. For these reasons it is probably not adequate to model the potential using statistically independent random variables.

A model potential with short-range order and long-range fluctuations is used here. It is generated using a generalization of Markov chains to higher dimensions called Markov fields. Results for the density of states are found which are qualitatively different from those obtained using independent random variables.

2. THE CORRELATED RANDOM POTENTIAL

Using the Gibbs representation of Markov random fields it has been shown that fields of correlated random variables can be generated site by site on a lattice [6]. Taking the Gibbs potential

$$\theta(w) = \sum_{\langle ij \rangle} \theta_{ij}(w) = -\sum_{\langle ij \rangle} \sigma_i(w)\, \sigma_j(w) , \tag{1}$$

where $\sigma_i(w)=w_i \in [-½, ½]$ is the value of the random variable on site i. The Gibbs distribution is, for the field or configuration of random variables w,

$$P(w) = e^{-K\theta(w)}/Z , \tag{2}$$

where Z is the partition function. This gives for the conditional probability

$$P(w_n < \Delta | w^0) = \frac{e^{\Delta(K \sum_{i \in N_n} \sigma_i)} - e^{-\frac{1}{2}(K \sum_{i \in N_n} \sigma_i)}}{2 \mathrm{sh}(\frac{1}{2} K \sum_{i \in N_n} \sigma_i)} \quad . \tag{3}$$

Given a configuration w^0 on the old graph this is the probability that the random variable on the new site n takes a value between $-\frac{1}{2}$ and $\Delta \leq \frac{1}{2}$. N_n is the set of neighbors of n in the old graph. Because the conditional probability depends only on the set of neighbors of n, the field of correlated random numbers can be generated iteratively. Examples are shown in Fig. 1 for values K=0.0, 1.5, 5.0 and 8.0. A square is completely black if the random variable takes a value $\frac{1}{2}$ and completely white if the value is $-\frac{1}{2}$. Intermediate values are partially filled. K=0 is the uncorrelated case and the correlations increase with K. As the process is symmetric in $\pm$ values of the random variables, the lack of symmetry in the pictures is due to the fact that the samples shown are much smaller than the correlation length. These random variables are all multiplied by a constant W, which gives the range of the potential. This potential is incorporated in the Schrödinger equation as shown in the next section.

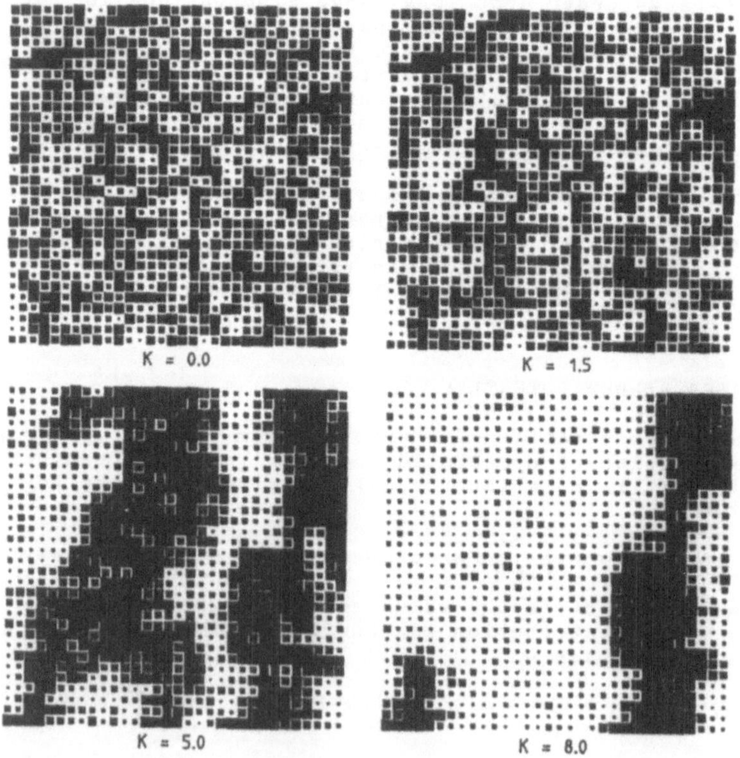

Fig. 1

3. MODEL AND METHOD

The model Hamilton operator describing the behaviour of an electron moving on a two-dimensional lattice with an applied perpendicular magnetic field is [7]

$$H = \sum_{lm} \varepsilon_{lm} |lm\rangle\langle lm| + \sum_{lm,l'm'} V_{lml'm'} |lm\rangle\langle l'm'| , \qquad (4)$$

where ε_{lm} are the correlated random site energies. The matrix elements $V_{lml'm'}$ are given by

$$V_{lml'm'} = \begin{cases} V, \text{ if } m=m' \text{ and } l=l'\pm 1 \\ V \exp(\pm 2\pi i \alpha l), \text{ if } l=l' \text{ and } m=m'\pm 1 , \\ 0, \text{ otherwise} \end{cases} \qquad (5)$$

where α is the magnetic phase factor: $\alpha = eBa^2/h$, and V is taken to be 1.

For rational $\alpha = p/q$, the tightbinding band splits into q magnetic subbands. At the band edge the energy spectrum becomes equivalent to the Landau spectrum of free electrons in the small magnetic field limit.

The density of states D(E) is calculated iteratively [7], and for a finite system of length L and width M the recursion relations are

$$D_L(E,\gamma) = N_L(E)/(\pi LM), \qquad (6)$$

$$N_L(E) = N_{L-1}(E) - \mathrm{Jm}\,\{\mathrm{Sp}\{g_L(VB_{L-1}V+I)\}\} , \qquad (7)$$

$$B_L = g_L(VB_{L-1}V+I)g_L , \qquad (8)$$

where I is the unit matrix, γ the imaginary part of the energy and g_L denotes the one-electron Green function

$$g_L = (E+i\gamma-H_L-Vg_{L-1}V)^{-1}, \qquad (9)$$

where H_L is the submatrix of the Hamiltonian of the Lth slice. For large enough systems and $\gamma\to 0$, $D_L(E,\gamma)$ converges to the density of states D(E).

4. RESULTS AND DISCUSSION

In Fig. 2 the density of states D(E) of the first Landau band is shown for K=0.0 (uncorrelated), K=1.5, K=5 and 8. It is seen that strong correlations produce a split of the density of states into two peaks. Since the two-peak structure has not been observed in experiments, it is concluded that correlations with $K\geq 5$ are too strong. The sample inhomogeneities, which are conjectured to be the cause of the background DOS, cannot be modelled by such a strongly correlated random potential. For weaker correlations K=1.5, the shape of the density of states is changed so that more weight is given to the wings, however, a constant background DOS as is inferred from the experiments is not obtained. It would be useful to have more information on

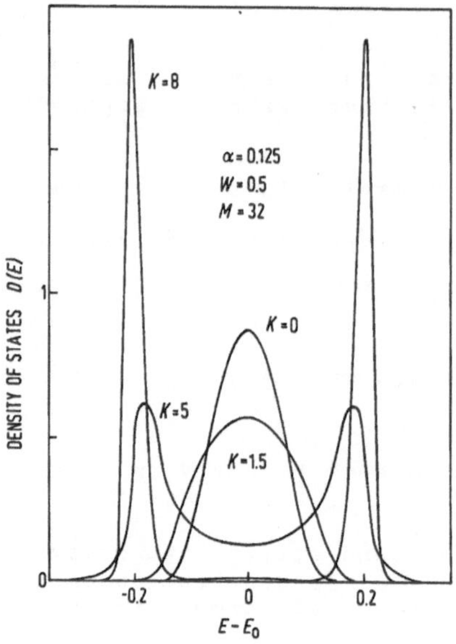

Fig. 2: Calculated density of states for correlation parameter K=0, 1.5 and 8

the sample inhomogeneities, their structure, range and amplitude, which would allow the choice of the appropriate parameters for the model.

5. REFERENCES

1. F. Wegner, Z. Phys. B51, 279 (1984)
2. B. Kramer, L. Schweitzer and A. MacKinnon, Z. Phys. B56, 297 (1984)
3. Brezin, Itzykson and Gross, Nuclear Phys. B235, 24 (1984)
4. T.P. Smith, B.B Goldberg, P.J. Stiles, and M. Heiblum, Phys. Rev. B32, 2696 (1985)
5. V. Mosser, D. Weiss, K.v.Klitzing, K. Ploog and G. Weimann, Sol. State Comm. 58, 5 (1986)
6. R. Johnston, PTB-Bericht, PG-2 (1986)
7. L. Schweitzer, B. Kramer and A. MacKinnon, J. Phys. C17, 4111 (1984)

Frequency- and Temperature-Dependent Conductivity in the Quantum Hall System

R. Joynt

Theoretische Physik, ETH-Hönggerberg, CH-8093 Zürich, Switzerland

1. Introduction

In the experimental understanding of localization, the most powerful tool has been the measurement of conductance as a function of several parameters. In the disordered insulator, a system similar to the quantum Hall system, temperature and frequency are the most interesting variables, and the verification of the well-known predictions of Mott bear out our overall picture of the physics of these systems.

It makes sense to try the same approach for the two-dimensional system in a strong magnetic field. We know already that this system has some amusing peculiarities in its ground state properties, namely that in the famous plateaux all states at the Fermi energy are localized in one direction which gives the quantized Hall conductance. The question is whether this insulating system ($\sigma_{xx} = 0$) also has behavior different from the Anderson insulator at finite temperature and driving frequency. Or, what is $\sigma_{xx}(T,\omega)$?

Let us start by giving the most simple-minded answer to this question. Take the case where the impurities are very dilute, that is, impurity density much less than $1/a^2$, where $a = \sqrt{(\hbar c/eB)}$ is the magnetic length. It is not hard to solve certain model potentials, such as a delta function or a $1/r^2$ potential, and the bound state wave functions always fall off as $\exp(-r^2/2a^2)$. One can then modify the arguments of Mott concerning resonance and hopping between localized states for this type of asymptotic behavior [1]. This leads to

$$\sigma_{xx}(T) \sim (1/T)\exp[-(T_0/T)^{1/2}] \text{ and } \sigma_{xx}(\sigma) \sim [\omega a N_0]^2 \ln(\gamma/\omega). \qquad (1)$$

In these fomulas $kT_0 = \lambda/(a^2N_0)$, where λ is a constant of order 1 and N_0 is the density of states at the Fermi level. γ is the Landau level width. On the other hand, if the impurities are dense: impurity density much greater than $(1/a^2)$, then it seems likely from numerical work [2] that the asymptotic behavior is $\exp(-r/a)$. This leads to the same results as in (1) except for the exponents: $(T_0/T)^{1/2} \rightarrow (T_0/T)^{1/3}$ and $\ln(g/\omega) \rightarrow \ln^3(\gamma/\omega)$. The main point is that both the temperature and the frequency-dependent conductivities reflect a single energy scale $(a^2 N_0)$. In the plateaux this will always be larger than the cyclotron energy eB/mc, where m is the effective mass. In frequency this is in the infrared, or, in temperature, about 1-10K for fields of a few tesla. In fact, however, at such high frequencies or temperatures, we are well out of the low energy asymptotic regime in which the Mott formulas are valid. Indeed we expect direct transitions between extended states at the center of Landau levels to dominate $\sigma_{xx}(\omega)$ and activated conduction to dominate $\sigma_{xx}(T)$ at these high energies. This measurement will then not tell us very much about the properties of the localized states. Is there any way out of this dilema?

I think that there is, at least if it is true that there are semiclassical states in the system [3]. These can be detected in various ways experimentally since their behavior is quite different from Anderson localized states. What makes them different is the existence of very strong correlation between the position and energy of the eigenstates. The point of this paper is to explore the experimental consequences of this correlation.

The plan is as follows. In Sect. 2 I review work on the frequency dependent conductivity. In Sect. 3 I present some new results on the temperature-dependent conductivity. Sect. 4 contains a few remarks about the effects of the electron-electron interaction, which is ignored until that point. Sect. 5 is the conclusion.

2. Frequency-Dependent Conductivity

The semiclassical picture of the quantum Hall system is the picture which is valid in the true $B \rightarrow \infty$ limit. All impurity potentials have a finite range, not a zero range as is sometimes assumed for analytical convenience. (This is rigorously true for the effective two-dimensional potentials in real MOSFETS or heterostructures.) When $a \sim B^{1/2}$ is much less than this range, then the states in the lowest Landau level have the simple form[4]

$$\psi(r,s) \sim \exp[-(r^2/2a^2)] \exp[i\phi(s)]; \qquad (2)$$

r and s are coordinates chosen so that r is everywhere perpendicular to an equipotential contour and s is always parallel to it. The reason we know that the states must look like this is simple. The kinetic energy is quantized in units of the cyclotron energy. For an electron to mix two levels, its wave function must be nonzero in two regions whose potential energy differs by the cyclotron energy. This is only possible if $\nabla V \cdot a \sim \hbar eB/mc$, where V is the potential energy, since a is the spatial extent of the wave function. By our hypothesis $B \to \infty$, however, the potential gradient is always too small to satisfy this condition. Hence the wave function always belongs to a single Landau level and potential energy is conserved, i.e., ψ is nonzero only on a contour of constant potential V. The states, whether localized on a closed contour or extended on an open one, nest nicely and low energy transitions are possible between them. We will use a simple approximation for $\phi(s)$, namely $\phi(s) = 2\pi sn/L$, where L is the length of the contour and n is an integer. The next state, which encloses this one, has one additional oscillation.

What we would now like to do is to calculate the frequency-dependent conductivity when the Fermi level lies in a region of localized states, i.e., when the quantum Hall effect is observed and $\sigma_{xx}(\sigma=0) = 0$. At finite frequencies we can have absorption of energy because of the 'sloshing of puddles' - transitions across the Fermi energy between neighboring states belonging to the same potential minimum or maximum.

To calculate the conductivity, we need first to understand the statistics of these states, which is equivalent to understanding the statistics of equipotential lines. Here we can use an analogy to percolation theory[4]. We observe that a state of energy E is located on the boundary of a region where V is less than E. (We concentrate on the less than half-filled case.) If we think of a lattice of empty or full 'sites' separated by a distance greater than ξ, where ξ is the correlation length of the potential, then our contour surrounds a cluster of full sites. The occupation probability in the site problem is

$$p(E) = 2\pi a^2 \int_{-\infty}^{E} \rho(E')\, d(E'), \tag{3}$$

where ρ is the density of states. The function p is energy dependent since a filled state sits on a line having $V < E$. Actually we only need to know about the properties of states with large L, since we are interested in low frequencies and the classical period of the orbit is $2\pi a^2 F/L$, where F is the average value of $|\nabla V|$, which we treat as constant. The trick then, when doing the sum over states, is to classify them according to their length L, and treat each class of states as having the same average properties. This means we need the function g(E,L), defined as the conditional density of states of length L, but only for large L. If $L \gg \xi$, it is clear from what we have said that

$$g(E,L) \sim [\xi\rho(E)/a^2] \exp[cL\ \ln(p/\xi)], \tag{4}$$

where c is a numerical constant of order one which is not easy to determine. For low frequencies we can now use a simple formula for the conductance[5]

$$\sigma_{xx}(\omega) \sim (e^2\omega^2\hbar/A) \sum_{\alpha\beta} |x_{\alpha\beta}|^2 \ \delta(E_\alpha - E_\beta + \hbar\omega)\ \delta(E_\alpha - \mu). \tag{5}$$

Here E_α and E_β are the energies of the eigenstates α and β, and $x_{\alpha\beta}$ is the dipole matrix element between these states; A is the total area of the sample; μ is the Fermi energy. Now for our trick. α is taken to be a state of length L, and the sum over β becomes a sum over the direction perpendicular to α. As β increases, we go along the potential gradient, eventually meeting a state of the right energy. We ignore transitions between different puddles. The Gaussian factor in the overlap matrix element will always suppress transitions of this kind relative to the ones we have considered. For transitions between our neighboring states, on the other hand, this factor is equal to one, since the transverse separation of these states is $a^2/L \ll L$. Performing the sum over α, we are left with an average over L, and the transverse summation:

$$\sigma_{xx}(\omega) \sim (e^2\omega^2/a^2)\nu \int dL\ \exp(cL\ \ln \nu/\xi)\ \Sigma\ \delta(\mu - E_\beta + \hbar\omega)|x_{L,\beta}|^2. \tag{6}$$

We have used $p(\mu) = \nu$, the filling factor for the Landau level. The interesting point now is that, setting $k = 2\pi\beta/L$ with β integral,

$$|x_{L,\beta}|^2 = (1/L^2) \int \langle x(s)x(s')\rangle \exp[ik(s-s')]\ ds\ ds'. \tag{7}$$

The integral is taken around the contour. Therefore the dipole matrix element just measures the noise spectrum of the system, as it should. This is a nice illustration of the classical limit of the fluctuation-dissipation theorem. If we now adopt a random-walk picture of the contour at large L, then we can compute the asymptotic form:

$$|x_{L,\beta}|^2 \sim \xi^2(L/\xi)^{2\rho}\ (k\xi)^{-1-2\rho}, \tag{8}$$

essentially from dimensional analysis, once we have assumed that $R \sim L^\rho$, where R is the root-mean-square radius of a cluster of perimeter L; ρ is known to be approximately 2/3 from numerical work on the percolation problem[6]. The calculation is now essentially complete. The result in terms of incomplete gamma functions, as well as a more complete derivation, may be found in reference [7]. We are most interested in the small-frequency limit, where we obtain

$$\sigma_{xx}(\omega) \sim \omega^{8/3}\exp(\omega_0/\omega), \quad \text{where} \tag{9}$$

$$\omega_0 = c \ln(1/\nu)\ (2\pi a^2\gamma/\xi^2 h). \tag{10}$$

This form for the conductivity is <u>universal</u>. The point is that if we probe only the states which are very long compared with the correlation length of the random potential, the paths will not know about the details of the distribution of the potential. At higher frequencies, this will cease to be true, and the frequency dependence of the conductivity offers information about the details of the potential, as discussed by Apenko and Lozovik[8]. Equations (9) and (10) together show that the characteristic frequency in the semiclassical picture is lower by a factor of a^2/ξ^2 compared to that of (1), and that the threshold at which this conductivity turns on may be quite a sharp one.

The Hall conductivity may be calculated in a similar fashion. Since it is not, strictly speaking, a dissipative quantity, one must be careful about applying the Kubo formula in this case[9]. There is an adiabatic part from motion of orbit centers, which is quantized and independent of frequency, as long as there are no extended states with energies near the Fermi level. There is also a contribution from the processes we have just discussed. In fact the only difference in the calculation of $\delta\sigma_{yx} = \sigma_{yx}(\omega) - \sigma_{yx}(0)$ and $\sigma_{xx}(\omega)$ is the correlation function. For $\delta\sigma_{yx}$ we need

$$\int \sin[k(s-s')]\ \langle x(s)y(s')\rangle ds\, ds'. \tag{11}$$

If we ignore the fact that our random contours must close, then this vanishes, since it measures the tendency of the contours to turn to the right or the left. Thus σ_{xx} turns on at a lower frequency than $\delta\sigma_{yx}$.

By concentrating on the very longest contours, we have only gotten at the asymptotically low-frequency behavior. When $\omega \simeq \omega_0$, then the paths cross over from random walk to circular shape. The conductivities are no longer universal functions, but certain things can still be said about them. One amusing result is that when $\nu > 1/2$, $\delta\sigma_{yx} > 0$ and when $\nu < 1/2$, $\delta\sigma_{yx} < 0$. That is because the electron orbits change handedness. When $\nu < 1/2$, the motion is around a minimum in the direction $\nabla V \times B$, but when $\nu > 1/2$, the electron is circling a maximum, and $\nabla V \times B$ has the other chirality. This also implies that the system is <u>optically</u> <u>active</u>: its refractive indices are different for left and right circularly polarized light. This is of practical importance because it is now possible to do microwave transmission experiments on structures having a dozen or so layers[10]. Investigation of the transmission coefficients as a function of field in a plateau should show an interesting complementarity between left and right polarization if there are semiclassical orbits. For a less than half filled band, (low field side of a plateau), and the direction of propagation parallel to the field direction, absorption will be strongest for left-handed polarization. I define this as light with counterclockwise circulation of the electric field vector, viewed along the direction of propagation. The reader can easily work these results out, using the criterion that the electric field of the microwaves be parallel to $\nabla V \times B$, and can therefore give up energy to the electrons.

Perhaps this polarization dependence would in fact be the most clear-cut way to test for semiclassical orbits. If the localization length is really a, and the characteristic energy is the cyclotron energy, then none of these interesting effects would show up at microwave frequencies.

There are certainly others which would test for the presence of long orbits. The ones which originally inspired this work were two-terminal measurements of the conductivity[11]. These do not directly measure σ_{xx} σ_{yx} but rather some combination of the two. They did show frequency dependence setting in sharply at $\omega \sim 10^5$ or 10^6 Hz. These effects are at present the subject of controversy. Since that situation will be reviewed by Dr. Kuchar in his contribution, I won't go into it further.

Another sort of experiment which seems to indicate long orbits is the measurement of current distributions by placing voltage probes at various places in the sample[12]. The results show that there are macroscopic inhomogeneities in the distributions which could well be due to the presence of semiclassical states. The fact, for example, that there are no plateaux for the Hall voltage if it is measured between the edge and the center of the sample is, in this theory, because there are orbits which intersect the central contact. The fact that the deviations change sign in the middle of the plateaux is another indication of the correctness of this viewpoint, for this is exactly, as we have already seen, the place where the orbits change their handedness. Thus their contribution to the Hall voltage changes sign, since only one side of the orbit, but always the same side, is actually being measured.

Another suggestive fact is that there seems to be no time-dependent effects in these two-dimensional systems[13]. If the external magnetic field is changed, and almost all states are strongly localized, how do they find their new positions in new Landau levels so quickly? The tunneling probabilities would seem to be very small unless the states have very long orbits. I know of no systematic study of relaxation times after a sudden change in magnetic field, but it would be well worth doing. Ideally, one could have a step-like change in the field from an 'in-between' to a 'plateau' value and monitor the approach to their final values of σ_{xx} and σ_{yx} as a function of time. In a highly inductive circuit, this is not necessarily easy, but nonetheless something like it should be tried. In gated systems, one could try similar things with the density, but physically it seems less likely that interesting effects will result.

3. Temperature-Dependent Conductivity

If there are long orbits, what are the consequences for the DC conductivity at finite temperature? We face here a topologically distinct problem from the AC case, since a DC current requires that the electrons follow paths that span the whole system. Internal vibrations do not count. It is clear that the crucial transitions, which must be assisted by phonons, are those which take an electron from one 'puddle' to another. The puddles themselves may be considered to have very large conductivity.

We can then proceed by following the argument of Ambegaokar, Halperin, and Langer[14]. We treat the system as a network of resistors, whose conductance is $\sigma_{ij}(T)$, each one connecting a pair of puddles i and j. In general σ_{ij} depends on the distance of closest approach of the 'shores' (of states at the Fermi energy) of puddles i and j. We then begin to thin out the network by eliminating a resistor if σ_{ij} is less than some chosen value. We continue this process until we come to value where the network fails to percolate across the sample. This value is then our answer for the conductance of the whole system, since the conductances vary over a wide range of magnitudes, and resistors with conductances greater than the critical value will act as short circuits. The temperature dependence is $\sigma_{ij} \sim |E_i-E_j|/kT$, where E_i is the energy of a site. As a consequence, as the temperature decreases, the electrons must look further and further away to find a site of similar energy. The conductance goes down quickly as a result of the small overlap. Here we shall assume that each puddle has many states within kT of the Fermi energy. Hence the conductance will be dominated by paths which connect only nearest neighbors. Longer hops will be killed off in the early stages of the thinning out process because of the Gaussian factor. The result is[15]

$$\sigma(T) \sim \exp[-(T_0^s/T)^{1/2}], \qquad (12)$$

as in (1), except that T_o^s is given by $z/a_{eff}{}^2N_o$. z is a pure number of order one which is characteristic of the percolation in the random network. The only difference between this result and the usual one is the appearance of a_{eff} instead of a, or, to put it another way, the renormalization of the overlap matrix element. As we saw before, it is the distance of closest approacy of two paths which counts, not the separation of their centers. For nearest neighbors, a reasonable approximation for a_{eff} is $a_{eff} \simeq a|\nu-1/2|$. When the level is half-filled σ diverges. This is correct, since current can percolate without hops in that limit. Of course, the formula is not quantitatively valid in that case.

Our assumption of nearest-neighbor hopping certainly breaks down at very low temperature: $T \ll T_o^s$. Eventually hops over distances greater than ξ will be necessary to complete the percolating network, and we will recover the result (1). It is unlikely that this temperature regime is reached in experiments.

A second approximation is the neglect of fluctuations. Rare very long paths of the sort discussed in the last section may be of inordinate importance. This will make some quantitative differences in the results in the direction of reducing T_o^s.

Experiments [16] have in fact seen the form (12) and a downward renormalization of T_o which could well have arisen from these physical effects. In these experiments T_o is about two or three orders of magnitude less than $1/a^2N_o$ in the denter of a plateau. This suggests that a_{eff} is about a factor of 20 larger than a, which is reasonable when the filling factor is rather small. Real experimental confirmation could only come if the dependence $T_o^s \sim |\nu - 1/2|^{+2}$ were verified. Since ν is difficult to measure independently, this is not easy, but it should be possible.

One must also take into account the fact that in the center of a plateau, two Landau levels are important and will in general have different values of ν. We have here taken the approach where interlevel hopping is neglected, in which case the conductances simply add, but this may no longer be a good approximation in the very center of the plateau.

4. Interactions

The first question is: what is the interaction? The answer is not trivial. The system is really three-dimensional, and in our semiclassical picture, the wave functions are actually sheets of charge e and width t, the latter being perhaps 100 A. The Coulomb energy of two such sheets is a linear function of their separation r as long as r is much less than t. When r is much greater than t, then the energy of course becomes e^2/r. There is a crossover regime where the energy depends on the detailed shape of the wave functions.

The implication is that there is no Coulomb gap for single-particle excitations, since the existence of this gap depends on the singular behavior of the interaction energy near the origin[17]. Therefore, for short hops, i.e., for $\sigma_{xx}(\omega)$, we do not need to worry about interaction effects. The same does not hold for $\sigma_{xx}(T)$, however. Here we deal with longer hops and the Coulomb interaction has its usual form. Application of the Efros-Shklovskii theory gives once again a result having the form (12) but with T_0^s replaced by T_0^c, where

$$T_0^c \simeq e^2/\kappa a \simeq 100K. \tag{13}$$

Here κ is the dielectric constant. This is a much higher temperature than is actually observed. In addition, the result of (12) can be distinguished experimentally from this one by its dependence on the filling factor.

Lastly, we would like to understand the more fundamental question of whether the single-particle picture used in deriving the semiclassical states remains valid in the presence of interactions. As long as we deal only with charge densities, as in the Hartree approximation, then there is clearly no problem. No length scales smaller than the correlation length of the potential can show up in the effective potential to ruin the approximation. In higher approximations there is no guarantee of this. Even in Hartree-Fock the phase of the wave functions enters the effective potential. Since these may vary on the scale of a, there is some reason to suppose that the WKB-like assumptions may ultimately break down when interactions are considered. On the other hand, the smoothness of the effective potential at short distances may work in the opposite direction.

5. Conclusion

The discovery of the quantum Hall effect is a dramatic chapter in the history of localization. The difficulty of making measurements other than the conductance has somewhat slowed down progress in finding out exactly what the wave functions look like, and what is the role of interactions. The particular question we have asked here is: how can we tell if the localization is semiclassical, forcing some of the states to be nearly extended? We have seen that there are a number of experiments, specifically current distribution, polarized microwave, time-resolved conductance, and frequency-, temperature-, and density-dependent conductance. The evidence is at present suggestive, but nowhere near conclusive, of the existence of long localized orbits. I hope that experimental work will continue, and see no reason that we shouldn't have answers fairly soon.

Acknowledgements

I am grateful to R. E. Prange and T. M. Rice for enlightening discussions.

References

1. Y. Ono: J. Phys. Soc. Japan 51, 237 (1982) and J. Phys. Soc. Japan 51, 2055 (1982)
2. H. Aoki, T. Ando: Phys. Rev. Lett. 54, 831 (1985
3. S. V. Iordansky: Sol. St. Comm. 43, 1 (1982)
4. S. A. Trugman: Phys. Rev. B 27, 7539 (1983) and R. Joynt, R. E. Prange: Phys. Rev. B29, 3303 (1984)
5. N. F. Mott and E. A. Davis: Electronic Processes in Non-crystalline Materials (Oxford Univ. Press, Oxford, 1971)
6. D. Stauffer: Phys. Rep. 54, 1 (1979)
7. R. Joynt: J. Phys. C18, L331 (1985)
8. S. M. Apenko, Yu.E. Lozovik: J. Phys. C18, 1197 (1985)
9. R. Kubo, S. J. Miyake, N. Hashitsume in Solid State Physics, vol. 17. ed. by F. Seitz and D. Turnbull (Academic Press, New York 1965)
10. F. Kuchar, R. Meisels, G. Weimann, W. Schlapp: Phys. Rev. B33, 2965 (1986)
11. M. Pepper, K. Wakabayashi: J. Phys. C16, L113 (1983)
12. G. Ebert, K. von Klitzing, G. Weimann: J. Phys. C18, L257 (1985).
13. I am grateful to T. M. Rice for pointing this out to me.
14. V. Ambegaokar, B. I. Halperin, J. S. Langer, Phys. Rev. B4, 2612 (1971)
15. R. Joynt, to be published
16. G. Ebert, K. von Klitzing, C. Probst, K. Ploog, G. Weimann: Sol. St. Comm. 45, 625 (1983)
17. A. L. Efros, B. I. Shklovskii: J. Phys. C8, L49 (1975)

Ground-State Energy of Two-Dimensional Electron Solids in High Magnetic Fields

G. Meissner and U. Brockstieger

Theoretische Physik, Universität des Saarlandes, D-6600 Saarbrücken, Fed. Rep. of Germany

The ground-state energy of a system of interacting electrons in a uniform charge-compensating background being subjected to a strong magnetic field has been studied. Many-body techniques have been employed under the assumption that the expectation values of the guiding center coordinates of the cyclotron motion of the electrons form a two-dimensional quantum solid.

Since the experimental discovery of the fractional quantized Hall effect in a strong-magnetic-field, low-temperature conductivity of high-mobility electron layers [1], correlation and exchange effects have been considered to be of significance in these two-dimensional (2D) electron systems of fractional filling factors $\nu = 2\pi r_L{}^2 n_e < 1$ of the lowest Landau level. Apparently, the ratio of two lengths

$$r_L/a = (g/2\pi)^{1/2}\ \nu^{1/2}\ , \tag{1}$$

i.e., the Larmor radius $r_L = (c\hbar/eB)^{1/2}$ of electrons of charge e in a magnetic field of amplitude B and their mean distance $a = 1/g\sqrt{n_e}$ at areal density n_e, respectively, provides a small parameter in these systems. Employing many-body techniques in exploring theoretically novel condensed phases, presumably underlying the experimental findings, will therefore be useful, e.g., in order to determine a critical filling factor $0 \leq \nu_c < 1$ separating the correlated quantum liquid proposed by LAUGHLIN [2] with $\nu > \nu_c$ and a 2D quantum solid of electrons [3] with $\nu \leq \nu_c$. In this paper, therefore, the ground-state energy of a system of interacting electrons in a uniform charge-compensating background is studied in the presence of high magnetic fields.

Our system has further been specified, by assuming the expectation values

$$\underset{\sim}{R}(\ell) = \underset{\sim}{X}(\ell) - \underset{\sim}{u}(\ell) = \underset{\sim}{a}_1\ell_1 + \underset{\sim}{a}_2\ell_2 \qquad (\ell = 1,2,..,N) \tag{2}$$

of position operators $\underset{\sim}{X}(\ell)$ for guiding center variables of the cyclotron motion to form a 2D lattice with primitive translation vectors $\underset{\sim}{a}_1$ and $\underset{\sim}{a}_2$. Many-body techniques have then been applied to the effective Hamiltonian H_{eff} obtained after eliminating the fast cyclotron motion. Pairwise local exchange corrections in our approach are still contained in the effective interaction $v_{eff}(\underset{\sim}{X}(\ell)-\underset{\sim}{X}(\ell'))$ between two position operators $\underset{\sim}{X}(\ell)$ and $\underset{\sim}{X}(\ell')$. Asymptotically, in the limit of high magnetic fields ($r_L \to 0$) it takes the form

$$v_{eff}(\underset{\sim}{R}) = \frac{r_L}{|\underset{\sim}{R}|}\left\{1 + \frac{r_L{}^2}{2|\underset{\sim}{R}|^2} + \,..\right\}(e^2/\varepsilon r_L)\ , \tag{3}$$

where ε denotes the dielectric constant. The dynamical behavior of our many-body system is determined by equal-time commutation relations for Cartesian components of the displacement operators of the guiding center variables, i.e.,

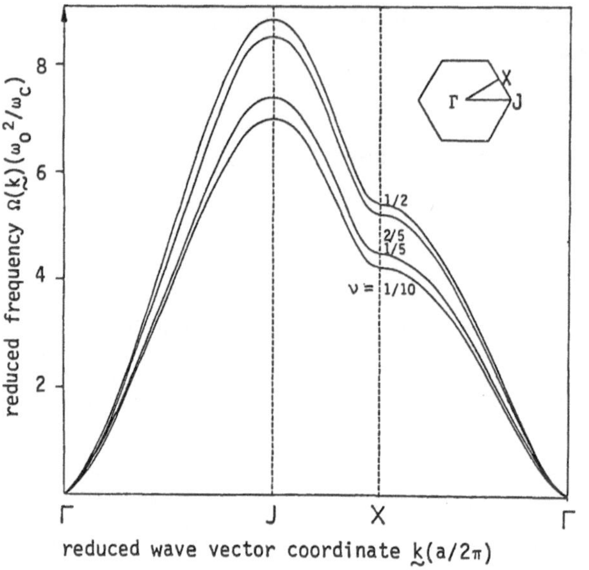

Fig. 1: Dispersion curve of magneto-phonon frequencies $\Omega(\underset{\sim}{k})$ versus wave vectors $\underset{\sim}{k}$ in a triangular lattice ($g = \sqrt{3}/2$) for various filling factors ν using reduced units where $\omega_c = \hbar/mr_L^2$, $(r_L/a) = (g/2\pi)^{1/2}\nu^{1/2}$, and $\omega_o = (e^2/ma^3)^{1/2}$ with m denoting the electron mass.

$$[u_\alpha(\ell),u_{\alpha'}(\ell')] = ir_L^2\delta_{\ell\ell'}\varepsilon_{\alpha\alpha'} \qquad (\alpha,\alpha' = 1,2)\ , \tag{4}$$

where $\varepsilon_{\alpha\alpha'}$ denotes the antisymmetric tensor. In the high-magnetic-field limit ($r_L \to 0$) the commutator of (4) becomes identical to zero, and thus our many-body system behaves classically. The dispersion relation obtained, e.g., in harmonic approximation, using reduced units (Fig. 1) for the magneto-phonon frequency

$$\Omega(\underset{\sim}{k}) = [\phi_{11}(\underset{\sim}{k})\phi_{22}(\underset{\sim}{k}) - \phi_{12}(\underset{\sim}{k})\phi_{21}(\underset{\sim}{k})]^{1/2} \equiv \omega_+(\underset{\sim}{k})\omega_-(\underset{\sim}{k}) \tag{5}$$

as the collective vibrational mode, shows the well-known transverse-phonon behavior, with $\Omega(\underset{\sim}{k}) \propto k^{3/2}$ in the long-wavelength limit [4]. Classically, the second-order coupling constants $\phi_{\alpha\alpha'}(\underset{\sim}{k})$ in (5) according to (3) take the limiting form

$$\phi^o_{\alpha\alpha'}(\underset{\sim}{k}) = \sum_{\ell\neq 0} [1 - \cos(\underset{\sim}{k}\cdot\underset{\sim}{R}(\ell))] \frac{\partial^2}{\partial R_\alpha \partial R_{\alpha'}} \frac{1}{|\underset{\sim}{R}(\ell)|}\ . \tag{6}$$

In this short communication we can only present some implications to be drawn from a microscopic expression, $E(\nu)/N$, having been derived with our effective Hamiltonian for the ground-state energy per electron [5]. In particular, the physical meaning of expressions obtained for the coefficients of the first three terms appearing in an expansion into powers of the filling factor ν, i.e.,

$$E(\nu)/N = \{-a_1\nu^{1/2} + a_2\nu^{3/2} + a_3\nu^{5/2} + ..\}\ (e^2/\varepsilon r_L) \tag{7}$$

will briefly be discussed emphasizing physical processes having to be regarded in a consistent calculation of a_3. Numerical estimates obtained for a triangular lattice in the high-magnetic-field limit ($r_L \to 0$) will also be given.

The coefficients of the first and second term related to the static lattice energy and to the vibrational zero-point energy, respectively, may clearly be worked out already in harmonic approximation. In reduced units one finds from the static lattice energy

$$-a_1 = \left(\frac{g}{2\pi}\right)^{1/2} \frac{1}{2} \left\{ \sum_{\ell \neq 0} \frac{1}{|\underset{\sim}{R}(\ell)|} - \frac{1}{g} \int d^2r \frac{1}{|\underset{\sim}{r}|} \right\} = -0.78213 \ , \tag{8}$$

where the numerical value 0.78213 following for a triangular lattice corresponds precisely to the value $2\sqrt{2}\ a_1 = 2.2122$ as calculated already by MEISSNER, NAMAIZAWA, and VOSS [6] and to $2\sqrt{2\pi}\ a_1 = 3.9210$ having been obtained by BONSALL and MARADUDIN [7]. From the vibrational zero-point energy of the guiding center variables one obtains the expression

$$a_2 = \left(\frac{g}{2\pi}\right)^{3/2} \frac{1}{4N} \sum_{\underset{\sim}{k}} \sum_{\alpha\alpha'} \phi_{\alpha\alpha'}(\underset{\sim}{k},-\underset{\sim}{k})\{\Omega(\underset{\sim}{k})\phi^{-1}_{\alpha'\alpha}(\underset{\sim}{k}) + \delta_{\alpha\alpha'}\} = 0.24101 \ . \tag{9}$$

A concise representation of results obtained for the coefficient a_3, being related to scattering of magnetophonons, can best be achieved, by extending an evident diagrammatic form of second-order coupling constants

$$\phi_{\alpha\alpha'}(\underset{\sim}{k}) \equiv \phi_{\alpha\alpha'}(\underset{\sim}{k},-\underset{\sim}{k}) \ := \tag{10a}$$

to higher order terms. Particularly important in the derivation of a consistent expression for that coefficient

$$a_3 = a_3^{\,in} - a_3^{\,ex} = 0.0586 \tag{11}$$

is the inclusion of contributions due to intermediate phonon exchange, $-a_3^{ex} < 0$, reducing the value, $a_3^{in} > 0$, as being obtained from scattering processes due to instantaneous anharmonicities only. In terms of the fourth-order coupling constants

$$\phi_{\alpha\alpha'\beta\beta'}(\underset{\sim}{k},-\underset{\sim}{k},\underset{\sim}{q},-\underset{\sim}{q}) \ := \tag{10b}$$

one obtains from instantaneous contributions of direct scattering

$$a_3^{\,in} = \left(\frac{g}{2\pi}\right)^{5/2} \frac{1}{16N^2} \sum_{\underset{\sim}{k},\underset{\sim}{q}} \sum_{\alpha\alpha'} \sum_{\beta\beta'} \phi_{\alpha\alpha'\beta\beta'}(\underset{\sim}{k},-\underset{\sim}{k},\underset{\sim}{q},-\underset{\sim}{q})$$

$$\times \{\Omega(\underset{\sim}{k})\phi^{-1}_{\alpha\alpha'}(\underset{\sim}{k}) + \delta_{\alpha\alpha'}\}\{\Omega(\underset{\sim}{q})\phi^{-1}_{\beta\beta'}(\underset{\sim}{q}) + \delta_{\beta\beta'}\} = 0.0861 \ . \tag{11a}$$

From dispersive anharmonicities [8] one finds to same order

$$-a_3^{\,ex} = \left(\frac{g}{2\pi}\right)^{5/2} \frac{1}{48N^2} \sum_{\underset{\sim}{k},\underset{\sim}{q}} \sum_{\alpha\alpha'} \sum_{\beta\beta'} \phi^{ex}_{\alpha\alpha'\beta\beta'}(\underset{\sim}{k},-\underset{\sim}{k},\underset{\sim}{q},-\underset{\sim}{q})$$

$$\times \{\Omega(\underset{\sim}{k})\phi^{-1}_{\alpha\alpha'}(\underset{\sim}{k}) + i\varepsilon_{\alpha\alpha'}\}\{\Omega(\underset{\sim}{q})\phi^{-1}_{\beta\beta'}(\underset{\sim}{q}) + i\varepsilon_{\beta\beta'}\} = -0.0275 \tag{11b}$$

with effective fourth-order coupling constants of intermediate phonon exchange

$$\phi^{ex}_{\alpha\alpha'\beta\beta'}(\underset{\sim}{k},-\underset{\sim}{k},\underset{\sim}{q},-\underset{\sim}{q}) \ := \ - \tag{10c}$$

Dispersive anharmonicities and resulting intermediate phonon exchange have not been taken into account in a previous self-consistent phonon approach [9], with the value 0.1600 reported for a_3. Our numerical estimates for the expansion coefficients of a triangular lattice in the limit of high magnetic fields have been obtained in using the asymptotic forms of the coupling constants to fourth order

$$\phi^{o}_{\alpha\alpha'\beta\beta'}(\underset{\sim}{k},-\underset{\sim}{k},\underset{\sim}{q},-\underset{\sim}{q}) = \sum_{\ell\neq 0}[1 - e^{i\underset{\sim}{k}\cdot\underset{\sim}{R}(\ell)}][1 - e^{i\underset{\sim}{q}\cdot\underset{\sim}{R}(\ell)}]\frac{\partial^4}{\partial R_\alpha \partial R_{\alpha'} \partial R_\beta \partial R_{\beta'}}\frac{1}{|\underset{\sim}{R}(\ell)|}$$

and to third order

$$\phi^{o}_{\alpha\beta\gamma}(\underset{\sim}{k},\underset{\sim}{q},-\underset{\sim}{k}-\underset{\sim}{q}) = \sum_{\ell\neq 0}[e^{i\underset{\sim}{k}\cdot\underset{\sim}{R}(\ell)} + e^{i\underset{\sim}{q}\cdot\underset{\sim}{R}(\ell)} + e^{-i(\underset{\sim}{k}+\underset{\sim}{q})\cdot\underset{\sim}{R}(\ell)}]\frac{\partial^3}{\partial R_\alpha \partial R_\beta \partial R_\gamma}\frac{1}{|\underset{\sim}{R}(\ell)|},$$

respectively. Finally, the effective fourth-order coupling constants of (10c)

$$\phi^{o\,ex}_{\alpha\alpha'\beta\beta'}(\underset{\sim}{k},-\underset{\sim}{k},\underset{\sim}{q},-\underset{\sim}{q}) = \sum_{\gamma\gamma'} \phi^{o}_{\alpha\beta\gamma}(\underset{\sim}{k},\underset{\sim}{q},-\underset{\sim}{k}-\underset{\sim}{q})\phi^{o}_{\alpha'\beta'\gamma'}(-\underset{\sim}{k},-\underset{\sim}{q},\underset{\sim}{k}+\underset{\sim}{q})$$
$$\times \{\Omega(\underset{\sim}{k}+\underset{\sim}{q})\phi^{-1}_{\gamma\gamma'}(\underset{\sim}{k}+\underset{\sim}{q}) + i\varepsilon_{\gamma\gamma'}\}\{\Omega(\underset{\sim}{k}) + \Omega(\underset{\sim}{q}) + \Omega(\underset{\sim}{k}+\underset{\sim}{q})\}^{-1} .$$

In conclusion, our quantum many-body theory of 2D electron solids in high magnetic fields gives consistent results and the correct classical limit for the ground-state energy. Various extensions of the presented results are possible in this approach to the ground-state energy $E(\nu)/N$, e.g., by taking the exchange interaction of the electrons into account [5] which, of course, has been suppressed through the limiting evaluations discussed in this paper.

We are grateful to Dr. V.J. Emery for useful conversations.

References

1. D.C. Tsui, H.L. Störmer, and A.C. Gossard, Phys. Rev. Lett. 48, 1599 (1982).
2. R.B. Laughlin, Phys. Rev. Lett. 50, 1395 (1983).
3. See, e.g., G. Meissner in: Series in Solid State Sciences, 51, Eds. W. Eisenmenger, K. Lassmann, and W. Güttinger, Springer, Berlin, Heidelberg, New York (1984), p. 263; K. Maki and V. Zotos, Phys. Rev. B28, 4349 (1983).
4. A.V. Chaplik, Zh. Eksp. Teor. Fiz. 62, 746 (1972) [Sov. Phys.-JETP 35, 395 (1972)]; G. Meissner, Z. Physik B23, 173 (1976).
5. G. Meissner and U. Brockstieger, to be published.
6. G. Meissner, H. Namaizawa, and M. Voss, Phys. Rev. B13, 1370 (1976).
7. L. Bonsall and A.A. Maradudin, Phys. Rev. B15, 1959 (1977).
8. For the role of dispersive anharmonicities in consistent approximations see: G. Meissner, Phys. Rev. Lett. 21, 435 (1968), and G. Meissner in: Lecture Notes in Physics 177, Ed. G. Landwehr, Springer, Berlin, Heidelberg, New York (1983), p. 70.
9. P.K. Lam and S.M. Girvin, Phys. Rev. B30, 473 (1984).

Energy Spectra of Interacting Two-Dimensional Electrons in a Magnetic Field

P.A. Maksym

Department of Physics, University of Leicester, Leicester, LE1 7RH, UK

The eigenstates of interacting two dimensional electrons in a magnetic field are studied theoretically for the case when the electron wave functions are subject to periodic boundary conditions. Classifying the states according to the irreducible representations of the appropriate magnetic space group enables them to be computed efficiently and leads to a detailed picture of their degeneracies. Numerical results for small systems are presented to show how the energies of the lowest few states depend on the number of flux quanta present and on the aspect ratio of the system. The energy spectra at various filling factors are found to behave differently when either the number of flux quanta or the aspect ratio are changed. In particular, when the aspect ratio is small the ground state energy has cusplike minima at 1/n fractional filling but the minima at even order fractional filling tend to disappear when the aspect ratio is increased. The energies of the low lying excited states also depend on aspect ratio but do not always have minima at 1/n filling. These effects are explained in terms of a competition between direct and exchange interactions which favours clustered configurations of electrons when the aspect ratio is large. The finite size scaling of the energy levels is briefly discussed.

1. Introduction

The discovery of the fractional quantum hall effect has stimulated great interest in the properties of interacting 2D electrons in a magnetic field. So far two theoretical approaches have been proposed to study the eigenstates of the electrons and their energy spectra. One approach, due to LAUGHLIN /1/, consists of minimizing the ground state energy with the aid of a variational wave function. The second approach is to numerically diagonalize the hamiltonian for a small number of electrons. This was pioneered by YOSHIOKA, HALPERIN and LEE /2/ and YOSHIOKA /3/ for electrons with periodic boundary conditions - the 'toroidal' geometry, and extended to other geometries by GIRVIN and JACAH /4/ (disk) and HALDANE and REZAYI /5/ (Spherical). The present work is also concerned with the numerical approach using the toroidal geometry. It consists of two parts. First, the symmetry properties of the hamiltonian are studied to investigate the possible degeneracies of the energy levels. Secondly, the dependence of the energy levels on the strength of the magnetic field and the aspect ratio of the periodic cell is studied in detail. A preliminary report of some of the results obtained is given here.

2. The Hamiltonian and its Symmetry Group

The n electrons interact with each other and with a uniform positive background charge. Periodic boundary conditions are imposed with an integral number, m, of magnetic flux quanta per periodic cell, so that the electrons also interact with their periodic images. All the electrons are taken to have the same spin, thus the filling factor, ν, is n/m. The hamiltonian consists of the following terms /2/:

$$H = \frac{1}{2m_o} \sum_{i=1}^{n} (\underline{p}_i + e\,\underline{A}_i)^2 + \frac{1}{2} \sum_{i \neq j}^{n} \sum_{l,k} \frac{e^2}{4\pi\varepsilon_o} \frac{1}{|\underline{r}_i - \underline{r}_j + l\underline{a} + k\underline{b}|} + H_{eb} + H_{bb} \; . \tag{1}$$

The first term is the kinetic energy of the electrons in the presence of the magnetic field, the second term is the electron-electron interaction energy and the last two terms are, respectively, the electron-background interaction and the interaction of the background with itself. The vectors $\underline{a}$ and $\underline{b}$ are the primitive vectors of the periodic array of cells. The eigenstates of the hamiltonian can be classified according to the irreducible representations of the symmetry group of the hamiltonian. In the absence of the magnetic field this group would simply be the space group constructed from the translational and point symmetry operators of the periodic array of cells. However in the presence of the magnetic field, the hamiltonian is not invariant under the action of these operators. Instead they have to be combined with other operators namely, gauge transformations or complex conjugation. BROWN /6/ has shown how to construct magnetic translation operators by combining the geometric translations with suitable gauge transformations. These operators form a sub-set of the full set of symmetry operators of H. The full set is constructed by combining the magnetic translation operators with operators derived from rotations or mirror operations /7,8/. In the symmetric gauge H is invariant under rotations, but in other gauges the geometric rotations may need to be combined with gauge transformations. For example, in the Landau gauge H is not invariant under the 90° rotation R, but is invariant under the operation exp $(-ie\,Bxy/\hbar)$ R where the exponential factor comes from a gauge transformation. The hamiltonian is not invariant under mirror operations in any gauge, however it is invariant under mirror operations combined with a suitable anti-unitary operator. Usually this is taken to be the time reversal operator, however in the case when all the electrons have the same spin it is more convenient to use the complex conjugation operator. Now denote the magnetic translation operators by $(0|\underline{\tau})$ where $\underline{\tau}$ is a geometric translation and $(0|\underline{0})$ is the identity operator; also let $(\alpha|\underline{0})$ denote one of the composite point operators, where α stands for a geometric rotation or mirror operation. The full set of symmetry operators of H is constructed by combining these operators in the same way that their geometric counterparts are combined to form a space group. However the rule for combining the resulting 'magnetic space operators' is

$$(\alpha|\underline{\tau}_1)(\beta|\underline{\tau}_2) = \exp\{-\frac{ien}{2\hbar}(\underline{\tau}_1 \times \alpha\underline{\tau}_2).\,\underline{B}\}(\alpha\beta|\underline{\tau}_1 + \alpha\underline{\tau}_2) \, . \tag{2}$$

Because of the phase factor in (2) the magnetic space operators form a ray representation of the geometric space group. Thus the eigenstates of H can be classified according to the appropriate irreducible matrix ray representations (alternatively it is possible to extend the order of the group to eliminate the phase factor in (2) /9/). This leads to a detailed picture of the possible degeneracies of the energy levels. Manipulating the full matrix representations is rather tedious because of the presence of the anti-unitary mirror operations which require special treatment /10/. Fortunately the results can easily be verified by considering the representations of the translational sub-group alone. These representations are all m/p dimensional where p is the highest common factor of m and n. Each can be labelled by a $\underline{k}$ vector in a suitable Brillouin zone and, physically, the y component of $\underline{k}$ is related to the momentum of the centre of mass. There are p^2 $\underline{k}$ points in the zone. For the case of a rectangular zone the $\underline{k}$ points all lie in the interior of the zone when p is odd, but when p is even there are $\underline{k}$ points on two if its edges (Fig.1). The degeneracy of the energy levels is a

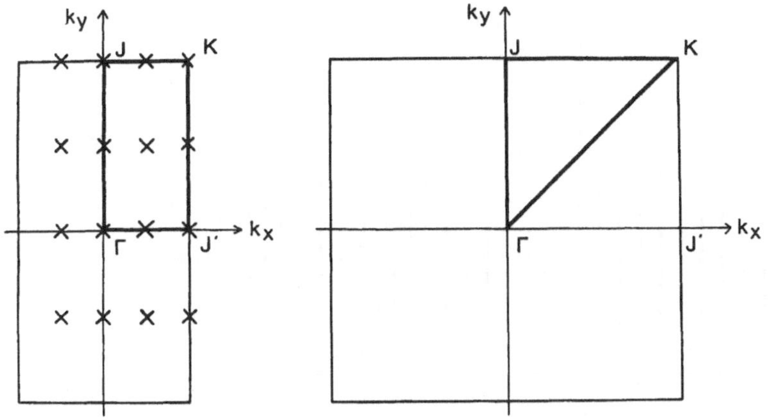

Fig. 1:
Rectangular and square Brillouin zones, showing irreducible segment and points and lines of high symmetry. The crosses indicate the allowed $\underline{k}$ points in the rectangular zone when p = 4.

multiple of m/p and the maximum possible degeneracy for electrons in a rectangular cell is 4 m/p. This corresponds to the 4 equivalent $\underline{k}$ vectors associated with the interior of the irreducible rectangle. The degeneracy is lower for levels associated with the edges of the irreducible rectangle, because the allowed $\underline{k}$ points occur only on two edges of the full zone. Thus the lines ΓJ', J'K, $\overline{KJ}$, JΓ are associated with a degeneracy of 2m/p and the points Γ, J', K, J are associated with a degeneracy of m/p. For electrons in a square cell the possible degeneracies are higher because of the higher symmetry: 8 m/p for $\underline{k}$ points equivalent to the interior of the irreducible triangle: 4 m/p for the lines ΓK, KJ, JΓ; 2 m/p at J; m/p at Γ and K. Together with an electron-hole symmetry at ½ filling, as described in the next section, these considerations explain the degeneracies found in the results of finite size numerical calculations.

3. Energy Spectra

The energy spectra are computed by numerically diagonalizing the hamiltonian, taking into account only the zeroth Landau level. In earlier work /2,3/ this was done by expanding the eigenstates in terms of Slater determinants of one electron Landau orbitals. However there is some advantage in choosing states which transform according to the irreducible representations of the appropriate magnetic space group: choosing states which transform according to the irreducible representations of the translational sub-group reduces the size of the hamiltonian matrix by a factor of 1/p and a further reduction by a factor of ∿ 1/2 can be obtained by utilizing the remaining symmetries. In the present work only the translational sub-group is used and the basis states are generated as follows. First a complete set of Slater determinants is found and sorted according to their values of $J = \sum_{i=1}^{n} j_i$ mod m where the centres of the one electron Landau orbitals are $X_j \equiv |\underline{a}|j/m$. This fixes k_y via the relation $k_y = 2\pi(p-J)/|\underline{b}|$. The next step is to construct states with a definite value of k_x. This is done by operating on the Slater determinants with projection operators constructed from the irreducible representations of the magnetic translation group. After some algebra it is found that the required basis states have the form

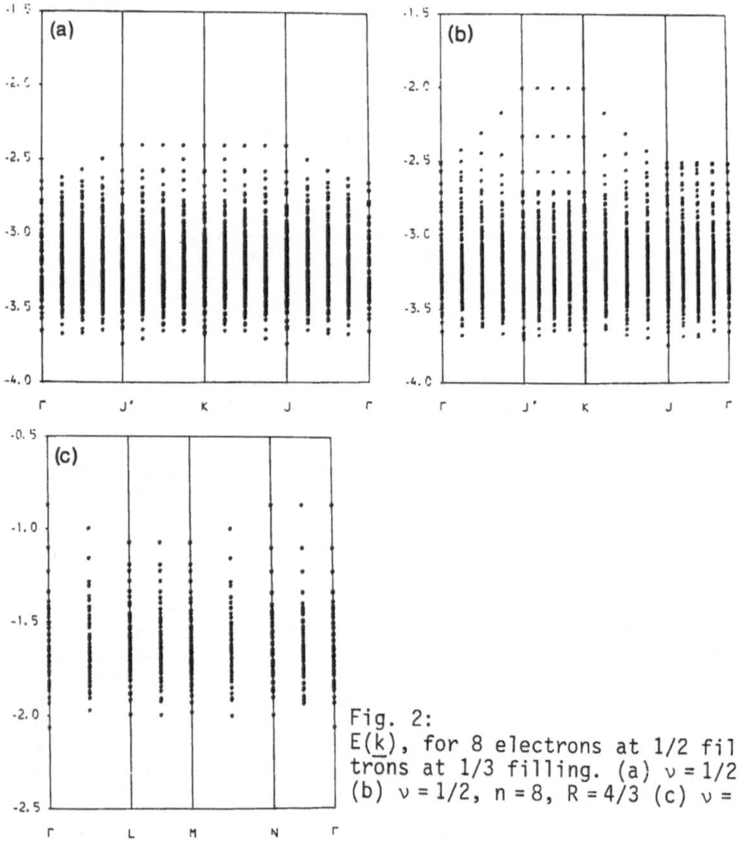

Fig. 2:
$E(\underline{k})$, for 8 electrons at 1/2 filling and 5 electrons at 1/3 filling. (a) $\nu = 1/2$, $n = 8$, $R = 1$ (b) $\nu = 1/2$, $n = 8$, $R = 4/3$ (c) $\nu = 1/3$, $n = 5$, $R = 4/3$

$$\sum_{q=0}^{p-1} \exp\{-ik_x|\underline{a}|q/p\} \mid j_1 + qm/p, \ldots . j_n + qm/p\rangle \ , \tag{3}$$

where $|j_1 \ldots j_n\rangle$ denotes a Slater determinant and sums are taken modulo m. A complete set of basis states is constructed by combining the complete set of Slater determinants according to (3).

Energy spectra for 8 electrons at $\frac{1}{2}$ filling and 5 electrons at 1/3 filling are shown in Fig. 2. The energies shown are total energies in units of $e^2/4\pi\varepsilon_0 l$ where $l^2 = \hbar/eB$ and the aspect ratio, R, is defined as $|\underline{b}|/|\underline{a}|$. Each frame shows $E(\underline{k})$ for a circuit around the irreducible segment of the Brillouin zone. For 8 electrons this includes the edges of the zone; for 5 electrons there are no $\underline{k}$ points on the edges and the circuit is around the interior points closest to the edges, the points L, M, N being adjacent to J', K, J respectively. The extra degeneracy for electrons in a square cell is clearly visible in the figures (compare (a), (b)). Detailed inspection shows that when the number of electrons is even there is a further degeneracy at Γ for $\nu = \frac{1}{2}$. In this case each level is 2m/p fold degenerate instead of m/p fold and this is thought to be a consequence of electron-hole symmetry. For n = 4 and m = 8 this can be verified by examining the elements of the hamiltonian matrix.

Overall the energy spectra for ½ and 1/3 filling look similar, but differences can be found by examining the low lying energy levels. Various authors have commented on the presence of a 'roton-like' minimum at 1/3 filling /5/, /11/. Here the properties of the ground state are considered in detail. For $\nu = 1/3$ the ground state is always found at K when n is even and at Γ when n is odd. In contrast, for $\nu = \frac{1}{2}$, the position of the ground state is very sensitive to the values of n and R. For n = 8 it is at $\underline{k} = (2\pi/|\underline{a}|, 2\pi/|\underline{b}|)$ when R = 1 and at K when R = 4/3. This $\underline{k}$ dependence is the origin of the degeneracy effects discussed by SU/12/. Further insight into the behaviour of the ground state energy, E_o can be obtained by studying its m dependence. This is shown in Fig. 3, together with the energy of the first excited state, E_1 for the n = 5 case with R = 0.1 and 1.0. The energy unit is $e^2/4\pi\varepsilon_o(2\pi/A)^{\frac{1}{2}}$ where A is the area of the cell. Clearly, there are cusp-like minima in $E_o(m)$ at $\nu = \frac{1}{2}$, 1/3, 1/4 when R = 0.1 but only at $\nu = 1/3$ when R = 1.0. The minima reflect the m dependence of the diagonal elements of H. These consist of the difference between a direct term, corresponding to the Coulomb repulsion, and a short-ranged exchange term /13/. When R = 0.1 the exchange term is negligible and the smallest diagonal elements of H correspond to regularly spaced configurations which minimize the Coulomb repulsion. The configuration $|\ 3\ 6\ 9\ 12\ 15\rangle$ is an example for n = 5, m = 15. This situation occurs when ν = 1/integer, which accounts for the cusps. At larger values of R the effect of the exchange term is to partly cancel the direct term for electrons whose orbit centres are sufficiently close. As a result the smallest diagonal elements of H correspond to configurations containing regularly spaced clusters of electrons (for example $|\ 1\ 2\ 7\ 8\ 12\rangle$ for n = 5, m = 15) and the ground state is dominated by these configurations. The clustering favours the cusps at $\nu = 1/3$ because all inequivalent clustered configurations then have the same J value. In contrast, at $\nu = \frac{1}{2}$, various inequivalent clustered configurations have different J values and this acts against the cusps. Note that the electron density remains substantially uniform despite the clustering: according to (3) all translated equivalents of any clustered configuration occur with equal intensity. Details are given in /13/.

Finally, it is natural to enquire how the energy levels scale with system size. The following intriguing observation sheds some light on this question. Suppose that the system size is increased by a factor s by placing s cells adjacent to each other. Then n, m and A increase by a factor s but R decrea-

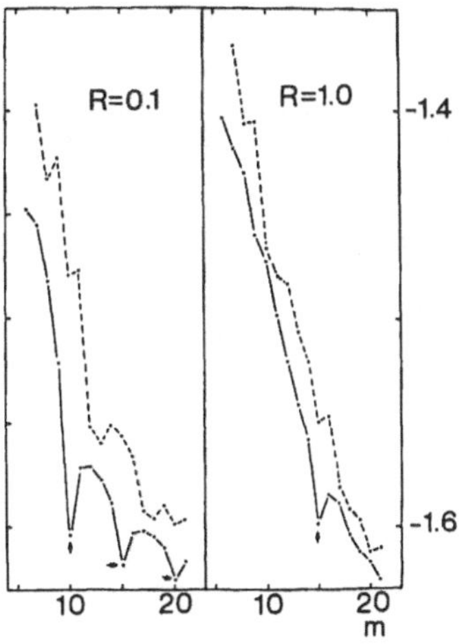

Fig. 3:
$E_o(m)$ and $E_1(m)$ for 5 electrons at $R = 0.1$ and 1.0. The points show the energy levels and the lines are to guide the eye. Arrows indicate the minima at 1/integer filling

ses by a factor s. This suggests comparison of E_0/n for n electrons at aspect ratio R with E_0/ns for ns electrons at aspect ratio R/s, with due allowance for the $1/s^{\frac{1}{2}}$ change in the energy unit $e^2/4\pi\varepsilon_0(2\pi/A)^{\frac{1}{2}}$. Results are shown in table 1 with energies expressed in these units, with A taken to be the area of the n electron system. The values of E_0/n and E_0/ns are remarkably close, the difference being < 3%. A similar comparison can be done for E_1. In this case the scaling is not so accurate, but still holds to within 10%.

Table 1: Comparison of energies per electron for various scaled system sizes

n/m	R	s	$E_0(n/m,R)/n$	$E_0(ns/ms, R/s)/ns$
2/4	1	2	- 0.975	- 0.948
3/6	1	2	- 1.14	- 1.16
3/9	1	2	- 1.236	- 1.235
2/4	0.9	3	- 0.983	- 0.961
2/6	0.9	3	- 1.029	- 1.013
2/4	1	4	- 0.975	- 0.946

Acknowledgement

I would like to thank Professor J.L. Beeby for a helpful discussion on the scaling of the energy levels.

References

1. R.B. Laughlin: Phys. Rev. B28, 4506 (1983)
2. D. Yoshioka, B.I. Halperin and P.A. Lee: Phys. Rev. Lett. 50, 1219 (1983)
3. D. Yoshioka: Phys. Rev. B29, 6833 (1984)
4. S.M. Girvin and T. Jacah: Phys. Rev. B28, 4506 (1983)
5. F.D.M. Haldane and E.H. Rezayi: Phys. Rev. Lett. 54, 237 (1985)
6. E. Brown: Solid State Physics 22, 313 (1968)
7. P.A. Maksym: J. Phys. C. Solid State Phys. 18, L433 (1985)
8. H. Overhof and U. Rössler: Phys. Status Solidi 26 461 (1968)
   W.G. Tam: Physica 42, 557 (1969)
9. J. Zak: Phys. Rev. 134A, 1602 (1964)
10. It is intended to give details in a subsequent publication; see also J.O. Dimmock and R.G. Wheeler in: The Mathematics of Physics and Chemistry Vol. 2, eds: H. Margenau and G.M. Murphy (Van Nostrand, Princeton, 1964)
11. D. Yoskioka: J. Phys. Soc. Japan 55, 885 (1986)
12. W.P. Su: Phys. Rev. B 30, 1069 (1984)
13. P.A. Maksym: J. Phys. C.: Solid State Phys. 19, L247 (1986)

High-Frequency Conductivity in the Quantum Hall Effect Regime

F. Kuchar[1], *R. Meisels*[1], *K.Y. Lim*[1], *P. Pichler*[2], *G. Weimann*[3], *and W. Schlapp*[3]

[1]Institut für Festkörperphysik, Universität, A-1090 Wien, Austria
and
Ludwig Boltzmann Institut für Festkörperphysik, Kopernikusgasse 15, A-1060 Wien, Austria

[2]Institut für Physik, Montanuniversität, A-8700 Leoben, Austria

[3]Forschungsinstitut der Deutschen Bundespost, D-6100 Darmstadt, Fed. Rep. of Germany

Quantum Hall effect (QHE) behavior of the Hall conductivity σ_{xy} is observed in multiple (MHS) and single heterostructures (SHS) of GaAs/$Al_xGa_{1-x}As$ at microwave frequencies (Ka band). Integer plateaus with even filling factor i are observed in σ_{xy} of the MHS as well as of the SHS with 2D electron and hole gas layers. In the case of the 2DEG in a SHS also the i=3 plateau is resolved. With this SHS it is clearly shown that the σ_{xy} plateaus occur equidistant on a 1/B scale and that the σ_{xy} values on the plateaus are multiples of a fundamental value.

At submillimeter frequencies of about 1000 GHz σ_{xy} measured on one of the MHS's is a smooth function of the magnetic field with weak structure above field values corresponding to ω_c, which cannot be related to the Hall plateaus of the d.c. experiments. This implies that a high-frequency breakdown of the IQHE occurs at frequencies between about 1/30 and 1/3 of ω_c.

1. Introduction

Most of the Quantum Hall effect experiments on MOSFET's or heterostructures are performed with d.c. /1/. For a complete understanding of the effect also the high-frequency behavior should be known. Experiments performed in the range 100 kHz - 50 MHz are discussed in /2/ and /3/. Experiments with 100 ns pulses /4, 5/ and a.c. experiments at frequencies up to 45 MHz /6/ in structures without gates showed no difference to the d.c. behavior. Some of the a.c. experiments /7 - 10/ - on gated structures - showed a so-called low-frequency breakdown of the integer Quantum Hall effect (IQHE). As an explanation of this observation an effective delocalization of semiclassical orbits in the sample was suggested /8, 10/. Experiments at microwave frequencies /2, 3/ (~ 30 GHz), however, clearly showed that the integer plateaus in GaAs/AlGaAs multiple heterostructures are not destroyed even at frequencies several orders of magnitude higher than used in /7 - 10/.

Experiments at high frequencies are therefore of great interest regarding the role of the localization in the case of the IQHE. Two frequency ranges seem to be particularly interesting: (a) Much higher than MHz but still small compared with the cyclotron resonance frequency ω_c, (b) close to ω_c. At ω_c inter-Landau-level transitions should destroy the IQHE. Due to the broadening of the Landau levels this might occur already at $\omega < \omega_c$. Apparently, it does not occur at 30 GHz /2, 3/ which is one to two orders of magnitude below ω_c under usual experimental conditions.

In this paper we report new experiments on GaAs/AlGaAs heterostructures in the microwave as well as in the submillimeter range. For measuring the Hall conductivity σ_{xy} in the microwave range a crossed waveguide arrangement is used /2/, in the submillimeter range an equivalent arrangement with linear grid polarizers and an optically pumped laser system.

2. Experimental Techniques

The contactless technique applied to measure directly the Hall conductivity σ_{xy} of the 2D carriers in GaAs/Al$_x$Ga$_{1-x}$As heterostructures uses a crossed polarizer-analyser arrangement in Faraday geometry. As shown in /3/ the electric field component of the transmitted radiation in this arrangement is proportional to σ_{xy} and hence the intensity proportional to σ_{xy}^2. An estimate of the magnitude of the intensity relative to the incident one gave a value of about 10^{-5} for a single GaAs/Al$_x$Ga$_{1-x}$As heterostructures at the i=2 plateau. With multiple heterostructures the intensity is multiplied by the square of the number of the 2DEG layers.

For the two spectral ranges investigated, two different designs of the polarizer-analyser system were used. In the microwave range (Ka band) it essentially consisted of crossed waveguides (cross section each 7.12 x 3.56 mm^2). As the microwave source a Hewlett-Packard 8690 B sweep oscillator with a 8697 A rf unit was used. The microwave power at the position of the sample was of the order of 1 mW. A reduction by a factor of 10 did not change the shape of the experimental curves. The intensity transmitted through the crossed waveguide (length 20 cm) was measured with a liquid-helium cooled bolometer. The sample was positioned in a flange (clear cross section 7.12 x 7.12 mm^2) in between them. In this context it is important to mention that the sample areas varied between 2.8 x 2.8 and 7 x 7 mm^2. In the first case the sample edges are inside the common cross section (3.56 x 3.56 mm^2) of the crossed waveguides, in the second case well outside of it.

In the submillimeter range an optical cryostat with a split-coil magnet was used. Linear-grid polarizers were used outside the cryostat in crossed positions to select again the signal proportional to σ_{xy}^2. Apertures smaller than the sample areas were used. Measuring temperatures were 4.2K for the submillimeter experiments, for the microwave experiments also 2.1K.

In both types of experiment the samples were shielded against visible and room temperature radiation by cold black polyethylene foils. Parameters of the samples used are given in Table I. The samples were grown by molecular beam epitaxy /11/. As concluded from the growth and doping conditions, the quantum wells are essentially rectangular in sample 1408 /3,11/. In sample 1325 the doped AlGaAs regions are situated asymmetrically so that well separated, single-sided heterostructures are produced in the GaAs wells. The 2DEG layers mainly form at the GaAs/AlGaAs interfaces (with re-

Table I. Parameters of the samples at T=4.2K. For further details see /11/. N_s is the total carrier density calculated from low-field Hall effect data.

Sample No.	Number of GaAs layers	Type of the 2D carriers	N_s (cm^{-2})	μ (cm^2/Vs)	Sample areas (mm^2)
1408a,b	15	n	$8.0x10^{12}$	$2.0x10^4$	7x7, 2.8x2.8
1325	10	n	$7.4x10^{12}$	$1.7x10^5$	7x5
1320	1	n	$5,2x10^{11}$	$3.0x10^5$	2.8x2.8
1360	1	n	$2.8x10^{11}$	$6.1x10^5$	7x7
1456	1	p	$4.0x10^{11}$	$5.0x10^4$	7x7

spect to the growth direction). Also at the AlGaAs/GaAs interfaces 2DEG layers exist, however, with a much lower electron density. This was concluded from d.c. ρ_{xx} measurements where a second, weak oscillation was observed. The period was approximately 12 times larger than the one corresponding to the main 2DEG layers. For the latter ones a carrier density of $6.9 \times 10^{11} cm^{-2}$ per layer was deduced from the period of the oscillations. For sample 1408 the QHE data showed that one of the 15 layers is depleted.

3. Experimental Results and Discussion

3.1 Microwave Experiments: Cyclotron Resonance

Figure 1 shows results for the single heterostructure 1360 in the crossed waveguide arrangement. Drawn is the part at very low magnetic fields, up to 0.2 Tesla. There is the prominent peak of the cyclotron resonance. The magnetic field is below the quantum limit, so this experiment can be described in classical terms.

The bolometer signal is, as mentioned in Chapter 2, proportional to the absolute square of the Hall-conductivity σ_{xy}. Using a classical Drude-type model we can write

$$\sigma_{xy} = \frac{n_s e^2}{m^*} \frac{\omega_c \tau^2}{(1-j\omega\tau)^2+\omega_c^2\tau^2} , \qquad (1)$$

where n_s is the sheet carrier concentration, ω_c and ω are the cyclotron and microwave angular frequencies, respectively. τ is the mean collision time. The measured signal S is proportional to $|\sigma_{xy}|^2$. The maximum signal is situated at a magnetic field corresponding to $\omega_{c,max}=\sqrt{\omega^2+\tau^{-2}}$, somewhat above ω_c. In the multiple heterostructures (Chapter 3.2) $\omega\tau$ is smaller than unity and $\omega_{c,max}$ is dominated by τ, hence $\omega_{c,max}\ \tau \approx 1$. For the determination of the effective mass from the microwave transmission in the crossed

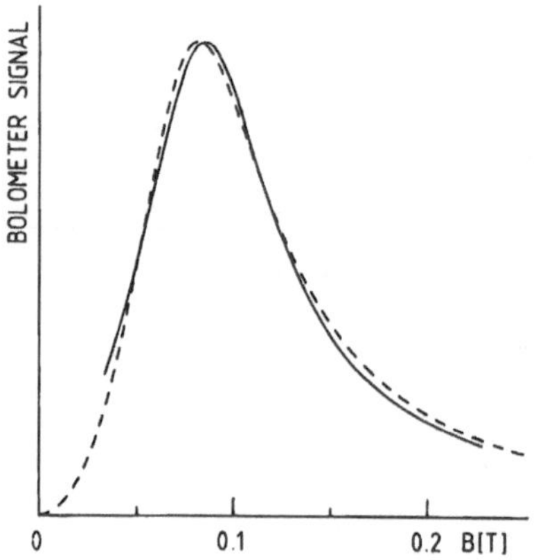

Fig.1: Cyclotron resonance of the 2D electrons in sample 1360 observed in the bolometer signal proportional to σ_{xy}^2. ν=28 GHz, T=4.2K. Dashed curve: fit according to (1)

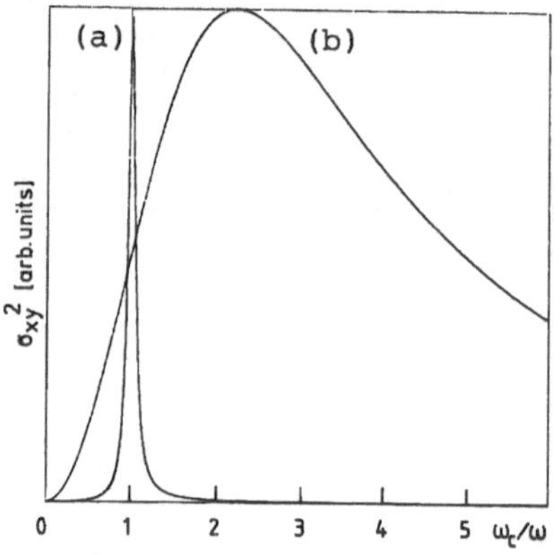

Fig.2: σ_{xy}^2 in the cyclotron resonance regime calculated according to (1); normalized to maximum. (a) weak damping, $\omega\tau=20$, (b) strong damping, $\omega\tau=0.2$

waveguide arrangement, one has to fit (1) to the experimental plot of the bolometer signal vs. magnetic field. Fit parameters are m*, entering via ω_c, and τ.

The ratio of full width at half maximum to the magnetic field at the maximum is $2/\sqrt{1+\omega^2\tau^2}$, giving a good starting value for the fit. From the fit a value of $m^*=0.067\ m_o$ was obtained, which is very close to values determined recently for 3D and 2D electrons in GaAs at far-infrared frequencies /12, 13/. The same value was also found in the usual transmission experiments, where the waveguides in front and behind the sample were not crossed.

For comparison with the experimental data of Fig. 1 , Fig. 2 shows calculated transmission-versus-field curves, according to (1). ω is held fixed, while τ is varied. For high ωτ values the peak is fixed at $\omega_c=\omega$, while for $\omega\tau \ll 1$ it has no relation to cyclotron resonance, and appears at $\omega_c\tau=1$.

3.2 Microwave Experiments: Quantum Hall Effect Regime

Figure 3 shows data recorded for the sample 1408b and 1325. The bolometer signal is proportional to σ_{xy}^2 as described in Chapter 2. In both samples structure is observed which at high magnetic fields clearly shows plateaus of σ_{xy}. These observations can be interpreted (a) by using the classical expression for the Hall conductivity σ_{xy} as in paragraph 3.1, and (b) by using the d.c. data of the resistivity components of ρ_{xx} and ρ_{xy}.

The results of a fit of (1) to the experimental data of sample 1408b is shown as the dashed curve in Fig. 3a. As for sample 1408a (7x7 mm^2) of /2/ the overall shape of the experimental curve is

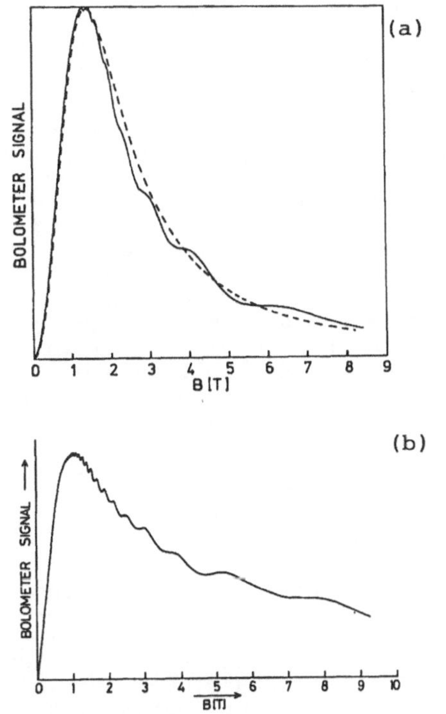

Fig.3: Bolometer signal ($\sim\sigma_{xy}^2$): (a) sample 1408b, T=4.2K, ν=33.4 GHz, dashed curve fit according to (1), details see text; (b) sample 1325, T=2.1K, ν=32.7 GHz. On an enlarged scale oscillations can be seen up to i=48. i=4 is at 5.8 and 7.5T, resp.

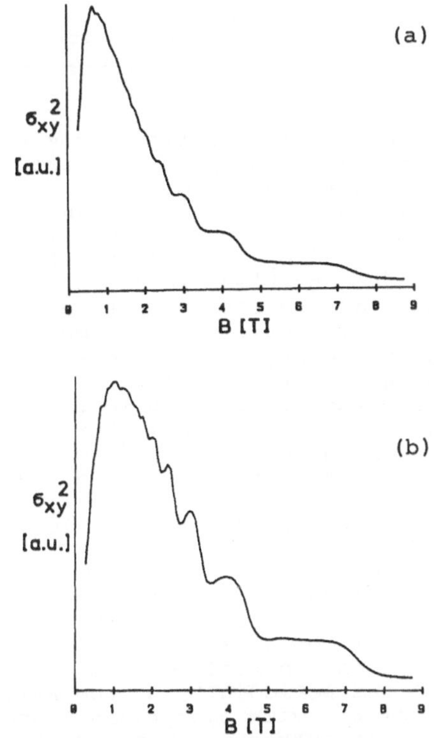

Fig.4: Model calculation of σ_{xy}^2 from the approximate values of ρ_{xx} and ρ_{xy} of sample 1408 (a). Diagram (b) qualitatively corresponds to sample 1325 where the ρ_{xx} contribution is enhanced compared with (a)

nicely reproduced by the calculation. Even so, the single-parameter (τ) Drude-type fit (fixed m*) is somewhat problematic. Actually two different τ values had to be used in $\omega_c\tau$ and $\omega\tau$, respectively. $\omega_c\tau$ determines the position of the peak of σ_{xy}, $\omega\tau$ the shape, in particular the width, of the curve. τ calculated from $\omega_c\tau=1$ is 2.8×10^{-13}s, τ from $\omega\tau$ is 6.6×10^{-12}s. For sample 1408a of /2/ the fit with a single $\tau=3.4\times10^{-13}$ could be applied. Since $\omega\tau$ was much smaller than unity, the fitted σ_{xy} curve corresponded to the classical linear dependence of σ_{xy} on magnetic field.

The plateau behavior is somewhat better pronounced in sample 1408b than in 1408a of /2/. We do not believe that this is due to the smaller area (fitting into the common cross section of the crossed waveguides) but due to small variations of the quality of the 2D layers across the wafer.

For sample 1325 the fit yielded too low values at high field strengths when adjusted to the maximum of the curve. This is possibly due to the existence of a second type of 2DEG layers as mentioned in Chapter 2.

We could also qualitatively reproduce the microwave data of σ_{xy} of the two samples from the d.c. values of ρ_{xy} and ρ_{xx} by using $\sigma_{xy}=\rho_{yx}/(\rho_{xx}^2+\rho_{xy}^2)$. This equation is valid only if $\omega\tau<<1$ i.e. if the cyclotron resonance does not contribute to σ_{xy}. Although this is not exactly fulfilled, the σ_{xy} values calculated from ρ_{xx} and ρ_{xy} data (Fig. 4) show the essential features that were also observed experimentally in the microwave experiment (Since for the d.c. measurements only van-der-Pauw samples on very small pieces of the same structures could be made no quantitative comparison is presented): In the high mobility sample 1325 the quantum oscillations of ρ_{xx} and the Hall plateaus of ρ_{xy} are visible down to lower magnetic fields, i.e. up to higher filling factors, than in the low mobility sample 1408. As an interesting detail, the more pronounced ρ_{xx} oscillations of sample 1325 have the effect of producing minima in σ_{xy} at high filling factors where ρ_{xx} is not negligible compared with ρ_{xy}. At low filling factors (high magnetic fields) σ_{xy} is practically equal to $1/\rho_{xy}$.

The microwave data of sample 1320 are analyzed in a different way as follows. Since the bolometer signal BS of Fig.5a is proportional to σ_{xy}^2 a plot of the square root of the bolometer signal versus the reciprocal magnetic field (1/B) yields information about the periodicity of the Hall-plateau structure. For obtaining this plot a weak background signal, BKG, (not contributed by the 2DEG), is extracted in a plot of BS versus $1/B^2$. The pure signal from the 2DEG, $(BS-BKG)^{1/2} \sim \sigma_{xy}$, versus 1/B is shown in Fig.5b. Clearly, three interesting features can be observed: (i) The centers of the plateaus are equidistant in 1/B. (ii) The plateau va-

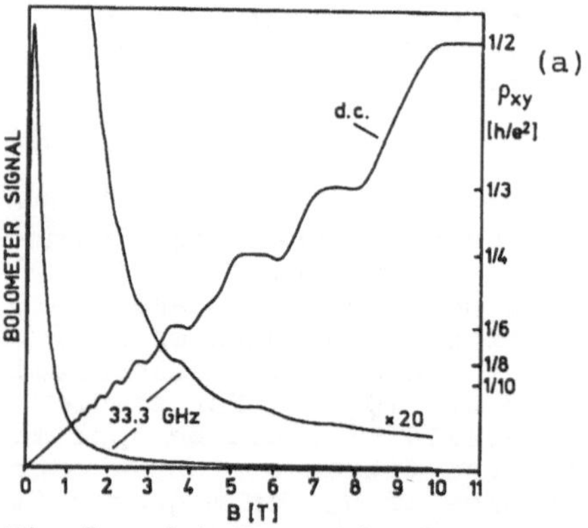

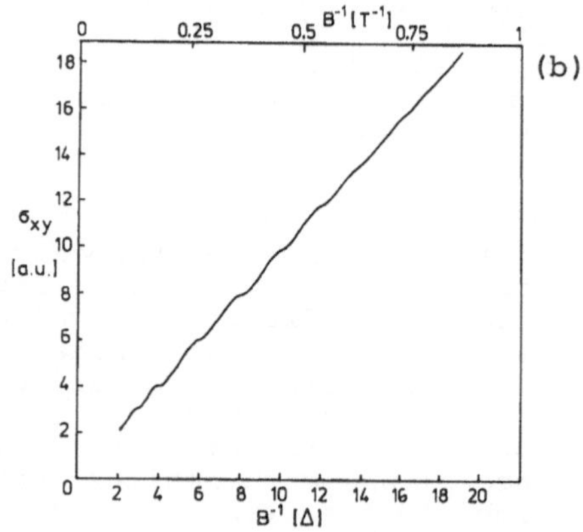

Fig.5: (a) Magnetic field dependence of the bolometer signal at 33.3 GHz and of ρ_{xy} (d.c.) for sample 1320. T=2.1K. (b) σ_{xy} vs. B^{-1}. The lower B^{-1} scale is given in multiples of the period $\Delta(B^{-1})$, the σ_{xy} scale in multiples of a fundamental value. Both were calculated from the corresponding values of the i=4 plateau

lues of the microwave Hall conductivity σ_{xy} are multiples of a fundamental value. (iii) Also the i=3 plateau is clearly resolved as in the d.c. data. This shows that the g factor enhancement /14/ is observable in d.c. as well as microwave transport experiments.

At first sight thc tilted plateaus of sample 1320 at high filling factors might lead to the conclusion that these plateaus are distorted as a consequence of using microwave frequencies. An analysis of (1) shows, however, that the steep σ_{xy} behavior close to the cyclotron resonance frequency causes the tilting of the plateaus. At our highest magnetic fields the influence of the cyclotron resonance is negligible and flat plateaus in σ_{xy} (corresponding to the d.c. ρ_{xy} plateaus) appear.

Also shown in Fig. 5a are d.c. ρ_{xy} data. These d.c. data were measured on a different piece of the same epitaxial layer, but showed a slightly different carrier concentration (by a factor of 1.063 smaller). In order to facilitate a comparison with the microwave results, the magnetic field scale was multiplied by the same factor. The minima on the high field side of the plateaus occur due to a ρ_{xx} admixture. Nevertheless, the width of the plateaus is clearly observable. The plateaus seem to be somewhat narrower in the microwave experiment. However, (i) one has to consider the remark on the tilting of the plateaus made above and (ii) that the measurement of a radiation intensity will never be as accurate as a d.c. resistance measurement. Therefore, we have to leave open at present whether the breakdown of the IQHE begins in the outermost regions of the plateaus already at about 30 GHz. The main part of the plateaus certainly still exists at this frequency.

Results regarding the microwave Quantum Hall effect of the two-dimensional hole gas (2DHG) are shown in Fig.7. The i=2 plateau is well developed at T=2.1K. Since the mobility of 2D holes still increases at lower temperatures /15/, plateaus at higher filling factors are expected to appear there.

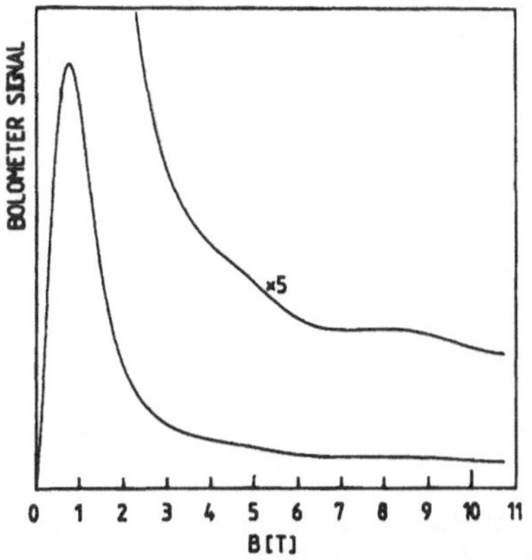

Fig.6: Bolometer signal for the single heterostructure 1456 with a 2D hole gas. T=2.1K, ν=30.4 GHz

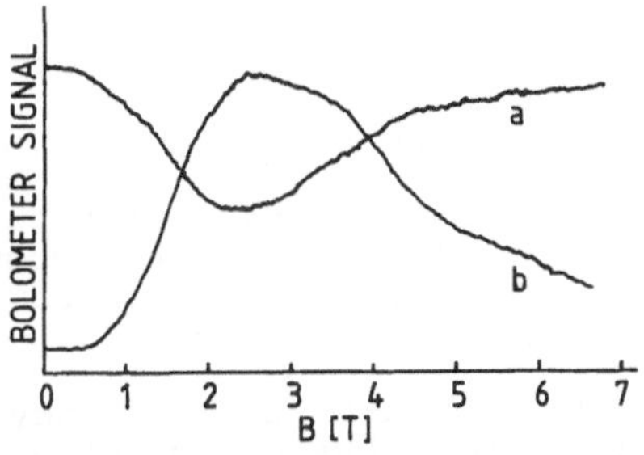

Fig.7: Submillimeter signal of sample 1408b observed without (a) and with (b) a crossed analyser. λ=305 μm, T=4.2K

For all of our samples with 2DEG layers the field values of the IQHE plateaus of the microwave experiments coincide with those of d.c. experiments. For the 2DHG layer we have not taken d.c. data.

3.3 Submillimeter Experiments: Cyclotron Resonance and Quantum Hall Effect Regime

In the submillimeter range the 305 μm (984 GHz) line of the CO_2-laser-pumped CH_3OD laser was used. In the 2DEG the cyclotron resonance occurs at about 2.7T in a SHS (sample 1320). Fig. 7 shows the signal of the Germanium bolometer with and without the crossed analyser. The cyclotron resonance is very broad in both types of experiments. There is no clear evidence of the plateau behavior so well developed at frequencies about a factor of 30 lower. The temperature of 4.2K, which was the lowest achievable with the split-coil magnet, is not too high to obscure the plateaus; in the microwave experiments they were clearly observable also at this temperature.

4. Conclusions

The "σ_{xy} spectroscopy" technique turned out to be extremely useful to investigate the high-frequency conductivity of 2DEG and 2DHG layers of GaAs/AlGaAs heterostructures in the microwave and submillimeter ranges. The main conclusions of our work are:

(i) The cyclotron resonance of 2D electrons can be observed in σ_{xy} at microwave frequencies and yields an effective mass value of 0.067 m_o, very close to the band edge values obtained for 2D and 3D electrons from far-infrared experiments.

(ii) Plateaus of the Hall conductivity σ_{xy} appear at microwave frequencies as high as the Ka band in the 2DEG and 2DHG of single and multiple heterostructures at high magnetic fields. This reflects the relation $\sigma_{xy} \approx 1/\rho_{xy}$ valid there. At low fields also the influence of ρ_{xx} on σ_{xy} could be observed in the multiple heterostructure with high electron mobility.

(iii) As in d.c. data on the 2DEG-SHS,the i=3 plateau and henc the spin splitting is clearly resolved. This is a strong indication that the g factor enhancement /14/ is independent of the measuring frequency, even if it is very close to the free-carrier spin resonance frequency /16/ as in the case of our sample 1320.

(iv) Edge effects seem to play no important role for the observation of the IQHE at microwave frequencies: Corresponding results were obtained with sample dimensions smaller and larger than the common cross section of the two crossed waveguides.

(v) As regards the accuracy of the IQHE, the microwave data cannot compete with d.c. data. Even so, the quantized behavior of σ_{xy} is demonstrated quantitatively in sample 1320: The σ_{xy} plateaus occur equidistant on a 1/B scale, their absolute values (square root of the bolometer signal) are multiples of a fundamental value. This was also reported recently by Volkov et al. /17/ for a GaAs/AlGaAs SHS in magnetic fields up to 3T.

(vi) At submm frequencies, the plateaus are not observed when ω_c is up to a factor of 3 higher than ω. This means that a high-frequency breakdown of the integer quantization occurs at frequencies somewhere between 1/30 and 1/3 of the cyclotron resonance frequency. It further means that disorder potentials play an important role for the high-frequency breakdown of the IQHE. Unfortunately, the range of the disorder potential cannot be determined from our experiments, since fluctuations of the density of the relevant centers are responsible for it. An interpretation within the model of semiclassical orbits /10,18,19/ would mean that the closed contours (closed orbits of localized states) cannot be much longer than several cyclotron orbits, which limits the validity of such a model. Certainly the orbits are much smaller than usual sample dimensions.

Acknowledgement

The authors thank Professors K. Seeger and G. Bauer for valuable discussions and their interest in this work. It was partly supported by "Fonds zur Förderung der wissenschaftlichen Forschung", Austria, project no. P5247.

1. See e.g. papers in Surface Sci. 131, Nos.1-3 (1982); 142, Nos.1-3 (1984).
2. F. Kuchar, R. Meisels, G. Weimann, and W. Schlapp: Phys.Rev. B 33, 2965 (1986).
3. R. Meisels, K.Y. Lim, F. Kuchar, G. Weimann, and W. Schlapp: In Springer Ser. Solid-State Sci. 67, 184 (Springer, Berlin, Heidelberg, 1986).
4. R. Woltjer, M. Mooren, J. Wolter, and J.P. André: Solid State Commun. 53, 331 (1985).
5. F. Kuchar, R. Meisels, G. Weimann, and H. Burkhard: Proc.17th Int. Conf. Phys.Semincond., San Francisco, eds. J.D. Chadi and W.A. Harrison, Springer, 1985.
6. B.B. Goldberg, T.P. Smith, M. Heiblum and P.J.Stiles: Proc. Yamada Conference XIII on Electronic Properties of Two-Dimensional Systems (to be published).
7. A.P. Long, H.W. Myron, and M. Pepper: Proc. 17th Int.Conf. Phys.Semicond., San Francisco, eds. J.D. Chadi and W.A. Harrison, Springer, 1985, p.279.
8. T.G. Powell, R. Newbury, C. McFadden, H.W. Myron, and M. Pepper: J.Phys. C 18, L 497 (1985).
9. M. Pepper and J. Wakabayashi: J. Phys. C 16, L 113 (1983).
10. R. Joynt: J. Phys. C 18, L 331 (1985).
11. G. Weimann and W. Schlapp: Appl. Phys. A 37, 3057 (1985).
12. W. Seidenbusch, G. Lindemann, R. Lassnig, J. Edlinger, and E. Gornik: Surf. Sci. 142, 375 (1984).
13. G. Lindemann, R. Lassnig, W. Seidenbusch, and E. Gornik: Phys.Rev. B 28, 4693 (1983).
14. Th. Englert, D.C. Tsui, A.C. Gossard, and Ch. Uihlein: Surface Sci. 113, 295 (1982).
15. G. Weimann and W. Schlapp: In Springer Ser. Solid-State Sci. 67, 33 (Springer, Berlin, Heidelberg, 1986).
16. D. Stein, K.v.Klitzing, and G. Weimann, Phys.Rev. Letters 51, 130 (1983).
17. V.A. Volkov et al.: JETP 43, 255 (1986) (in Russian).
18. S.M. Apenko and Yu.E. Lozovik: J.Phys.C 18, 1197 (1985).
19. R.F. Kazarinov and Serge Luryi: Phys.Rev. B 25, 7626 (1982); B 27, 1386 (1983).

A New Approach to the Quantum Hall Effect

R. Woltjer, R. Eppenga, and M.F.H. Schuurmans
Philips Research Laboratories, NL-5600 JA Eindhoven, The Netherlands

A new model is proposed to explain important experimental results on the Quantum Hall effect. The existence of plateaus in the Hall resistance and broad minima in the magnetoresistance can be explained without explicit reference to localized electron states. The main assumption is a moderate inhomogeneity in the electron density across the width of the sample. Earlier experiments on the distribution of the Hall voltage across the sample can be explained when a gradient in the electron density is assumed.

1. INTRODUCTION

After the discovery of the Quantum Hall effect by VON KLITZING et al. [1] an enormous effort started to explain the width of the plateaus and the exact value of the Hall resistance at these plateaus. The exact value $R_H = h/ie^2$ has convincingly been explained by LAUGHLIN [2] with important additions by HAJDU [3], but the width of the plateaus is still a controversial subject. Flat plateaus in the Hall resistance and broad minima in the magnetoresistance are commonly explained by localisation of the carriers in the tails of the density of states in each Landau level caused by the high magnetic field [4]. To explain the broad plateaus at low temperatures in low mobility samples almost all electrons (95%) must be localized [5].

A number of different mechanisms of localisation has been proposed, but only few of them are able to explain the large degree of localisation needed in the magnetic field. The semi-classical localisation due to a smoothly varying potential is the most studied one [6,7,8]. Without inelastic scattering the cyclotron orbits drift along the equipotential lines under influence of the Lorentz force. Closed equipotential lines (around a local potential minimum or maximum) correspond to the localized states; electrons on such closed lines cannot move from one current contact to the other and thus cannot contribute to the d.c. current through the sample. A.C. electric fields at high frequencies will give the electrons the possibility to contribute to the a.c. current by travelling up and down their equipotential orbits. This gives rise to delocalisation at higher frequencies, as calculated by JOYNT [9]. Experiments at high frequencies set a severe limit on the spatial extension of these closed equipotential lines [10,11], because no frequency dependence was found up to 30 GHz in high mobility samples.

Up to now localisation has only been used for a qualitative description of the observed magnetotransport effects. Quantitative theoretical predictions for the Hall resistance and the magnetoresistance as a function of the magnetic field (or the electron density) are not available and detailed comparison with experimental results is therefore not possible.

2. QUANTUM HALL EFFECT IN THE ABSENCE OF LOCALISATION

We present a new model for the explanation of the Quantum Hall effect. In particular we show that a moderate inhomogeneity in the electron density across the sample already leads to the Quantum Hall effect without invoking the localisation corresponding to such an inhomogeneity. Thus localisation seems not to be essential for the explanation of the broad and flat plateaus in the Hall resistance. The existence of inhomogeneity has been demonstrated in several experiments [12,13,14] and a systematic inhomogeneity can be created by using a tilted gate on a sample [15].

In our model measured voltages (V_{xx} and V_H) are determined by integrals over the local resistivity tensor of a two-dimensional electron gas. The Hall voltage in a rectangular Hall bar geometry is then given by

$$V_H(x) = \int [\, \rho_{xy}(x,y) J_x(x,y) + \rho_{xx}(x,y) J_y(x,y) \,]\, dy, \tag{1}$$

where ρ_{xx} $(= \rho_{yy})$ and ρ_{xy} $(= -\rho_{yx})$ are the xx and the xy elements of the isotropic local resistivity tensor and J_x and J_y are the x and y components of the current density. The coordinates x and y refer to the long and the short axis of a Hall bar, respectively. The total current flowing in the x-direction through the sample is given by

$$I = \int J_x(x,y)\, dy. \tag{2}$$

It is independent of x because of current conservation. Combination of {1} and {2} yields the Hall resistance

$$R_H(x) = \frac{V_H(x)}{I} = \frac{\int [\, \rho_{xy}(x,y) J_x(x,y) + \rho_{xx}(x,y) J_y(x,y) \,]\, dy}{\int J_x(x,y)\, dy}. \tag{3}$$

The distribution of the current over the sample is determined by Kirchhoff's laws. A component of the current in the y-direction will lead to a mixing of ρ_{xx} in the Hall resistance as often observed in experiments. When the resistivity tensor is homogeneous across the sample {3} leads to the well known result

$$R_H(x) = \rho_{xy} \left[1 + \cot\!an\, \Theta_H \frac{\int J_y(x,y)\, dy}{\int J_x(x,y)\, dy} \right], \tag{4}$$

where $\Theta_H = \arctan(\rho_{xy} / \rho_{xx})$ is the Hall angle and the part in square brackets describes the geometrical correction factor [15]. For x far from the ends of a long Hall bar we have $J_y(x,y) = 0$ and therefore $R_H = \rho_{xy}$.

To get some feeling for the influence of inhomogeneities on magnetotransport we have carried out calculations in a simplified situation. These model calculations without invoking localisation, and with a number of simplifications, yield flat plateaus in the Hall resistance as a function of the magnetic field. By assuming a systematic inhomogeneity (a few percent) of the electron density across the sample we have been able to reproduce the distribution of the Hall voltage over the sample as measured by WOLTJER et al. [13] and EBERT et al. [16].

3. THE LOCAL RESISTIVITY TENSOR

In a homogeneous two-dimensional electron gas of density n_e, and in the absence of scattering, we have, in a perpendicular magnetic field, $\rho_{xy} = B/en_e$ and $\rho_{xx} = 0$. For ρ_{xy} we ignore the small effects of scattering and take $\rho_{xy} = h/ie^2$, where $i = hn_e/eB$ is the filling factor of the Landau levels. For ρ_{xx} we derive a simple expression, that incorporates the effects of scattering, starting from

$$\rho_{xx} \sim \sum_k \sum_l \iint A_{kl} D_k(E) f(E) D_l(E')\{1-f(E')\} dE dE', \qquad \{5\}$$

where A_{kl} is the scattering probability from Landau level k to Landau level l, $D_k(E)$ is the density of states in Landau level k and f(E) the Fermi-Dirac distribution. We only consider elastic scattering in the Born approximation thereby eliminating localisation, i.e. $A_{kl} \sim (k + 1/2)\delta_{kl}$. When we ignore the spin-splitting and the broadening of the Landau levels, the density of states in Landau level k is given by

$$D_k(E) = (eB/h)\delta(E - E_k), \qquad E_k = (k + 1/2)\hbar\omega_c , \qquad \{6\}$$

where $\delta(\ldots)$ is the Dirac delta function. The Fermi-Dirac distribution f(E) is used to calculate the occupation of the Landau levels. Inserting these approximations in {5} gives

$$\rho_{xx} \sim \sum_k (k + 1/2)\ (eB/h)^2 f(E_k)\{1 - f(E_k)\}. \qquad \{7\}$$

The chemical potential in $f(E_k)$ is determined by the requirement that the sum of $f(E_k)$ over k equals the filling factor i. We now have a simple $\rho_{xx}(B/n_e)$ with thermally activated minima at integral filling factors. In figure 1, ρ_{xx} as well as the chemical potential are given for three temperatures as a function of the reciprocal filling factor.

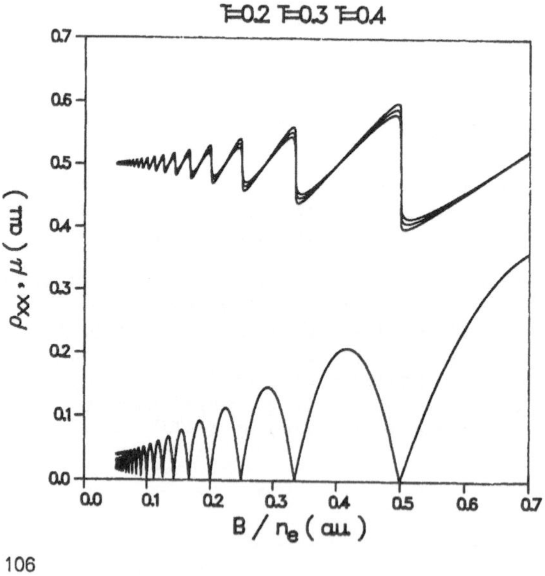

1: Calculated magnetoresistivity and chemical potential as a function of the reciprocal filling factor at three different temperatures (at B = 0.5 these temperatures correspond to $\hbar\omega_c/k_BT$ = 24, 32 and 48, respectively). This simple ρ_{xx} is used as input for our model calculations.

For the inhomogeneous two-dimensional electron gas we take the expressions for ρ_{xx} and ρ_{xy} as obtained for the homogeneous gas, but with $n_e = n_e(x,y)$. The spatial dependence of the resistivity tensor originates from the spatial distribution of the electron density only. We stress that the resistivity tensor as chosen does not contain the effects of localisation of electron states. The Hall resistivity as such does not show the staircase-like structure and the magnetoresistivity does not show broad minima around the integral filling factors. Nevertheless, in combination with the inhomogeneity of the electron density it will be shown to lead to the phenomena observed in the Quantum Hall regime.

We now return to the interpretation of magnetotransport properties by evaluating {3} with the ρ_{xx} and the ρ_{xy} as described before. If $\rho_{xx}(B/n_e(x,y))$ (almost) goes to zero for integral filling factors i_0, Kirchhoff's laws tell us that (almost) all current will flow along paths with $hn_e(x,y)/eB = i_0$. In view of the inhomogeneity in $n_e(x,y)$ there can be a range of magnetic field values B at which such current carrying paths exist between the current contacts. The Hall resistance as described by {3} is then dominated by contributions from these paths and $R_H = \rho_{xy}$ is evaluated at filling factor i_0 for this range of magnetic field values. We obtain plateaus in the Hall resistance! The magnetoresistance is (almost) zero over the same range of B-values. We note that if ρ_{xx} has deep minima at fractional filling factors in a many-electron description as proposed by Laughlin [17] we will in the same way find plateaus at fractional filling factors.

Because it is difficult to evaluate the current distribution according to Kirchhoff's laws in two dimensions we take a more simple geometry to be able to see some of the effects of inhomogeneities in a quantitative way. The consequences of this effective one-dimensional approach for the two-dimensional case will be discussed later in this paper.

4. THE ONE-DIMENSIONAL APPROACH

We consider an infinitely long Hall bar with an electron density that only varies over the width of the sample: $n_e(x,y) = n_e(y)$ and thus $J_y(x,y) = 0$. We will not specify the detailed variation of n_e as a function of y. Instead we consider the distribution of n_e values as they may appear across the sample. We take this distribution to be Gaussian

$$G(n_e) = \frac{1}{\sqrt{\pi}\,\delta n} \exp\left\{-\left(\frac{n_e - n_0}{\delta n}\right)^2\right\}, \qquad \{8\}$$

where n_0 is the mean electron density and δn its spread. Correspondingly we modify {3} by replacing the integration with respect to y by an integration with respect to n_e, using $dy/w = G(n_e)dn_e$, where w is the width of the sample. In view of $J_y = 0$ we have

$$J_x(n_e/B) = E_x(B)/\,\rho_{xx}(n_e/B), \qquad \{9\}$$

where $E_x(B)$ is the local electric field in the x-direction; it is independent of x and y. We finally obtain the Hall resistance across the sample to be given by

$$R_H(B) = \frac{\int [\, G(n_e)\rho_{xy}(n_e/B)/\rho_{xx}(n_e/B)\,]\, dn_e}{\int [\, G(n_e)/\rho_{xx}(n_e/B)\,]\, dn_e} \qquad \{10\}$$

and the magnetoresistance per square to be given by

$$R_{xx}(B)/\square = \frac{\int E_x \, dx}{\int J_x(x,y)\, dy} = \left\{ \int [\, G(n_e)/\rho_{xx}(n_e/B)\,]\, dn_e \right\}^{-1} . \qquad \{11\}$$

5. RESULTS OF THE CALCULATIONS

The results for $R_{xx}(B)$ and $R_{xy}(B)$ with a relative spread $\Delta = \delta n/n_0$ of 3% in the electron density of {8} are given in figure 2 for three different temperatures. Around integral filling factors the Hall resistance $R_H(B)$ shows plateaus which are absent in $\rho_{xy}(B)$. These plateaus broaden when the temperature is lowered as observed in experiments. The magnetoresistance $R_{xx}(B)$ shows broad minima at the magnetic fields where the Hall resistance shows plateaus. To give an indication of the flatness of the plateaus obtained in our model we give some results for a sample with 3% inhomogeneity. In a magnetic field of 5 Tesla at a temperature of 6 Kelvin the plateau is flat within 0.3 % over 4 % change in magnetic field. When the temperature is lowered to 2 Kelvin the plateau is flat on a level of 10^{-8} over the same range. The minima in the magnetoresistance R_{xx} tend to zero exponentially when the temperature reaches absolute zero. This is consistent with the thermally activated behaviour observed in experiments [19]. If inelastic scattering due to variable range hopping had been taken into account in ρ_{xx} the magnetoresistance R_{xx} would tend to zero more slowly at low temperatures.

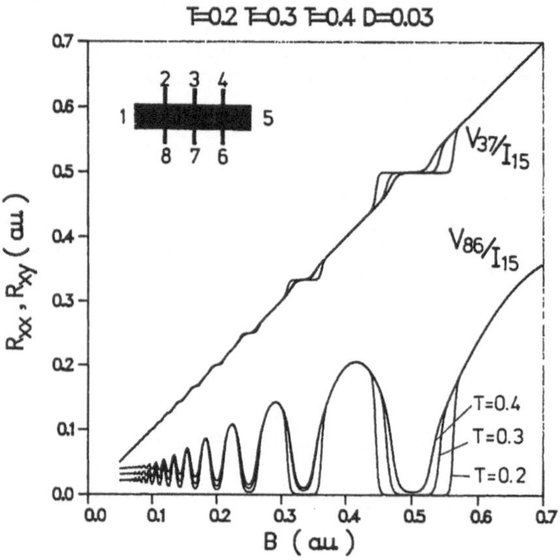

2: Calculated magnetoresistance and Hall resistance as a function of the perpendicular magnetic field at the same three temperatures as in figure 1. To obtain these results a spread Δ of 3% in a Gaussian distribution of the electron density was assumed. Localisation, spin-splitting and Landau level broadening are not taken into account in our calculations.

The behaviour of R_{xx} with B as displayed in figure 2 is not fully correct due to the simplifications in our description of ρ_{xx} and due to the assumption that the local electron density is independent of the magnetic field. This last assumption gives unphysical fluctuations of the electro-chemical potential over the width of the sample due to the steps of the chemical potential at integral filling factors shown in figure 1. The electro-chemical potential can be flattened by allowing a redistribution of the electron density over the width of the sample. This compensates for the steep change in chemical potential at integral filling factors. While sweeping the magnetic field the redistribution of the electron density causes $n_e(x,y,B)$ to remain longer in the neighbourhood of an integral filling factor than it would do without the redistribution. For more discussions on the influence of the electron density on the chemical potential we refer to GERHARDTS et al.[18]. Clearly, we should take into account the effects of the redistri-bution, but until now we have not succeeded in solving this problem.

One further result from a series of calculations at different temperatures is a linear relation between the minima in the R_{xx} and the slope of the R_{xy} at the magnetic field values where the minima occur. A typical result of such calculations is given in figure 3 and is comparable with the linear dependence found by TAUSENDFREUND et al.[19].

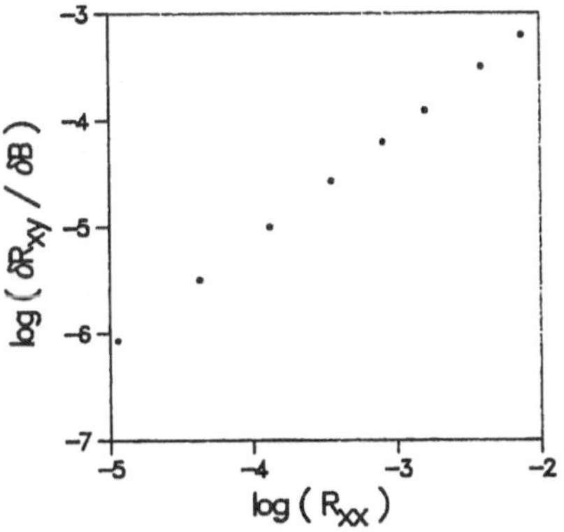

3: Calculated slope of the plateau in the Hall resistance as a function of the magnetoresistance. The points (at different temperatures) are calculated at the magnetic field where the magnetoresistance reaches its minimum value.

6. THE EFFECTS OF SYSTEMATIC INHOMOGENEITIES

To simulate the effects of a systematic inhomogeneity in the electron density across the width of a sample we have carried out calculations with a spatial dependence of the mean electron density n_0 in {8}, keeping the relative spread $\Delta = \delta n/n_0$ constant. We then have a systematic inhomogeneity with a statistical fluctuation superimposed on it. The distribution function is now given by:

$$G(n_e,y_0) = \frac{1}{\sqrt{\pi}\,\Delta n_0(y_0)} \exp\left\{-\left(\frac{n_e - n_0(y_0)}{\Delta n_0(y_0)}\right)^2\right\}, \qquad \{12\}$$

where $n_0(y_0)$ is a systematic inhomogeneity as a function of the y_0-coordinate across the sample. The integrations over the y-coordinate in {3} must be replaced by integrations over n_e and y_0 by using: $dy = G(n_e,y_0)\,dn_e dy_0$. The Hall resistance over a part of the Hall bar width is obtained by the appropriate limitation of the y_0 integration interval.

In order to explain the results of EBERT et al. [16] we use a gradient of 6% in $n_0(y_0)$ across the width of the sample in addition to a relative spread Δ of 2% in {12}. The results of our calculations, given in figure 4 for the voltages between the different probes, compare very well with their measured voltages. The fraction of the current that flows through the middle of the sample as a function of the magnetic field is shown in figure 5. Not all the

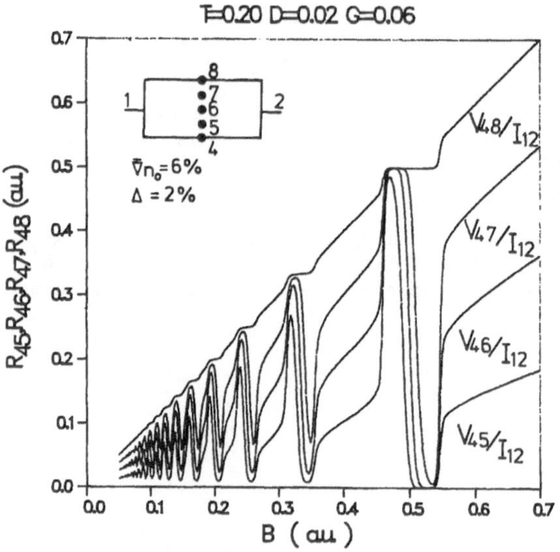

4: Calculated Hall resistances between the potential probes in the Hall direction of a Hall bar geometry as a function of the perpendicular magnetic field. A linear gradient of 6% in $n_0(y_0)$ in the Hall direction is assumed in addition to a relative spread Δ of 2% in the electron density in {12}.

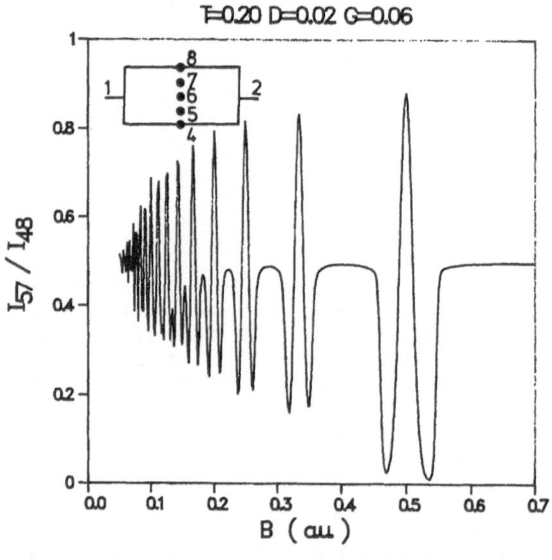

5: Fraction of the total current flowing between probe 5 and 7 as calculated with the same parameters as in figure 4.

structure present in the experimental results is reproduced in the calculations because we have not taken into account the spin-splitting. Our results lead to the following picture: The current through the sample flows through a fraction of the total width at integral filling factors. The position of the current-carrying paths changes with the magnetic field. At different magnetic fields different paths through the sample have integral filling factors and thus will carry most of the current because of their low resistivity [20].

7. DISCUSSION

One of the consequences of our model is the concentration of the current transport in current carrying filaments. This will have strong influence on the break-down of the Quantum Hall effect as observed by SIMON et al. [12]. The experimental determination of the current density at which break-down takes place is difficult because the part of the width of the sample that carries the current is not known. Calculating the current density by using the overall width of the sample will give an underestimate of the actual current density as demonstrated by BLIEK et al. [21].

More elaborate calculations on our model should include a better derivation of ρ_{xx} by taking into account Landau level broadening, spin-splitting and inelastic scattering. Incorporation of this last effect will probably result in the break-down of the Quantum Hall effect at high currents in our model. The large electric Hall field over a current filament can open the possibility of nearly elastic scattering between two different Landau levels on both sides of the current filament. When the difference in potential energy eV_H between both sides compensates for the difference in cyclotron energy $\hbar\omega_c$ elastic scattering can take place between both sides.

How do these calculations in our simplified one-dimensional approach compare with a realistic two-dimensional electron gas? Our most important assumption is the existence of a local resistivity tensor. This tensor can only be defined when there are large enough areas in the sample with nearly homogeneous electron density. Close to integral filling factors we need paths with a homogeneous electron density between the two current contacts to give paths with very low resistivity along the sample. Two effects work in favour of this situation. Firstly, the redistribution of charge over the width of the sample that causes the electron density to prefer an integral filling factor as described in section 5. Secondly, the localisation present in real samples will give a range of electron densities around an integral filling factor with a low resistivity. Both of these effects enlarge the probability to find a path with a low resistivity between the current contacts, compared with the percolation-like situation of finding a path with a homogeneous electron density through a random distribution of electron densities.

Another effect that cannot be neglected in the two-dimensional case is a type of semiclassical localisation near a maximum or a minimum in the electron density. Around such extrema a closed loop with low resistivity can be formed excluding the area in the loop from participating in d.c. current transport through the sample; an electric field cannot exist between the edges of such an area.

8. CONCLUSIONS

In conclusion we have shown that it is possible to explain (i) the Quantum Hall effect and (ii) the distribution of the Hall voltage across a sample in terms of a local resistivity tensor which does not exhibit the effects of localisation. We stress that this means that the Quantum Hall effect can exist in the absence of localisation. On the other hand inhomogeneities in the electron density, that are necessary for our explanation, will imply localisation. Thus excluding localisation is an ad hoc and more or less artificial operation. However, if localisation is disregarded in the local description of an inhomogeneous sample, the plateaus in the Hall resistance and the broad minima in the magnetoresistance can be explained. Moreover earlier measured distributions of the Hall voltage over a sample can be understood in detail. In a particular experimental situation it remains to be decided to what extent the various effects are related to localisation or to inhomogeneity of the electron density.

Acknowledgements. We want to thank D.v.d.Marel, J.M.Lagemaat and J.A.Pals for stimulating discussions.

REFERENCES

1. Klitzing K. von, Dorda G. and Pepper M., Phys.Rev.Lett. 45 (1980) 494
2. Laughlin R.B., Phys.Rev.B. 23 (1981) 5632
3. Hajdu J., Proceedings of the summerschool on "The physics of the two-dimensional electron gas" held 2-14 June 1986 in Antwerpen
4. Ando T., Surface Science 113 (1982) 182
5. Störmer H.L., Festkörperprobleme XXIV (1984) 25
6. Joynt R. and Prange R.E., Phys.Rev.B. 29 (1984) 3303
7. Iordansky S.V., Solid State Comm. 43 (1982) 1
8. Kazarinov R.F. and Luryi S., Phys.Rev.B. 25 (1982) 7626
9. Joynt R., Festkörperprobleme XXV (1985) 413
10. Meisels R., Lim K.Y., Kuchar F., Weimann G. and Schlapp W., in Two-dimensional systems: Physics and devices, Springer Ser.Solid-State Sci., Vol 67 (Springer, Berlin, Heidelberg 1985)
11. Woltjer R.,Mooren J. and Wolter J., Solid state comm. 53 (1985) 331
12. Simon Ch., Goldberg B.B., Fang F.F., Thomas M.K. and Wright S., Phys.Rev.B. 33 (1986) 1190
13. Woltjer R., Eppenga R., Mooren J., Timmering C.E. and André J.P., to be published in Europhysics Letters
14. Zheng H.Z., Tsui D.C. and Chang A.M., Phys.Rev.B. 32 (1985) 5506
15. Lippmann H.J. and Kuhrt F. Zeitschrift Für Naturforschung 13A (1958) 474
16. Ebert G., Klitzing K.von and Weimann G., J.Phys.C. 18 (1985) L257
17. Laughlin R.B., Phys.Rev.B. 27 (1983) 3383
18. Gerhardts R.R. and Gudmundsson V., to be published
19. Tausendfreund B. and Klitzing K.von, Surface Science 142 (1984) 220
20. Pudalov V.M. and Semenchinskii, JETP Lett. 42 (1985) 232
21. Bliek L., Braun E., Hein G., Kose V., Niemeyer J., Weimann G. and Schlapp W., to be published

A New Quantum Effect in the Transverse Magnetoresistance of Two-Dimensional Conductors with a Narrow Constriction in the Conducting Channel

L. Bliek[1], G. Hein[1], V. Kose[1], J. Niemeyer[1], G. Weimann[2], and W. Schlapp[2]

[1]Physikalisch-Technische Bundesanstalt, Bundesallee 100, D-3300 Braunschweig, Fed. Rep. of Germany
[2]Forschungsinstitut der DBP beim FTZ, Postfach 5000, D-6100 Darmstadt, Fed. Rep. of Germany

A new type of quantum steps has been observed in the transverse magnetoresistance of a two-dimensional conductor, having a narrow constriction in the width of the conducting channel. The quantized resistance values are integer multiples of $h/(2n_W e^2)$, where n_W is related to the number of quanta $h/(2eB)$ of magnetic flux through the constriction.

INTRODUCTION

To study a possible influence (IMRY /1/, LINDELOF et al /2/) of the penetration of a small number of quanta of magnetic flux into a Quantum Hall Effect sample, we studied two devices with narrow constrictions in the middle. The effects to be described here were clearly pronounced only for the sample with the narrowest constriction. The length of this constriction was 10.2 μm. Its effective width, as determined from the resistance at zero magnetic field, amounted to 1.0 μm. In normal QHE samples the number of flux quanta is far too large for our purpose, and even in this constriction we have 5000 quanta h/2e of magnetic flux per Tesla of magnetic flux density. The sample was made out of MBE-grown GaAs-(GaAl)As heterostructure material with 3.0×10^{15} electrons per m^2 at 1.2 K. Consequently there were 3.2×10^4 electrons in the constriction. The samples were provided with two current leads and eight potential probes, as indicated schematically in the insert of Fig. 1. This allowed measurements of the Hall voltage on either side of the constriction and of the potential drop in current direction both over the constriction and on either side of it.

RESULTS

The measurement results for the wide part of our samples show the usual behaviour: We observe plateaux in the Hall voltage measured on the contact pairs 1-5, 2-6, 3-7 and 4-8 accompanied by a zero voltage drop in the direction of the current, measured using the contact pairs 1-2, 3-4, 5-6, and 7-8, both over a considerable part of the magnetic field sweep. This holds true for all measuring currents used. At the same time, the voltage drop over any part of a sample including its constriction, i.e. between the contact pairs 2-3, 1-3, 2-4, 1-4, 6-7, 5-7, 6-8, and 5-8, behaves quite differently. Typical results for the sample described above are shown in Fig.1. The results are perfectly reproducible on sweeping the magnetic field up and down and on using different sets of contacts. Instead of a wide range of zero voltage, we now have, depending on the current, minima or narrow ranges of zero voltage, at the edges of which the voltage rises step-wise.

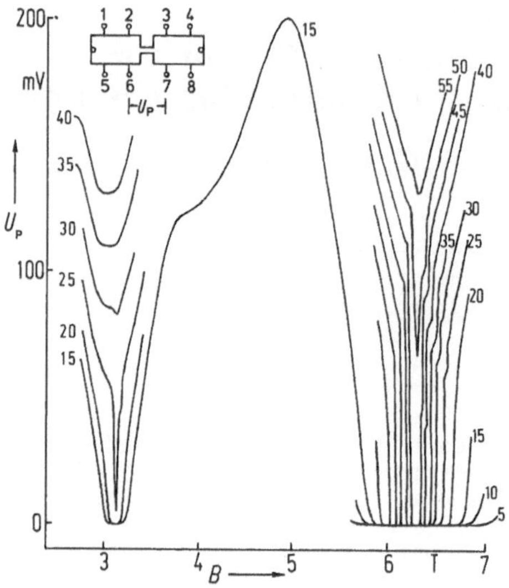

Figure 1: Potential drop in the direction of current flow as a function of magnetic flux density for a GaAs-(GaAl)As heterostructure with a narrow constriction. The parameter numbers on the curves give the measuring current in μA. The temperature of the liquid He bath was 1.2 K. The resistance minima near 3 T and 6 T correspond to the quantized Hall resistance steps $h/(4e^2)$ and $h/(2e^2)$, respectively

Obviously, those steps are due entirely to the presence of a constriction, since the rest of the device still has a vanishing resistance. Dividing the measured voltages by the measuring currents almost makes corresponding steps coincide, indicating that the steps are a feature in the resistance of the constriction.

Figs. 2 and 3 show data, taken on the same sample, after it had been stored at room temperature for one month. As indicated by the position of the resistance minimum on the B scale, the carrier density had become slightly smaller than it was during the first measurement. Comparison of Figs. 1, 2 and 3 reveals the influences of temperature and magnetic field.

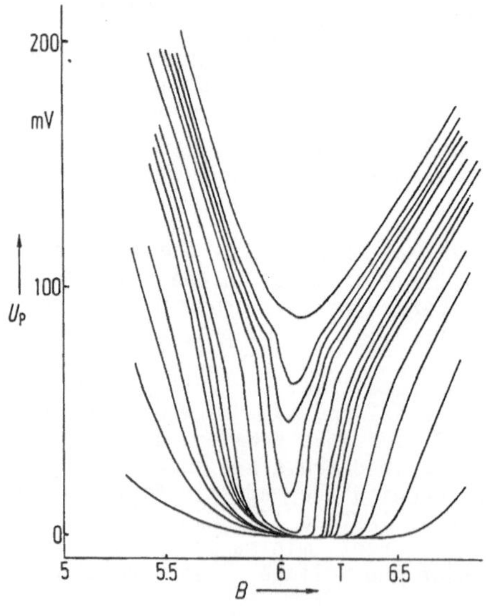

Figure 2: Similar data as represented in Fig. 1 for the same sample, but taken at a later date and for a bath temperature of 4.2 K. Measuring currents used are 5, 10, 13, 15, 18, 19, 20, 21, 23, 25, 26, 27, 28 and 30 μA

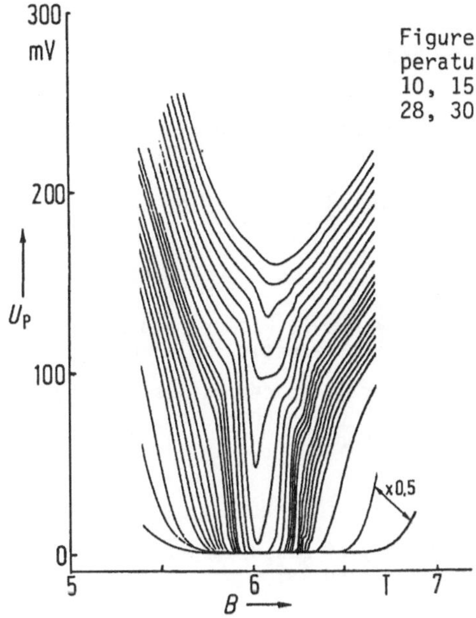

Figure 3: Data as in Fig. 2, but for a temperature of 1.2 K and measuring currents 5, 10, 15, 18, 19, 20, 21, 22, 23, 25, 26, 27, 28, 30, 32, 34, 36, 38, 40, 42, 44 and 46 μA

As usual in magneto-quantum effects, the steps are more pronounced at higher magnetic fields and at lower temperatures.

On the B scale the steps are almost evenly spaced, with a slight decrease in the spacing towards higher magnetic inductions. Trying to relate this spacing to the magnetic flux in the constriction, one has a choice of three possible units of magnetic flux density:

$$\frac{h}{2eLw}, \quad \frac{h}{2eL^2} \quad \text{and} \quad \frac{h}{2ew^2},$$

where L and w are the length and the width of the constriction respectively.

2eBLw/h is the number of flux quanta in the constriction and also the number of electrons that can be accommodated in the constriction in the two lowest Landau levels.

Thinking of a regular lattice of flux quanta or of centers of cyclotron orbits, $L(2eB/h)^{1/2}$ is the number of flux quanta in a strip, one flux quantum wide and with the length of the constriction or the corresponding number of electrons in the two-lowest Landau levels. $w(2eB/h)^{1/2}$ is the equivalent number in the perpendicular direction. Denoting the magnetic flux density values, corresponding to the middle of a step in the resistance of the constriction, by B_S, we find that of the three possible quantities only

$$n_L = L\cdot(2eB_S/h)^{1/2} \qquad (1)$$

calculated for each step gives a series of numbers in which the difference is one unit for adjacent steps. The experimental values for n_L are integers, within the accuracy of measurement. For the results represented in the figures, n_L ranges from 528 to 607 in the region near 6 T and from 393 to 402 near 3 T. For the points, where the resistance rises steeply to the next step, a similar equation holds, but with an additional constant on the right-hand side.

At first sight, the resistance values R_S, corresponding to mid-points of steps, appear to have an even spacing.

They can approximately be represented as integral multiples of a constant K_R:

$$R_S \approx n_R \cdot K_R \ , \qquad (2)$$

where n_R is an integer between 2 and 19 and $K_R \approx 200\Omega$. Closer inspection reveals that the spacing decreases with increasing magnetic field and that R_s has to be multiplied by $B_s^{1/2}$ to obtain uniformly spaced values. We have expressed B_S in each of the three "natural units" given above and R_s in units of the resistance quantum $h/(e^2)$ and obtained

$$R_s \cdot \sqrt{B_s} = n_R \frac{\sqrt{h/(2e)}}{w} \cdot \frac{h}{2e^2} \ , \qquad (3)$$

n_R being an integer number within the uncertainty of measurement. As mentioned above, we observed values of n_R between 2 and 19.

Equation (3) holds true for all the 155 steps we evaluated as well as for the resistance minima. After cooling the sample that had been at room temperature between measurements down to e.g. 1.2 K, new steps were frequently observed, while ones previously present had disappeared. Yet, eq. (3) was found to hold in all cases. Hence, this equation obviously describes a fundamental property of the resistance steps, whereas the approximate validity of eq. (2) may be just fortuitous and due only to the validity of eq. (3) and the limited range of B-values for which steps are observed.

Making use of eq. (1), eq. (3) can be rewritten as:

$$\rho_s = \frac{n_R}{n_L} \cdot \frac{h}{2e^2} \ , \qquad (4)$$

where ρ_s is the resistivity of the constriction, or as

$$R_s = n_R \cdot \frac{n_L}{n_o} \cdot \frac{h}{2e^2} \ , \text{ where} \qquad (5)$$

$$n_o = \frac{2eB_s}{h} \cdot L \cdot w \qquad (6)$$

is the number of flux quanta in the constriction or the number of electrons in the two lowest Landau levels.

We conclude, therefore, that we have observed a new type of quantization in the magnetoresistance of a twodimensional conductor with a narrow constriction in the conducting channel. The quantized resistance values are proportional firstly to the ratio of the number of flux quanta in a strip with the length of the constriction and one flux quantum wide and the total number of flux quanta in the constriction and secondly to $h/(2e^2)$. The resistance assumes its quantized values at magnetic flux densities, for which the number of flux quanta in the above-mentioned strip is an integer.

REFERENCES

1. Imry, Y.
   J. Phys. C: Solid State Phys. 15, L221 (1982)

2. Lindelof, P.E. and Hansen, O.P.
   Appl. of High Magn. Fields in Semiconductor Phys., ed. G. Landwehr p. 45 (Springer, Berlin 1983)

A Model for the Quantum Effects Observed in the Transverse Magnetoresistance of Two-Dimensional Conductors with a Constricted Channel

L. Eaves and F.W. Sheard

Department of Physics, University of Nottingham, Nottingham, NG7 2RD, UK

A model is proposed to explain the three main features of recent experiments by Bliek et al. on the dissipative transverse magnetoresistance of the two-dimensional electron gas in GaAs/(AlGa)As heterostructures in which there is a narrow constriction in the conducting channel. These features are the high breakdown current densities observed, the magnetic field values of the quantum steps and the quantised values of the magnetoresistance.

Recently, Bliek et al. [1,2] reported a remarkable quantum effect in the transverse magnetoresistance of a two-dimensional electron gas with a narrow constriction in the conducting channel. This new effect occurs at currents and magnetic fields close to the region of breakdown of the dissipationless state of the Quantum Hall Effect (QHE). The devices were fabricated from MBE-grown GaAs/(AlGa)As heterostructure material with areal electron density $\nu_0 = 3 \times 10^{15}\ \mathrm{m}^{-2}$. The width and length of the constriction were $w = 1.0\ \mu\mathrm{m}$ and $L = 10.2\ \mu\mathrm{m}$ respectively. The device was provided with two current- and eight voltage-probes so that the voltage drop along the current direction could be monitored on either side of the constriction and also across the constriction itself. The key experimental results reported in [1] can be summarised as follows. (a) The threshold current densities J_c for the breakdown of the dissipationless state on the $h/2e^2$ and $h/4e^2$ Hall plateaux are much higher than those reported previously (e.g. [3-5]). Bliek et al. reported values of J_c around 30 A m^{-1} on the $h/2e^2$ plateau at 6T compared to values of typically 0.3 to 2 A m^{-1} in the earlier work. (b) At currents around breakdown, the transverse magnetoresistance increases in a step-like fashion. Bliek et al. found that the magnetic field values B_s at these steps obeyed the relation

$$L(2eB_s/h)^{\frac{1}{2}} = n_L\ , \qquad (1)$$

where n_L takes on large integer values from 528 to 607 on the $h/2e^2$ plateau and from 393 to 402 on the $h/4e^2$ plateau. (c) Bliek et al. observed that the magnetoresistance steps are "quantised" according to the equation

$$R_s = \frac{n_R}{w}\left(\frac{h}{2e}\right)^{\frac{1}{2}}\left(\frac{h}{2e^2}\right)B_s^{-\frac{1}{2}} = \frac{n_R n_L}{n_0}\,\frac{h}{2e^2}\ , \qquad (2)$$

where $n_0 = (2eB_s/h)Lw$ is the number of flux quanta in the constriction and n_R takes on low integer values (between 2 and 19).

In this article we show that the solution of the one-electron Schroedinger equation in crossed electric ($\underline{E}$) and magnetic ($\underline{B}$) fields using the Landau gauge gives a valuable insight into these observations. For $\underline{B} \parallel \underline{z}$ and $\underline{E} \parallel \underline{x}$ the eigenfunctions and energy eigenvalues are

$$\psi_{n,k_y}(x,y) = \phi_n(x - x_0)e^{ik_y y} \qquad (3)$$

and

$$\varepsilon_{n,k_y} = (n + \tfrac{1}{2})\hbar\omega_c + eEx_o + \frac{e^2E^2}{2m^*\omega_c^2} \quad , \tag{4}$$

where ϕ_n are simple-harmonic oscillator solutions, $-x_o = \ell_B{}^2k_y + eE/m\omega_c{}^2$, $\ell_B = (\hbar/eB)^{\frac{1}{2}}$, $\omega_c = eB/m^*$ and $k_y = 0, \pm 2\pi/L, \pm 4\pi/L$ .... etc. The y-axis corresponds to the direction of current flow down the length L of the constriction. For simplicity we neglect spin splitting. The spatial extent X of the simple harmonic states $\phi_n(x - x_o)$ is given by the classical limit of simple harmonic motion so that $X = 2(2n + 1)^{\frac{1}{2}}\ell_B$. The separation Δx_o between the centres x_o of adjacent $\phi_n\ (x - x_o)$ is given by $\Delta x_o = (2\pi/L)\ell_B{}^2$. For normal macroscopic samples (mm-size) and magnetic fields $\sim$10T, Δx_o is on the sub-atomic length scale $\sim 10^{-12}$ m. However, Δx_o is $\sim 10^{-10}$ m in the experiment of Bliek et al. if we take L to correspond to the length of the constriction.

Let us now consider the three key features of the Bliek et al. experiment. Firstly, the critical electric fields E_c and current densities J_c are given quite accurately by the expressions

$$eE_c\ell_B\{(2n + 1)^{\frac{1}{2}} + (2n + 3)^{\frac{1}{2}}\} = \hbar\omega_c \quad \text{and} \quad J_c = E_c/\left(\frac{h}{pe^2}\right) \tag{5}$$

where n = 0 and p = 2 for the $h/2e^2$ plateau and n = 1 and p = 4 for the $h/4e^2$ plateau. Equation (5) corresponds to the electric field at which the classical limits of two eigenstates $\phi_n(x - x_o)$ and $\phi_{n+1}(x - x_o{}')$ of equal energy just touch $((x_o - x_o{}') = \hbar\omega_c/eE_c)$. Taking $m^*/m = 0.07$, equation (5) gives $J_c = 31$ A m^{-1} and 15 A m^{-1} for the $h/2e^2$ and $h/4e^2$ plateau respectively, which agree to within 10% with the threshold current densities reported by Bliek et al. for breakdown. Equation (5) corresponds to the effective onset of elastic or quasi-elastic inter-Landau level scattering. This scattering process requires spatial overlap between $\phi_n(x - x_o)$ and $\phi_{n+1}(x - x_o{}')$. A related effect has been investigated in the high field magnetoresistance of n-GaAs and n-InP [6-9] and its possible role in the breakdown of the dissipationless state of the QHE has been discussed previously [5,8-11]. Inter-Landau level scattering can occur by the emission of long wavelength acoustic phonons whose momentum component $q_y = (x_o{}' - x_o)/\ell_B{}^2 \sim \ell_B{}^{-1}$ by momentum conservation. However, other inter-Landau level scattering mechanisms between the eigenstates e.g. due to impurities, charge inhomogeneities or many-body effects can be envisaged. Note that for the conditions of the experiment, the emitted phonon energy $\varepsilon_q \simeq \hbar v_s q << \hbar\omega_c$, where v_s is the sound velocity (for this reason we term the effect "quasi-elastic"). The fact that the high breakdown current density reported in [1] agrees so closely with equation (5) is strong evidence that (quasi-)elastic inter-Landau level scattering is the intrinsic mechanism for the breakdown of the dissipationless QHE. The lower threshold current densities observed in previous experiments on larger samples ($\sim$mm size) suggest inhomogeneous current flow [3].

Secondly, let us consider equation (2) governing the remarkable quantised values of the dissipative resistance. The following argument shows that this equation implies that <u>each</u> current-carrying electron passing through the constriction dissipates energy in quantised units of $\hbar\omega_c$. Let us suppose that after passage through the constriction ν_1 electrons per unit area are excited by inter-Landau level transitions into the n = 1 Landau level. The total areal density is ν_o. On entering the region beyond the constriction where $E_x = 0$, the ν_1 electrons can dissipate energy $\hbar\omega_c$ to the lattice by making transitions back to the lowest Landau level. The power dissipation is therefore given by

$P = \nu_1 v_d w \hbar\omega_c$, where $v_d = E/B$. By definition, the transverse magnetoresistance $R_y = P/I_y^2 = \nu_1 \hbar\omega_c/\nu_o^2 e^2 v_d w$. This can be rewritten by noting that energy and momentum conservation for the acoustic phonon emitted in the inter-Landau level transition require that $q_y = k_y - k_y' = 2\pi n_y/L = \omega_c/(v_d - v_s) \simeq \omega_c/v_d$, where n_y is the integer number of eigenfunctions traversed. The relation is exact if the scattering involves no phonons and is purely elastic. Hence the transverse magnetoresistance can be written as

$$R_y = \frac{\nu_1}{\nu_o^2} n_y \frac{2\pi}{L\omega} \frac{\hbar}{e^2} = 2\left(\frac{\nu_1}{\nu_o}\right) \frac{n_y}{n_o} \frac{h}{2e^2} \tag{6}$$

since $n_o = \nu_o L w$. If we set $\nu_1 = \nu_o$, corresponding to the case when all the current-carrying electrons emit one quantum of energy $\hbar\omega_c$, equation (6) reduces to

$$R_y = 2 \frac{n_y}{n_o} \frac{h}{2e^2} \, . \tag{7}$$

This is formally identical to equation (2) given in [1] if we set the quantum number $n_R = 2$ (the lowest integer value observed) and identify n_y with n_L. Higher <u>even</u> integer values of n_R would correspond to each current-carrying electron emitting energy in quantum $2\hbar\omega_c$, $3\hbar\omega_c$ etc. Note that on the $h/2e^2$ plateau at 30 A/m, $V_x = (h/2e^2)\, I_y = 0.4\text{V} \simeq 40\ \hbar\omega_c$.

Before discussing further the implication of this result, let us consider ar alternative view of the dissipative process in the constriction. This gives a more physical picture of how the voltage drop down the constriction arises. If ν_1 electrons per unit area undergo inter-Landau level scattering which shifts their eigenfunction centre by $x_o' - x_o = \ell_B^2(k_y - k_y')$ during the time $\Delta t = L/\text{v}$ to travel through the constriction, then the corresponding lateral current is given by $J_x = \nu_1 e(x_o' - x_o)/\Delta t = 2\pi\nu_1 e v_d n_y \ell_B^2/L^2$. The current density along th y-axis is $J_y = \nu_o e v_d$. Since the net current must flow parallel to the length the constriction, we must regard the axis of the constriction as being tilted slightly by an angle θ to the y-axis if J_x is non-zero. Setting the current perpendicular to the axis of the constriction equal to zero gives $J_x\cos\theta = J_y s$ Hence $\tan\theta = J_x/J_y = 2\pi\ell_B^2\nu_1 n_y/\nu_o L^2$. This tilting slightly rotates the Hall field and produces a component of electric field parallel to the axis of the constriction given by $E_y' = E\sin\theta \simeq E\tan\theta = (2\pi\ell_B^2/L^2)(\nu_1/\nu_o)n_y E$. The dissipative resistance is then $R_y = LE_y'/I_y$, yielding a result identical to equation (

Setting $n_R = 2$ in equation (2) for the $h/2e^2$ plateau implies that passage through the constriction drives <u>all</u> the current-carrying electrons from the n = into the n = 1 Landau level from which they subsequently de-excite to give rise to the dissipative resistance. This is a surprising and quite unexpected conclusion but is clearly indicated by the large and quantised values of R_s observed in the experiment. We suggest that quasi-elastic inter-Landau level scattering in the presence of the large Hall electric field is the driving mechanism for this process. However the transition process must be a many-electron phenomenon in which the inter-electron Coulomb energy ($\sim e^2/4\pi\varepsilon_o\varepsilon_r\ell_B$) plays a role.

Finally, let us examine the implications of the discrete magnetic field values B_s for the quantum magnetoresistance steps. This result also seems to require that <u>all</u> of the electrons make an inter-Landau level transition during their traversal of the constriction. For a single electron, the transition rate between the n = 0 and n = 1 eigenstates centred at x_o and x_o' respectively is dominated by the overlap integral and is proportional to $\ell_B^2 q_y^2 \exp(-\frac{1}{2}\ell_B^2 q_y^2)$ [9]. This function is negligibly small for $\ell_B q_y \gg 1$ and has a broad maximum at $\ell_B q_y = \sqrt{2}$ corresponding to $n_y = L/\ell_B\pi\sqrt{2} \sim 220$ on the $h/2e^2$ plateau at 6T.

In order to explain the occurrence of magnetoresistance steps at particular values of magnetic field B_s given by equation (1), we require a mechanism that for a particular B_s will strongly select a single value of $n_y = n_L$. We postulate that in an excitation in which all electrons passing through the constriction are excited into the n = 1 Landau level, the many-electron transition is governed by an expression of the general form $F(\ell_B q) = [f(\ell_B q_y)\exp(-\ell_B{}^2 q_y{}^2/2)]^N$ where N is a large integer corresponding to the number of electrons involved and $f(\ell_B q_y)$ is a polynomial. Such a function will be very sharply peaked at a critical value of $\ell_B q_y = \alpha$ giving

$$\left(\frac{\hbar}{eB_s}\right)^{\frac{1}{2}} \frac{2\pi}{L} n_y = \alpha \qquad \text{or } L\left(\frac{2eB_s}{h}\right)^{\frac{1}{2}} \left(\frac{\alpha}{2\pi^{\frac{1}{2}}}\right) = n_y .$$

This is the same as equation (1) if we again identify n_y with n_L and set $\alpha = 2\pi^{\frac{1}{2}}$. Note that if all of the electrons scatter from the n = 0 to n = 1 Landau level, Pauli's principle requires that n_y is the same for all the transitions.

In conclusion, therefore, we believe that the remarkable new quantum magnetoresistance effect reported by Bliek et al. points to a coherent, many-electron inter-Landau level scattering process in the breakdown of the dissipationless state of the QHE. We have accounted for even values of n_R in our model. Odd integers could arise if the transitions only involved one spin species.

References

1. L. Bliek, G. Hein, V. Kose, J. Niemeyer, G. Weimann and W. Schlapp, Int. Conf. "The Application of High Magnetic Fields in Semiconductor Physics", Wurzburg, Springer Series in Solid State Sciences (1986a).
2. L. Bliek, E. Braun, G. Hein, V. Kose, J. Niemeyer, G. Weimann and W. Schlapp Semicond. Sci. Technol. 1 (1986b) 110.
3. M.E. Cage, R.F. Dziuba, B.F. Field, E.R. Williams, S.M. Girvin, A.C. Gossard, D.C. Tsui and R.J. Wagner, Phys. Rev. Lett. 51, (1983) 1374.
4. G. Ebert, K. von Klitzing, K. Ploog and G. Weimann, J.Phys.C 16 5441 (1983).
5. H.L. Stormer, A.M. Chang, D.C. Tsui and J.C.M. Hwang, Proc. Int. Conf. Phys. Semiconductors (San Francisco) publ. Springer, 1985, p.267.
6. L. Eaves, P.S.S. Guimaraes, J.C. Portal, T.P. Pearsall and G. Hill, Phys. Rev. Lett. 53, 608 (1984).
7. P.S.S. Guimaraes, L. Eaves, J.C. Portal and G. Hill, Proc. Int. Conf. Phys. Semiconductors (San Francisco) publ. Springer, 1985, p.459.
8. L. Eaves, P.S.S. Guimaraes and J.C. Portal, J. Phys. C 17, (1984) 6177.
9. P.S.S. Guimaraes, L. Eaves, F.W. Sheard, J.C. Portal and G. Hill, Physica 134B (1985), 47.
10. D.C. Tsui, G.J. Dolan and A.C. Gossard, Bull. Am. Phys. Soc. 28, 365 (1983).
11. O. Heinonen, P.L. Taylor and S.M. Girvin, Phys. Rev. B30, 3016 (1984).

Photoexcited Quantum Hall Behavior in Modulation Doped GaAs/AlGaAs Heterostructures: Parallel Conduction and Subthreshold Photoeffects

W.P. Kirk[1], *P.S. Kobiela*[1], *H.D. Shih*[2], *and M.A. Reed*[2]

[1]Physics Department, Texas A&M University, College Station, TX77843, USA
[2]Central Research Laboratories, Texas Instruments, Inc., Dallas, TX75265, USA

1. INTRODUCTION

Although the modulation doping technique has been instrumental in achieving high electron mobilities [1] it is also responsible for a radiation sensitive effect known as persistent photoconductivity (PPC). [2] This effect is characterized by a light-induced conductivity enhancement that persists for long times (in some systems, > 10^8 sec) at low temperatures. The PPC arises from the excitation of electrons out of deep donor-related traps in the AlGaAs, known as DX centers, which suppress recapture due to a large lattice relaxation.[3] In the GaAs/$Al_xGa_{1-x}As$ system these excited electrons are able to maintain quasiequilibrium with the two-dimensional electron gas (2-DEG) layer in the GaAs, forming a parallel conduction path in the $Al_xGa_{1-x}As$.

The PPC has been studied in the GaAs/$Al_xGa_{1-x}As$ system by a number of workers, [4] but none of these studies have explored the effects of PPC on the behavior of the 2-DEG in the quantum limit. In this paper we present a study of PPC effects on quantum transport processes in high mobility GaAs/$Al_xGa_{1-x}As$ modulation doped structures. [5] Transport coefficients are compared at high and low fields to derive information concerning density, mobility, and the distribution of carriers in the 2-DEG and parallel conduction paths. From these measurements we determine quantitatively how parallel conduction affects the accuracy of quantized Hall resistance standards. Also, data showing subthreshold photoeffects will be presented.

2. TWO-CARRIER MODEL

The application of a two-carrier analysis is helpful for understanding the behavior of a 2-DEG in equilibrium with a parallel conduction layer when Hall conditions apply. In the limits of high and low magnetic field B, expressions for the diagonal resistivity ρ_{xx} and Hall resistivity ρ_{xy} are found as follows: in low field limit ($\omega_{c_1}\tau_1$ and $\omega_{c_2}\tau_2 \ll 1$),

$$\rho_{xx} = \frac{1}{e(n_1\mu_1 + n_2\mu_2)} \quad (1)$$

$$\rho_{xy} = \frac{n_1\mu_1^2 + n_2\mu_2^2}{e(n_1\mu_1 + n_2\mu_2)^2}\, B\ ; \quad (2)$$

in high field limit ($\omega_{c_1}\tau_1$ and $\omega_{c_2}\tau_2 \gg 1$),

$$\rho_{xx} = \frac{(n_1/\mu_1) + (n_2/\mu_2)}{e(n_1 + n_2)^2} \quad (3)$$

$$\rho_{xy} = \frac{1}{e(n_1 + n_2)} B, \quad (4)$$

where e is the carrier charge and n_1 and μ_1 are the carrier density and mobility respectively for the 2-DEG, and n_2 and μ_2 correspond to similar quantities for parallel conduction in the AlGaAs layer.

In the quantum limit, the development of Landau levels causes the carrier density in the 2-DEG to take on discrete values, viz. $n=in_0$, where i is the integral number of filled Landau levels and $n_0=eB/h$ is the density of a singly filled Landau level. Thus, Eq. (4) becomes

$$\rho_{xy} = \frac{1}{e(n_2 + i\,\frac{eB}{h})} B . \quad (5)$$

Clearly when $n_2=0$, we recover the quantized Hall resistance $\rho_{xy}=R_H=h/ie^2$. The fact that n_2 appears in Eq. (5) obviously indicates under these circumstances that ρ_{xy} will depart from the expected R_H value, implying a source of error in the measurement of R_H.

3. EXPERIMENTAL PROCEDURE

The samples were modulation doped GaAs/$Al_{0.3}Ga_{0.7}As$ heterostructures grown in a Riber 2300 MBE on a Cr-doped GaAs substrate. The minimum channel width used in these studies was 150 µm, to exclude localization effects due to short channel effects. The samples were cooled slowly (≈30 hrs) from room temperature to millikelvin temperatures in a light-tight container to eliminate residual PPC and to minimized thermally induced strain. Low field values of the mobility and carrier density were measured at T=1 K, and found to be $\mu=10^5$ $cm^2/V{\cdot}s$ and $n=2.8\times10^{11}$ cm^{-2}, respectively. Light excitation was done by direct illumination from a GaAsP/GaAs red LED. Light dose was controlled by varying the time the LED was activated by a constant (20 mA) current. Dose quantities refer to the calculated number of photons arriving at the sample. For our particular experimental configuration, there were approximately 7.8×10^{11} photons/sec striking the sample surface. No attempt has been made to correct for reflection of photons at the sample surface nor for absorption in the sample; thus absolute intensity figures must be viewed cautiously. However, relative dose values reported here were easy to control and are thus highly precise.

4. RESULTS AND DISCUSSION

In Fig. 1 we show ρ_{xy} as a function of magnetic field B for various photoexcitation doses, while Fig. 2 shows ρ_{xx} as a function of B for the same photon doses. The measurements were taken sufficiently long after excitation and at approximately the same time after photoexcitation to eliminate possible transient and nonexponential decay effects of the PPC. The photon doses ranged from a minimum of 3.9×10^9 to 2.5×10^{13} photons. A systematic shift of the quantum Hall plateaus (and the accompanying ρ_{xx} minima) toward higher magnetic fields is seen as the density of the carriers in the 2-DEG increases. Upon reaching a critical photon dose ($>7.8\times10^{11}$ photons), quantum transport apparently breaks down, and the plateaus deviate from the expected values. At the same time the ρ_{xx} minima depart significantly from zero.

The density of carriers as a function of light dose was determined in two ways: (1) from high-field values of the Hall resistance ($n=B/e\rho_{xy}$) and (2) from the periodicity of the Shubnikov-de Haas (SdH) oscillations in

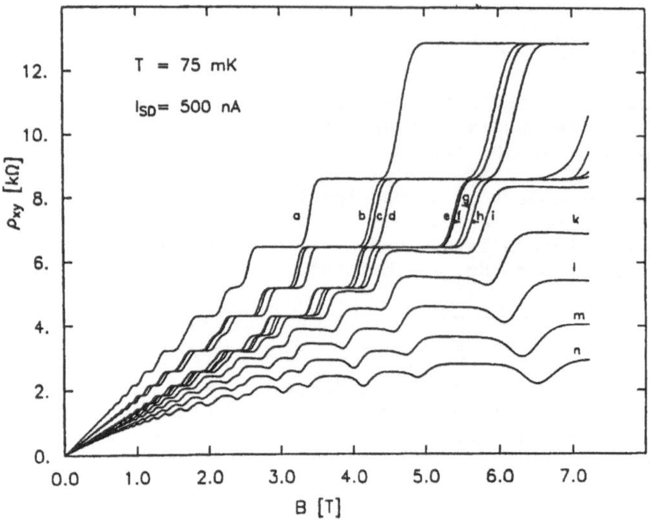

FIG. 1. ρ_{xy} vs B for GaAs/$Al_{.3}$ $Ga_{.7}$As sample. Photon dose ranges:

(a) 0
(b) 3.9×10^{9}
(c) 7.8×10^{9}
(d) 1.2×10^{10}
(e) 1.6×10^{10}
(f) 3.1×10^{10}
(g) 3.9×10^{11}
(h) 7.8×10^{11}
(i) 1.6×10^{12}
(k) 3.1×10^{12}
(l) 6.2×10^{12}
(m) 1.3×10^{13}
(n) 2.5×10^{13}

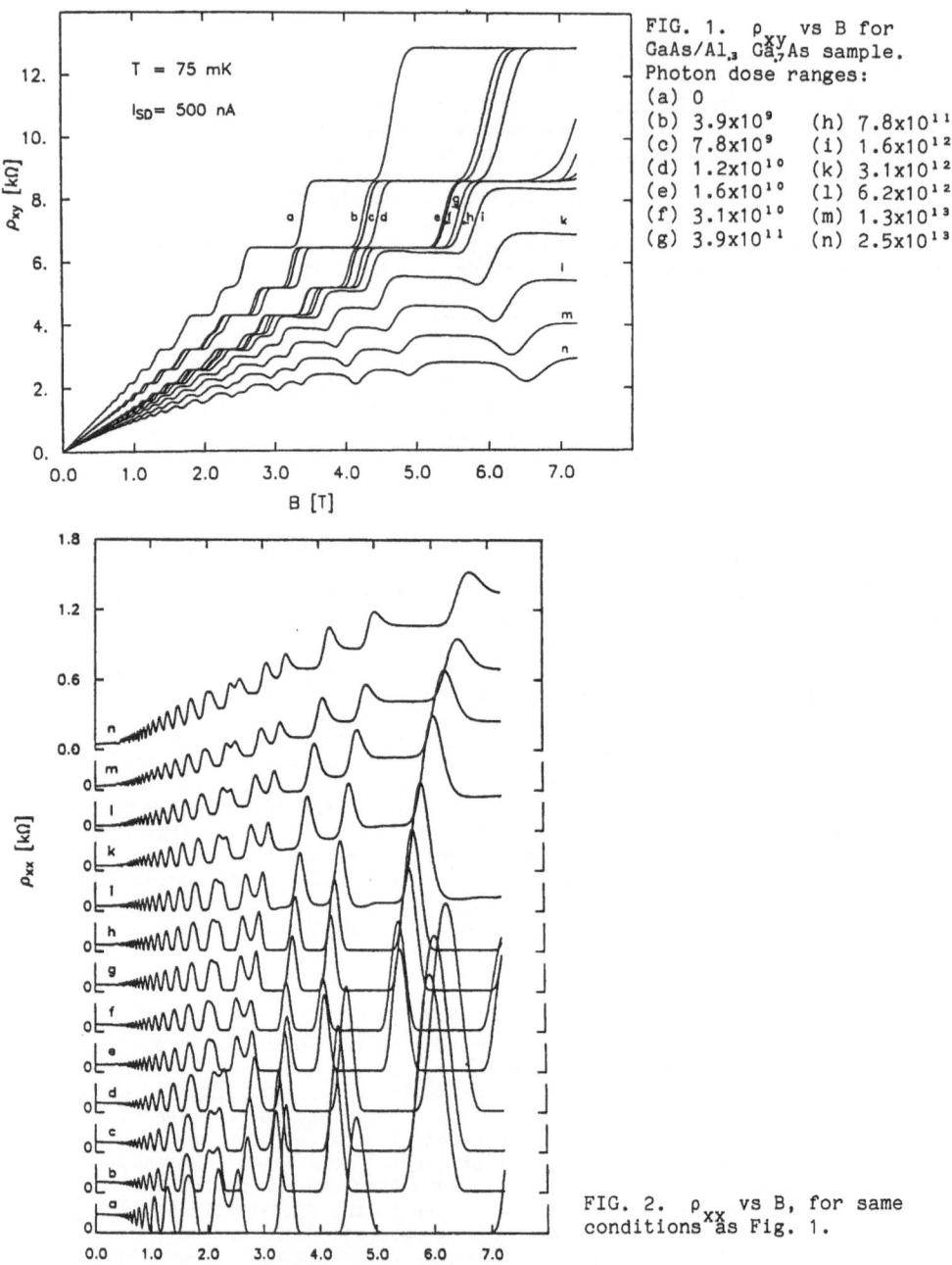

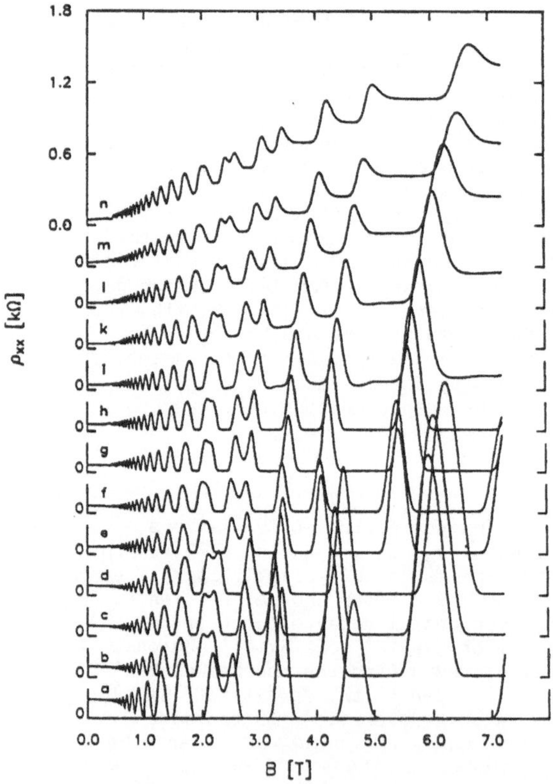

FIG. 2. ρ_{xx} vs B, for same conditions as Fig. 1.

ρ_{xx}($n=2e/h\Delta(1/B)$). The results of these two methods are shown in Fig. 3(a). At a threshold dose of 1.6×10^{12} photons, where the quantum transport clearly deviates, these two methods of determining carrier density depart. Since the high-field Hall resistance method gives the

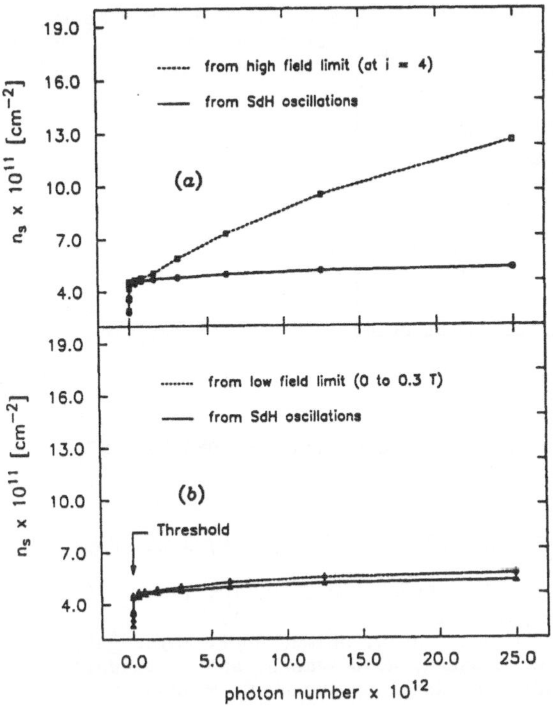

FIG. 3.(a) Carrier density vs photon dose from SdH and i = 4 Hall plateau. (b) Carrier density vs photon dose from SdH and slope of ρ_{xy} between 0 and 0.3 T.

combined carrier density in both the 2-DEG and the AlGaAs, whereas the SdH oscillations are essentially measuring the 2-DEG carrier density, then the carrier density determined from low-field Hall resistance values as predicted by Eq. 2 should agree well with the method using SdH oscillations. Excellent agreement is shown in Fig. 3(b). Having determined the densities in the two regions, the mobility of the electrons in the AlGaAs can be calculated. For example, at a photon dose of 2.5×10^{13}, $n_1 = 5.5 \times 10^{11}$ cm^{-2} and $n_2 = 7.5 \times 10^{11}$ cm^{-2}. Assuming that μ_1 does not change appreciably, then $\mu_2 = 0.19 \times 10^4$ $cm^2/V \cdot s$. If we define the limits of conduction in the AlGaAs by observing the minima in ρ_{xx} and use the value derived for the mobility in the AlGaAs along with our resolution of 0.05 Ω for ρ_{xx}, then the carrier density in the AlGaAs, for photon doses up to 7.8×10^{11}, is found to be $<4 \times 10^6$ cm^{-2}. Bridge techniques used by other workers [6] have measured minimum resistances in these regions to be $<10^{-7}$ Ω. From this value an upper limit can be placed on the number of carriers (<10 cm^{-2}) in the AlGaAs region, if no deviation is observed. This limit implies, using Eq. 5, a deviation of ≈1 part in 2×10^9 in the quantized Hall resistance due to parallel conduction.

At subthreshold levels, we observed a nonlinear step-like structure in the carrier density as a function of photon number, as shown in Fig. 4. Conduction through the next higher subband can be ruled out because of the well-behaved SdH oscillations. This anomalous behavior may indicate more than one type of trap (DX center) is responsible for degrading effects, such as light-induced pinch-off voltage shifts and high-frequency degradation. The dynamics of this process deviates from present understanding [7] of photoexcitation mechanisms and needs further investigation.

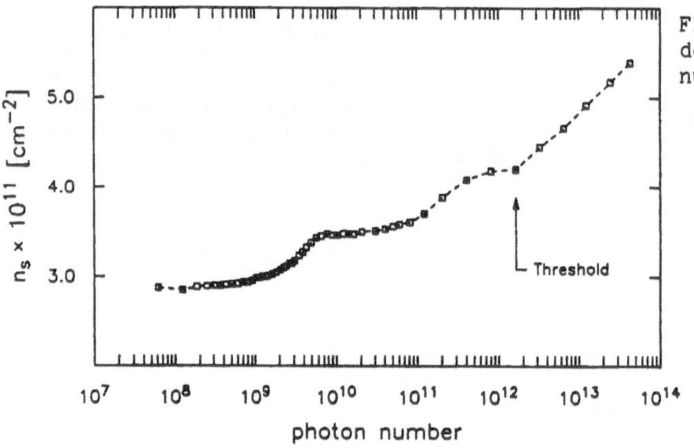

FIG. 4. Carrier density determined by SdH vs photon number on logarithmic scale.

This work was supported by National Science Foundation: DMR 8405197.

REFERENCES

1. R. Dingle, H. L. Stormer, A. C. Gossard, and W. Weigmann, Appl. Phys. Lett. 33, 665 (1978).
2. M. K. Sheinkman and Y. Ya. Shik, Sov. Phys. Semicond. 10, 128 (1976).
3. D. V. Land and R. A. Logan, Phys. Rev. Lett. 39, 635 (1977).
4. E. F. Schubert, K. Ploog, H. Dambkes, and K. Heime, Appl. Phys. 33A, 63 (1984); H. L. Stormer, A. C. Gossard, W. Weigmann, and K. Baldwin, Appl. Phys. Lett. 39, 912 (1981); A. Katalsky and J. C. M. Hwang, Appl. Phys. Lett. 44, 333 (1984).
5. Additional details found in M. A. Reed, W. P. Kirk, and P. S. Kobiela, IEEE. J. Quantum Electronics (in press).
6. D. C. Tsui, A. C. Gossard, B. F. Field, M. E. Cage, and R. F. Dziuba, Phys. Rev. Lett. 48, 3 (1982).
7. H. J. Queisser, Phys. Rev. Lett. 54, 234 (1985).

Quantum Hall Effect in Bicrystals of p-$Hg_{0.75}Cd_{0.23}Mn_{0.02}Te$

G. Grabecki[1], *T. Suski*[2], *T. Dietl*[1], *T. Skośkiewicz*[1], *and M. Gliński*[3]

[1]Institute of Physics, Polish Academy of Sciences, PL-02-668 Warszawa, Polanc
[2]High Pressure Research Center, Polish Academy of Sciences, PL-01-142 Warszawa, Poland
[3]International Laboratory for High Magnetic Fields and Low Temperatures, PL-53-529 Wroclaw, Poland

Magnetoresistance and Hall coefficient of the inversion layer formed at the grain boundary in p-$Hg_{0.75}Cd_{0.23}Mn_{0.02}Te$ have been measured. The quantized values of the Hall resistivity have been observed either at millikelvin temperatures or under hydrostatic pressure.

1. INTRODUCTION

It is presently well established that defects at the grain boundary (GB) plane in n-Ge /1/, p-InSb /2/, and narrow-gap p-$Hg_{1-x}Mn_xTe$ /3/ lead to formation of a two-dimensional (2d) inversion layer. In those systems the carrier gas resides in a symmetric v-like potential well which quantizes the motion in the direction perpendicular to the GB plane into a set of electric subbands. In many respects, the GB is similar to yet another 2d system, namely, δ-doped layer /4/. We describe here our magnetoresistance and Hall effect studies of a natural GB in p-$Hg_{0.75}Cd_{0.23}Mn_{0.02}Te$, carried out as a function of temperature and hydrostatic pressure. In particular, our work specifies experimental conditions for the Quantum Hall Effect (QHE) to be observed in this system. The material we study differs from that in which the QHE has so far been found, by the presence of localized spins which reside in the d-shell of Mn atoms. Though spin-effects will not be discussed here, it is worth to note that their role in systems of reduced dimensionality begin to attract recent attention /5/.

2. SAMPLES AND EXPERIMENT

The crystals of $Hg_{0.75}Cd_{0.23}Mn_{0.02}Te$ were grown by A.Mycielski and B.Witkowska using the solid state recrystallization method. The energy gap was estimated from an absorption measurement at 4.2K to be 210 meV. Typical ingots consist of several single crystalline grains with the net acceptor concentration of the order of $10^{16}cm^{-1}$. The grain boundaries can be clearly seen after a preferential etching /3/. The as-grown ingots were cut into Hall bars with the GB parallel to the surface of typical dimensions 5x0.2 mm^2. The samples were mounted either in a dilution refrigerator or in a high pressure cell. A liquid was used as a pressure transmitting medium. The InSb pressure gauge was applied to determine pressure values in the whole temperature range.

3. RESULTS

The total 2d electron concentrations at atmospheric pressure were typically of the order of $10^{12}cm^{-2}$. A careful analysis of the Hall effect and periods of the Shubnikov-de Haas (SdH) oscillations demonstrated that at least three electric subbands were occupied. In the case of the sample for which the re-

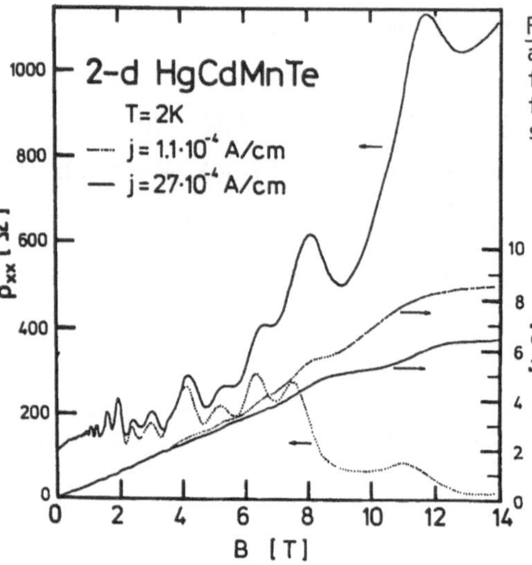

Fig.1. Grain boundary diagonal and Hall resistivities as a function of the magnetic field for two different current densities

sults will be presented here, the concentrations at subsequent electric subbands were found to be $5.1\cdot10^{11}$, $3.4\cdot10^{11}$, and $1.15\cdot10^{11}cm^{-2}$ The resulting total electron concentration is somewhat smaller than that deduced from the low-field Hall coefficient R_H, partly because R_H appears to be affected by quantum corrections to the conductivity of disordered systems /1,6/. The SdH oscillations from the electrons in the uppermost subband appear in a magnetic field of 0.2T pointing to a high mobility value of $\sim5\cdot10^4 cm^2/Vs$. The mobility of electrons in the ground state subband is about one order of magnitude smaller, as it could have been expected /4/.

Figure 1 presents the diagonal and Hall resistivity, ρ_{xx} and ρ_{xy}, measured under atmospheric pressure at 2K. For a low current density (dashed curve) a deep minimum in ρ_{xx} is observed in $\sim$13T. At the same time ρ_{xy} tends towards a quantized value, $h/3e^2$. The high field resistivities were found to be very sensitive to the current density j (compare solid and dashed curve in Fig.1). The range of j where the non-Ohmic behavior begins is similar to that found in other systems /7/. The results plotted in Fig.2 demonstrate that the QHE can be observed in this multisubband system already in $\sim$5T, providing that the temperature is lowered down to the millikelvin range. In particular, in $\sim$5.5T and at 35mK, $\rho_{xy}=h/7e^2$, within our present experimental accuracy of 0.5%.

Under a hydrostatic pressure the electron concentration and thus the number of occupied subbands decreases. Furthermore, for not too high pressures, the electron mobility was found to increase. The above facts make conditions for the observation of QHE more favourable and, indeed, as shown in Fig. 3, we have observed well-developed Hall plateaux and deep minima in ρ_{xx} at a temperature as high as 4.2 K in magnetic fields below 7 T. The results of Fig. 3a and 3b were obtained for hydrostatic pressures p set at room temperature at 4.3 and 5 kbar, respectively. The actual pressure at 4.2 K is, however, much lower (because of the thermal collapse of the pressure transmitting medium). It is measured to be as low as 0.2±0.1 and (0.3±0.1) kbar, respectively. This implies an unexpectedly large pressure coefficient of the GB electron concentration, difficult to reconcile with the known pressure coefficients of the band structure parameters. One explanation is based

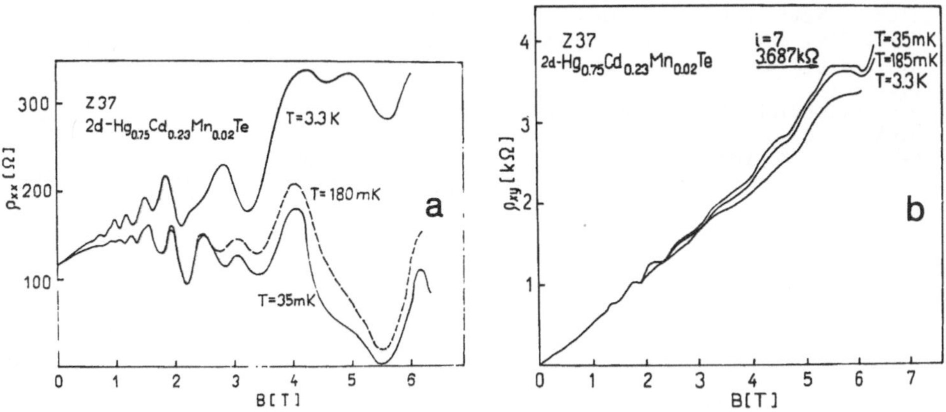

Fig.2.Plots of ρ_{xx} and ρ_{xy} vs. magnetic field for different temperatures.

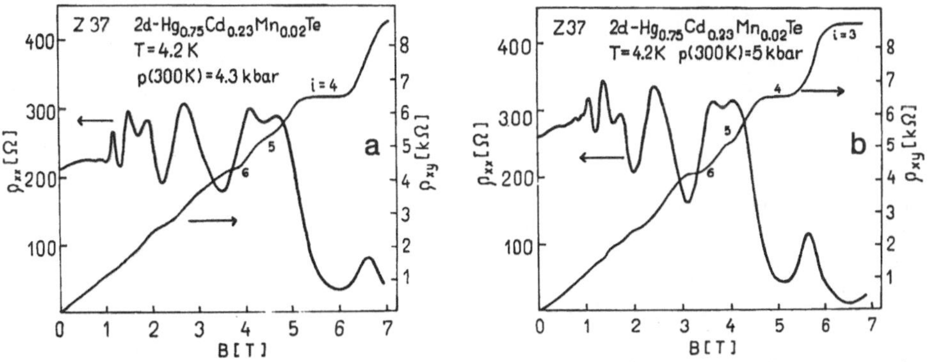

Fig.3.Plots of ρ_{xx} and ρ_{xy} vs. magnetic field for two values of hydrostatic pressure.

on the conjecture that high pressure causes a metastable change in the charge state of the GB defects, which persists despite of the pressure drop at low temperatures. This would place our system in the category of materials which exhibit an interesting effect of high-pressure freeze out /8/.

REFERENCES

1. G.Landwehr, E. Bangert, S. Uchida: Solid State Electron. 28, 171 (1985), and references therein
2. R. Herrmann, W. Kraak, G. Nachtwei, Th. Schurig: phys. stat. sol. (b) 135, 423 (1986), and references therein
3. G. Grabecki T. Dietl, P. Sobkowicz, J. Kossut, W. Zawadzki: Appl. Phys. Lett. 45, 1214 (1984)
4. F. Koch, A.Zrenner, M. Zachau: in Two-Dimensional Systems and New Devices ed. by G. Bauer at al. (Springer, Berlin 1986) p.175
5. G. Bergman: Phys. Rev. Lett. 49, 162 (1982); F. Komori, S. Kobayasahi, W. Sasaki: J. Mag. Mater. 35, 74 (1983);
G.Grabecki, T. Dietl, J. Kossut, W. Zawadzki: Surface Sci. 142, 588 (1984)
M.Chmielowski, T. Dietl, P. Sobkowicz, F. Koch, in Proc. Int. Conf. Phy-

sics of Semiconductors, Stockholm 1986, to be published; M. von Ortenberg: Phys. Rev. Lett. 49, 1041 (1982); A.V. Nurmikko, X.-C. Zhang, S.-K.Chang, L.A. Kolodziejski, R.L. Gunshor, S. Datta: J. of Luminescence 34, 89 (1985)
6. B.L. Altshuler, A.G. Aronov, D.E. Khmelnitzkii, P.A. Lee: Phys. Rev. B22, 5142 (1980)
7. see, e.g., K. von Klitzing: Physica 126B+C, 242 (1984)
8. see, e.g., S. Porowski, W. Trzeciakowski: phys.stat.sol.(b) 128, 11 (1985).

High-Precision Quantum Hall Effect Experiments at the National Institutes

E. Braun

Physikalisch-Technische Bundesanstalt, Bundesallee 100,
D-3300 Braunschweig, Fed. Rep. of Germany

This paper gives a review of the most recent high-precision measurements in various national institutes.

One of the basic duties of the national institutes is the realization and maintenance of the unit of resistance, the ohm. Therefore since the discovery of the quantum Hall effect these institutes have great interest in high-precision measurements of quantized Hall resistances in order to prove if they will provide a more effective maintenance of the ohm.

Up to now results of high-precision measurements of quantized Hall resistances have been presented by the national institutes of Australia /1/, Canada /2/, France /3/, Germany /4/, Great Britain /5/, Japan /6/, Netherlands /7/, Switzerland /8/, USA /9/ and by the Bureau International des Poids et Mesures (BIPM) in Paris /10/.

All experiments show that a sample is suitable for high-precision measurements only, if the longitudinal resistivity (on both sides) ρ_{xx} is lower than a certain value and if the resistances measured between any two contacts by a two-terminal method show equal values which are close to the quantized value (quantized magnetoresistance effect /11/). In GaAs samples this limit is about 1 mΩ, it is higher in silicon.

Due to a general shortage of high-quality samples experimental results obtained with samples having higher values of ρ_{xx} were sometimes corrected by assuming a linear dependence of $\Delta\rho_{xy} = f(\rho_{xx})$ where $\Delta\rho_{xy}$ is the deviation from the quantized value ρ_{xy}. In these cases ρ_{xx} and ρ_{xy} are measured as a function of temperature and ρ_{xy} is obtained from the extrapolation $\rho_{xx} = 0$.

The two-dimensional electron gas has been realized in Si-MOSFETs, in GaAs-$Ga_{1-x}Al_xAs$ heterostructures and in InP-$In_xGa_{1-x}As$ heterostructures.

The heterostructures had been grown by molecular beam epitaxy (MBE) or by metal organic vapor deposition (MOCVD). No dependence on the material nor on the step number (i=4 or i=2) was found.

While the relative uncertainty of the reproducibility of the ratio of a quantized Hall resistance and a reference resistance is in the order of 10^{-8}, the relative uncertainty of the quantized Hall resistance R_H itself is usually ten times larger. This is due to the fact that the realization of the SI-ohm in an institute is very difficult. Only a few laboratories have set up the so-called Thomson Lampard capacitor (TLC). The other laboratories are linked to the TLC at the NML in Australia via intercomparisons of 1 Ω standard resistances with the BIPM. The NML is the only institute, which has done a continuous realization of the ohm and intercomparisons with the BIPM during the past 22 years.

The mean value of 12 different determinations of R_H (/1/ to /10/) is

$$R_H = 25\ 812.8070 \pm 0.0049\ \Omega.$$

From this a value for the fine structure constant can be derived:

$$\alpha^{-1} = 137.035997 \pm 0.000037$$

This is in excellent agreement with the value for α^{-1} recently obtained by KINOSHITA /12/ from the anomalous magnetic moment of the electron:

$$\alpha^{-1} = 137.0359898 \pm 0.0000027$$

In conclusion, the consistency of the results - some minor discrepancies have not been mentioned in this paper - favour the individual monitoring of the as-maintained ohm,as well as an international agreement on an adopted value for R_H for this pupose. The already mentioned shortage of suitable samples, however, is a practical problem.

References:

1. B.W. Ricketts and M.E. Cage: Proc. of CPEM86, to be published in IEEE-IM
2. B.M. Wood and M. D´Iorio: Proc. of CPEM 86, to be published in IEEE-IM

3. F.Delahaye, D. Dominguez, F. Alexandre, J.P. Andre, J.P. Hirtz and M. Razeghi: Metrologia 22,103 (1986)
4. L. Bliek, E. Braun, H.-J Engelmann, H. Leontiew, F. Melchert, W. Schlapp, B. Stahl, P. Warnecke and G. Weimann: PTB-Mitteilungen 93,21 (1983)
E. Braun et al. to be published.
5. A. Hartland, R.G. Jones, B.P. Kibble and D.J. Legg: Proc. CPEM 86, to be published in IEEE-IM
6. K. Yoshihiro, J. Kinoshita, K. Inagaki, Ch. Yamanouchi, T. Endo, Y. Murayama, M. Koyanagi, A. Yagi, I. Wakabayashi and S. Kawaji: Phys Rev. B 33,6874 (1986)
7. W. van der Wel, K.J.P.M. Harmans, R. Kaarls and I.E. Mooij: IEEE-IM 34,314 (1985)
8. W. Schwitz, L. Bauder, H.-J. Bühlmann, M. Py and M. Illegems: Proc. of CPEM 86, to be published in IEEE-IM
9. M. Cage, R. Dziuba, B.F. Field, T.E. Kiess and C.T. van Degrift: Proc. of CPEM 86, to be published in IEEE-IM
M.E. Cage, B.F. Field, R.F. Dziuba, S.M. Girvin, D.C. Tsui and A.C. Gossard: Phys.Rev. B30,2286 (1984)
10. T.J. Witt, T.Endo and D.Reymann: Proc. CPEM 86 to be published in IEEE-IM
11. F.F.Fang and P.J.Stiles: Phys.Rev. B27,6487 (1983)
12. T. Kinoshita: Proc. CPEM 86, to be published in IEEE-IM

Part II

Fractional Quantum Hall Effect

Fractional Quantum Hall Effect in Silicon MOSFETs

I.V. Kukushkin and V.B. Timofeev

Institute of Solid State Physics, Academy of Sciences of the USSR, Chernogolovka 142432, USSR

In Si MOSFET structures with high mobility of 2D electrons in a strong perpendicular magnetic field ($H \lesssim 20$ T) the thermoactivated conductivity is used for determination of activation gaps W in the energy spectrum of interacting electrons at the fractional filling factors: $\nu = 1/3$, $2/3$, $4/3$, $7/3$ and $4/5$. These gaps increase proportionally to the reciprocal magnetic length under the condition that the electron mobility μ_e is fixed. The dependence of the activation gaps on H and μ_e can be factorized within a simple relation. A destruction of FQHE due to disorder is discussed. The magnitudes found for W, and their dependence on H and μ_e, correlate with the theoretical conception concerning the condensation of 2D electrons into quantum incompressible Fermi fluid. An optical spectroscopy method for the determination of the gaps for quasi-particle excitations in the Fermi fluid energy spectrum is suggested.

The phenomenon of the integer quantum Hall effect (QHE) discovered by von Klitzing et al. /1/ is a result of the gaps in the energy spectrum of 2D electrons in a high perpendicular magnetic field and is due to the existence of localized states within those gap regions /2/. Later in GaAs/AlGaAs heterojunctions with very high mobility of carriers in the 2D space charge layer, it was discovered that quantization of the Hall resistance ρ_{xy} and simultaneous vanishing of the diagonal resistivity ρ_{xx} occur also at fractional values of the filling factor $\nu = N_s\, h/eB$ with odd denominators /3,4/. Fractional quantization (FQHE) occurs for rational $\nu = 1/q$ $(q=3,5,7,...)$ and all their multiples $\nu = p/q$ $(p=1,2,3,...)$ /5,6/. The FQHE is observed at temperatures much lower than the characteristic temperatures necessary for the observation of integer QHE. It indicates the existence of a new kind of gaps at fractional ν caused by creation of a new correlated many-particle state with a finite gap in its excitation spectrum. According to Laughlin's conception /7/ this many-particle ground state is an incompressible electron liquid existing at rational fractions of ν. In this model the appearance of an electron is equivalent to the creation of q excitations (namely quasi-electrons) and the disappearance of an electron is equivalent to the creation of q quasiholes. These excitations are separated from the ground state by corresponding gaps which in principle are different for quasielectrons (Δ_e) and quasiholes (Δ_h) /8,9/. Up to now the gaps in the spectrum of the incompressible Fermi-liquid have been determined with the use of the thermoactivated conductivity method /6/. Under thermoactivation of excitations to mobility edges the transport coefficients ρ_{xx}, σ_{xx} at corresponding fractional ν are proportional to exp (- W/kT), which enables one to get the activation energy W and the resulting gap $\Delta = \Delta_e + \Delta_h \approx 2W$ /10/. For small Coulomb gaps 2D-electron gas condensation occurs at such low temperatures that the activation processes are strongly masked by the variable range hopping conductivity /11/. In that case the dependences of $\rho_{xx}(T)$ and $\sigma_{xx}(T)$ are not described by the simple Arrhenius law and the determination of Coulomb gaps by this method becomes less reliable, so that other methods for independent measurement of these gaps are to be found.

This paper consists of two parts. In the first part we present the results concerning the determination of the activation gaps at fractional ν in Si MOSFET'S, with the use of the activated conductivity (or magnetoresistivity) method and the analysis of the dependence of these gaps on magnetic field and electron mobility / 12-13 /. In the second part the optical spectroscopy method for the measurement of Coulomb gaps in the energy spectrum of incompressible Fermi liquid existing at rational fractions of ν will be discussed /14-16/.

Two samples of long geometry having maximum mobility, 3.6×10^4 and $2.8\ 10^4$ cm^2/v s, at T=1.5 K and one sample of Corbino geometry with maximum mobility 3.2×10^4 cm^2/v s at T=1.5 K and $4.1\ 10^4$ cm^2/v s at T=0.35 K have been investigated. The behaviour of the mobility μ_e on the electron density N_s for these MOSFET'S is shown in Fig.1. In the case of high quality samples the mobility demonstrates a strong temperature dependence within the region (1.5-0.35)K and increases when the temperature is lowered. All measurements have been performed under AC regime with 20 Hz frequency and have demonstrated ohmic behaviour in the whole dynamical range from 5.0 to 5 nA.

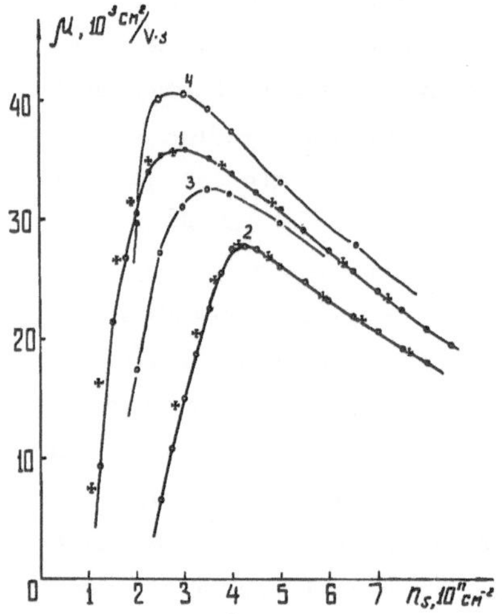

Fig. 1. The dependence of electron mobility on concentration. Curves (1) and (2) correspond to two long MOSFET'S N1, N2 at T=1.5 K, curves (3) and (4) correspond to Corbino MOSFET N3 measured at 1.5 K and 0.35 K respectively. (0) and (+) - measurements of the conductivity at H=0 and Hall conductivity respectively.

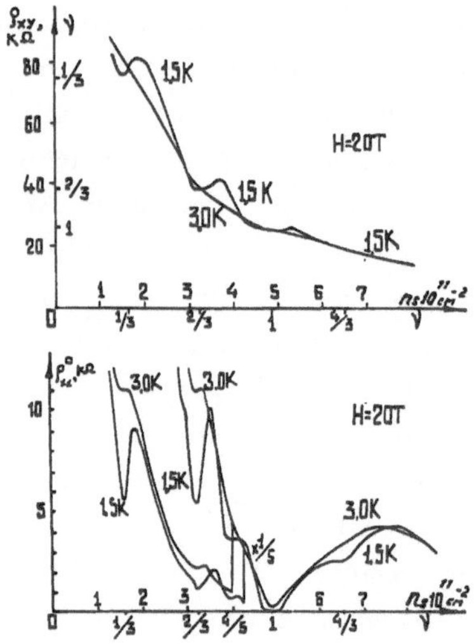

Fig. 2 The behaviour of the diagonal ρ_{xx} and Hall ρ_{xy} components of the magnetoresistivity tensor in the region of fractional filling factors ν =1/3,2/3,4/5,4/3 at H=20 T, T=1.5 K and T=3 K for long MOSFET N1.

Fig.2 illustrates the dependences of the diagonal and Hall resistivity $\rho_{xx}(\nu)$ and $\rho_{xy}(\nu)$, respectively, in the region of ν =2/3 measured at H=20 T and T=1.5 and T=3 K for sample 1. It is evident that the plateau of ρ_{xy} and the minimum of ρ_{xx} at ν =2/3 disappear when the temperature increases. The same behaviour of $\rho_{xy}(\nu)$ and $\rho_{xx}(\nu)$ was observed at ν =1/3, 2/3, 4/3 and 4/5.

Fig. 3 illustrates the dependences of the diagonal conductivity $\sigma_{xx}(\nu)$ measured at H=8 T and T=0.35 and T=1.5 K for Corbino geometry (sample N 3). Minima of σ_{xx} at fractional filling factors ν =4/3, 5/3, 7/3, and 8/3 are clearly seen, which are associated with the condensation of 2D-electrons into an incompressible Fermi-liquid. When the temperature increases the minima of $\sigma_{xx}(\nu)$ gradually disappear but the peculiarities of $\sigma_{xx}(\nu)$ at the above-mentioned fractions still remain up to T=1.7 K. The activation behaviour of ρ_{xx} at ν =2/3 is shown in Fig.4 (the left straight line in Fig.4 corresponds to the dependence $\rho_{xx}^{min} \sim \exp(-W/kT)$). An analogous

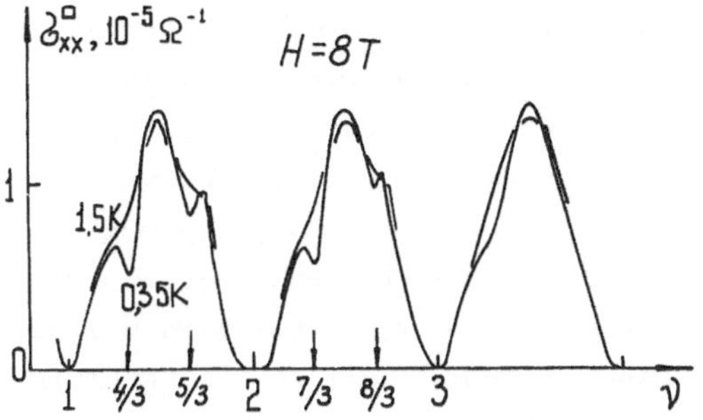

Fig.3. The behaviour of the conductivity σ_{xx} at H=8 T, and T=0.35 K and T=1.5K for Corbino MOSFET N3.

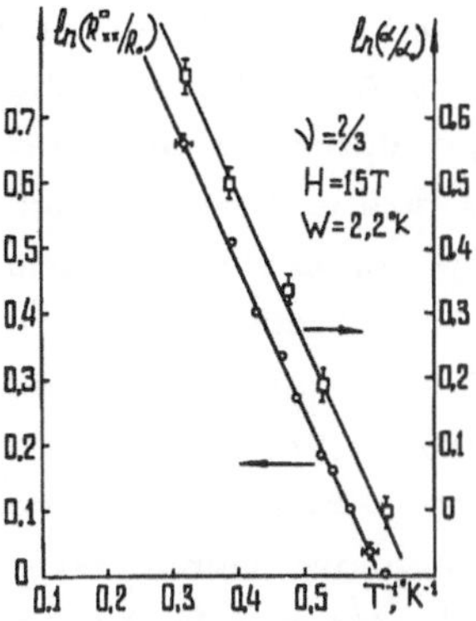

Fig.4. The temperature dependence of ρ_{xx} and derivative $1/\rho_{xy} \cdot d\rho_{xy}/d\nu \equiv \alpha$ at ν =2/3 and H=15 T for the long MOSFET N1. ρ_0 and α_0 correspond to the values of ρ_{xx} and α measured at T=1.5 K.

temperature dependence was observed for the derivative $d\rho_{xy}/d\nu$ in the corresponding plateau region at the above fractions (shown by the straight line in the right part in Fig.4). The activation energies found in this way, with less accuracy, were approximately the same. When the temperature is lowered the onset of breakdown of the simple Arrhenius description is observed. In Fig.5 the temperature dependence of σ_{xx} in coordinates $\ln \sigma_{xx}$ on T^{-1} measured at ν =4/3 and H=8 T down to 0.35 K for Corbino geometry (sample N3) is represented. One can see that this dependence is not described by a simple linear law. It means that the magnetoconductance is not only

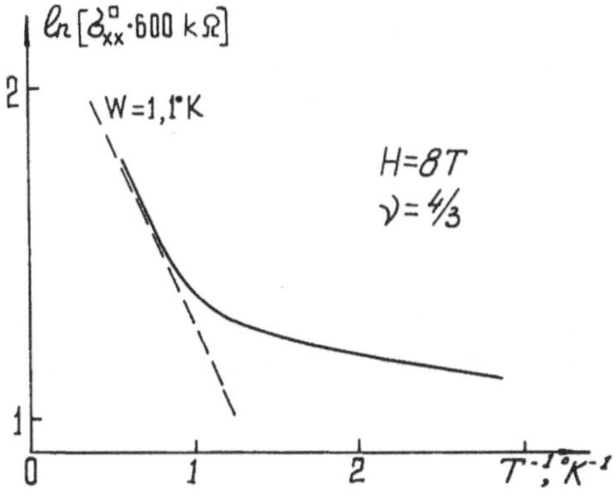

Fig.5. The temperature dependence of σ_{xx} at ν =4/3 and H=8 T MOSFET N3).

due to thermoactivated processes within the investigated temperature region. A similar deviation from the linear dependence at very low temperature has been observed in GaAs–AlGaAs heterojunctions /11/. It is due to the fact that at sufficiently low temperature the variable range hopping mechanism of the conductivity is more effective than the thermoactivation mechanism. Therefore the slope of the curve in $\ln\sigma - T^{-1}$ coordinates directly corresponds to the activation energies W only in the relatively high temperature region (the left part of the curve in Fig.5). For example the activation energy found with the use of the above-mentioned procedure from Fig.5 is equal $W \approx (1.1 \pm 0.05)$K at $\nu = 4/3$ and H=8 T.

It has been shown for different Si MOSFET structures that the activation gaps at fractional $\nu = 1/3$, 2/3, 4/3 and 4/5 increase proportionally to the reciprocal magnetic length in the region H=8–20 T: $W \sim l_0^{-1} \sim H^{1/2}$. In this region, under variation of the electron density N_s, the Hall mobility was fixed with the accuracy about 10%. This dependence shown in Fig.6 reflects just the effect of Coulomb manyparticle interaction, responsible for FQHE in accordance with the theory.

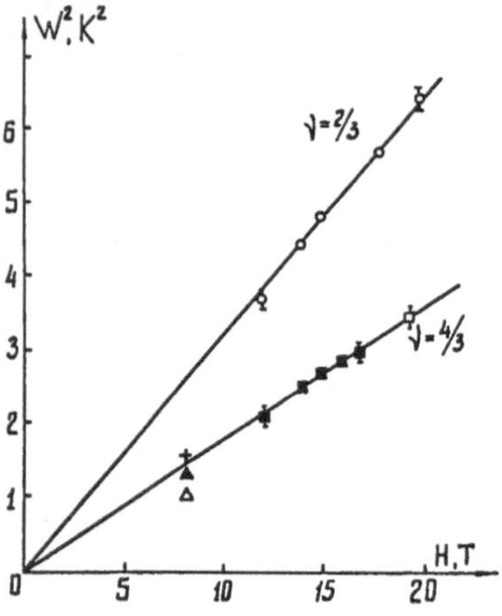

Fig.6. The dependence of the activation gap W on H at $\nu = 2/3$ and 4/3 measured for two long MOSFET'S (dark and open symbols) with fixed mobility: $\mu_e = (3.5 \pm 0.1) \times 10^4$ cm²/V s (circles) and $\mu_e = (2.7 \pm 0.1)\ 10^4$ cm²/V s(squares). The symbols ▲, △ correspond to W measured by means of the activated conductivity method at $\nu = 4/3$ and $\nu = 7/3$ for Corbino MOSFET. The symbol (+) corresponds to Coulomb gap measured by means of optical spectroscopy method at $\nu = 7/3$.

It was found that the magnitudes of the activation gaps are sensitive to disorder and depend on the mobility. At a given magnetic field the magnitudes of the gaps W increase as the mobility grows. The scale of W for fractional ν at different μ_e and under different magnetic fields in various structures can be factorized within the simple expression (see also Fig.7):

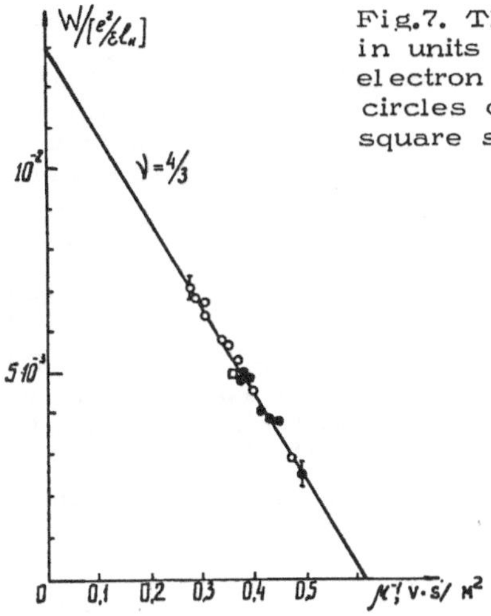

Fig.7. The dependence of the activation gaps in units of Coulomb energy on the reciprocal electron mobility at $\nu = 4/3$. The open(dark) circles correspond to two long MOSFET'S, the square symbol corresponds to Corbino MOSFET.

$$W = G_\nu^\infty \cdot (1 - \mu_0/\mu_e) \cdot e^2/\varepsilon l_0 \, .$$

Here G_ν^∞ is the activation gap made dimensionless by the Coulomb energy when $\mu_e \to \infty$, μ_0 is the minimal mobility when the activation energy approaches zero (FQHE is not observed when $W \to 0$), and ε is the dielectric constant. It follows from experimental observation that μ_0 does not depend on H. The magnitudes of G_ν^∞ ($2G_\nu^\infty = \tilde{\Delta} = \tilde{\Delta}_e + \tilde{\Delta}_h$, $\tilde{\Delta}$ – is the dimensionless Coulomb gap) found with the use of the above procedure are equal to 0.018 (for $\nu = 1/3$), 0.013 (for $\nu = 2/3$), 0.013 (for $\nu = 4/3$) and 0.006 (for $\nu = 4/5$). These magnitudes are about a factor of two less than those calculated theoretically /8,9,17,18/.

It should be mentioned that the experimentally measured activation energy W corresponds to the half energy width between mobility edges of the extended states to which elementary excitations of quasi-electrons and quasiholes with fractional charges occur. When the broadening of the corresponding quantum state, due to disorder, approaches the magnitude of the Coulomb gap, one may expect the destruction of FQHE. This destruction due to disorder has been considered theoretically in /19/. If one accepts that the quantum level broadening is equal to $\Gamma \approx \hbar\omega_c/\sqrt{\mu H}$, according to references /20/, a semiempirical criterion for the minimal mobility can be introduced in the following form:

$$\mu_0 \approx (h/e)^3 \cdot (\varepsilon/m)^2 ,$$

where m is the effective mass of 2D-carriers. This criterion works

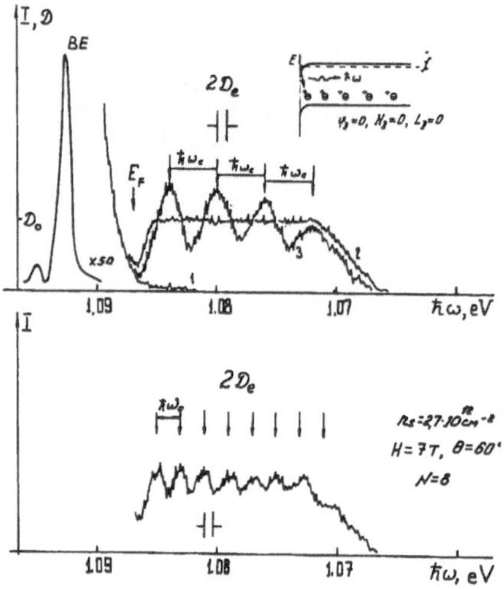

Fig.8. The insert represents schematically the band bending under the condition of nonequilibrium electron-hole excitation near the interface region in Si MOS-structure. BE- the luminescence of excitons bound with boron atoms from crystal volume. Spectrum(1)-the long wave tail of BE-line at $V_G=V_T$, $N_s=0$. Spectrum(2) and (3) correspond to the recombination of 2D-electrons with nonequilibrium holes at $T=1.6K$, $N_s=2.7\ 10^{11}cm^{-2}$, $H=0$ and $H=7$ T respectively. $D_o=1.6\ 10^{11}cm^{-2}$ is the density of states at H=0. The spectrum (b) in the lower part of the figure is measured at T=1.6K, $N_s=2.7\ 10^{11}cm^{-2}$. H=7 T and $\varphi=60^o$.

satisfactorily in the cases of electron inversion layers in Si MOSFET and electron and hole 2D-layers in Ga/AlGaAs heterojunctions.

Now we discuss an optical spectroscopy method for measurements of the gaps in the energy spectrum of quasiparticle excitations of incompressible Fermi-fluid under the condition of FQHE. This method is founded on the analysis of radiative recombination (RR) spectra caused by the recombination of 2D-electrons with holes bound to boron atoms when nonequlibrium electron-hole excitation near SiO_2-Si interface region of MOSFET structure occurs (see the insert of Fig.8). Under the condition of e-h excitation the band bending behind the 2D-electron layer disappears and the radiative recombination process is due to the wave functions overlapping of 2D-electrons and nonequilibrium holes bound to boron atom cores just close to the interface. Because the inhomogeneous energy broadening of nonequlibrium holes is not large (< 0.8 meV) the radiative recombination spectra reflect the density of states of 2D-electrons. Fig.8 represents RR spectra measured at H=0 (spectrum 2) and H=7T(spectrum 3) when 2D-electron concentration was equal to N_s= =2.7 10^{12} cm^{-2} and T=1.6 K. Without magnetic field the RR spectra have the steplike distribution which directly reflects the constant 2D-electron density of states. The spectral full width at half maximum corresponds to Fermi-energy of 2D-electrons. In magnetic field the RR spectra demonstrate Landau level quantization and $\hbar\omega_c$ scale is sensitive to the magnitude of H projection relatively to 2D-plane due to two-dimensionality of electrons (compare spectra (3) and (b) in

Fig. 8). It should be emphasized that 2D-electrons recombine only with holes having the momentum J_z =-3/2. Therefore the optical transitions for electrons with S_z =-1/2 are forbidden and these electrons are discriminated in optical RR spectra. It means that with the use of the optical spectroscopy method we have opportunity to investigate phenomena in Si MOSFET at filling factors $\nu > 2$.

Fig. 9 illustrates the RR spectra of 2D-electrons measured at H=8 T and T=1.6 K under the variation of ν in the region close to ν =7/3 for Corbino MOSFET N 3 (the corresponding dependence of $\sigma_{xx}(\nu)$ was shown in Fig.3). It should be mentioned that under conditions corresponding T=1.6 K and $2.27 < \nu < 2.40$ a considerable part of 2D-electrons are delocalized and the Landau level width is about 0.3 meV /15/. Due to that the shape of these RR spectra reflects the distribution of nonequilibrium holes (< 0.8 meV) and is not sensitive to 2D-electron density of states. Actually one can see from Fig. 9 that the shape of 2De-line does not practically change but the energy position of this line is non-monotonous under the variation of ν within the range $2.27 < \nu < 2.40$. With the aim to show the effect of 2D-electron condensation, it is useful to compare the found nonmonotonous dependence of the energy position of RR line

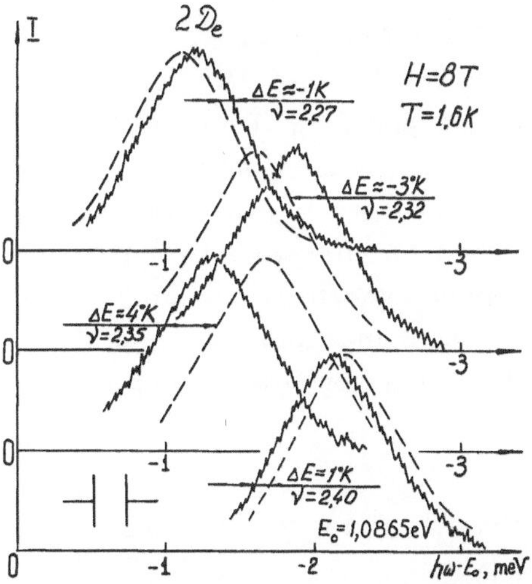

Fig. 9. The radiative recombination spectra of 2D-electrons with non-equilibrium holes measured at H=8 T in the region $2.27 < \nu < 2.40$ for Corbino MOSFET. Solid and dashed curves correspond to T=1.6 K and T=4 K respectively.

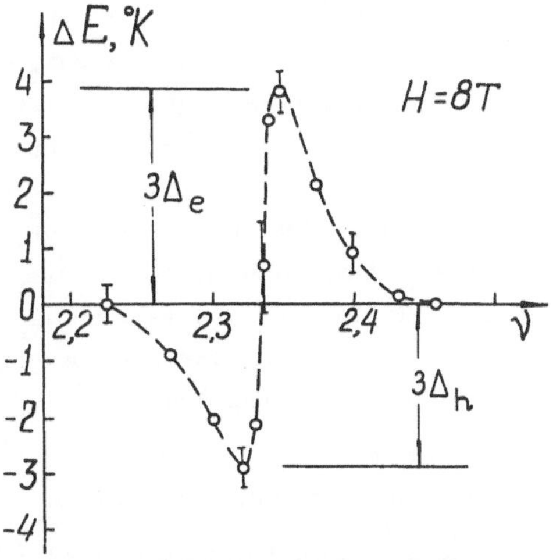

Fig.10. The dependence of the 2De-line energy position difference measured at T=1.6 K and T=4 K on filling factor ν at H=8 T in the region ν =7/3.

on ν with that one obtained under the condition when incompressible Fermi-fluid does not exist (at temperature $T > \Delta$). The energy position of 2De-line measured at T=4 K $> \Delta$ monotonically changes on ν and does not exhibit any peculiarities at fractional ν =7/3. In our opinion the energy difference ΔE for the spectral positions of 2De-lines measured at T=1.6 K and T=4 K characterizes the effect of 2D-electron interaction when the condensation into incompressible Fermi-fluid occurs. Fig.10 demonstrates that the dependence $\Delta E(\nu)$ measured at H=8 T exhibits the peculiarity just at ν =7/3. Namely, the magnitude ΔE is negative and minimal at ν which are a little bit lower than ν =7/3 than changes the sign and reaches the maximum at ν which are a little larger than ν =7/3. The dependence $\Delta E(\nu)$ can be explained if one takes into consideration that under any recombination act a one 2D-electron disappears and with respect to the theoretical Fermi-fluid model it is equivalent to creation of three quasiholes with the charge 1/3 e at $\nu < 7/3$ and to absorption of three quasielectrons at $\nu > 7/3$. It means that when three quasielectrons are absorbed, a phonon energy $\hbar\omega$ increases on $3\Delta_e$ (analogously a photon energy decreases on $3\Delta_h$ when three quasiholes are created). The corresponding conservation laws, for radiative recombination, can be written in the following way:

$$\text{at } \nu > 7/3 : \; E_e + E_h + 3N\Delta_e \longrightarrow \hbar\omega + 3(N-1)\Delta_e ,$$
$$\text{at } \nu < 7/3 : \; E_e + E_h + 3N\Delta_h \longrightarrow \hbar\omega + 3(N+1)\Delta_h .$$

From Fig.10 it follows that $3\Delta_e = (4 \pm 0.3)$K and $3\Delta_h = (3 \pm 0.3)$K at H=8 T and ν =7/3. The resulting Coulomb gap Δ measured with the use of the optical spectroscopy method agrees with that one found by the activated conductivity method for the same Corbino MOSFET. Therefore with the use of optical spectroscopy method the possibility for the separate measurements of the gaps Δ_e and Δ_h appears as well as for the investigation of a temperature dependence of these Coulomb gaps.

References

/1/ K. von Klitzing, G.Dorda and M.Pepper, Phys.Rev.Lett., 45 (1980) 494.
/2/ R.B.Laughlin, Phys.Rev., B23 (1981) 5632.
/3/ D.C.Tsui, H.L.Stormer and A.C.Gossard, Phys.Rev.Lett., 48 (1982) 1559.
/4/ H.L. Stormer, D.C.Tsui, A.C.Gossard and J.C.M.Hwang, Physica 117B/118B1 (1983) 688.
/5/ H.L.Stormer, Surface Sci., 142 (1984) 130.
/6/ D.C.Tsui in: Proc. 17th Int.Conf.Phys.Semicond., San Francisco 1984, Eds.D.J.Chadi and W.A.Harrison (Springer,Berlin,1985) 247.

/7/ R.B.Laughlin, Phys.Rev.Lett., 1983, 50, 1395.
/8/ R.B.Laughlin, Surf.Sci., 1983, 142, 163.
/9/ R.Morf, B.I.Halperin, Phys.Rev., 1986, B33, 2221.
/10/ A.M.Chang, M.A.Paalanen, D.C.Tsui, H.L.Stormer, J.C.M.Hwang, Phys.Rev., 1983, B28, 6133.
/11/ S.Kawaji, J.Wakabayashi, J.Yoshino, H.J.Sakaki, Phys.Soc.Jap., 1984, 53, 1915.
/12/ I.V.Kukushkin, V.B.Timofeev, Zh.Eksp.Teor.Fiz, 1985,89,1692.
/13/ I.V.Kukushkin, V.B.Timofeev, Surf.Sci., 1986, 170, 148.
/14/ I.V.Kukushkin, V.B.Timofeev, Pis'ma Zh.Eksp.Teor.Fiz., 1984, 40, 413.
/15/ I.V.Kukushkin, V.B.Timofeev, Pis'ma Zh.Eksp.Teor.Fiz.,1986, 43, 387.
/16/ I.V.Kukushkin, V.B.Timofeev, Pis'ma Zh.Eksp.Teor.Fiz.,(in press).
/17/ A.H.MacDonald, Phys.Rev., 1984, B30, 2550.
/18/ S.M.Girvin, A.H.MacDonald,P.M.Platzman.Ph.Rev.Lett.,1985,54,581.
/19/ R.B.Laughlin, M.L.Cohen. J.M.Kosterlitz, H.Levine, S.B.Libby, A.M.M.Pruisken, Ph.Rev.,1985, B32, 1311.
/20/ T.Ando, Y.Uemura, J.Phys.Soc.Japan, 1974, 36, 959.

Recent Advances in the Study of the Fractional Quantum Hall Effect

R.J. Nicholas[1], R.G. Clark[1], A. Usher[1], J.R. Mallett[1], A.M. Suckling[1], J.J. Harris[2], and C.T. Foxon[2]

[1]Clarendon Laboratory, Parks Road, Oxford, OX13PU, UK
[2]Philips Research Laboratories, Redhill, Surrey, UK

A description is given of the occurrence of fractional quantum effects, and how these are influenced by the presence of disorder. It is shown that, when electrons are photoexcited into a GaAs-GaAlAs heterojunction, the fractional state existing at a Landau level occupancy of 7/5 can dominate over that at 4/3. Activation energy measurements of the resistivity show how the energy gap of the fractional states is reduced by the presence of disorder. A new method of analysis of the resistivity minima is presented which is used to give information on the correlation length of the many-body ground state.

1. INTRODUCTION

The existence of the fractional quantum Hall effect (FQHE) is taken to be evidence for the formation of a new highly correlated ground state of a two-dimensional electron gas. This occurs at very low temperatures, in high magnetic fields, and in systems where there is only a very small amount of disorder present. The main experimental observations are that minima are observed in the electrical resistivity component ρ_{xx}, at fractional Landau level occupancies $\nu = nh/eB = p/q$, where p is an integer and q is an odd integer (1-9); while corresponding Hall plateaus are seen at quantized Hall resistivity values of $h/\nu e^2$. To date, fractional states have been reported at $\nu = p/q = 1/3$, 1/5, 2/5, 2/7, 3/7 and 4/9, and the equivalent 'hole' analogues of these states have been observed at occupancies $\nu = 1 - (p/q)$. These states occur when all of the electrons lie in the lowest spin split Landau level, but it has recently been shown that they can all exist in a similar manner in the upper spin state at occupancies of the form $\nu = 1 + (p/q)$. Once $\nu > 2$ the electrons occupy the second Landau level. At this point the experimental position becomes less clear, with some reports of the observation of 7/3 and 8/3 states (3,4), and some suggestions that even denominator fractions may occur (8,9). The significance of these results is that the existence of minima in the resistivity and quantized Hall plateaus may be shown, by using the gauge invariance arguments of LAUGHLIN (10), to result from the formation of a mobility gap in the density of states. In other words the degeneracy of the individual Landau levels for isolated electrons has been lifted by the residual Coulomb interactions, leading to the formation of an energy gap between the ground and excited states of the system.

The theoretical treatments of the phenomenon break into three main groups. Firstly there is the original quantum fluid picture developed by Laughlin (11-13), which is based on a postulated trial ground state wave-function. This possesses quasi-particle excitations which have fractional charge of unit e/q. The higher fractional states are thought to result from a hierarchy (12,14,15) in which the ground state of each succeeding fraction (say 2/5), is the result of a condensation of the quasi-particles associated with the preceding level (1/3). These quasi-particles exhibit fractional charge, corresponding to the denominator of the fraction concerned. This picture implies that no 'daughter' state may exist unless its parent state exists - a prediction which is brought into question below. Further extensions of this theory include the calculations by MACDONALD et al. (16) for the N = 1 Landau level, in which it is found that the fractional states repeat themselves, but that the quasi-particle energy gap associated with the 1/5 states is comparable with that of the 1/3 state. Several calculations have been made of the energy gap for the 1/3 state, i.e. the energy needed to create a separated quasi-electron - quasi-hole pair. These give numbers of order $Ce^2/4\pi\varepsilon l_B$, with $C \sim 0.1$ (17-19), where l_B is the cyclotron radius ($l_B = \sqrt{\hbar/eB}$). If the quasi-particles may be approximated to point charges, then the energy gap will scale as $q^{-2.5}$ (15). Calculations of the dispersion relation for the quasi-particle pair excitation (20,21) suggest that there may be a rather smaller 'indirect gap' at finite wavevectors, close to the reciprocal lattice constant for a Wigner crystal.

The second approach is that originated by TAO and THOULESS (22,23), who used the Landau gauge as a starting point, in which the single particle states may be thought of as a set of parallel tracks. They suggested that a collective state could be formed by an ordered filling of the tracks such that every qth single particle state is filled. Quasi-particle excitations of this state then consist of a 'defect' with one extra track filled or unfilled, while a quasi-particle pair excitation would consist of the translation of a single occupied track by one electron state. This model also suggests that even denominator fractions may occur, but calculations of the energy gaps for both the odd and even fractions give values considerably greater than are found from the Laughlin approach, or by experiment.

More recently there have been several calculations made based on a Wigner crystal ground state (24,25), which predict a lower energy ground state than that found by Laughlin. KIVELSON et al (24) have found large contributions to the energy from cooperative ring exchange, in which electrons move coherently along a closed path in the crystal lattice. These contributions are enhanced at rational filling factors, again leading to an energy gap and quasi-particle excitations. The energy gap values are comparable with those predicted for the Laughlin ground state.

2. ELECTRON CONCENTRATION DEPENDENCE

Experimental results are described here which were taken using Hall bridge specimens with channel widths of 50 - 150 μm. These were modulation doped $GaAs-Ga_{0.68}Al_{0.32}As$ heterojunctions grown by M.B.E.* (26), using spacer layers of 400 and 800Å, and with resulting electron concentrations in the range $0.6 - 4 \times 10^{11} cm^{-2}$. The sample mobilities ranged from 0.1 to $2.1 \times 10^6 cm^2/Vs$, depending upon the sample and electron concentration. For any one sample it was possible to change the electron concentration by factors of 2 - 3 by excitation of persistent photoconductivity using a red L.E.D. The samples were mounted on a laminated copper cold finger in a dilution refrigerator, and were cooled to temperatures as low as 20 mK. The temperature was measured by a calibrated carbon resistor. All electrical connections were carefully screened and filtering was employed both inside and outside the cryostat. Low frequency A.C. currents were used at densities in the range 10^{-2} to 10^{-6} A/m, in order to determine the

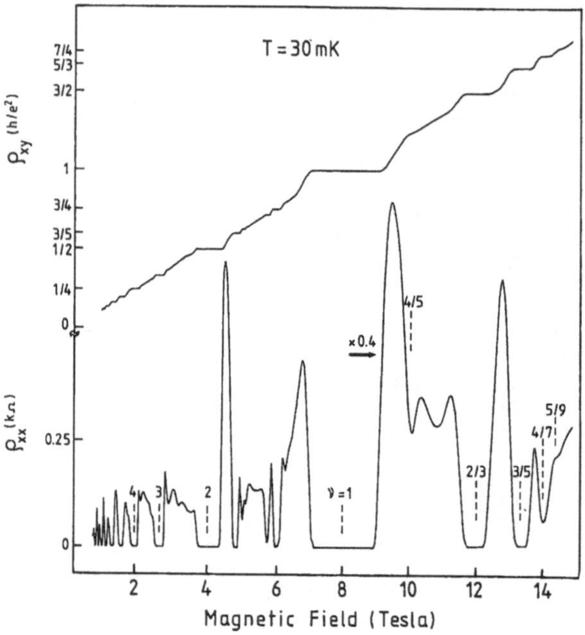

Fig. 1: The resistivity and Hall voltage (upper trace) for sample G63 at 30 mK. The fractional occupancies for $\nu < 1$ are indicated on the trace. The Hall resistivity is given in units of (h/e^2).
The current is 100 nA, corresponding to a current density of 6.6×10^{-4} A/m.

onset of electron heating effects. The magnetic field was produced by a Nb_3Sn superconducting solenoid.

Fig. 1 shows a typical recording of the resistivity ρ_{xx} and Hall component ρ_{xy}, for the highest mobility sample G63 at an electron concentration of $1.9 \times 10^{11} cm^{-2}$, following the photoexcitation of the majority of the carriers. The current density is 6.6×10^{-4} A/m. This shows what is probably the most comprehensive set of fractional states observed to date. The region above 8T corresponds to the incomplete filling of the lower spin state of the N = 0 Landau level, and clear features can be observed at 2/3, 3/5, 4/7, with a weak feature at 5/9. There is also a weaker minimum

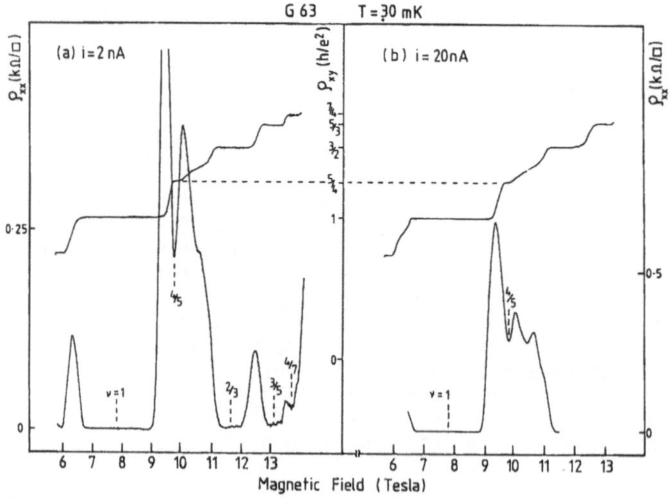

Fig. 2: Experimental recordings of ρ_{xx} and ρ_{xy} for sample G63 for two different measuring currents.

at 4/5, which is shown more clearly in fig. 2, where measurements are shown at current densities of 1.3×10^{-4}, and 1.3×10^{-5} A/m. At the lower current densities there is also a Hall plateau associated with $\nu = 4/5$, the first time that this has been observed, and the furthest into the Landau level tail that a fractional state has been detected. The conductivity between the 2/3 and 3/5 minima shows a substantial reduction relative to the results taken with higher current levels, and this is not understood, although it may be related to the very low measuring fields which can lead to contact problems. The fractional features can also be seen to repeat themselves in the region $1 < \nu < 2$, with a similarly regular set of q = 3, 5, 7 fractions, which is obviously symmetric about the half-occupied level. This behaviour is entirely consistent with the picture of a series of states becoming progressively more bound as the denominator decreases or as they move towards the centre of each level, where the influence of disorder is least. Considerable changes from this picture can be seen however when the electron concentration is varied and more disordered samples are studied.

Figure 3 shows the electron concentration dependence of the FQHE in a slightly more disordered sample G62, from 1.5 to $2.3 \times 10^{11}\mathrm{cm}^{-2}$. The q = 7 states have all disappeared, and the 3/5 state is considerably weaker, as judged by the depth of its associated resistivity minimum. The 7/5 state is only seen at the highest carrier concentration. The only anomalous feature is that the 5/3 minimum rapidly increases in strength as more carriers are photoexcited, with much less change for $\nu = 4/3$. By the highest trace the width of the 5/3 Hall plateau is clearly greater than that for 4/3. This tendency becomes much more pronounced for the next sample described, G29, which is shown in fig. 4, varying from 1.9 to $3.4 \times 10^{11}\mathrm{cm}^{-2}$. The mobility variation in this sample is extremely rapid ($\alpha\ n^{4.5}$; FOXON et al (26)), which is thought to be due to the presence of long-range potential fluctuations. At the lowest concentration the 5/3 and 4/3 states are hardly even visible as minima in the resistivity, but the 5/3 feature rapidly evolves to be a well-defined Hall plateau and resistivity minimum. In contrast the 4/3 state only ever gives rise to a weak resistivity minimum, but still shows a Hall plateau. On closer inspection however, the Hall resistivity of this plateau is found to be a few per cent too high. The very surprising feature is the appearance of a stronger minimum at $\nu = 7/5$, which does have an accurately quantized Hall plateau, obviously dominating the 4/3 features. A final and even more pronounced example of this trend is shown in fig. 5, for sample G71. Following full photoexcitation of the sample, the electron concentration reaches $3.8 \times 10^{11}\mathrm{cm}^{-2}$, by which time the 5/3 minimum, and its corresponding Hall plateau, have become well-resolved features. The 4/3 minimum is almost completely annihilated, and 7/5 takes over both the resistivity and Hall plateau.

It would thus appear that there is a systematic change in the character of the phenomenon as the influence of disorder becomes more pronounced. Further evidence for this comes from the earlier report of CLARK et al (27), who were using more disordered samples grown in a completely different growth kit. Following photoexcitation a deep resistivity minimum and accurately quantized plateau appeared at 7/5, with the complete absence of a fractional state at 4/3, but with well-behaved 5/3 structures. The common feature in all of these studies is that the inversion in strength of the 7/5 and 4/3 states is brought about by the photoexcitation of additional carriers into the 2DEG. At the same time the 5/3 state becomes more strongly favoured in even the highest mobility samples (for G63 the activation energy for 5/3 is substantially larger than for 4/3 (8)). It

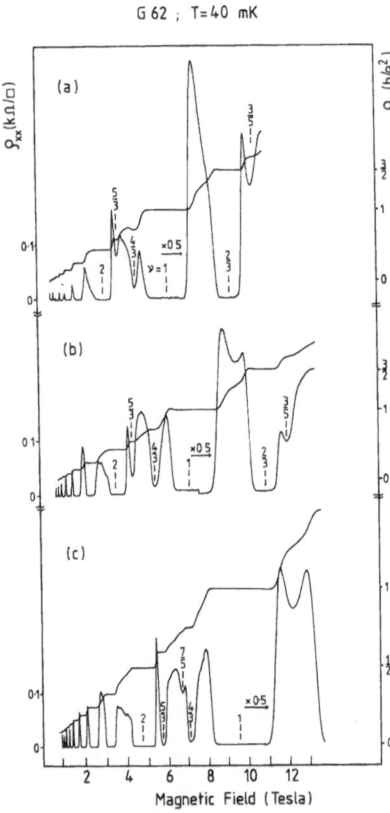

Fig. 3: ρ_{xx} and ρ_{xy} traces for sample G62 at 40 mK. The first trace (a) is before illumination, (b) is in an intermediate state, and (c) has the fully saturated electron concentration.

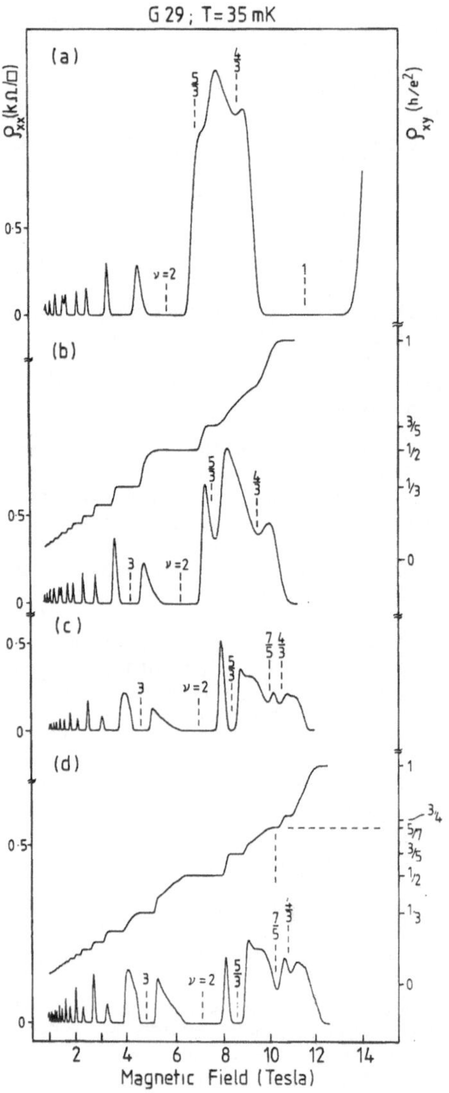

Fig. 4: ρ_{xx} and ρ_{xy} traces for sample G29 at 35 mK. Four different carrier concentrations are shown, as produced by photo-excitation.

appears to be the photoexcitation process which is critical, rather than the increase in electron concentration, as earlier results on samples of comparable concentration do not show such behaviour. In a recent measurement in which the electron concentration was increased by the use of a back electrode, BOEBINGER et al (6) observed the opposite behaviour; at high concentrations a very weak 7/5 state was completely suppressed by a broad

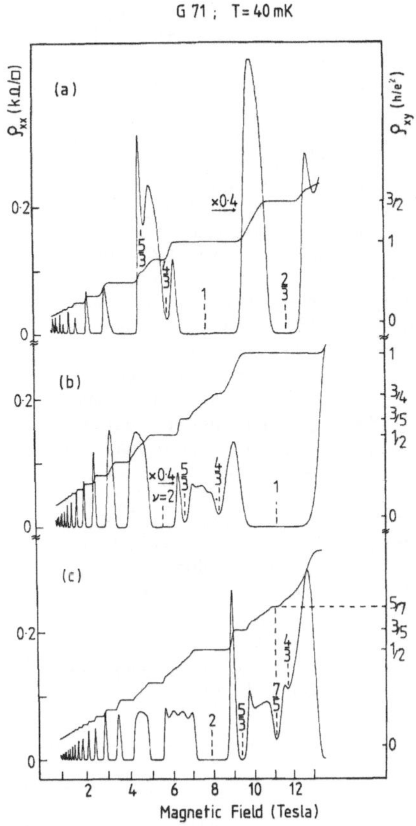

Fig. 5: ρ_{xx} and ρ_{xy} traces for sample G71 at 40 mK, for three different electron concentrations.

4/3 resistivity minimum at low temperatures, although the high electron concentrations were again found to favour the 5/3 state. The photo-excitation mechanism is associated with the presence of deep traps in the GaAlAs doped layer, and therefore once these have been excited they will act as positively charged remote scattering centres which alter the disorder in the system. The predominant sign of these scattering centres would seem to be the most likely cause of the systematic difference in behaviour between the results for the region $\nu < 1.5$, where the Landau level is electron-like and therefore under the influence of an attractive potential, and for $\nu > 1.5$ where the hole-like states will be repelled from the scattering centres. This may also be the reason why the q = 3 states, with larger charge units, are suppressed relative to q = 5 while they are electron-like. The final, and probably the most significant, conclusion is that these results pose considerable difficulties for the interpretation of the FQHE in terms of a heirarchical model, in which the existence of fractional states at a lower order is a necessary prerequisite for the formation of successive orders.

3. ENERGY AND MOBILITY GAPS

The most clear predictions from the theoretical descriptions of the FQHE described above concern the magnitude of the energy gap associated with the formation of the collective ground state. We have attempted to measure this by measurements of the activated behaviour of the resistivity minima associated with the formation of the fractional states. Once an energy gap has been formed the density of states may be represented schematically as shown in fig. 6, with a mobility gap separating the conducting states of each type of particle, which will begin at the 'mobility edges' E_C and E_C'. When the effects of disorder become small, this mobility gap will be close to the energy required to form a quasi-electron quasi-hole pair. At temperatures below the gap energy one would expect an exponentially activated resistivity, with an energy (Δ) equal to half the mobility gap. Such behaviour is shown in fig. 7, for the samples G63, G62 and G71, which are in order of increasing 'dirt', as judged by mobility and quality of the FQHE states observed. The data are taken for the resistivity minimum associated with $\nu = 2/3$, and with electron concentrations adjusted by photoexcitation so that the minima all occur at approximately the same magnetic field. It can be seen that the effect of disorder is to cause a significant decrease in the magnitude of the activation energy. This may be due to either one or both of the following effects: the energy gap of the many body state may have been reduced, and in addition a broadening of the density of states may bring the two mobility edges closer together, thus reducing the measured mobility gap. The maximum value of Δ is 2.3 K, for G63, which is larger than other values reported for similar fields (5,7), but still approximately a factor of three smaller than most theoretical predictions (17-19). This is probably due to some residual disorder left in the system, to the possibly rather lower 'indirect gap' for quasi-particle excitation (20,21) and to the influence of the finite extent of the wavefunction out of the plane. ZHANG and DAS SARMA (28), have shown recently that there is an almost two-fold reduction in the gap energy when the finite z-extent of the electrons is considered.

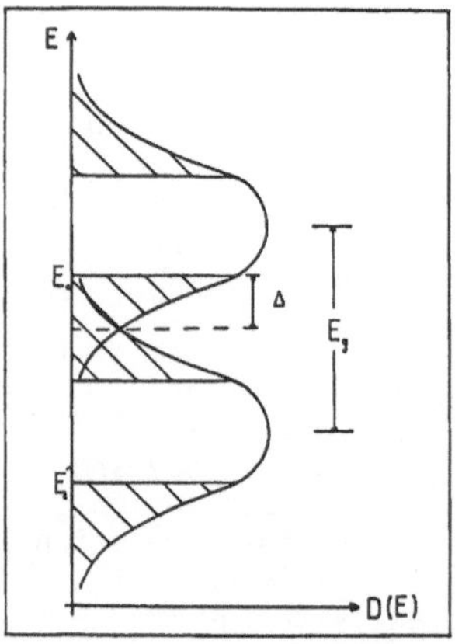

Fig. 6: A schematic picture of the density of states for the many-body ground state. The mobility edges E_C are shown, as is the activation energy Δ.

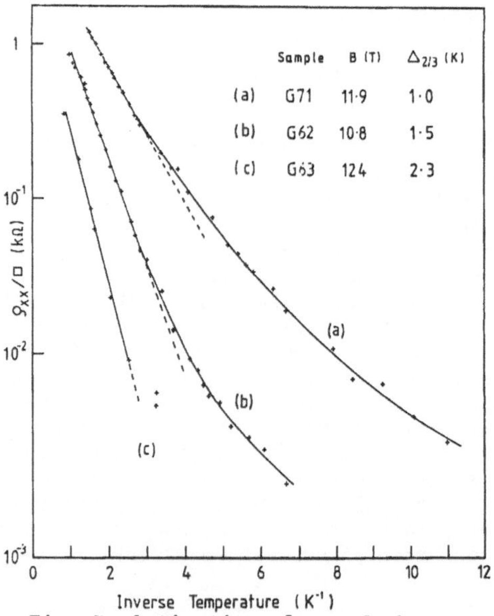

Fig. 7: Activation plots of the resistivity for the 2/3 minimum in the samples shown in figs. 1, 3 and 5.

At still lower temperatures there is a deviation from the Arrhenius plots, as hopping conduction sets in. An approximate fit to the temperature dependence gives $\rho \sim \exp(T^{\alpha})$, with α close to 1/4, as expected for bulk materials, rather than 1/2 as predicted for 2-D systems in high magnetic fields (29). This may be due to the presence of some leakage through the doped layers in the structure.

It is interesting to study the widths of the Hall plateau and resistivity minima in order to try to analyse the localisation effects associated with the quasi-particle states. The widths at absolute zero should measure the proportion of these states directly, however the activation energies are particularly small in disordered samples and even 30 mK is not sufficiently low to be a good approximation to zero. Instead we adopt the following procedure: the conductivity is known to follow an activated behaviour when E_F lies between the mobility edges. This gives a thermally excited quasi-particle population of

$$n \sim D(E_c)\ kT \exp(E_F - E_c)/kT \tag{1}$$

for one level. This is dominated by the exponential term, so we have for the conductivity:

$$\sigma \approx \sigma_c \exp(E_F - E_c)/kT \tag{2}$$

for conduction at the mobility edge. If we now choose a fixed value of conductivity, then this will occur for a Fermi energy of

$$E_F(\sigma) = E_c - kT \ln(\sigma_c/\sigma). \tag{3}$$

Provided that the energy dependence of D(E) is relatively weak, we may now equate this to the occupancy of the system (as $\nu \propto n/eB$) to give:

$$\nu(\sigma) = \nu_c - AkT \ln(\sigma/\sigma_c), \tag{4}$$

where A is a proportionality factor related to the density of states. This says that if we measure the width of any σ_{xx} minimum at a fixed σ, we would expect a linear plot as a function of temperature extrapolating to $\Delta\nu$ (T = 0), with a gradient proportional to the value of σ. The results of such an analysis for various values of σ are shown in fig. 8 for the 2/3 minimum in sample G71. These are in good agreement with (4), corresponding to a fan of lines all extrapolating back to a value of $\Delta\nu = 0.09$. Similar analyses for the 5/3 minimum (30) also give values around 0.09. The data for the other samples is rather less complete, but for G63, for example, analyses of the 2/3, 4/3 and 5/3 minima all suggest widths of order $\Delta\nu = 0.04$.

This analysis implies that for a given sample there is a critical level of 'occupational disorder', which will destroy fractional states of a given denominator. Purer samples with less random disorder can tolerate less of this before becoming de-localised - i.e. they have a smaller proportion of localised states present. This also gives some idea of the correlation length of the quantum fluid, since the electron system may still be regarded as having an exact occupancy, provided that it is only observed over a finite length scale (i.e. throughout the resistivity minimum the effective occupancy remains at the exact value). The ordering length is then given approximately by $l_B\sqrt{\pi/\Delta\nu}$, which is of order 10 l_B for the samples studied, consistent with the calculations of LAUGHLIN (11-13), who estimates ± 5 l_B. The influence of the disorder will be to

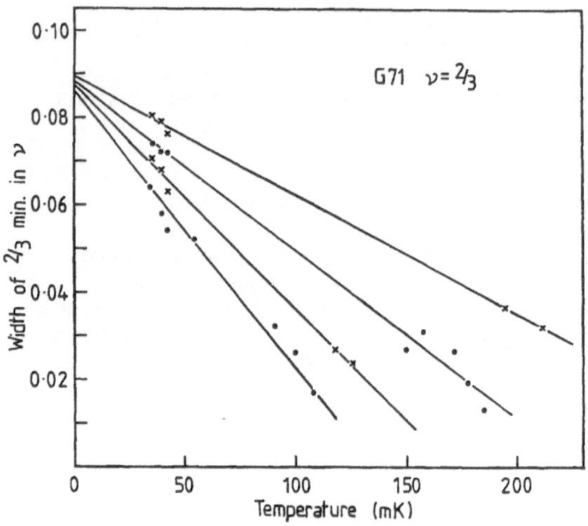

Fig. 8: A plot of the temperature-dependent width of the 2/3 minimum for sample G71 at 12T, using the procedure described in the text.

reduce the correlation length of the many-body wavefunction. This will then reduce the binding energy as the electron-electron interactions will be suppressed. As a result the more disordered samples will be less sensitive to the 'occupational disorder' and will give rise to wider plateau and resistivity minima, but will not show higher order fractions as these correspond to longer distance correlations.

REFERENCES

1. D.C. Tsui, H.L. Stormer and A.C. Gossard, Phys. Rev. Lett. 48, 1559 (1982)
2. A.M. Chang, P. Berglund, D.C. Tsui, H.L. Stormer and J.C.M. Hwang, Phys. Rev. Lett. 53, 997 (1984)
3. E.E. Mendez, L.L. Chang, M. Heiblum, L. Esaki, M. Naughton, K. Martin and J. Brooks, Phys. Rev. B30, 7310 (1984)
4. G. Ebert, K. von Klitzing, J.C. Maan, G. Remenyi, C. Probst, G. Weimann and W. Schlapp, J. Phys. C17, L775 (1984)
5. G.S. Boebinger, A.M. Chang, H.L. Stormer and D.C. Tsui, Phys. Rev. Lett. 55, 1606 (1985)
6. G.S. Boebinger, A.M. Chang, H.L. Stormer and D.C. Tsui, Phys. Rev. B32, 4268 (1985)
7. J. Wakabayashi, S. Kawajii, J. Yoshino and H. Sakaki, Surf. Sci. 170, in press (1986)
8. R.G. Clark, R.J. Nicholas, A. Usher, C.T. Foxon and J.J. Harris Surf. Sci. 170, 141 (1986)
9. R.J. Nicholas, R.G. Clark, A. Usher, C.T. Foxon and J.J. Harris, Solid State Commun. in press (1986)
10. R.B. Laughlin, Phys. Rev. B23,5632 (1981)
11. R.B. Laughlin, Phys. Rev. Lett. 50, 1395 (1983)
12. R.B. Laughlin, Surf. Sci. 142 163 (1984)
13. R.B. Laughlin, in Solid State Sciences (Springer Verlag) 53 p.279 (1984)
14. F.D.M. Haldane, Phys. Rev. Lett. 51, 605 (1983)
15. B.I. Halperin, Phys. Rev. Lett. 52, 1583 (1984)
16. A.H. MacDonald and S.M. Girvin, Phys. Rev. B33, 4414 (1986)
17. S.M. Girvin, Phys. Rev. B30, 558 (1984)

18. A.H. MacDonald, G.C. Aers and M.W.C. Dharma-wardana, Phys. Rev. B31, 5529 (1985)
19. R. Morf and B.I. Halperin, Phys. Rev. B33 2221 (1986)
20. S.M. Girvin, A.H. MacDonald, and P.M. Platzman, Phys. Rev. B33, 2481 (1986)
21. F.D.M. Haldane and E.H. Rezayi, Phys. Rev. Lett. 54, 237 (1985)
22. R. Tao and D.J. Thouless, Phys. Rev. B28, 1142 (1983)
23. R. Tao, Phys. Rev. B29, 635 (1984)
24. S. Kivelson, C. Kallin, D.P. Arovas and J.R. Schrieffer, Phys. Rev. Lett. 56, 873 (1986)
25. R. Keiper and O. Zeip, Phys. Stat. Solidi 133b, 769 (1986)
26. C.T. Foxon, J.J. Harris, R.G. Wheeler and D.E. Lacklison, J. Vac. Sci. and Technol. B4, 511 (1986)
27. R.G. Clark, R.J. Nicholas, M.A. Brummell, A. Usher, S. Collocott, J.C. Portal and F. Alexandre, Solid State Commun. 56, 173 (1985)
28. F.C. Zhang and S. Das Sarma, to be published (1986)
29. Y. Ono, J. Phys. Soc. Japan 51, 237 (1982)
30. R.J. Nicholas, R.G. Clark, A. Usher, J.R. Mallett, A.M. Suckling, J.J. Harris and C.T. Foxon, Solid State Sciences (Springer Verlag) 67, p. 194 (1986)

* Grown at Philips Research Laboratories, Redhill, Surrey, UK.

Activation Energies of the Fractional Quantum Hall Effect in GaAs/AlGaAs Heterostructures

J. Wakabayashi

Department of Physics, Gakushuin University, Mejiro, Toshima-ku, Tokyo 171, Japan

1. INTRODUCTION

There has been considerable progress in understanding the fractional quantum Hall effect (FQHE) since the first observation by TSUI et al. [1] in a two-dimensional electron system of a GaAs/AlGaAs heterostructure. The phenomenon is characterized by formation of a plateau in the Hall resistivity ρ_{xy} and concurrent vanishing of the diagonal resistivity ρ_{xx} similar to the (integral) quantum Hall effect [2] except that the filling factor becomes a fractional number p/q, where q is an odd integer and p is an integer prime to q. The fractional filling vactor ν is defined by $\nu = N_s h/eB$ where N_s is the electron number density and eB/h is the degeneracy of a Landau level. The temperature dependence of ρ_{xx} and ρ_{xy} observed experimentally at fractional filling suggests to us the existence of a finite energy gap between the ground state and the lowest excited state.

Basic understanding of the phenomenon is considered to be established [3] based on Laughlin's theory of an incompressible quantum fluid [4]. Although there are some other theoretical approaches to the FQHE [5,6], Laughlin's theory describes most successfully the experimentally observed phenomena. His trial wave function for $\nu = 1/q$ describes the ground state, which is a new quantum liquid state originating from the strong Coulomb interaction between the electrons. The theory also shows that the 1-1/q state is an electronic conjugate of the 1/q state, which is called electron-hole symmetry [7,8]. Laughlin's wave function for the lowest excited state at $\nu = 1/q$ reveals that the excitation is associated with the creation of fractionally charged quasi-electrons (-e/q) and quasi-holes (+e/q) and that there is an energy gap for the excitation. Successive theoretical works [7,9,10] have shown that the higher order states of $\nu = p/q$ arise from the formation of a new quantum liquid state by quasi-particles originating from the Coulomb interaction between the quasi-particles. All p/q fractional numbers, except 1/5 and 4/5, consist of two hierarchies of continued-fraction filling factors derived from 1/3 and 2/3. Therefore, understanding the 1/3 and 2/3 effect is a basis for understanding the FQHE.

The activated temperature dependence of the resistivity ρ_{xx} or the conductivity σ_{xy} at each fractional filling is thought to originate from the excitation of quasi-particles which carry the current. Therefore, the comparison between the theoretical excitation energy and the experimental activation energy is another important study besides the comparison between theory and experiment of the simple fractional numbers.

In the following, a brief review of the theory of the excitation energy of the FQHE will be given in the next section, followed by the experimental results and discussion of the activation energies at $\nu = 1/3$ and $\nu = 2/3$.

2. THEORY OF THE EXCITATION ENERGY

Several theoretical calculations have been done for the excitation energy of a pair of one free quasi-electron and one free quasi-hole, $2\Delta_{1/q}$, in various approximations. The pair excitation energy $2\Delta_{1/q}$ is given by $\alpha_{1/q}\cdot e^2/\varepsilon\ell$ where e is the electronic charge, ε is the dielectric constant of GaAs and ℓ is the radius of the ground Landau orbit. The constant $\alpha_{1/q}$ depends on the theoretical approximations. The hypernetted chain method was employed by LAUGHLIN [7] and CHAKRABORTY [11] to give $\alpha_{1/3} = 0.056$ and $\alpha_{1/3} = 0.0526$, respectively. These results, however, seem to be too small when compared with the experimental results as discussed in Sect.3.3. HALDANE and REZAYI [12] performed numerical diagonalization of finite systems of up to 8 electrons. Their extrapolation of the results to an infinite number of electrons gives $\alpha_{1/3} = 0.105 \pm 0.005$. GIRVIN et al. [13] calculated the collective excitation spectrum in analogy with Feynman's theory for the excitation spectrum of superfluid ^{4}He [14]. They estimated the excitation energy at infinite wave vector by using the asymptotic exciton dispersion and obtained $\alpha_{1/3} = 0.106$. Quite recently, MORF and HALPERINI [15] performed elaborate calculations using the Monte Carlo method for systems of up to 72 electrons and obtained $\alpha_{1/3} = 0.099 \pm 0.009$. Judging from these results, $\alpha_{1/3} \simeq 0.1$ seems to be conclusive theoretically.

These results, however, are for an ideal 2D system in an infinite magnetic field; that is, the distribution of the electron wave function normal to the interface is a delta function, only the lowest Landau level is considered, and no disorder (including impurities) is assumed. The observed activation energies are expected to be reduced in magnitude from the theoretical excitation energy for the following reasons: the effect of the finite thickness of inversion layers (three-dimensional effect), the effect of mixing of higher Landau levels, and the effect of disorder including impurities.

The effect of finite thickness of the inversion layer was first investigated on the ground state energy by MacDONALD and AERS [16]. When the mean electron-electron distance is comparable with the thickness of the inversion layer, the effect of finite thickness of the inversion layer causes a reduction of the Coulomb interaction and, hence, the reduction of the cohesive energy of the ground state and the excitation energy. The effect of the finite thickness on the excitation energy was investigated by YOSHIOKA [17] and by ZHANG and DAS SARMA [18] based on numerical calculations for the finite systems. Their results give a similar reduction of the excitation energy $2\Delta_{1/3}$ by 25% and by 29%, respectively, of that at $(b\ell)^{-1} = 0$ when $(b\ell)^{-1} = 0.5$, where $3/b$ is the inversion layer thickness [19]. YOSHIOKA [17,20] investigated the effect of mixing of higher Landau levels on the ground state energy and the excitation energy as a function of $\lambda = (e^2/\varepsilon\ell)/\hbar\omega_c$, where ω_c is the cyclotron frequency. According to his results, the excitation energy $2\Delta_{1/3}$ decreases approximately linearly by 16% when λ increases from 0 to 1. It is most difficult to treat the effect of disorder theoretically and also experimentally. Recently, ZHANG et al. [21] and REZAYI and HALDANE [22] investigated the effect of a single charged impurity for finite systems and found a similar reduction of the excitation energy $2\Delta_{1/3}$ as a function of the strength of the charge of the impurity. However, both theoretical and experimental study of the effect of disorder including impurities on the excitation energy is very primitive at present.

3. EXPERIMENTAL RESULTS AND DISCUSSIONS

3.1 Temperature Dependence of ρ_{xx}

A crude experiment on the temperature dependence of ρ_{xx} at ν = 1/3 and 2/3 was first carried out by TSUI et al. [23]. They reported the activated behaviour of ρ_{xx} at temperatures above 0.4 K. CHANG et al. [24] made a systematic study of the temperature dependence of ρ_{xx} and ρ_{xy} around ν = 2/3 by varying the electron density with a backside gate bias in magnetic fields between 6.6 T and 10.6 T and at low temperatures between 770 mK and 65 mK. They observed a single activated temperature dependence of ρ_{xx} and ρ_{xy} at temperatures between 400 mK and 65 mK. KAWAJI et al. [25] performed the first quantitative experiment on the temperature dependence of ρ_{xx} at ν = 1/3 in very high mobility samples. They also measured the temperature dependence of ρ_{xx} at ν = 2/3 in the same sample. The activation energies at ν = 1/3 and ν = 2/3 showed no electron-hole symmetry when they were scaled by the magnetic field. Their results at both ν = 1/3 and ν = 2/3 showed that there were two activation energies which appear at high temperature and at low temperature, respectively, as shown in Fig.1. They considered that their results showed an interchange of two activated conduction processes which appear at high temperature and at low temperature, respectively. However, their data points are not enough to give a definite conclusion that the two processes interchange with each other.

Recently, WAKABAYASHI et al. [26] have performed systematic experiments on the temperature dependence of ρ_{xx} in several samples having a wide range of mobility. Typical traces of ρ_{xx} against the magnetic field B for each class of sample are shown in Fig.2. The mean value of the electron density

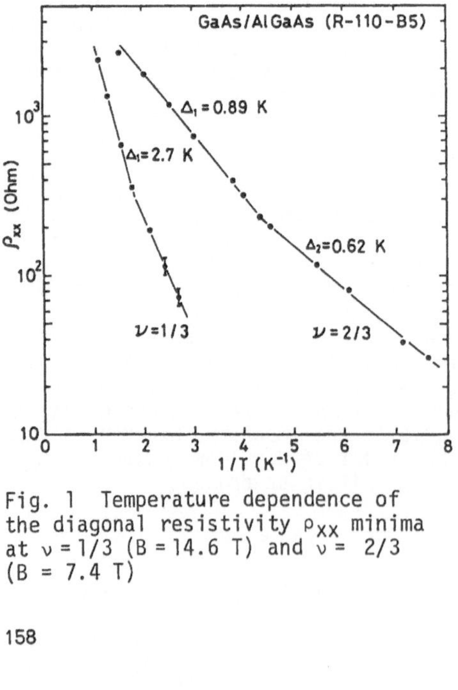

Fig. 1 Temperature dependence of the diagonal resistivity ρ_{xx} minima at ν = 1/3 (B = 14.6 T) and ν = 2/3 (B = 7.4 T)

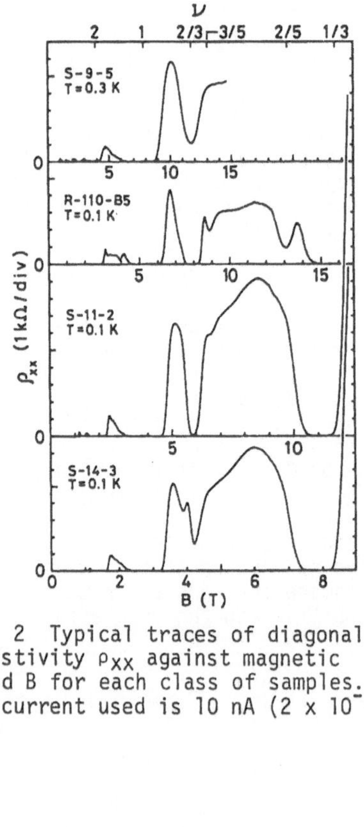

Fig. 2 Typical traces of diagonal resistivity ρ_{xx} against magnetic field B for each class of samples. The current used is 10 nA (2 x 10^{-6} A/m)

in units of cm^{-2} and the mobility in units of $cm^2/V.s$ is 1.2×10^{11} and 1.1×10^6 for R-110s, 1.9×10^{11} and 0.32×10^6 for S-9s, 0.94×10^{11} and 0.21×10^6 for S-11s, and 0.62×10^{11} and 0.26×10^6 for S-14s, respectively. Figure 3 shows typical results of temperature dependence of ρ_{xx} minima at $\nu = 1/3$ for each class of sample. The results were all well expressed by a sum of two activated conduction processes as

$$\rho_{xx}(T) = \rho_{01} \exp(-W_1/T) + \rho_{02} \exp(-W_2/T) \tag{1}$$

and not by the interchange of two activated conduction processes. Each solid line in Fig.3 represents a fitted curve obtained by (1). We denote the activation energy in (1) as W_1 (measured in kelvin) for the high temperature region and W_2 for the low temperature region in order to distinguish them from Δ_1 and Δ_2 which are obtained straightforwardly from a log ρ_{xx} versus 1/T plot. The results at $\nu = 2/3$ were also expressed as a sum of two activated conduction processes as shown in Fig.4. The result of current dependence of ρ_{xx} at $\nu = 2/3$ of the sample S-14-3 at 78 mK indicates that the deviation observed in the sample S-14-3 at low temperature is due to Joule heating. Therefore, the activation energy W_1 was determined directly from the slope of the log ρ_{xx} versus 1/T plot in this case. The single activated conduction at $\nu = 2/3$ observed by CHANG et al. [24] may be due to very small values of ρ_{02} of their high mobility sample, similar to the result of the sample R-110-B11 in Fig.4 which has a mobility of 1.1×10^6 $cm^2/V.s$.

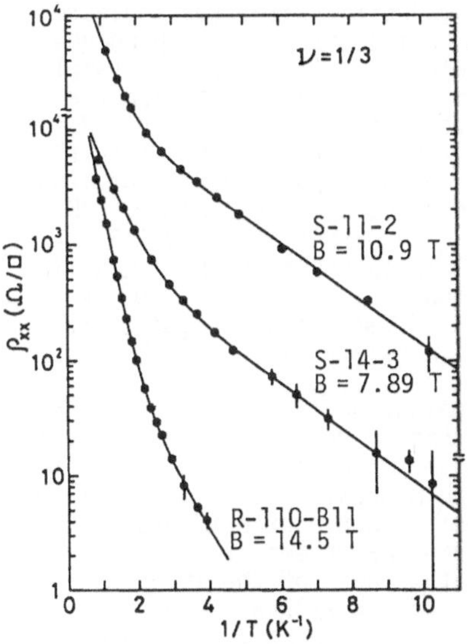

Fig. 3 Examples of temperature dependence of ρ_{xx} minima at $\nu = 1/3$. The current used is 5 nA except for the sample R-110-B11, 10 nA. Each solid line represents a fitted curve for $\rho_{xx}(T) = \rho_{01} \cdot \exp(-W_1/T) + \rho_{02} \cdot \exp(-W_2/T)$.

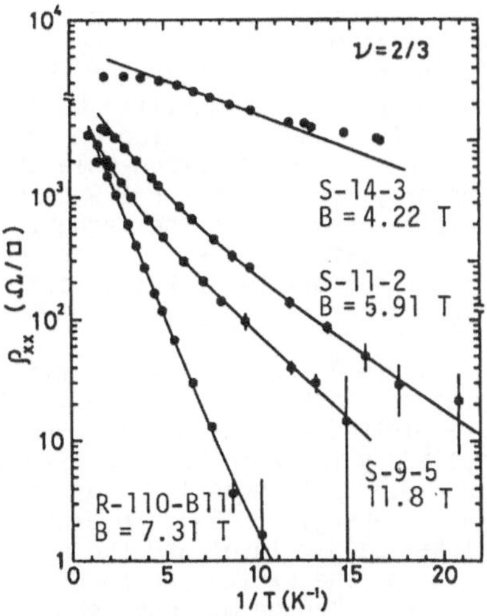

Fig. 4 Examples of temperature dependence of ρ_{xx} minima at $\nu = 1/3$. The current used is 1 nA for S-14-3, 2 nA for S-11-2, 5 nA for S-9-5, and 10 nA for R-110-B11. Solid lines are the same as in Fig. 3 except for sample S-14-3.

The results in Figs.3 and 4 have a large error bar (one standard deviation)at low temperatures and it was also possible to fit an equation similar to (1) but with the second term replaced by $C/T \cdot \exp[-(T_0/T)^{1/2}]$ to the data, where C and T_0 are numerical constants. The formula for the second term is derived by ONO [27] for 2D hopping conduction at low temperatures in a strong magnetic field. It was difficult, however, to achieve a good fit to some of their data using Mott's variable range hopping formula of $C/T^{2/3} \exp[-(T_0/T)^{1/3}]$ [28] for the second term of (1). Quite recently, however, WAKABAYASHI et al. [29] carried out careful experiments on the temperature dependence of ρ_{xx} at $\nu = 1/3$ and $\nu = 2/3$. Their data confirm that the temperature dependence at low temperature is also activated. The origin of the W_2 process will be discussed briefly in Sect.3.4.

In the following, we will discuss the activation energy W_1 and compare it with the theoretical excitation energy for the 1/3 and 2/3 effects. It should be noted that W_1 and Δ_1 show similar dependence not only on the filling factor [30] but also on the magnetic field.

3.2 Scaling of Activation Energy

According to the theoretical results, the excitation energy is expected to show $B^{1/2}$ dependence because ℓ has $B^{-1/2}$ dependence. The magnetic field dependence of the activation energy W_1 for $\nu = 1/3$ and $\nu = 2/3$ is shown in Fig.5. The data for the R-110 samples show a mean value and one standard deviation of six measurements for $\nu = 1/3$ and seven measurements for $\nu = 2/3$, respectively, of three samples. Other data points show results of each measurement. It is clear in Fig.5 that the activation energies $W_1(1/3)$ and $W_1(2/3)$ show no systematic dependence on the magnetic field. (We indicate the filling factor in parentheses.) The similar magnetic field dependence was observed in $\Delta_1(1/3)$ and $\Delta_1(2/3)$. The results of $W_1(1/3)$ of S-11 samples at B = 11.0 T and $W_1(2/3)$ of S-9 samples at B = 11.6 T differ by about a factor of three in spite of the similar magnitude of the magnetic field and the difference of mobility being only about a factor of 1.6. This result means that the activation energy for $\nu = 1/3$ and $\nu = 2/3$ is different, that is, there is no electron-hole symmetry as discussed by KAWAJI et al. [25,31]. Further, $W_1(2/3)$ of the S-9 samples is smaller than that of the R-110 samples even though the magnetic field of S-9 is larger than that of R-110. The S-9 samples have, however, much smaller mobility, about 0.32×10^6 $cm^2/V.s$, than the mobility of R-110 (1.1×10^6 $cm^2/V.s$). These results suggest that the activation energy depends on the electron mobility [24].

Figure 6 shows the activation energy W_1 against μB. The activation energies for both $\nu = 1/3$ and $\nu = 2/3$ show a systematic dependence when they are plotted against μB. This result confirms that the activation energy is scaled by the electron mobility as well as the magnetic field. It is clear in Fig.6 that there is no electron-hole symmetry in the experimental activation energy [25,31]. YOSHIOKA [17] considered the effect of mixing of the higher Landau level and the effect of finite thickness of the inversion layer on the excitation spectrum for $\nu = 1/3$ and $\nu = 2/3$. He found that the energy at "roton minimum" showed the same tendency between $\nu = 1/3$ and $\nu = 2/3$ as observed experimentally. The quantitative agreement, however, is not satisfactory.

Both activation energies for $\nu = 1/3$ and $\nu = 2/3$ in Fig.6 show saturation at high μB. A possible explanation for this behaviour is the effect of finite thickness of the inversion layer. Increasing the magnetic field causes a decrease of the radius of the Landau orbit, which results in a re-

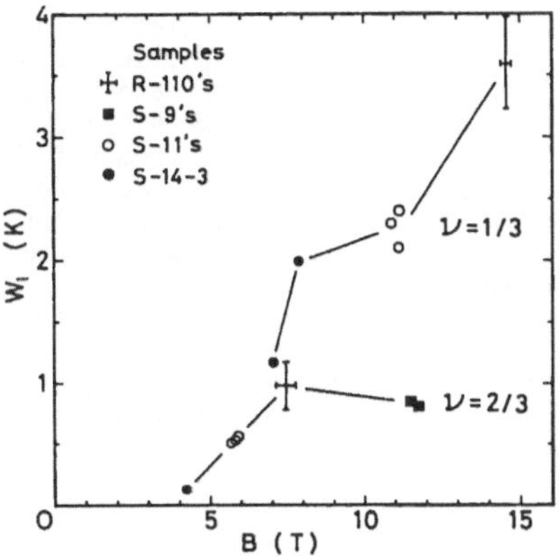

Fig. 5 Activation energies for ν = 1/3 and 2/3 against magnetic field. The data for the R-110 sample show the mean value and one standard deviation.

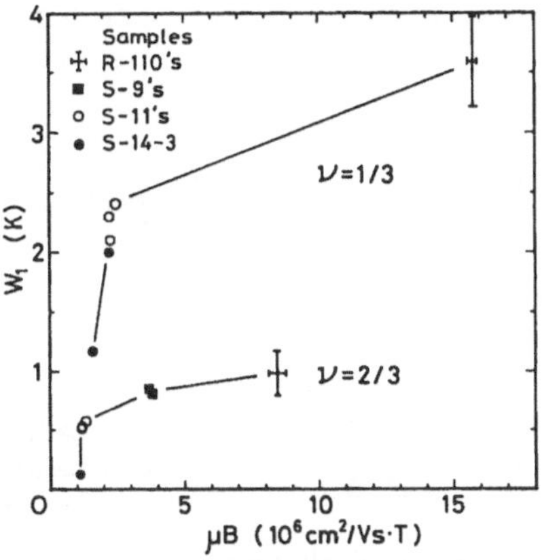

Fig. 6 Activation energies for ν = 1/3 and 2/3 against μB.

duction of the mean electron-electron distance. When the electron-electron distance becomes comparable with the thickness of the inversion layer, the effect of finite thickness causes a reduction of excitation energy, as discussed in the previous section. Another possible origin of the saturation behaviour is increasing mobility. If increasing mobility means a decreasing effect of disorder, the "ideal" activation energy will be given at infinite mobility. Recently KUKUSHKIN and TIMOFEEV [32] reported the μ^{-1} dependence of the activation energy at $\nu = 4/3$ observed in a silicon MOSFET. It is difficult in the present case to get a true mobility dependence of the activation energy because the data have been obtained at different magnetic fields. Another interpretation of the μB dependence is that decreasing μB increases the broadening of excited energy levels and consequently decreases the excitation energy gap. The broadening of energy levels is considered to occur in the excited states for quasi-particles because the ground state of the present system is degenerate.

Recently, BOEBINGER et al. [33] measured the magnetic field dependence of the activation energies Δ_1 for $\nu = 1/3$ and $\nu = 2/3$ up to 28 T. Their results show a rather smooth magnetic field dependence up to 20 T as shown in Fig.7. They claimed that the activation energies Δ_1 for $\nu = 1/3$ and $\nu = 2/3$ lay on a single line when they were plotted against the magnetic field. This means that the electron-hole symmetry is valid. However, they used two types of samples; a Hall bar type and a quasi-Corbino type. It is considered that we cannot rely upon the temperature dependence of σ_{xx} obtained in quasi-Corbino samples for the following reason. The source-drain current in quasi-Corbino samples which have no closed electrode cannot be free from the contribution of the Hall current. Therefore, one cannot measure σ_{xx} by using quasi-Corbino samples. Thus their overlapping data for ν =1/3 and 2/3 around B $\sim$20 T are not reliable. Although the only point at B $\sim$ 26 T for $\nu = 1/3$ appears to lie on the extrapolated line of the activation energies for $\nu = 2/3$, this may be because they ignored the mobility as a parameter. Therefore, we conclude that the activation energy cannot be scaled by the magnetic field only.

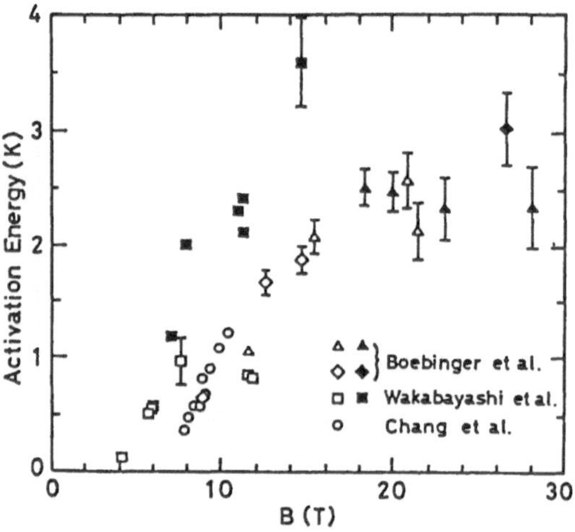

Fig. 7 Activation energies for ν = 1/3 and 2/3 against magnetic field. Closed symbols indicate data for ν= 1/3 and open symbols for ν = 2/3. (△ ▲) and (◇ ◆) are from ref. 33, (□ ■) are from ref. 26, and (○) are from ref. 24. Triangles are data from $\sigma_{xx}(T)$ of quasi-Corbin samples. All other data are from $\rho_{xx}(T)$ of Hall bar samples. (Note that the definition of the activation energy of ref. 33 differs from others by a factor of two.)

3.3 Quantitative Comparison Between Experiment and Theory

In Fig.6 the R-110 samples have 3.6 K as a mean value of the activation energy $W_1(1/3)$ at B = 14.6 T. If we assume $3/b$ = 145 A [19,34] then $(b\ell)^{-1}$ = 0.72 and this results in a reduction of the excitation energy by about 35% according to Yoshioka's calculation [17]. Also, his calculation shows that the effect of mixing of the higher Landau level at B = 14.6 T (λ = 0.683) causes a reduction of the excitation energy by about 11%. Then, the total reduction caused by these two effects is 43%, which allows us to estimate the "bare" activation energy as 6.2 K. This value leads to 0.062 for the excitation energy $2\Delta_{1/3}$ in units of $e^2/\varepsilon\ell$. Theoretical results by LAUGHLIN, 0.056 [7], and by CHAKRABORTY, 0.0526 [11], seem to be unreasonable because the value 0.062 is the lower limit of the activation energy if the present estimate of the reduction by the two effects is reasonable. On the other hand, the theoretical results of $2\Delta_{1/3} \sim 0.1e^2/\varepsilon\ell$ by other authors [12,13,15] seem to be reasonable when they are compared with the estimated "bare" activation energy, 0.062, because we have not yet taken the effect of disorder into account in the above estimate of the reduction. It is worth noting that the maximum activation energy $W_1(1/3)$ obtained in the R-110 sample is 4.2 K at B = 14.4 T. If we use this maximum value in the above calculation, the excitation energy becomes 0.073 in units of $e^2/\varepsilon\ell$. This result means that if we assume a reduction of about 25% caused by a disorder effect, the experimental results and the theoretical results are consistent with each other. However, it is difficult at present to include the theoretical results by ZHANG et al. [21] and by REZAYI and HALDANE [22] in the estimate of reduction of the experimental activation energy.

It is necessary to investigate the effect of disorder, including impurities, both experimentally and theoretically for a further quantitative comparison between experiment and theory. At present we have only mobility as a parameter which represents the effect of a certain amount of disorder. However, the mobility at zero magnetic field does not represent all the transport properties in a strong magnetic field as reported in Fig.1 of [25] and [31]. Recently, MENDETZ [35] reported activation energies of Δ_1 of 2D hole systems, which are not directly correlated with low temperature mobility. It is necessary for experiment to find a parameter representing the effect of disorder by controlling them systematically.

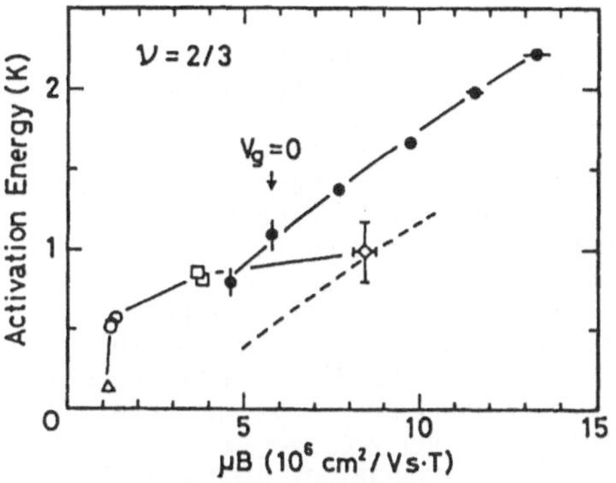

Fig. 8 Activation energies for ν = 2/3 against μB. The broken line shows calculated results from the data of [24]. Open symbols are data from Fig. 6. Closed symbols are data from Fig. 9.

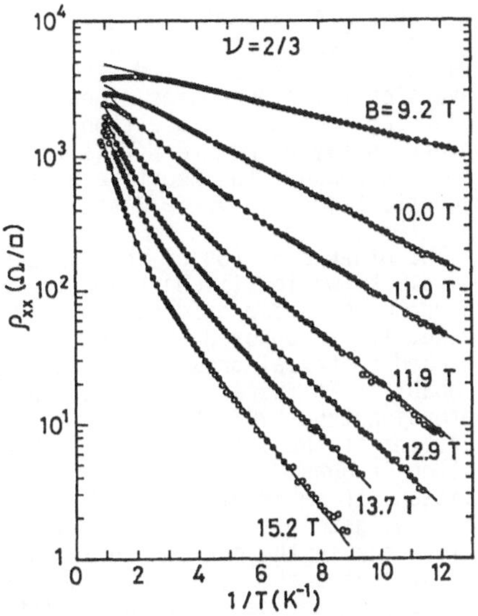

Fig. 9 Temperature dependence of ρ_{xx} minima at $\nu = 2/3$ in a single sample with a backside gate. Each solid line represents a fitted curve for $\rho_{xx}(T) = \rho_{01}\cdot\exp(-W_1/T) + \rho_{02}\cdot\exp(-W_2/T)$.

3.4 Results on a Sample with a Backside Gate

So far we have discussed mainly the experimental results obtained from the sample with no gate electrode. CHANG et al. [24] measured the activation energy at ν = 2/3 altering the magnetic field and mobility with a bias on the backside gate. Although they mentioned that the activation energy depends on the electron mobility, they plotted the activation energy against the mobility "or" the magnetic field (see Fig.3 of [24]). The broken line in Fig.8 shows the μB dependence of the activation energy calculated from their data. Chang et al.'s results show stronger μB dependence than the results obtained from the sample with no gate electrode in Fig.6. Quite recently, WAKABAYASHI et al. [36] studied the temperature dependence at ν = 2/3 of a sample with a backside gate. Their results show that there were two activated conduction processes, as in the sample with no gate electrode, as shown in Fig.9 in contrast to Chang et al.'s results. The other important result in Fig.9 is that the activation energies measured in a sample at high resistivity with a negative bias on the backside gate is W_2 and not W_1 which is considered to correspond to the theoretical excitation energy [29]. The activation energy W_1 obtained from Fig.9 also shows stronger μB dependence than the resutls in Fig.6, as shown in Fig.8. We note that the activation energy obtained with no bias on the backside gate of this sample shows good agreement with the results obtained from the sample with no gate electrode. The results in Fig.8 suggest that there is another parameter other than the magnetic field and the mobility which scales the activation energy.

A possible parameter to give such an effect is the surface field in the inversion layer normal to the interface. When the backside gate is negatively biassed, the surface field strength increases and the thickness of the inversion layer decreases. Then the activation energy is expected to increase according to the consideration on the effect of finite thickness of the inversion layer. However, this is contrary to the experimental fact shown in Fig.8 because the negative bias on the backside gate decreases both the magnetic field for $\nu = 2/3$ and the mobility.

The other function of the surface field is controlling the disorder. It is well known in silicon inversion layers [37] that a strong surface field increases disorder originating from the interface roughness. In the present system it is supposed that the interface is smooth owing to the lattice-matched epitaxial growth of GaAs and AlGaAs crystal. However, it is highly probable that there is residual interface roughness at the GaAs-AlGaAs heterojunction [38]. Further, it is also plausible to expect residual impurities at the interface [39], which are considered to produce the localization of quasi-particles and the finite width of the fractional Hall plateau. If we assume this disorder at the interface, the effect of the backside gate bias on the activation energy W_1 is considered to be reasonable, because the positive bias pulls the electrons away from the interface and decreases the disorder which electrons feel, and vice versa for the negative bias. The electron mobility, that is, does not describe all the effects of disorder, as indicated in the difference of μB dependence of the activation energy between the samples with and without the gate bias in Fig.8.

Results in Fig.9 show that the W_2 process becomes dominant when a large negative bias is applied to the backside gate. This suggests that the W_2 process is highly correlated with the disorder [40] discussed above. The origin of the W_2 process, however, will be discussed elsewhere [29,41].

ACKNOWLEDGEMENTS

The author is indebted to Professor S. Kawaji for continuous stimulation and for supporting all his work. He also thanks Professor H. Sakaki and Dr. J. Yoshino for cooperating and providing him with samples. This work is supported by a Grant-in-Aid for Special Distinguished Research from the Ministry of Education, Culture and Science.

REFERENCES

1) D. C. Tsui, H. L. Stormer, A. C. Gossard: Phys. Rev. Lett.48,1559(1982)
2) S. Kawaji: Proc. Int. Symp. Foundation of Quantum Mechanics, Tokyo,1983 p. 327 (Physical Society of Japan, 1984)
3) D. Yoshioka: Prog. Theor. Phys. Suppl. No. 84, p. 97 (1985)
4) R.B. Laughlin: Phys. Rev. Lett. 50, 1395 (1983)
5) R. Tao: Phys. Rev. B29, 636 (1984)
6) S. Kivelson, C. Kallin, D. P. Arovas and J. R. Schrieffer: Phys. Rev. Lett. 56, 873 (1986)
7) R. B. Laughlin: Surf. Sci. 142, 163 (1984)
8) S. M. Girvin: Phys. Rev. B29, 6012 (1984)
9) F. D. M. Haldane: Phys. Rev. Lett. 51, 605 (1983)
10) B. I. Halperin: Phys. Rev. Lett. 52, 1583 (1984)
11) T. Chakraborty: Phys. Rev. B31, 4026 (1985)
12) F. D. M. Haldane and E. H. Rezayi: Phys. Rev. Lett. 54, 237 (1985)
13) S. M. Girvin, A. H. MacDonald and P. M. Platzman: Phys. Rev. Lett. 54, 581 (1985), and Phys. Rev. B33, 2481 (1986)

14) R. P. Feynman and M. Cohen: Phys. Rev. 102, 1189 (1956)
15) R. Morf and B. I. Halperin: Phys. Rev. B33, 2221 (1986)
16) A. H. MacDonald and G. C. Aers: Phys. Rev. B29, 5976 (1984)
17) D. Yoshioka: J. Phys. Soc. Jpn. 55, 885 (1986)
18) F. C. Zhang and S. Das Sarma: Phys. Rev. B33, 2903 (1986)
19) F. F. Fang and W. E. Howard: Phys. Rev. Lett. 16, 797 (1966)
20) D. Yoshioka: J. Phys. Soc. Jpn. 53, 3740 (1984)
21) F. C. Zhang, V.Z. Vulovic, Y. Guo and S. Das Sarma: Phys. Rev. B32, 6920 (1985)
22) E. H. Rezayi and F. D. M. Haldane: Phys. Rev. B32, 6924 (1985)
23) D. C. Tsui, H. L. Stormer, J. C. M. Hwang, J. S. Brooks and M. J. Naughton: Phys. Rev. B28, 2274 (1983)
24) A. M. Chang, M. A. Paalanen, D. C. Tsui, H. L. Stormer and J. C. M. Hwang: Phys. Rev. B28, 6133 (1983)
25) S. Kawaji, J. Wakabayashi, J. Yoshino and H. Sakaki: J. Phys. Soc. Jpn. 53, 1915 (1984)
26) J. Wakabayashi, S. Kawaji, J. Yoshino and H. Sakaki: J. Phys. Soc. Jpn. 55, 1319 (1986)
27) Y. Ono: J. Phys. Soc. Jpn. 51, 237 (1982)
28) N. F. Mott and E. A. Davis: "Electronic Properties in Non-Crystalline Materials" (Clarendon, Oxford, 2nd ed.)
29) J. Wakabayashi, S. Sudou, S. Kawaji, K. Hirakawa, J. Yoshino and H. Sakaki: Submitted to J. Phys. Soc. Jpn.
30) J. Wakabayashi, S. Kawaji, J. Yoshino and H. Sakaki: Surf. Sci. 170, 136 (1980)
31) J. Wakabayashi, S. Kawaji, J. Yoshino and H. Sakaki: Proc. 17th Int. Conf. Physics of Semicond., San Francisco, 1984, p. 283 (Springer-Verlag, New York, 1985)
32) I. V. Kukushkin and V. B. Timofeev: Surf. Sci. 170, 148 (1986)
33) G. S. Boebinger, A. M. Chang, H. L. Stormer and D. C. Tsui: Phys. Rev. Lett. 55, 1606 (1985)
34) T. Ando: J. Phys. Soc. Jpn. 51, 3900 (1982)
35) E. E. Mendetz: Surf. Sci. 170, 561 (1986)
36) J. Wakabayashi, S. Sudou, S. Kawaji, K. Hirakawa and H. Sakaki: Proc. 18th Int. Conf. Phys. Semiconduc., Stockholm, 1986, to be published
37) T. Ando, A. B. Fowler and F. Stern: Rev. Mod. Phys. 54, p.502 (1982)
38) B. A. Joyce: Proc. Int. Conf. on Modulated Semicond. Structures, Kyoto, p. 1 (1985)
39) J. C. M. Hwang, A. Kastalsky, H. L. Stormer and V. G. Keramidas: Appl. Phys. Lett. 44, 802 (1984)
40) J. Ihm and J. C. Phillips: J. Phys. Soc. Jpn. 54, 1506 (1985)
41) J. Wakabayashi, S. Sudou, S. Kawaji, K. Hirakawa, J. Yoshino and H. Sakaki: in preparation

Fractional Quantum Hall Effect of p-Type GaAs-(GaAl)As Heterostructures in the Millikelvin Range

G. Reményi[1], *G. Landwehr*[1], *W. Heuring*[1], *G. Weimann*[2], *and W. Schlapp*[2]

[1]Physikalisches Institut der Universität Würzburg, D-8700 Würzburg, Fed. Rep. of Germany
[2]Forschungsinstitut der Deutschen Bundespost, D-6100 Darmstadt, Fed. Rep. of Germany

1. INTRODUCTION

The magnetoresistance components ρ_{xx} and ρ_{xy} of modulation doped p-type GaAs-(GaAl)As heterojunctions were investigated between 50 mK and 4.2 K in magnetic fields B up to 12.5 T and in the temperature range of 400mK to 4.2K in fields up to 21 T. The high quality samples had hole mobilities up to 2.2×10^5 cm^2/Vs and a carrier concentration around 2.3 - 2.5 $\times 10^{11}$/cm^2 at the lowest temperature employed. Pronounced Shubnikov-de Haas oscillations and well-developed Hall plateaus were observed, especially at temperatures below 300 mK. Both the integral and the fractional quantum Hall-effect showed up with well-resolved steps in ρ_{xy} at 50 mK. Plateaus with fractions with odd and even denominators were seen.

In order to obtain information about energy gaps and the density of states, the resistivity of various ρ_{xx} minima was measured as a function of temperature with B as a parameter. At constant magnetic field two activation energies were found for various occupation numbers.

The orbital degeneracy at the top of the GaAs valence band is expected to lead to a complex, nonparabolic set of two-dimensional (2D) bands describing holes confined to GaAs-(GaAl)As interfaces /1/. The high hole mobilities in GaAs-(GaAl)As heterostructures allow the possibility of very detailed experimental studies that should reveal many new aspects and provide further tests of our understanding of 2D hole transport. We report here experiments on the integral and fractional quantum Hall effect and on Shubnikov-de Haas oscillations.

2. EXPERIMENT

The samples investigated were produced by molecular beam epitaxy with Be as a dopant. Their structure and the specimen geometry are shown in Figs. 1 and 2. Ohmic contacts were produced by an In (Zn)-alloy.

Table 1: Electrical properties of samples at T = 0.5 K.

Sample /Reference/	Mobility μ in cm^2/Vs	Concentration h_S 10^{11}/cm^2
1394-2	60×10^3	2.00
1458-1	185×10^3	2.55
1458-2	90×10^3	2.45
1458-3	140×10^3	2.53

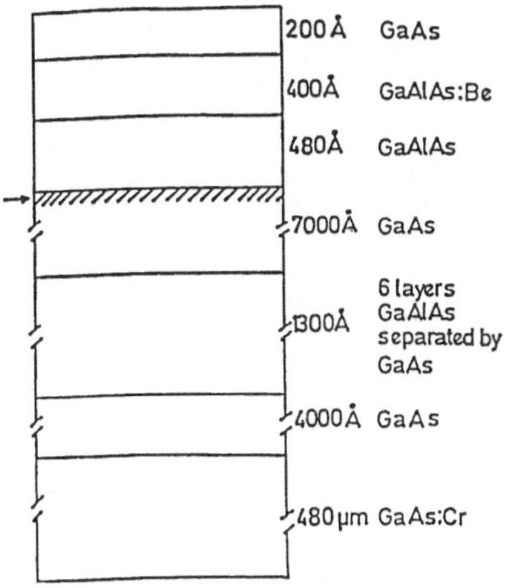

Fig. 1: Schematic of the p-type GaAs-GaAlAs heterojunctions

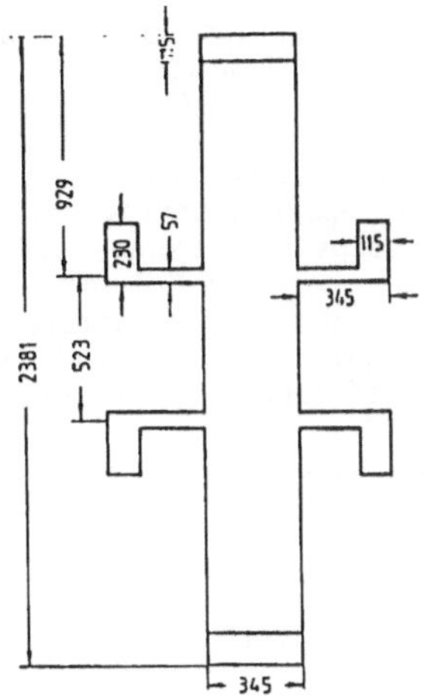

Fig.2: Sample geometry, dimensions in µm

The hole mobilities of 4 specimens cut from two chips measured at 0.5 K are shown in table 1.

The hole mobility increased significantly when the temperature was lowered into the Millikelvin range. The highest mobility was measured on sample 1458-1, it was 2.2×10^5 cm^2/Vs at T = 38 mK.

Resistivity ρ_{xx} and Hall resistance ρ_{xy} were measured with a cryogenic four-point a.c. resistance bridge or with a.c. lock-in technique. The dilution refrigerator was equipped with an epoxy mixing chamber which allowed a direct immersion of the samples in the cryogenic liquid. The high-field experiments were performed in a 10 MW Polyhelix-Bitter-type magnet with He3-cryostat which allowed us to produce temperatures down to 0.4 K. The result of a high-field experiment between 3 T and 22 T on sample 1458-1 is shown in Fig. 3. One can recognize well-developed Hall plateaus. The one between 9 T

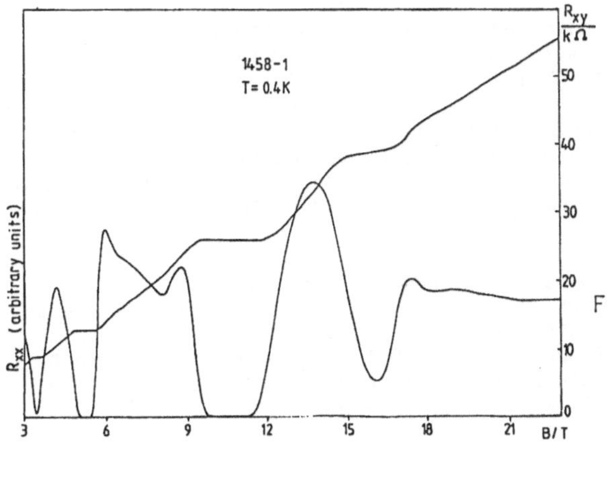

Fig. 3: Hall resistance R_{xy} and magnetoresistance R_{xx} at T = 0.4 K for magnetic fields B between 3 T and 23 T

and 12 T belongs to the occupation number $\eta = 1$. The slightly tilted plateau around 16.5 T which is accompanied by a dip in the magnetoresistance R_{xx} belongs to $\eta = 2/3$. The structure around 18 T can be attributed to $\eta = 3/5$.

Data obtained for the same sample at 50 mK in the cryostat equipped with a 12.4 T superconducting coil are shown in Fig. 4. One can recognize that the Hall steps have become steeper and that considerable structure has become visible between 6 T and 8 T in R_{xx}. The dip seen at 0.4 K around 7.5 T has no corresponding structure in R_{xy}. However, additional features show up when the sample is cooled to 50 mK. This will be discussed in conjunction with the fractional Hall effect subsequently.

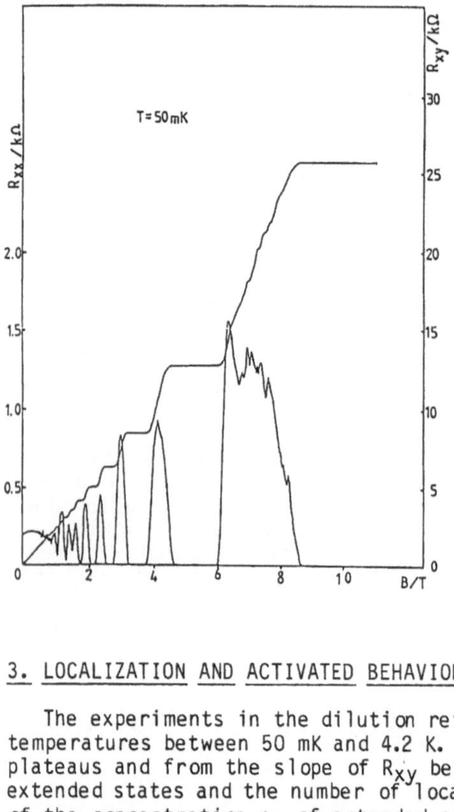

Fig. 4:
Magnetoresistance R_{xx} and Hall resistance R_{xy} of GaAs-(GaAl)As heterostructure 1458-1 as a function of magnetic field B at T = 50 mK. Well-developed Hall plateaus can be recognized. The structure between 6 T and 8 T is caused by the fractional QHE.

3. LOCALIZATION AND ACTIVATED BEHAVIOR

The experiments in the dilution refrigerator were performed at various temperatures between 50 mK and 4.2 K. From the varying widths of the Hall plateaus and from the slope of R_{xy} between adjacent plateaus the number of extended states and the number of localized states were estimated. The ratio of the concentration n_e of extended states to total concentration as deduced from the classical Hall line indicates the degree of localization in a Landau level. The results for temperatures between 50 mK and 0.8 K are shown in Fig. 5 by crosses. One sees that number of extended states is decreasing with decreasing temperature.

Extrapolation to T = 0 yields a finite number of extended states, in agreement with predictions by AOKI and ANDO /9/. Further information can be expected from a study of the widths of the Hall plateaus, which are correlated with the localization. If one assumes that the ratio of the width b of a plateau to its maximum width b_0 (which is given by the difference between adjacent R_{xx} peaks) is given by the relation

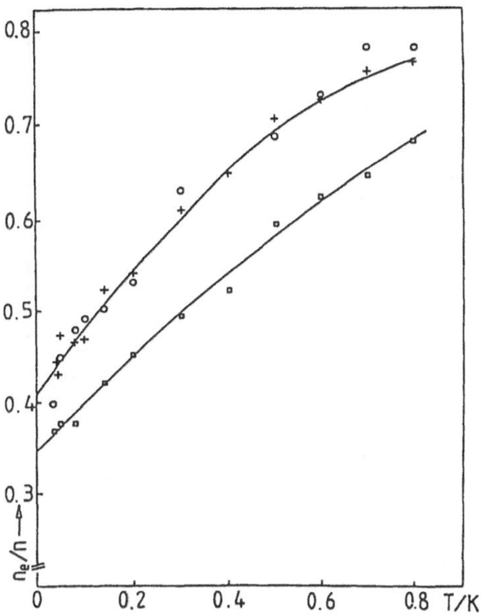

Fig. 5:
Ratio of the concentration of extended states to the total hole concentration as a function of temperature. The crosses were deduced from the analysis of the slope of R_{xy} between two Hall plateaus. The squares and circles were derived from the analysis of the temperature dependence of the Hall plateaus.

$$n_e/n = 1 - b/b_o$$

one obtains for plateau 1 (squares) and plateau 2 (circles) the results shown in Fig. 5. One can see that the results obtained by the different approaches are compatible, although the coincidence of the upper two curves deduced from the slope should be considered accidental.

In order to get information about energy gaps, the magnetoresistance was measured as a function of temperature at different magnetic fields. One expects activated behavior if the Fermi level is in the localized region between two Landau levels. The experiments in the Millikelvin range for filling factors 7, 8 and 10 yielded activated behavior in two temperature ranges. The data are shown in Figs. 6, 7 and 8, they can be represented by two straight lines in a logarithmic plot against 1/T. The behavior of this kind was also found in other samples at large filling factors. It should be mentioned that the minima in R_{xx} are still well developed at these high η -values. Data for small filling factors were also taken at higher temperatures above 300 mK. In these cases only a single activation energy could be defined. The two activation energies E_{a1} and E_{a2} for sample 1458-1 have been plotted in table 2.

Table 2: Activation energies deduced from the temperature dependence of Shubnikov-de Haas minima at filling factors η = 7, 8, 10.

η	B/T	E_{a1}/mK	E_{a2}/mK
7	1.50	244	22
8	1.33	580	17
10	1.05	330	12

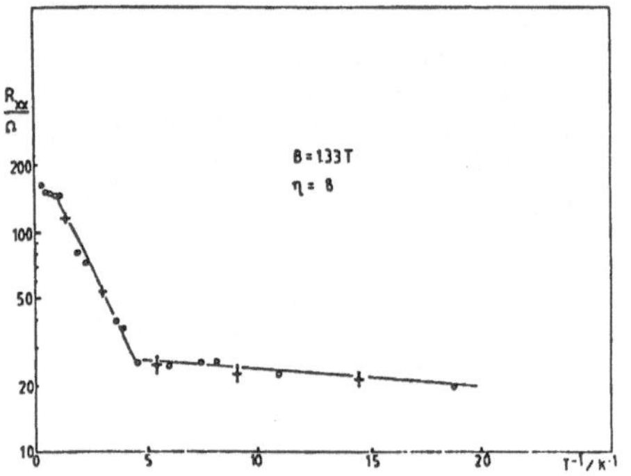

Fig.7

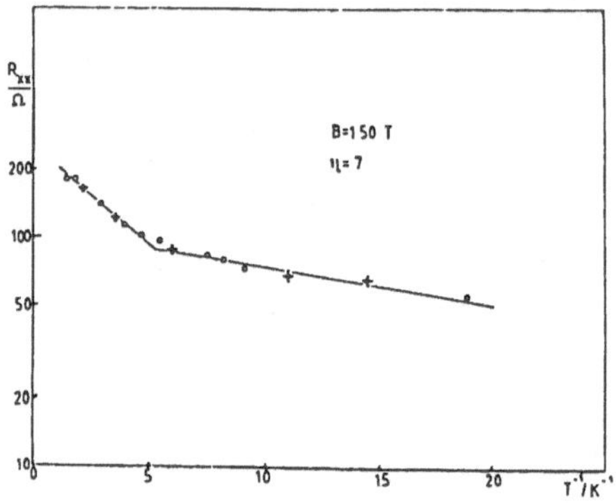

Fig.6

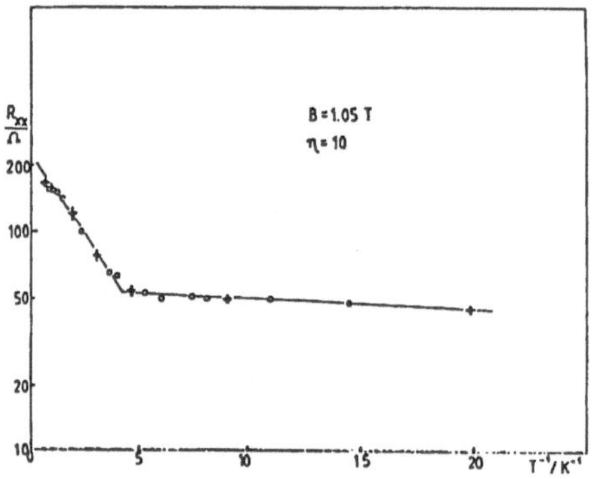

Fig.8

Fig. 6, 7, 8: Magnetoresistance R_{xx} at magnetic fields of 1.5 T, 1.33 T and 1.05 T corresponding to filling factors of 7, 8 and 10 respectively as a function of 1/T for sample 1458-1

4. FRACTIONAL QUANTUM HALL EFFECT

The Hall plateaus obtained in sample 1458-1 have been mentioned already. For the Shubnikov-de Haas minimum which can be attributed to $\eta = 2/3$ activation measurements were performed at 16 T. A logarithmic plot of R_{xx} as a function of T^{-1} yielded data shown in Fig. 9. Two activation energies of 0.09 meV for the low temperature region and 0.05 meV for higher temperatures can be deduced. Behavior of this kind is predicted by a model proposed by IHM and PHILIPPS /4/, who suggested the existence of a second activation energy in the fractional Quantum Hall Effect due to excitation of carriers which, due to the presence of potential fluctuations, have not condensed into the Laughlin quantum liquid ground state.

For another sample 1394-1 structures were seen between $\eta = 1$ and 2 at temperatures of 50 and 38 mK. The minimum around 6T at 50mK can be formally assigned to $\eta = 3/2$. However, the minimum disappears when the temperature is lowered to 38 mK and structures appear at $\eta = 14/9$ and 7/5. These large variations suggest that a reliable assignment of quantum numbers is difficult because there seems to be a superposition of various fractionals. No distinct features could be distinguished in the Hall resistance between quantum numbers 1 and 2 (see Fig. 10).

The sample with the highest mobility, 1458-1, shows very sharp structures in R_{xx} between 6 and 8 Tesla. However, the minima shift their position with

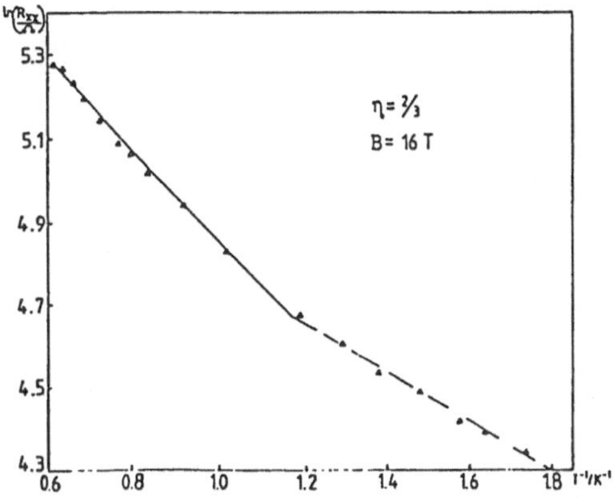

Fig. 9: Activated behavior of the FQHE at a filling factor 2/3 (sample 1458-1)

Fig. 10: Magnetoresistance and Hall effect of sample 1394

temperature and change their magnitude, so that it was not possible to assign quantum numbers. The structures in R_{xy} are better defined, especially at very low temperatures. In nearly all data Hall plateaus (with accompanying minima in the magnetoresistance) showed up at filling factors of 10/7, 13/9 and 1.37(≈4/3). The accompanying R_{xy} values varied somewhat however with temperature. On the other hand the analysis of several data showed η-values of 1.25 ± 0.05 (5/4), 1.20 ± 0.03 (6/5) and 1.17 ± 0.03 (7/6). R_{xy} and B were measured with a precision of about 1 %. During the experiment the sample was warmed up to 77K and subsequently cooled to 50 mK. The results were not changed by this procedure.

It is well known that at low temperatures slight inhomogeneities in the carrier concentration can cause structures in the Quantum Hall regime, therefore it is necessary to investigate a relatively large number ofsamples in order to be sure that no artifacts are observed. Therefore we are at present not in a position to claim to have reliably identified the FQHE with even fractions.

ACKNOWLEDGEMENT

Part of the data presented were taken at the Hochfeldmagnetlabor Grenoble, Max-Planck-Institut für Festkörperforschung.

LITERATURE

1. E. Bangert, G. Landwehr: Proceedings EP 2 DS VI, p. 590, Kyoto (1985)
2. H.L. Stoermer, A.M. Chang, z. Schlesinger, D.C. Tsui, A.C. Gossard, W. Wiegmann: Phys. Rev. Lett. 51, 126 (1983)

3. G.S. Boebinger: PhD. Thesis (1986, M.I.T.)
4. J. Ihm. J.C. Philipps: J. Phys. Soc. Japan, 54, 1506 (1985)
5. G. Reményi, G. Landwehr, W. Heuring, E. Bangert, G. Weimann W. Schlapp: Proc. of the 18th Int. Conf. on the Semiconductor Physics, Stockholm 1986
6. M. Razeghi, J.P. Duchemin, J.C. Portal, L. Dmowski, G. Reményi, R.J. Nicholas, A. Briggs: J. Applied Physics Letters 48, (11), p. 712 (1986)
7. G. Ebert, K. von Klitzing, J.C. Maan. G. Reményi, C. Probst, G. Wiegmann: J. Phys. C17, L 775 (1984)
8. R.G. Clark, R.J. Nicholas, A. Usher, J.J. Harris: Proc. EP2 DS VI, p. 233, Kyoto (1985).
9. A. Aoki, T. Ando: Solid State Commun.38, 1079 (1981)
10. E.E. Mendez: Surface Science 170, 564 (1986)

The Two-Dimensional Density of States at Fractional Filling Factors

T.P. Smith III[1], *W.I. Wang*[1], *and P.J. Stiles*[2]

[1]IBM Thomas J. Watson Research Center, P.O. Box 218, Yorktown Heights, NY 10598 USA

[2]Department of Physics, Brown University, Providence, RI 02912, USA

Magnetocapacitance measurements have yielded the density of states of a two-dimensional electron gas from the weak-field limit to the extreme quantum limit. Quantitative information about the density of states at filling factors of 1/3 and 2/3 has been obtained.

The fractional quantum Hall effect (FQHE) [1] is a result of many-body interactions in two-dimensional electronic systems (2DES). Thus far, experimental studies of the FQHE have primarily focused on transport properties associated with this effect. However, the density of states (DOS) should also exhibit gaps or minima at fractional filling factors.

In principle, the DOS is directly related to the capacitance. However, if contact is made through the 2DES, conductivity effects influence the measured capacitance in a complicated way [2] and the DOS is not easily extracted. HICKMOTT [3] has measured the capacitance of accumulation layers in GaAs-AlGaAs capacitor relying on transport perpendicular to the 2DES but series resistance effects in his samples are not negligible. In order to measure the DOS both in-plane and series resistance effects must be small.

To accomplish this, GaAs heterostructure capacitors were fabricated on conducting substrates. A 2DES is located at the interface between an undoped GaAs layer and a modulation doped (AlGa)As layer. The 2DES is separated from the n^+ GaAs by 50 nm of undoped GaAs. The proximity of the conducting substrate to the heterojunction allows electrons to flow in and out of the 2DES with negligible resistive loss even in the extreme quantum limit. To form a capacitor an aluminum electrode was deposited on top of the structure. The DOS at fractional filling factors was studied in two samples. From analysis of the weak-field capacitance oscillations the mobility of sample A is at least 200,000 cm^2/Vs and the mobility of sample B is at least 400,000 cm^2/V-2. Both samples have a carrier concentration of 1.9 x $10^{11}/cm^2$. The area of the capacitors on sample A is 7.4 times larger than that of the capacitors fabricated on sample B.

Analysis of the weak-field and integer quantum Hall regime allows comparison of the results from our new samples with previous results and provides a basis for interpreting our results at fractional filling factors. The measured magnetocapacitance for sample A is shown in Fig. 1. The decrease in the capacitance between Landau levels is relatively small (4% at ν = 2) indicating that the DOS in the gap between Landau levels is large (40% of the zero-field DOS at ν = 2). We modeled our capacitance data using a DOS of the form given by GERHARDTS [4] with $\Gamma = \Gamma_0 B^{\alpha}$. To take into account the effects of spin splitting (observed at B = 8 Tesla) we used a self-consistent oscillatory g-factor used previously by ENGLERT et al. [5]. The best fit for this model

FILLING FACTOR, ν

DATA:

f = 989 Hz

T = 0.4 K

$N_s = 1.9 \times 10^{11} e^-/cm^2$

FIT: $\Gamma = \Gamma_0 B^\alpha$

$\Gamma_0 = 1.06\,meV,\ \alpha = 0.81$

$g = g_0 + E_{ex}/\mu_B B \Sigma(n_\uparrow - n_\downarrow)$

$g_0 = 0.52$

$E_{ex} = 12.8 \times 10^{-12} meV\text{-}cm^2$

MAGNETIC FIELD, B (TESLA)

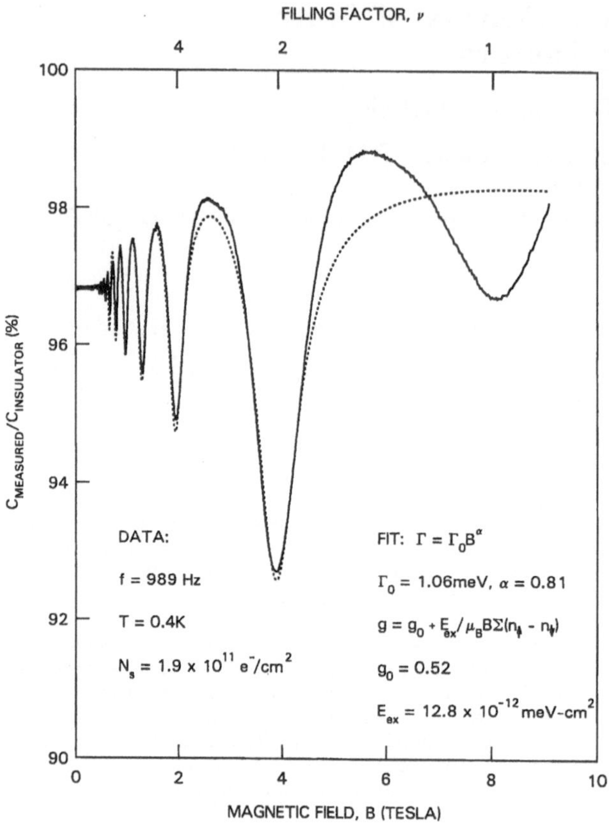

1. (a) The measured and calculated magnetocapacitance below the extreme quantum limit. The dotted line is the calculated capacitance using an oscillatory g-factor.

is also shown in Fig. 1. The fit is very good up to about 5 Tesla. The magnetic field dependence of the Landau level broadening is not too far from the $\sqrt{B}$ dependence predicted by ANDO and UEMURA [6] but the level width for our fit is almost an order of magnitude larger than predicted (3.2 meV vs 0.6 meV at 5 Tesla). However, our results agree fairly well with magnetization [7, 8] (Γ = 4.4 meV at 5 Tesla) and specific heat [9] (Γ = 2.1 meV with a 20% background at 5 Tesla) experiments.

Figure 2 shows the high field magnetocapacitance and DOS for samples A and B. The DOS has the form shown by either the upper or lower curve in Fig. 2b and Fig. 2d. However, the magnitude lies somewhere between these two curves. Consistent with the results at low magnetic fields, the structure at fractional filling factors is much stronger in sample B which has a higher mobility Nonetheless, the minima in the capacitance at filling factors of 2/3 and 1/3 ar relatively weak for both samples. Although these small changes in the capacitance reflect large changes in the density of states, the DOS remains larger than the zero-field DOS for $\nu < 1$.

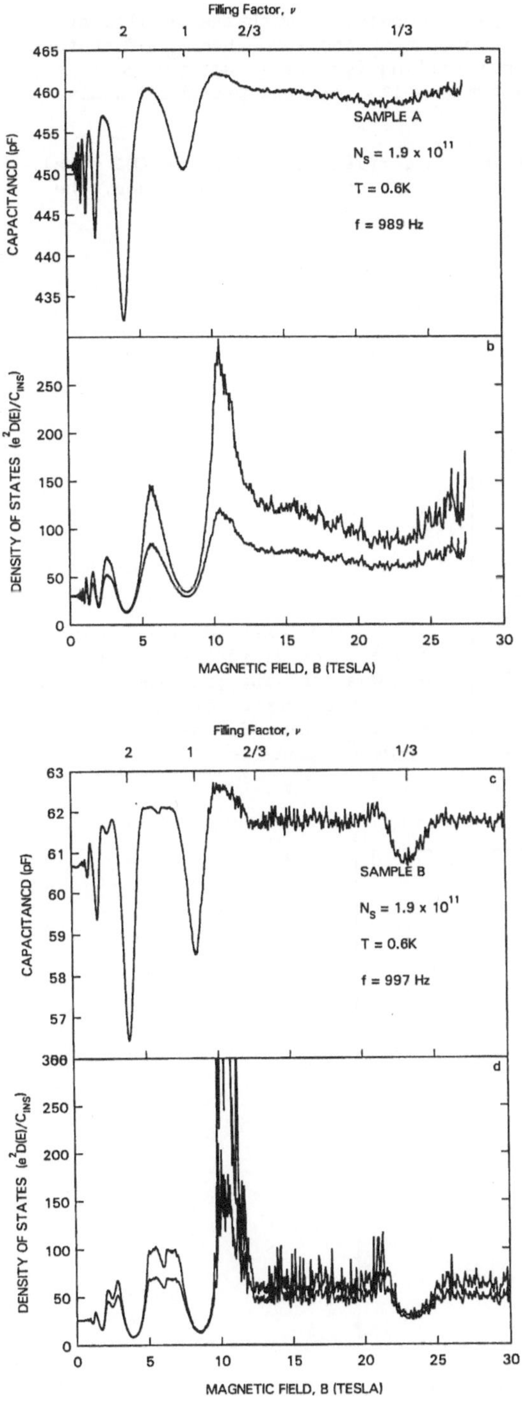

2. (a) The measured magnetocapacitance of sample A and (b) the measured density of states. (c) The measured magnetocapacitance of sample B and (d) the corresponding density of states.

The large DOS in the fractional gaps is consistent with the results at integer filling factors. However, the origin of the large number of states between Landau levels and at fractional filling factors is not yet well understood. Disorder, scattering, inhomogeneity, are all probably involved but this has not been confirmed.

Figure 3 shows the dependence of the DOS on carrier concentration (DC bias). As the magnetic field strength at which $\nu = 1/3$ occurs becomes larger, the

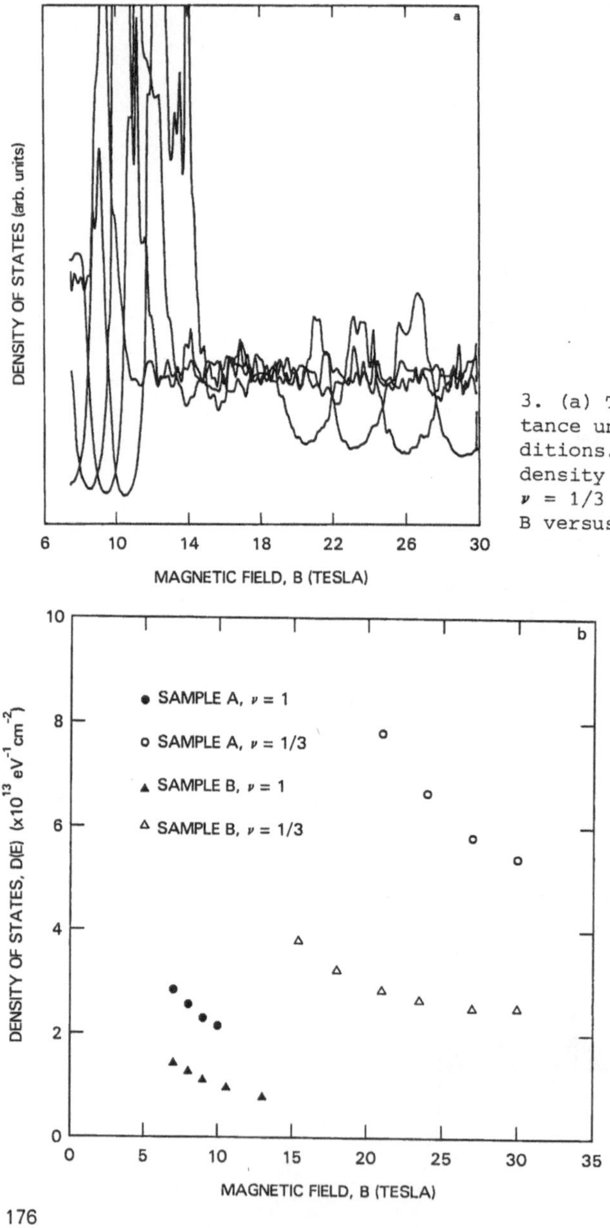

3. (a) The measured magnetocapacitance under different bias conditions. (b) The magnitude of the density of states at the $\nu = 1$ and $\nu = 1/3$ minima for samples A and B versus magnetic field.

reduction in the DOS at $\nu = 1/3$ saturates. This indicates that localization effects are no longer present and the magnetic length is shorter than the localization length. Both samples exhibit this behavior (see Fig. 3b) with the saturation occuring at higher magnetic fields for sample A which has the lower mobility.

In summary, we have measured the DOS of a 2DES from the weak-field limit to the extreme quantum limit. At the lowest fields the density of states can be modeled with a Gaussian DOS. Spin splitting is observed in the lowest Landau level and the g-factor appears to be enhanced by more than an order of magnitude over the free electron value. At fractional filling factors the DOS is drastically reduced but still very large. At high magnetic fields the reduction in the density of states at a filling factor of 1/3 saturates indicating that the magnetic length has become shorter than the localization length.

Acknowledgements

The authors are grateful to L. Rubin, B. Brandt and the staff at the Francis Bitter National Magnet Laboratory for their support. We would also like to thank M. Christie and L. Alexander for fabricating the capacitors and D. A. Syphers, E. E. Mendez, D. H. Lee, L. L. Chang, L. Esaki, F. F. Fang, and T. Jackson for their help and advice. This work was supported in part by the Army Research Office.

References

1. D. C. Tsui, H. L. Stormer, and A. C. Gossard, 48, 1559 (1982).
2. F. Stern, unpublished IBM internal report.
3. T. Hickmott, private communication
4. R. R. Gerhardts, Surf. Sci. 58, 227 (1976).
5. Th. Englert, D. C. Tsui, A. C. Gossard, and Ch. Uihlein, Surf. Sci. 113, 295 (1982).
6. T. Ando and Y. Uemura, J. Phys. Soc. Jap. 36, 959 (1974).
7. T. Haavasoja, H. L. Stormer, D. J. Bishop, V. Narayanamurti, A. C. Gossard and W. Wiegmann, Surf. Sci. 142, 294 (1984).
8. J. P. Eisenstein, H. L. Stormer, V. Narayanamurti, A. Y. Cho, A. C. Gossard, and C. W. Tu, Phys. Rev. Lett. 55, 875 (1985).
9. E. Gornik, R. Lassnig, G. Strasser, H. L. Stormer, A. C. Gossard, and W. Wiegmann, Phys. Rev. Lett. 54, 1820 (1985).

Possible Fractional Filling Factors in the Quantum Hall Effect at Temperatures up to 4.2 K

G. Hein[1], *G. Weimann*[2], *and W. Schlapp*[2]

[1]Physikalisch-Technische Bundesanstalt, Bundesallee 100, D-3300 Braunschweig, Fed. Rep. of Germany
[2]Postfach 5000, D-6100 Darmstadt, Fed. Rep. of Germany

Data of the quantum Hall effect and the voltage U_p parallel to the current I_p through $GaAs$-$Ga_xAl_{1-x}As$ heterostructures show reproducible structures. These structures are observed by adding or subtracting the data taken with either reversed sample current or magnetic field direction, they seem to be correlated to fractions of the filling factor ν. We observed structures at values of the filling factor of $\nu = p/q$ with $q = 2,3,4,5,7,9,11$ in samples with charge carrier mobilities of nearly 450 000 $cm^2/(Vs)$ at temperatures of 4.2 K and 1.2 K and 2 μA to 15 μA sample current. Data of the second derivative of the Hall voltage with respect to the magnetic flux density show structures at filling factors up to 19/2 at a temperature of 1.2 K.

Experimental and Results

We measured the quantum Hall effect and the voltage drop U_p parallel to the current I_p of $GaAs$-$Ga_xAl_{1-x}As$ heterostructures reversing both the direction of the sample current and of the magnetic field relative to each other. The data, as well as those for the magnetic field, were digitized and recorded with a microcomputer. One example for the possible combinations of the direction of the magnetic field and the sample current, designated α, β, γ and δ, is shown in Fig. 1. Because of a possible misalignment of the Hall contacts relative to each other, one obtains a voltage U_{pa} in addition to the Hall voltage U_H. The additional voltage U_{pa} can vary because reversing either the field or the current direction causes the current to flow in different areas or edges of the sample /1/.

The measured data of the quantum Hall effect are described by the following equations:

$$U_\alpha = U_{pa} + U_H ,$$

$$U_\beta = -U'_{pa} - U_H ,$$

$$U_\gamma = U'_{pa} - U_H ,$$

$$U_\delta = -U_{pa} + U_H .$$

Adding the measured voltages U_α and U_γ, i.e. $U^+_{\alpha\gamma} = U_\alpha + U_\gamma$, yields a residual voltage of $U_{pa} + U'_{pa}$ which in the ideal case of an exact alignment of the Hall contacts should be zero and otherwise is proportional to the voltage U_p. A typical experimental result is shown in Fig. 2 together with one original measurement of U_α for comparison. The apparently noisy structure in addition to the residual voltage $U_{pa} + U'_{pa}$, shown in Fig. 2, was found in measurements on several samples differing in electron mobility and concentration, and is obviously caused by physical properties of the samples.

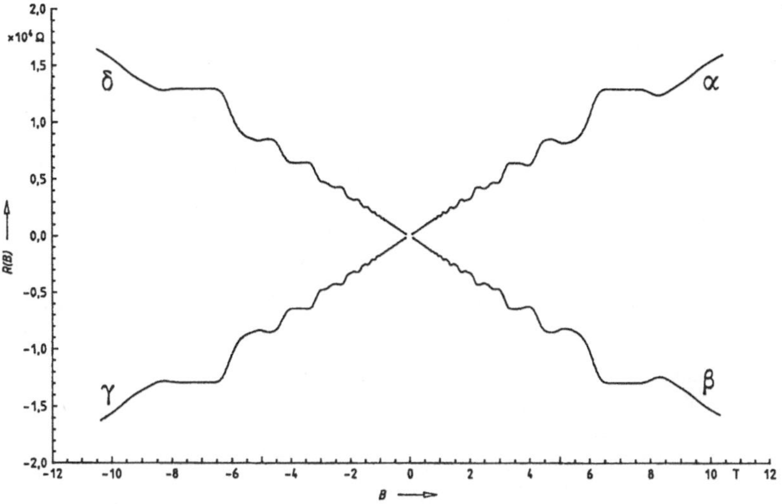

Figure 1: Measured Hall voltages for all four possible combinations of the sample current and the magnetic field directions. The resistance R_H is shown negative when a negative Hall voltage is divided by a positive sample current I_p. The negative B-axis represents the reversed field direction

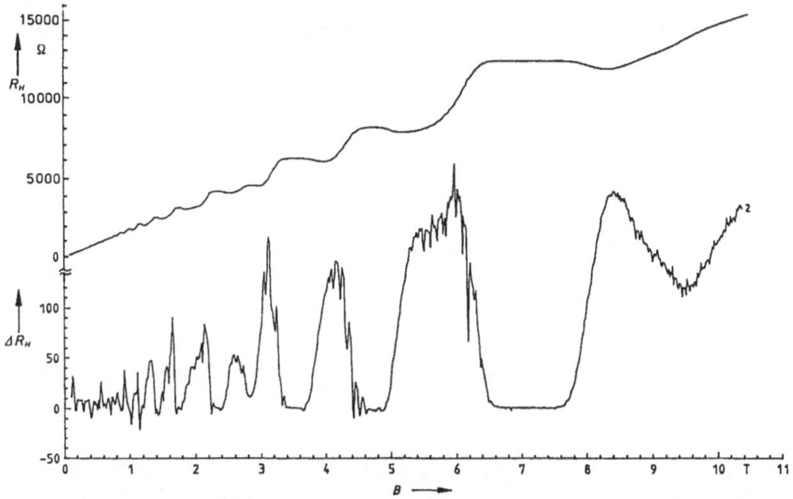

Figure 2: Residual resistance $\Delta R_H = (U_\alpha + U_\gamma)/I_p$ (trace 2) and $R_H = U_H/I_p$ (1)

The structures are reproducible for numerous measurements with different sample currents in the range 2 μA to 15 μA for temperatures between 1.2 K and 4.2 K. Summing the voltages U_p measured parallel to the sample current I_p yields similar structures with peaks and minima at the same values of the magnetic field as found for the Hall measurements.

In Fig. 3 the residual voltages $U^+_{\alpha\gamma} = U_\alpha + U_\gamma = U_{pa} + U^-_{pa}$ for four different samples are plotted against the filling factor ν which is given by the following relation:

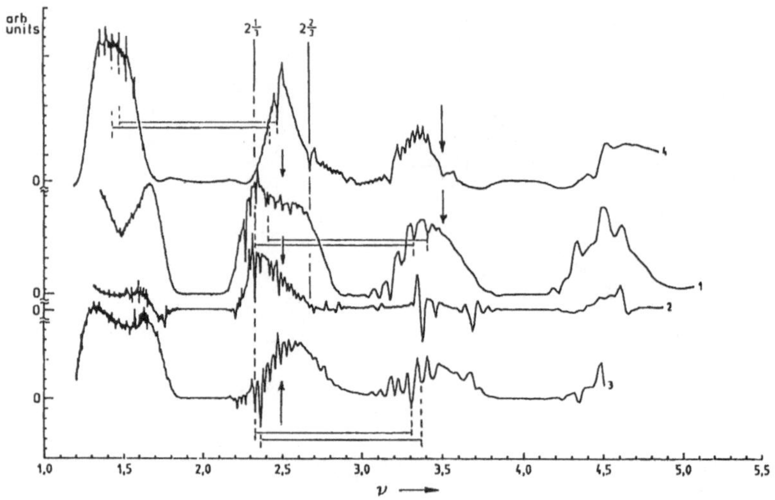

Figure 3: Residual voltages $U^{+}_{\alpha\gamma}$ for four different samples normalized to the filling factor ν with the help of equ. 1 of the text. Bars connect equivalent minima

Trace	carrier density n_s	mobility μ
1	$n_s = 3.4 \cdot 10^{11}\ cm^{-2}$	$\mu = 428\ 000\ cm^2/(Vs)$
2	$n_s = 3.25 \cdot 10^{11}\ cm^{-2}$	$\mu = 486\ 000\ cm^2/(Vs)$
3	$n_s = 3.13 \cdot 10^{11}\ cm^{-2}$	$\mu = 443\ 000\ cm^2/(Vs)$
4	$n_s = 2.95 \cdot 10^{11}\ cm^{-2}$	$\mu = 190\ 000\ cm^2/(Vs)$

$$\nu = 1/(B \cdot P), \qquad (1)$$

where P is the period related to the low-field Shubnikov-de Haas oscillations of U_p. At the center of a Hall plateau the filling factor becomes an integer n (n = 0, 1, 2, ...) and one obtains

$$n = 1/(B_n \cdot P),$$

where B_n is the value of the magnetic flux density at the center of the plateau n. The period P is then given by

$$P = 1/(n \cdot B_n),$$

which can be used to calculate the filling factor with eqn. 1. Taking differe values of n to determine the period P, one finds a dependence of the period P the magnetic field which when plotting the measured data of U_p against the fi ling factor and using the n = 2 plateau to calculate P causes a shift of the low-field minima of U_p against the integer values of $\nu = n$.

Figure 3 shows clearly the occurrence of minima at fractional values of $\nu = q/p$ with p = 2,3,4,5,7,9,11 and also equivalent minima at the same denominator p but different q, corresponding to a change in ν by 1 or 2. In Fig. such equivalent minima are connected by a bar.

In a subsequent experiment we investigated the second derivative of the Hall voltage U_α against the magnetic flux density, i.e. d^2U_α/dB^2. The dif-

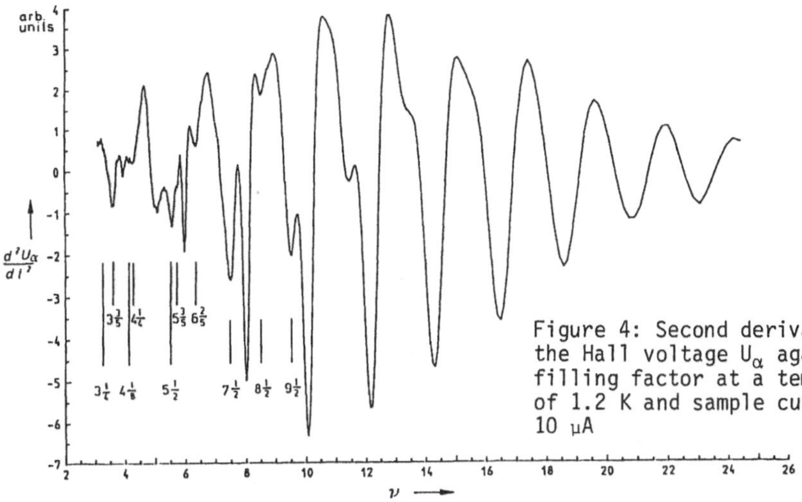

Figure 4: Second derivative of the Hall voltage U_α against the filling factor at a temperature of 1.2 K and sample current of 10 μA

ferentiation was carried out prior to digitizing with an electrical differentiation amplifier. In order to increase the signal-to-noise ratio the data of 15 subsequent measurements were averaged. The result is shown in Fig. 4 for a flux density range of 5 T and a temperature of 1.2 K. Again we observed structures which when plotted against the filling factor at values of $\nu < 5$ become very similar to those of Fig. 3. The additional minima are related to the filling factors as indicated in Fig. 4. For the calculation of the filling factor we used the period P determined from the Hall plateau of n = 4. The shift of the minima at low magnetic fields ($\nu > 10$) against the integer values of ν is then again caused by the dependence of the oscillation period on the magnetic field as mentioned above. Fig. 4 shows the existence of minima at filling factors of integer + 1/2 (e.g. 13/2, 15/2, 17/2, 19/2).

Although structures corresponding to fractional filling factors are clearly seen, related plateaux in the voltage U_H are not observed. The reason for this might be the high temperature of the sample, which prevents the parallel resistance from becoming zero.

We conclude that by the method of adding or subtracting the measured Hall voltages or the voltages U_p respectively, we resolved small structures hidden in both voltages. A slight inhomogeneity in the carrier concentration and the possible different path for the sample current when reversing either the current or the magnetic field direction causes a small shift of the additional structures on the magnetic field axis. This then causes peaks and minima to show up after adding or subtracting the measured voltages. The correlation to fractional values of the filling factor ν suggests that the anomalous quantum Hall effect /2/ observed at very high magnetic fields and extremely low temperatures might play an important role in our observations. Therefore further experiments are in preparation.

Acknowledgment

We wish to thank Dr. L. Bliek and Dr. L. Schweitzer for valuable discussions and Mr. H.J. Engelmann for preparing the samples.

1. Ch. Simon, B.B. Goldberg, F.F. Fang, M.K. Thomas, S. Wright, Phys. Rev., B33, 1190 (1986)
2. D.C. Tsui, R.L. Störmer, A.C. Gossard, Phys. Rev. Letters, 48, 1559 (1982)

Part III

Heterostructures and Superlattices: Optics

Tunneling Cyclotron Resonance in Semiconductor Superlattices

S.J. Allen, Jr., T. Duffield*, R. Bhat, M. Koza, M.C. Tamargo, J.P. Harbison, F. DeRosa, D.M. Hwang, P. Grabbe, and K.M. Rush***

Bell Communications Research, Inc., Redbank, NJ 07701, USA

1. INTRODUCTION

One of the more remarkable achievements in the field of materials science has been the growth of semiconductor single crystals atomic layer by atomic layer. By varying the composition along the growth direction a very great degree of control can be exercised over semiconductor doping and bandgaps. Although the scientific understanding and technological development of these materials has been unusually rapid, the early vision[1] that focused on the engineering of bulk, 3-dimensional, bandstructures has been largely forsaken in favor of properties that emerge due to confinement of the electron states to the two dimensions of a single layer.

Experiments that address the question of quantum transport along the superlattice growth direction have been relatively scarce. The original work of DINGLE et al.[2] revealed splittings in excitonic features that were caused by tunneling between quantum wells, but in the limit of a large number of coupled wells the optical absorption spectra became featureless. More recently SAKAKI et al.[3,4] have reported measurements of the Fermi surface geometry in some superlattice systems with extended states along the growth direction. SOLLNER et al.[5] have observed strong resonant tunneling features in double barriers separated by a thin quantum well. Most relevant to the following discussion is the recent theoretical and experimental work of MAAN et al.[6,7] that examines the effect of a magnetic field, perpendicular to the growth direction, on the mini-bands in the superlattice. In this configuration, interband magneto-luminescence[7] shows a well-defined set of Landau levels that disappear when the cyclotron energy exceeds the mini-band width.

There are many issues that can be addresssed by performing cyclotron resonance in the Voight geometry in semiconductor superlattices. We enumerate some of them here.

- Electron mass tunneling in the growth direction

---

* Visiting guest scientist at the Francis Bitter National Magnet Laboratory, Massachusetts Institute of Technology, Cambridge Mass.

** Present Address: University of Maryland, College Park, MD 20742.

- Coherence length in the growth direction
- Quantitative measure of the tunneling probability as a function of barrier composition and thickness
- Localization
- Magnetic breakdown
- Miniband breakdown
- Magnetic states in periodic potentials

Here we describe cyclotron resonance measurements on extended states in GaAs/(Al,Ga)As semiconductor superlattices. The magnetic field is oriented perpendicular to the growth direction and the electrons are forced to execute cyclotron motion by tunneling through the periodic barriers. The resonance frequency is directly related to the transport mass along the growth direction. At the outset, we wish to point out that cyclotron resonance is rather unforgiving in the sense that any scattering event, be it small angle or large angle, will contribute to the linewidth, so that the observation of a well-defined resonance implies coherent tunneling through many barriers. That such conditions are obtained in these superlattices, especially with alloy barriers, was not at all obvious before these experiments were performed.

We then describe qualitatively features that develop as the superlattice period exceeds the magnetic length scale. The miniband concept is no longer useful, and the cyclotron resonance is inhomogeneously broadened by the superlattice structure.

2. EXPERIMENT

The $Al_xGa_{1-x}As$/GaAs superlattice samples used in this study were grown in an atmospheric pressure organo-metallic chemical vapor deposition, (OMCVD), reactor[8] and by molecular beam epitaxy, (MBE). Approximately 5-6 microns of superlattice was grown directly on a GaAs substrate. The period of the superlattice was varied from 10 nm to 50 nm and the thickness of the $Al_xGa_{1-x}As$ barrier was nominally 2 nm. Transmission electron micrographs revealed perfect crystallization and uniform layered structure. However, the barriers in the OMCVD material were not perfectly sharp and had sloping walls tending to produce a triangular-shaped barrier with a full width at half maximum of 2 nm. The fractional Aluminum content, x, in the barrier was varied between nominal values of .10 and .75 .

The carrier concentration was in the low 10^{15} cm^{-3} range. Two considerations led to the choice of low donor concentration. Cyclotron resonance, with the magnetic field in the plane of the sample, does not occur at $w_c = eB/m_c$, where e is the electron charge, B the magnetic field, and m_c the cyclotron mass. Rather it occurs at the combined cyclotron and plasma resonance given by $w_o^2 = w_c^2 + w_p^2$. Here w_o is the observed resonance frequency and

w_p is the plasma frequency, which we can estimate from the simple 3D result, $w_p^2 = ne^2/(\ m)$. Here n is the volume density of carriers, the dielectric constant and m the electron mass in the growth direction. The exact collective plasma modes of the superlattices are known to exhibit a rich and varied behavior.[9] But, we expect that this 3D result will be a valid measure of the importance of these corrections. By keeping n in the 10^{15} range they can be kept much smaller than the cyclotron frequency and m_c can be accurately measured. Low doping levels also reduce impurity scattering. At liquid nitrogen temperatures the transport mobility was of the order of 40,000 cm^2/volt·sec in our samples. This should be compared with the mobilities of the order of 2,000 cm^2/volt·sec obtained in the heavily doped superlattices used in the Fermi surface studies of SAKAKI et al.[3,4]. The low doping levels do require us to perform the resonance experiments at elevated temperatures to avoid carrier freeze out, but this does not seriously compromise the experiments. All the data reported below was obtained at liquid nitrogen temperatures.

3. RESULTS

3.1 Transport Mass in Small Period Superlattices

We first describe results for the superlattices that have a nominal period of 10 nm and barriers of 2nm but varying Al content in the barrier. Here, at least for low Al content, we can characterize the results in terms of a transport mass along the growth direction.

Cyclotron resonance experiments were performed in magnetic fields to 8 Tesla.[10] The resonance was detected by doing swept frequency spectroscopy with a Fourier transform spectrometer. For every sample, resonances were recorded with the magnetic field in the plane of the sample, and perpendicular to it. Typical data are shown in Fig. 1 a. The resonance frequency, w_o, was squared and plotted versus B^2. In this way the plasma resonance, that alters the result when the field is in the plane of the sample, appears as a non-zero intercept, (in close agreement with the estimates made above), but does not effect the slope of the line which we take as a measure of the cyclotron mass, m_c. This mass is the geometric mean of the transport mass along two orthogonal directions. To extract and display a measure of the transport mass along the growth direction we evaluate $m = m_c^2/m_o$. The mass determined from these measurements is plotted in Fig. 2. as a function of Al concentration, for superlattices with a 10nm period.

With the magnetic field perpendicular to the surface the electrons execute cyclotron motion in the plane of the barriers and wells and the resonance is expected to be close to the GaAs mass. Indeed, the results in Fig. 2. show a mass, m_o, which differs from the bulk GaAs mass by an amount that can be readily explained by band non-parabolicity.

The mass along the growth direction shows a strong increase with Al content or barrier height. It is apparent however that

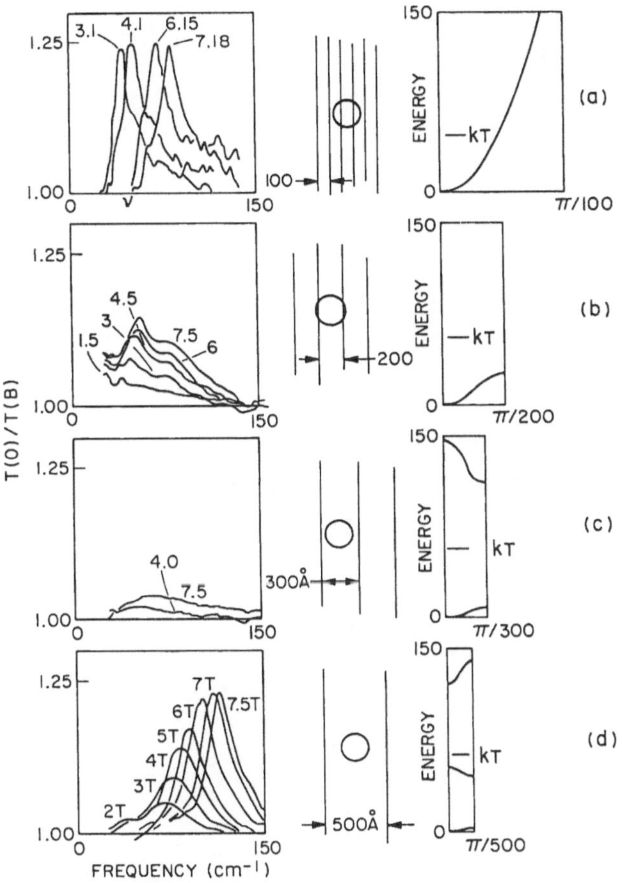

Fig. 1. Cyclotron resonance for four different superlattice periods. kT indicates the temperature at which the data was obtained. At the right is shown the miniband structure in zero magnetic field with energy units of cm^{-1}. In the center is shown the relative size of the period and cyclotron orbit at 8 Tesla.

there is scatter in the experimental data that lies outside the experimental uncertainties in determining the mass and Al content in the barrier. This indicates that there are material parameters, beyond our control or understanding, such as charge trapped in the barriers or barrier shape, that influence transport through the barriers. On the other hand it does demonstrate that these measurements are sensitive and specific to transport through the barrier.

It is important to point out that in the largest magnetic field used in these experiments the electrons still execute a cyclotron motion with a diameter that exceeds the superlattice period. (At 8 Tesla the quantum limit zero point motion is of the order of 18 nm.) As a result, for the 10nm superlattices, we

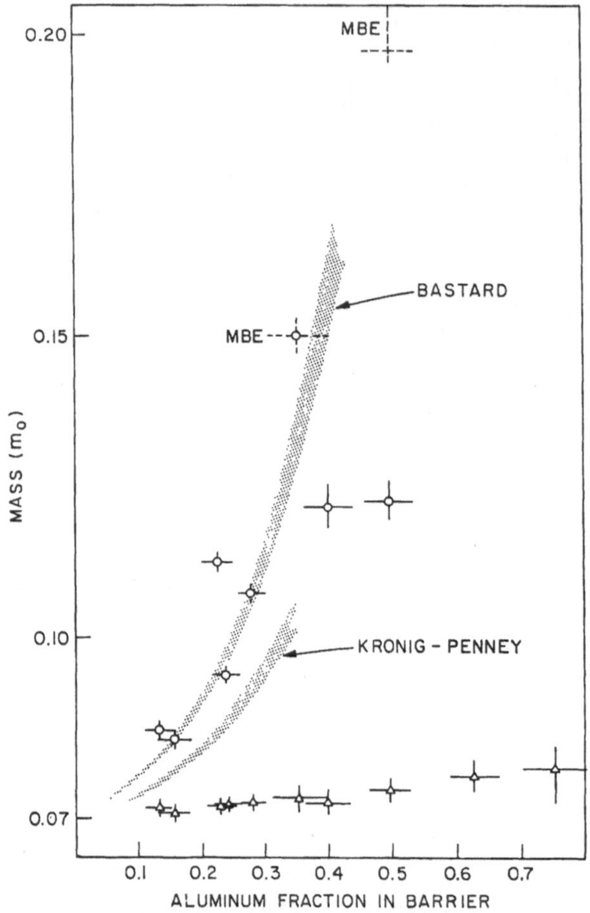

Fig. 2. Electron mass along and perpendicular to the growth direction. All the data with the exception of the two indicated points are taken from OMCVD material. No mass can be measured along the growth direction for x > .5.

choose not to consider, the interesting complications described by MAAN[6] as the cyclotron orbit shrinks inside the period of the superlattice. There may well be subtle changes in the CR as the number of superlattice periods embraced by the cyclotron orbit is changed but they are not apparent in the existing data.

In marked contrast to the lack of experimental studies of extended states in semiconductor superlattices, there are a number of theoretical discussions that address mini-band structure.[1,11-16] Here we take a simple Kronig-Penney potential to model the superlattice and calculate a cyclotron mass, m_c, at an energy k_bT. In a superlattice, in addition to conduction and valence band discontinuities, there is also an effective mass discontinuity at the interfaces between the quantum wells and the barriers. BASTARD[16] has shown that this changes the continuity condition for stationary states with probability current, F, to 1/m dF/dz rather than dF/dz as in the original Kronig-Penney model. We have calculated the mini-band structure for our samples using both the Kronig-Penney and Bastard solutions for the potential. We assume that the electron mass in the $Al_xGa_{1-x}As$ barriers, required for the Bastard solution, is m_o^* multiplied by the ratio of the direct gap in the barrier to that in GaAs, where m_o is the mass measured with the field parallel to the growth direction. We also assume that the barrier height is related to the Al concentration by $E_c = (.63 \pm .02)\ E_g$[17], where E_c is the conduction band offset and E_g is the difference in energy gap between the barrier and GaAs.

The range of magnetic fields used in these experiments spans the classical and quantum limits. At the highest fields the cyclotron energy exceeds k_bT. Experimentally, however, the mass exhibits no field dependence and we choose to compare experiment with a "classical" calculation of the cyclotron mass for an electron with energy k_bT. The results of the calculations for the mass along the growth direction are shown as the shaded areas in Fig. 2. For a given Al concentration, the limits represent the uncertainty in E_c with the upper limit given by $E_c = .65\ E_g$. Despite the scatter in the data the calculation that includes the change in the band-structure as parameterized by Bastard is significantly better than the simple Kronig-Penney model.

Equally important is the fact that the scattering rate inferred from the cyclotron resonance linewidth implies a coherence length that is greater than ten superlattice periods. Electrons coherently tunnel through many superlattice periods in the structures with 10 nm period.

3.2 Mini-band Breakdown

The effective mass in the mini-band ceases to be a useful parameter to describe cyclotron resonance if the cyclotron orbits are comparable to or smaller than the superlattice period. Here we experimentally explore the evolution of the cyclotron resonance as we expand the superlattice period. The superlattices used in this study were comprised of 2 nm barriers with Al fraction of .3 but with periods of 10, 20, 30 and 50 nm.

Figure 1. shows the cyclotron resonance spectra for each of these samples. In the center of the figure the size of the cyclotron orbit at 8 Tesla is indicated with respect to the superlattice period, and at the extreme right the mini-bands in the absence of a magnetic field are shown.

As the period of the superlattice expands, the mini-band width shrinks until it is smaller than the cyclotron frequency. It is not surprising that no resonance can be detected under these conditions. Rather, only a broad resonance develops that grows in strength but which can not be characterized by a mass. Finally, at 50 nm period a well-defined resonance develops whose intensity grows and whose frequency shifts with magnetic field.

The extremes of the behavior are well understood. At 10 nm period the mini-band picture is useful and we have a resonance that can be characterized with a mass derived from the mini-band structure. At 50 nm we have a sub-band-like transition between quantum well-like states which develops intensity and shifts frequency as the cyclotron frequency approaches and exceeds the sub-band transition frequency. Indeed, the low magnetic field transition approaches 67 cm-1, which is in rough agreement with the calculated energy separation of the first two minibands for this superlattice.

HARPER[18] pointed out some time ago that the eigenstates of an electron in a periodic potential and in a magnetic field can be described by Landau states, with an effective mass appropriate to the periodic potential, only in the limit that the magnetic length is much larger than the period of the potential. Otherwise, the Landau levels have an energy that depends on the position of the cyclotron orbit, and the cyclotron resonance will suffer "Harper" broadening.[19-21] MAAN[6] has recently calculated the effect of a periodic potential on Landau levels in just this limit.

We show schematically the energy levels of such a calculation in Fig. 3. The position of the cyclotron orbit is a good quantum

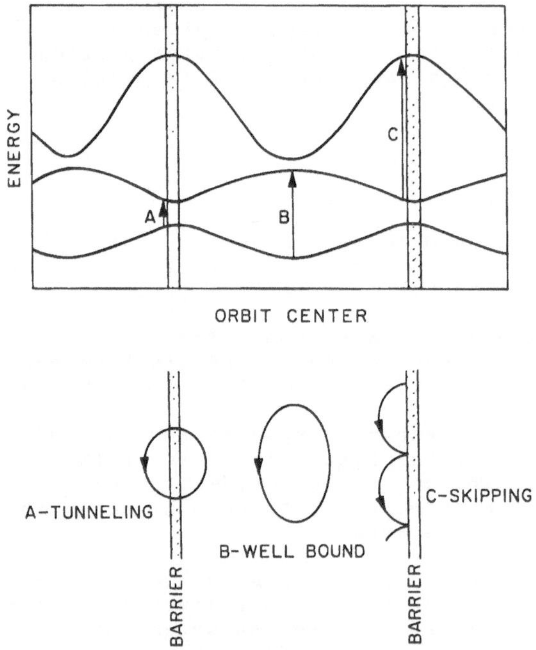

Fig. 3. Schematic diagram of the Landau level energies as a function of orbit position in the limit of superlattice period large compared with cyclotron orbit.

number. We can identify three types of cyclotron resonance that represent the extrema in Fig. 3. Although the states are quantum mechanical, we graphically describe them by the following classical analogues. Transitions at the well center are "well-bound" resonances and correspond to the resonance seen in Fig. 1 d, for the 50nm period. The alternative interpretation, given above, as a magnetic field-assisted sub-band transition is equally valid.

The other two limiting cases are " barrier bound" resonances. The low-frequency resonance corresponds to a tunneling mode or splitting. It is possible that the low-frequency features induced by a magnetic field in the 20nm and 30nm period superlattices ,(Figs. 1 b,c) , are related to these tunneling resonances,but further experiments and modeling are required to confirm or deny this interpretation. The high-frequency transition at the barrier has as its classical analogue skipping orbits which for an orbit centered right on the barrier should occur at approximately twice the cyclotron frequency. We have not seen a resonance as yet that resembles this particular mode.

4. CONCLUSION

We have performed cylotron resonance in semiconductor superlattices with the magnetic field oriented in such a way that the electrons must tunnel through the barriers. For relatively short superlattice period and Al fraction less than .5 the resonance is well characterized by a mass which may be used to determine the transport mass along the growth direction. Comparisons with model calculations suggest that the mini-band structure be calculated with the boundary conditions prescribed by Bastard. Most important is the fact that the electrons appear to propagate coherently through many barriers in these superlattices.

When the superlattice period exceeds the magnetic length the effective mini-band mass is no longer useful in characterizing the resonance. Cyclotron resonance transitions now depend on the position of the orbit center with respect to the superlattice barriers and wells. Well-bound resonances have been seen and are qualitatively understood as sub-band transitions in a strong magnetic field. Barrier-bound resonances that appear as tunneling and skipping modes have yet to be clearly identified.

ACKNOWLEDGEMENT

We wish to thank Larry Rubin and Bruce Brandt for their support of the experiments carried out at the Francis Bitter National Magnet Laboratory.

1. L. Esaki and R. Tsu, IBM J. Res. Develop. 14, 61 (1971).

2. R. Dingle, A.C. Gossard, and W.Wiegmann, Phys. Rev. Lett. 34, 1327 (1975).

3. J. Yoshino, H. Sakaki and T. Furata, Proceedings of the XVIIth International Conference on the Physics of Semiconductors, edited by J.D. Chadi and W.A. Harrison, (1984) p. 519.

4. L.L. Chang, H. Sakaki, C.A. Chang and L. Esaki, Phys. Rev. Lett. 38, 1489 (1977).

5. T.C.L.G. Sollner, W.D. Goodhue, P.E. Tannenwald, C.D. Parker, and D.D. Peck, Appl. Phys. Lett. 43, 588 (1983).

6. J.C. Maan, Springer Series in Solid State Sciences, 53 183 (1984).

7. G. Belle, J.C. Maan, and G. Weimann, VIth International Conference on the Electronic Properties of Two-Dimensional Systems, Kyoto, Japan (1985); —— –, Solid State Commun. 56, 65 (1985).

8. H.M. Manasevit, Appl. Phys. Lett., 12, 156 (1968).

9. A.C. Tselis and J.J. Quinn, Phys. Rev., B29, 3318 (1984), and refernces contained therein.

10. T. Duffield, R. Bhat, M. Koza, F.DeRosa, D.M. Hwang, P. Grabbe and S.J. Allen, Jr., Phys. Rev. Lett., 56, 2724 (1986).

11. R. Dingle, in Festkorperprobleme, Vol XV, edited by H.J. Queisser, Pergamon/Vieweg, Braunschweig (1975), p. 21.

12. G.A. Sai-Halasz, Proceedings of the XIVth Internationsl Conference on the Physics of Semiconductors, Edinburgh, edited by B.L.H. Wilson, (1978) p. 21.

13. G.A. Sai-Halasz, L. Esaki and W.A. Harrison, Phys. Rev. B18, 2812 (1978).

14. D. Mukherji and B.R. Nag, Phys. Rev., B12, 4338 (1975).

15. G.A. Sai-Halasz, R. Tsu and L. Esaki, Appl. Phys. Lett., 30, 651 (1977).

16. G. Bastard, Phys. Rev., B24, 5693 (1981).

17. T.J. Drummond and I.J. Fritz, Appl. Phys. Lett. 47 284 (1985).

18. P.G. Harper, Proc. Phys. Soc., 68, 879 (1955).

19. R. Tsu and J.Janak, Phys. Rev. B9, 404 (1974).

20. P.L. Taylor, Phys. Rev., B15, 3558 (1977).

21. R. Rammel, J. Phys., 46, 1435 (1985).

Thermodynamic and Magneto-Optic Investigations of the Landau Level Density of States for 2D Electrons in GaAs

E. Gornik, W. Seidenbusch, and G. Strasser

Institut für Experimentalphysik, Universität Innsbruck,
A-6020 Innsbruck, Austria

In this report different experimental results relevant to determine the density of states in a high magnetic field are summarized: Measurements of the specific heat of GaAs/GaAlAs multilayers reveal clear evidence for a Gaussian-like density of states superimposed on a constant background. Temperature-dependent measurements of the resistivity in the regime of the Hall plateaus confirms the existence of a flat, mobility-dependent background between Landau levels. Magnetization data give evidence of a magnetic field dependence of the Gaussian-like density of states.

From cyclotron resonance transmission a variation $\Gamma_{CR} \sim \sqrt{1/\mu}$ of the linewidth Γ_{CR} with the zero-field mobility μ is obtained for integer filling factors. The same dependence on mobility is obtained for the amount of background states. Evidence is found that the observed Gaussian and flat part of the density of states is correlated with the number of ionized impurities present,which also determine the zero-field mobility.

INTRODUCTION

In a 2-dimensional electron system (2DES) a magnetic field perpendicular to the plane of electrical confinement leads to full quantization of the electron motion. The energy spectrum consists of sharp Landau levels (LL) separated by the cyclotron energy $\hbar\omega_c$.

In a real system the LL are broadened due to scattering by impurities, phonons or other scattering mechanisms. In the simplest approximation the levels are described by a level width Γ. In the case of high magnetic fields, where $\hbar\omega_c >> \Gamma$, real gaps appear between the LL. This leads to an oscillatory structure of practically all physical quantities as a function of the magnetic field.

The most fundamental quantity underlying all these physical properties of the system is the form of the density of states D(E). A complete theoretical description of D(E) has not been performed until now. Diagrammatic /1,2/ as well as path integral /3,4,5/ techniques have been used to calculate D(E). In the self-consistent Born approximation /2,6/ a semi-elliptic form of D(E) without background is obtained. Using a path integral technique within lowest order cumulant expansion GERHARDTS/4/ obtained a pure Gaussian D(E) for long-range potentials. Exact results have been obtained by WEGNER and BREZIN et al. /5/ for some restricted types of short-range scattering distributions for the lowest LL. For a white noise potential a Gaussian D(E) is obtained. A Poisson distribution of scatterers (with nonzero higher order correlations) yields a peak density of states at the center of the LL and a weakly decaying D(E) towards the next LL.

Experimental investigations on the D(E) of 2D electrons in GaAs have first been performed with thermodynamic techniques: From specific heat /7/ measurements a Gaussian density of state on a flat background was determined. From the analysis of the magnetization /8/ an increasing Gaussian levelwidth with magnetic field was found. Temperature-dependent resistivity /9/ measurements reveal also a magnetic field-dependent Gaussian density with a flat density of states between LL. Similar results were obtained from capacitance experiments /10/.

It is the aim of this paper to give some new information on the origin of the form of D(E). We will first describe the results from the specific heat experiments and compare them critically with the other techniques. Additional information on D(E) is obtained from cyclotron resonance studies /11/. A correlation between the cyclotron linewidth and the background density of states is found.

1. Specific Heat

The most direct method to determine D(E) is the measurement of the electronic specific heat given by

$$C_V = \frac{dU}{dT} = \frac{d}{dT}\int_0^\infty E\, D(E)\, f(E,E_F)\, dE\,, \qquad (1)$$

where $f(E,E_F)$ is the Fermi distribution function. An externally induced temperature change leads to a reordering of the electrons. The heat capacity is proportional to D(E) at the Fermi energy, sampling localized and delocalized states.

The first calculation of the specific heat in 2D systems was performed by ZAWADZKI and LASSNIG /12/. They assumed a Gaussian density of states $D(E) \sim e^{-E^2/2\Gamma_G^2}$ independent of the magnetic field. Two contributions to the specific heat are found: intra- and inter-Landau level contributions. Results for a levelwidth $\Gamma_G = 0.25$ meV are shown in Fig. 1 for two temperatures. The intra LL contributions lead to an oscillatory behavior with a vanishing specific heat at integer filling factors. (The filling factor is defined as $\nu = n_s/\pi l^2$ neglecting spin splitting with n_s the electron concentration).

The inter-LL contributions appear as sharp spikes at the position of integer ν-values. These spikes are only present at low magnetic fields and "high" temperatures where kT is comparable with the LL splitting. At the lower temperature (dashed curve) the inter-LL peaks have disappeared. The intra-LL contributions for a given filling factor depend on kT/Γ_G. A maximum is found for $kT/\Gamma_G \sim 0.2$, which means that the specific heat is not sensitive to Γ_G in this range.

A heat pulse technique was applied to determine the electronic specific heat: In this technique a short heat pulse has to heat the sample adiabatically. The thermal time constant of the system has to be considerably longer than the heating pulse. The thermal isolation was achieved with thin (5 to 10 µm) superconducting wires. The change in sample temperature was measured with a Au:Ge film, a thin Ni-Cr film served as a heater. A detailed description of the experimental set-up is given in Ref. /13/.

The experiments were performed on two different multilayer materials. We want to concentrate here only on the higher mobility sample consisting of

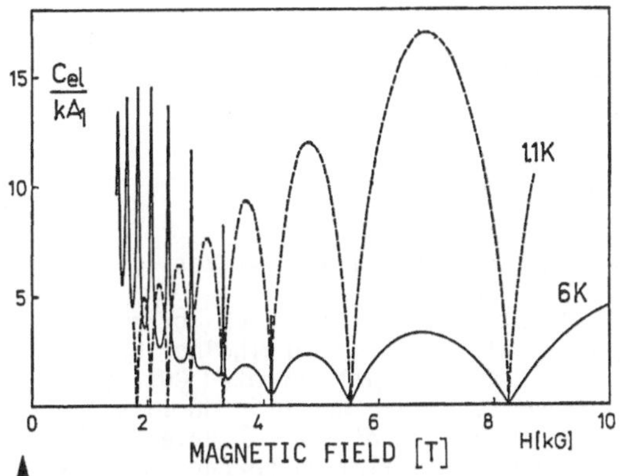

Fig. 1: Specific heat of 2D electrons versus magnetic field for two temperatures after Ref. 12.
Full curve: T=6K; dashed curve: T=1.1 K
$n_s = 8.0 \times 10^{11} cm^{-2}$, $\Gamma_G = 0.25$ meV,
$A_1 = 1/\pi \cdot (eB/\hbar)$.

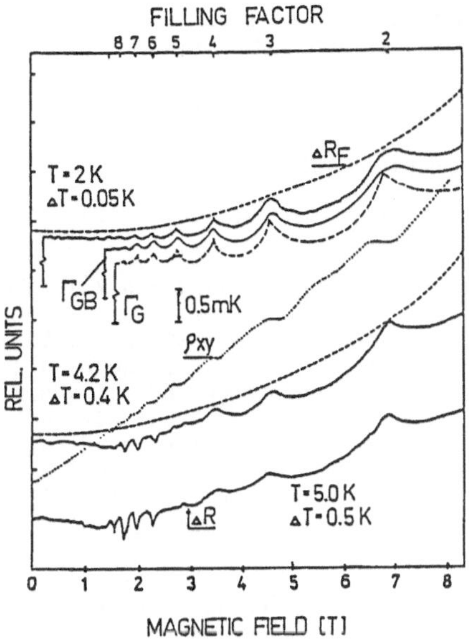

Fig. 2: Temperature change measured with the Au:Ge film versus magnetic field (curves ΔR), for a heat pulse raising the sample temperature by ΔT (ΔR_F changes in d.c. detector resistance). Theoretical calculations for a pure Gaussian levelwidth (Γ_G) and a Gaussian with background (Γ_{GB}) are also shown for 2 K.

94 layers of 220 Å GaAs and 500 Å GaAlAs. The mobility at 4.2 K was 80.000 cm²/Vs, and $n_s = (7.7 \pm 0.3) \times 10^{11} cm^{-2}$. The total sample thickness was 20 µm.

Fig. 2 shows the observed temperature change expressed as curves ΔR versus the magnetic field for three temperatures as obtained from averaging over 10 runs. The applied heat pulse raised the sample temperature by the values indicated as ΔT. The dashed curves ΔR_F show the background dc-resistance variation of the detector film on an extended scale. Oscillations of the sample temperature are clearly observed. The temperature changes are most pronounced for integer ν-values which represent the number of fully occupied LL. The ν-values are determined from the oscillatory conductivity measured on the same sample before being thinned down, shown as dotted curve ρ_{xy}.

The size of the ΔR signal is proportional to the rise in sample temperature. The data show that the sample temperature is higher for all integer ν-values at T = 2 K and ν-values up to $\nu = 4$ for the higher temperature. At lower fields ($\nu < 4$) peaks with opposite sign are observed for T = 4.2 and 5.0 K. From a qualitative comparison with the calculations we can identify the behavior at higher field due to intra LL contribution and the peaks at lower field and higher temperature due to inter LL contribution (see Fig. 2).

From the experimental data the form of the density of states can be determined by comparing the observed temperature change with calculations of ΔT. The main influence on ΔT comes from the form of the electric specific heat $C_{el}(T,B,\Gamma)$ which is calculated using different types of model density of states. In Fig. 2 the T = 2 K data are compared with a) a Gaussian density of states

$$D_G(E) = (\pi e^2)^{-1} \sum_n (2/\pi)^{1/2} \Gamma_G^{-1} \exp[(E - \lambda_n)/2\Gamma_G^2]^2$$

and b) a Gaussian with background

$$D_{GB}(E) = D_G(E)(1 - x) + (\pi e^2)^{-1}(\frac{x}{\hbar\omega_c}) \Theta(E)$$

with $\lambda_n = \hbar\omega_c(n + 1/2)$ and x is the percentage of background states.

The levelwidth Γ is defined as total width at half maximum which correlates to Γ_G as: $\Gamma = 2.4 \times \Gamma_G$. The resulting temperature change (including the behavior of the background) is shown in Fig. 2 by the full (denoted GB) and dotted (denoted G) curves. It is clearly evident that the curve GB fits the data considerably better than curve G. The flat part of the ΔT curve at T = 2 is the direct evidence for the existence of a flat background density of states. The pure Gaussian $D_G(E)$ results in a sharp spike-like change in ΔT even if we increase Γ_G considerably. It has been shown previously, that a Lorentzian density of states does not fit the data /7/.

The next question is whether Γ_G is constant with B or not. Other experimental techniques give evidence for a magnetic field-dependent level width /8,9,10/. The influence of different level widths on C_{el} is shown in Fig. 3. The experimental result (curve 1 for T = 4.2 K) is compared first with a magnetic field-independent Gaussian width Γ_G = 0.75 meV (curve 3) and Γ_G = 1.5 meV (curve 4). It is evident that inter LL peaks are very sensitive to Γ_G, while intra LL peaks at high field are not. As a consequence the data are consistent with a magnetic field-dependent Γ_G. The best linewidth fit to the data is achieved at $\sim$ 2 T where inter LL contributions are dominant, giving Γ_G = 0.75 meV. A good fit over the whole magnetic field range can be achieved with Γ_{GB} = 0.6 meV x $\sqrt{B}$ (B in T) and a background of x = 0.2 (curve 2). The $\sqrt{B}$ dependence was taken as used in Ref. /8/. If we try to fit the sample with 172 double layers described in Ref. /7/ with a magnetic field-dependent width we obtain Γ_{GB} = 0.9 meV x $\sqrt{B}$ and x = 0.30.

Summarizing the specific heat data we can state that there is clear evidence for a Gaussian density of states at the positions of the LL, sitting on a flat background. The analysis of the experiments is consistent with Γ_G which increases according to a $\sqrt{B}$ law. However, due to the rather weak sensitivity of C_{el} to Γ_G at higher magnetic fields, this result relies more on other experimental techniques.

2. Other techniques to determine D(E)

2.1 Magnetization

Another thermodynamic technique to determine D(E) is the measurement of the magnetization of the 2D electron gas. The magnetization is given by M = - dF/DB where F is the free energy. At absolute zero, both, the Fermi level and the magnetization exhibit a saw-tooth oscillation periodic in inverse field, with discontinuities at integer-filling factors. The magnetization oscillations are of constant amplitude $M_o = n_s A \mu_B^*$, where A is the sample area and μ_B^* the effective Bohr magneton. The first calculation of M for a Gaussian density of states was performed by ZAWADZKI and LASSNIG /14/.

Successful measurements of the magnetization were performed by EISENSTEIN et al. /8/. An extremely sensitive capacitive technique was developed to measure the torgue of the sample due to the change in magnetic moment /15/. In this paper we will only discuss the results: Fig. 4 shows the maximum amplitudes of the de Haas van Alphen oscillations normalized by the ideal amplitude M_0 versus magnetic field. The solid lines represent calculations with a magnetic field-independent Gaussian $D_G(E)$ with Γ_G = 1 meV and 2 meV. The data are plotted for two samples: A GaAs/GaAlAs multilayer (50 periods, $\mu = 8.0 \times 10^4 cm^2/Vs$, $n_s = 5.4 \times 10^{11} cm^{-2}$) and a single layer heterostructure ($\mu = 2.85 \times 10^5 cm^2/Vs$ and $n_s = 2.7 \times 10^{11} cm^{-2}$). It is evident that a constant Γ_G cannot account for the observed behavior. The best fit for the multilayer sample is obtained with $\Gamma_G = 1$ meV $\sqrt{B(T)}$ shown as dashed curve. No background density of states is assumed.

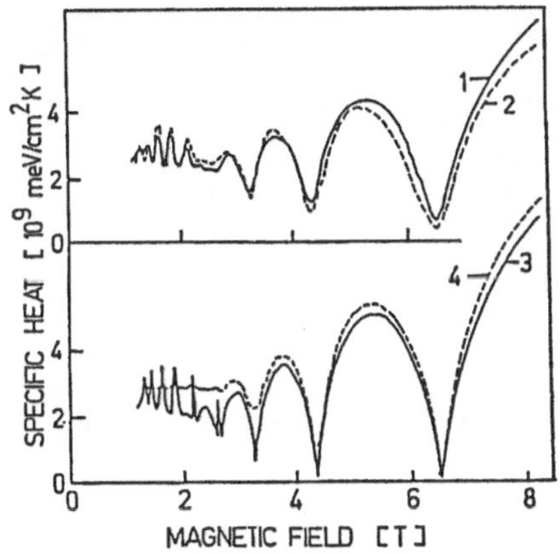

Fig. 3: Comparison of calculated and experimental C_{el} for a sample with 94 layers (described in the text) vs. magnetic field at 4.2 K:
Curve 1: experiment,
curve 2: $\Gamma_G = 0.6$ meV $\sqrt{B}$, $x = 0.2$
curve 3: $\Gamma_G = 0.75$ meV, $x = 0$
curve 4: $\Gamma_G = 1.5$ meV, $x = 0$

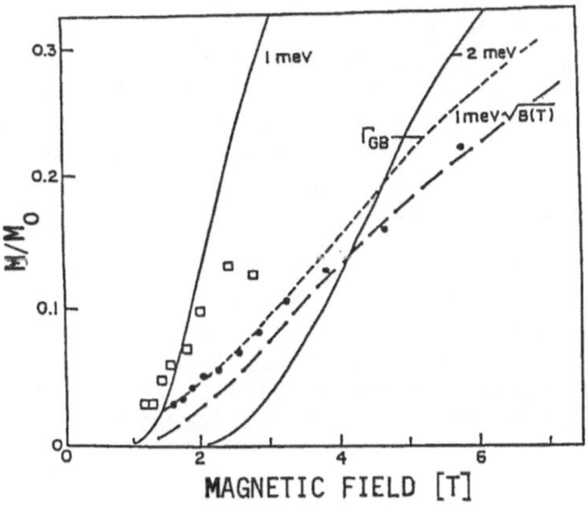

Fig. 4: Normalized dHvA oscillation amplitude vs. magnetic field after Ref. 8; multilayer: full circles, single layer: squares.
The curves are calculations with Γ_G-values indicated and $\Gamma_{GB} = 0.8$ meV $\cdot \sqrt{B(T)}$, $x = 0.2$

We have tried to fit these data with a reduced $\Gamma_{GB} = 0.8$ meV $\sqrt{B(T)}$ and a background density of $x = 20$ % shown as curve denoted Γ_{GB}. The fit is quite good up to 4 T but deviates for higher fields. The single heterolayer sample has a somewhat smaller width. We were able to fit the data with $x = 10$ % and $\Gamma_{GB} = 0.65$ $\sqrt{B(T)}$. However, the fit with background is not as good as without background. The magnetization technique seems to be quite sensitive to determine the linewidth in the whole magnetic field range. Unfortunately the method has only been applied in a rather narrow magnetic field range. It should also be noted that this technique is most sensitive when LL are half filled. Under this condition the maxima in amplitude appear.

2.2 Temperature dependence of ρ_{xx}

For a 2D electron system the resistivity ρ_{xx} exhibits strong oscillation with magnetic field (Shubnikov-de Haas oscillation). Minima of ρ_{xx} occur at integer filling factors which are strongly temperature dependent. As the

Fermi level moves towards the center of a LL the temperature dependence of ρ_{xx} becomes weaker. The analysis of this effect was used by WEISS et al. /16/ to determine the density of states between LL:

The analysis assumes that nearly all states are localized except states in the center of LL. By fitting the activated behavior of ρ_{xx} the density of states for integer-filling factors is determined. In the analysis a Gaussian width dependent on $\sqrt{B}$ and a constant background is assumed. The main result of this technique is a quite accurate determination of the background density of states, which varies significantly with sample mobility. The Gaussian width Γ_G is rather weakly sample dependent, which might be due to the assumption of a very narrow range of extended states in the center of LL.

A summary of the determined background values from the activated resistivity measurements for a large number of samples together with the results from specific heat is given in Fig. 5. The x-values are calculated from Ref. /16/ by the ratio of the midpoint density to the density at B = 0 given by $D(E)_{B=0} = 28 \times 10^9 cm^2 meV^{-1}$. The x-values from the two techniques are in good agreement. It is evident that the background density of states increases very steeply for very low mobility samples while it decreases slowly for high mobilities. We have not included the x-data from magnetization since the data can also be fitted without background.

From the above results we can conclude that there exists a flat background density of states with Gaussian peaks at the LL positions. The amount of background states depends on the sample mobility. The width of the Gaussian peaks does not show a consistent behavior as a function of mobility from the above techniques. Similar results for the density of states have been obtained from capacitance techniques /10/.

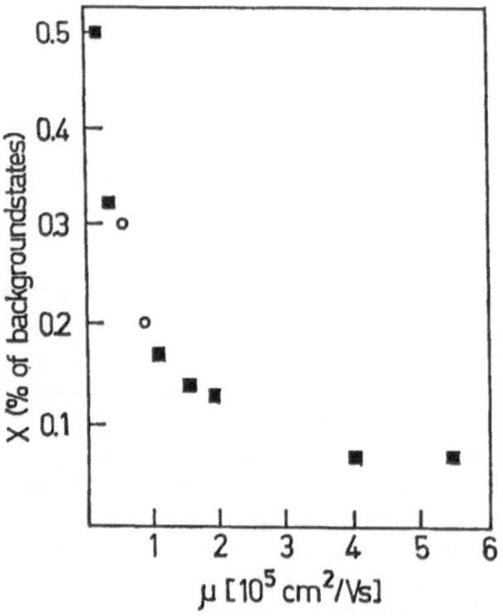

Fig. 5: Summary of determined background density of states in % of the B = 0 density.
(■) - from temperature dependent ρ_{xx};
(o) - from specific heat;

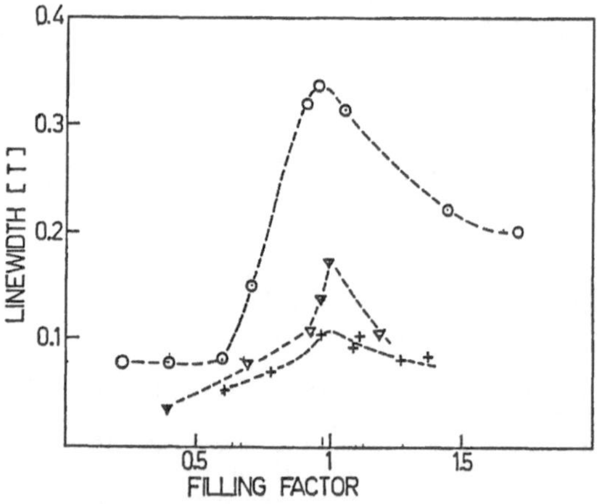

Fig. 6: Measured cyclotron resonance transmission linewidth as a function of filling factor for three different samples:
(∇) $n_s = 1.2 \times 10^{11} cm^{-2}$, $\mu = 4.5 \times 10^5 cm^2/Vs$
(o) $n_s = 1.2 \times 10^{11} cm^{-2}$, $\mu = 1.2 \times 10^5 cm^2/Vs$ data taken from Ref. 17
(+) $n_s = 2.3 \times 10^{11} cm^{-2}$, $\mu = 1.0 \times 10^6 cm^2/Vs$

3. Cyclotron Resonance Spectroscopy

While there is consistent information on the background, the value of the Gaussian width Γ_G and its dependence on the zero-field mobility has not been derived yet in a satisfactory way. One technique which should give information on changes of Γ_G is the analysis of the cyclotron resonance linewidth. In cyclotron resonance (CR) the observed linewidth of the transmission spectrum is determined by the broadening of both, the initial and final LL. The linewidth contains information on the individual level width Γ but not in a trivial way. In the techniques described in the previous chapters the pure LL width was extracted from experiments which were not influenced by scattering processes or transitions between different LL; in the analysis the level width was always assumed to be the same for all LL. It is clear that cyclotron resonance will only give good values of Γ_G after a careful theoretical fit of the data. However, tendencies of Γ_G as a function of certain external parameters for a situation where no significant physical quantity changes can be determined.

Previous experimental /17-20/ and theoretical /2,6/ investigations have revealed a filling factor-dependent CR linewidth due to screening. Maxima of the linewidth occur at integer values of the filling factor and minima in between.

Fig. 6 shows a plott of the measured CR linewidth Γ_{CR} as a function of filling factor for three different samples at a temperature of 4.2 K. All samples show a clearly defined maximum of the linewidth for $\nu = 1$. A systematic behavior of the linewidth value at $\nu = 1$ with the sample mobility is evident, while the linewidth for a filling factor of 0.5 seems not to be correlated clearly to the mobility. There is also clear evidence for a temperature dependence of the linewidth but we only want to compare here data at one temperature.

To get information on D(E) from these data we have to make the assumption that Γ_{CR} at integer filling factors is directly correlated to the Gaussian level width Γ_G. This correlation is demonstrated in Fig. 7 where the Γ_{CR} at $\nu = 1$ is plotted as a function of mobility. The linewidth decreases according to a slope of 1/2 with increasing mobility. If we plot the percentage of background states from Fig. 6 on the same double logarithmic scale we find a very interesting result: The background shows the same dependence on mobility as the linewidth maxima in the CR.

At integer filling factors Γ_{CR} is due scattering on ionized impurities /2,6/. The influence of screening is weak therefore we believe that we can correlate Γ_{CR} with the mobility and thus with the number of ionized impurities Ni: the zero-field mobility is limited by single scattering events at the impurities /21/, therefore the scattering time τ ($\mu = e\tau/m^*$) is inversely proportional to Ni for a given carrier concentration. We believe that the plot over μ in Fig. 7 is meaningful, since the samples we investigated had low carrier concentration only varying between 1.2×10^{11} and $2.4 \times 10^{11}\,cm^{-2}$. In addition we find that the data points which obey most precisely the slope (-1/2) are from samples with the same density and considerably different mobilities (f.e. $n_s = 1.2 \times 10^{11}\,cm^{-2}$ and $\mu = 1.0 \times 10^5\,cm^2/Vs$ and $4.0 \times 10^5\,cm^2/Vs$).

From Fig. 7 we can draw therefore the conclusion that the Gaussian peaks Γ_G and the amount of background density of states x are both correlated with the number of impurities in the sample. A dependence $\Gamma_G \propto \sqrt{Ni}$ and $x \propto \sqrt{Ni}$ can be derived from Fig. 7. A behavior $\Gamma_{CR} \propto \sqrt{Ni}$ has been found in high

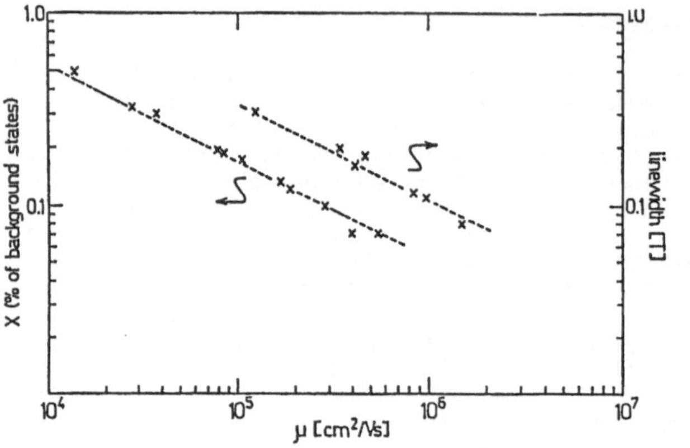

Fig. 7: Cyclotron resonance linewidth at filling factor ν = 1 and percentage of background density of states as a function of sample mobility (at zero magnetic field)

purity bulk semiconductors where long-range impurity potentials dominate /11/. Since we find the same dependence in the 2D system, long-range impurity potentials have to be responsible for Γ_{CR}, Γ_G and x as well in our situation, where screening is weak (ν.. integer). The origin of these potentials is in our opinion ionized impurities in the high-doped GaAlAs separated from the 2D gas by the spacer. However, the impurities in the spacer and in the GaAs (close to the channel) may also influence the level width since they scatter more effectively. It is therefore difficult to determine the impurities which play the dominant role.

What can we say about the form of D(E) for half-filled LL? From Fig. 6 a strong decrease in Γ_{CR} is found which does not correlate with mobility. According to Ref. /2/ and /6/ the linewidth is determined for half-filled LL by the residual doping in the GaAs. The long-range impurities are completely screened. We mentioned before that the specific heat and $\rho_{xx}(T)$ data do not give useful information in this situation. However, we have found that the magnetization results, which are analysed at non integer-filling factors, do not give evidence for a background density of states. Combining the magnetization and CR results we can speculate that the background density of states has to be influenced by screening for half-filled LL in the same way as Γ_{CR} and thus Γ_G. This means that we expect the background density of states and the Gaussian level width to oscillate with filling factor. This has not been proven experimentally yet. A good method might be the capacitance technique which is most sensitive at half-filled LL /10/.

ACKNOWLEDGEMENT

The work was partly supported by the Stiftung Volkswagenwerk (Projekt I/61 840). We would also like to thank Dr. R. Lassnig and Prof. W. Zawadzki for helpful and critical discussions.

REFERENCES

1. T. Ando, Y. Uemura: J. Phys. Soc. Jap. 36, 959 (1974)
2. R. Lassnig, E. Gornik: Solid State Commun. 47, 959 (1983)

3. R.R. Gerhardts: Z. Phys. B21, 275 (1975) and Surf. Sci. 58, 227 (1976)
4. F. Wegner: Z. Phys. B51, 279 (1983)
5. E. Brezin, D.I. Gross, C. Itzykson: Nuclear Phys. B235, 24 (1984)
6. T. Ando, Y. Murayama: J. Phys. Soc. Jap. 53, 693 (1985)
7. E. Gornik, R. Lassnig, G. Strasser, H.L. Störmer, A.C. Gossard, W. Wiegmann: Phys. Rev. Lett. 54, 1820 (1985)
8. J.P. Eisenstein, H.L. Störmer, V. Narayanamurti, A.Y. Cho, A.C. Gossard: Phys. Rev. Lett. 55, 875 (1985)
9. E. Stahl, D. Weiss, G. Weimann, K. v. Klitzing, K. Ploog: J. Phys. C18, L783 (1985)
10. V. Mosser, D. Weiss, K. v. Klitzing, K. Ploog, G. Weimann: Solid State Commun. 58, 5 (1986)
11. E. Gornik: Physica 127B, 95 (1984)
12. W. Zawadzki and R. Lassnig: Solid State Commun. 56, 537 (1984)
13. E. Gornik, R. Lassnig, G. Strasser, H.L. Störmer, A.C. Gossard: Surf. Sci. 170, 277 (1986)
14. W. Zawadzki, R. Lassnig: Surf. Sci. 142, 225 (1984)
15. J.P. Eisenstein: Appl. Phys. Lett. 46, 695 (1985)
16. D. Weiss, K. v. Klitzing, V. Mosser: Springer Series in Solid States Sciences 67, 204 (1986)
17. Th. Englert, J.C. Maan, Ch. Uihlein, D.C. Tsui, A.C. Gossard: Physica 117B & 118B, 631 (1983)
18. W. Seidenbusch, R. Lassnig, E. Gornik, W. Weinmann: Physica 134B,314(1985)
19. R. Lassnig, W. Seidenbusch, E. Gornik, G. Weimann: Proc. Int. Conf. on the Physics of Semiconductors, Stockholm (1986)
20. G.L.I.A. Rikken, H.W. Myron, P. Wyder, G. Weimann, W. Schlapp, R.E. Horstman, J. Wolter: Surf. Sci 170 (1986)
21. W. Walukiewicz, H.E. Ruda, J. Lagowski and H.C. Gatos, Phys. Rev. B30, 4571 (1984)

Magneto-Luminescence in Modulation-Doped AlGaAs-GaAs Multiple Quantum Well Heterostructures

C.H. Perry[1]*, *J.M. Worlock*[2], *M.C. Smith*[3], *and A. Petrou*[4]

[1]Physik-Department der Technischen Universität München, D-8046 Garching, Fed. Rep. of Germany

[2]Bell Comm. Research, Holmdel, NJ 07733, USA

[3]Sandia National Laboratories, Albuquerque, NM 87185, USA

[4]Physics Department, SUNY at Buffalo, NY 14260, USA

The interband magneto-photoluminescence spectrum of n-type modulation-doped multiple quantum well heterostructures of $Al_xGa_{1-x}As$-GaAs has been studied in magnetic fields up to 19 Tesla. The fields were applied parallel to the growth axis. Free carrier Landau level transitions have been identified with Landau indices up to $\ell = 10$. The peaks are strictly linear with magnetic field at high temperatures and for high Landau states at all temperatures. From these patterns hole effective masses can be deduced.

At low temperatures, energy anomalies develop in the $\ell = 0$ and $\ell = 1$ interband transitions. Associated with these are pronounced intensity oscillations. Both are periodic in 1/B and appear to be empirically related to Landau level crossings with the lowest unoccupied conduction subband state and not the Fermi level. Above threshold fields of $\sim$ 5 Tesla excitonic Landau level transitions have been identified and become dominant over the free carrier transitions at higher magnetic fields. Binding energies of 3 - 4 meV of two-dimensional excitons are estimated for 250 Å well widths.

At energies of approximately 25 meV and 105 meV below the band gap, magneto-luminescence replicas of the Landau band-to-band transitions are observed. These are attributed to transitions from the Landau free carriers to residual acceptors of carbon and manganese respectively. Transitions to both ground and excited states have been studied as a function of magnetic field and the binding energies determined.

I. Introduction

Photoluminescence is known to be a powerful technique for the study of excitonic recombination in doped semiconductors. Semiconductor quantum wells formed from GaAs between AlGaAs have long been known to show strong excitonic behaviour in interband spectra /1,2/; however, until recently no interband magneto-optical studies have been carried out on these systems /3-5/. Application of a magnetic field in optical experiments not only allows excitonic recombination to be distinguished from free carrier recombination, but can also provide information on electron non-parabolicity, hole effective mass, and the Coulomb binding between electrons and holes etc.

Our samples were grown by MBE. Alternating layers of GaAs and Si-doped AlGaAs were laid down to form quantum wells with varying layer widths, aluminium content, carrier concentration and mobility. Resonant Raman scattering in high magnetic fields, utilizing the spin-split-off valence band has been the subject of earlier investigations by this group /6/. They

*) Permanent address: Physics Department, Northeastern University, Boston, Ma 02115, USA

yielded a conduction band effective mass, $m^*_{CB} = 0.068 \pm 0.002\ m_o$, in agreement with other published work.

Exciton recombination is prevalent in undoped wells of high purity /7/ as contrasted with bulk GaAs, in which intrinsic (impurity involved) excitation dominates. When excess electrons are introduced into the quantum wells by doping the AlGaAs, as in our samples, very few electrons remain in the barriers. These electrons screen the electron-hole interaction, so that in zero field the exciton recombination is suppressed and direct band-to-band (BB) recombination due to the free carriers dominates the luminescence spectrum.

When doped quantum wells are subjected to magnetic fields in the 0 - 20 Tesla range, the main BB luminescence breaks into discrete components /8-10/. The observed peaks correspond to radiative transitions between conduction band (CB) and valence band (VB) Landau levels of equal quantum number ℓ, where $\ell = 0, 1, 2...$ Their energy variation with magnetic field is strictly linear and in the ratio 1 : 3 : 5 : .... the slopes directly give the inverse reduced mass $(1/m^*_{CB} + 1/m^*_{VB})$.

We also observe replicas of the BB luminescence; these appear at lower energies ($\Delta E \simeq 25$ meV and $\Delta E \simeq 105$ meV) and have been interpreted as electronic transitions (BA) from CB to residual carbon and manganese acceptor states, respectively /11,12/. In a magnetic field the "BA replicas" break into two sets of lines associated with transitions from CB Landau levels to the ground state (A^o) and an excited state (A^{o*}) of the acceptor.

In the last part we report our findings of the effect of magnetic fields on the photoluminescence (PL) intensity or efficiency /13/. An overall increase in luminescence is observed with increasing axial field. Large oscillations in the intensity occur whose peaks and valleys are approximately periodic in 1/B. In addition, above $\sim$ 5 Tesla and under particular excitation conditions, another set of luminescence peaks become visible. These occur at energies slightly lower than the transitions between the lowest three Landau states by 2 to 5 meV. In many cases the entire PL intensity transfers from the $\Delta\ell = 0$ Landau transitions to these excitonic or weakly bound states above the threshold field.

II. Experimental

The photoluminescence in the MQW heterostructures was excited using discrete laser lines from Ar^+ and He-Ne lasers or a tunable IR dye laser in the 7000 - 9000 Å range. The samples were immersed in superfluid helium ($T \simeq$ 1.8 K) and were subjected to axial magnetic fields up to 19 Tesla in a conventional Bitter magnet. The emitted light was collected through the bottom window of the cryostat and focussed onto the entrance slit of a Spex double grating monochromator. A thermo-electrically cooled RCA3134A GaAs photomultiplier tube and the associated photocounting electronics were used for signal detection. Both the spectrometer and the dye laser output could be scanned independently by stepping motors under control of a DEC minicomputer, which allowed either emission or excitation spectroscopy to be undertaken as a function of magnetic field.

Two techniques were used to study the luminescence. The first involves scanning the spectrometer while keeping the magnetic field fixed; the second approach is to ramp the magnetic field while fixing the spectrometer at a constant energy. The former is more appropriate for observing transitions between Landau levels of low quantum number ℓ, while the latter proved particularly useful in studying high Landau transitions. The PL signal could alternatively be directed to a Spex triple monochromator equipped with an optical multichannel OMA detector. A single field ramp from 0 - 15 Tesla served to obtain a complete set of data between 1,507 - 1,544 eV; the result could be displayed as an intensity contour map.

III. Results and Discussion

a) Band-to-band Landau transitions

In this section we describe our PL results in an axial magnetic field, concentrating on the energetics of the main band-to-band recombination. We observed well-defined transitions between states of high Landau index ℓ whose energies fall in the ratio (1:3:5:7) as well as interesting anomalies from the lowest Landau levels.

Landau levels will decay radiatively with photon energies

$$E_{\ell\to\ell} = E_0 + ehB(\ell + \frac{1}{2})/\mu^* c \,, \qquad (1)$$

where $1/\mu^* = (1/m^*_{CB} + 1/m^*_{VB})$.
High index transitions (ℓ = 3 and above) are fitted using Eq.(1) with the band gap E_0 and the reduced effective mass μ^* treated as adjustable parameters.

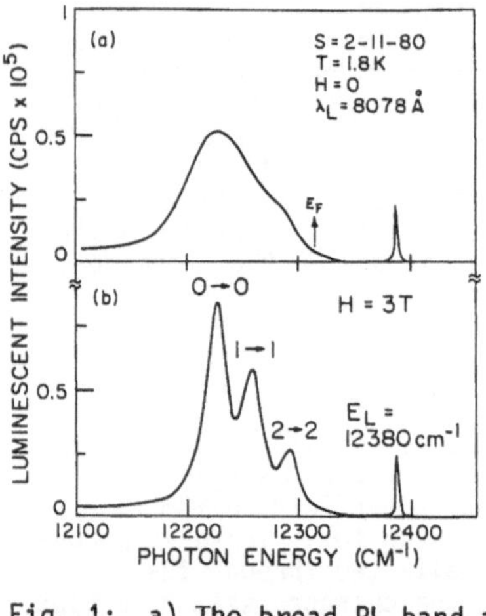

Fig. 1: a) The broad PL band at B = 0. b) Three peaks are observed at B = 3 Tesla corresponding to band-to-band transitions between Landau levels.

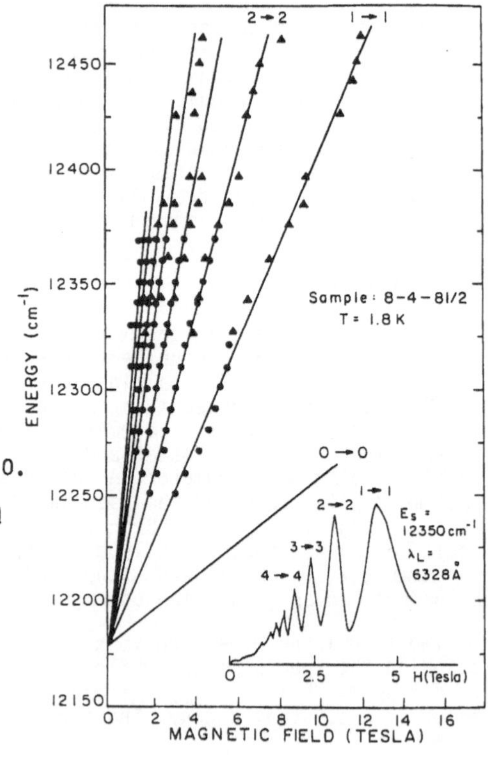

Fig. 2: Emission (●) and excitation (▲) energies of interband Landau transitions. The solid lines are a least squares fit with slopes in the ratio 1:3:5... An example of an emission-field scan is shown in the inset.

Fig. 1 shows the PL spectra for sample 2-11-80 at B = 0 and B = 3 Tesla. The broad BB zero-field spectrum breaks into definite peaks at finite fields corresponding to $\Delta\ell = 0$ transitions. Two different techniques (see experimental section) were used to obtain the $\ell = 1$ to $\ell = 8$ transitions displayed in Fig. 2 for sample 8-4-81/2. The solid lines were derived from a least squares fit to the data using Eq.(1), from which an effective hole mass $m^*_{VB} = 0.44\ m_0$ and a band-gap $E_0 = 1.410$ eV were derived. These parameters differed from sample to sample. For example the band gap of high purity GaAs is 1.5188 eV (12,250 cm^{-1}), whereas the origin of the Landau fans in our doped MQW heterostructures varies from 1.505 to 1.517 eV. Qualitatively this red shift could be similar to that observed in highly doped bulk crystals /15/. If m^*_{CB} is assumed constant /5/, the calculated values of m^*_{VB} vary from 0.32 to 1.0 m_0; this spread could be due to differences in electron concentration, well size and many-body effects (exchange and correlation energy).

At 70 K, only the lower two Landau levels are observed in sample 2-11-80 as shown in Fig. 3a). Their behaviour is linear over the entire range of fields investigated. Fig. 3b) shows data taken with the same excitation source (6328 Å He-Ne) but at T ≈ 2 K. Here we show only the lowest $\Delta\ell = 0$ transitions as compared with the plot in Fig. 2. The energy versus field fans show significant deviations from linear behaviour. In

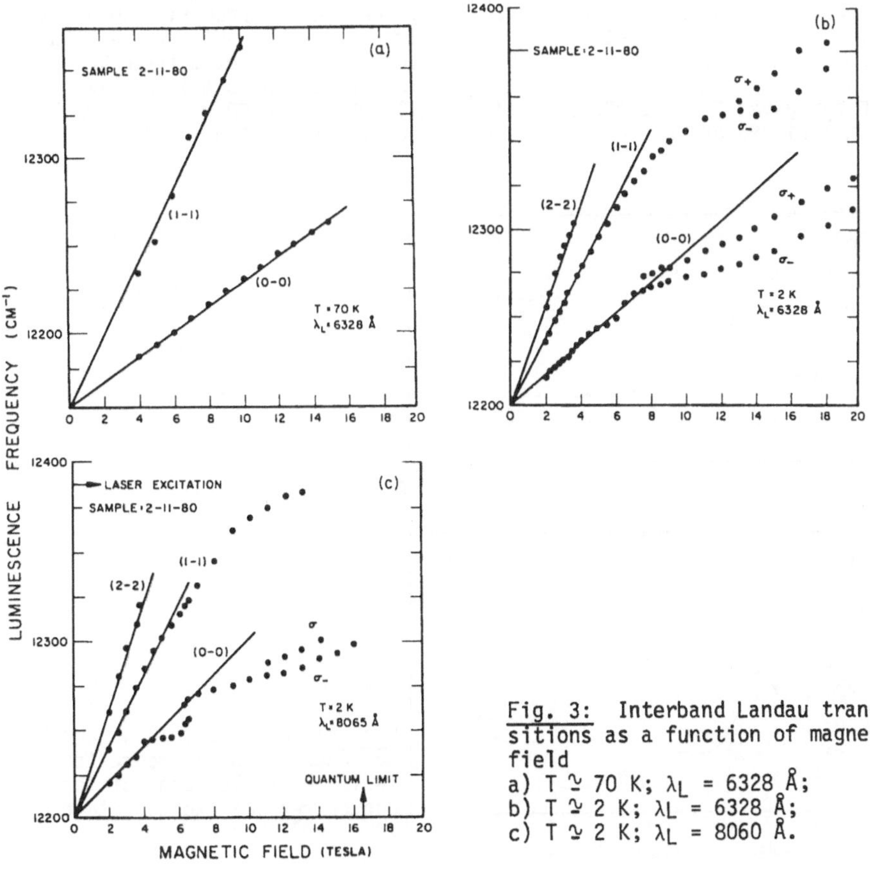

Fig. 3: Interband Landau transitions as a function of magnetic field
a) T ≈ 70 K; λ_L = 6328 Å;
b) T ≈ 2 K; λ_L = 6328 Å;
c) T ≈ 2 K; λ_L = 8060 Å.

addition the lines split into circularly polarized components, σ_+ and σ_-, due to the enhanced spin splitting of the two spin states in each Landau level /16/. If the laser excitation energy is decreased to just above the Fermi level (Fig. 3c)), the step-like discontinuities become more pronounced. At the quantum limit, where all the electrons are in the lower spin state, the σ_+ transition disappears. This non-linear behaviour in the lower Landau transitions is discussed in a later section as it is related to field-dependent intensity oscillations and the appearance of excitonic states /14/.

b) Magneto-luminescence due to acceptor states

All the modulation-doped MQW samples that we have studied exhibit features in the PL below the main BB peak; they are attributed to band-to-acceptor (BA) recombination. The intensity of the BB and BA transitions are comparable when the excitation is only a few meV above the Fermi level. At higher excitation energies the replica is characteristically smaller, generally only a few per cent of the intrinsic BB luminescence. Since the electrons involved in the transition are presumably the same, the intensity ratio reflects the relative population of holes in the valence band vis-a-vis the acceptor. The spectrum of sample 8-4-81/3, which has only one major peak at zero field, is shown in a field of 7 Tesla in Fig. 4a); a swept field scan is shown in Fig. 4b). Here transitions to A^{0*} from five different conduction band Landau levels are visible. Transitions to A^0 were considerably broader and only $0 \rightarrow A^0$ could be resolved. A summary plot from BB, BA^{0*}, and BA^0 in a magnetic field is shown in Fig. 5. The origin of the CB → VB fan gives E_0 = 12,190 cm^{-1} (1.511 eV), while the CB → A^{0*} is at 12060 cm^{-1} (1.495 eV) and CB → A^0 is estimated to be at 11980 cm^{-1}

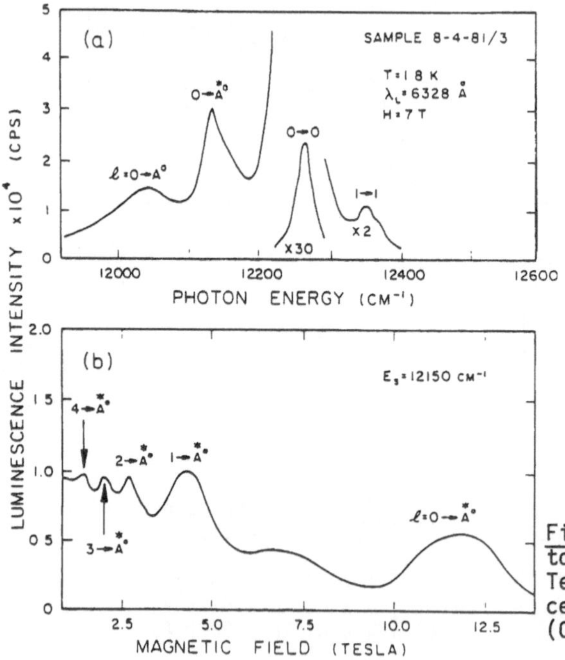

Fig. 4: a) Band-to-band and band-to-acceptor luminescence for B = 7 Tesla. b) Field scans of the acceptor replica excited state (CB - A^{0*}).

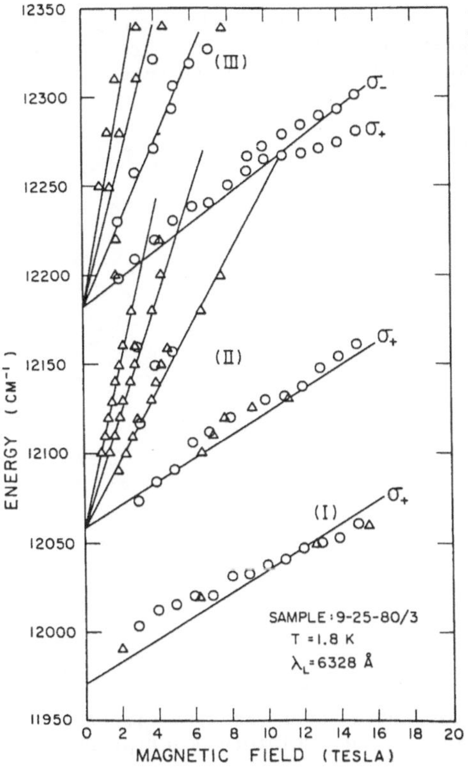

Fig. 5: Landau fans I and II are associated with the graphite acceptor for (CB → A^0) and (CB → A^{0*}) respectively; fan III is due to inter-band Landau transitions. O: spectrometer scanned at fixed magnetic field. Δ: B-field scans at fixed luminescence energy. Solid lines show 1:3:5... ratios.

(1.486 eV). From these values the ground state and excited state binding energies were found to be ∿ 25 meV and ∿16 meV, respectively. Miller et al. /17/ have observed BA photoluminescence in undoped quantum wells. We follow them in interpreting our shallow acceptor state to be residual carbon in the GaAs layer.

It is interesting to note that at high fields, while the (0 → 0) BB luminescence splits into two circularly polarized components σ_- and σ_+ as shown in Fig. 5, both 0 → A^0 and 0 → A^{0*} show only one component polarized as σ_+. From intensity studies of CB → A^{0*} and CB → A^0 as a function of laser excitation energy E_L, we conclude that hot electrons interact with the acceptor and populate the excited state A^{0*}.

We also observe transitions to a deep acceptor state located approximately 105 meV below E_0. In some samples, the BA replica associated with this impurity is also comparable to the BB luminescence. Both CB → A^0 and CB → A^{0*} transitions are observed; their relative intensities again depend on E_L. A composite plot incorporating the CB-acceptor (A^0 and A^{0*}) Landau transitions is shown in Fig. 6. On the basis of a least squares fit (solid lines) each fan has transitions in the ratio (1:3:5:...). The origins of the fans are at 11650 cm^{-1} (1.445 eV) for A^{0*}, 11340 cm^{-1} (1.406 eV) for A^0, and 12180 cm^{-1} (1.510 eV) for E_0 (not shown in Fig. 6). The binding energy (BE) for A^0 is ∿ 104 meV and ∿ 65 meV for A^{0*}. Previous work /18/ on manganese-doped bulk GaAs gives BE(A^0) = 113 meV. This acted as a basis for

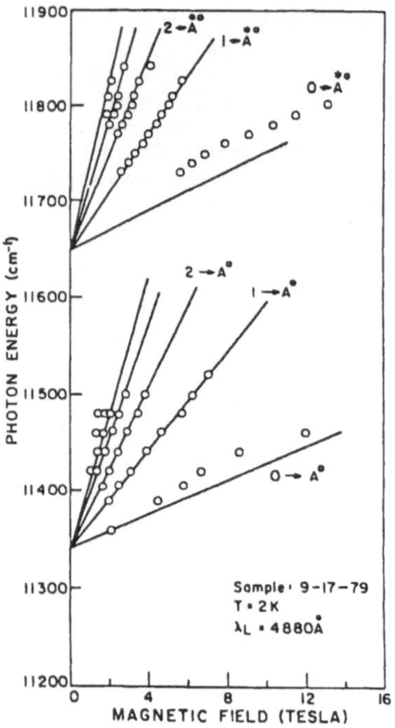

Fig. 6: PL peaks due to band-manganese acceptor transitions, obtained using field scans. The solid lines are fits to the data.

identifying the deep impurity as manganese and was confirmed by circular polarization studies in the magnetic field. A more complete description is given in the work by Petrou et al. /12/.

c) Magneto-Luminescence Intensity Oscillations

More recently we have concentrated on the effect of magnetic fields on the PL intensity or efficiency /13/. We observed that the total luminescence increased with field and exhibited strong oscillations, especially when the electron temperature was low. For these studies we used a variety of laser sources between 4880 - 8200 Å and the luminescence was detected using the optical multichannel analyzer system described in the experimental section. A typical PL-intensity contour map with magnetic field is shown in Fig. 7. Such data provided the peak intensities of the various $\ell \rightarrow \ell$ transitions as well as the total integrated intensity. An example is shown in Fig. 8.

The intensities were found to vary in a periodic manner with the maxima (minima) being periodic in 1/B for all the samples investigated, with the $0 \rightarrow 0$ transition being the major contributor. Plots of $1/B_i$ versus i are shown in Fig. 9, where the index i labels the intensity peak and B_i is the field at which the ith peak occurs. The parallel resistivity, ρ_{xx} shows a different 1/B periodicity from the luminescence oscillations even though the optical and electrical measurements were made simultaneously. Comparison between the two results can be seen in Fig. 10. From the Shubnikov-de Haas result (slope of 1/B plot) the derived Fermi energy E_F is in good agreement with calculated value from the known carrier concentration. An equivalent energy E_E can be obtained from the intensity data

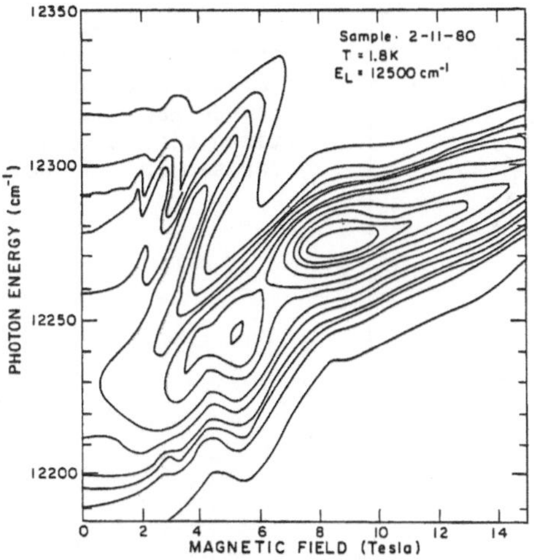

Fig. 7: Intensity contours for sample 2-11-80 at T $\simeq$ 1.8 K, with E_L = 12,500 cm^{-1}. Bulges to the low energy side of the 0 → 0 are seen where the energy step was observed in Fig. 3c). Peaks in the intensity, indicated by island-shaped contours, occur at the same field values.

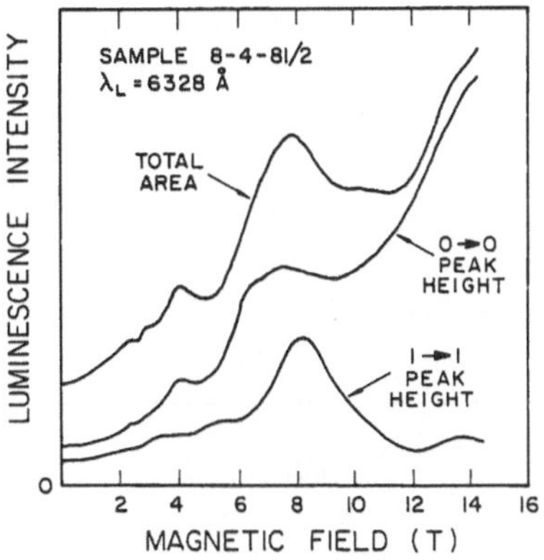

Fig. 8: Variation of the total integrated intensity and peak heights for the 0 → 0 and 1 → 1 BB Landau transitions with magnetic field for sample 8-4-81/2.

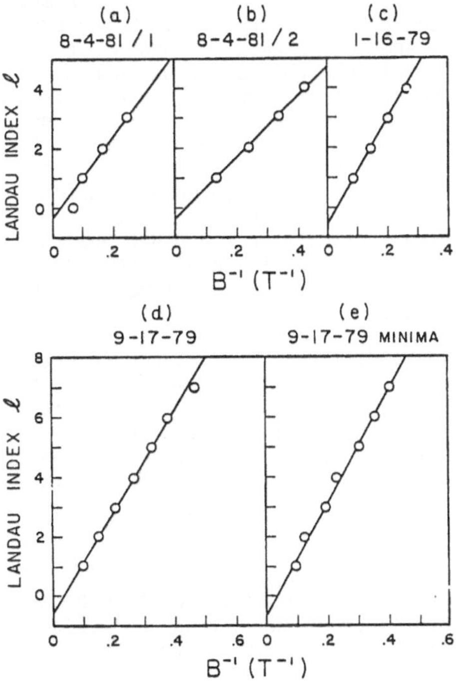

Fig. 9: Plots of the intensity peak index i versus 1/B for five samples.

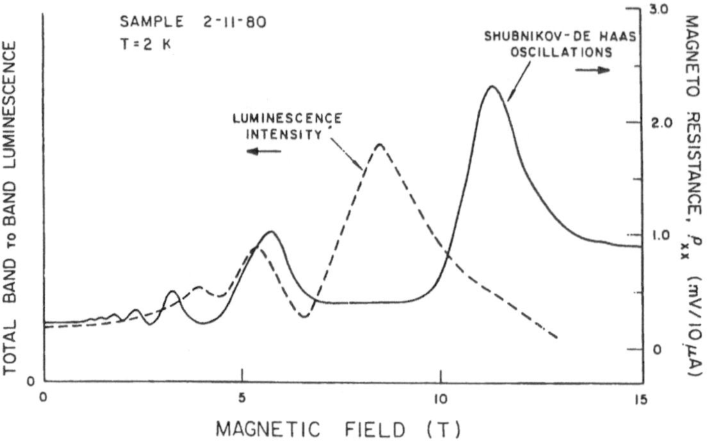

Fig. 10: Comparison of intensity oscillations with Shubnikov-de Haas data taken simultaneously for sample 2-11-80.

in a similar manner by multiplying the slope of the 1/B plots by $he/c\mu^*$.

There is no correspondence between E_E and E_F for any of the samples investigated. However, there appears to be an empirical relation between E_E and the lowest unoccupied conduction subband state $E_{n,o}$ /19,20/. The mechanism by which the magnetic field exerts a cylcic effect on the PL efficiency or on other competing non-radiative recombination processes remains to a large part unexplained.

Related to the $\ell \rightarrow \ell$ intensity oscillations are the appearance of a new series of Landau transitions. These are observed at energies of 2-5 meV, below $\ell = 1 \rightarrow 1$, $2 \rightarrow 2$, and $3 \rightarrow 3$ free Landau states. From their behaviour in a magnetic field we conclude that they are due to weakly bound donor states or evolve from subband-heavy hole excitonic recombination, normally only expected in undoped wells. Above threshold fields of $\sim$ 5 Tesla it is possible to shift the entire PL intensity from the free Landau transitions to these excitonic transitions. An example of field scans taken at two different emission energies is shown in Fig. 11. A composite plot of the complete Landau fan with the band-to-band ($\Delta\ell = 0$) and the hydrogenic states (n = 1, 2, 3) superimposed is given in Fig. 12. The complexity of the system makes quantitative conclusions quite difficult, but further analysis should provide the binding of excitons as a function of Landau level and magnetic field. In zero field the exciton binding energy is estimated to be 3-4 meV for samples with $\sim$250 Å well-widths.

Based on the binding energies alone it is difficult to distinguish between donor-bound state recombination and recombination with excited states of the exciton. The cyclic variation of the overall luminescence intensity with magnetic field appears to favor modulation of the rate of decay of highly excited excitons; this subject continues to be explored both theoretically and experimentally (now up to 40 Tesla) by this group in modulation-doped MQW heterostructures.

IV. Summary

Photoluminescence spectroscopy in high magnetic fields has proved to be an informative tool for investigating the behaviour of two-dimensional electrons in modulation-doped, n-type GaAs-$Al_xGa_{1-x}As$ MQW heterostructures.

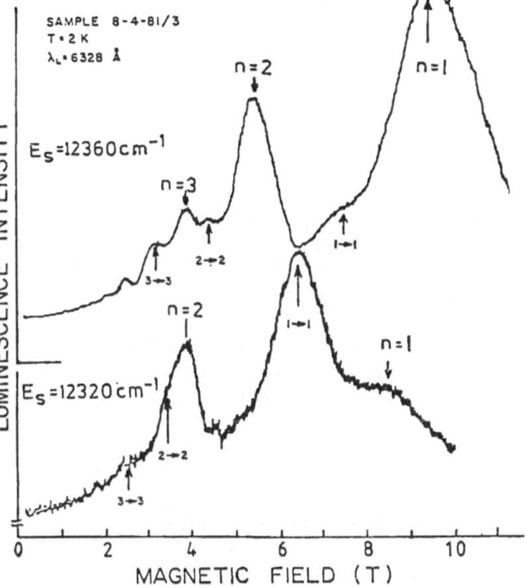

Fig. 11: The intensity variation with swept magnetic field for E_S = 12320 cm^{-1} and E_S = 12360 cm^{-1}. In the top spectrum, the excitonic excitation dominates the $\Delta\ell$ = 0 band-to-band transitions.

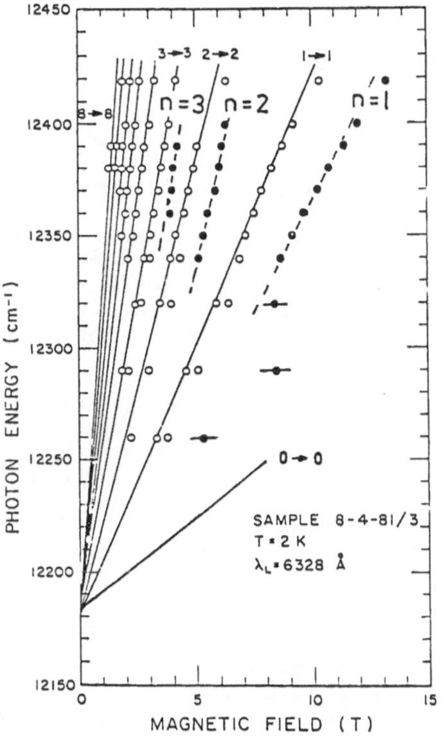

Fig. 12: The observed $\Delta\ell$ = 0 free Landau transitions for ℓ = 1,2,3,... and the superimposed n = 1,2, and 3 excitonic transitions observed above ∿ 5 Tesla for sample 8-4-81/3.

Three types of field-dependent excitations are observed: One corresponds to transitions from Landau levels in the conduction band to Landau levels in the valence band (band-to-band transitions with $\Delta\ell$ = 0). From these patterns, hole-effective masses m^*_{VB} and band-gap values E_0 could be deduced. Anomalies are observed in the ℓ = 0 and ℓ = 1 transitions at low electron temperatures. The second type involves evidence for the existence of weakly bound states or excited excitonic states that exist alongside the free carrier luminescence. Intensity studies indicate that they can dominate the spectrum at high fields. Thirdly, magneto-photoluminescence associated with residual acceptor impurity states has been observed. Landau fans emanating from transitions between conduction band Landau levels and the ground and excited states were used to derive the binding energies and the chemical species. Finally, variation of the luminescence intensity with a periodic 1/B dependence occurs in all the transitions observed which are related to competing radiative and non-radiative mechanisms.

Acknowledgements: We wish to thank A.C. Gossard and W. Wiegmann and G. Weimann for the MQW samples and their continued interest in this work. We are pleased to acknowledge helpful discussions with D.M. Larsen and J. Wornock and to thank L. Rubin and the staff of the National Magnet

Laboratory for their hospitality and cooperation. Finally CHP benefited from interactions with F. Koch, G. Abstreiter and their group while at TU München. This work was supported by NSF Grant DMR-8121702.

References:

1. R. Dingle, W. Wiegmann, and C.H. Henry, Phys.Rev.Lett. 33, 827 (1974)
2. R.C. Miller, A.C. Gossard, D.A. Kleinmann, and O. Munteau, Phys. Rev. B 29, 3740 (1984)
3. S. Tarucha and H. Okamoto, Solid State Comm. 52, 815 (1984)
4. J.C. Maan, G. Belle, A. Fasolino, M. Alterelli, and K. Ploog, Phys. Rev. B 30, 2253 (1984)
5. W. Ossau, B. Jäkel, E. Bangert, G. Landwehr, and G. Weimann, 2nd Int. Conf. on Modulated Semiconductor Structures, Kyoto, 1985, p. 120
6. J.M. Worlock, A.C. Maciel, C.H. Perry, R.L. Aggarwal, A.C. Gossard, and W. Wiegmann: In Application of High Magnetic-Fields in Semiconductor Physics, ed. by G. Landwehr (Springer, Berlin, Heidelberg 1983) p.186
7. C. Weisbuch, R.C. Miller, R. Dingle, A.C. Gossard, and W. Wiegmann, Solid State Comm. 37, 219 (1981)
8. J.M. Worlock, A.C. Maciel, A. Petrou, C.H. Perry, R.L. Aggarwal, M. Smith, A.C. Gossard, and W. Wiegmann, Surf. Sci. 142, 486 (1984)
9. C.H. Perry, A. Petrou, M.C. Smith, J.M. Worlock, and R.L. Aggarwal, J. Luminescence 31&32, 491 (1984)
10. M.C. Smith, A. Petrou, C.H. Perry, J.M. Worlock, and R.L. Aggarwal, 17th Int. Conf. on Physics of Semiconductors, ed. by J.D. Chadi and W.A. Harrison (Springer, New York 1984), p. 547
11. A. Petrou, M.C. Smith, C.H. Perry, J.M. Worlock, and R.L. Aggarwal, Solid State Comm. 52, 93 (1984)
12. A. Petrou, M.C. Smith, C.H. Perry, J.M. Worlock, J. Warnock, and R.L. Aggarwal, Solid State Comm. 55, 865 (1985)
13. M.C. Smith, A. Petrou, C.H. Perry, and J.M. Worlock, 2nd Int. Conf. on Modulated Semiconductor Structures, Kyoto 1985, p. 56
14. M.C. Smith, C.H.Perry, A. Petrou, and J.M. Worlock, Bull.Am.Phys.Sol. 31, 558 (1986)
15. V.A. Vilkotskü, D.S.Domanevskii, R.D. Kakanakov, V.V. Krasovskii, and V.D. Tkachev, phys.stat.solidi(b) 91, 71 (1979)
16. Th.Englert, D.C.Tsui, A.C.Gossard, and Ch.Uihlein, Surf.Sci. 113. 295 (1982)
17. R.C.Miller, A.C.Gossard, W.T.Tsang, and O.Munteau, Phys.Rev. B 25, 3871 (1982)
18. M. Ilegems, R. Dingle, and L.W. Rupp, Jr., J.Appl.Phys. 46, 3059 (1975)
19. Z.-J. Tien, PhD Thesis, Northeastern University, 1981 (unpublished)
20. M.C. Smith, PhD Thesis, Northeastern University, 1985 (unpublished)

Magneto-Luminescence in GaAs-(GaAl)As Superlattices

W. Ossau, B. Jäkel, and E. Bangert

Physikalisches Institut der Universität Würzburg,
Röntgenring 8, D-8700 Würzburg, Fed. Rep. of Germany

1. INTRODUCTION

During the past few years it has become possible to grow systems consisting of alternate layers of different semiconductors with controlled thicknesses and relatively sharp interfaces. These one-dimensional periodic structures, generally referred to as superlattices, are most extensively studied in the system built with alternate layers of GaAs and (GaAl)As. It has been reported by many authors that due to the quantum size effect excitons have a character different from that in bulk material /1/. The exciton becomes quasi-two dimensional, its binding energy is enhanced, the degeneracy of heavy- and light-hole band is removed, resulting in two exciton systems.

Recent studies /2,3,4/ have revealed discrepancies between theory and experiment, suggesting that a more rigorous exciton model is required to describe exciton states in quantum wells. It is necessary to take into account:

the off-diagonal elements of the exciton-Hamiltonian,
the penetration of the wavefunctions into the barrier and - resulting from that - the mismatch between well- and barrier masses of electrons and holes.

This paper is organized as follows:
In Section 2 we report our experimental results obtained by photoluminescence measurements at low temperatures and high magnetic fields. Section 3 illustrates the calculations performed with the aim to deduce the hole-subband dispersion relation as a function of the well width. In addition we present variational calculations for the binding energy of the exciton ground state as well as for the transverse extension of the exciton in a quantum-well. Finally Section 4 is devoted to the comparison between theory and experiment.

2. EXPERIMENTAL RESULTS

Low-temperature photoluminescence experiments are performed on several GaAs-quantum wells with thicknesses between 27 and 300 A in high magnetic fields up to 9.5 T. The superlattices were grown by molecular beam epitaxy and consist of up to 110 GaAs layers, sandwiched between (GaAl)As barriers. The width of the barriers is rather large, in order to ensure decoupling of the individual GaAs wells. The high quality of the samples, which were provided by Dr. G. Weimann of the FTZ of the Deutsche Bundespost, made it possible to observe excitonic lines with linewidth less than 0.2 meV .

Fig. 1 shows a typical luminescence spectrum of a quantum well with 180 A thickness at B = 0 T. We observe excitons formed between the lower lying

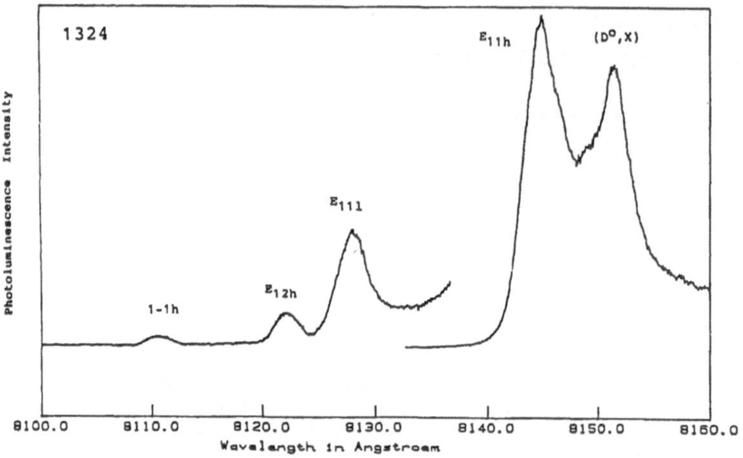

Fig. 1: Luminescence spectrum at $T = 1.8$ K and $B = 0$ T for a sample with well width $L_z = 180$ Å.

electron subbands and the heavy, light and excited hole subbands. In addition we identify interband transitions and excitons bound to defects /5/. All these lines have an individual magnetic field dependence which can be recognized in fig. 2. The excitonic lines E(11h), E(111) and E(13) show a nonlinear energy shift and magnetic field splitting, which is small for the heavy-hole exciton and observable only at temperatures below 2 K. Furthermore we observe lines which are more affected by the magnetic field, they are interpreted as interband (Landau) transitions. Several authors /2,6/ have fitted these linear magnetic field shifts employing an equation usually applied to interband transitions between nondegenerate bands:

$$E_{1,1} = E_o + e\hbar B(1 + \tfrac{1}{2})/\mu \,, \qquad 1/\mu = (1/m_e^* + 1/m_{VB}^*) \,. \qquad (1)$$

In these works the reduced mass is used as a fitting parameter leading to drastically enhanced hole masses greater than one m_o. In order to interpret our data - as well as to get insight into the mechanism of excitonic effects on lower lying transitions - we have calculated the hole-subband levels as a function of magnetic field (Sec. 3/A).

We have systematically studied the diamagnetic shift of the heavy-hole exciton and its dependence on the well width. In three dimensions the energy shift of an isotropic exciton ground state has been calculated for a wide range of field strengths /7/. By a change from three to two dimensions the ground-state shift is drastically reduced,as will be shown for the low-field case subsequently.

Generally the change in energy of the ground state is given by:

$$\Delta E_{hh}(1s) = \frac{e^2B^2}{8\mu} \langle \Psi | \varrho^2 | \Psi \rangle \,, \qquad \varrho^2 = x^2 + y^2 \,. \qquad (2)$$

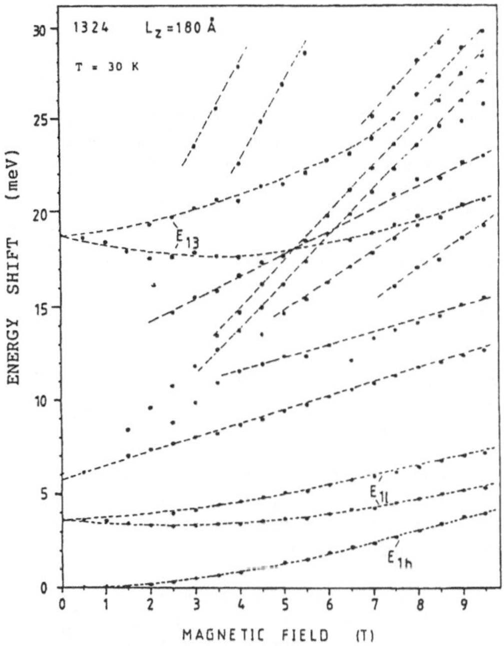

Fig. 2: Relative shift of photon energy as function of magnetic field. The sample temperature is 30 K and L_z = 180 Å.

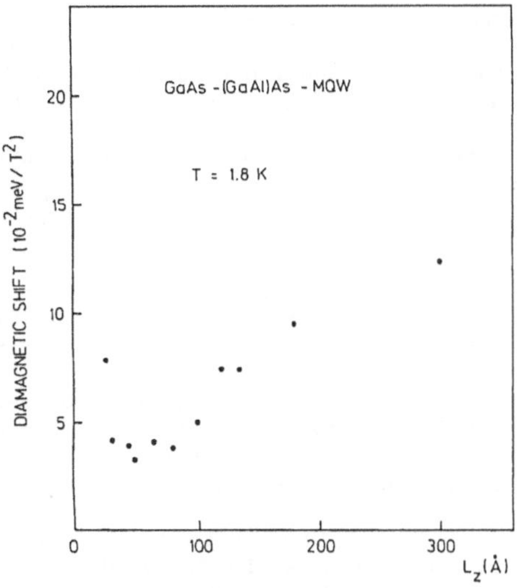

Fig. 3: Diamagnetic shift of the heavy-hole exciton in dependence on the well thickness.

All the symbols have their usual meaning. The term in brackets indicates the extension of the exciton parallel to the interface. In three dimensions this value is 2 a_B^2 , where a_B is the effective Bohr radius of the exciton. By reducing the dimensionality to two its value becomes 3/8 a_B^2 . This means that the diamagnetic shift of two-dimensional excitons is 16/3 times smaller than that of 3D ones.

In fig. 3 we have plotted the diamagnetic shift of the heavy-hole exciton as a function of the well width L_z. As expected the diamagnetic shift becomes smaller with decreasing well thickness. However for L_z about 80 Å the data reach a minimum and then increase rapidly with further decreasing L_z. We have explained this increase of the diamagnetic shift by the penetration of the exciton wavefunction into the adjacent barriers /4/. This means that the exciton progressively loses its quasi-two dimensional character. The decreasing of ΔE_{hh}(1S) for large well widths is explained by several groups /3,4/ employing three-dimensional wavefunctions for the exciton confined in barriers of infinite height. To take into account the penetration of the wavefunctions into the barriers we have extended the variational calculations of the binding energy of the ground state exciton done by Bastard et. al /8/ to the case of quantum wells with finite barrier height (Sec. 3/B).

3. THEORY

A) Hole subbands and Landau-levels

To compare the interband transitions between the hole and electron subbands we need realistic calculations of the magnetic field dependence of these subbands. As we are interested only in small energy shifts of the interband transitions, we treat the Γ_6-conduction band and the Γ_8-valence bands separately and neglect interactions with the Γ_7-spin-split-off valence band.

The energy shift of the first electron subband is given by

$$\Delta E_c = (n + \tfrac{1}{2}) \frac{\hbar e}{m_e} B \pm g_e \mu_B B \, . \qquad (3)$$

All symbols have their usual meaning. Although the split-off band is not included, its effect on the effective mass m_e and g-factor can be introduced following the treatment by Roth et al. /9/.

The degeneracy of the valence band and the effect of confinement in a one-dimensional potential well produce interesting features of the hole subbands. It has been shown that the dispersion of the subbands in the direction parallel to the well is strongly nonparabolic /10/.

In the neighbourhood of the Γ-point the upper valence band is described by the well-known Luttinger Hamiltonian /11/. In the basis of the four Γ_8 states with $J = 3/2$ and $J_z = +3/2, +1/2, -1/2, -3/2$ respectively it reads:

$$H = \begin{pmatrix} a_+ & b & c & 0 \\ b^* & a_- & 0 & c \\ c^* & 0 & a_- & -b \\ 0 & c^* & -b^* & a_+ \end{pmatrix} , \qquad (4)$$

where

$$a_\pm = E_v - \tfrac{1}{2}(\gamma_1 \mp 2\gamma_2) k_z^2 - \tfrac{1}{2}(\gamma_1 \pm \gamma_2)(k_x^2 + k_y^2) + J_z \varkappa \mu_B B,$$

$$b = \sqrt{3}\, \gamma_3 (k_x - i k_y) k_z \, ,$$

$$c = \frac{\sqrt{3}}{2} \left[\gamma_2 (k_x^2 - k_y^2) - 2i \gamma_3 k_x k_y \right] ;$$

$\gamma_1, \gamma_2, \gamma_3$ are material parameters describing the hole effective masses. These parameters have different values in GaAs and in (GaAl)As. The energy-gap difference between the two materials is fairly well known and expressed as /12/:

$$\Delta E_g = 1.155\, x + 0.37\, x^2 \ \text{eV}.$$

A recently much debated question is how this difference is distributed between valence- and conduction-band edges. Values for the band offset of the valence band between 15 % and 40 % of ΔE_g have been discussed.

The Luttinger-Hamiltonian is modified to include the external magnetic field in z-direction in the following way:

1. Terms are added which arise from the Zeeman hole-spin coupling

$$\varkappa \mu_B J_z B + q \mu_B J_z^3 B ,$$

where J_z is the spin 3/2 matrix $\varkappa$ and q are material parameters. Actually q is very small and therefore neglected.

2. The harmonic oscillator representation for the motion in the (x,y)-plane is included based upon the commutation rules of the k_x and k_y components /13/.

Then the wavefunctions can be written:

$$\psi_I = \sum_{i=1}^{4} \begin{pmatrix} A_{1i} \cos(k_{z_i} z) & \varphi_n \\ A_{2i} \sin(k_{z_i} z) & \varphi_{n+1} \\ A_{3i} \cos(k_{z_i} z) & \varphi_{n+2} \\ A_{4i} \sin(k_{z_i} z) & \varphi_{n+3} \end{pmatrix} ,$$

where φ_n are harmonic oscillator functions. The wavefunctions in the barrier are expressed with

$$\psi_{II} = \sum_{i=1}^{4} \begin{pmatrix} B_{1i} \, e^{-k_i z} & \varphi_n \\ B_{2i} \, e^{-k_i z} & \varphi_{n+1} \\ B_{3i} \, e^{-k_i z} & \varphi_{n+2} \\ B_{4i} \, e^{-k_i z} & \varphi_{n+3} \end{pmatrix} .$$

The boundary conditions are $\psi_I = \psi_{II}$ and $\frac{\partial H_I}{\partial k_z} \psi_I = \frac{\partial H_{II}}{\partial k_z} \psi_{II}$ at the interface.

B) Variational calculations

As explained above, the bandstructure of the valence band in zinc-blende structures is degenerate, therefore the Hamiltonian of a Wannier exciton is quite complicated. The additional confinement of the exciton in a superlattice makes the exciton problem much more difficult, so that an exact solution is not accessible at present. Many calculations of exciton binding energies in quantum wells have been performed in the last years / 8,14 /. Most of these calculations have treated the hole as particle with either light hole or heavy hole mass.

The current treatment is the following:

Due to the reduction in symmetry along the growth axis the degeneracy of the valence-band is removed as a consequence of the band discontinuities. In addition, contributions of the off-diagonal elements of the Luttinger-Hamiltonian are neglected. This leads to the formation of two separate exciton systems i.e. the heavy-hole ($J_z = \pm 3/2$) and the light-hole ($J_z = \pm 1/2$) exciton.

There are a lot of experimental data (including our own ones of the diamagnetic shift of the heavy-hole exciton), which cannot be explained in a satisfying fashion if the mixing of heavy- and light-hole states is neglected. Recently, few calculations of the exciton binding energy taking this mixing into account have been published / 15,16,17 /. However, the wavefunctions employed are very complicated. To understand the diamagnetic shift of the heavy-hole exciton we will follow the approach of two separate exciton systems and discuss corrections due to heavy- and light-hole mixing in Sec. 4.

For nondegenerate bands the exciton Hamiltonian reads

$$H = \frac{p_{z_e}^2}{2m_e} + \frac{p_{z_h}^2}{2m_h} + \frac{P_X^2 + P_Y^2}{2(m_e + m_h)} + \frac{p_x^2 + p_y^2}{2\mu} - \frac{e^2}{\kappa[x^2+y^2+(z_e-z_h)^2]^{1/2}} + V_{conf}(z_e) + V_{conf}(z_h) \tag{5}$$

with

$$(m_e + m_h)\vec{R} = m_e\vec{r}_e + m_h\vec{r}_h \ , \ \vec{r} = \vec{r}_e - \vec{r}_h \ ,$$

where $m_e, \vec{r}_e, m_h, \vec{r}_h$ are the effective masses and positions of electrons and holes respectively, κ is the relative dielectric constant, P the center-of-mass momentum and μ the reduced electron hole mass in the transverse direction. To obtain the energies of the excitons we apply a variational approach with the following trial function

$$\psi = N \cos\frac{\pi z_e}{L} \cos\frac{\pi z_h}{L} \exp\left[-\frac{1}{\lambda}\left[\varrho^2 + (z_e - z_h)^2\right]^{1/2}\right],$$

where N is a normalization constant and λ the trial parameter.

This nonseparable wavefunction has the advantage of only one variational parameter and provides a suitable interpolation between thin and thick wells. For barriers of infinite height the trial function appproaches the exact form when L_z goes to 0 and ∞ respectively. We minimize the energy as a function of λ and calculate the diamagnetic shift making use of the obtained wavefunction.

4. RESULTS AND DISCUSSION

The values of the reduced mass μ and the heavy-hole mass m_h used for the variational calculation are usually expressed in terms of the Luttinger parameters γ_i. If the Luttinger matrix is written in the basis of the eigenstates of J^2 and J_z and diagonalizes it for small $k_{\parallel} = (k_x, k_y)$ one will obtain for the heavy-hole anisotropic mass

$$\frac{1}{m_h} = \frac{1}{m_o}(\gamma_1 - 2\gamma_2) \ , \qquad \frac{1}{\mu} = \frac{1}{m_e} + \frac{1}{m_o}(\gamma_1 + \gamma_2) \ . \tag{6}$$

The values of γ_1, γ_2 and m_e give $m_h = 0.45\ m_o$, $m_{h\parallel} = 0.10\ m_o$ and $\mu = 0.040\ m_o$. We used these parameters for the variational calculation of the exciton ground state. Assuming an aluminium content of 40 % and a ratio of the band-gap discontinuities of 85/15 we obtained a diamagnetic shift drawn in fig.4 as a full line. The qualitative agreement between theory and experiment is obvious. The experimental findings - decreasing diamagnetic shift with decreasing L_z, a minimum near 70 A and subsequently increasing diamagnetic shift with further decreasing L_z due to the penetration of the wavefunctions into the barriers - are well described. However the quantitative agreement between theory and experiment is rather poor. Including the nonparabolicity of the conduction-band only produces a small improvement for narrow wells (broken line). For example, to describe the diamagnetic shift observed for a well with 180 A thickness quantitatively a reduced mass of 0.052 instead of 0.040 m_o is needed. This means that the parallel mass of the heavy-hole has to be 0.24 m_o instead of 0.1 m_o. Such drastically enhanced hole masses are employed in a variety of studies /2,3,6/ to explain the experimental data.

We try to overcome this discrepancy as follows:
Instead of the hole mass $m_o/(\gamma_1+\gamma_2)$, obtained from the diagonal terms of Luttinger's Hamiltonian only, we insert the hole mass of the 2D-hole

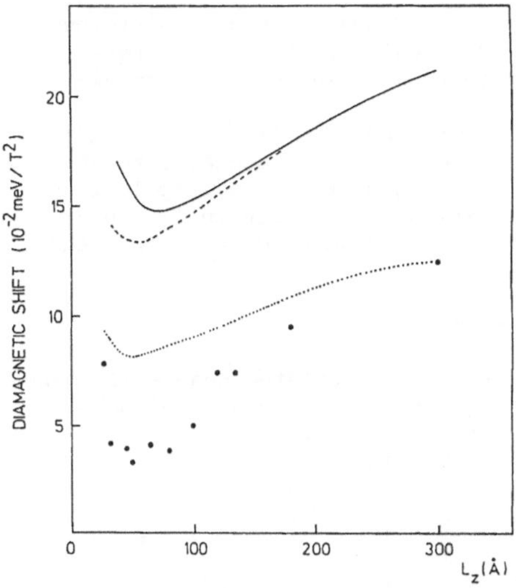

Fig. 4: Diamagnetic shift of the heavy-hole exciton as a function of the well width. The lines are calculated.

subbands. Since the subband dispersions result from the full 4*4 Hamiltonian, effects of the light-hole-heavy-hole coupling are incorporated into the exciton states by this procedure. This coupling between both bands gives rise to strong nonparabolicities /10/. To obtain numerical results for the heavy hole mass we used the relation

$$m_{h\|} = \frac{h^2}{2\pi}\left[dA(E)/dE\right], \qquad (7)$$

where A(E) is the area in the 2D-k-space enclosed by the contour of constant energy E. For simplicity we replaced γ_2 and γ_3 by $\bar{\gamma} = (3\gamma_3 + 2\gamma_2)/5$.

Fig. 5 shows the result for well widths of 180 Å. We find strong nonparabolicities which are more pronounced when the well width is large and the fraction of the band-gap discontinuity for the valence band is small. This

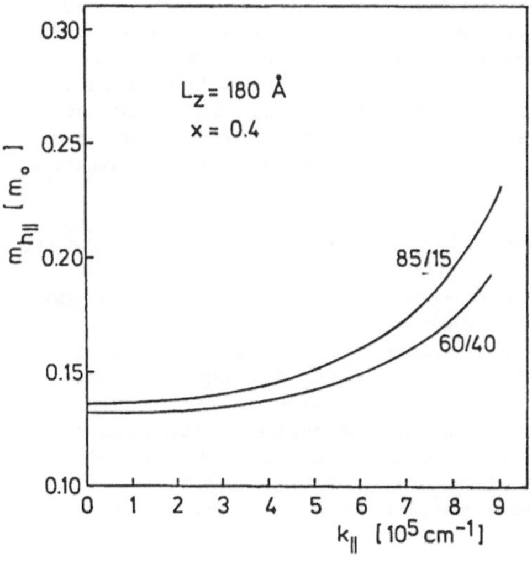

Fig. 5: Calculated heavy-hole mass m_h as a function of k .

fact is easy to explain, because for these conditions the energy difference between heavy- and light-hole subband is small, and therefore the interaction large. From fig. 5 it is easy to see that the value of the band-gap splitting is a crucial input parameter.

Another aspect of our calculation is that the heavy-hole mass for small $k_{\parallel}$ is enhanced compared with the value arising from the diagonal term of the Luttinger Hamiltonian. The mass values obtained from an extrapolation of $k_{\parallel}$ to 0 show a pronounced L_z dependence which can be seen in fig. 6. For very large well thicknesses only $m_{h\parallel}$ approaches the value deduced from the diagonal terms.

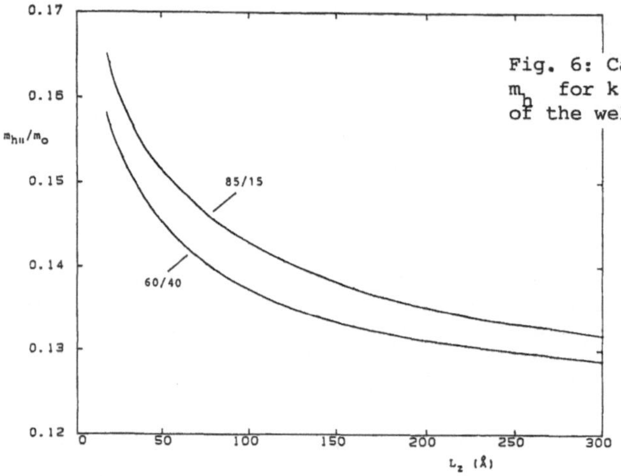

Fig. 6: Calculated heavy-hole mass m_h for $k = 0$ as a function of the well width.

In the framework of these numerical results the experimental data are discussed for a sample with 180 A well width. Assuming a 15 % fraction of the band-gap discontinuity for the valence band we calculate $m_{h\parallel}(k_{\parallel}=0) = 0.136\ m_o$ for $L_z = 180$ A, corresponding to a reduced mass $\mu = 0.045\ m_o$, which is still too small to describe the observed diamagnetic shift in a satisfying way. Of course $hk_{\parallel} = 0$ is not the relevant momentum for the exciton. If the center-of-mass momentum is zero, as assumed usually, the squared momentum of the hole is equal to the squared momentum of the relative motion in the layer plane. This means that the mean value of $k_{\parallel}^2$ is equal to $(p_x^2 + p_y^2)/\hbar^2$, which can be obtained from the variational calculations (equ. 5). With this rough estimate we calculate for the wavevector $k = 0.74\ 10^6\ cm^{-1}$ and the hole mass $m_{h\parallel} = 0.182\ m_o$ leading to a reduced mass $\mu = 0.049\ m_o$ which gives a better agreement with the diamagnetic shift observed for this sample.

Nonparabolic effects produce enhanced electron masses in the narrow well regime - heavy-hole masses, however, are larger for wide quantum wells. This results in a nearly constant reduced mass of the exciton for the whole L_z range studied ($0.048 \pm 0.002\ m_o$ for 60/40 and $0.050 \pm 0.002\ m_o$ for 85/15 splitting). We have calculated the diamagnetic shift as a function of the well width, considering the relevant wavevector for the heavy-hole exciton (dashed-dotted line in fig. 4). The dotted line in fig. 4 represents the values calculated taking into account the nonparabolicities of the conduction band. The quantitative agreement is better than that obtained with the reduced mass estimated from only the diagonal terms of the Luttinger-Hamiltonian. However, the very small diamagnetic shift near $L_z = 70$ A is not well described either.

There are several reasons for this discrepancy:

1) We take into account the mixing between heavy- and light-hole band by the use of a changed hole mass, but still consider heavy- and light-hole excitons as separate systems.
2) We have determined the heavy-hole mass using a spherical approximation for the Luttinger-Hamiltonian and an estimate for the relevant hole-momentum. Both treatments may produce incorrect mass values.
3) Besides the reduced mass of the exciton, its transverse extension is another important factor determining the quantity of the diamagnetic shift. The expectation value of ϱ^2 is dependent on the choice of the wavefunction. Therefore another trial wavefunction may lead to smaller diamagnetic shifts.
4) The nonparabolicity of the conduction band is more pronounced than assumed usually /12/.

We conclude that the incorporation of the mixing between heavy- and light-hole band states results in a better agreement between theory and experiment. However, it has become obvious, that a more rigorous exciton model is required for a quantitative description of exciton states in quantum wells.

In order to interpret the observed interband-transitions (fig. 2) we have calculated the magnetic field dependence of the hole-subband energies as illustrated above. Adding the energy variation of the lower-lying conduction-band-Landau levels (equ.3), we finally get the energy shift of the interband transitions as a function of the magnetic field.

Comparing the energies of the calculated transitions with the experimental observations,we find a surprisingly good agreement between theory and experiment (fig. 7). Because of the complex valence-band structure it is rather obvious that we observe more lines than predicted with the over simplified theory mentioned above (equ.1). Indeed it is surprising that there are no excitonic effects influencing the energy variation of these low-lying Landau transitions. Comparing fig. 2 with fig.7 there are lines which do not coincide with calculated interband transitions. Therefore these

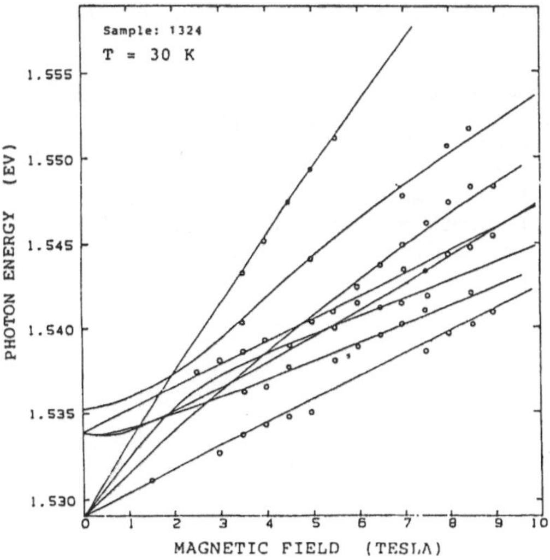

Fig. 7: Calculated interband transitions as a function of the magnetic field. The circles represent the experimentally observed transitions.

lines are attributed to excitons formed between the lowest electron subband and excited hole-subbands (e. g. E(12h) and E(13h)).

ACKNOWLEDGEMENTS We wish to thank G. Weimann and W. Schlapp for the MQW samples. This work is financially supported by the Deutsche Forschungsgemeinschaft.

REFERENCES

1. R. Dingle, Festkörperprobleme, Vol. XV, (Pergamon Braunschweig 1975) p. 21
2. J.C. Maan, G. Belle, A. Fasolino, M. Altarelli and K. Ploog, Phys. Rev. B30 (1984) p. 2253
3. N. Miura Y. Iwasa S. Tarucha and H. Okamoto, Proc 17th Intern. Conf. on the Physics of Semiconductors, San Francisco, 1984, p. 360
4. W. Ossau, B. Jäkel, E. Bangert, G. Landwehr and G. Weimann, 2nd Int. Conf. on Modulated Semiconductor Structures, Kyoto 1985,p. 120
5. Y. Nomura, K. Shinozaki and M.Ishii, J. Appl. Phys. 58 (1985), p. 1864
6. M.C. Smith, A: Petrou, C.H. Perry, J.M. Worlock and R.L. Aggarwal Proc. 17th Intern. Conf. on the Physics of Semiconductors, 1984, p. 547
7. D. Cabib, E. Fabri and G. Fioro, Nuovo Cimento 10B, 1972, p. 185
8. G. Bastard, E.E. Mendez, L.L. Chang and L. Esaki, Phys. Rev. B26, 1982 p. 1974
9. L.M. Roth, B. Lax and S. Zwerdling, Phys. Rev.114, 1959, p. 90
10. E. Bangert and G. Landwehr, Superlattices and Microstructures 1, 1985, p. 363
11. J.M. Luttinger, Phys. Rev. 102, 1956, p. 1030
12. H.J. Lee, L.Y. Yuraval J.C. Wooley and A.J. Springthorpe, Phys. Rev. B21, 1980, p. 659
13. C.R. Pidgeon and R.N. Brown, Phys. Rev. 146, 1966, p. 575
14. R.L. Greene and K.K. Bajaj, Solid State Commun. 45, 1983, p. 831
15. G.D. Sanders and Y.C. Chang, Phys. Rev. B. 32, 1985, p. 5517
16. K.S. Chan, J. Phys. C 19, 1986, p. L125
17. L.J. Sham, Proceedings of the Yamada Conf. on Mod. Semiconductor Structures, Kyoto 1985, p. 573

Note added in proof: Following the completion of this manuscript we received a preprint from G.E.W. Bauer and T. Ando (Proc. of the ICPS, Stockholm 1986). These authors take into account the full Luttinger Hamiltonian and calculate energies variationally with a basis consisting of the square well eigenfunctions normal to the well and two-dimensional hydrogenic wavefunctions in the plane. For a well width of 180 Å they find gratifying agreement with theory and our data for the diamagnetic shift.

Magneto-Optics in GaAs-GaAlAs Quantum Wells

D.C. Rogers[1], *J. Singleton*[1], *R.J. Nicholas*[1], *and C.T. Foxon*[2]

[1]Clarendon Laboratory, Parks Road, Oxford, OX1 3PU, UK
[2]Philips Research Laboratory, Redhill, Surrey, UK

The inter-band photoconductivity of GaAs-(Ga,Al)As quantum wells has been studied in steady magnetic fields up to 16T. Landau level transitions with indices up to 14 are visible, at energies up to 400 meV above the energy gap of GaAs. The Landau levels persist down to ∿ 2T, enabling an accurate extrapolation to B = 0 to yield exciton binding energies, which are substantially lower than those deduced from previous high-field measurements. Several diamagnetically-shifted subband excitons are observed, including one associated with a bound state in the spin-orbit-split-off band. The Landau levels are fitted using a calculation of the binding energy of 2D excitons as a function of B /1/: the resulting electron effective masses are checked against pump and probe cyclotron resonance measurements.

In this work we report interband photoconductivity experiments performed on a selection of GaAs(Ga,Al)As multiquantum well (MQW) samples.* Each consists of a nominally undoped GaAs buffer layer grown on a semi-insulating GaAs substrate, followed by 60 periods of thick ($\gtrsim$ 150Å) $Ga_{0.65}Al_{0.35}As$ barriers and GaAs wells, and finally 1300Å (Ga,Al)As. Samples with wells of thickness 22Å, 55Å, 75Å and 110Å were available. The experiments were performed with the sample at 55K in a 16T superconducting magnet /2/.

The photoconductive response of the samples is dominated by the GaAs buffer layer, and the absorption spectrum of the MQW is superimposed on this relatively smooth background /3/; typical results at B = 0 are shown in fig. 1. Several excitonic transitions between quantum well (QW) hole subbands (heavy hole [HHN] and light hole [LHN]) and electron (EN) subbands are visible as minima, and both allowed ($\Delta N = 0$) and forbidden ($\Delta N = 1,2$) transitions are seen, enabling the subband confinement energies to be deduced directly /3/. Of especial interest is the highest energy subband transition in the two widest QW, which cannot be explained consistently with the levels deduced from the other features. This fits well to the energy calculated for a transition from the first spin-orbit split-off band subband to the E1 subband, and it is also observed as a very strong feature at 300 K, where only allowed ($\Delta N=0$) transitions are observed from the light hole and heavy hole subbands.

Fig. 2 shows the effect of applying a magnetic field perpendicular to the QW planes in the 55Å sample: the traces are dominated by a series of Landau level transitions evolving from the HH1-E1 exciton, and by the excitons shifting up diamagnetically. The small shoulder-like features a and b in fig. 2 are visible in all the wider wells and are identified in the following way: at low magnetic fields, the HH1-E1 Landau level transitions evolve from bound states of the exciton ($\ell = 1$ from $2p^+$, $\ell = 2$ from $3d^{2+}$ etc.). Careful extrapolations of the low-field (B<6T) Landau level transitions down to B = 0 show that the transitions evolve from points close to, but just below the feature corresponding to a /2/.

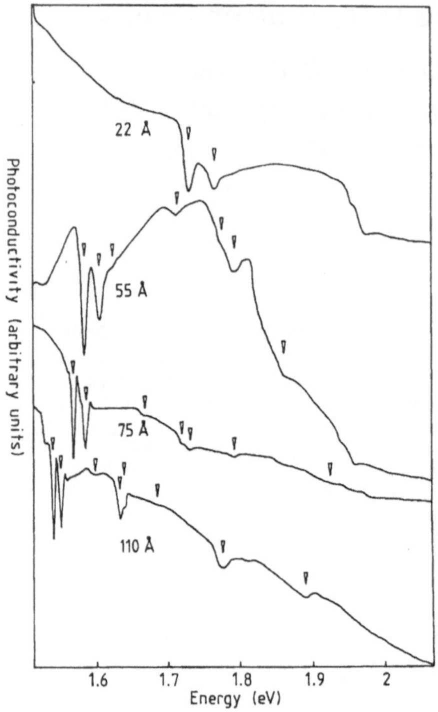

Figure 1: Photoconductive spectra of MQW at B=0 and T=55 K. Excitonic subband transitions are indicated.

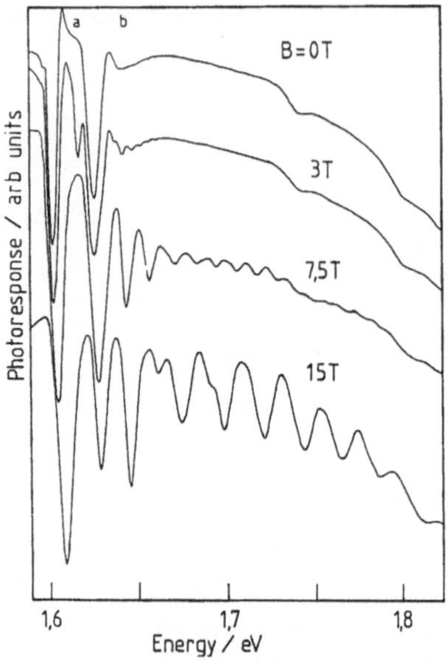

Figure 2: Typical spectra for 55 Å QW sample, at 55 K and various magnetic fields.

Thus a, b represent the series limits for the HH1 - E1 and LH1 - E1 excitonic transitions respectively, allowing the exciton binding energy to be deduced directly in the 55Å, 75Å and 110A wells.

We now attempt to fit the magnetooptical data. Previous authors /4,5/ have interpreted the higher (ℓ>1) Landau level transitions as free carrier transitions: this is not true in bulk GaAs, and so we base our analysis on a model /1/ of excitonic states in the high-field limit in highly anisotropic systems. The exciton binding energy is given by:-

$$E_B \simeq 3\ \left[(\hbar eB)/4(\ell+\tfrac{1}{2})\mu^* R^*\right]^{\frac{1}{2}} R^* D \qquad (1)$$

with μ^* the exciton reduced mass, R^* the excitonic Rydberg and D_1 a parameter related to the dimensionality of the exciton, taking values from 0.25(3D) to 1(2D). The B = 0 exciton binding energy is given by $E_B(0) = 4R^*D$. To complete the theory, we need dispersion relationships for motion of the electrons and holes in the plane of the layers. The electrons are described by the simple k.p perturbation theory of /6/, which, after some rearrangement /7/ yields the following Landau level energies:-

$$E_e(B,\ell,K_2) = (\ell+\tfrac{1}{2})(\hbar eB/m^*_{sb})(1 + (K_2/E_g)(m^*_{sb}/m^*_o)(\ell+\tfrac{1}{2})(\hbar eB/m^*_{sb})), \qquad (2)$$

where m_o^* is the GaAs band-edge effective mass, m_{sb}^* is the "subband-edge" effective mass, and K_2 is the parameter expressing the non parabolicity of the conduction band (-0.83 in bulk GaAs). As (2) contains two adjustable parameters, m_{sb}^* and K_2, pump and probe cyclotron resonance (CR) measurements were performed on the samples to check m_{sb}^* /7/. Cyclotron effective masses of the 22Å well are plotted as a function of $\hbar\omega_c$ in fig. 3: the very large apparent non-parabolicity ($K_2 \approx 2.3$) is thought to be due to the polaron effect, and this will be dealt with in a subsequent paper /7/. For the present purposes, we choose m_{sb}^* to lie within the limits $m_{int}^*/(1+\pi\alpha/8) \leq m_{sb}^* \leq m_{int}^*/8/$, where m_{int}^* is the intercept of the line fitted to the CR data (fig. 2). The valence band is complicated by interactions between heavy and light hole levels, and the heavy hole mass, m_{hh}^* is treated as an adjustable parameter. The complete transition energy is thus:-

$$E_T = E_g + E_{E1} + E_{HH1} + E_e + (L+\tfrac{1}{2})(\hbar eB/m_{hh}) - E_B,$$

where $E_B(1)$ is corrected for electron non parabolicity, E_{E1} and E_{HH1} are the B = 0 subband confinement energies, and E_e is given by (2). K_2, m_{hh}^* and D_1 are varied to get the best fit possible: an example is shown in fig. 4.

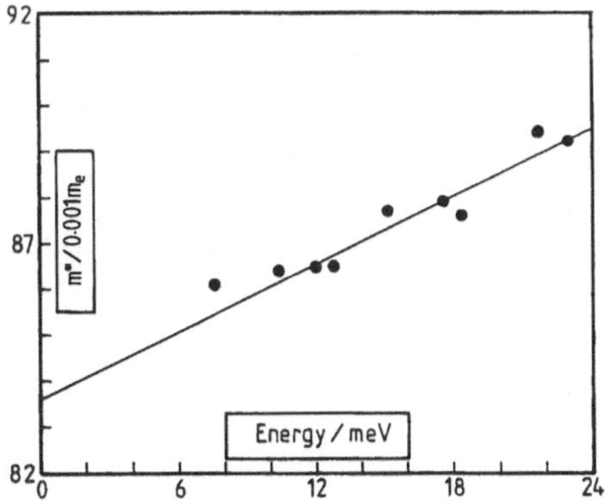

Figure 3: E1 subband effective mass deduced from pump and probe CR experiments. The points are data: the line is a least-squares fit.

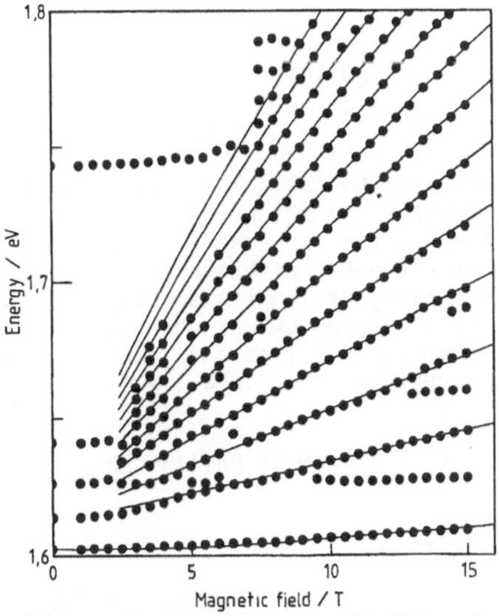

Figure 4: Theoretical fit (lines) to HH1-E1 Landau level transitions in 55Å well. Data are shown as points.

Values of the fitting parameters for all the wells are shown in the table. The fitted values of the exciton binding energies are in good agreement with the more reliable direct measurements, and the HH1-E1 binding energy can be seen to peak at ∿ 55Å and fall for narrower wells: this is at variance with theory /10/ which predicts a maximum value of the binding energy ∿ 10 meV at ∿ 25Å. For all the wells K_2 was found to be -1.3±0.1: this is greater than the value -0.83 which describes the GaAs conduction

Table: Fitting parameters for MQW interband magnetooptical data (EBE = exciton binding energy).

Well	22 Å	55 Å	75 Å	110 Å
M^*_{sb}/M_e	0·082	0·0715	0·069	0·0675
K_2	-1·3	-1·3	-1·3	-1·3
M^*_{hh}/M_e	0·36	0·55	0·85	0·85
EBE HH1-E1 (fitted)	11 meV	12·5 meV	10 meV	9 meV
EBE HH1-E1 (direct)	-	12·5 meV	10 meV	8 meV
EBE LH1-E1 (direct)	-	13·5 meV	11 meV	9 meV

band reliably /11/. Recent CR measurements on GaAs-(Ga,Al)As heterojunctions, in which the polaron effect is screened out also reveal K_2 = -1.3 /12/ and the increase is thought to be due to confinement. The heavy hole mass decreases with increasing well width, consistent with progressive decoupling of light and heavy hole subbands. Reduction of the heavy hole mass in narrow QW was recently observed using hole CR /9/, in qualitative agreement with this work: however, a heavy hole mass equal to the bulk value was reported in wide wells.

* Grown by MBE at Philips Research Laboratories, Redhill, Surrey, UK.

References

1. O. Akimoto and H.Hasegawa: J.Phys.Soc.Jpn. 22 181 (1967)
2. D.C. Rogers, J.Singleton et al.: Phys.Rev. B (in press)
3. D.C. Rogers and R.J. Nicholas: J.Phys. C 18 L891 (1985)
4. N. Miura et al.: Proc.17th Int.Conf.Phys.Semicond. 359 (Springer, New York 1984)
5. J.C. Maan et al.: Phys.Rev B 30 2253 (1984)
6. E.D. Palik et al.: Phys.Rev 122 475 (1961)
7. J. Singleton, R.J.Nicholas and D.C. Rogers: to be published.
8. W. Xiaoguang et al.: Phys.Stat.Sol. b 133 229 (1986)
9. Y. Iwasa et al.: Surface Science 170 401 (1986)
10. K.S. Chan: J.Phys.C 19 L125 (1986)
11. Q.H.F. Vrehen: J.Phys.Chem.Solids 29 129 (1968)
12. M.A. Hopkins and R.J. Nicholas: to be published.

Inelastic Light Scattering in Two-Dimensional Systems in High Magnetic Fields

A. Pinczuk[1] *and D. Heiman*[2]

[1]AT&T Bell Laboratories, Murray Hill, NJ 07974, USA
[2]Francis Bitter National Magnet Laboratory, Massachusetts Institute of Technology, Cambridge, MA 02139, USA

We review recent resonant inelastic light scattering research carried out in modulation doped GaAs-(AlGa)As quantum well heterostructures. The valence subband states are probed as a function of applied magnetic field by measurements of the transitions of high mobility two-dimensional hole gases. In the case of two-dimensional electron gases in high magnetic fields normal to the plane, we observe combined Landau level-intersubband excitations, Landau level transitions and quasi-3D magnetoplasmons.

1. *INTRODUCTION*

Inelastic light scattering is a powerful experimental probe of electronic excitations of semiconductors. In recent years the method has been widely applied to studies of systems of lower dimensionality in semiconductor quantum wells and heterojunctions. Interest in light scattering by two-dimensional electron systems in semiconductors was stimulated by a proposal of BURSTEIN et al.[1]. The proposal pointed out that sensitivity to observe the excitations of two-dimensional (2D) systems is achieved in *resonant* light scattering. These enhanced spectra are obtained with photon energies close to one of the optical transitions to states occupied by the free carriers. The proposal led to the observations of resonant light scattering by 2D electron systems in modulation doped GaAs-(AlGa)As heterostructures [2,3]. The first observations of resonant inelastic light scattering by 2D electrons in high magnetic fields were reported in 1981 [4,5].

The light scattering method is a spectroscopic tool that yields the energies of the electronic excitations. By means of polarization selection rules it is possible to obtain *separate* spectra of single particle and collective excitations [6-8]. These are unique features that allow determination of the structure of energy levels and collective electron-electron interactions.

Light scattering examinations of 2D electron systems have been reviewed in a number of publications (see Refs. [9-11] and references therein). In this paper we consider recent work on high mobility systems in modulation doping GaAs-(AlGa)As heterostructures in high magnetic fields [12-14]. In Section 2 we present a brief discussion of light scattering by 2D systems. In Section 3 we consider 2D hole gases and in Section 4 the 2D electron gases.

2. *KINEMATICS AND SELECTION RULES*

BURSTEIN et al. [1,15] have discussed the mechanisms and selection rules that apply to resonant inelastic light scattering by 2D electron systems in

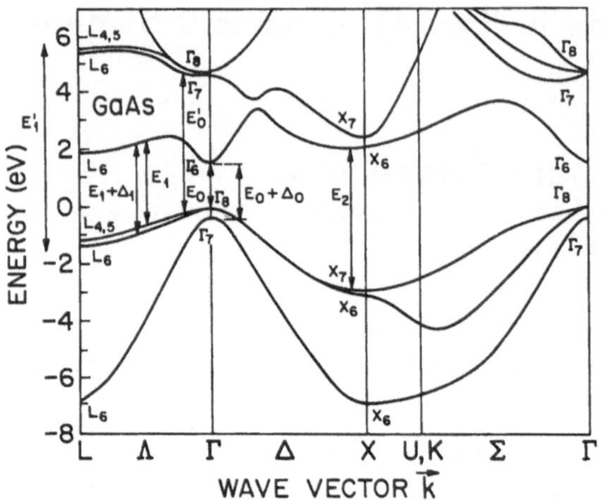

Figure 1. Band structure of GaAs, showing the main interband optical transitions that contribute to resonant inelastic light scattering [after J. R. Chelikowsky and M. L. Cohen: Phys. Rev. Vol. B14, p. 556, 1976]

semiconductors. Within the framework of effective-mass theory light scattering mechanisms are similar to those of the bulk semiconductors [1,15-17]. The resonant enhancements occur for photon energies close to optical gaps associated with the states of the free carriers (examples of such optical transitions can be seen in Figs. 3 and 10). We show in Fig. 1 the electron energy band structure of GaAs and several of its direct optical energy gaps. In the case of free electrons in conduction band states the relevant resonances occur at the fundamental, E_0, and spin-orbit split-off, $E_0 + \Delta_0$, gaps. In GaAs the energies are $E_0 \simeq 1.5$ eV and $E_0 + \Delta_0 \simeq 1.9$ eV. Two types of spectra are measured. *Polarized* spectra are obtained with polarizations of incident and scattered light that are parallel to each other. In the *depolarized* spectra the two polarizations are orthogonal. In the case of the $E_0 + \Delta_0$ resonance the polarized spectra have been interpreted as *charge density fluctuations* and the depolarized spectra as *spin-density fluctuations* [6,7,15]. Spectra obtained with photon energies near the fundamental E_0 gap indicate more complex selection rules [18]. BURSTEIN et al. [1,15] also predicted that light scattering by charge density fluctuations coupled to optical phonons should show strong resonant enhancement at the E_1 optical gap. This effect was demonstrated in InAs metal-insulator-semiconductor devices [19]. For light scattering by free holes in a direct gap semiconductor the most important resonance is at the E_0 gap [1,15]. A new approach to resonant light scattering by free holes in GaAs quantum wells has been recently demonstrated [20]. This approach has been used to carry out spectroscopic studies of 2D hole gases [12,13,20].

The nearly backscattering geometry shown in Fig. 2 is frequently used. Light propagates inside the sample along directions close to the normal to the plane of the quasi 2D system. It is often convenient to set $\theta + \phi = 90°$. In this case the in-plane and normal components of the scattering wavevector are given by [21]

$$k = \frac{2\pi}{\lambda_L} (\cos\theta - \sin\theta) \qquad (1.a)$$

and

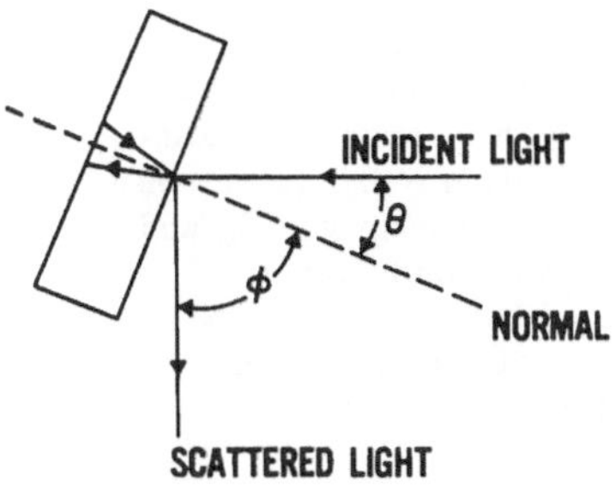

Figure 2. Nearly backscattering geometry used in resonant light scattering experiments

$$k_z = \frac{4\pi}{\lambda_L} \eta(\lambda_L) \left\{ 1 - [1/2\eta(\lambda_L)]^2 \right\}, \tag{1.b}$$

where λ_L is the laser wavelength and $\eta(\lambda_L)$ is the refractive index. The in-plane component k can be varied from a small value, at $\theta = 45°$, up to a maximum of $k \sim 10^5\ cm^{-1}$ for $\theta \simeq 0°$. The more conventional backscattering geometry corresponds to $\theta + \phi = 0$. In this case $k = (4\pi/\lambda_L)\sin\theta$. Values of k are larger by a factor of two in this configuration.

The kinematics of the light scattering processes follows from the condition of conservation of wavevector. For the in-plane component this condition is simply: $\vec{k} = \vec{q}$, where $\vec{q}$ is the wavevector of the elementary excitation. Breakdown of this conservation was found in low mobility systems [22]. It has been assigned to wavevector relaxation due to electron scattering by the ionized impurities. In *periodic multilayer* 2D systems there is an additional conservation condition on k_z. To see how this arises consider an elementary excitation that in each plane has a phase factor $e^{iq_z \ell d}$, where ℓ is an integer that labels the planes and d is the period. The light scattering intensity is proportional to

$$I(q_z, k_z) \sim \left| \sum_{\ell=1}^{N} e^{i(q_z - k_z)\ell d} \right|^2 = \frac{1 - \cos N(q_z - k_z)d}{1 - \cos(q_z - k_z)d} \tag{2}$$

with a maximum for $k_z = q_z$ proportional to N.

3. *TWO-DIMENSIONAL HOLE GASES*

Optical studies of GaAs-(AlGa)As heterostructures have revealed basic properties of the states near the energy gap [23]. Resonant inelastic light scattering by 2D hole gases in p-type modulation doped GaAs-(AlGa)As quantum wells allows direct spectroscopic determinations of the energy spacings between valence subband states [12,13,20]. The spectral lineshapes are very different from the relatively narrow bands of the 2D electron gases. They are determined by the complexities of the dispersions of the valence subbands. This behavior is a consequence of the degeneracy of the valence band states in zincblende-type semiconductors. It leads to sets of separate, but coupled, light and heavy subbands [24-30].

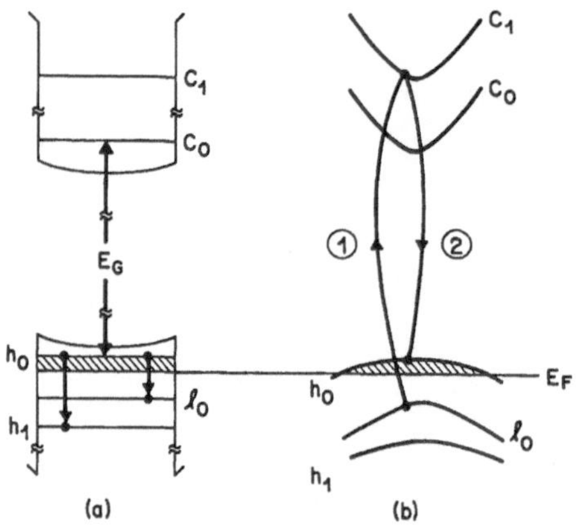

Figure 3. Schematic representation of energy levels and subband dispersion in p-type GaAs-(AlGa)As quantum wells. (a) shows the fundamental energy gap E_G and indicates two intersubband transitions of holes. (b) shows the two optical transitions in resonant light scattering by $h_o \rightarrow \ell_o$ excitations

Our studies are carried out on p-type modulation-doped GaAs quantum wells. The light scattering spectra are excited with a tunable infrared dye laser operating at photon energies (1.6 – 1.7 eV) in resonance with optical transitions of the higher lying states derived from the fundamental E_0 gap. Intersubband transitions of interest are represented in Fig. 3(a). h_0 is the ground heavy subband, ℓ_0 the lowest light subband and h_1 the first excited heavy subband. Fig. 3(b) shows the optical transitions that take part in resonant light scattering by $h_0 \rightarrow \ell_0$ excitations. This procedure has the advantage of large resonant enhancements and at the same time minimizes the background due to optical emission at the fundamental gap.

Figure 4 shows low-temperature resonant light scattering spectra measured in a sample with a relatively low free hole density and high mobility. d_1 is the thickness of the GaAs wells, d_2 that of the Be-doped center of the barriers and d_3 of the undoped spacers introduced to enhance carrier mobility. The structures labeled $h_0 \rightarrow \ell_0$ and $h_0 \rightarrow h_1$ are assigned to excitations of the corresponding intersubband transitions.

The differences in spectral lineshapes of $h_0 \rightarrow \ell_0$ and $h_0 \rightarrow h_1$ excitations are direct evidence of the complex dispersions of the valence subbands. The $h_0 \rightarrow h_1$ excitations appear as relatively narrow bands because the two subbands are nearly parallel near the Brillouin zone center. The $h_0 \rightarrow h_1$ excitations appear as broad structures because the two subbands have very different dispersions. The measured $h_0 \rightarrow h_1$ energies are in excellent agreement with the spacings of levels of a square well [20]. However, the model gives energies that are higher than the measured $h_0 \rightarrow \ell_0$ transitions. The discrepancy has been explained by envelope function calculations that predict *electron-like* dispersions for the ℓ_0 subband [24]. The electron-like dispersion of the ℓ_0 subband is a major prediction of effective-mass theories [24-28] that is supported by our light scattering experiments.

Only minor differences are observed in the positions of $h_0 \rightarrow h_1$ bands in $z(y'y')\bar{z}$ (polarized) and $z(y'x')\bar{z}$ (depolarized) spectra. This is evidence of weak depolarization field effects. The absence of a phonon-like mode coupled to free holes and the presence of the LO phonon is also evidence of weak depolarization

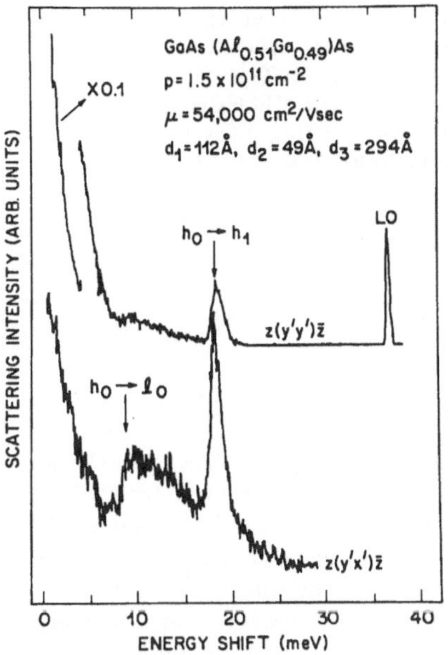

Figure 4. Polarized and depolarized spectra of 2D holes confined in p-type modulation doped GaAs-(AlGa)As quantum wells

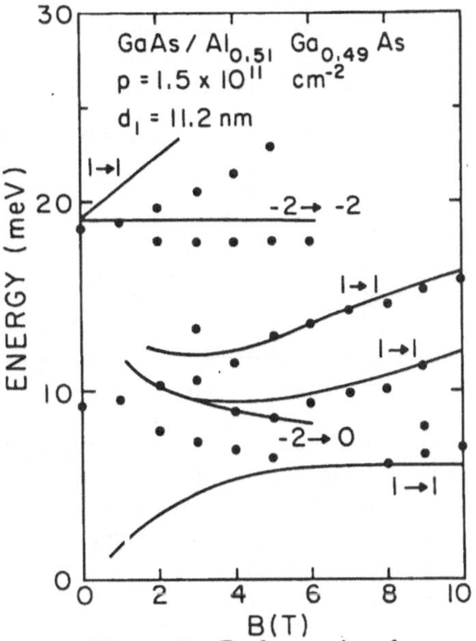

Figure 5. Peak energies of intersubband transitions of 2D holes in high magnetic fields

field effects. In the case of 2D hole systems in Si space charge layers depolarization field effects are absent in spectra from (110) and (111) surfaces [31].

A high magnetic field applied normal to the layers splits the $h_0 \rightarrow \ell_0$ and $h_0 \rightarrow h_1$ transitions into complex multiplets [13]. Figure 5 displays the energies of the strongest peaks observed in the light scattering spectra. The solid lines are the calculated transition energies with the envelope-function approach of Ref. 24. The *allowed* transitions originate from one of the filled levels and terminate in an empty level with a change of Landau quantum number $\Delta m = 0, \pm 2$. The field-independent $h_0 \rightarrow h_1$ excitation at 18 meV is assigned to the $-2 \rightarrow -2$ transition. In agreement with this assignment, the transition disappears above $\sim$ 6 Tesla. At this field the $m = -2$ level, the first excited state of the h_0 multiplet, is no longer occupied by holes. In the $h_0 \rightarrow \ell_0$ multiplet we find that the transition assigned as $-2 \rightarrow 0$ decreases in energy with magnetic field until it disappears, as expected, for $B \gtrsim 6$ Tesla. This behavior is further verification of the electron-like mass curvature of the light hole valence subband.

The $B = 0$ polarized $z(y'y')\overline{z}$ spectrum in Fig. 4 has an intense broad structure at energies below those of conventional intersubband excitations. Figure 6 shows that this low-energy scattering has a maximum at an energy shift of $\sim$ 1.2 meV. We have determined that this spectral feature has no dependence on the angle θ (see Fig. 2). This shows that it is independent of the magnitude of

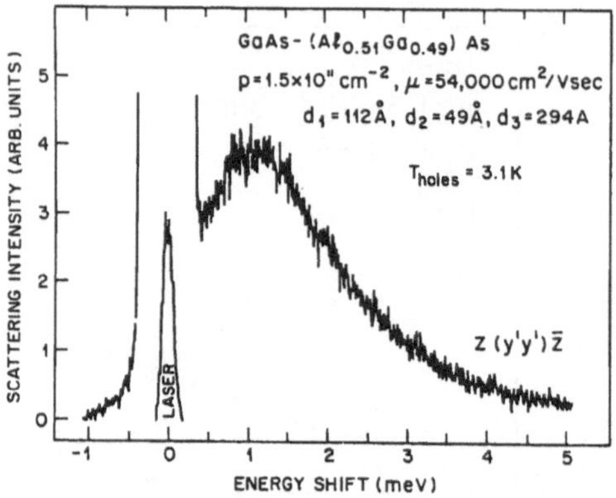

Figure 6. Low energy part of a polarized spectrum like that in Fig. 4. The temperature of holes is determined from Stokes-antiStokes intensity ratios

the in-plane component of scattering wavevector [see Eqs. (1)]. Therefore, the low energy excitations are not multilayer plasmons of the type observed by OLEGO et al. [21] in the n-type superlattices. The results of Fig. 7(a) show that a high magnetic field normal to the layers causes striking changes in the spectra, while in-plane fields have no effect. This behavior confirms the quasi-2D character of the low-energy excitations [12]. The energies of the two best resolved low-energy peaks are given in Fig. 7(b). For B $\geq$ 4 Tesla there is linear dependence with hole effective masses of 0.35 and 0.55. These values are in agreement with cyclotron resonance determinations [32-34]. The energies of the strongest peaks in the spectra are consistent with the calculated 1 $\rightarrow$ 2 cyclotron resonance transitions. Further work is required to reach a better understanding of these low energy excitations.

4. *TWO-DIMENSIONAL ELECTRON GASES*

We have recently reported the observation of new transitions of the 2D electron gas of modulation doped GaAs-(AlGa)As quantum wells in high magnetic fields [14]. In these experiments the photon energies are close to resonance with states

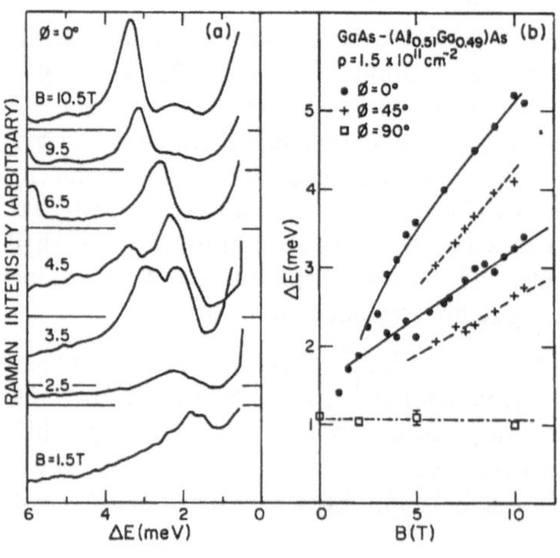

Figure 7. (a) Magnetic field dependence of the polarized spectrum like that in Fig. 4.
(b) Energy as a function of magnetic field of two peaks. ϕ is the angle between the field and the normal to the layers

of the fundamental optical gap E_0. The intermediate states in the light scattering processes are the higher valence subbands derived from the Γ_8 heavy and light states of GaAs. This approach allows light scattering by transitions that were not observed in previous light scattering work [4,5]. The most unexpected are those in which the electrons undergo a combined change of Landau level and subband state.

Combined transitions were previously measured in far infrared optical absorption experiments in which *tilted* magnetic fields cause the coupling between in-plane and normal to the plane motions required for a simultaneous change in subband index and Landau level number [34]. In the experiments reported here the combined transitions are observed with magnetic fields *normal* to the plane. We attribute the new selection rules to the mixing between subband and in-plane motions in the valence states that participate in the two virtual transitions of the resonant light scattering processes (see Fig. 10).

Figure 8 shows spectra measured in a modulation doped sample that consists of 15 GaAs quantum wells of thickness $d_1 = 270Å$. The superlattice period is 980Å and the electron mobility is $\mu = 1.3 \times 10^5$ cm^2/V sec. The sample is immersed in superfluid He and the magnetic fields are along the normal to the layers. The spectra were excited with photon energies in the range 1.54-1.58 eV. They are 30-50 meV above the fundamental gap. The $z(y'x')\bar{z}$ spectrum for $B = 0$ shows the peak at $E_{01} = 8.0$ meV, the spacing between the two lowest conduction subbands [9]. The $z(x'x')\bar{z}$ spectrum shows two additional peaks labeled I_- and P. I_- is assigned to the lowest collective intersubband excitation [9,15] and P to the plasmon of the multilayers [21]. The spectra for $B \neq 0$ show

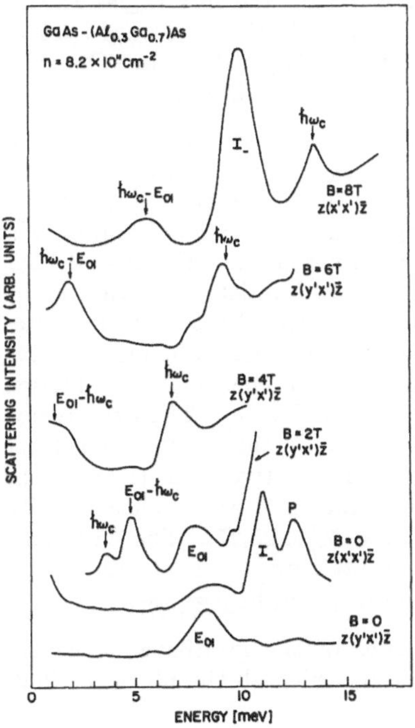

Figure 8. Light scattering spectra of 2D electrons for different values of magnetic field

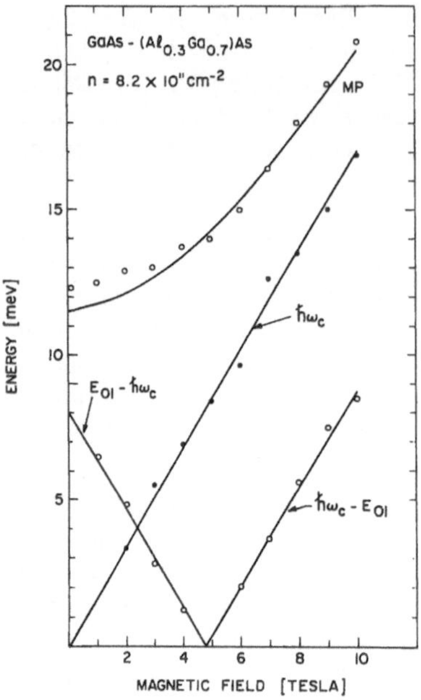

◄ Figure 9. Energies of combined transitions, Landau level transitions and quasi-3D magnetoplasma modes as function of magnetic field

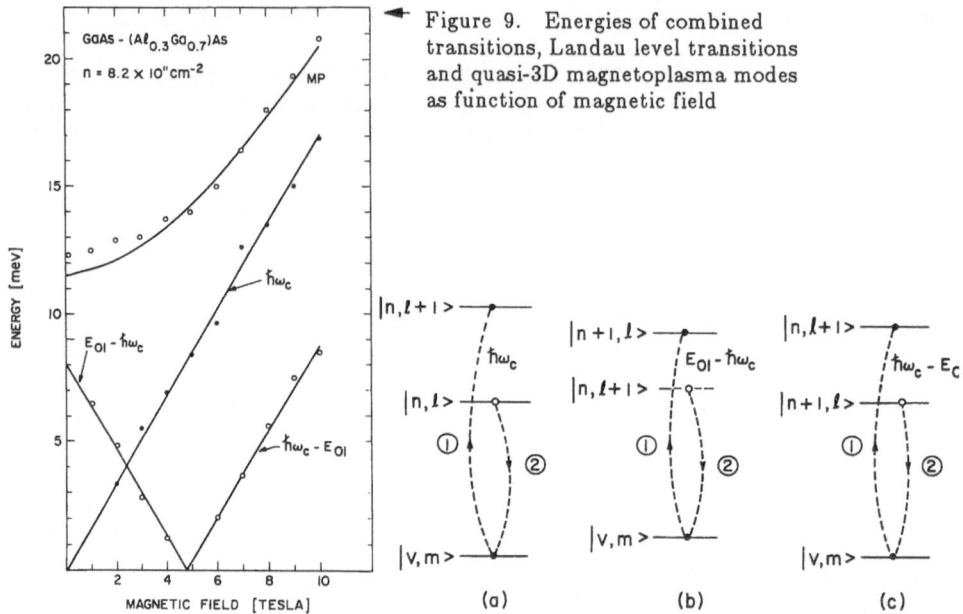

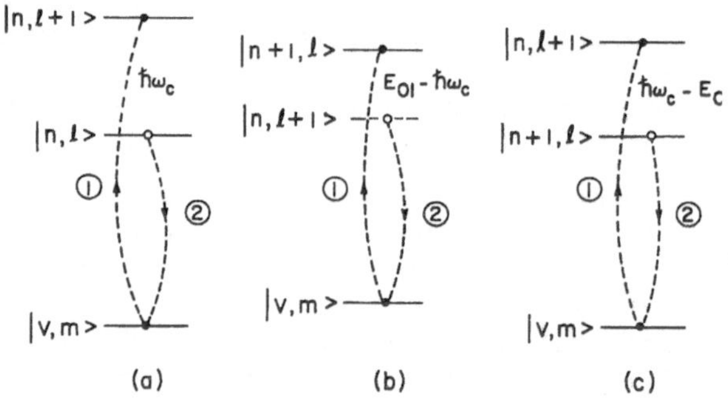

Figure 10. The two optical transitions in resonant inelastic light scattering by 2D electrons in high magnetic fields. The numbers are the order of the transitions

two new bands. For $B \leq 4$ Tesla they are at $E_{01} - \hbar\,\omega_c$ and $\hbar\omega_c$ ($\omega_c = eB/m^*$ is the cyclotron frequency of electrons). For $B > 4$ Tesla the new bands are at $\hbar\omega_c - E_{01}$ and $\hbar\omega_c$. Figure 9 shows the positions for $B \leq 10$ Tesla. The solid lines are calculated energies with $E_{01} = 8.0$ meV and $m^* = 0.068\ m_o$.

Figure 10 shows the two virtual optical transitions in resonant inelastic light scattering at energies $\hbar\omega_c$, $E_{01} - \hbar\omega_c$ and $\hbar\omega_c - E_{01}$. $|n, \ell>$ are conduction states with subband index n and Landau level quantum number ℓ. $|v, m>$ are the valence states derived from the fourfold degenerate Γ_8 states of GaAs. v is the subband index and m the Landau level quantum number. For $B = 0$ Figs. 10(b) and 10(c) describe a new mechanism for stokes and antistokes scattering by intersubband excitations. It exists because away from the Brillouin zone center there is extensive coupling and mixing between the valence subbands [24-29]. The mixing allows, *within the dipole approximation*, a change in conduction subband index in the processes shown in Figs. 10(b) and 10(c). The observation of combined transitions for $B \neq 0$ can be interpreted by the additional valence band mixing introduced by the magnetic field. In effective-mass theories the states $|v,m>$ consist of admixtures of harmonic oscillator wavefunctions with quantum numbers in the range $m - 1 \leq \ell' \leq m + 2$ [24,25,27]. The additional field-dependent mixing allows resonant light scattering by new transitions, including the excitations with energies $\hbar\omega_c$, $E_{01} - \hbar\omega_c$ and $\hbar\omega_c = E_{01}$.

In Fig. 9 MP labels the magnetoplasma modes of the multilayer structure. Since $k_z d \approx 6 \sim 2\pi$ we should observe the mode in which all the planes oscillate in phase [35]. The solid line is the magnetoplasma energy calculated from $\omega_p(B) = [\omega_p^2(0) + \omega_c^2]^{1/2}$, with $\omega_p^2(0) = 4\pi ne^2/\epsilon_s m^* d$ and $\epsilon_s = 13$ is the background dielectric constant. For $B > 4$ Tesla there is good agreement between measured and calculated energies. The discrepancies at lower field may be due to the interaction between the magnetoplasmon and the collective intersubband excitation I_- [36,37]. The interaction could be strong even for $kd < 0.1$ because, as can be seen in Fig. 8, the energies of the two modes are very close for $B = 0$.

5. *CONCLUDING REMARKS*

We have presented a discussion of recent inelastic light scattering experiments in GaAs quantum wells in high magnetic fields. This work demonstrates several applications of the method to spectroscopic studies of two-dimensional systems in semiconductors. Carrier temperatures determined by stokes- antistokes ratios from samples immersed in superfluid He are ~ 3 K in the case of holes (see Fig. 6) and ~ 7 K in the case of electrons. Carrier heating by the incident laser beam, unavoidable in resonant light scattering, is less severe in the experiments reported here because the photon energies are closer to the fundamental gap. The use of smaller incident powers in conjunction with optical multichannel detection would make possible light-scattering investigations of the two dimensional systems with even lower carrier temperatures.

The authors wish to acknowledge the collaboration of M. Altarelli, J. H. English, A. Fasolino, A. C. Gossard, R. Sooryakumar, H. L. Störmer and W. Wiegmann. Discussions with L. J. Sham have given insight into the structure of the valence subband states. The Francis Bitter National Magnet Laboratory is supported by the National Science Foundation through its Division of Materials Research.

1. E. Burstein, A. Pinczuk, S. Buchner: In *Physics of Semiconductors 1978, Proc. of the 14th Int. Conf. on Physics of Semiconductors*, ed. by B. L. H. Wilson, (The Institute of Physics, London 1979) p. 1231.
2. G. Abstreiter, K. Ploog: Phys. Rev. Lett. 42, 1308, 1979.
3. A. Pinczuk, H. L. Störmer, R. Dingle, J. M. Worlock, W. Wiegmann, A. C. Gossard: Solid State Commun. 32, 1001 (1979).
4. J. M. Worlock, A. Pinczuk, Z. J. Tien, C. H. Perry, H. L. Störmer, R. Dingle, A. C. Gossard, W. Wiegmann, R. L. Aggarwal: Solid State Commun. 40, 867 (1981).
5. Z. J. Tien, J. M. Worlock, C. H. Perry, A. Pinczuk, R. L. Aggarwal, H. L. Störmer, A. C. Gossard, W. Wiegmann: Surface Sci. 113, 89 (1982).
6. A. Pinczuk, J. M. Worlock, H. L. Störmer, R. Dingle, W. Wiegmann, A. C. Gossard: Solid State Commun. 36, 43 (1980).
7. A. Pinczuk, J. Shah, A. C. Gossard, W. Wiegmann: Phys. Rev. Lett. 46, 1341 (1981).
8. Ch. Zeller, G. Abstreiter, K. Ploog: Surface Sci. 113, 85 (1982).
9. A. Pinczuk, J. M. Worlock: Surface Sci. 113, 69 (1982).

10. J. M. Worlock, A. C. Maciel, C. H. Perry, Z. J. Tien, R. L. Aggarwal, A. C. Gossard, W. Wiegmann: In *Application of High Magnetic Fields in Semiconductor Physics*, ed. by G. Landwehr (Springer-Verlag, Heidelberg, Berlin 1983), p. 186.
11. G. Abstreiter, R. Merlin, A. Pinczuk: IEEE J. Quantum Electron QE-22, 1771 (1986).
12. A. Pinczuk, D. Heiman, R. Sooryakumar, A. C. Gossard, W. Wiegmann, Surface Sci. 170, 573 (1986).
13. D. Heiman, A. Pinczuk, A. C. Gossard, J. M. English, A. Fasolino, M. Altarelli: In *Proc. of the 18th Int. Conf. on the Physics of Semiconductors* (to be published).
14. A. Pinczuk, D. Heiman, A. C. Gossard, J. H. English: ibid.
15. E. Burstein, A. Pinczuk, D. L. Mills: Surface Sci. 98, 451 (1980).
16. A. Mooradian: In *Festkoerperprobleme* vol. IX, 74 (Pergamon-Vieweg, Braunschweig 1969).
17. G. Abstreiter, M. Cardona, A. Pinczuk: In *Light Scattering in Solids*, vol. IV, 5 (Springer-Verlag, Berlin, Heidelberg 1984).
18. D. Olego, A. Pinczuk, A. C. Gossard, W. Wiegmann: Bull Am. Phys. Soc. 28, 447 (1983).
19. L. Y. Ching, E. Burstein, S. Buchner, H. H. Wieder: J. Phys. Soc. Japan 49 (Suppl. A), 951 (1980).
20. A. Pinczuk, H. L. Störmer, A. C. Gossard, W. Wiegmann: In *Proc. of the 17th Int. Conf. of Semiconductors*, ed. by J. D. Chadi and W. A. Harrison, (Springer-Verlag, New York 1985), p. 329.
21. D. Olego, A. Pinczuk, A. C. Gossard, W. Wiegmann: Phys. Rev. B25, 7867 (1982).
22. A. Pinczuk, J. M. Worlock, H. L. Störmer, A. C. Gossard, W. Wiegmann: J. Vac. Sci. Technol. 19, 561 (1981).
23. Reviews of much of the research on the field can be found in a recent Special Issue of the IEEE Journal on Quantum Electronics on "Semiconductor Quantum Wells and Superlattices: Physics and Applications: IEEE J. Quantum Electron. vol. QE-22, number 9 (1986).
24. M. Altarelli, U. Ekenberg, A. Fasolino: Phys. Rev. B32, 5138 (1985); M. Altarelli: Festkoerperprobleme 25, 381 (1985).
25. G. Landwehr, E. Bangert: Superlattices and Microstructures 1, 363 (1985).
26. Y. C. Chang and J. N. Schulman: Superlattices and Microstructures 1, 357 (1985).
27. D. A. Broido, L. J. Sham: Phys. Rev. B31, 888 (1985); S. E. Yang, D. A. Broido, L. J. Sham: Phys. Rev. B32, 6630 (1985).
28. T. Ando: J. Phys. Soc. Japan 54, 1528 (1985).
29. R. Sooryakumar, D. S. Chemla, A. Pinczuk, A. C. Gossard, W. Wiegmann, L. J. Sham: Solid State Commun. 54, 859 (1985).
30. R. C. Miller, A. C. Gossard, G. D. Sanders, Y. C. Chang, J. N. Schulman: Phys. Rev. B32, 8452 (1985).
31. M. Baumgartner, G. Abstreiter, E. Bangert: J. Phys. C17, 1617 (1984).
32. H. L. Störmer, Z. Schlesinger, A. Chang, D. C. Tsui, A. C. Gossard, W. Wiegmann: Phys. Rev. Lett. 51, 126 (1983).
33. Z. Schlesinger, S. J. Allen, Jr., Y. Yafet, A. C. Gossard, W. Wiegmann: Phys. Rev. B32, 5231 (1985).

34. Y. Iwasa, N. Miura, S. Tarucha, H. Okamoto, T. Ando: Surface Sci. 170, 587 (1986).
35. W. Beinvogl, J. F. Koch: Phys. Rev. Lett. 40, 1736 (1978).
36. A. C. Tsellis, J. J. Quinn: Phys. Rev. B29, 2334 (1984).
37. S. Das Sarma: Phys. Rev. B29, 2334 (1984).
38. J. K. Jain, S. Das Sarma: Work in progress.

High Magnetic Field Studies of Confined Impurities in Semiconductors

B.D. McCombe, N.C. Jarosik, and J.-M. Mercy*

Department of Physics and Astronomy, SUNY at Buffalo, NY 14260, USA

The combination of far infrared spectroscopy and high magnetic fields has proven to be a powerful tool in the study of the electronic properties of bulk semiconductors. Recent advances in growth techniques have permitted the reproducible fabrication of repeated semiconductor heterostructures with interface abruptness on an atomic scale and a very high degree of doping control. Simple considerations show that the electronic states of shallow impurities in such confining structures can depend strongly on location with respect to confining barriers. Recent far infrared magneto-spectroscopy studies of shallow donors doped in the GaAs wells of GaAs/AlGaAs multiple quantum well structures are reviewed. Multiple quantum well samples with quantum well widths between 80Å and 450Å and doped in the well centers, well edges, and simultaneously in the barrier and well centers have been investigated. Results are in generally good agreement with recent theoretical calculations. Possible use of such measurements to determine the impurity distribution along the growth direction is discussed.

1. Introduction

Far infrared (FIR) spectroscopic studies of semiconductors in high magnetic fields have proven to be extremely useful in sorting out many of the complex electronic properties of bulk semiconductors, and in recent years have contributed substantially to our understanding of electronic systems of reduced dimensionality. Recent developments in materials growth techniques have permitted the growth of semiconductor heterojunctions, superlattices and quantum wells (QW), with abrupt interfaces on the scale of the unit cell and that can be selectively doped on the scale of a few lattice constants. This has led to new devices, as well as making possible the observation of new physical phenomena. Knowledge of the effect of confining potential barriers on the electronic states of the impurities, particularly binding energies, is becoming increasingly important.

In these lecture notes recent FIR studies in high magnetic fields of the electronic states of shallow (hydrogenic) donors confined in multiple QW structures are reviewed. Hydrogenic donors are particularly well suited for initial investigations directed at understanding the effects of confinement due to their rather simple electronics structure in the builk in the materials of interest. The GaAs-AlGaAs heterostructure system grown by molecular beam epitaxy (MBE) was chosen as the experimental vehicle for these studies because of the high degree of materials, interface, and doping control that has been developed. Both absorption and photoconductivity measurements are described.

Related experimental studies have been carried out by photoluminescence [1,2] and Raman Scattering [3] at zero magnetic field. The FIR experiments allow a more precise determination of the impurity energies and, the application of a high magnetic field provides a number of benefits including the possibility of detailed line shape studies.

2. Background

Impurities in semiconductors have received considerable attention, both experimentally and theoretically, for a number of years. Many shallow impurities in bulk semiconductors are well-described by a simple hydrogenic model. For a donor, the extra electron is weakly bound to the net positively charge impurity center through the Coulomb attraction. The semiconductor host crystal is taken into account in the one-band effective mass approximation through the use of an effective mass for the electron and a Coulomb interaction screened by the appropriate dielectric constant of the host crystal, ε_s. For a simple parabolic and isotropic conduction band the known results for the hydrogen atom can simple be taken over by replacing the free electron mass by the effective mass, m*, and $-e^2/r$ by $-e^2/\varepsilon_s r$. This yields the following well-known allowed energy states,

$$E_n = -\frac{m^* e^4}{2\varepsilon_s^2 \hbar^2 n^2}, \quad n = 1, 2, 3, \ldots, \tag{1}$$

where the energy is measured with respect to the conduction band edge. The effective Bohr radius is

$$a_o^* = \frac{\hbar^2 \varepsilon_s}{e^2 m^*} . \tag{2}$$

For donors in GaAs the binding energy, Ry*(3D), is approximately 5.8 meV, and the effective Bohr radius is ~100Å.

The effects of confinement are expected to become important when the impurity ion is within a few Bohr radii of some confining potential barrier. There are two limiting cases that provide a qualitative picture of these effects. In the first case, the width of the confining potential is allowed to go to zero resulting in a 2 dimensional (2D) situation. In this case the binding energy becomes Ry*(2D) = 4Ry*(3D), and the effective Bohr radius is $a_o^*(2D) = a_o^*(3D)/2$. The other interesting limit is that of an impurity ion located at an infinite potential discontinuity. In this case the envelope wave function of the impurity is excluded completely from the half-space where the potential is infinite, and by symmetry, only a subset of the 3D hydrogen atom solutions are allowed in the other half space (all s-like solutions are excluded) [4]. In this case the binding energy is Ry*(edge) = Ry*(3D)/4. Thus it is clear that confinement leads to large changes in the binding energies (a factor of 16 between the 2 limits for infinite potential barriers).

Recently, several workers have considered the problem of a hydrogenic impurity in an isolated semiconductor quantum well (formed by the energy band discontinuity) in the effective mass approximation. The Hamiltonian for this problem can be written

$$H = \frac{p^2}{2m_i^*} - \frac{e^2}{\varepsilon_{si}[\rho^2+(z-z_i)^2]^{1/2}} + V(z) , \tag{3}$$

where $V(z) = V_o$, $i = 2$, for $|z| > L/2$

$= 0$, $i = 1$, for $|z| < L/2$,

$\rho^2 = x^2+y^2$; z_i is the z-coordinate of the impurity ion; the origin of the coordinate system has been taken to be at the center of the QW with the z-axis parallel to the growth axis; L is the width of the well; $m_1^*(m_2^*)$ and $\varepsilon_{s1}(\varepsilon_{s2})$ are the electron effective mass and dielectric constant of the smaller (larger) gap semiconductor; V_o is the conduction band discontinuity (≡barrier height), and image contributions have been ignored. The Hamiltonia are not separable in general, and are not amenable to analytic solution. Theoretical work has made use of variational aproaches with infinite barrier height [5] and finite barrier heights [6,7] for an isolated QW with parameters appropriate to the GaAs/AlGaAs system. These calculations differ in the complexity of the trial functions and the treatment of boundary conditions. However, all results are in qualitative agreement; and the finite barrier height calculations agree quantitatively. Substantial shifts in binding energy are found as a function of well width and of position of the impurity within the well. The results are qualitatively in accord with expectations from the limiting cases outlined above. The binding energy increases as the QW width is decreased, while it decreases as the impurity ion is moved from the center to the edge of the well at constant well-width. Very recently coupling between QWs in MQW structures has been included [8]. Results indicate a reduction in the binding energy of center impurities compared to the isolated-well case due to the spreading out of the electron wavefunction in adjacent wells, and a peak in the density of states for electrons in the wells bound to impurity ions in the barriers.

An external magnetic field along the QW axis (the z-direction) has also been included in the theoretical calculations, and the dependence on magnetic field of the ground state (m=0) and the lowest lying excited state (m=±1) energies have been determined [9]. Here, the usual hydrogen atom notation for states in the low magnetic field limit, with m the azimuthal quantum number, has been used. The results are qualitatively very similar to results for hydrogenic impurities in bulk semiconductors.

The continuum states consist of quantized confinement subbands in the z-direction, with the x-y motion of each subband quantized into Landau levels whose energies are given for parabolic bands by (neglecting spin)

$$E_n = (N+\tfrac{1}{2})\hbar\omega_c \ , \qquad (4)$$

where N is the Landau quantum number, and $\omega_c = eB/m^*c$, with B the magnetic induction, and m_c^* the cyclotron effective mass (m_1^* for the QW of Eq. (3)). In the low field (Zeeman) limit ($\gamma = \hbar\omega_c/2Ry^*(3D)$) the m = ±1 states are split apart linearly by the field ($E(m{=}{+}1) - E(m{=}{-}1) = \hbar\omega_c$), while the ground state (m=0) is unaffected in lowest order. At higher magnetic fields the ground state begins to move up in energy with slope approaching that of the N=0 Landau level ($\hbar\omega_c/2$) for $\gamma \to \infty$. The lowest excited state (m=-1), after initially decreasing, also begins to move up in energy with a similar slope. Both states in the high field limit, $\gamma >> 1$, are bound states associated with the N=0 Landau level. The lowest m=+1 state moves up more rapidly with a slope approaching that of the N=1 Landau level ($3\hbar\omega_c/2$); this becomes a bound state associated with the N=1 Landau level for $\gamma>>1$. The energy separation between m=+1 and m=-1 is $\hbar\omega_c$ at all values of B.

For impurities at the well centers and the magnetic field along the growth axis electric dipole transitions are governed by the parity selection rule ($\Delta\ell = \pm1$, with ℓ the orbital quantum number) and angular momentum conservation along the magnetic field direction, ($\Delta m = \pm1$ for light propagation along the magnetic field). This results in dominant allowed transitions 1s to 2p(m=±1) for light propagation along the magnetic field with electric field vector polarized perpendicular to the magnetic field. For impurities away from the well center similar selection rules hold.

3. Experimental Details

The MQW GaAs-AlGaAs samples were grown by molecular beam epitaxy. The GaAs QWs (widths between 80Å and 450Å) were selectively doped with Si donors at $5 \times 10^{15} cm^{-3}$ (for the 450Å well-width sample) or $1 \times 10^{16} cm^{-3}$ (for all others) over the central 1/3 of the well for the center-doped samples, or the "top" 1/3 of the wells for the edge-doped sample. The number of wells varied from 12 to 45; barrier widths were 125Å to 150Å to minimize interactions between adjacent QWs. The molar fraction of Al in the $Al_xGa_{1-x}As$ barriers was $x \approx .22$ (450Å wells) or $x \approx 0.3$ (all other samples). The composition and well-width were determined by photoluminescence measurements at NRL.

The FIR spectroscopic measurements were carried out in a light pipe system in conjunction with a Fourier Transform Interferometric Spectrometer with various cryogenic detectors, and a 9T superconducting magnet system. Related measurements have been made with a FIR laser spectrometer [10]. Transmission data were obtained at low temperatures ratioed either to background spectra from a reference MQW sample with no intentional donor doping, to backgrounds from the same sample immediately after cool down in the dark, or in some cases to zero magnetic field reference spectra. In most samples studied immediately after cool-down in the dark, the confined donor impurity absorption was not observable; donors in the wells are ionized under these circumstances. After illumination for a few seconds at low temperatures with red light, donors in the QWs are neutralized and clear intra-impurity absorption lines are observable as long as the sample is maintained at temperatures below ~140K. This persistent photoeffect is illustrated in Fig. 1. With the

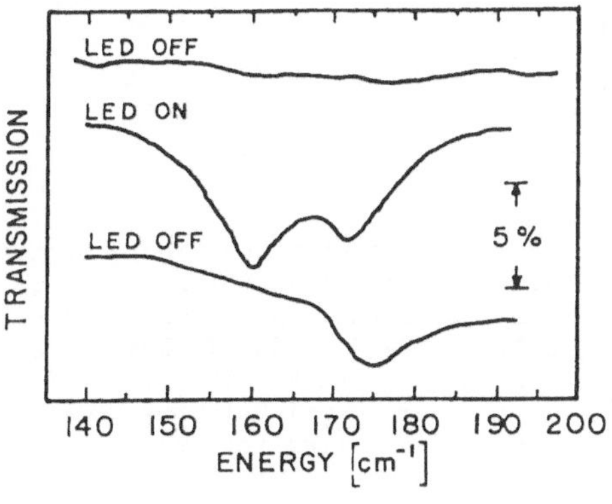

Fig. 1: Transmission spectra at 9T and 4.2K for a center-doped 210Å well-width MQW sample illustrating the effects of illumination by a red LED. top-before: center-during: bottom-after

LED off after cool down, no structure is observed. With the LED on, both confined impurity and bulk impurity (from the GaAs buffer layer) 1s-2p(m=+1) transitions are observed. When the LED is turned off again, only the confined impurity 1s-2p(m=+1) line remains (with a "tail" to lower frequencies). The assignment of the high-energy feature to confined impurities was verified by tilting the magnetic field. The confined impurity line position scales approximately with the normal component of the magnetic field, while the feature attributed to bulk impurities is independent of the angle between the magnetic field and the normal to the sample surface.

4. Absorption Measurements

4.1. Center-Doped Donors

A series of 4 center-doped samples with QW widths of 450Å, 210Å, 138Å, and 80Å were investigated to determine the effects of varying confinement on the electronic states of the donor impurities. The use of high magnetic fields in these studies was found to be very beneficial in several ways: 1) It permits the unambiguous identification of features that are electronic in nature; 2) The lines become narrowed at high fields, permitting much higher precision in the transition energy determinations; and 3) Tilting the magnetic fields permits the separation of confinement-related features from bulk features. A compilation of the results is shown in Fig. 2. Experimental points for both 1s-2p(m=±1) are compared with theoretical calculations [9] (light face lines) and the corresponding transitions observed in a "bulk" MBE Si-doped epitaxial layer (heavy dashed lines). Substantial shifts to higher energy for the confined impurity transitions are clearly evident, and the agreement with theory is good over the whole range of magnetic fields investigated. For the 80Å QW the observed 1s-2p(m=+1) transition energies show slightly more deviation (to lower energies) from the theoretical values.

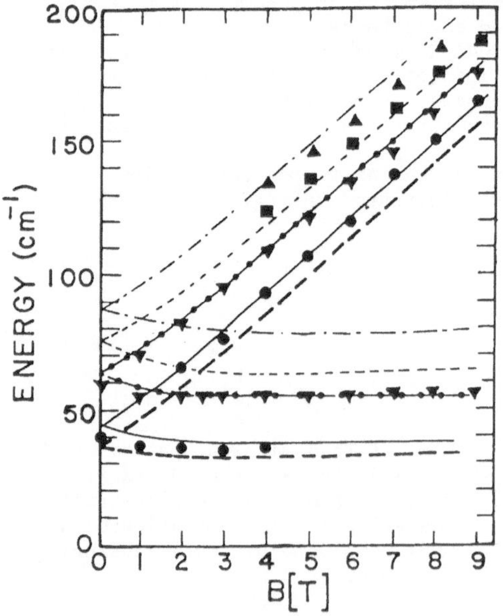

Fig. 2: Plot of transition energies vs. magnetic field for several MQW samples. Experimental results: ▲ - 80Å; ■ -140Å; ▼ -210Å; ● -450Å; thick dashed line-"bulk". Theory: —·—·-80Å; - - - -140Å; -●-●-●- 210Å; ——— 450Å (after ref. 9)

4.2. Edge-Doped Donors

The edge-doped sample used in these studies consisted of twelve 375Å GaAs QWs separated by 125Å barriers of $Al_{0.3}Ga_{0.7}As$. A transmission spectrum for this sample is shown in Fig. 3. Also shown are the calculated positions of the transition from the ground state with m=0 to the first excited state with m=+1 for an impurity at the edge of the QW (lower energy arrow) and at the center (higher energy arrow). These are frequently referred to as 1s(m=0)-2p(m=+1) transitions, although for edge impurities they more closely resemble 2p(m=0) to 3d(m=+1) transitions. Note the strongly asymmetric line shape, broadened to higher energies. The experimental transmission minimum in this case occurs measureably below ($\sim 4 cm^{-1}$) the calculated transition energy for an impurity at the edge. There is no perceptible absorption at the energy corresponding to an impurity at the center of the well. The experimentally measured energies of the transmission minima are consistently below the calculated energies for impurities at the edge of an isolated QW. There are several possible explanations for this discrepancy: 1) Electrons in the well bound to positive ions in the barrier would skew the line profile in this direction; 2) Coupling of adjacent wells tends to lower the transition energy for edge impurities; and 3) Calculations for impurities at the edge of the well are more sensitive to the choice of wave function matching conditions at the boundary (than impurities at the center); and the wave function matching conditions may not be adequate.

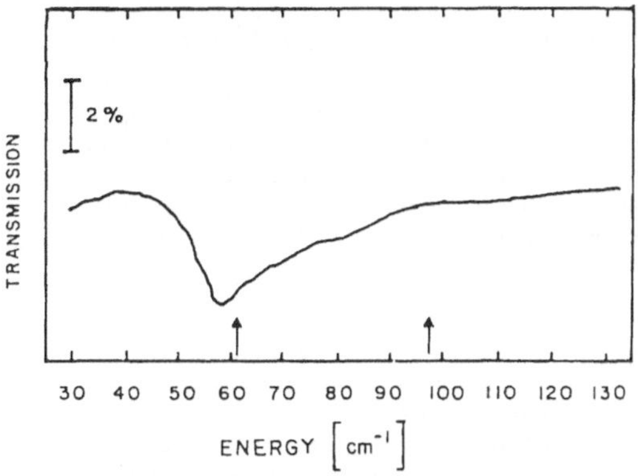

Fig. 3: Transmission spectra at 4T and 4.2K for an edge-doped, 375Å well-width MQW sample. Arrows indicate the calculated ground state with m=0 to 1st excited state with m=+1 transition energy for an impurity at the edge of the QW (lower energy) and at the center of the QW (higher energy)

Finally, the lack of any discernible absorption in this sample corresponding to impurities at the well center means that there is no substantial redistribution of impurities in the GaAs on the scale of ~60Å under these growth conditions (QWs doped just prior to growth of the AlGaAs barriers).

4.3. Additional Features

Under certain experimental conditions some center-doped MQW samples exhibit additional absorption peaks at lower frequencies in the vicinity of the 1s-2p(m=+1) center-impurity transition [10]. In an attempt to determine the origin of these absorption lines MQW samples doped both in the GaAs wells and the AlGaAs barriers have been studied.

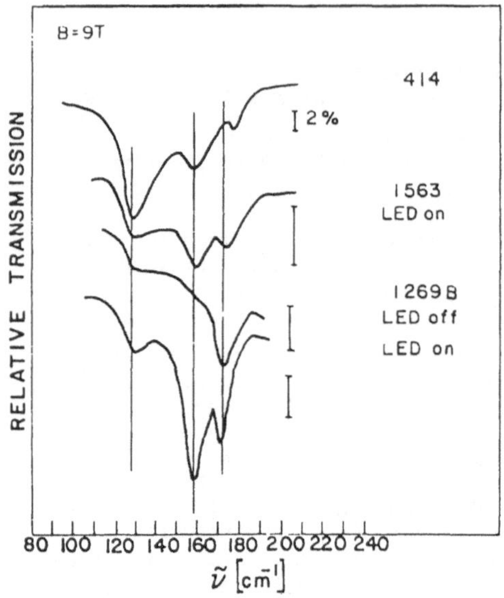

Fig. 4: Transmission spectra at 4.2K for 3 samples with nominal well-widths of 210Å. 414 amd 1563 are doped with donors in both the barrier and well centers. 1269B is doped only in the well centers

Examples of these absorption features at 9T in three samples, all having the same nominal QW widths (210Å), are shown in Fig. 5. Samples 414 and 1563 are doped with Si donors in the central 1/3 of the GaAs wells ($10^{16}cm^{-3}$) and in the central 1/3 of the barriers ($2x10^{16}cm^{-3}$ in sample 414 and $1x10^{16}$ in sample 1563). Sample 1269B is the center doped sample of Figs. 1 and 2. Three transmission minima are observed with varying relative intensities depending on the sample and experimental conditions. The lowest frequency line is dominant in sample 414, comparable in intensity to the center-impurity transition in sample 1563, and appears as a weak line (under optically pumped conditions) or a shoulder, or onset, in 1269B.

Studies of sample 414 for various angles of the magnetic field with respect to the surface normal demonstrate that all three features are confinement related. Temperature dependence of the line intensities show clearly that all features originate in impurity states, and that the lowest frequency line corresponds to a considerably smaller binding energy than the center impurity (highest frequency) line. This evidence, as well as the fact that the lowest frequency line occurs at a frequency below the calculated position of the m=0 to m=+1 transition for edge impurities, favors the assignment of this line to a transition involving electrons in the wells bound to positive ions (ionized donors) in the AlGaAs barriers. The origin of the middle line in 414 is presently uncertain.

5. Photoconductivity Measurements and Line Profiles

Since the transition energies depend strongly on the position of the impurity ion in a QW, the line profile of a particular intra-impurity transition contains information about the distribution of impurities along the growth direction. Recent theoretical work has focussed on calculations of the absorption line profile in a magnetic field and its dependence on impurity distribution along the QW axis [12]. In order to

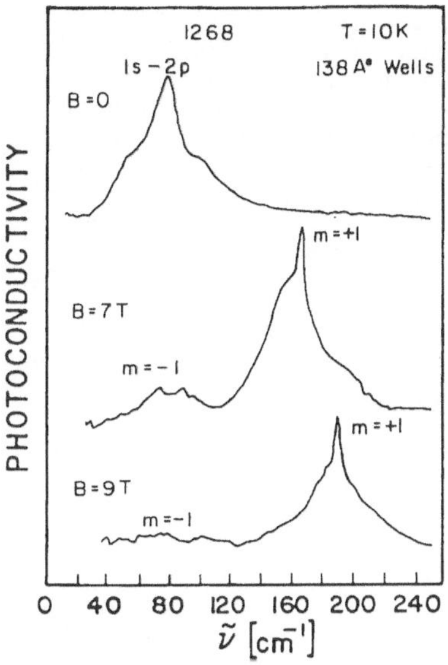

Fig. 5: Capacitively coupled photoconductivity spectra for a center-doped MQW sample at 9-10K and several magnetic fields

investigate details of the line profile, it is desirable to use a technique that avoids the necessity of dividing two spectra to extract weak (typically 1-4%) absorption features. Photothermal ionization spectroscopy has been a very useful technique in studying shallow impurities in bulk materials, since it typically has enhanced sensitivity and improved signal to noise and does not suffer from the background problems associated with transmission measurements. A capacitive coupling technique was used in the present experiments to study photothermal ionization spectra of the confined donors [13].

An example of results for one center-doped sample is shown in Fig. 5. The dominant feature at zero field is the 1s-2p hydrogenic transition, with a series of overlapping transitions to higher excited states contributing to the high-frequency "tail". The evolution of these lines w magnetic field clearly shows the advantages of the photoconductive technique, as well as the beneficial effect of high magnetic fields in separating the various overlapping transitions into clear and well-defined lines. The various transitions are identified in the figure.

Although the photoconductive signal at zero magnetic field is of excellent quality, due to the large number of overlapping transitions, it is not suitable for a detailed lineshape analysis. On the other hand, the 1s-2p(m=+1) transition at high field is well separated from other transitions and is free from such complications. Greene and Bajaj [12] have calculated the 1s-2p(m=-1) transition energy as a function of positive impurity ion position at high magnetic field for various well widths for isolated QWs. These authors have also calculated the line profile for certain impurity distributions. For a uniform distribution the line profile is proportional to $d\Delta(z_i)/dz_i)^{-1}$ where $\Delta(z_i)$ is the lowest m=0 to m=+1 transition energy, and z_i is the position of the impurity ion in the QW, with $z_i=0$ at the center of the well.

Figure 6 shows a comparison of the 1s→2p(m=+1) line in photoconductivity with the same line in absorption at a field of 7T. Both

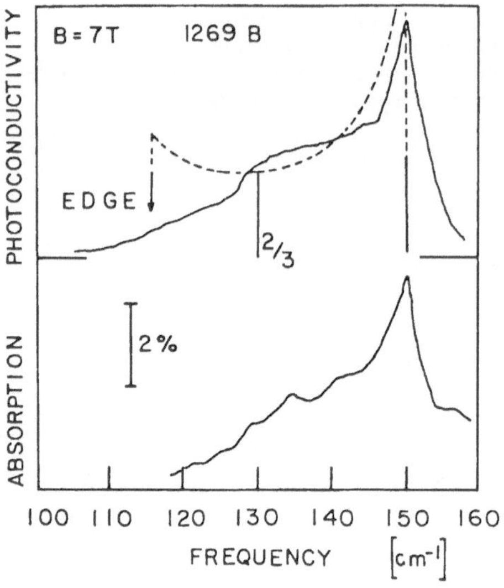

Fig. 6: Comparison of the photoconductivity (top panel) and absorption (bottom panel) line shapes for a center-doped 210Å well-width sample. A theoretical line profile is indicated by the dashed line for uniform doping, as described in the text. Upper trace -T~10K; lower trace-T~15K

measurements were taken at elevated temperatures where effects of barrier impurities are minimal. Both exhibit similar lineshapes, a sharp peak corresponding to impurities near the center with an asymmetric tail extending to lower energies resulting from impurities distributed toward the interface. The photoconductive signal exhibits a clear onset at $128 cm^{-1}$ with a weak tail extending to even lower energies. The signal-to-noise advantage of photoconductivity is clearly evident in the data, and the qualitative similarity in the lineshapes indicates that the photoconductive signal in this case reflects the absorption profile reasonably well. A theoretical profile neglecting lifetime broadening of the individual transitions has been obtained from the results of reference [12] for a uniform distribution extending 70Å to either side of the well center (dashed line in Figure 7). This produces an onset close to that observed and otherwise reproduces the qualitative features of the line profile. This agreement provides evidence of substantial impurity redistribution 30-35Å beyond the intended doping profile. The line profile in narrower wells should be even more sensitive to redistribution on this scale since the transition energy changes more rapidly with position for impurities near the edge.

Acknowledgements

We are grateful to a number of collaborators who were involved in various aspects of this work; B.V. Shanabrook, R.J. Wagner and J. Furneaux of NRL for photoluminescence measurements, FIR laser magnetospectroscopy, and numerous enlightening discussions; G. Wicks and J. Ralston of Cornell University, and J. Comas and W. Beard of NRL for MBE samples. This work was supported in part by ONR through grant #N0001483K0219 and NSF through grant #ECS-8200312 to NRRFSS at Cornell University.

†Present Address: AT&T Bell Laboratories, Murray Hill, NJ.

References

1. R.C. Miller, A.C. Gossard, W.T.Tsang, and O. Munteaunu, Phys. Rev. B 25, 3871 (1982).
2. B.V. Shanabrook and J. Comas, Surf. Sci. 142, 504 (1984).
3. B.V. Shanabrook, J. Comas, T.A. Perry and R. Merlin, Phys. Rev. B 29, 7096 (1984).
4. J.D. Levine, Phys. Rev. 140, A568 (1965).
5. G. Bastard, Phys. Rev. B 24, 4714 (1981).
6. C. Mailhiot, Yia-Chung Chang, and T.C. McGill, Phys. Rev. B 26, 4449 (1982).
7. R.L. Greene and K.K. Bajaj, Solid State Commun. 45, 825 (1983).
8. P. Lane and R.L. Greene, Phys. Rev. B33, 5871 (1986) and private communication.
9. R.L. Greene and K.K. Bajaj, Phys. Rev. B 31, 913 (1985).
10. R.J. Wagner, B.V. Shanabrook, J.E. Furneaux, J. Comas, N.C. Jaroskik, and B.D. McCombe, Proc. of the 11th Int'l Symposium on GaAs and Related Compounds, Biarritz, France, 1984 (Conference Series #74, Adam Hilger Ltd., Bristol and Boston, 1985) p. 315.
11. N.C. Jarosik, B.D. McCombe, B.V. Shanabrook, J. Comas, John Ralston, and G. Wicks, Phys. Rev. Letters 54, 1283 (1985).
12. R.L. Greene and K.K. Bajaj, Phys. Rev. B34, 951 (1986).
13. J.M. Mercy, N.C. Jarosik, B.D. McCombe, J. Ralston and G. Wicks, Proc. of PCSI-13, Pasadena, CA, Jan. 1986, to be published in JVST.

Determination of the Electric Subband Energies and Wavefunctions in GaAs-AlGaAs Heterostructures Using Resonant Subband-Landau-Level Coupling

H. Sigg, C.J.G.M. Langerak, and J.A.A.J. Perenboom*

High Field Magnet Laboratory and Research Institute for Materials, NL-6525 ED Nijmegen, The Netherlands
*Also at: Max-Planck Institut für Festkörperforschung, D-7000 Stuttgart 80, Fed. Rep. of Germany

The resonant coupling of the subband-Landau-levels in the two-dimensional electron gas (2DEG) of high mobility GaAs-AlGaAs heterojunctions has been studied using cyclotron resonance (CR). When the applied magnetic field has a component in the plane of the 2DEG, the electron motion parallel and perpendicular to the interface is coupled and the degeneracy of Landau levels belonging to different electric subbands will be lifted; this coupling will lead to hybridisation of the subband-Landau-levels. A splitting of the CR is observed when $\hbar\omega_c$ coincides with the transition energy E_{10} between the two lowest subbands [1]. This effect was observed some time ago by SCHLESINGER, HWANG and ALLEN [2].

CR measurements were performed on high mobility, MBE grown, modulation doped GaAs-$Al_{0.3}Ga_{0.7}As$ heterojunctions [3] using a far-infrared Michelson interferometer. The substrate was 300 µm thick and wedged to eliminate interference effects, and a 6 µm thick Mylar foil with a 15 nm thick Cr layer served as a semi-transparant backgate. The electron concentration, determined each time from the low-field Shubnikov-de Haas oscillations, was of the order of 2×10^{11} cm^{-2} and could be varied over approximately 20% with this backgate; it was also varied by means of illumination with a red light emitting diode (LED). The samples were immersed in a pumped liquid helium bath and could be rotated over an angle θ up to 15 degrees around an axis perpendicular to the magnetic field.

In the far-infrared spectra we observed a broadening and splitting of the CR. For low and high values of the magnetic field only one resonance line is well developed. As the field approaches the hybridisation regime, this resonance becomes less pronounced and a second resonance starts to develop. Around 12.2 T, the field reaches a value for which the unperturbed CR energy would have been equal to E_{10} and then the two resonances are of equal strength. The splitting of this doublet is proportional to the tilt-angle θ between the applied magnetic field and the normal to the interface.

Figure 1 shows the energies of the CR doublet (solid circles) for the case of $\theta = 3.8°$, the open circles represent the average value. The unperturbed CR energy $\hbar\omega_c = (e\hbar/m^*)B$ is represented by the straight solid line and was determined from the CR far away from the subband-Landau-level crossing. The subband transition energy E_{10} is now obtained from the intersection of the dashed line through the

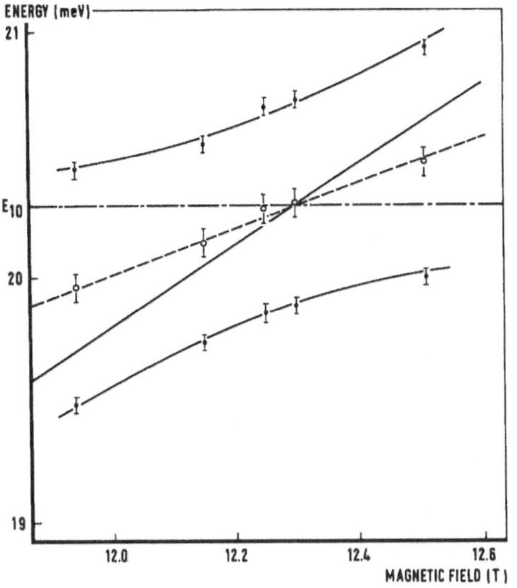

Fig. 1. Measured CR energies as a function of magnetic field (solid circles). The mid-points of the energy doublets are represented by the open circles. The solid line is the unperturbed CR determined from measurements far away from the subband-Landau-level crossing.

mid-points of the CR doublets and the unperturbed CR. ΔE_{10}, the splitting between the two branches around E_{10}, is a measure for the coupling strength: $\alpha_{10} = (1/\theta)(\Delta E_{10}/E_{10})$ and can be calculated from the expressions given by ANDO [4]. The term in the Hamiltonian, responsible for the observed hybridisation, is proportional to z and the component of the applied magnetic field in the plane of the 2DEG, $zB\sin\theta$. The coupling strength is proportional to $(\sin\theta/\theta) \int \zeta^*_1(z) z \zeta_0(z) dz$ and can be easily evaluated for the case of a simple triangular well potential. We obtain $\alpha_{10} = 0.87$. In Ref. [2] a much larger value $\alpha_{10} = \sqrt{2}$ is obtained for the case of a harmonic potential well.

E_{10} and the coupling strength were determined for several samples under the conditions listed in the inset of Fig. 2. The coupling strength becomes smaller with increasing backgate voltage. This can be understood as follows: The interface potential will bend off at its backgate side when a positive backgate voltage is applied and the wavefunction $\zeta_1(z)$ will then extend further into the GaAs layer than $\zeta_0(z)$; in the limit $\sin\theta << 1$, α_{10} is proportional to the overlap matrix element $<z>$, and this matrix element, and consequently the coupling strength, will be reduced. When the sample is illuminated with a red LED however, α_{10} is found to stay approximately constant, implying that in this case the shape of the potential is much less affected.

Figure 3 gives the subband transition energy as a function of the 2D electron concentration n and shows a decrease of E_{10} when the electron concentration is increased with an applied backgate voltage. In contrast, an increase of the carrier concentration through illumination leads to an increase of E_{10}. With photoexcitation one produces a transfer of charge from the AlGaAs layer to the GaAs layer, which leads to an increase of the electric field at the interface: The

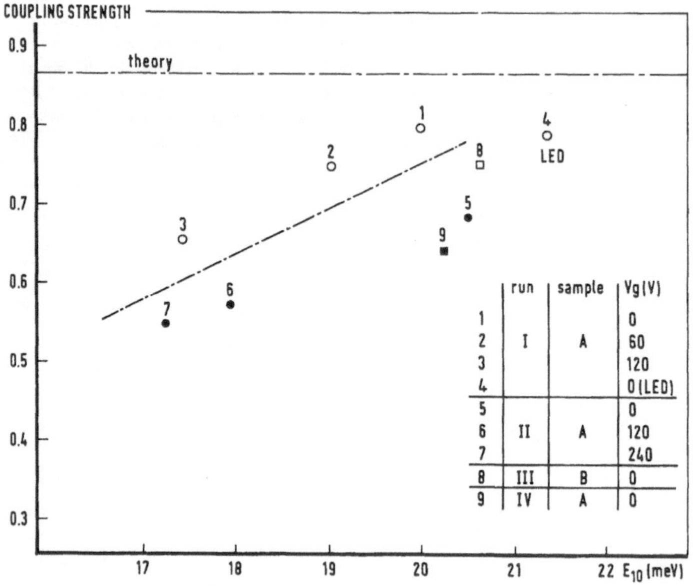

Fig. 2. The coupling strength α_{10} as a function of E_{10}. In the inset the references to sample and backgate voltages are given.

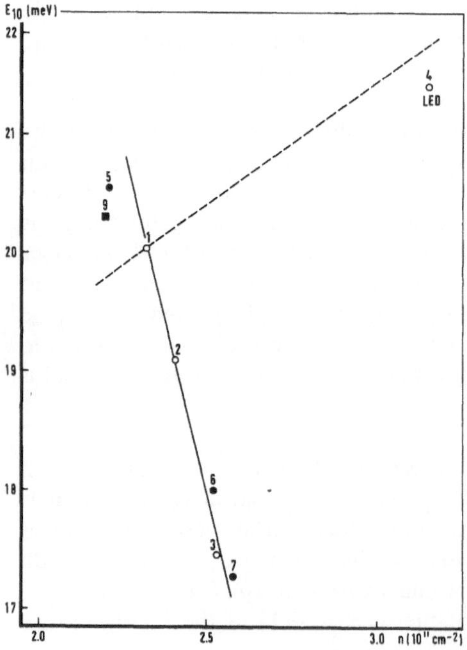

Fig. 3. Subband transition energy E_{10} as a function of the electron density n. The numbers refer to the list in Fig. 2. The solid line and the dashed line are calculated from Ref. [6].

dashed line represents the result of selfconsistent potential well calculations of STERN and DAS SARMA [6], assuming a value for the unintentional background doping of 2×10^{14} cm^{-3}. On the other hand, with a backgate voltage an additional electron concentration Δn can be introduced into the 2DEG, but with no charge flow through the interface: using Poisson's equation, one finds an induced electric field between the channel and the backgate of $\Delta F_{BG} = -e\Delta n/\varepsilon$, where ε is the dielectric constant for GaAs. The "slope" of the confining potential is herewith reduced, and again with the results of Ref. [6] as a function of depletion field and carrier concentration, in this case the solid line is obtained, in remarkably good agreement with the experimental data.

With resonant subband-Landau-level coupling one can determine the subband transition energy E_{10} and study its variation with the field at the interface. The coupling strength depends on the extension of the subband wavefunctions and gives valuable information on the shape of the interface potential. It is therefore a technique to sensitively test interface potential calculations.

Acknowledgement

Part of this work was suppported by the Stichting voor Fundamenteel Onderzoek der Materie (FOM) with financial support from the Nederlandse Organisatie voor Zuiver Wetenschappelijk Onderzoek (ZWO), and by a grant from the European Economic Community.

References

1 G.L.J.A. Rikken, H.W. Myron, C.J.G.M. Langerak and H. Sigg, Surf. Sci. **170**, 160 (1986).
2 Z. Schlesinger, J.C.M. Hwang and S.J. Allen, Phys. Rev. Lett. **50**, 2098 (1983).
3 G. Weimann and W. Schlapp, Appl. Phys. Lett. **46**, 411 (1985).
4 T. Ando, Phys. Rev. B **19**, 2106 (1979).
5 T. Ando, A.B. Fowler and F. Stern, Rev. Mod. Phys. **54**, 437 (1982) and references therein.
6 F. Stern and S. Das Sarma, Phys. Rev. B **30**, 840 (1984).

Cyclotron Resonance in n-GaAs/GaAlAs Heterojunction

F. Thiele[1], *W. Hansen*[1], *M. Horst*[1], *J.P. Kotthaus*[1], *J.C. Maan*[2], *U. Merkt*[1], *K. Ploog*[3], *G. Weimann*[4], *and A.D. Wieck*[1,2]

[1]Institut für Angewandte Physik, D-2000 Hamburg 36, Fed. Rep. of Germany
[2]Hochfeldlabor des Max-Planck-Institut für Festkörperforschung, F-38042 Grenoble, France
[3]Max-Planck-Institut für Festkörperforschung, D-7000 Stuttgart 80, Fed. Rep. of Germany
[4]Forschungsinstitut der Deutschen Bundespost, D-6100 Darmstadt, Fed. Rep. of Germany

1. Introduction

We report on cyclotron resonance experiments of inversion electrons in high mobility GaAs/GaAlAs heterojunctions in a wide range of magnetic fields B and temperatures T with special emphasis on saturation effects and subband nonparabolicity. The experiments have been performed with a Fourier spectrometer (B<12T) and a far-infrared laser (B>12T).

2. Saturation effects

In high mobility samples one must take into account saturation effects when mobilities are extracted from experimental linewidths. To demonstrate this, we have calculated cyclotron resonance profiles assuming classical Drude conductivity for the inversion electrons [1]. In Fig.1(a) the ratio of

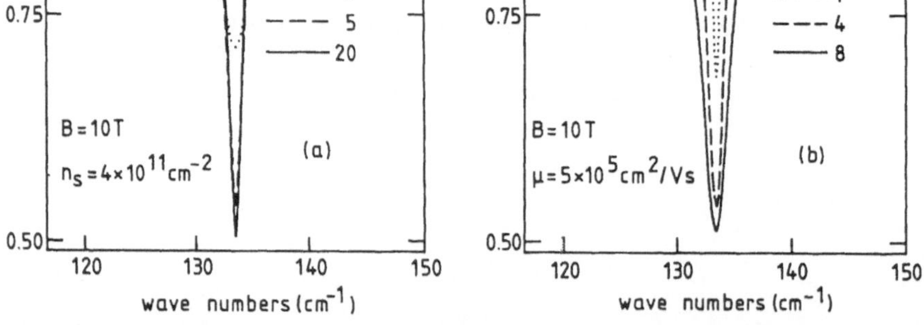

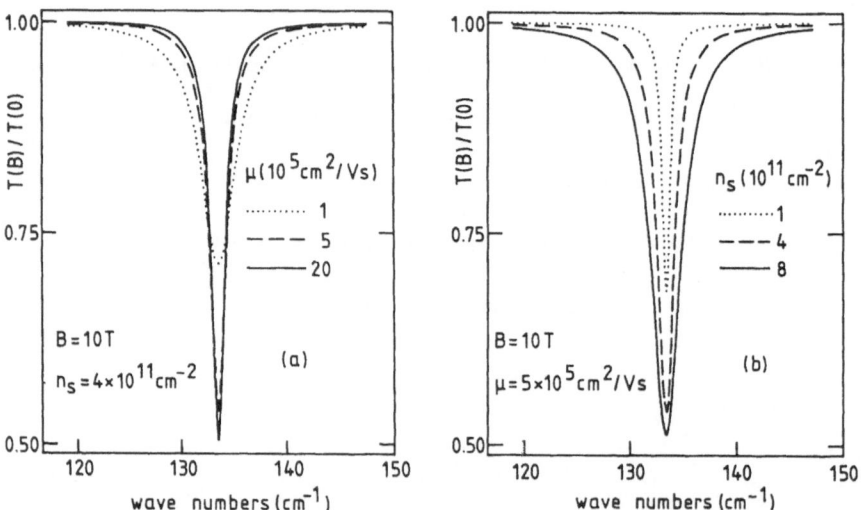

Fig. 1. Calculated cyclotron resonance profiles demonstrating saturation effects. The ratio of the transmittance in the presence T(B) and in the absence of the magnetic field T(0) is shown for (a) various mobilities μ and (b) various electron densities n_s assuming classical Drude conductivity and an effective mass $m^*=0.07\ m_o$.

transmittance T(B) for linearly polarized light in a magnetic field and transmittance T(0) in the absence of a magnetic field is depicted for various mobilities μ and electron densities n_s. When the mobility is increased at a fixed density the ratio T(B)/T(0) cannot become less than 0.5 and the mobility no longer can be determined from the halfwidth $\Delta\omega$ (FWHM) using the simple relation $\mu=2\omega_c/\Delta\omega\cdot B$ that is valid in the limit of small changes of transmittance. Saturation effects are also evident when the density n_s is changed at a fixed mobility as is shown in Fig.1(b). To extract realistic values of mobilities we have compared our experimental spectra with calculated line profiles [see Fig.3(b)].

3. Cyclotron masses and linewidths

Figure 2(a) shows cyclotron masses $m^*=eB/\omega_c$ of two samples with different electron densities at liquid helium temperatures versus magnetic field B or filling factor $\nu=n_s\cdot h/e\cdot B$. As a result of band nonparabolicity [2,3] the mass continuously increases at filling factors $\nu<2$ in both samples. For the sample with electron density $n_s=5.0\cdot10^{11}cm^{-2}$ oscillations of the mass are apparent. We interpret this as quantum oscillations caused by nonparabolicity. The higher Landau transitions $n\rightarrow n+1$, which are observed at lower magnetic fields, have lower transition energies, i.e., correspondingly higher masses. It is also clear from the figure that the subband masses which are

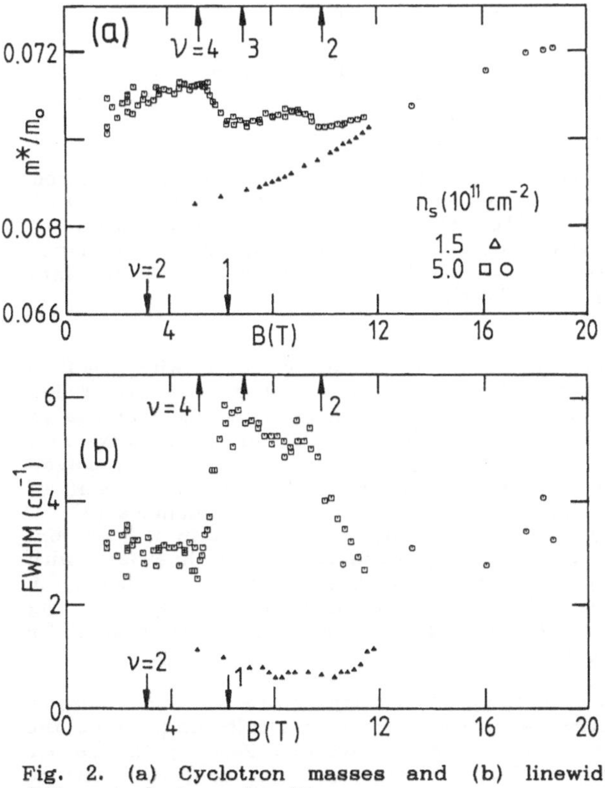

Fig. 2. (a) Cyclotron masses and (b) linewidths for two samples with different electron densities n_s.

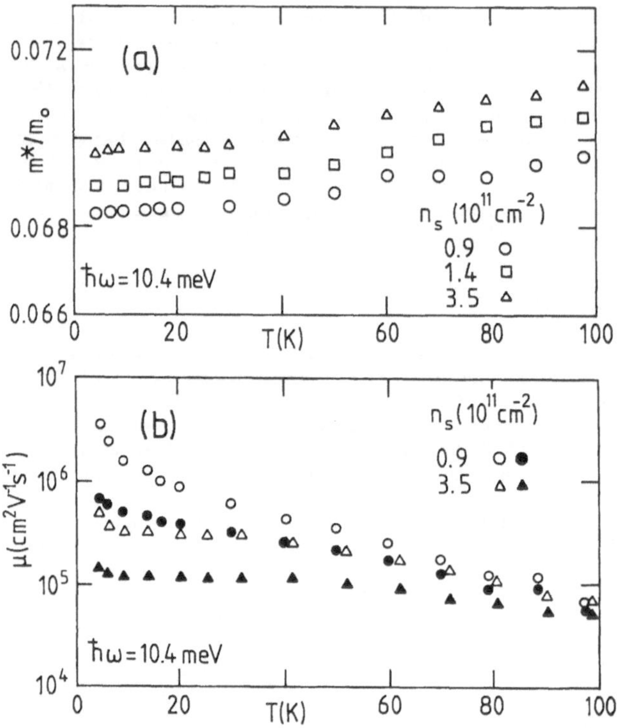

Fig. 3. (a) Temperature dependence of cyclotron masses measured at a fixed laser energy $\hbar\omega$ for three samples with spacer thicknesses 630 Å (n_s=0.9·10^{11}cm^{-2}), 360 Å (n_s=1.4·10^{11}cm^{-2}), and 200 Å (n_s=3.5·10^{11}cm^{-2}). (b) Temperature dependence of mobilities extracted from cyclotron resonance linewidths using the simple relation $\mu=2/\Delta B$ (closed symbols) and taking into account saturation effects (open symbols).

obtained in the limit of vanishing magnetic fields increase with electron density. Polaron effects were found to be unimportant at magnetic fields B<16T [2-4]. Linewidths are shown in Fig.2(b) where a strong increase at filling factor $\nu\sim3$ is observed for the sample with the higher density. This effect we interpret as being caused by a superposition of unresolved Landau transitions $0^-\to1^-$ and $1^+\to2^+$ which are expected, again as a result of nonparabolicity, to occur at slightly different resonance frequencies [3], i.e., inhomogenous broadening. It is interesting to note that in the present samples we do not observe an oscillatory behavior of the linewidth with maxima at even filling factors. This effect has been observed previously in samples with lower mobilities [5] and has been explained by screening effects [6,7]. Also no $\sqrt{B}$-dependence is found which is characteristic for short-range scatterers only [6].

Figure 3(a) shows masses as a function of temperature for three samples with different electron densities. At present we do not fully understand the reason for the increase of the masses with temperature. Possibly the masses increase because the subband energies increase [8] and nonparabolicity is stronger. On the other hand, polaron effects may become effective at elevated temperatures similar as under hot electron conditions [9]. Figure 3(b) shows apparent mobilities $\mu=2/\Delta B$ as extracted directly from the observed linewidths (full symbols) and values corrected for the saturation effect (open symbols). Due to the interaction of the inversion electrons with optical phonons the mobility decreases at higher temperatures [10]. At the highest temperatures saturation effects become unimportant and the mobility reaches a value which is independent of mobility at low temperatures. In fact, the sample showing the highest mobility at low temperatures has a rather thick spacer layer (d=630 Å) which strongly reduces ionized impurity scattering. This mechanism is dominant at low temperatures [10] but has no effect on the polar optical scattering. If we evaluate the cyclotron width on this sample at low temperatures and $\hbar\omega$=10.4meV taking into account saturation we extract a mobility from cyclotron resonance that exceeds the value of $\mu_{dc}=4\cdot 10^5 cm^2/Vs$ obtained from transport experiments.

We thank U. Rössler and G. Lommer for valuable discussions and acknowledge financial support of the Deutsche Forschungsgemeinschaft.

References

1. K. W. Chiu, T. K. Lee, J. J. Quinn: Surf. Sci. 58, 182 (1976).
2. M. Horst, U. Merkt, W. Zawadzki, J. C. Maan, K. Ploog: Solid State Commun. 53, 403 (1985).
3. G. Lommer, F. Malcher, U. Rössler: Superlattices and Microstructures, in press.
4. G. Lindemann, W. Seidenbusch, R. Lassnig, J. Edlinger, E. Gornik: Physica 117B&118B, 649 (1983).
5. Th. Englert, J. C. Maan, Ch. Uihlein, D. C. Tsui, A. C. Gossard: Solid State Commun. 46, 545 (1983).
6. T. Ando, Y. Murayama: J. Phys. Soc. Jpn. 54, 1519 (1985).
7. R. Lassnig, E. Gornik: Solid State Commun. 47, 959 (1983).
8. F. Stern, S. Das Sarma: Phys. Rev. B 30, 840 (1984).
9. E. Gornik: private communication.
10. W. Walukiewicz, H. E. Ruda, J. Lagowski, H. C. Gatos: Phys. Rev. B 30, 4571 (1984).

Cyclotron Resonance in InSb in Crossed Electric and Magnetic Fields

U. Merkt

Institut für Angewandte Physik, Universität Hamburg, Jungiusstr. 11, D-2000 Hamburg 36, Fed. Rep. of Germany

In electron inversion layers of metal-oxide-semiconductor structures on InSb cyclotron resonance in crossed electric and magnetic fields ExB is studied. The crossed-field configuration is established by the surface electric field of the space-charge potential and by a magnetic field that is applied parallel to the layers. Destruction of the Landau quantization is observed when the ratio E/B exceeds a critical value.

1. Introduction

Semiconductor electrons in crossed electric and magnetic fields ExB represent an illustrative system to study the motion of lattice electrons in external fields [1,2] and to test the validity of the effective-mass approximation (EMA) that is commonly employed in the theoretical description. When one wants to include the case of purely electric fields one must account for the presence of a potential barrier [3]. Otherwise the electron motion would be unlimited in space with continuous acceleration along the electric field direction. In fact, it turns out that the barrier is also needed in the crossed-field configuration at finite magnetic fields since there is a transition from a magnetic-type of motion, which is restricted in space by the cyclotron radius l, to an electric-type of motion when the field ratio E/B exceeds a critical value [4]. This transition can only be described by the two-band model and is analogous to the behavior of an electron in vacuum, as described by the relativistic equation of motion [5].

We have recently demonstrated the use of inversion layers on InSb to study cyclotron resonance in the crossed-field configuration [6-8]. In these experiments the above transition is observed through the destruction of the Landau quantization itself when the electric field of the space-charge potential results in field ratios that exceed the critical value u which is the maximum velocity possible in the conduction band of a narrow-gap semiconductor. The maximum velocity u is a consequence of the highly non-parabolic conduction band: The energy dispersion is parabolic only in the vicinity of the conduction band minimum but becomes linear at higher wave vectors. This linear increase defines the maximum velocity $u=(\varepsilon_g/2m_0^*)^{1/2}$ with the gap energy ε_g and the band edge effective mass m_0^*.

2. Theory

2.1 Classical trajectories

It is illuminating to discuss the classical trajectories of electrons that run along a barrier [8,9] in the presence of crossed fields ($E \parallel z$, $B \parallel x$) before treating the quantum mechanical description in terms of eigenenergies and

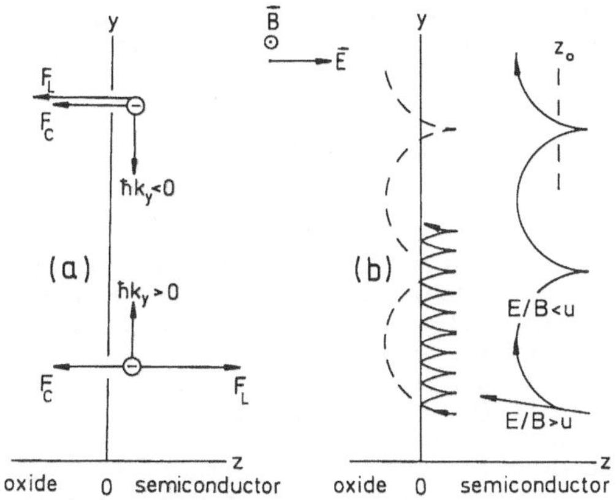

Fig.1. Electrons in crossed electric and magnetic fields near a barrier. (a) Coulomb force F_C and Lorentz force F_L for positive and negative momentum $\hbar k_y$. (b) Classical trajectories for two center coordinates z_0. The motion of the bulklike electron ($z_0/l \gg 1$) changes from a magnetic-type (E/B<u) to an electric-type (E/B>u).

quantum numbers. The electrons are subjected to the Coulomb force F_C, which is directed toward the interface and tries to bind all electrons to the barrier, and to the Lorentz force F_L, which pulls electrons with positive momentum $\hbar k_y>0$ into the interior of the semiconductor against the action of the Coulomb force [see Fig.1(a)]. On InSb with its small effective mass the Lorentz force may win the trial of strength and the electron can complete a Landau cycloide that lies completely inside the semiconductor [see Fig.1(b)]. Such bulklike electrons will show cyclotron resonance in crossed fields that is not affected by the presence of the barrier. Electrons with a negative momentum are bound to the barrier by the combined action of both forces and behave similar to electrons in purely electric subbands. The optical excitations of such electrons are diamagnetically shifted intersubband resonances [10].

The behavior of the bulklike electrons is rather unphysical if strong electric or vanishing magnetic fields are considered. In such fields an electric type of motion with continuous acceleration along the electric field direction is expected. For an electron in free space this is indeed obtained from the relativistic equation of motion [5]. The drift velocity E/B has a physical meaning only if it is less than the light velocity c and, in fact, there are two drift velocities in the theory of special relativity: A magnetic drift velocity $v_{dm}=E/B$ and an electric drift velocity $v_{de}=c^2B/E$ for field ratios less and higher than the light velocity, respectively. In coordinate frames moving with the proper drift velocities we observe Landau circles in the magnetic case (E/B<c) and continuous acceleration along the electric field direction in the electric case (E/B>c). In the magnetic case the trajectories in the laboratory frame are cycloides and in the electric case we have straight lines tilted by the influence of the magnetic field [see Fig.1(b)]. In the latter case, the Landau quantization is destroyed and the electrons will hit the barrier as they do in the purely electric case. For semiconductors it has been realized some time ago [1,2] that the validity of the EMA implies that the E/B ratio is not too large. The EMA predicts magnetic-type of motion for all field ratios E/B and as such is analogous to a nonrelativistic theory. Then the two-band model for crossed fields was developed which correctly accounts for both types of behavior, as does the relativistic theory for electrons in free space [1,11].

2.2 Effective-mass approximation

Before we describe the quantization in crossed electric and magnetic fields we first consider the purely electric case. It is well-known that the energy spectrum in an electric field is continuous as long as there is no barrier in the direction of the electric field ($E \parallel z$) [11]. In any realistic system, however, there is a potential barrier and the electrons become quantized into electric subbands: the motion is bound in the z-direction and is described by discrete subband energies $\mathcal{E}_i$ (i=0,1,...), whereas the motion is free in the plane parallel to the barrier, i.e., perpendicular to the electric field [12]. This free motion is described by the quasi continuous wave vectors k_x and k_y and we have the well known quasi two-dimensional subbands

$$\mathcal{E}_{i,k_x,k_y} = \mathcal{E}_i + \frac{\hbar^2(k_x^2 + k_y^2)}{2m_0^*} \quad . \tag{1}$$

In order to obtain a simple description for the subband energies $\mathcal{E}_i$ we assume a constant electric field E, i.e., we use the triangular-well approximation [12]. This potential provides only a rough approximation to a self-consistent potential but it has the merit that the energies can be calculated analytically:

$$\mathcal{E}_i = \left(\frac{9\pi^2}{8m_0^*} \right)^{\frac{1}{3}} (e\hbar E)^{\frac{2}{3}} \left(i + \frac{3}{4}\right)^{\frac{2}{3}} \quad . \tag{2}$$

For inversion layers, a suitable effective field E can be derived from a comparison of the Fang-Howard variational model [12] and the triangular-well approximation. When one puts equal the average distances of electrons $\langle z \rangle$ in the ground subband calculated in the two approaches, an effective field

$$E = \frac{\pi^2}{12} \frac{e}{\varepsilon_0 \kappa} \left(n_{depl} + \frac{11}{32} n_s \right) \tag{3}$$

is obtained with dielectric constants ε_0 and κ of vacuum and semiconductor, respectively. In the following we use this equation to relate the electron density n_s to the strength E of the surface electric field. This procedure will allow a fairly accurate description of the experiments as long as only lower subband states (i=0,1) are involved.

In case of a purely magnetic field applied parallel to the interface we obtain magnetic surface levels [8]

$$\mathcal{E}_{i,k_x,z_0} = \frac{\hbar^2 k_x^2}{2m_0^*} + \left[\nu_i(z_0) + \frac{1}{2} \right] \hbar\omega_c \tag{4}$$

which are depicted for the ground magnetic level ($i=0, k_x=0$) in Fig.2(a). The center coordinate $z_0=k_y l^2$ is the most important quantum number and denotes the center of oscillation of the electron as long as it is positive, i.e., away from the interface. It has a somewhat formal meaning if it is negative. Then, it describes strong binding of the electron to the interface due to the action of the Lorentz force [8]. In general, one needs the indices $\nu_i(z_0)$ of the parabolic cylinder functions to calculate the magnetic energies according to (4). In the bulk of the semiconductor ($z_0 \to +\infty$) we have the common highly degenerate Landau levels since the indices become independent of the center coordinate z_0. The eigenenergies in crossed fields [8,11] are obtained from the purely magnetic solution by adding the straight line $\hbar^2 k_d^2/2m_0^* + eEz_0$ with momentum $\hbar k_d = m_0^*(E/B)$ to the magnetic surface levels:

$$\varepsilon_{i,k_x,z_0} = \frac{\hbar^2 (k_x^2 + k_d^2)}{2m_0^*} + [\ \nu_i(z_0) + \frac{1}{2}\ ]\ \hbar\omega_c + eEz_0 \quad . \tag{5}$$

This is demonstrated in Fig.2(a,b). There is straightforward interpretation for this: the first term corresponds to the kinetic energy of the classical ExB drift, the second is the potential energy of the electron in the electric field. Again, the electron position is given by its center coordinate z_0 which now obeys the relation $z_0=(k_y-k_d)l^2$.

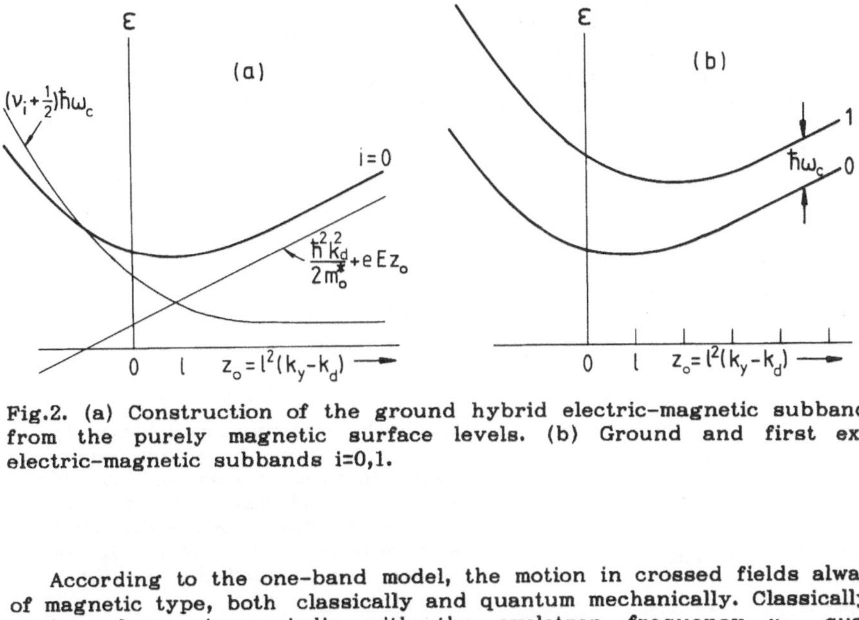

Fig.2. (a) Construction of the ground hybrid electric-magnetic subband i=0 from the purely magnetic surface levels. (b) Ground and first excited electric-magnetic subbands i=0,1.

According to the one-band model, the motion in crossed fields always is of magnetic type, both classically and quantum mechanically. Classically the motion always is periodic with the cyclotron frequency ω_c, quantum mechanically we always start from the purely magnetic energies and the bands run parallel separated by the cyclotron energy $\hbar\omega_c$ for the center coordinates which are more than about one cyclotron radius inside the semiconductor [see Fig.2(b)]. Electrons with such center coordinates will show cyclotron resonance in crossed fields that is not affected by the presence of the electric field. This may be regarded as the main result of the EMA.

2.3 Three-level k·p theory

The eigenenergies of electrons in crossed fields have been calculated taking into account an InSb-type bandstructure [6]: a Γ_6 conduction band at energy $+\mathcal{E}_g/2$, a Γ_8 valence level at energy $-\mathcal{E}_g/2$, and a Γ_7 spin-orbit split-off level at energy $-\mathcal{E}_g/2-\Delta$ with spin-orbit energy Δ. It turns out that the most important parameter in this description is the ratio of the two relevant velocities $\delta = v_{dm}/u$. Only in the regime $0<\delta<1$ we have magnetic quantization. As the electric field becomes stronger ($\delta>1$) there is quantization similar to the one of purely electric subbands. In other words, a sufficiently strong transverse electric field destroys the Landau levels. The energies of the Landau levels ($k_x=0$) for bulklike electrons ($z_0/l \gg 1$) in crossed fields ($0<\delta<1$) are

$$\mathcal{E}_{i,\pm} = v_{dm}\hbar k_y + (1-\delta^2)^{\frac{1}{2}} \left[\left(\frac{\mathcal{E}_g}{2}\right)^2 + \mathcal{E}_g D_{i,\pm}\right]^{\frac{1}{2}} \quad , \tag{6}$$

$$D_{i,\pm} = (1-\delta^2)^{\frac{1}{2}} \hbar\omega_c \left(i+\frac{1}{2}\right) \quad \pm \quad \frac{1}{2} g_0^* \mu_B B \tag{7}$$

with cyclotron frequency $\omega_c = eB/m_0^*$ and effective spin factor g_0^* taken at the conduction band edge. For vanishing electric fields ($\delta=0$) we recover the well-known formulas for Landau levels in narrow-gap semiconductors [13]. Except for the spin term which is of atomic origin in semiconductors and as such does not have correspondence in the free-electron case, the Landau levels in crossed fields can be derived using the semirelativistic analogy [6] (see also Fig.3). The factors $(1-\delta^2)^{1/2}$ in (6) and (7) correspond to the relativistic Doppler shift and to a Lorentz transformation of the magnetic field, respectively.

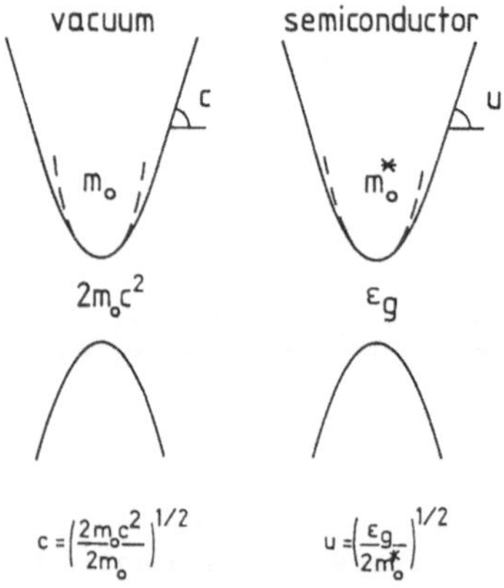

Fig.3. The relativistic analogy between electrons in vacuum and in semiconductors according to the two-band model. The maximum velocity u in the conduction band of the semiconductor is derived by its correspondence to the light velocity c.

The transition matrix elements for the fundamental and harmonic cyclotron transitions have been calculated ignoring effects of electron spin [7]. In view of our experiments we consider light with amplitude of vector potential A_0 incident parallel to the electric field direction ($E \parallel z$) and polarized perpendicular to the magnetic field direction ($B \parallel x$). We also assume ratios $\hbar\omega_c/\mathcal{E}_g \ll 1$ as it is justified in view of our experiments and evaluate the first nonvanishing order of the matrix elements analytically. We normalize the results with the matrix element $M_{01}(\delta=0)=A_0 e l \omega_c/\sqrt{2}$ of the fundamental transition in the absence of an electric field and thus obtain for the fundamental transition

$$M_{01}(\delta) = (1-\delta^2)^{\frac{1}{4}} \frac{2\delta^4-4\delta^2+4(1-\delta^2)^{\frac{3}{2}}}{(2-\delta^2)^2 + \delta^2} , \tag{8}$$

and for the first harmonic transition

$$M_{02}(\delta) = \sqrt{8} \left(\frac{\hbar\omega_c}{\mathcal{E}_g}\right)^{\frac{1}{2}} \delta\left[1-\delta^2\right]^{\frac{1}{2}} \frac{2\delta^4-5\delta^2+1+4(1-\delta^2)^{\frac{3}{2}}}{(2-\delta^2)^2 + \delta^2} . \tag{9}$$

The transition strengths, i.e., squared matrix elements are depicted in Fig.4.

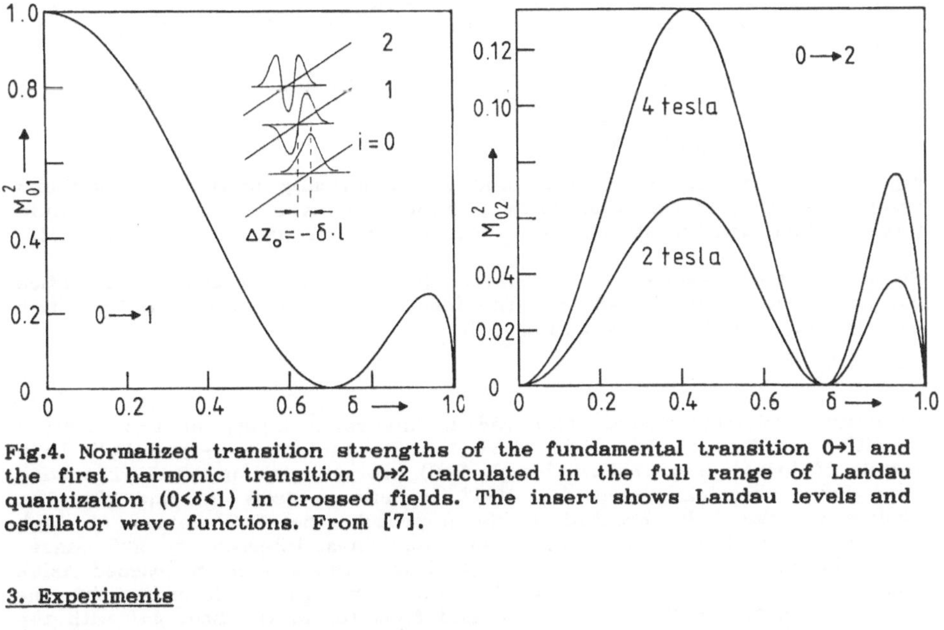

Fig.4. Normalized transition strengths of the fundamental transition 0→1 and the first harmonic transition 0→2 calculated in the full range of Landau quantization (0<δ<1) in crossed fields. The insert shows Landau levels and oscillator wave functions. From [7].

3. Experiments

3.1 The divergence of the cyclotron mass

Fourier spectra taken for various magnetic fields B at a fixed inversion electron density n_s are shown in Fig.5. The sharp resonances are cyclotron resonances 0→1 (CR), the broad and weak ones are diamagnetically shifted intersubband resonances 0→1 (ISR). Laser spectra taken at a fixed photon energy $\hbar\omega$ for various electron densities n_s are shown in Fig.6. In this figure, we observe at low densities 0→1 cyclotron transitions and at higher densities a second peak which is due to 1→2 cyclotron transitions. This peak appears as a result of the inhomogeneity of the surface electric field [6,8]. As the density n_s, i.e., the surface electric field E increases, the 0→1 resonances shift to higher magnetic fields and finally the line vanishes.

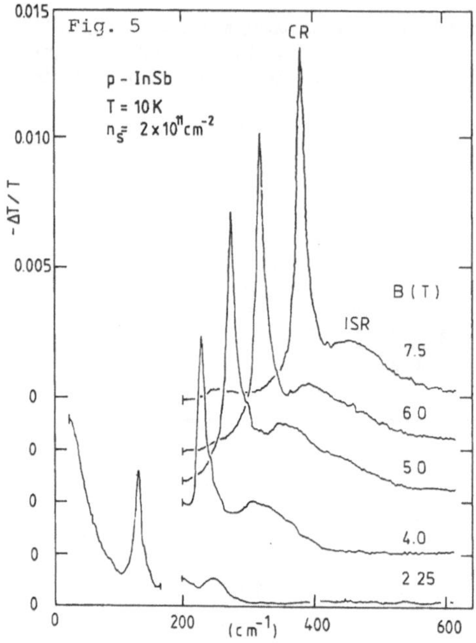

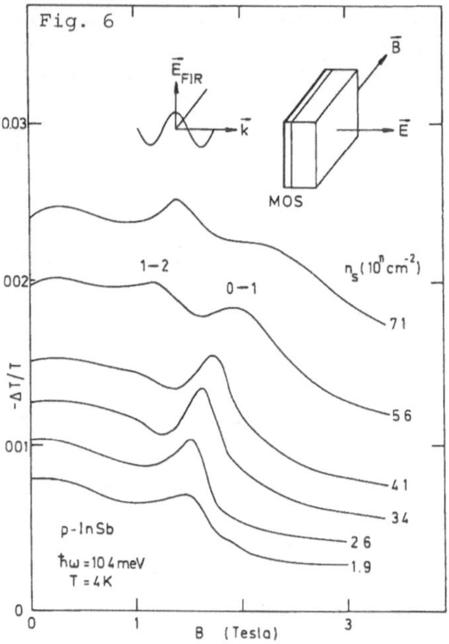

Fig.5. Cyclotron resonances (CR) and diamagnetically shifted intersubband resonances (ISR) in magnetic fields B parallel to inversion layers on p-InSb (crossed-field configuration). All resonances are 0→1 transitions. From [14].

Fig.6. Cyclotron resonances in crossed fields at various electron densities n_s, i.e., electric field strengths. The inset shows the sample configuration with the incident light wave. From [6].

Apparent cyclotron masses measured in inversion layers at two electron densities n_s, i.e., electric field strengths $E \neq 0$ and in a n-type bulk InSb sample of low doping ($n = 6 \times 10^{13} cm^{-3}$, $E = 0$) are depicted in Fig.7. The solid lines have been calculated for $0^+ \rightarrow 1^+$ transitions from (6) and (7). The cyclotron masses in crossed fields approach the E=0 values at high magnetic fields but strongly differ at lower ones: Whereas the E=0 masses extrapolate to the band edge mass $m_0^* = 0.014 m_0$, the masses in crossed fields show a steep increase. Theoretically, they diverge as described by the relation $m^* \propto m_0^*(1-\delta^2)^{-1}$ which is obtained from (6) in the limit $\delta \rightarrow 1$ with the usual definition of the cyclotron mass. The divergence is indicated by the arrows for the two electric field strengths in Fig.7. Experimentally the disappearance of the cyclotron maximum 0→1 in Fig.6 corresponds to the condition $\omega\tau < 1$ since we have a finite electron relaxation time τ.

The divergence of the cyclotron mass and the disappearance of cyclotron resonance are spectacular manifestations of the relativistic analogy [4,6]. The factor $(1-\delta^2)$ in the denominator of the apparent mass ($\delta \rightarrow 1$) is a consequence of two relativistic effects: the magnetic field has to be Lorentz transformed to the system moving with the magnetic drift velocity E/B which gives one factor $(1-\delta^2)^{1/2}$. Since the energy has to be transformed back to the laboratory system we have another factor $(1-\delta^2)^{1/2}$ that corresponds to the relativistic transverse Doppler shift.

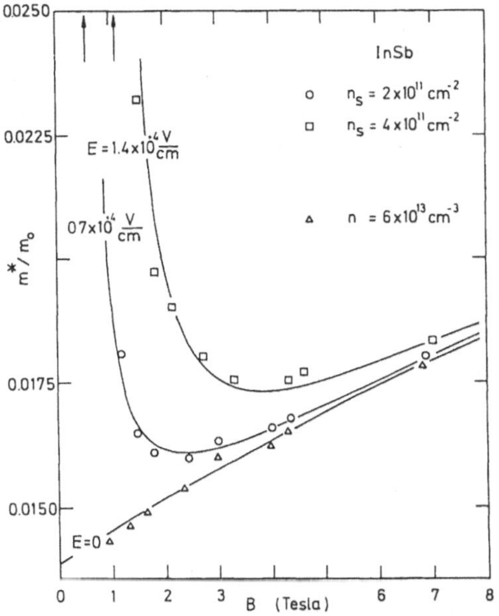

Fig.7. Cyclotron masses in crossed fields (circles and squares) and in the absence of an electric field (triangles). The solid lines are calculated for $0^+ \to 1^+$ transitions, the arrows indicate the critical magnetic field strengths where cyclotron resonance vanishes and the mass diverges in the absence of scattering for the two electric field strengths $E \neq 0$. From [6].

3.2 Breakdown of the selection rules

There is still more evidence for the destruction of Landau quantization in a transverse electric field: the well-known oscillator selection rules break down [15] and harmonic cyclotron resonance can become comparable in strength to the fundamental one (see Fig.4). In the bulk of InSb the harmonics $0 \to 2$ are rather weak with oscillator strengths that are only about 10^{-3}–10^{-4} of the one for the fundamental resonance and they are induced by different mechanisms: band warping, inversion asymmetry, or impurities [13]. Figure 8 shows spectra in crossed fields measured at a constant photon energy $\hbar\omega$ versus magnetic field B at various electron densities n_s. At the lower densities the predominant maxima at $B \simeq 1.8$T are due to $0 \to 1$ cyclotron transitions. At higher densities this resonance shifts to higher magnetic fields and vanishes as was described above. At densities $n_s > 3.0\times10^{11}\,cm^{-2}$ $1 \to 2$ resonances and new resonances occur that are marked by arrows. If their resonance position is extrapolated to electron density $n_s = 0$ it agrees with the one of the harmonic $0 \to 2$ transition in bulk n-type InSb [7].

Qualitatively, the observation of strong harmonic transitions can be explained by our k·p model. Direct quantitative comparison, however, is not possible since the relation between electron density and electric field strength given in (3) no longer is a good approximation for the final state i=2. With the field calculated according to (3) we obtain the electron density $n_s \simeq 3\times10^{11}\,cm^{-2}$ that should correspond to maximum excitation of the $0 \to 2$

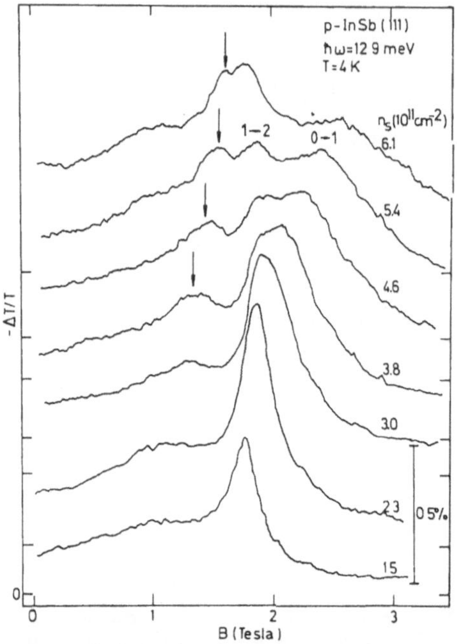

Fig.8. Fundamental cyclotron resonances (0→1, 1→2) and harmonic cyclotron resonance (0→2, see arrows) in the crossed-field configuration. From [7].

transition [$\delta \approx 0.4$, see Fig.(4)] whereas in the experiments we find densities $n_s \approx 4\text{-}5 \times 10^{11} cm^{-2}$. In view of the simplifications we have employed in our description we consider this as satisfactory agreement.

An intuitive picture why harmonics arise in crossed fields is presented in the inset of Fig.4. Landau levels i=0,1,2 are depicted away from the interface together with oscillator wave functions. In the two-level model there is a shift Δz_0 of the center coordinate when the energy of the electron is changed [8]. This shift is $\Delta z_0 \approx -\delta \cdot l \cdot \Delta n$ in the limit $\delta \ll 1$. Intuitively this means that the Landau electron steps into the direction of the electric field during a cyclotron transition. It is then clear that the usual selection rules for the harmonic oscillator no longer are valid. However, this simplified picture does not use the full four-component wave function and thus cannot account quantitatively for the results given in (8) and (9).

4. Conclusion

Inversion layers on InSb in magnetic fields applied parallel to the layers represent an interesting system for magneto-optical experiments. The externally applied magnetic field and the internal surface electric field of the space-charge potential provide a crossed-field configuration with strong electric fields. This way, cyclotron resonance in narrow-gap semiconductors can be studied experimentally in crossed electric and magnetic fields. The destruction of the Landau quantization itself by the strong transverse electric field that has been predicted a long time ago [4,15] could be demonstrated by a divergence of the apparent cyclotron mass and by a

breakdown of the oscillator selection rules. These observations not even in principle can be described within the framework of the one-band effective-mass approximation (EMA) that successfully describes the dynamics in purely electric or in purely magnetic fields but becomes incorrect at large field ratios E/B, i.e., magnetic drift velocities. On the other hand, a fairly good description is obtained with the two-band model of the semiconductor bandstructure. This is analogous to the motion of an electron in free space: only if its velocity is less than the light velocity c can it be described by nonrelativistic mechanics which is a one-band model. In a semiconductor the limiting velocity that corresponds to the light velocity is the maximum velocity u possible in the conduction band according to the two-band model.

This review is based on work which in course of time I did together with J.H. Crasemann, M. Horst, S. Klahn, and W. Zawadzki. I thank J.P. Kotthaus for his continuous encouragement and the Deutsche Forschungsgemeinschaft for financial support.

References

1. J. Zak, W. Zawadzki: Phys.Rev. 145, 536 (1966)
2. A.G. Aronov, G.E. Pikus: Sov. Phys. JETP 24, 339 (1967)
3. J.H. Crasemann, U. Merkt, J.P. Kotthaus: Phys. Rev. B 28, 2271 (1983)
4. W. Zawadzki, B. Lax: Phys. Rev. Lett. 16, 1001 (1966)
5. J.D. Jackson: Classical Electrodynamics, 2nd. ed. (Wiley, New York, 1975) pp.522 and 582
6. W. Zawadzki, S. Klahn, U. Merkt: Phys. Rev. Lett. 55, 983 (1985); Phys. Rev. B 33, 6916 (1986)
7. S. Klahn, M. Horst, U. Merkt: In Proceedings of the 18th International Conference on the Physics of Semiconductors, Stockholm, 1986, in press
8. U. Merkt: Phys. Rev. B 32, 6699 (1985)
9. F. Koch: In Physics in High Magnetic Fields, Proceedings of the Oji International Seminar, Hakone, Japan, 1980, ed. by S. Chikazumi and M. Miura (Springer, Berlin, 1981)
10. S. Oelting, U. Merkt, J.P. Kotthaus: Surf. Sci. 170, 402 (1986)
11. W. Zawadzki: Surf. Sci. 37, 218 (1973)
12. T. Ando, A.B. Fowler, F. Stern: Rev. Mod. Phys. 54, 437-672 (1982)
13. C.R. Pidgeon: In Handbook on Semiconductors, ed. by T.S. Moss (North Holland, Amsterdam, 1980), Vol.2, Chap.5, pp.223-328
14. S. Oelting: 15th School on the Physics of Semiconducting Compounds (Jaszowiec, Poland, 1986), unpublished
15. W. Zawadzki: in Proceedings of the 9th International Conference on the Physics of Semiconductors, Moscow, 1968 (Nauka, Leningrad, 1968), Vol.1, p.312

Electronic Excitations in Laterally Microstructured AlGaAs-GaAs Heterojunctions

W. Hansen[1], *J.P. Kotthaus*[1], *A. Chaplik*[1,3], *and K. Ploog*[2]

[1]Institut für Angewandte Physik, D-2000 Hamburg 36, Fed. Rep. of Germany
[2]Max-Planck-Institut für Festkörperforschung, D-7000 Stuttgart 80, Fed. Rep. of Germany
[3]Institute of Semiconductor Physics, 630090 Novosibirsk 90, USSR

The far infrared transmission of laterally periodical microstructured GaAs heterojunctions is investigated at low temperature (T=2K) and in high magnetic fields (B<12T). At B=0T a well-defined resonance is observed that increases in frequency with increasing magnetic field. Position and strength of this resonance are discussed in terms of depolarization and quantization in wire grid structures.

Introduction

Cyclotron resonance of laterally unstructured two-dimensional electron systems (2DES) has been studied extensively on various substrates [1]. Here we study the behaviour of electrons at a heterojunction interface that are confined in narrow channels of submicron width. A first study of the infrared response of laterally structured heterojunctions has been carried out on rather macroscopic discs of 4μm diameter and has been discussed in terms of macroscopic depolarization alone [2]. If the structures become smaller, one expects in addition to depolarization noticeable effects of quantization in the lateral confining potential.

Sample Preparation

The heterojunction wire grids studied here are prepared by combination of holographic lithography and etching processes on laterally unstructured heterojunctions [3]. The original samples are grown by MBE and consist of a 40nm thick Si-doped AlGaAs layer and a 6 nm thick undoped AlGaAs-spacer grown on top of an undoped 2 μm thick GaAs buffer layer. At T=77K electron density and mobility of the unstructured sample are $n_s=1.0\cdot10^{11}$ cm^{-2} and μ=70.000 cm^2/Vsec, respectively. The photoresist etching mask is defined holographically by an Argon-laser beam with a period of l=1.4 μm. Wet etching is used to define stripes of width 2a=0.5 μm and depth of about 70nm. The surface geometry was controlled by SEM on GaAs monitor samples which were subjected to the same preparation procedure.

Experimental Results

The far infrared transmission ($20cm^{-1}<\tilde{\nu}<200cm^{-1}$) and the magnetic field dependence of the microwave (f≈30GHz) transmission and reflection have been measured at temperature T=2K on three different samples. Figures 1a and b show typical spectra of one sample in relatively low magnetic fields (0.5T<B<4.2T) with different polarizations of the incident radiation field vector with respect to the heterojunction stripes. In Fig. 1a the electric field vector is parallel to the stripes, whereas in Fig.1b it is normal. We present the ratio $\Delta T/T = (T(B_2)-T(B_1)) / T(B_2)$ of the transmission T(B) at magnetic fields $B_2>B_1$. The field B_2 is sufficiently large that the corresponding resonance is outside the spectral regime shown. The field

Supported by the Stiftung Volkswagenwerk

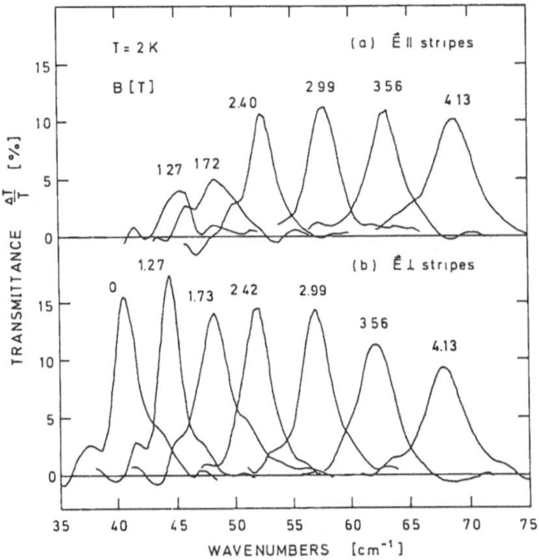

Fig.1: Far infrared transmission spectra of a laterally microstructured heterostructure at different magnetic fields (a) parallel polarization, (b) normal polarization

values B_1 are specified above the corresponding resonance peaks. The $\Delta T/T$ signal thus directly reflects the infrared response of the 2DES at magnetic field B_1. At low magnetic fields all samples show significantly different oscillator strengths for the two different polarizations. With the magnetic field decreasing towards B=0T the resonances diminish for the parallel polarization and become stronger for the perpendicular polarization. At higher fields (B>4T) the resonances have about the same strength for the two polarizations. The full width at half maximum (FWHM) of the resonances varies nonmonotonically between $3cm^{-1}$ and $5cm^{-1}$ as the amplitude does.

As demonstrated in Fig.2 the square of the resonance frequency increases with B^2 with a starting point at finite frequency. The resonance position does not depend on polarization. The observed magnetic field dependence is well described by $\omega_{res}{}^2=\omega_0{}^2+(eB/m^*)^2$, with $\hbar\omega_0$=5meV and m^*=0.072, a value comparable to CR masses found on unstructured samples. The behaviour of the wire grid structure is quite different from the usual 2DES, where the cyclotron resonance frequency is directly proportional to the magnetic field.

In the microwave response of the sample for magnetic fields B>0.8T Shubnikov de Haas (SdH) oscillations are clearly resolved. Fan charts of the SdH extrema show linear behaviour as expected for the case of an unstructured 2DES. The extracted two-dimensional electron density of the sample shown above is $n_s = (4.8\pm0.2)\cdot10^{11}$ cm^{-2}. The density could be raised up to $n_s = (7.7\pm0.3)\cdot10^{11}$ cm^{-2} by illumination with bandgap radiation.

The width of the stripes of heterojunctions on the sample shown above is determined from SEM on a monitor sample to be (0.45 ± 0.04) μm.

Discussion

In order to explain the positions and polarization dependence of the resonances we consider the Maxwell Garnett theory employed in [2] for the case of electron stripes. Furthermore, we include the possibility of non-

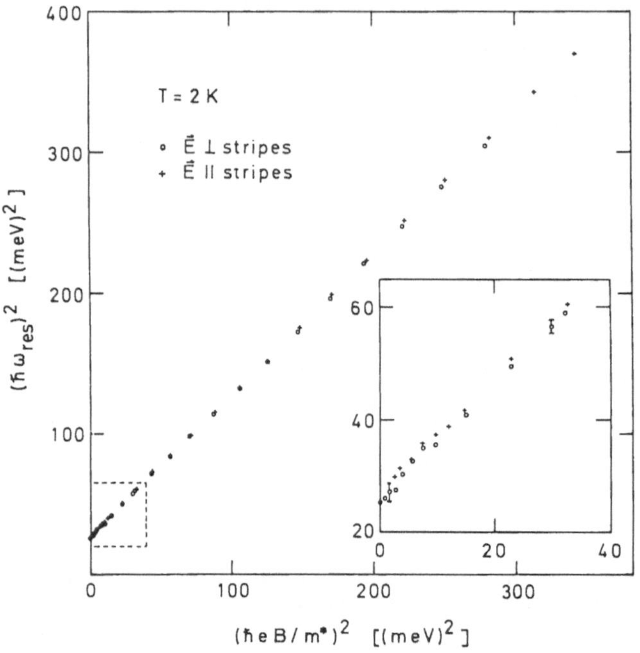

Fig.2: Square of the resonance position versus B^2. Here the units are photon energy and CR energy, respectively, $m^*=0.072m_e$ is chosen so that the line connecting the points has a slope of unity.

classical conductivity due to quantization of the electron system by the confining potential. This is motivated by the fact that the true width of the electron channels may be much less then the etched width due to large band-bending at the boundaries of the stripes. For sake of simplicity we assume a parabolic confining potential $U(x) = m\Omega^2x^2/2$ with a characteristic frequency Ω as it is used in [4].

In this model we calculate a high-frequency resonance position of $\omega^2=\omega_d{}^2+\Omega^2+\omega_c{}^2$, where ω_c is the cyclotron resonance frequency and ω_d is the depolarization frequency. In our case of a stripe geometry the depolarization frequency is given by

$$\omega_d{}^2 = \frac{e^2\, n_s}{\bar{\varepsilon}\, \varepsilon_0\, a\, m^*}$$

with $\bar{\varepsilon}$ an appropriately averaged dielectric constant. If we use $\bar{\varepsilon}=(\varepsilon_s+1)/2$ with $\varepsilon_s=12.53$ as is proposed in [2] and 2a=460 nm, $n_s = 4.8\cdot10^{11}\mathrm{cm}^{-2}$, then ω_d is much higher than the measured resonance frequency. Assuming $\bar{\varepsilon}=\varepsilon_s$ gives values of ω_d which are close to the observed value ω_0. Since the discrepancy between the observed ω_0 and the calculated one would increase if we include quantization, we do not need a finite quantization frequency Ω to explain our observations.

For the normal polarization we calculate the oscillator strength to be

$$\mathrm{Re}\ (\sigma_{xx}) = \sigma_0\ [\ \omega^2\ (\ 1 + \omega^2\tau^2 + \omega_c^2\tau^2)\ /\ DN\ ]$$

and for the parallel polarization

$$\mathrm{Re}\ (\sigma_{yy}) = \mathrm{Re}\ (\sigma_{xx}) + \sigma_0\ [\ \omega_0^2\tau^2\ (\omega_0^2-2\omega^2)\ /\ DN\ ]$$

with $DN = \omega^2 (1 + \omega_0^2\tau^2 + \omega_c^2\tau^2 - \omega^2\tau^2) + \tau^2(\omega_0^2 - 2\omega^2)^2$
$\sigma_0 = e^2 n_s \tau / m^*$ and $\omega_0^2 = \omega_d^2 + \Omega^2$.

With constant scattering rate τ and increasing ω_c the resonance amplitude for normal polarization is expected to decrease monotonically from σ_0 to $\sigma_0/2$. On the other hand, for the parallel polarization the amplitude at the same frequencies increases with B from zero to $\sigma_0/2$ (assuming $\omega\tau >> 1$). At low magnetic fields (B<4T) the experimental results show more complex behaviour than the above prediction, which describes the overall variation only qualitatively. Consequently, the assumption of constant and isotropic scattering rate seems to be too simple.

In order to compare the absolute resonance amplitude with the calculated oscillator strength, we have to introduce a constant f=2a/l which describes the coverage of the sample with stripes and have $\Delta T/T \propto f \cdot Re(\sigma)$. From resonance position ($\propto (n_s/a)^{1/2}$) and the resonance amplitude ($\propto n_s \cdot a$) we can extract values for n_s and a independently. Taking the resonance position at B=0T and the amplitude at high magnetic fields (B>4T), where the amplitudes are almost constant, we obtain $n_s = (0.40 \pm 0.02) \cdot 10^{11} cm^{-2}$ and $2a = (490 \pm 20)$nm. This n_s-value is only slightly smaller than the one extracted from SdH analysis at microwave frequency. The value of a is about the same as the one extracted from SEM pictures. This is surprising, since it implies that the electrons are distributed over the full width of the heterojunction stripe. Note that the discrepancy becomes even larger if one assumes a finite frequency Ω due to quantization of the system.

A further hint that quantization still plays a less imporant role is given by the linear dependence of the microwave SdH extrema plotted versus 1/B. In case of quantization in the wire nonlinear behaviour is expected in the low-field regime, as can be seen by modifying the theory in [4] for the two-dimensional case.

In conclusion, well-defined far infrared resonances with low linewidth demonstrate homogeneity of the wire grid. The high-frequency mobility extracted from the resonance linewidth indicates that the elastic mean free path is about 700nm and exceeds the dimension of the artificial structure. A simple theory used to describe the resonance position yields the result, that the electrons are distributed uniformly over the whole width of the etched geometry and one-dimensional quantization still seems to be negligible. We finally wish to note that experimental observations similar to ours have recently been reported for a 2DES on the surface of liquid helium, but are interpreted in a more complex manner [5,6].

References

1. see for example T. Ando, A. B. Fowler, F. Stern: Review of Modern Physics 54 (1982)
2. S. J. Allen, Jr., H. L. Störmer, J. C. M. Hwang: Phys. Rev. B 28, 4875 (1983)
3. E.Batke et al.: In Semiconductors Quantum Well Structures and Superlattices, ed. by K. Ploog and N.T. Linh (Edition de Physique, Les Ulis, 1986) p.155
4. D. Childers, P. Pincus: Phys. Rev. 177, 1036 (1969)
5. D.C. Glattli et al.: Phys. Rev. Lett. 54, 1710 (1985)
6. V.A. Volkov, S.A. Mikhaĭlov: JETP Lett. 42, 556 (1985)

Landau Level Width and Cyclotron Resonance in 2D Systems

R. Lassnig

Institut für Experimentalphysik, Universität Innsbruck, A-6020 Innsbruck, Austria

The density of states of quasi 2D electrons at high magnetic fields splits into Landau levels, which are separated by the cyclotron energy $\hbar\omega_c$. These levels (with Landau index n) are broadened due to material inhomogeneities, impurities and phonons. In GaAs-GaAlAs heterostructures at low temperatures, the most important scattering mechanism is ionized impurity scattering, which is strongly influenced by screening /1,2/.

In this contribution a self-consistent screening theory is presented for the Landau level broadening and the cyclotron resonance in GaAs-GaAlAs heterostructures at high magnetic fields. With respect to previous models, the following improvements are introduced:

(1) For the first time a theory of temperature dependent screening at high magnetic fields is developed. It is shown that, even at temperatures of 4°K, the temperature suppresses strongly the so-called resonant screening.
(2) The local field ("Hubbard") correction to the electronic polarisability is derived from self-consistency equations for the vertex corrections. This electron-electron interaction can also reduce the screening strongly.
(3) The interference of long-range scattering potentials is treated within a correlation function technique, thereby getting rid of long-range divergences.
(4) Cyclotron resonance transmission spectra are calculated within self-consistent current relaxation theory, introducing a new expression for the memory function.

The calculation is restricted to parabolic bands and zero spin splitting, and Landau level widths are calculated within the self-consistent Born approximation. The resulting semi-elliptic density of states represents the main features of the system, and the considered effects can be treated analytically over a wide field of parameters, although details will be evidently altered by level mixing and higher order theory.

The broadening parameter Γ_n of the Landau levels is given by /1/:

$$(1) \quad \Gamma_n^2 = 4 \sum_q U^2(q) |\langle n| e^{iqx} |n\rangle|^2 / \varepsilon^2(q)$$

where U(q) denotes the unscreened electron-impurity interaction. This effective interaction includes the form factors for the impurities /1/ and the long-range interference of the scattering potentials in a Yukawa-type approximation ($q \to \sqrt{q^2+q_c^2}$, where q_c is determined from the x-space correlation function of the impurity distribution /3/). Especially in the case of weak screening, the interference of the scattering potentials is a necessary requirement for reliable theoretical results.

For partially filled levels the dielectric function ε depends essentially on Γ_n (via the intra-Landau level polarisability of the electron gas), and Γ_n depends in turn on ε; so the problem is self-consistent. The fact that a smaller level width results in stronger screening, and this in turn in a smaller level width, is often referred to as "resonant screening" /4/ and is most pronounced at zero temperature.

On the other hand, screening becomes very weak if the Fermi energy lies in between two Landau levels: In that case only virtual transitions to different Landau levels are possible, which are suppressed as $1/\hbar\omega_c$ (in contrast to $1/\Gamma_n$ for the resonant screening). Thus, in an experimental sweep of the Fermi energy through a system of well-separated Landau levels an oscillatory screening of the ionized impurities will be observed, resulting in oscillatory Γ_n and cyclotron resonance linewidth Γ_{CR}. Maxima are expected when the filling factor $\nu = 2\pi l^2 N_{el}$ is an even integer (N_{el} and l are the electron density and the Landau radius). Experimentally, this was first observed by Englert et al./5/ and later confirmed by other groups /6-8/.

Finite temperatures, however, represent a limit to this resonance effect, since a level width Γ_n much smaller than k_BT will not result in a stronger polarisability: The electrons are then quite uniformly distributed over the Landau level and cannot screen any more. This leads to a hot electron effect, even at 1°K due to the small level widths.

Fig.(1a) shows the calculated temperature dependence of the level width: Γ_0 is plotted versus the filling factor (neglecting spin splitting). N_{el} is kept constant, so the drawing corresponds to a magnetic field sweep. N_i and d_{sp} are the bulk impurity concentration and the spacer thickness. The curves correspond to temperatures of 0.1, 1 and 10°K; the difference is most pronounced if the Landau levels are half filled, and especially for $\nu<2$ the temperature effect can change the results by orders of magnitude.

In Fig.(1b) the cyclotron resonance linewidth is plotted for the same experimental situation. We extend the self-consistent current relaxation theory of Götze /9/ and Gold /10/

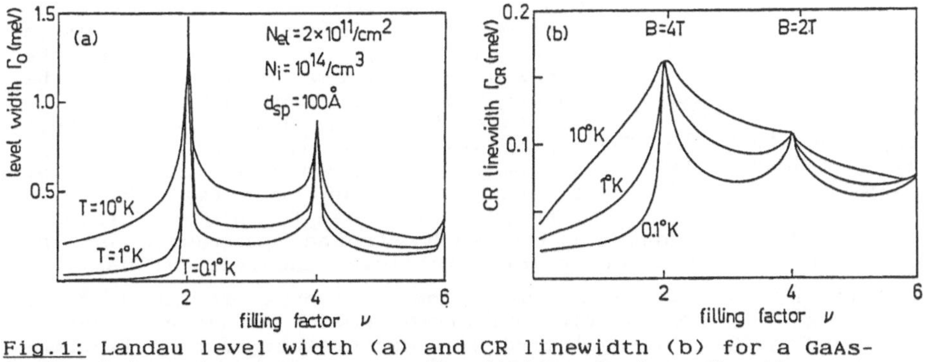

Fig.1: Landau level width (a) and CR linewidth (b) for a GaAs-GaAlAs heterostructure versus the filling factor, for three different temperatures.

to the dynamic magnetic field case and obtain the dynamic conductivity $\sigma(z)$ via the memory function M(z):

$$(2)\quad \sigma(z)=\frac{ie^2N_{el}}{m}\,\{z-\omega_c+M(z)\}^{-1},$$

$$(3)\quad M(z)=\frac{1}{2mN_{el}z}\sum_q\frac{U^2(q)}{\varepsilon^2(q)}\cdot q^2\cdot\Pi(q,z+M(z))$$

where the dynamic polarisability $\Pi(q,z+M)$ must be calculated for the unscattered electron gas. This is quite essential in order to obtain consistent results. For $\nu<2$, the dynamic polarisability is proportional to $\nu/(z-\omega_c+M(z))$, and the memory function is obtained in the simple form:

$$(4)\quad M(z)=(\omega_c-z)/2+i\sqrt{\gamma-(\omega_c-z)^2/4}$$

$$(5)\quad \gamma=\frac{1}{2mN_{el}z}\,\frac{\nu}{2\pi l^2}\sum_q\frac{U^2(q)}{\varepsilon^2(q)}\cdot q^2\cdot e^{-q^2l^2/2}\cdot(q^2l^2/2)$$

We find that this type of memory function is most suited for weakly perturbed systems, whereas for CR close to the metal-insulator transition Gold's expression /10/ is more appropriate. From Fig.(1b) it is seen that the oscillation of the screening strength is much less pronounced in Γ_{CR}, compared to the level width. This behaviour is due to the fact that CR represents a local excitation and thus probes microscopic properties of the sample. Long-range potentials contribute weaker to Γ_{CR} than to Γ_n; mathematically this is expressed by vertex corrections (in Kubo theory) or the factor q^2 in the integral Equ.(3) of Mori-type theory.

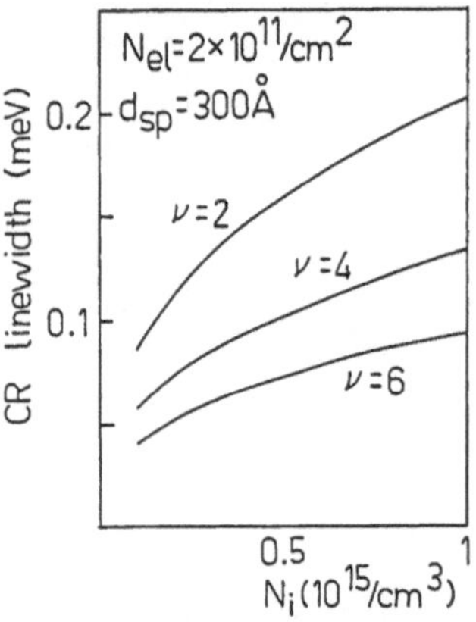

Fig.2: CR linewidth vs. impurity density for filled Landau levels.

Due to the high degeneracy of the system, the specific choice of the perturbation expansion is always problematic, meaning that a quantitative description of the experiments is rather doubtful. However, if a Landau level is completely filled, then the screening is weak and can be treated quite accurately. Of course, there exists the trade-off that then, solutions are very sensitive to the treatment of long - range scattering interferences. Fig.(2) shows the CR linewidth for filled Landau levels (ν=2,4,6) as a function of N_i. It is seen that the linewidths increase roughly with the square root of the impurity concentration. This behaviour has been previously found in the bulk /11/. An important observation is that, at constant electron density, the CR linewidths decrease for higher filling factors. This is due to the much stronger influence of the nonresonant screening at (relatively) lower magnetic fields.

Further, we have calculated the influence of local field corrections on the static polarisability. The Dyson Equation shown in Fig.(3a), which includes electron-electron and electron-impurity interaction, can be treated numerically on the computer.

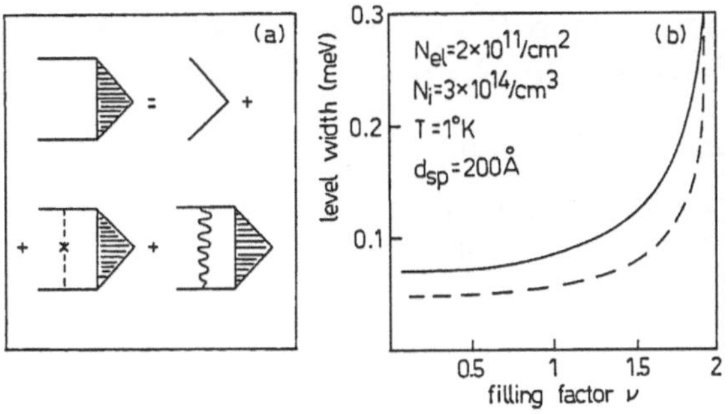

Fig.3: Self - consistency equation for the vertex corrections leading to the local field corrections to the polarisability, (a).
(b) Landau level width vs. filling factor with (full line) and without (broken line) local field corrections.

Here, we give an analytical approximation which represents a lower bound for the strength of the effect:

$$(6)\quad \Pi(q) = \Pi_0(q) / \left(1 + \Pi_0(q) \sum_p \frac{V_c(p)}{\varepsilon(p)} e^{iqp_y} |\langle n| e^{ipx} |n\rangle|^2\right),$$

where V_c is the Coulomb interaction and Π_0 is the polarisability without local field corrections. Fig.(3b) shows that the level width including local field corrections (full line) is considerably higher than without this effect (broken line).

This discussion is intended only to show the dominant trends that are found in a Landau level system. For specific samples (experiments), the detailed form of the results will depend on the scattering center distribution, on k_BT/Γ_n, on N_{el} and on ν; further overlapping lines due to nonparabolicity can play a role for $\nu>1$. Zero temperature approximations are often not justified and we have demonstrated that whenever the level width is of the order of k_BT or smaller, then hot electron effects on the screening are strongly present. The complexity of the calclations for partially filled levels (due to resonant screening) is reduced for filled levels, but then the long-range interference of the scattering potentials requires a careful handling. Local field corrections to the polarisability have been derived for the first time and have a substantial influence on the level widths.

Acknowledgements: This work was partially supported by the Stiftung Volkswagenwerk. I would like to acknowledge stimulating discussions with E.Gornik, J.Hajdu and A.Gold.

/1/ R.Lassnig, E.Gornik; Solid State Comm. 47, 959 (1983)
/2/ T.Ando, Y.Murayama; J.Phys.Soc.Japan 54, 1519 (1985)
/3/ R.Lassnig, to be published
/4/ S.Das Sarma; Phys.Rev.B 23, 4592 (1981)
/5/ Th.Englert, J.C.Maan, Ch.Uihlein, D.C.Tsui, A.C.Gossard; Physica 117B&118B, 631 (1983)
/6/ W.Seidenbusch, G.Lindemann, R.Lassnig, J.Edlinger, E.Gornik; Surf.Science 142, 375 (1984)
/7/ W.Seidenbusch, R.Lassnig, E.Gornik, G.Weimann; Physica 134B, 314 (1985)
/8/ G.Rikken, H.W.Myron, P.Wyder, G.Weimann, W.Schlapp, R.R.Horstman, J.Wolter; Surf.Science 170 (1986)
/9/ E.Gornik; Physica 127B, 95 (1984)
/9/ W.Götze; Solid State Comm. 27, 1393 (1978)
/10/ A.Gold; Phys.Rev.B 32, 4014 (1985)
/11/ E.Gornik; Physica 127B, 95 (1984)

Hole Cyclotron Resonance in p-Type GaAs-AlGaAs Superlattices in High Magnetic Fields

Y. Iwasa, N. Miura, S. Takeyama, and T. Ando

Institute for Solid State Physics, University of Tokyo,
7-22-1 Roppongi, Minato-ku, Tokyo 106, Japan

We report two aspects of holes in GaAs-AlGaAs multiple quantum wells, namely the energy band structure and excitonic effects. The valence band structure is studied by cyclotron resonance of two-dimensional holes at high magnetic fields. The heavy hole cyclotron mass strongly depends upon the well size and magnetic field. This behavior is explained by the valence band Hamiltonian. We also discuss the exciton state in a hole plasma. Experimental evidence for the exciton stabilization by magnetic fieldsis presented.

1. Introduction

In semiconductor superlattices, the band structures are known to be modified by the quantum size effect/1/. The most remarkable feature of these structures is the subband formation due to restricting the carrier motion along one direction. However, if the band edge structure is not simple, the dispersion relation perpendicular to the quantizing axis is also affected. The four-fold degenerate valence band of GaAs is a fundamental example/2-5/. We have studied this problem by cyclotron resonance of p-type modulation doped quantum wells. The energy structure of two-dimensional (2D) holes is discussed as a function of well size and magnetic field. Each subband is highly non-parabolic because of the coupling with the other subbands. This causes a complicated behavior of the cyclotron mass.

Excitonic effects are expected in the 2D hole gas system. It is known that at low temperature, the heavy hole exciton is quenched by charged carriers of sufficient concentration/6/. We report here that this peak reappears at high magnetic field. The exciton is thought to be stabilized by the magnetic field.

Six GaAs-AlGaAs samples are grown by the MBE technique. These samples have different hole concentrations p and well layer thicknesses L_z, but have almost the same barrier size ($L_B \simeq 500$Å) and Al mole fraction ($x \simeq 0.5$). They have 40 or 50 periods. We have estimated p using Shubnikov-de Haas (SdH) oscillations.

Cyclotron resonance measurements were carried out up to 120T. An optically pumped molecular gas laser and a H_2O laser were used as far infrared (FIR) light source. We also measured the exciton transmission spectra by use of an optical multi-channel analyzer up to 35T. The magnetic field is applied perpendicular to the 2D plane.

2. Subband Structure

Figure 1 shows the magnetic field dependence of the FIR transmission intensity at various wavelengths. Many absorption peaks are observed, and

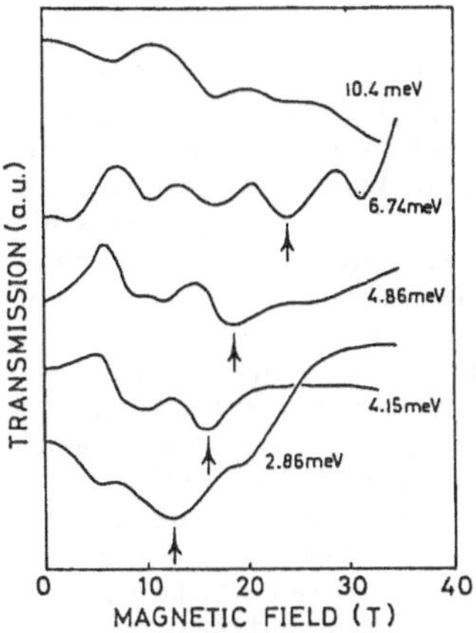

Fig.1 Experimental traces of FIR magneto-absorption spectra at various wavelengths for the sample with L_z=108Å L_B=583Å, x=0.53, p=5.4x10^{11}cm^{-2}. T≃6K.

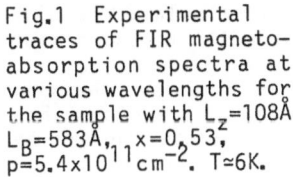

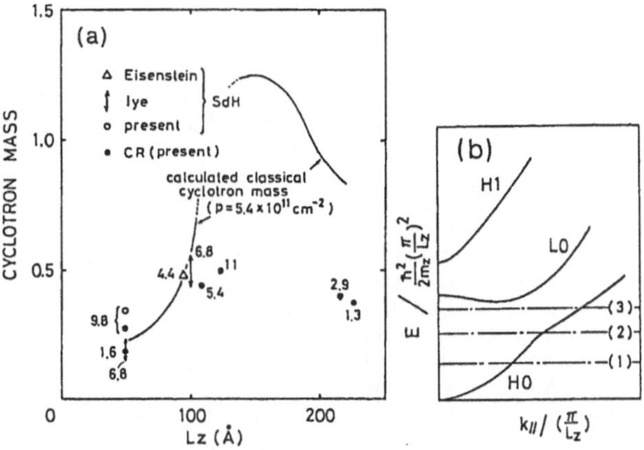

Fig. 2 (a) L_z dependence of cyclotron mass of the lowest heavy hole subband obtained at low magnetic fields. The mass is in units of free electron mass. Small numbers are carrier concentrations (10^{11}cm^{-2}/sheet). Solid lines are the calculated classical cyclotron mass. (b) Schematic diagram of energy dispersion relations of 2D holes. The lowest three subbands are shown. At fixed carrier concentration relative positions of E_F are also indicated. (1) L_z=50Å, (2) L_z=100Å (3) L_z=200Å

resonance peak energies do not necessarily depend linearly upon magnetic field. These characteristics indicate that the Landau levels are highly non-linear against the magnetic field and are not equally spaced. At higher temperatures, all peaks disappear except the one indicated by the arrow. This peak is assigned to the main transion 1a→2a of the lowest heavy hole subband. At low magnetic fields, this peak is thought to correspond to the classical cyclotron resonance which is determined from the dispersion relation at zero magnetic field.

We estimated the low-field cyclotron mass. Figure 2(a) shows the L_z dependence of the mass in several samples. The masses obtained from the temperature dependence of the SdH oscillations are also shown. The small numbers are hole concentrations in units of 10^{11}cm^{-2}/sheet. Although samples have different carrier concentrations, we can see the general tendency that the mass is quite small in samples with small L_z, increases with increasing L_z, and becomes smaller again when L_z is about 200Å. The mass is thought to approach the bulk value ~0.4 in the large L_z limit.

We can understand this behavior as follows. The L_z dependence of the cyclotron mass is determined by the position of the Fermi level (E_F) and the subband structure which is largely affected by the coupling with the other subbands. For simplicity, we assume that p is constant. When L_z is varied, the relative position of E_F changes as shown in figure 2(b). In the case of small L_z, E_F lies at (1) where the coupling with the other subbands is not significant. Therefore, the diagonal part of the Kohn-Luttinger Hamiltonian is emphasized, and the mass is very small. For L_z=100Å, E_F lies at (2), where the repulsion by the light hole subband is strong so that the mass is fairly large. For L_z=200Å, E_F increases like (3) where the cyclotron mass becomes smaller again.

Following ref.7, we have calculated the L_z dependence of the classical cyclotron mass for a fixed carrier concentration of $p=5.4\times10^{11}cm^{-2}$/sheet as shown in figure 2(a). Although p is not constant in the experiments, the qualitative behavior is in good agreement. When L_z is small, the lowest heavy hole subband near E_F is almost parabolic, therefore, the experimental result is well explained by classical calculations. However, in case of large L_z, where the subband near the Fermi level is highly non-parabolic, the classical mass is not a good approximation, because experimental values are obtained at finite magnetic fields.

Since the subband structure is complicated, as shown in Figure 2(b), the magnetic field dependence of the cyclotron resonance energy is expected to show a complicated behavior. We have performed a higher field experiment and plotted the resonance energies against magnetic field in figure 3. In this sample, only the main transition 1a→2a is observed at a low temperature. When the temperature is increased, another peak appears, which is assigned to the transition -2b→-1b. The initial state -2b is unoccupied at low temperature. The main transition shows a distinct warp as a function of magnetic field. We also plotted the calculated result. 40% valence band discontinuity is in better agreement with the experiment than 15%.

Detailed comparison beween experiments and theory on other samples will be reported elsewhere. The qualitative agreement is satisfactory. For samples with $L_z \simeq 200$Å, the character of the main transition is changed by the magnetic field. Namely, at low magnetic fields, the transition is heavy hole-like, however, it begins to have a light hole character at high magnetic fields. This results from the large coupling of heavy and light hole subbands.

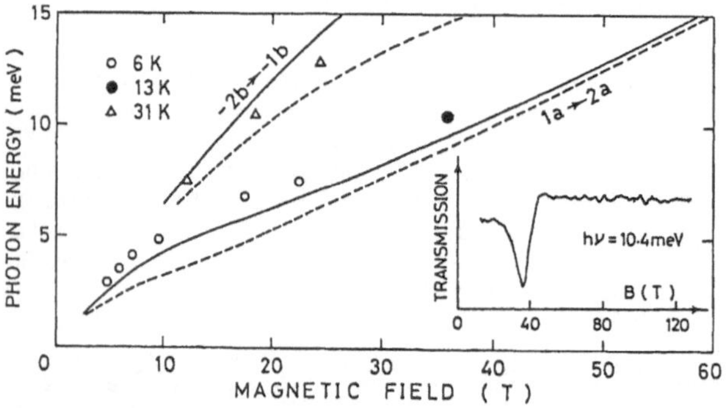

Fig. 3 Plots of photon energies of the absorption peaks versus magnetic field. L_z=50Å, L_B=517Å, x=0.47, $p=1.6\times10^{11}cm^{-2}$. The inset shows the experimental trace for hν=10.4meV. The solid lines are calculated result assuming 40% valence band offset, and the broken lines are after Dingle's rule.

3. Excitons in 2D Hole Plasma

In p-type modulation doped quantum wells, it is known that only the exciton associated with the lowest heavy hole subband is quenched in the case of sufficiently high carrier concentrations/6/. Since higher subband excitons still retain significant strength, it is suggested that the screening effect is not so important but heavy hole subband filling prevents exciton formation.

In order to study the effect of free carriers, we have measured exciton spectra in the doped system. Figure 4 shows interband absorption spectra of a sample with L_z=50Å, $p=9.8\times10^{11}cm^{-2}$/sheet. Small structures below the absorption edge are considered to be due to impurities. At 4.2K the lowest heavy hole exciton is absent at zero magnetic field. When the field is applied, an extra peak indicated by the arrow appears, in addition to the

Landau level transition peaks/8/. This peak is due to the heavy hole exciton which is stabilized by the magnetic field. The intensity of the extra peak increases from a threshold field of 14T.

We also measured the temperature effect on the same sample in the absence of a magnetic field. As the temperature is increased up to 200K, an extra peak also appears at the absorption edge. This peak is also assigned to the heavy hole exciton.

According to these results it can be said that at low magnetic field and low temperature, there exists a region where the heavy hole exciton is unstable. However, when we increase field or temperature, the exciton becomes stable. Modulation doped quantum wells are good samples to observe this phenomenon, because ionized impurities are spacially separated. It is not clear yet what is the most important factor to determine this behavior of the heavy hole exciton. Magnetic field and temperature have the same effect on the screening and the Pauli exclusion principle. Miller et al. stressed that the Pauli principle is essential for explaining the absence of only the lowest heavy hole exciton/6/.

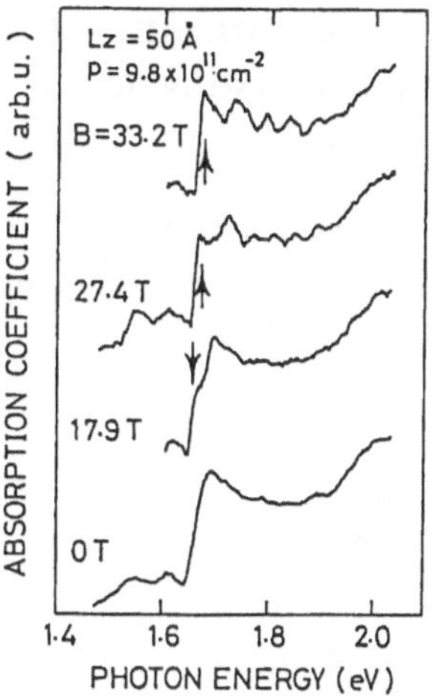

Fig. 4 Interband magnetooptical spectra of the sample with L_z=50Å, L_B=525Å, x=0.47 p=9.8x10^{11}cm^{-2} at 4.2K.

According to our calculation/3/, the third lowest Landau level begins to depopulate near the threshold field. The lowest Landau level is still fully occupied. If the Landau level mixing by the Coulomb potential is not significant, the appearence of the exciton is difficult to explain by band filling effects. We suggest that the magnetic field dependence of the screening effect on the exciton should also be considered.

Acknowledgement

We are grateful to S.Tarucha and H.Okamoto for kindly providing the samples and valuable discussions.

References

1. R.Dingle : Festkörperprobleme 15, 21 (1975).
2. A.Fasolino and M.Altarelli : Two-Dimensional Systems, Heterostructures and superlattices, ed. by G.Bauer, F.Kucher, H.Heinrich (Springer 1984) 176
3. Y.Iwasa, N.Miura, S.Tarucha, H.Okamoto and T.Ando : Proceedings of the 6th Int. Conf. on EP2DS ed. by T.Ando (Kyoto 1985) 587
4. E.Bangert and G.Landwehr : Proceedings of the 6th Int. Conf. on EP2DS ed. by T.Ando (Kyoto 1985) 593
5. Z.Schlesinger, S.J.Allen,Jr, Y.Yafet, A.C.Gossard and W.Wiegmann : Phys. Rev. B32, 5231 (1985)
6. R.C.Miller and D.A.Kleinmann : J. Lumines. 30 520 (1985)
7. T.Ando : J. Phys. Soc. Jpn 54, 1528 (1985)
8. N.Miura, Y.Iwasa, S.Tarucha, H.Okamoto : Proceedings of the 17th ICPS ed. by J.D.Chadi, W.A.Harrison (San Francisco 1984) 359

Magnetoplasma Resonance in the Dynamic Conductivity of a Type I Superlattice

*N.J.M. Horing, H.C. Tso, and X.L. Lei**

Department of Physics and Engineering Physics,
Stevens Institute of Technology, Hoboken, NJ 07030, USA

We have examined the effect of a magnetic field on the strong plasma resonance of the dynamic conductivity of a close-packed Type I superlattice using a magneto-hydrodynamic analysis of the superlattice memory function. We have determined explicitly the finite jump in the imaginary part of the memory function characterizing a plasma resonance shifted by the magnetic field to the magneto-plasmon frequency $(\omega_p^2 + \omega_c^2)^{\frac{1}{2}}$ where ω_c is the cyclotron frequency and $\omega_p = (4\pi e^2 n_{2D}/md)^{\frac{1}{2}}$ is the bulk plasmon frequency for a closely packed Type I superlattice of period d and 2D electron sheet density n_{2D}.

This report is concerned with our investigation of the effect of a magnetic field on the plasma resonance in the memory function, and in the dynamic conductivity of a Type I semiconductor superlattice. Our work follows the usual procedure of expanding the high frequency magneto-resistivity to lowest order in the impurity scattering potentials (phonons are neglected here). It is well known that for a bulk three dimensional (3D) solid state plasma, the memory function $M(\omega)$ is determined by the inverse dielectric function $K(\vec{p},\omega)$ for the 3D system in wavenumber-$\vec{p}$, and frequency-ω representation as [1abc,2]

$$M(\omega) = \frac{-N}{4\pi nm\omega} \int \frac{d^3p}{(2\pi)^3}\, p_x^2\, p^2 |u(\vec{p})|^2 \,[K(\vec{p},\omega) - K(\vec{p},o)] \,, \qquad (1)$$

where N is the 3D density of randomly smeared impurities, n is the 3D density of electron carriers, and $u(\vec{p})$ describes the impurity scattering potential in $\vec{p}$-representation. Equation (1) predicts a rather mild plasma resonance effect in the structure of the memory function for three dimensions.[1]

We have generalized equation (1) to accommodate the intrinsic spatial inhomogeneity of a Type I superlattice of interacting two dimensional (2D) planar sheets of electron carriers which are equally spaced in the z-direction (spacing d). The electron sheets (parallel to x-y plane) are taken to have identical polarization properties, and our generalization also includes a magnetic field in the z-direction perpendicular to the planes. The explicit construction of the spatially inhomogeneous inverse dielectric function $K(z_1\ z_2;\ \bar{p},\omega)$ for the Type I superlattice[3] has been presented in ref. 3 $[\vec{p} = (\bar{p},\ p_z)\ ;\ \bar{p} = (p_x,p_y)\ ;\ p = |\bar{p}| = (p_x^2 + p_y^2)^{\frac{1}{2}}]$. Our analysis of plasma resonance [4] in the high frequency conductivity of a Type I superlattice has been presented in ref. 5, and we found a strong resonance effect around the plasma frequency for a closely packed superlattice[5] in the absence of a magnetic field. We now extend this to include a magnetic field.

Our determination of the superlattice conductivity tensor element $\sigma_+ = \sigma_{xx} + i\sigma_{yx}$ (electric field is in x-direction)

$$\sigma_+ = i(n_{2D}e^2/m)/[\omega - \omega_c + M(\omega)] \tag{2}$$

in terms of the memory function $M(\omega)$ expanded to lowest order in the impurity scattering potentials due to remote impurity sheets, dynamically screened by the Type I superlattice response function $K(z_1,z_2;\ \bar{p},\omega)$, leads to the result

$$M(\omega) = \frac{2\pi e^2 N_r Z_r^2}{n_{2D} m\omega} \int \frac{d^2\bar{p}}{(2\pi)^2}\ \frac{p_x^2}{|\bar{p}|}\ [e^{-pc} + [S(p,o) - 1]\cosh pc]^2 A(\bar{p},\omega,o), \tag{3}$$

where N_r is the remote impurity sheet area density and Z_r is its charge number, n_{2D} is the electron sheet area density, ω_c is the cyclotron frequency and c is the separation of a remote impurity sheet from the nearest electron sheet of the superlattice. In obtaining $M(\omega)$ above we averaged over a random distribution of impurities on each remote impurity sheet. Furthermore,

$$S(p,x/d) = \text{Sinh}\ pd/[\cosh pd - \cos x] \tag{4}$$

is the Type I superlattice structure factor and $A(\bar{p},\omega,x)$ is given by

$$A(\bar{p},\omega,x) = - \frac{4\pi\alpha_o^{2D}(\bar{p},\omega)}{1+4\pi\alpha_o^{2D}(\bar{p},\omega)S(p,x/d)} + \frac{4\pi\alpha_o^{2D}(\bar{p},o)}{1+4\pi\alpha_o^{2D}(\bar{p},o)S(p,x/d)} , \tag{5}$$

where $4\pi\alpha_o^{2D}(\bar{p},\omega)$ is the wavenumber and frequency dependent 2D polarizability of the planar 2D electron sheets.

An accurate determination of $M(\omega)$ calls for the use of the random phase approximation (RPA) for the 2D polarizability[6] $4\pi\alpha_o^{2D}(\bar{p},\omega)$, augmented with a finite electron linewidth[7]. However, our concern here is focused on the principal magnetoplasmon resonance alone, and the hydrodynamic model of 2D dielectric response suffices in this matter. To this end, we have generalized Fetter's[8] hydrodynamic model of 2D dielectric response to include a magnetic field, with the result

$$4\pi\alpha_o^{2D}(\bar{p},\omega) = - (2\pi e^2 n_{2D}/m)/[\omega(\omega+i\delta) - \omega_c^2 - \beta^2 p^2], \tag{6}$$

where $\delta > o$ is an infinitesimal shift and the parameter β is chosen here as $\beta=\sqrt{3}V_f/2$ (V_f = Fermi velocity). For close packing of the (strongly interacting) superlattice planes, $pc < pd << 1$, the magnetohydrodynamic model yields

$$M(\omega) = \frac{-N_r Z_r^2 e^2}{2n_{2D} md\omega} \frac{\omega_p^2}{\beta^2} [\ln \frac{\omega_p^2+\omega_c^2}{|\omega^2-\omega_p^2-\omega_c^2|} + i\pi\eta_+(\omega^2-\omega_p^2-\omega_c^2)], \tag{7}$$

where $\eta_+(x)$ is the Heaviside step function describing a jump discontinuity in Im $\{M(\omega)\}$ associated with pronounced plasma-resonant behavior at the magnetoplasma resonance $\omega = (\omega_p^2 + \omega_c^2)^{\frac{1}{2}}$, where $\omega_p = (4\pi e^2 n_{2D}/md)^{\frac{1}{2}}$ is the effective bulk plasmon frequency for a closely packed Type I superlattice[4]. This generalization of our earlier prediction of strong plasma-resonant behavior for a closely packed Type I superlattice in a magnetic field would not be altered if one were to employ an RPA description of 2D magnetoplasma dynamics. The jump discontinuity in Im $\{M(\omega)\}$ at the magnetoplasma resonance, and the location of the "close-packed" resonance at $\omega^2 = \omega_p^2 + \omega_c^2$ should be the same in the RPA as they are in our 2D magneto-hydrodynamic model.

Acknowledgement: This work was partially supported by the U.S. Army, Electronics Materials Research Division at the Electronics Technology and Devices Laboratory, Fort Monmouth, N.J.

REFERENCES AND FOOTNOTES

1. (a) Dawson, J., Oberman, C.: Phys. Fluids 5, 517 (1962); ibid. 6, 394 (1963)
   (b) Ron, A., Tzoar, N.: Phys. Rev. 131, 1943 (1963)
   (c) Götze, W., Wölfle, P.: Phys. Rev. B6, 1226 (1976)
2. Horing, N.J.M., Lei, X.L., Cui, H.L.: Phys. Rev. B33, 6929 (1986)
3. Horing, N.J.M., Fiorenza, G., Cui, H.L.: Phys. Rev. B31, 6349 (1985)
4. Das Sarma, S., Quinn, J.: Phys. Rev. B25, 7603 (1982); ibid. B27, 6516 (1983)
5. Lei, X.L., Horing, N.J.M., Zhang, J.Q.: Phys. Rev. B33, 2912 (1986)
6. (a) Horing, N.J.M., Orman, M., Yildiz, M.: Phys. Lett. 48A, 7 (1974)
   (b) Horing, N.J.M., Yildiz, M.: Ann. Phys. (NY) 97, 216 (1976)
7. Platzman, P.M., Horing, N.J.M., Tzoar, N.: Proc. 2nd Int. Conf. "Electronic Properties of 2D Systems" at Berchtesgaden, Germany, ed. by Koch, J.F. and Landwehr, G. (Phys. Dept. der Tech. Univ. München, Munich 1977) p. 1039
8. Fetter, A.L.: Ann. Phys. (NY) 81, 367 (1973)

Radiative Recombination of Two-Dimensional Electrons with Nonequilibrium Holes in Si MOS-Structures

I.V. Kukushkin and V.B. Timofeev

Institute of Solid State Physics, Academy of Sciences of the USSR, 142432 Chernogolovka, USSR

It has been shown that radiative recombination spectra of 2D-electrons with nonequilibrium holes in p-(001) Si MOS-structures can be used as an effective tool for the investigation of the 2D-density of states under applied magnetic field. Oscillations of the Landau level width with electron occupation have been found.

In this report we present results concerning the investigations of the radiative recombination (RR) spectra of 2D-electrons with nonequilibrium holes in p-(001)Si MOSFET's. In the considered case the optical transitions are indirect and the corresponding probability does not depend on the energy of the particles involved in the recombination process. It follows from magnetotransport investigations that under electron-hole excitation the band bending behind the 2D-electron layer disappears and the RR process is due to a wave function overlap of 2D-electrons and injected holes bound to boron atom cores close the the interface (see insert of Fig. 1). Then - because the inhomogeneous energy broadening caused by nonequilibrium holes is not too large (<0,8 meV) - the RR spectra reflect the density of states of 2D-electrons /1-3/.

Without magnetic field ($H = 0$) the recombination spectra have a step-like distribution and the full width at half maximum (FWHM) of these spectra linearly increases with the 2D-electron concentration N_s (see the $2D_e$-spectra in Fig. 1). Such behavior of the 2D-electron spectra reflects the constant density of states of the electron system at $H = 0$.

The 2D-line has been observed not only as TO- and TA-phonon replicas but as well in the non-phonon (NP-) spectral region. It indicates, that when a recombination occurs the Brillouin momentum is transferred to an impurity center. Therefore it follows that the 2D-electron recombines with a nonequilibrium hole bound to a boron atom located at the interface region.

Due to the symmetry-lowering of the electron system one can expect that RR spectra of 2D-electrons measured in the direction parallel to the interface should be polarized. It was found that the degree of polarization P is constant within the whole $2D_e$ spectral region and that the magnitude of P found experimentally: $P_{exp} \approx 0.28$ is close to the theoretically calculated value $P_{th} \simeq 0.33$ under the assumption that all hole momentum projections corresponding to $J = 3/2$ participate in the recombination process /2/.

Fig. 2 illustrates the dependences of the FWHM on ΔE and the degree of polarization P of the $2D_e$-line on concentration N_s. The arrows shown in the figure indicate the critical concentration N_s^c corresponding to the transition from the region of strong localization to the state of metallic conduc-

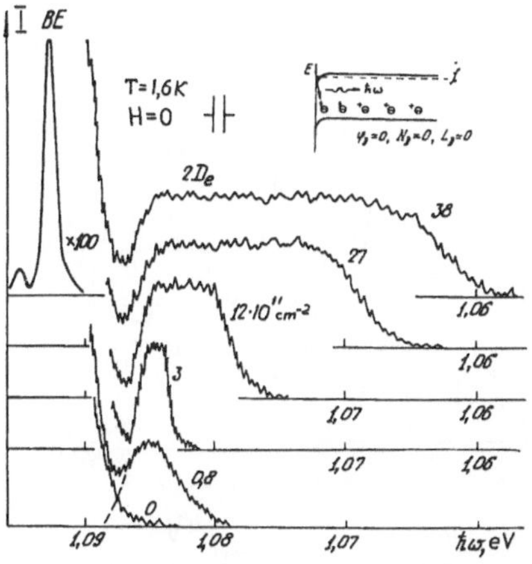

Fig. 1: TO-phonon component of the RR-spectra of Si MOSFET measured under Ar-laser excitation (the power was about 10^{-3} W/cm²) and different gate voltages V_g at T=1.6 K. 2D-electron densities are measured Shubnikov-de Haas oscillations of the conductivity and are indicated for each spectrum in units of 10^{11}cm⁻². The BE line corresponds to excitons bound to boron atoms.

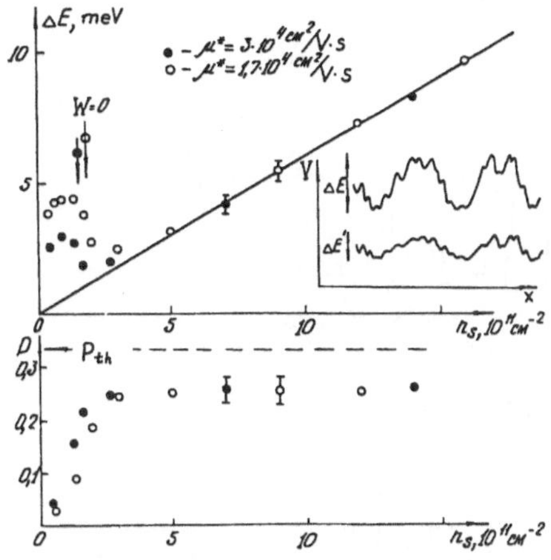

Fig. 2: The dependences of the 2D-luminescence band width (upper part) and polarization degree (lower part) on the concentration N_s at T=1.6 K. The dark and open circles correspond to electron mobilities $3\cdot10^4$ cm²/Vs and $1.7\cdot10^4$ cm²/Vs, respectively. The arrows correspond to the critical concentration N_s^0 when the transition from strong localization to metallic conductivity occurs.

tivity. The magnitude of N_s^0 has been determined for two samples with different electron mobilities by extrapolation of the activation energy dependence W on N_s in the limit $W\to0$. In the region of metallic conductivity the width ΔE depends linearly on N_s and the degree of polarization within the $2D_e$-line remains constant. When $N_s<N_s^0$ the width E reflects an amplitude of random potential and demonstrates a nonmonotonous behavior on N_s which can be explained in terms of the screening of random potential fluctuations. Under the condition of strong localization ($N_s<N_s^0$) the degree of polarization starts to drop due to the valley mixing of localized electronic states. Therefore by means of the optical spectroscopy method the regions of strong localization and metallic conductivity are clearly separated.

Under applied magnetic field the RR spectra of 2D-electrons demonstrate oscillatory structure due to Landau level quantization (Fig. 3). The magnitude of Landau splitting is sensitive only to the projection of H relative to the 2D-plane due to two-dimensionality of the electrons (lower part of the Fig. 3) /3/. With the use of Landau level fans constructed for various magnetic fields the Fermi energy and the energy E_0 at the bottom of the band can be easily found (see upper part of Fig. 3).

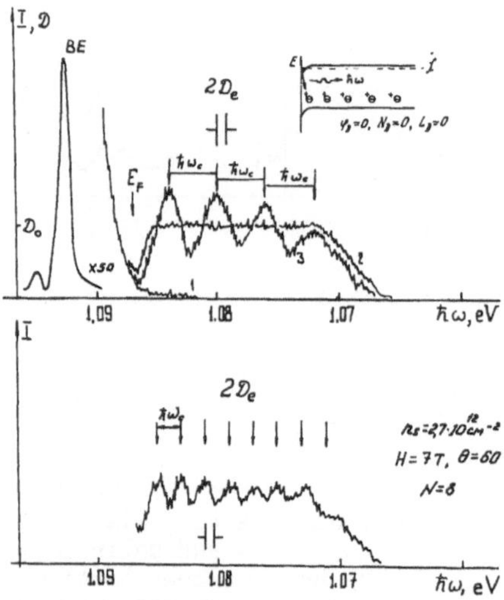

Fig. 3: The RR spectra of 2D-electrons measured without (H=0) and with magnetic field (H=7T) at T=1.6 K. At $N_S = 2.7\cdot10^{12}$ cm^{-2} and H=7T four Landau levels are completely occupied (the filling factor ν =16). Landau level fans are shown in the upper part of the figure. E_0 and E_F are the bottom of the energy band and the Fermi energy. Γ_F is a broadening due to the damping of one-particle excitations in the 2D-electron Fermi sea. For the lower spectrum the angle of H with respect to the 2D-plane is 60°.

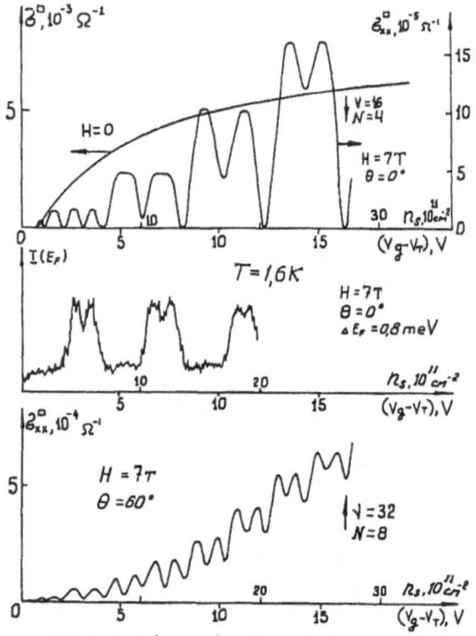

Fig. 4(a,c)-the dependences of conductivity and magnetoconductivity (H=7T) on V_g for a Corbino MOSFET having an electron mobility of $3\cdot10^4$ cm²/Vs at N $=4\cdot10^{11}$cm^{-2} and T=1.6 K.
(a) - Θ = 0°, (c) - 60°.
(b) - the dependence of the luminescence intensity measured at the blue limit of the 2D-line ($\hbar\omega$ =1.0865 eV) on N_S. The spectral slit is equal to 0.08 meV, H=7 T and T=1.6 K.

Because the blue limit of the RR spectra connected with the Fermi energy of 2D-electrons is fixed it allows measurements of RR intensity oscillations on N_S just at the spectral position corresponding to the Fermi-level. Fig. 4 represents these oscillations measured at $\hbar\omega$ = 1.0865 eV, H = 7 T, T =1.6 K, with the use of a spectral width δE = 0.8 meV. One can see that:

1. the RR intensity oscillations at the Fermi-level position are periodic on the V_g (or N_S) scale,

2. the intensity oscillations and the Shubnikov-de Haas oscillations of the conductivity are completely correlated,

3. in the RR oscillations the valley-orbit splitting is resolved,

4. due to the selection rules the transitions with the participation of the electron spin projection S_z = -1/2 are forbidden and this spin state is discriminated in the optical RR spectra at T = 1.6 K.

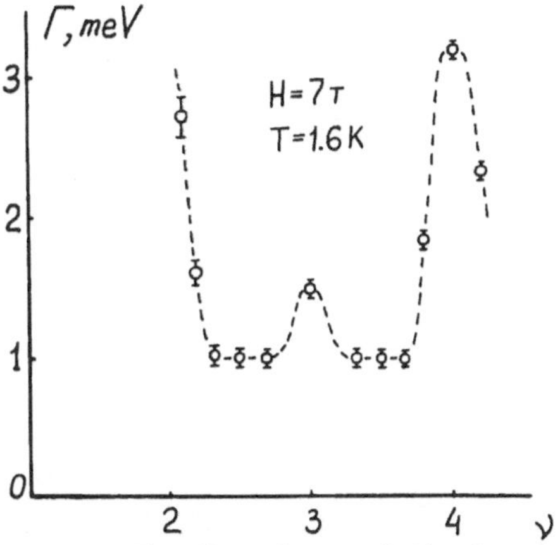

Fig.5: The dependence of the lower Landau level (N=0) width of filling factor ν measured with the use of the radiative recombination spectra of 2D-electrons with nonequilibrium holes at H=7 T and T=1.6 K.

With the use of the described spectroscopic method it was possible to investigate the energy distribution behavior of the density of states within the lowest Landau level on the filling factor ν. Here we represent only the result concerning the dependence of the Landau level width Γ on ν (see Fig. 5). One can see that the width Γ oscillates as a function of ν, namely: it reaches a maximum at ν-values, where the Fermi-level is located just in the middle of the energy gap, and has a minimum when the Fermi-level is in the center of level (the region of the extended states). Such a dependence of the Landau level width on ν can be explained in terms of selfconsistent screening of random potential fluctuations /4/.

REFERENCES

1. I.V. Kukushkin, V.B. Timofeev, Pis'ma Zh. Eksp. Teor. Fiz. (JEPT Lett.) 1984, 40, 413
2. I.V. Kukushkin, V.B. Timofeev, Zh. Eksp. Teor. Fiz. (JEPT), in press, 1986.
3. I.V. Kukushkin, V.B. Timofeev, Pis'ma Zh. Eksp. Teor. Fiz. (JEPT Lett.), 1986, 43, 387
4. T. Ando, Y. Murayama, Proc. of the 17th Int. Conf. on the Physics of Semiconductors, San Francisco 1984, J.D. Chadi and W.A. Harrison, Eds. Springer Verlag 1985, p. 317.

Part IV

Heterostructures and Superlattices: Transport, Bandstructure

Theory of p-Type Inversion Layers in Magnetic Fields

L.J. Sham

Department of Physics, B-019, University of California, San Diego, La Jolla, CA 92093, USA

Due to the bulk valence band degeneracy, Landau levels in the quasi-two dimensional hole systems have nonlinear magnetic field dependence and multiple cyclotron frequencies, features not existent in the conduction electron systems. It is shown how the Landau levels can be sorted out into subbands, classified according to the approximate nature of the spin up or down heavy or light holes. Calculations are used to interpret the observed cyclotron resonances. The warping effect is shown to play an important role.

1. Introduction

The four-fold degeneracy at the top of the bulk semiconductor yields interesting complexities for the holes confined in an inversion layer or in a quantum well. The first theories for the p-inversion layer are for the Si-MOSFET [1,2,3]. We shall concentrate here on the III-V compounds, in particular, the GaAs/AlGaAs heterojunction. Many of the properties of holes in these systems have been measured by experiments in high magnetic fields [4,5,6,7]. The corresponding theories [8,9,10,11,12] will be reviewed here and the understanding which they provide for the experimental observations will be discussed. The behavior of holes in a quantum well [13] is similar and will not be discussed here.

Despite the complications due to the degeneracy at the top of the bulk valence band, a simple physical picture which accounts for the salient features as consequences of the valence band complex is possible. As part of the job to achieve that here, we shall also point out the relationship between different sets of notations for the quantum numbers characterizing the Landau levels for holes. I hope that this effort will make it easier to relate theories of holes to experiments.

2. Subband k·p Method

Luttinger's formalism [14] is used for the valence bands. The basis set consists of the four degenerate states at the top of the bulk bands which transform like the eigenstates of the angular momentum J=3/2 and shall be ordered in the descending order of the eigenvalues of the z-component, m_J=3/2,1/2,-1/2,-3/2. We have chosen the z axis to be normal to the interface. The effective Hamiltonian can be separated into two terms, H_0+H_A, with the first term given by:

$$\begin{vmatrix} W-3K/2 & S & R & 0 \\ S^{\dagger} & U-K/2 & 0 & R' \\ R^{\dagger} & 0 & U+K/2 & S' \\ 0 & R'^{\dagger} & S'^{\dagger} & W+3K/2 \end{vmatrix} \quad \text{where,} \qquad (1)$$

$$W = -(\hbar^2/2m)\ (\gamma_1-2\gamma_2)k_z^{\ 2} + V(z) - (\hbar^2/2m)\ (\gamma_1+\gamma_2)k^2$$
$$U = -(\hbar^2/2m)\ (\gamma_1+2\gamma_2)k_z^{\ 2} + V(z) - (\hbar^2/2m)\ (\gamma_1-\gamma_2)k^2$$
$$R = (\sqrt{3}\hbar^2/4m)\ (\gamma_3+\gamma_2)k_-^{\ 2}$$
$$S = (\sqrt{3}\hbar^2/m)\ \gamma_3 k_z k_+$$
$$K = \kappa\hbar\omega_c \qquad (2)$$

V(z) is the self-consistent effective potential for the inversion layer. The prime on R and S denotes transpose when these are later expressed in matrix form using the subband basis set. The dagger denotes the usual Hermitian conjugate. m is the free electron mass. ω_c is the cyclotron frequency of the free electron. The Luttinger parameters are taken to have the values,

$$\gamma_1 = 6.85, \quad \gamma_2 = 2.1, \quad \gamma_3 = 2.9, \quad \kappa = 1.2 \qquad (3)$$

k_z is the wave-vector operator along the normal and k is the magnitude of the wave-vector components k_x and k_y in the plane of the interface.

$$k_+ = k_x+ik_y \quad , \quad k_- = k_x-ik_y \quad . \qquad (4)$$

The second term of the effective Hamiltonian, which may be called the anisotropy term, consists of zeros in the 4x4 matrix except where R appears in (1), replacing by

$$R_A = (\sqrt{3}\hbar^2/4m)\ (\gamma_3-\gamma_2)\ k_+^{\ 2} \quad . \qquad (5)$$

Note that, when the in-plane wave-vector is zero, the Hamiltonian is diagonal. Thus, the k_x=0 and k_y=0 subband states can be obtained by solving the two uncoupled Schrödinger equations for the motion normal to the interface, yielding two series of doubly degenerate series of heavy ($m_J=\pm3/2$) and light ($m_J=\pm1/2$) hole states. We shall use these as the basis set to transform the Hamiltonian into a larger matrix, and, hence, name the method the subband k·p method. In fact, the basis set has to be established self-consistently, since the potential depends on the total density of holes and, therefore, on the finite in-plane k states.

The k_x,k_y terms in the off-diagonal part of the Hamiltonian mix the heavy and light hole states in the subband but the name of each subband remains, by convention, the same as the k=0 state of the subband. In a finite normal magnetic field, the in-plane components of k are given by

$$k_\alpha \ = \ -i\ \partial/\partial x_\alpha + eA_\alpha/\hbar c \quad . \qquad (6)$$

Following Luttinger [14], the components are expressed in terms of the raising and lowering operators, a^+ and a:

$$k_+ = a^+\sqrt{2}/R_c \quad , \qquad k_- = a\ \sqrt{2}/R_c \qquad (7)$$

where R_c is the cyclotron radius, $\sqrt{(\hbar c/eB)}$. In terms of the k=0 states

$\psi_{\nu j}(z)$ with energy $\varepsilon_{\nu j}$, with ν denoting the subband and j the m_J for heavy or light hole states of both spin directions, the Hamiltonian retains the form of (1) with

$$W_{\nu\nu'} = \delta_{\nu\nu'}[\varepsilon_{\nu h} - (\gamma_1+\gamma_2)\hbar\omega_c(a^+a + 1/2)]$$

$$U_{\nu\nu'} = \delta_{\nu\nu'}[\varepsilon_{\nu \ell} - (\gamma_1-\gamma_2)\hbar\omega_c(a^+a + 1/2)]$$

$$R_{\nu\nu'} = [(\sqrt{3})/2]\ (\gamma_3+\gamma_2)\hbar\omega_c <\psi_{\nu h} \mid \psi_{\nu'\ell}> a^2$$

$$S_{\nu\nu'} = \sqrt{6}\ \gamma_3\hbar\omega_c\ R_c <\psi_{\nu h}\mid k_z \mid \psi_{\nu'\ell} > a$$

$$R_{A\nu\nu'} = [(\sqrt{3})/2]\ (\gamma_3-\gamma_2)\hbar\omega_c <\psi_{\nu h} \mid \psi_{\nu'\ell}> a^{+2} \tag{8}$$

where the suffixes h and ℓ are alternate notations of the heavy and light hole states.

The k=0 states are not the only possibility for the basis set. Refs. 8 and 10 have used other basis functions. However, the subband k•p method is not only a convenient method for computation, but is also a convenient conceptual device, as we shall see next.

3. Landau Levels under Cylindrical Symmetry Approximation

For convenience, we shall first neglect the anisotropy term R_A. The Hamiltonian H_0 is then symmetric about the normal to the interface.

This approximation is sometimes justified by observing that the values of γ_3 and γ_2 are nearly the same. If these two Luttinger parameters are set equal everywhere in the Hamiltonian, this leads to the so-called spherical approximation and has been studied in the context of the inversion layer and quantum wells [10]. This further approximation is unnecessary since it provides no simplification in computation and has the undesirable feature of changing the correct energy order of the top three subbands from 0h,0ℓ. (This happens in the square well as well as in the inversion layer.) Since the 0ℓ and 1h subbands are very close in energy, the wrong order has profound effect on the effective mass and Landau level spacing of the second highest subband. No further consideration of the spherical approximation will be given.

Consider first the diagonal terms of the subband k•p matrix for the Hamiltonian. The resulting Landau levels are grouped into subbands and each subband has two sets of Landau levels corresponding to two spin directions with the usual linear magnetic field dependence. The off-diagonal terms provide inter-subband mixing as well as intra-subband level mixing. These mixings give rise to nonlinear field dependence as well as uneven level spacings. The cylindrical symmetry approximation eliminates the intra-subband mixing and reduces the inter-subband mixing to sets of four levels with the wave function taking the form

$$\left[u_{n-2}\zeta_{1,n} \quad u_{n-1}\zeta_{2,n} \quad u_{n}\zeta_{3,n} \quad u_{n+1}\zeta_{4,n}\right] \tag{9}$$

where u_n are the usual harmonic functions for the cyclotron motion in the interface plane, i.e., eigenfunctions of a^+a, and ζ_{jn} are linear combinations of the k=0 basis functions for motion in the z direction. The staggered structure of the four components of the wave function fits the elements of H_0 which connect different m_J with the right number of raising or lowering operators.

The wave function (9) gives us a set of quantum numbers to specify a Landau level: $\{k,n,\nu,j\}$. k resolves the degeneracy of each Landau level: for example, with a special choice of the gauge, k can represent the wave-vector of the plane wave along the x-axis. Then n labels the levels within each subband ν,j, which is determined by the dominant component of the wave function. In practice, the determination of ν,j is done by tracing from low field values [11]. ν is the subband index, assuming integer values from 0 on. j takes on either the m_J values (3/2, 1/2, -1/2, -3/2) or, more expressively, $(h+,\ell+,\ell-,h-)$. The definition of n in (9) is arbitrary. Ref. 8 replaces n by n-1 and Ref. 10 replaces n by n+1. We might call this set of quantum numbers the theorists' convention. Its advantage lies in the same Landau level number n for levels in different subbands which interact through the off-diagonal terms and, thus, in having clearly the same set of selection rules, $n \rightarrow n+1$, for transitions within the valence subbands. Its disadvantage lies in the fact that for each different j, n takes on a different set of integers. For example, for j = -3/2 or h-, n starts from -1. The fact that the dominant component of the wave function has an harmonic wave function u_{n+1} is obscured and, thus, complicates the quasi-selection rule for transition from the valence band to the conduction band: $n \rightarrow n+1$ for the same example j=h-.

To remove that complication, we can use what we might call the experimentalists' convention. The Landau level index n is replaced by m, taking on integer values from zero to infinity uniformly for all subbands. Thus,

$$m = n - j - 1/2 \quad . \tag{10}$$

A strong transition from a valence subband level to a conduction subband level conserves the Landau index m. However, we have to remember now that levels which mix have the same m+j+1/2. I am going to adopt the compromise of using the theorists' convention for computing the valence subband levels and the experimentalists' convention for correlation with experiment and for exposition.

The top bundle of Landau levels belongs to a heavy-hole subband. The field dependence of all levels is nonlinear, except one: the top level of the spin-down bunch, i.e., j = -3/2 or h-, and n = -1 in the theorists' notation for m=0 in the experimentalists'. From (9), the wave function for that level has only the fourth component to be non-zero. From the corresponding single Hamiltonian matrix element, the energy is

$$E(k,0,0h-) = \varepsilon_{oh} - (\gamma_1+\gamma_2 + 3\kappa)\, \hbar\omega_c/2 \quad . \tag{11}$$

For the Luttinger parameter values quoted above, the slope of the Landau is -0.3 meV/Tesla. The closest level to it, the top heavy-hole spin-up level, i.e., j=3/2 or h+ and n=2 (theorists' notation) or m=0 (experimentalists'), has a steeper initial slope as the magnetic field increases but bends over to cross the m=0 straight line. Thus, the highest Landau level is spin-down at low fields and spin-up at high fields. All three computations [8-10] show the same qualitative feature but they differ in the value of the crossing magnetic field.

Because the spacing between two Landau levels with m differing by unity is dependent on m and is not linear in B, expressing the cyclotron frequency in terms of cyclotron mass is neither convenient nor instructive. It is better to compare the transition energy directly with experimental measurement. Since all three computations [8-10] have been done for the same experimental system [6] with the hole density at $5 \times 10^{11}/cm^2$, they can be directly compared. Qualitative features of the observed cyclotron frequencies can be understood from the theories. There is more than one cyclotron frequency since the level spacings are not uniform. The nonlinear field dependence of the Landau levels is reflected in the cyclotron frequencies. Strong transition lines are between completely full and completely empty levels with m differing by unity. Weak lines may be due to partially filled levels or m differing by more than 1 because of intersubband mixing.

Quantitatively, there is some disagreement among the calculations. Table I shows typical values for the cyclotron frequencies at 8 Tesla. Ref. 8-10 are results of cylindrical approximation and Ref. 11 shows the modification of Ref. 9 due to adding the anisotropy term, which will be discussed in the next section. Ref. 10 appears to give the best agreement with experiment. However, a number of factors, which could cause the variation in the calculated results, might also make this agreement fortuitous. The highest measured frequency could also be lined up with the 0+→1+ transition to give two lines in good agreement with Ref. 9 or 11. Alternatively, the middle measured frequency might be the 1+→2+ transition, in good agreement with Ref. 8. Further experiment, such as polarization measurement, is needed to sort out which transition is being measured.

Table I. Comparison of calculated and measured cyclotron frequencies in meV.

Transition	Ref. 8	Ref. 9	Ref. 10	Ref. 11	expt
0- → 1-	3.70	5.44	3.5	5.20	3.56
0+ → 1+	2.94	3.28	2.2	3.30	2.14
1+ → 2+	2.18	2.30	1.6	2.62	1.56

The factors which are taken into account in different ways by the theories or which can cause uncertainties are:

1. The number of basis functions needed for convergence. Ando [15] claimed that in the zero magnetic field calculation of the classical effective mass 100 terms are required. In that case, since the influence of very low valence subband basis states is important on the highest subband, the result is very sensitive to the band edge discontinuity and to the depletion density, neither of which are accurately known. Ref. 8, using only 12 terms could hardly claim convergence.
2. The Hartree approximation at zero magnetic field is used for the self-consistent potential. The effect of the magnetic field on the self-consistent potential has not been studied, The exchange-correlation part of the potential can be included in the local density approximation [16] and is presumed to be unimportant.
3. For the transition, there could be further many-body effects, which should not be confused with the exchange-correlation potential above, unlike the electron case [9].
4. The effect of anisotropy will be discussed next.

4. Effect of Cylindrical Anisotropy

In zero magnetic field, the anisotropic term gives a pronounced warping to the spin-up Fermi surface in the inversion layer. In finite fields, one expects that the effect of neglecting the anisotropy term is not so drastic because the cyclotron motion involves averaging around the orbit. Inspection of Table I, columns labeled Ref. 9 and Ref. 11 supports that. However, studies by Ref. 11 and 12 show some interesting anisotropy effects.

Since the R_A term differs from the corresponding R term in the Hamiltonian (1) by a factor of four raising operators, the H_A term connects Landau levels with the index n differing by 4. This has some effect on the Landau levels in general. More pronounced are the splittings where these levels would cross in the cylindrical symmetry approximation, for example, in the top heavy-hole subband, the two levels n=1, j=-3/2 and n=5, j=3/2. A consequence of these couplings of the Landau levels is the increased possibility of more weak transitions. In Ref. 12, the dipole elements are calculated. Some of the observed cyclotron resonances in Ref. 7 can thus be explained. There, however, remain [7] some transitions in high magnetic fields (>20T) not accounted for.

An elegant interpretation of the Shubnikov-de Haas oscillations [17] is given by Bangert and Landwehr [12] to demonstrate the effect of anisotropy. A minimum is missing if two Landau levels cross at the Fermi level, as is shown to be the case for n=1, j=-3/2 and n=6, j=3/2, the difference in n being not equal to 4. On the other hand, if the cylindrical symmetry approximation predicts the absence of a minimum due to the crossing of two levels at the Fermi level with n differing by 4, the anisotropy will split the levels and restore the missing minimum. Such a case is also shown to exist.

5. Conclusions

It seems to me that, as in the case of electron cyclotron mass in Si-MOSFET, the many-body effect will render the quantitative comparison of calculated cyclotron frequencies of holes in the one-particle approximation with measurement meaningless. The complexity created by the valence band complex, particularly in the many-body effect on the transition between Landau levels and in the cylindrical anisotropy, should be exploited to seek distinguishing qualitative features which could be observed. We have made a promising beginning. Since many of the problems we discussed occur in p-type quantum wells, further theoretical and experimental studies could be done in these better characterized systems as well.

6. Acknowledgement

I acknowledge with pleasure my collaboration with David Broido and Eric Yang and the support of the U.S. National Science Foundation under Grant. No. DMR 85-14195.

7. Literature

1. S.S. Nedorezov: Fiz. Tver. Tela 12, 2269 (1970). [English Trans: Sov. Phys.-Solid State 12, 1814 (1971).
2. E. Bangert, K. von Klitzing, G. Landwehr: Proc. 12th Int. Conf. on Phys. of Semiconductors, ed. M.H. Pilkuhn, (Teubner, Stuttgart, 1974) p. 714.
3. F.J. Ohkawa and Y. Uemura: Prog. Theor. Phys. Supp. 57, 164 (1975).
4. H.L. Stormer, Z. Schlesinger: A. Chang, D.C. Tsui, A.C. Gossard, W. Wiegmann, Phys. Rev. Lett. 51, 126 (1983).

5. J.P. Eisenstein, H.L. Stormer: V. Narayanamurti, A.C. Gossard, W. Wiegmann, Phys. Rev. Lett. 53, 2579 (1984).
6. Z Schlesinger, S.J. Allen: Y. Yafet, A.C. Gossard, W. Wiegmann, Phys. Rev. B32, 5231 (1985).
7. W. Erhardt, W. Staguhn, P. Byszewski, M. von Ortenberg, G. Weimann, L. van Bockstal, P. Janssen, F. Herlach, G. Landwehr: Proc. Second Int. Conf. on Modulated Semiconductor Structures (1985), to be published.
8. E. Bangert and G. Landwehr: Superlatt. Microstrc. 1, 363 (1985).
9. D.A. Broido and L.J. Sham: Phys. Rev. B31, 888 (1985).
10. U. Ekenberg: EP2dS VI (1985), Surf. Sci., to be published. U. Ekenberg and M. Altarelli, Phys. Rev. B32, 3712 (1985).
11. S.R. Eric Yang, D.A. Broido, L.J. Sham: Phys. Rev. B32, 6630 (1985).
12. E. Bangert and G. Landwehr: EP2DS VI (1985), Surf. Sci. to be published.
13. Y. Iwasa, N. Miura, S. Tarucha, H. Okamoto, T. Ando: EP2DS VI (1985), to be published.
14. J.M. Luttinger: Phys. Rev. 102, 1030 (1956).
15. T. Ando: J. Phys. Soc. Jpn 54, 1528 (1985).
16. W. Kohn and L.J. Sham: Phys. Rev. 140, A1133 (1965).
17. G. Remenyi, G. Landwehr, G. Weimann: to be published.

Transport Properties of p-Type GaAs-(GaAl)As Heterojunctions in High Magnetic Fields

G. Landwehr

Physikalisches Institut der Universität Würzburg,
Röntgenring 8, D-8700 Würzburg, Fed. Rep. of Germany

1. INTRODUCTION

In the last few years p-type inversion layers in GaAs-(GaAl)As heterojunctions have been investigated by various groups. The work was motivated by unusual physical properties, especially by the very high hole mobility at low temperatures. Actually, the interest in p-type inversion layers is not new. The first silicon field effect transistors had p-channels for technological reasons. When it was realized that n-channel MOSFETs were advantageous because of the considerably higher electron mobility, more and more attention was attributed to these devices. Nevertheless quantum effects in p-type silicon inversion layers were investigated in detail in the 70's because it turned out that they had interesting physical properties. In particular, it is possible to modify the subband structure by the surface electric field. The effective masses are strongly dependent on the orientation of the interface and on the magnitude on the electric field present. It was predicted that due to lacking inversion symmetry the Kramers-degeneracy was lifted at finite wave vectors /1, 2/. However, the calculated splitting turned out to be rather small and was never seen in Shubnikov-de Haas experiments. Only recently the splitting was seen in an optical experiment /3/. The previous work on p-type silicon inversion layers has been reviewed by ANDO, FOWLER and STERN /4/.

Work on two-dimensional electronic systems received a substantial impetus when modulation doped GaAs-(GaAl)As heterostructures were introduced /5/. Due to spatial separation of ionized impurities and mobile carriers very high mobilities at low temperatures were achieved. At present, the top electron mobility at helium temperatures is well above 10^6 cm^2/Vs /6/. When the technique of modulation doping was applied to generate p-type heterostructures it was demonstrated that hole mobilities could be significantly improved /7/. In the last few years low temperature hole mobilities above 2×10^5 cm^2/Vs have been achieved /8/. These high mobilities have allowed to perform very interesting transport and optical experiments which have yielded detailed information on the structure of electric subbands and their dependence on external fields. Due to the degeneracy of the top of the valence band of GaAs and due to the existing anisotropy of the constant energy surfaces the electric subbands are more complicated than in n-type material. Even for (100) symmetry planes the motion of holes parallel and perpendicular to the interface is strongly coupled.

Most experiments which can yield information on electric subbands of 2D-systems involve high magnetic fields. Usually it is assumed that the electric subbands are not substantially modified by the application of a transverse magnetic field, which means it is anticipated that the Peierls approximation holds. The influence of high magnetic fields is manifested by Landau quantization which is imposed on the band structure. In the case of p-type silicon inversion layers this approach seems to be justified, judging from the relatively good agreement between theory and experiment. One should recall, however, that the hole mobility in MOSFETs of good quality is a few thousand cm^2/Vs. The comparison of the Shubnikov-de Haas and cyclotron resonance data - which have been obtained on high mobility p-type GaAs-(GaAl)As heterostructures - with theoretical predictions derived from Landau-level calculations shows that the magnetic field has a significant influence on the subband structure. This has the consequence that it is no longer possible to distinguish between light and heavy holes, except in the quantum limit. The main purpose of this paper is to demonstrate this. Moreover, it is the intention to show that in spite of the relatively complicated structure of electric subbands in GaAs-(GaAl)As heterostructures a quite substantial understanding of many special features has been achieved in the meantime.

2. REVIEW OF EXPERIMENTAL DATA

During the last few years such a large amount of activity has been devoted towards the investigation of the physical properties of p-type GaAs-(GaAl)As heterostructures that it is not possible to give a comprehensive review here and to discuss all relevant results. Instead, only a few experiments will be selected in order to indicate the present state of our knowledge and to point out remaining problems.

The first experiments by STOERMER and coworkers on the transport properties of high mobility p-type modulation doped heterostructures in the helium 3 temperature range yielded well-resolved Shubnikov-de Haas oscillations and the integral quantum Hall effect /9/. Part of the data is reproduced in Fig. 1. A plot of the position of the maxima and minima of the oscillations

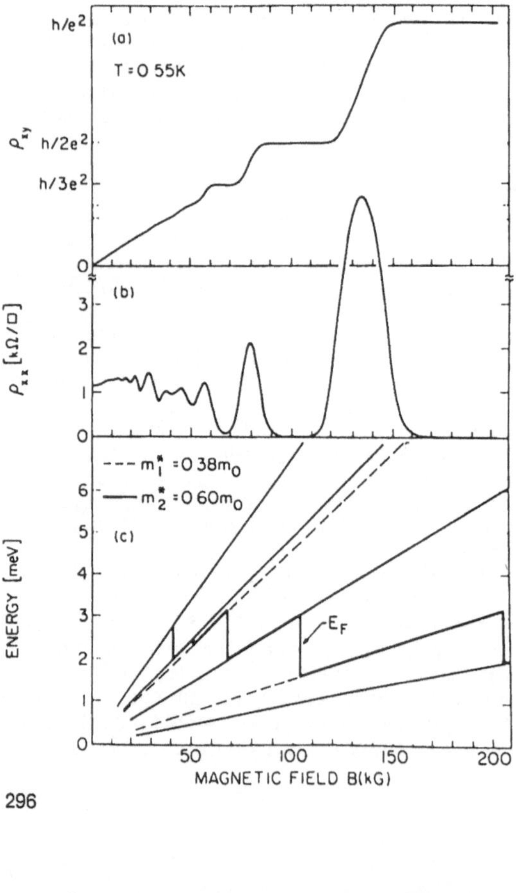

Fig. 1:
a) Hall resistance ρ_{xy};
b) magneto-resistance ρ_{xx};
c) Landau-level scheme for a GaAs-(GaAl)As heterostructure (after ref. 9)

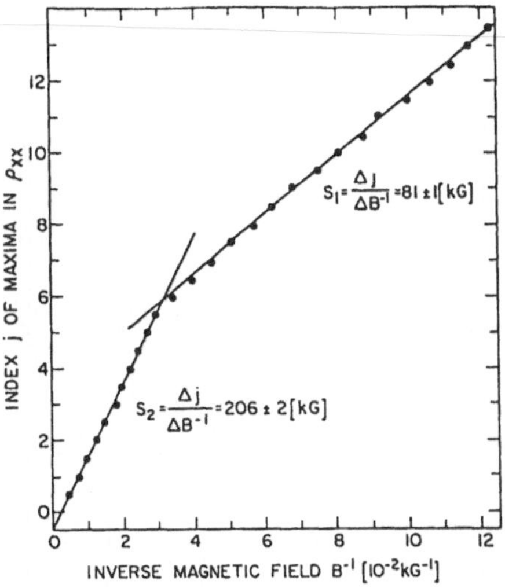

Fig. 2:
Positions of oscillatory extrema on a 1/B scale vs. whole numbers for the ρ_{xx} data shown in Fig. 1 (after ref. 9)

on an inverse magnetic field scale versus whole numbers gave two intersecting straight lines as can be seen in Fig. 2. In addition, cyclotron resonance experiments were performed with a Fourier-transform-spectrometer resulting in a two-line-spectrum which was attributed to two kinds of holes with effective masses of 0.38 m_0 and 0.60 m_0. It was concluded that two subbands exist which are degenerate at k = 0 and that the spin degeneracy is lifted at $k \neq 0$ as a consequence of the lacking inversion symmetry at the GaAs-(GaAl)As heterojunction interface.

Subsequently magneto-transport experiments on p-type heterostructures with low carrier concentration were extended to higher magnetic fields by MENDEZ and coworkers /10/. Besides the integral quantum Hall effect they observed plateaus in the Hall-resistance for fractional quantization accompanied by minima in the component ρ_{xx} of the magneto-resistance. An example of the data is given in Fig. 3. Again a plot of the Shubnikov-de Haas extrema in units of 1/B against whole numbers yielded 2 straight sections from which in conjunction with the Quantum-Hall-effect concentrations of two kinds of heavy holes were deduced. Attempts to derive the effective mass of holes from the temperature dependence of the Shubnikov-de Haas oscillations at relatively low magnetic fields yielded effective masses between 0.13 m_0 (taken at 0.4 T) and 0.19 m_0 (taken at 0.9 T). These results indicate that a straight forward interpretation of the data does not seem to be justified. The relaxation time as determined from the damping of the Shubnikov-de Haas oscillations turned out to be three times smaller than the scattering time deduced from the zero field conductivity /11/. This again indicates that the interpretation of the data needs special caution.

In the following selective data on high quality p-type GaAs-(GaAl)As heterojunctions will be presented which were grown by Dr. G. Weimann, Forschungsinstitut der Deutschen Bundespost, Darmstadt. The samples have a hole concentration between 2 x $10^{11}/cm^2$ and 2.5 x $10^{11}/cm^2$ and a high mobility, which varies between 60.000 cm^2/Vs and 185.000 cm^2/Vs at 0.5 K. On these specimens careful measurements were performed in the temperature range between 50 mK and 4.2 K in magnetic fields up to 12.5 T. The analysis of the data revealed features which were not seen in the early experiments. They clearly indicate that a proper interpretation without detailed theoretical calculations is not possible. Data, obtained for the sample with the highest mobility (220.000 cm^2/Vs at 50 mK) in a dilution refrigerator are shown in

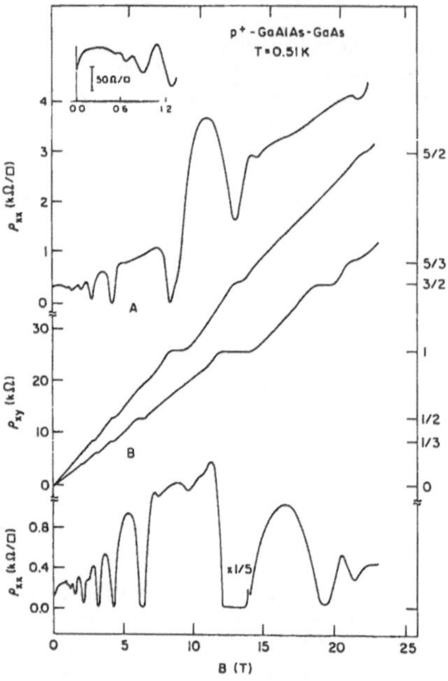

Fig. 3:
Hall resistance ρ_{xy} and magnetoresistance ρ_{xx} of 2 heterostructures at very high magnetic fields at T = 0.51 K (after ref. 10)

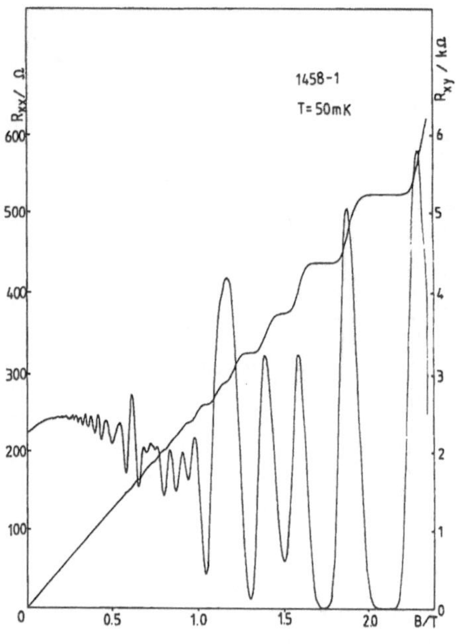

Fig. 4:
Hall resistance ρ_{xy} and magnetoresistance ρ_{xx} of a p-type heterostructure at T = 50 mK (after Remenyi and Heuring)

Fig. 4. In order to reveal the relatively complicated Shubnikov-de Haas pattern, magnetoresistance and Hall-resistance have been plotted in the low field range up to 2.5 T. When the temperature is raised, the oscillations broaden but still have a complicated appearance. This can be seen in Fig. 5 where data obtained at 500 mK for a sample with a somewhat lower mobility of 140.000 cm^2/Vs at 0.5 K are shown. A plot of the positions of the maxima and minima on a 1/B scale against whole numbers reveals that it is impossible to obtain straight line sections. Inspection of the experimental data shows that at the values whère the 1/B periodicity is broken, maxima appear instead of minima. This is indicated by arrows in Fig. 5. In order to obtain a straight line representation of the extrema, as found in previous experiments /9/, it is necessary to leave out that particular Landau-quantum number. Such a plot with two irregularities is shown in Fig. 6. As will be pointed out in the following, these findings can be understood on the basis of Landau-level calculations for the heterostructure which take into account the full anisotropy of the band structure.

Another hint that the heterostructure data need a rather elaborate interpretation came from cyclotron resonance experiments performed with far-infrared lasers /12/. The early experiments were performed with Fourier transform spectrometers /9, 13/. These have the advantage, that a wide frequency range can be covered, however, at the expense of resolution due to the small intensities in the far-infrared. A spectrum obtained at two particular frequencies is shown in Fig. 7. The observed multiline spectrum indicates that the Landau-level scheme must have a complicated structure and and that it is no longer possible to speak of effective masses in a naive

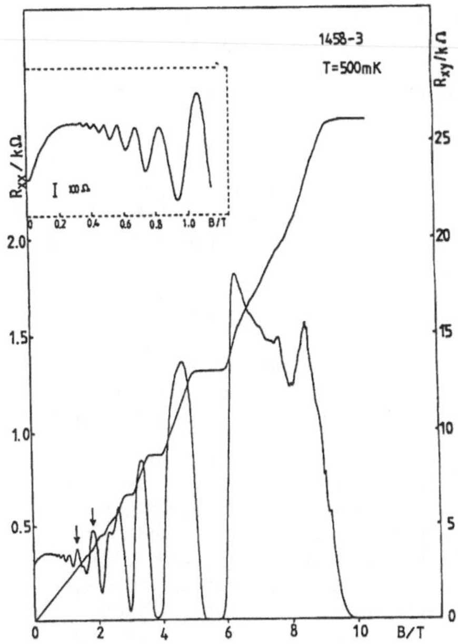

Fig. 5:
R_{xx} and R_{xy} for a different structure at T = 0.5 K. Insert:enlarged scale at low magnetic fields

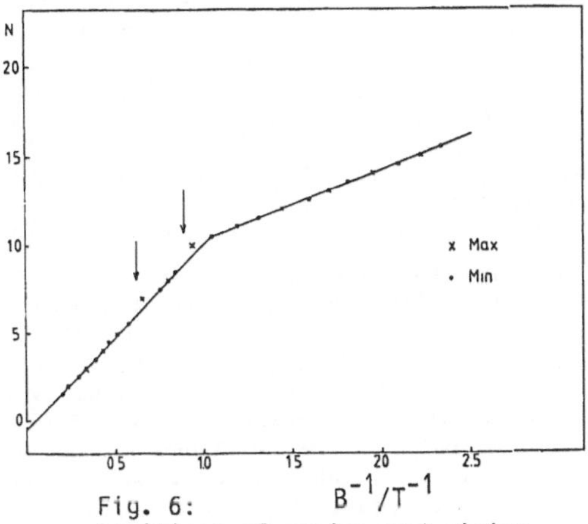

Fig. 6:
Positions of maxima and minima on a 1/B scale vs. whole numbers N for the data shown in Fig. 5

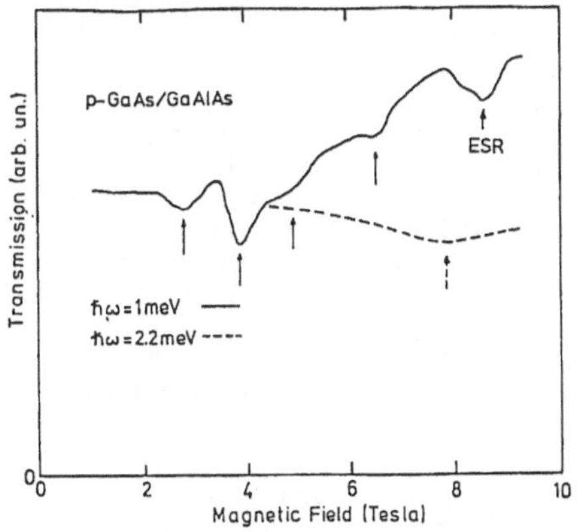

Fig. 7:
Cyclotron resonance at two fixed frequencies for a p-type heterostructure (after ref. 12)

sense. Instead it is necessary to interpret the cyclotron resonance data in terms of magneto-optical transitions.

3. LANDAU-LEVEL CALCULATIONS

During the recent past the electrical subbands of p-type GaAs-(GaAl)As heterostructures have been calculated by several groups /14 - 17/. This area will not be discussed in great detail because it is subject of a contribution by L. J. Sham in this issue. Instead, preferentially, new results will be presented which were recently obtained by E. Bangert and which allow a detailed interpretation of experimental data. It should be mentioned, however, that certain aspects still lack quantitative agreement between theory and experiment.

Our first attempt to calculate electric subbands was based on a self-consistent solution of both Schrödinger's and Poisson's equation employing the Luttinger 4 x 4 matrix representation of the valence band of GaAs. The calculations were performed in the self-consistent Hartree-approximation for zero magnetic field. It turned out, however, that it was necessary to incorporate a magnetic field into the calculations because the initial agreement between theory and experiment was not satisfying. Therefore, Luttinger's matrix was modified. To simplify the calculations, the in-plane anisotropy of the band structure was ignored in the beginning. Because still the agreement between theory and experiment left something to be desired, the full anisotropy was included /18/. The Landau-levels obtained for a hole concentration $2.3 \times 10^{11}/cm^2$ are shown in Fig. 8. Also the Fermi energy has been plotted in Fig. 8. It is evident that the calculation was done for T = 0 K which should not be serious drawback, however, because most experiments were done at very low temperatures. Two series of Landau-levels can be distinguished, the a-series can be derived from the heavy hole band with small curvature and b-series from the band with the smaller effective mass at k = 0. Details are explained in the original paper. The dashed lines refer to the original calculations without inclusion of the anisotropy in the interface plane. One can recognize around 7 T that the crossing of the levels -1b and 3a is suppressed. In order to allow a comparison with the low field Shubnikov-de Haas data the Landau-levels have been plotted on a larger scale for magnetic fields between 0.3 T and 2 T. One can see that there are numerous crossings between different Landau-levels. Whenever the Fermi-level passes such a crossing point or it is very close to it, a Shubnikov-de Haas minimum should be missing. This is actually observed. The anomalies in the plot of position of the extrema against whole numbers can be attributed to the Landau-level crossings in Fig. 8b around 1.1 T and 1.5 T. The coincidences of the Fermi-level at lower magnetic fields can no longer be resolved.

In order to get more detailed information on the Shubnikov-de Haas oscillations, especially on the positions of the maxima and minima, Bangert calculated the density of states assuming a Gaussian profile and a line width Γ proportional to $B^{1/2}$. The results, obtained with $\Gamma = 0.1 \times B^{1/2}$ meV (B in Tesla) are shown in Fig. 9, together with the Fermi energy as a function of the magnetic field. A plot of the maxima and minima against whole numbers reproduces the experimental data represented by Fig. 6 very well. The kink is predicted to occur at 0.69 T in good agreement with the experiments.

Finally, it should be mentioned that the theoretical cyclotron resonance transition data are in reasonable agreement with the experimental findings. One should note, however, that the agreement is not perfect. In Fig. 10 the theoretical predicted transition energies for cyclotron resonance absorption have been plotted against the magnetic field. The experimentally observed transitions are indicated by dots, the diameter of which indicate the strength of the observed lines. It is obvious that the transition 2a - 1a corresponding to an effective mass about 0.4 m_0 is dominant. Many of the theoretically predicted transitions at fields below 5 T have not been observed, although the calculated matrix elements indicate that they should be seen. The origin of this discrepancy has not yet been found out.

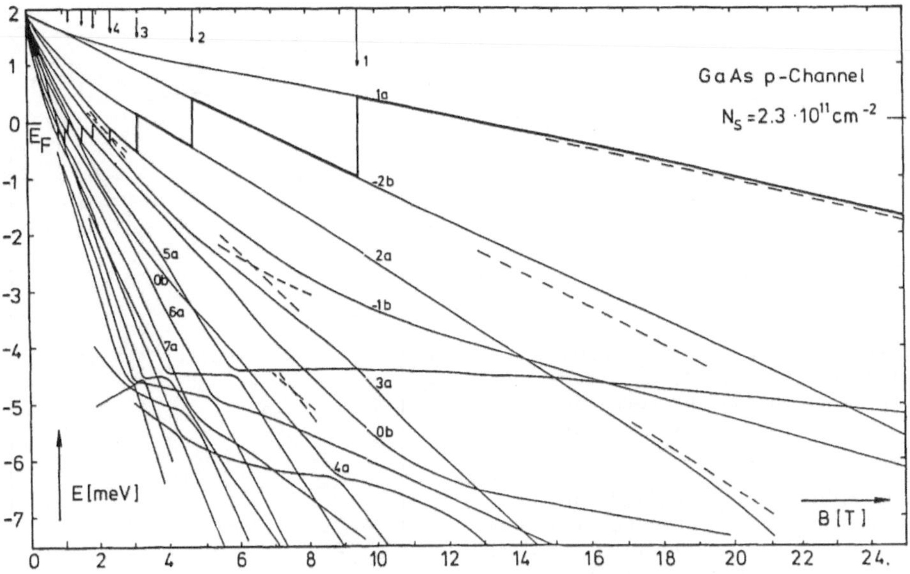

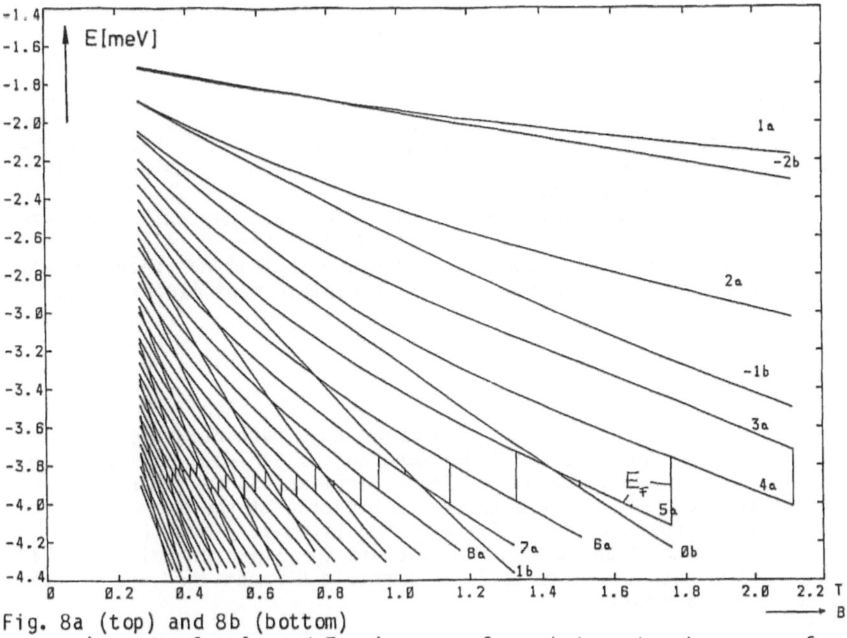

Fig. 8a (top) and 8b (bottom)

a) Landau-levels and Fermi energy for a heterostructure as a function of magnetic fields up to 25 T

b) Details of the same calculations on an enlarged scale between 0.3 T and 2.1 T (after ref. 18)

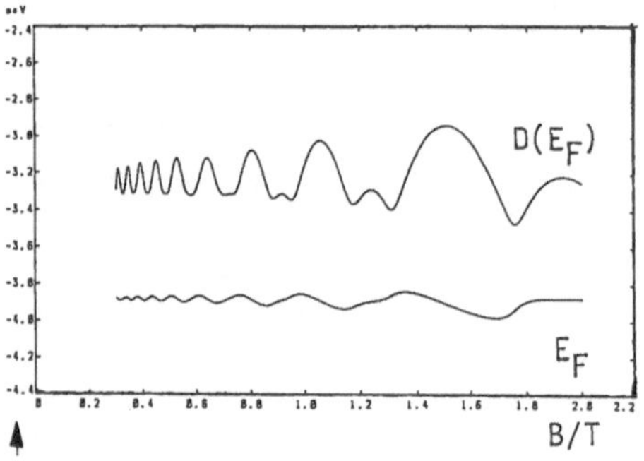

Fig. 9:
Density of states at the Fermi level for a p-type heterostructure (arb. units) and Fermi energy as function of B (E. Bangert)

Fig. 10:
Comparison of experimental and theoretical cyclotron resonance line positions. Dots: experimental results. Lines: theoretical calculations (after ref. 18)

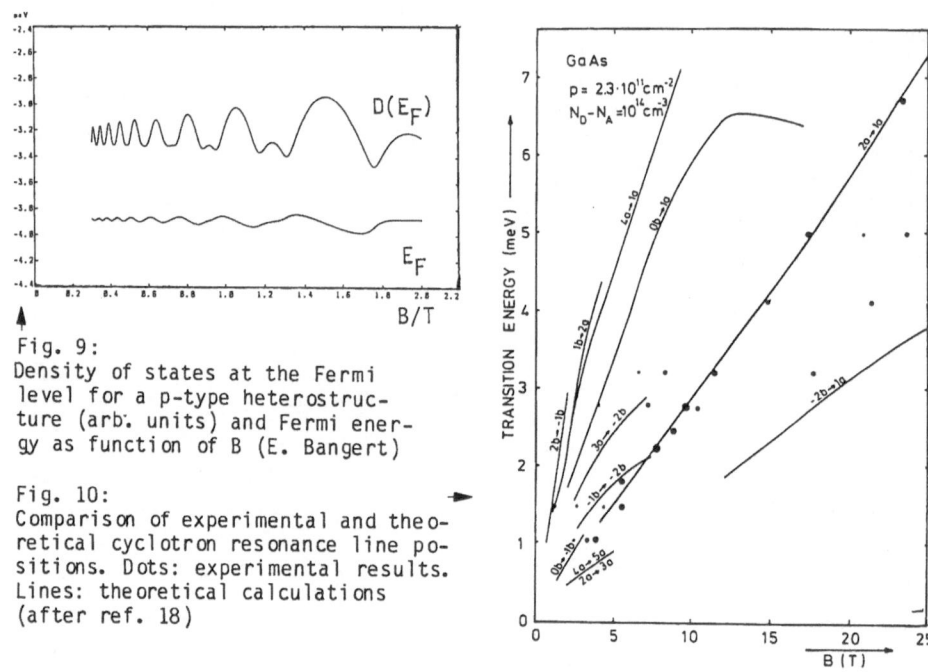

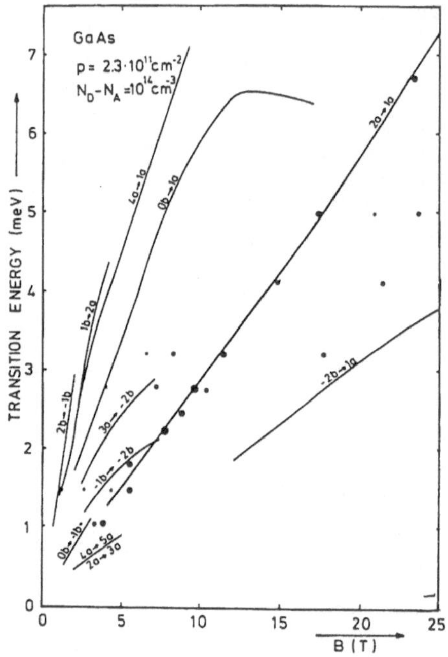

4. CONCLUSION

The brief review given indicates that p-type GaAs-(GaAl)As heterostructures are very intriguing systems. Especially appealing is the possibility to modify the band structure by an external magnetic field. The aspects in conjunction with the fractional quantum Hall-effect are discussed separately in this issue. It turned out that the subband structure is relatively complicated but, on the other hand, simple enough that it has been possible to perform theoretical calculations which can explain many experimental details.

5. ACKNOWLEDGMENT

The experimental work in Würzburg would not have been possible without the excellent samples provided by G. Weimann and W. Schlapp, FTZ Darmstadt, and without a very close cooperation. The experiments in the mK-range were performed at the Hochfeldmagnetlaboratorium Grenoble of the Max-Planck-Institut für Festkörperforschung. The work benefited very much from the dedication of G. Reményi and W. Heuring who did the experimental work. Last but not least the enthusiasm of E. Bangert should be mentioned who did the theoretical calculations.

6. LITERATURE

/1/ E. Bangert, K. von Klitzing and G. Landwehr: Proc. 12th Int. conf. Physics of Semiconductors, (M.H. Pilkuhn Ed.), Teubner Verlag, Stuttgart, p. 714 (1974)
/2/ F.J. Ohkawa and Y. Uemura: Progr. Theor. Physics, Suppl., 57, 164 (1975)
/3/ A.D. Wiek, E. Batke, D. Heitmann, J.P. Kotthaus and E. Bangert: Phys. Rev. Lett.

/4/ T. Ando, A.B. Fowler and F. Stern, Rev. of Modern Physics 54, 437 (1982)
/5/ H.L. Stoermer, R. Dingle, A.C. Gossard and W. Wiegmann: Inst. Phys. Conf. Ser. 43, 557 (1978)
/6/ H.L. Stoermer: Surface Science 142, 130 (1984)
/7/ H.L. Stoermer, A.C. Gossard, W. Wiegmann, R. Blondel and K. Baldwin: Appl. Phys. Lett. 44, 139 (1984)
/8/ E.E. Mendez and W.I. Wang: Appl. Phys. Lett. 46, 1159 (1985)
/9/ H.L. Stoermer, A.M. Chang, Z. Schlesinger, D.C. Tsui, A.C. Gossard and W. Wiegmann: Phys. Rev. Lett. 51, 126 (1983)
/10/ E.E. Mendez, W.I. Wang, L.L. Chang and L. Esaki: Phys. Rev. 30, 1087 (1984)
/11/ E.E. Mendez: Surface Science 170, 564 (1986)
/12/ W. Erhardt, W. Staguhn, P. Byszewski, M. von Ortenberg, G. Landwehr, G. Weimann, L. von bockstal. P. Janssen, F. Herlach and J. Witters: Surface Sc. 170, 581 (1986)
/13/ Z. Schlesinger, S.J. Allen, Y. Yafet, A.C. Gossard and W. Wiegmann: preprint
/14/ D.A. Broido and L.J. Sham: Phys. Rev. B31, 888 (1985)
/15/ E. Bangert and G. Landwehr: Superlattices and Microstructures 1, 363 (1985)
/16/ U. Ekenberg and M. Altarelli: Phys. Rev. B30, 3569 (1984)
/17/ T. Ando, J. Phys. Soc. Japan 54, 1528 (1985)
/18/ E. Bangert and G. Landwehr: Surface Science 170, 593 (1986)

Two-Dimensional Hole Gas in GaInAs/InP Heterojunction

L. Dmowski[1,*], *D. Gauthier*[1], *J.C. Portal*[1], *M. Razeghi*[2], *P. Maurel*[2], *and F. Omnes*[2]

[1]CNRS-INSA, F-31077 Toulouse and
CNRS-SNCI, 166X, F-38042 Grenoble, France
[2]Thomson CSF, BP 10, F-91041 Orsay, France
*also at High-Pressure Research Centre, PAS Unipress, Warsaw, Poland

Studies of a two-dimensional hole gas in a $Ga_{0.47}In_{0.53}As/InP$ heterojunction grown by metalorganic chemical vapour deposition are reported. In a sample with a total hole density $p_{tot} = 7.6 \times 10^{11}$ cm^{-2} a Hall mobility $\mu_H = 10500$ cm^2/V.s was measured at 4.2 K. Low temperature persistent photoconductivity was observed significantly increasing the hole density at the interface.

1. Introduction

Till now GaInAs/InP heterostructures were widely grown and studied with respect to the two-dimensional electron gas confined at the interface. However, we could not find any paper relating the existence of a 2 D hole gas in this system in contrast to GaAs/GaAlAs where both types of 2 D carriers have already been studied. In this paper we present magnetotransport data of p-type GaInAs/InP heterojunctions which confirm two-dimensionality of the hole gas and the lifting of the spin degeneracy due to the lack of the inversion symmetry at the interface as it was already demonstrated for GaAs/GaAlAs [1, 2].

2. Experimental results

The samples were grown in a $i\text{-}Ga_{0.47}In_{0.53}As/i\text{-}InP/p^+\text{-}InP$ configuration on Fe-doped semi-insulating InP substrates (as shown in Fig. 1) using a low pressure metalorganic chemical vapour deposition system. The p^+-InP layer was doped with Zn to a level $N_A - N_D = 10^{17}$ cm^{-3} while the GaInAs layer was not intentionally doped.

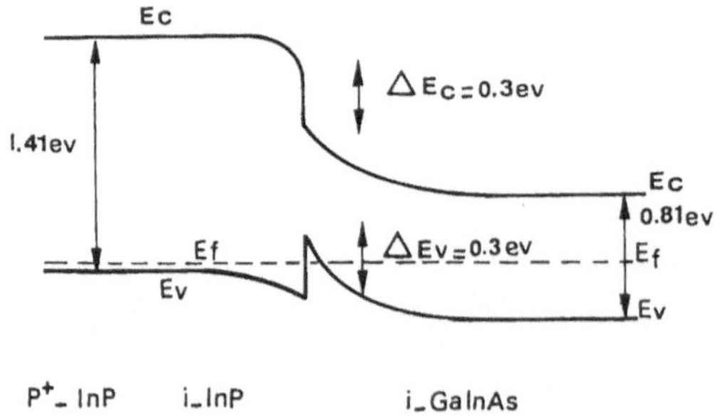

Fig. 1
Schematic configuration of the $i\text{-}Ga_{0.47}In_{0.53}As/i\text{-}InP/p^+$ InP heterojunction.

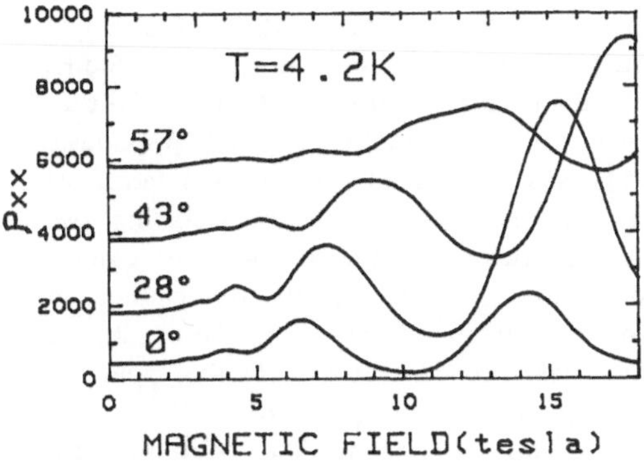

Fig. 2
Diagonal resistivity ρ_{xx} versus magnetic field for various angles at 4.2 K.

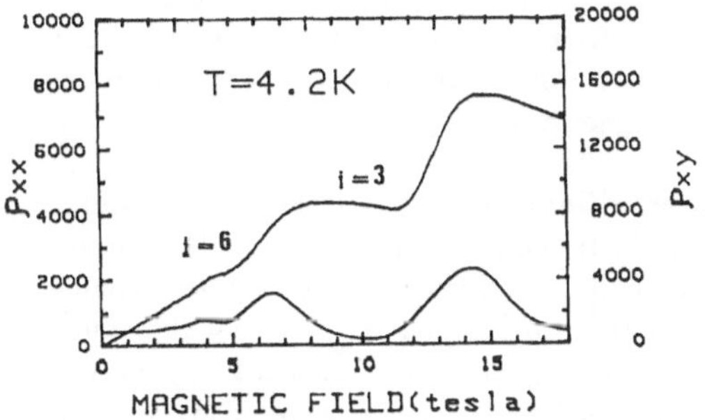

Fig. 3
Shubnikov-de-Haas and Hall effect as a function of magnetic field at 4.2 K.

Magnetotransport measurements (Shubnikov-de-Haas and Hall effect) were performed at 4.2 K in magnetic fields up to 18 T. Shubnikov-de-Haas experiments were done in the normal and tilted magnetic field orientations (Fig. 2). The ρ_{xx} extrema appeared at constant values of the perpendicular component of the magnetic field $B_{\perp} = B_o \cos\Theta$ demonstrating the bidimensionality of the holes at the interface. Figure 3 shows the diagonal resistivity ρ_{xx} and the Hall resistivity ρ_{xy} as a function of B at 4.2 K. Although this temperature is not low enough to obtain very high quality of QHE plateaus, one can distinguish the steps typical for QHE which also confirms the two-dimensionality of the carriers. It is worth pointing out the relatively large width of the step $\rho_{xy} = h/3\,e^2$ corresponding to three entirely occupied single Landau levels. In the case of electron systems, plateaus with odd indices occur at lower temperatures and develop into narrower structures than those having even indices. In contrast the plateaus observed in the hole system do not show this odd/even disparity [1]. This feature is attributed to the lifting of the spin degeneracy due to the lack of the inversion symmetry at the interface and the splitting of each hole subband into two "spin subbands" even at zero magnetic field. This splitting was already demonstrated experimentally [1, 2] as well as by calculations [3, 4] for p-type GaAs/AlGaAs single interfaces.

We have estimated the total hole density as $p_{tot} = 7.58 \times 10^{11}\ cm^{-2}$ taken from the QHE plateau with i = 3. The corresponding minimum of SdH effect belongs to a magnetic field B_3 necessary for the complete filling of 3 single Landau levels. The carrier concentration deduced from low field S.d.H. oscillations was significantly smaller than that determined from either classical or quantum Hall measurements. An observation reported previously for heterostructure GaAs/AlGaAs [1, 5] was attributed to the existence of two hole "spin subbands" degenerate at the Brillouin zone centre having two different effective masses and generating two independent sets of Landau levels. If we follow the interpretation that the hole density determined at low magnetic fields gives the population of the spin down heavy hole subband p_{hh^-} equal 2.16×10^{11} cm^{-2}, then the hole concentration of the spin up subband (with heavier effective mass) is $p_{hh^+} = p_{tot} - p_{hh^-} = 5.42 \times 10^{11}$ cm^{-2}. This gives a population ratio $P_{hh^+}/P_{hh^-} = 2.5$ similar to that obtained for a 2 DHG in GaAs/AlGaAs for similar total densities [6]. Recent calculations (see the contributions by Sham and Landwehr in this issue) have indicated, however, that the distinction between two kinds of holes is only a rough approximation in the presence of high magnetic fields.

In contrast to the 2 DHG in GaAs/AlGaAs [5, 7] we have observed a significant persistent variation of the 2 DH concentration after illuminating the sample at low temperature. The hole densities in both spin subbands increased to the values of $p_{hh^-} = 2.84.10^{11}$ cm^{-2} and $p_{hh^+} = 8.00\ 10^{11}$ cm^{-2} giving a higher ratio $p_{hh^+}/p_{hh^-} = 2.8$ for higher total concentrations. This tendency is in agreement with recent calculations [4]. We have also observed that the hole concentration p_H derived from low field Hall measurements always underestimates the total concentration $p_{tot} = p_{hh^+} + p_{hh^-}$. This discrepancy increases with increasing hole concentration. We attribute it to the conducting system consisting of two types of carriers with different mobilities. Solving the equations which describe the Hall mobility μ_H and the Hall density p_H of such a system and taking into account that for a 2 DH spin subband system $p_{hh^+} > p_{hh^-}$ and $\mu_{hh^+} < \mu_{hh^-}$ we obtain :

$$(1) \qquad \mu_{hh^+} = \frac{\mu_H\ p_H}{p_{hh^+} + p_{hh^-}} \left[1 - \sqrt{\frac{p_{hh^-}}{p_{hh^+}} \left(\frac{p_{hh^+} + p_{hh^-}}{p_H} - 1 \right)} \right]$$

$$(2) \qquad \mu_{hh^-} = \frac{\mu_H\ p_H}{p_{hh^+} + p_{hh^-}} \left[1 - \sqrt{\frac{p_{hh^+}}{p_{hh^-}} \left(\frac{p_{hh^+} + p_{hh^-}}{p_H} - 1 \right)} \right]$$

For the total hole density $p_{tot} = 7.58.10^{11}$ cm^{-2} and $\mu_H = 10450$ cm^2/V.s (before illuminating) we have obtained $\mu_{hh^+} = 6240$ cm^2/V.s, $\mu_{hh^-} = 14850$ cm^2/V.s and for $p_{total} = 10.84\ 10^{11}$ cm^{-2} and $\mu_H = 7320$ cm^2/V.s (after illuminating) $\mu_{hh^+} = 4100$ cm^2/V.s, $\mu_{hh^-} = 10770$ cm2/V.s.

3. Conclusion

The first observation of a two-dimensional hole gas in $Ga_{0.47}In_{0.53}As$/InP heterojunction is reported. Similarly to the GaAs/AlGaAs interface, the lowest heavy hole bound state consists of two subbands with different hole densities and different mobilities. The lifting of the Kramers degeneracy due to the lack of the inversion symmetry at the interface gives an approximate population ratio of the subbands $p_{hh^+} / p_{hh^-} = 2.5$ suggesting spin splitting at zero magnetic field at finite wave vectors as found in GaAs/AlGaAs.

Acknowledgements

We acknowledge the financial support from Conseil Régional Midi-Pyrénées and ESPRIT Programs and Prof. G. Landwehr for useful discussions.

References

1. H.L. Störmer, Z. Schlesinger, A. Chang, D.C. Tsui, A.C. Gossard and W. Wiegmann : Phys. Rev. Lett. 51, 126 (1983)
2. J.P. Eisenstein, H.L. Störmer, V. Narayanamurti, A.C. Gossard and W. Wiegmann : Phys. Rev. Lett. 53, 2579 (1984)
3. E. Bangert and G. Landwehr : In Superlattices and Microstructures, 1, 363 (1985)
4. U. Ekenberg and M. Altarelli : Phys. Rev., B32, 3712 (1985)
5. E.E. Mendez, W.I. Wang, L.L. Chang and L. Esaki : Phys. Rev., B30 1087, (1984)
6. L.L. Chang, E.E. Mendez, W.I. Wang, L. Esaki and P.M. Tedrow : In Proc. 17th Int. Conf. Phys. Semicond., San Francisco, 1984, ed. by J.D. Chadi and W.A. Harrison (Springer, New York, 1985), p. 299
7. H.L. Störmer, A.C. Gossard, W. Wiegmann, R. Blondel and K. Baldwin : Appl. Phys. Lett., 44, 139 (1984).

Magnetotransport in the δ-Doping Layer

F. Koch[1], A. Zrenner[1], and K. Ploog[2]

[1]Physik-Department der Technischen Universität München, D-8046 Garching, Fed. Rep. of Germany

[2]Max-Planck-Institut für Festkörperforschung, D-7000 Stuttgart, Fed. Rep. of Germany

The δ-layer consists of a sheet of Si-donor atoms embedded during MBE-growth in a single atomic plane of GaAs. We examine the electronic transport in the plane of the layer and for magnetic fields in the range 0 - 20 T.

I. Introduction

For the "brave new world" of MBE(molecular-beam-epitaxy)-growth of synthetic semiconductor materials, of band structure engineering and the many, proliferating types of superlattices, impurities don't just happen - they are caused. Donors and acceptor atoms can be deliberately positioned in precise numbers with atomic layer precision during crystal growth. The accurately controlled spacers for remote doping, donor states placed in a specific position of the quantum well and the present case of δ-function doping are examples of impurity engineering.

We consider here various magnetotransport phenomena in the plane of a layer of Si-atoms embedded in GaAs during MBE-growth. Densities N_D are of the order of a milli-monolayer of Si, but in any case sufficiently high to assure a degenerate gas of conduction electrons and metallic transport in the plane of the layer. The electrons occupy subband levels of a V-shaped potential.

The original publication on this novel 2-d system is found in Ref. /1/. A MES-FET device based on this doping layer is described in /2/. Parallel-field magnetotransport is discussed in the recent Refs. /3,4/. The energy level structure has been measured in tunneling experiments /5/, while central cell effects and valley-orbit splitting of the L-minima of the GaAs band structure are contained in Ref. /6/. The present paper provides an overview on the magnetotransport work with additional results on previously published work and with a brief look at work currently in progress.

II. The Subband Structure of the δ-Layer

The electronic levels of the N_D donors are calculated assuming the positive ion density $+N_De$ to be uniformly spread over a sheet at $z = 0$ and screened self-consistently by the subband charge of the $N_s = N_D$ electrons. For densities below $\sim 10^{13}$ cm^{-2} it has been shown that the number of metallic electrons in the doping sheet equals the design donor density N_D. The δ-sheet is an n^+-layer accumulation layer in a lightly doped n-type background material.

The simplest version of the calculation assumes a constant effective mass $m^* = 0.067\ m_0$ to describe subbands associated with the Γ-point of the GaAs

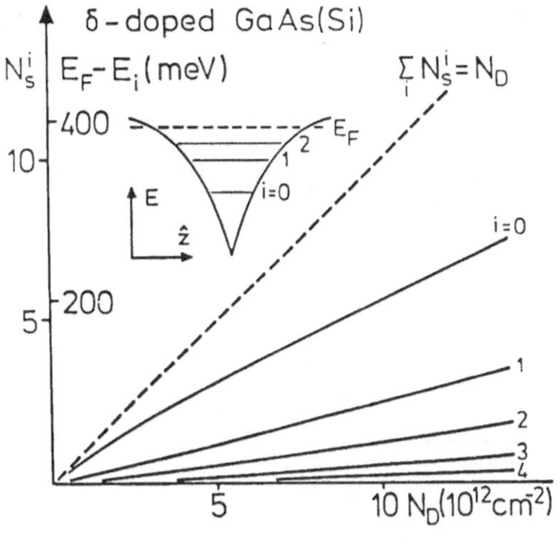

Fig. 1: Calculated subband levels and occupancies versus donor density in the δ-accumulation layer. A parabolic conduction band has been assumed.

band structure. The energies E_i and occupancies N_s^i of each of the filled levels are shown in Fig. 1 as a function of N_D. We note that the energies, in particular for carriers in the ground state E_0, lie significantly above the band edge for typical values of N_D. Thus, for example, for N_D = 7 x 10^{12} cm^{-2} the energy of parallel motion of the carriers rises to 147 meV above E_0, which itself in the sense of an average kinetic energy of the $\hat{z}$-directed motion lies about 30 meV above the conduction band minimum. From the nonparabolicity effects, as calculated in Ref. /7/, the mass at the Fermi energy in the E_0 subband would be ∿ 80% above the band-edge value. The error incurred in calculating energies and subband occupations in Fig.1 is only a small fraction of this amount. Nevertheless, Fig. 1 makes clear that for the interpretation of data with high N_D, a more extensive calculation is needed. Such work has been done and will appear elsewhere. For the present the simple, nonparabolic calculation is sufficient.

III. The $B_\perp$-Magnetoconductivity

The straightforward way to determine the partial occupations N_s^i of the subbands, and thus their energies E_i, is the oscillatory magnetoconductivity in a field applied perpendicular to the doping plane. This was first done in Ref. /1/.

In the accompanying Fig. 2 we show another example of such conventional Shubnikov-de Haas data. For a sample with a design density of 7.4 x $10^{12}$$cm^{-2}$ Si-atoms, a complex pattern of oscillations appears in a perpendicular field. High fields are a necessity and the data of Fig. 2 was taken at the Grenoble magnet facility. The Fourier spectrum in the insert shows that 4 periods in reciprocal field can be identified. The densities are 4.0, 1.75, 0.8, and 0.4, all in units of 10^{12} cm^{-2}. The sum is to be compared with the design density of 7.4 x 10^{12} cm^{-2}.

The figure also shows that the mobilities in the four subbands differ substantially. Whereas the groundstate subband has μ_0 ∿ 1400 cm^2/Vsec, the carriers moving in subbands whose wavefunctions extend far beyond the ions at z = 0, typically have μ-values of 5 - 10,000 cm^2/Vsec.

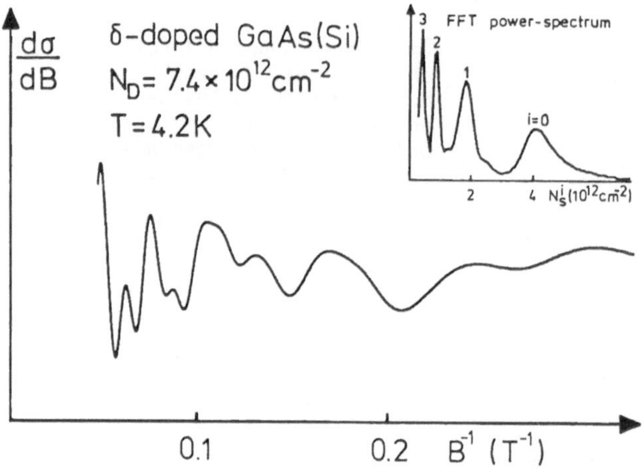

Fig. 2: Shubnikov-de Haas oscillations for a Si-doped layer with 7.4×10^{12} cm^{-2} donors. The corresponding fast Fourier transform (FFT) reveals four individual subband contributions.

The parabolic model calculation that we quoted in Fig. 1 for the present density of 7.0×10^{12} cm^{-2} gives N_s^i values that differ systematically by a small but significant amount from the experimental values in Fig. 2. The difference is explained satisfactorily by nonparabolicity.

IV. The $B_{\parallel}$-Magnetoconductivity

The doping-layer for typical N_D has a number of occupied, bound levels. In the previously discussed Fig. 2, four of these have been identified. The carriers move in the symmetric, V-shaped self-consistent potential. For a field $B_{\parallel}$ the levels shift in energy and become magnetoelectric subbands /3,4/. The bottom of the subband, the lowest energy state remains at $k_{\parallel} = 0$. For the latter the energy shift is a strictly diamagnetic term

$$\Delta E(B_{\parallel}) = \frac{e^2}{2m^*} \langle z_i^2 \rangle B_{\parallel}^2 .$$

Here $\langle z_i^2 \rangle$ is the average square of the extent of the wavefunction in the $\hat{z}$-direction for the ith subband.

The diamagnetic shift causes the bands to rise above E_F and empty the carriers, redistributing them among the remaining states. The effect is quite analogous to the better known Shubnikov-de Haas oscillations. In this "diamagnetic" version, the oscillation sequence is not simply periodic in 1/B .

The cutoff of occupancy of a given subband is registered as a peak in the conductivity derivative $\partial\sigma/\partial B$. This has been argued qualitatively and confirmed by comparison of theory and experiment in Ref. /4/. Fig. 3 is an example of the effect as it is observed for a δ-sheet doped with a Si-density of 7.4×10^{12} cm^{-2}. From the analysis in Ref. 4 the peaks labeled n = 2, 3, 4 represent the cutoff condition of the corresponding subband. 4 is also the number of occupied bands evident in the Shubnikov-de Haas data of Fig. 2.

Generally for the higher order subbands such as 3 and 4 the percentage occupancy is so low that the redistribution of the charge in the self-consistent potential with rising $B_{\parallel}$ does not change that potential. In this

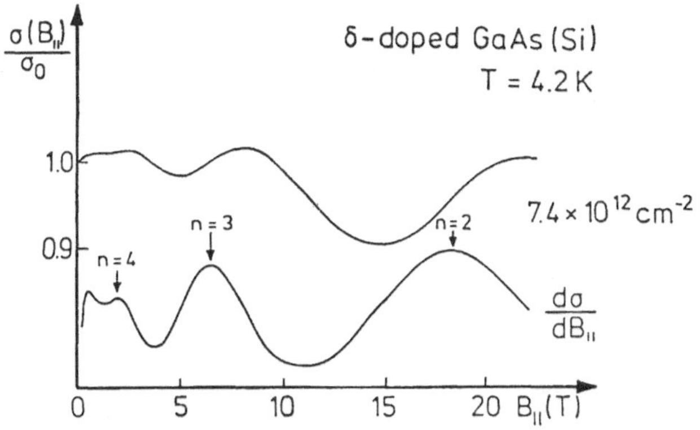

Fig. 3: The diamagnetic Shubnikov-de Haas effect for a Si-doped layer with 7.4×10^{12} cm^{-2} donors. The maxima in $d\sigma/dB_\parallel$ are labeled and correspond to the cutoff condition at which the nth subband level coincides with E_F.

sense $\langle z_i^2 \rangle$ is independent of B and the diamagnetic energy term as cited above is a good approximation. Cutoff implies $\Delta E(B_\parallel) = E_F - E_i$. This in turn allows us to proceed in a simple way to evaluate from a measured peak such as n = 3 at 6.6 T, the corresponding $\langle z_3^2 \rangle$. The experimental value for $E_F - E_3$ from the $B_\perp$-data comes to 12 meV. It follows that $\langle z_3^2 \rangle^{1/2}$ = 145 Å. This experimental value compares nicely with the calculated number.

There is yet another interesting aspect of the $B_\parallel$-data. Fig. 3 shows clearly an n = 4 peak. Thus, although the $B_\perp$ oscillations did not allow us to measure the occupation of such a subband, it does exist and is partially filled. The calculation in Fig. 1 predicts a density $N_s^4 \sim 1 \times 10^{11}$ cm^{-2}, but because of the nonparabolicity effect this number is relatively uncertain.

V. The Magnetic-Field-Induced Transition to an Insulating State

At the low end of the N_D scale, say in the range of 10^{11} dopant atoms per cm^2, it is not at all clear that the description of the layer as a 2-d sheet of metallic-conducting electrons is valid. The statistical nature of the randomly-placed Coulomb centers will become apparent when their average spacing in the plane is greater than the electron Bohr radius in GaAs ($a_B^* = 100$ Å). The question is "How much greater?". At really low density one has isolated, hydrogenic donors that represent an insulator.

There are a number of experimental approaches to this question of the transition from the metallic 2-d sheet to the insulating state. In principle one can examine a closely spaced sequence of samples with increasing density N_D, for each of which the electronic density N_s equals N_D. A plot of conductivity vs. N_D would show the expected transition at some density. $N_D \lesssim 1/(a_B^*)^2 \sim 10^{12}$ cm^{-2}. For a given insulating layer raising temperature T would give the transition. Alternatively, one could vary the charge N_s for a given N_D using a Schottky-barrier surface contact to search for the metal-insulator transition. Such work is in progress but is of no real concern to the present discussion of magnetotransport phenomena. We choose to show here how a perpendicularly applied magnetic field can tune the transition to the insulating state for a layer with a given N_D.

Samples with N_D significantly below $1/a_B^{*2}$ still conduct at T = 4.2 K. This is the case for the layer with $N_D = N_s = 2.9 \times 10^{11}$ cm^{-2} for which we show data in Fig. 4. The sample is in the form of a Hall bar with current and voltage contacts.

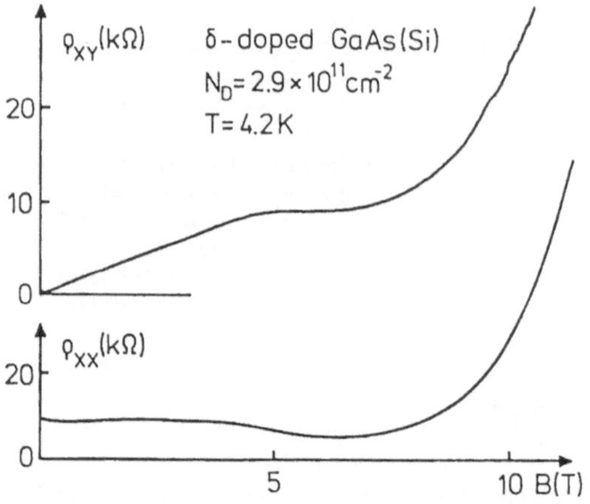

Fig. 4: Magnetic-field-induced metal-insulator transition in δ-doped GaAs

The Hall resistance ρ_{xy} in the figure rises linearly to reach a plateau in the 5-7 T range. The value of the slope, together with the measured ρ_{xx} value, gives the carrier density as $N_s = 2.9 \times 10^{11}$ cm^{-2} and mobility of ∿ 3000 cm^2/Vsec. This measures the doping value more precisely than the design doping estimate. For the measured density the average spacing of the Si donors is ∿ 185 Å.

With increasing $B_\perp$ the resistance ρ_{xx} falls to a broad minimum at 6.5 T and then rises steeply. We effectively lose contact to the voltage probes which themselves become insulating. We believe to have reached the expected magnetic-field-induced transition to an insulating state. The effect occurs as the Bohr radius decreases with applied field. In a 10 T field the lowest Landau level of a free electron has a width of wavefunction which is only 80 Å. The cyclotron energy $\hbar\omega_c$ is 2-3 times the impurity binding energy of 6 meV. It follows that the confinement of the electrons is dominated by the magnetic field. The transition to the insulating state is caused by the increasing shrinkage of the electronic wave function with rising $B_\perp$. The plateau in ρ_{xy} and the coincident minimum in ρ_{xx} represent the magnetic quantum limit. At a field of 6.5 T all of the ∿ 3 x 10^{11} cm^{-2} carriers condense into the lowest Landau level.

VI. The "Half δ"-Layer

Of course half a Dirac δ-function makes no mathematical sense, but the physical system of a sheet of N_D donor atoms at $z = 0$ can be engineered to have the related electronic screening charge all on one side. The potential becomes a triangular well instead of the symmetric V-shape. The case is analogous to carriers confined at the Si-SiO_2 interface by oxide charge at the boundary layer. This is what we refer to as a δ/2-layer.

The situation that we show in Fig. 5 has a design density of N_D = 7 x 10^{12} Si cm^{-2} placed in the GaAs terminating layer just prior to growing the $Al_xGa_{1-x}As$ ($x = 0.30$) heterostructure barrier. The oscillation spectrum allows us to identify two subbands with occupations $N_s^0 = 3.8 \times 10^{12}$ cm^{-2} and $N_s^1 = 0.9 \times 10^{12}$ cm^{-2}.

The numbers are significantly less than the design doping would have led us to expect. Possibly some of the Si atoms have become deep levels with tightly bound electrons.

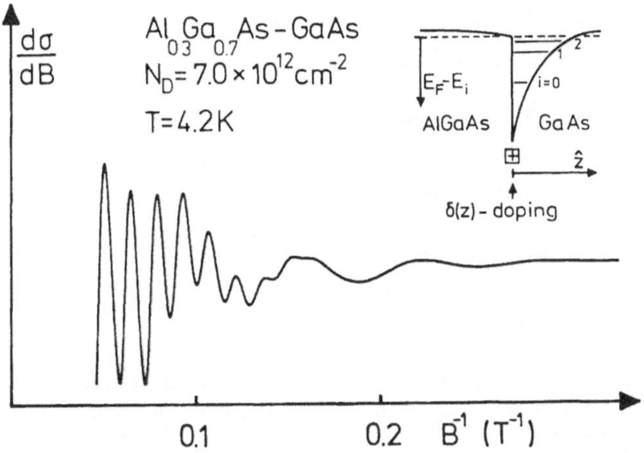

Fig. 5: Shubnikov-de Haas oscillations for a "δ/2"-layer with a design density of $7.0 \times 10^{12} cm^{-2}$ donors

If we now take the measured densities and the 70-30 rule of dividing up the band discontinuity, a model calculation gives a potential well which rises to within about 40 meV of the barrier height. The system acts as if the Fermi level were pinned at 40 meV below the $Al_xGa_{1-x}As$ conduction band edge.

There are a number of features of the δ/2-system still to be explored in future work.

VII. Acknowledgement

Work of the δ-doping layer has been supported by the Deutsche Forschungsgemeinschaft under the Schwerpunktsprogramm "Physikalisch-technische Grundlagen für die III-V-Halbleiter-Elektronik".

References:

/1/ A. Zrenner, H. Reisinger, F. Koch, and K. Ploog, Proc. of the ICPS San Francicsco (1984), eds. J.P. Chadi and W.A. Harrison, Springer Verlag, New York, p. 325
/2/ E.F. Schubert and K. Ploog, Jap.J. of Appl. Phys. 24, L 608 (1985)
/3/ H. Reisinger and F. Koch, 6th Intern. Conf. on Electronic Properties of Two-Dimensional Systems, Surface Sci., to be published
/4/ A. Zrenner, H. Reisinger, and F. Koch, Phys. Rev. B 33, 5607 (1986)
/5/ M. Zachau, F. Koch, K. Ploog, P. Roentgen, and H. Beneking, to be published
/6/ A. Zrenner and F. Koch, ICPS 18, to be published
/7/ U. Rössler, Solid State Commun. 49, 943 (1984)

Magnetotransport and Photoconductivity of IV–VI Compound Superlattices

A. Martinez[1,*], *T.K. Chu*[1], *and R.S. Allgaier*[1,2]

[1]Naval Surface Weapons Center, Silver Spring, MD 20903, USA
[2]Theodore Associates, Inc., 10510 Streamview Court, Potomac, MD 20854, USA
*also at Physics Department, The American University, Washington, DC 20016, USA

I. INTRODUCTION

The magnetotransport and optical properties of IV-VI compound semiconductor superlattices have been the subject of numerous theoretical and experimental investigations recently.[1-7] Several systems have been studied, among them PbSnTe/PbTe, PbSeTe/PbSnTe, and PbEuTe/PbTe. Superlattices of these materials have been grown using both, hot-wall[8,9] and molecular beam epitaxies.[10] This increased activity in the study of IV-VI semiconductor superlattices may be a sign of renewed interest in these family of semiconductors, particularly since these materials gain more technological importance as recent reports[11] point to interdiffusion problems in the more technologically popular HgTe/CdTe system.

In this paper, we report on the magnetotransport properties of both n- and p-type PbS/PbSe superlattices in magnetic fields up to 8 Tesla at several temperatures, and on their photoconductive behavior. This superlattice system displays a number of unusual properties, among them a non-linear dependence of the Hall resistance on magnetic field, negative magnetoresistance, and persistent photoconductivity. We propose that some of these phenomena are caused by the formation of at least two electrical sub-bands arising from the multivalley nature of the band structure in these materials.

The lead salts are narrow gap semiconductors with direct gaps occurring at the L points of the Brillouin zone. The surfaces of constant energy are four ellipsoids of revolution oriented with their major axes parallel to the <111>-crystallographic directions. The mass anisotropy ratio ($K=m_l/m_t$), and thereby the prolateness of these ellipsoids, varies, with K equal to 1.3, 1.8, and 10 for PbS, PbSe, and PbTe respectively.

II. EXPERIMENTAL PROCEDURE

The PbS/PbSe superlattices used in this study were grown on a modified hot-wall evaporator featuring two furnaces similar to the one used by CLEMENS et al.[8] to grow PbSnTe/PbTe superlattices. The samples were deposited on freshly cleaved [111]-oriented BaF_2 substrates, with and without depositing buffer layers of one of the two components before the superlattice was grown. A detailed description of the growth process and materials characterization of these structures has been reported and is being published elsewhere.[12] Hall bars were obtained from the grown superlattices by means of a photolithographic technique followed by chemical etching. Electrical contacts were made onto the samples after evaporation of gold pads using silver paste. The samples were studied in a variable temperature cryostat with magnetic fields of up to 8 T at several temperatures from 4.2 K to 250 K.

III. RESULTS and DISCUSSION

The Hall resistivity dependence with magnetic field of an n-type sample is shown in Fig. 1, at different temperatures. As seen there, a non-linear magnetic field dependence was observed for low temperatures, while at the higher temperatures this deviation from linearity disappeared. These data were fit to a two-mobility model [13] with excellent results. For the data shown in Fig. 1 at 4.2 K the parameters obtained from this fit were μ_1 = 1.0 m^2/Vs, μ_2 = 0.32 m^2/Vs, n_1 = 1.0 x 10^{23} m^{-3}, and n_2 = 2.2 x 10^{23} m^{-3}. The total carrier concentration computed from these results coincides, within the experimental error, with the carrier concentration derived from the slope of the Hall resistivity at 250 K which was 3.3 x 10^{23} m^{-3}. This behavior is similar to that observed by HEREMANS et al. [5] in a PbTe/PbEuTe quantum well. They attributed this non-linear dependence of the Hall resistance to the existence of two electric sub-bands with different mobilities in the quantum well, corresponding to a carrier distribution between the ellipsoid with its major axis oriented along the <111>-direction perpendicular to the film plane (transverse valley) and the three remaining [$\bar{1}$11] [1$\bar{1}$1], and [11$\bar{1}$] ellipsoids (oblique valleys) or the secondary heavy hole valence band of PbTe.

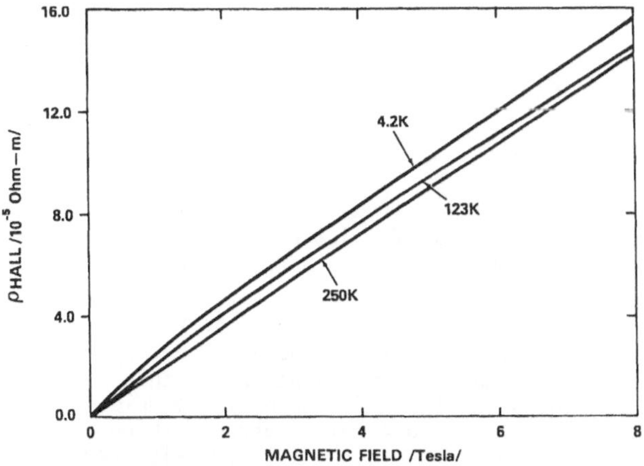

Fig. 1 Magnetic field dependence of the Hall resistivity for an n-type superlattice, 10 periods, each layer 150 Å thick grown on a PbSe buffer layer.

Recent photovoltaic spectral response studies indicate that PbS/PbSe superlattices form type I structures, so that both holes and electrons are confined in the PbSe layer.[14] Consequently, carriers confined in the PbSe layer will be distributed between the transverse and the oblique valleys which will generate sub-bands with different properties for two reasons. First, the proper projection of the 3-d Fermi surface onto the film plane[15] yields 2-d effective masses for motion in the film plane of $0.04m_0$ and $0.052m_0$ for the transverse and the oblique valleys respectively. Second, according to the same formalism the effective masses in the z-direction, m_z, are also different yielding values of $0.07m_0$ and $0.04m_0$ for the transverse and the oblique valleys respectively. A heavier m_z leads to a more confined wavefunction and hence to weaker interface scattering. Both the mass and scattering differences suggest that the carrier mobility in the transverse valley will be larger than that in oblique valleys. The fits of the present Hall data to a two-mobility model produced mobilities which at the most differed by a factor of 3, compared to a difference of two orders of magnitude reported for PbEuTe/PbTe. This may be explained by observing that the mass anisotropy ratio for the two semiconductors in which the carriers are confined are significantly different, viz. K_{PbSe} = 1.8 and K_{PbTe} = 10. In addition in the case of the PbTe/PbEuTe quantum well the carriers are not only scattered due to the presence of the interface but, in addition, due to the europium in the confining barriers which is a strong scatterer.

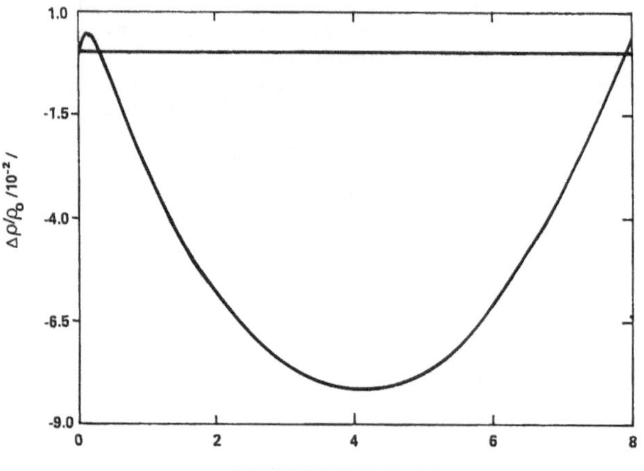

Fig. 2 Magnetic field dependence of $\Delta\rho/\rho_o$ at T = 4.2 K for an n-type superlattice, 10 periods, each layer 150 Å thick, no buffer layer

We also observed negative magnetoresistance in n-type PbS/PbSe superlattices at 4.2 K, but not in p-type samples. The magnetoresistance magnetic field dependence is presented in Fig. 2 for an n-type sample consisting of 10 periods, each layer being 150 Å thick, grown with no buffer layer. As shown there, a parabolic dip in the magnetoresistance occurred with a negative minimum at H = 4 Tesla. We observed similar behavior in other n-type samples grown with and without buffer layers. However, for the samples grown on PbSe buffers, which displayed higher mobilities, the negative dip occurred at fields below 2 Tesla; at higher fields, a positive effect approached a linear dependence on the magnetic field. The corresponding magnetic field dependence of the Hall resistivity also displayed departure from linearity. We do not fully understand the cause of this rather unusual effect in our samples. Similar effects have been observed in III-V semiconductors and usually have been explained in terms of scattering by localized spins of shallow donor impurities.[16] This explanation seems inappropriate for the case of IV-VI semiconductors since shallow impurities have not been observed in this type of semiconductors. We speculate that a magnetic field dependent carrier redistribution among the four ellipsoids may contribute to the negative magnetoresistance. Carrier transfer from the oblique valleys to the transverse one would cause such an effect since carriers would go from a lower to a higher-mobility ellipsoid. This kind of carrier redistribution is known to take place in uniaxially-strained lead salts.[17,18] Furthermore, the failure to observe negative magnetoresistance in p-type samples is consistent with this type of carrier transfer in which carriers would be transferred from the higher to the lower-mobility ellipsoid. This is suggestive of the occurrence of magnetostrictive effect which, as reviewed by KEYES[19], occurs in multivalley degenerate semiconductors. Magnetostrictive effects have been observed by BELSON et al.[20] and THOMSON et al.[21] in p-type PbTe. However, the deformation potentials for bulk PbS and PbSe are substantially smaller than those for PbTe, and even in the last case the typical amplitude of the strain is about 10^{-7}. The possibility remains that there may be an enhancement of the magnetostrictive effects in layered structures but more experimental and theoretical work is needed to validate this assumption.

Persistent photoconductivity was also observed in both n- and p-type PbS/PbSe superlattices as shown in Fig. 3. This behavior was observed in all the samples studied for temperatures below 100 K. The samples were cooled while enclosed in a copper can and then exposed to light from a red light-emitting diode. Figure 3 shows the resistivity as a function of time, including two intervals in which the sample was exposed to light as identified by the dashed lines. The effect saturated with a change in the resistivity in the proximity of 3%. The persistent photoconductivity displayed by these samples may be due to a combination of band bending effects and the presence of diffusion potential barriers. The photogenerated electron-hole pairs are spatially separated by the potential barrier caused by band bending due to electric fields induced by the charge transfer from the PbS layers (the potential barrier in these superlattices) to the PbSe layers. Interdiffusion of PbS into the PbSe layer may cause a spike in the band modulation at the interface which will produce a similar spatial separation effect on the electron-hole pairs that will hinder recombination, thus contributing to the persistent photoconductivity.

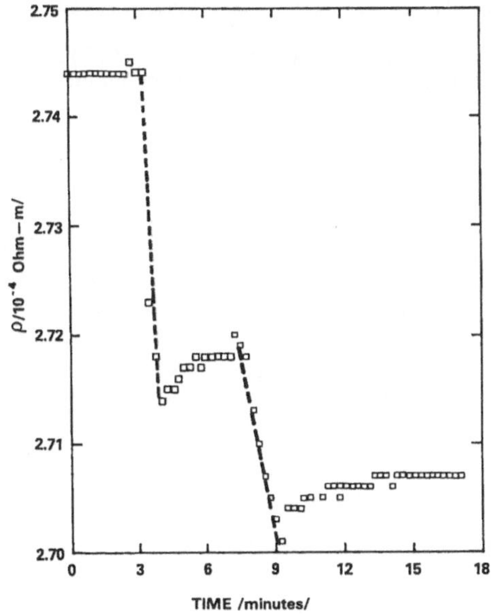

Fig. 3 Photoconductivity relaxation for an n-type superlattice

IV. ACKNOWLEDGMENTS

We thank Dr. J. R. Cullen of NSWC and Dr. H. D. Drew of the University of Maryland for helpful discussions on this work. This work was supported by the Naval Surface Weapons Center's Independent Research Fund and by the Office of Naval Research.

References:

1. S. Takaoka, T. Okumura, K. Murase, A. Ishida, and H. Fujiyasu: Solid State Commun. 58, 637 (1986).
2. H. Pascher, P. Pichler, G. Bauer, H. Clemens, E. J. Fantner, and M. Kriechbaum: Surface Sci. 170, 657 (1986).
3. K. Murase, S. Ishida, S. Takaoka, and T. Okumura: Surface Sci. 170, 486 (1986).
4. A. Ihsida, H. Fujiyasu, H. Ebe, and K. Shinohara: J. Appl. Phys. 59, 3023 (1986).
5. J. Heremans, D. L. Partin, P. D. Dresselhaus, M. Shayegan, and H. D. Drew: Appl. Phys. Lett. 48, 928 (1986).
6. G. Bauer: Surface Sci. 168, 462 (1986), and references therein.
7. P. P. Ruden, T. L. Reinecke, and F. Crowne: Superlatt. and Microstruc. 1, 197 (1985).
8. H. Clemens, E. J. Fantner, and G. Bauer: Rev. Sci. Instrum. 54, 685 (1983).
9. H. Kinoshita and H. Fujiyasu: J. App. Phys. 51, 5845 (1980).
10. D. L. Partin: J. Vac. Sci. Technol. 21, 1 (1982).
11. D. K. Arch, J. L. Staudenmann, and J. P. Faurie: Appl. Phys. Lett. 48, 1588 (1986).
12. A. Martinez, F. Santiago, C. R. Anderson, and T. K. Chu: Bull. Am. Phys. Soc. 31, 532 (1986).
13. R. A. Smith: In Semiconductors, 1st ed. (Cambridge University Press, Cambridge 1959).
14. T. K. Chu, D. Agassi, and A. Martinez: to be published.
15. H. Schaber and R. E. Doezema: Phys. Rev. B20, 5257 (1979).
16. Y. Toyozawa, J. Phys. Soc. Japan 17, 986 (1962).
17. A. Martinez and B. Houston: Appl. Phys. Lett. 43, 77 (1983).
18. J. R. Burke and G. P. Carver: Phys. Rev. B17, 2719 (1978).
19. R. W. Keyes: In Solid State Physics, ed. by F. Seitz, D. Turnbull, and H. Ehrenreich, Vol. 20, 37 (1967).
20. H. S. Belson, J. R. Burke, and E. Callen: Phys. Rev. B3, 4243 (1971).
21. T. E. Thomson, P. R. Aron, B. S. Chandrasekhar, and D. N. Langenberg: Phys. Rev. B4, 518 (1971).

Oscillatory Current-Voltage Characteristics and Magnetocapacitance Effects in Single Barrier $n^+GaAs/(AlGa)As/n\text{-}GaAs/n^+GaAs$ Heterostructures

L. Eaves, D.K. Maude, F.W. Sheard, and G.A. Toombs

Department of Physics, University of Nottingham, Nottingham, NG7 2RD, UK

The LO phonon-related oscillatory structure in the reverse-bias J(V) curves of single-barrier GaAs/(AlGa)As heterostructures is described and explained by the variation of the impedance of the undepleted section of the n-GaAs layer. The anomalous behaviour of the magnetocapacitance is also accounted for, without invoking magnetic freeze-out of carriers throughout the complete length of n^- layer. An analysis is given of new magnetocapacitance data including the oscillatory structure in C(V).

1. Introduction

In 1984, Hickmott et al [1,2] reported oscillatory structure in the low-temperature reverse-bias current-voltage characteristics J(V) of single barrier $n^+GaAs/(AlGa)As/n^-GaAs/n^+GaAs$ heterostructures. The period ΔV corresponded to the energy ($\hbar\omega_L$ = 36 meV) of the longitudinal optic phonon mode in GaAs. The band structure in reverse bias is shown in Figure 1. The current arises from quantum tunnelling of electrons through the (AlGa)As barrier from the n^+GaAs region into the n^- layer. Hickmott et al observed the oscillations only at high magnetic fields (B > 4T), though subsequently we demonstrated [3-5] that they can be observed even at B = 0. Several models [6-17] have emerged to explain the oscillatory structure and a lively controversy has arisen in the literature. This article reviews some of this work and explains how our model [5-8] accounts for the electrical behaviour of these devices and, in particular, the oscillatory structure in J(V). In addition, we present new magnetocapacitance data and compare them with measurements made by Hickmott et al [1,2,18]. A simple equivalent circuit derived from a small signal analysis of the device explains the magneto-capacitance data under reverse bias without invoking magnetic freeze-out throughout the entire length of the n^-GaAs layer as has previously been suggested [1,2,18]. We show that the frequency-, voltage- and magnetic field-dependence of the capacitance under reverse bias can be understood in terms of the large magnetoresistance variation of the undepleted (low electric field) region of the n^- GaAs layer at liquid helium temperatures. The model also explains the weak oscillatory structure in the reverse bias capacitance-voltage curves C(V).

2. Oscillatory Structure in J(V)

Our layers were grown by MBE as follows: 200 μm thick n^+GaAs substrate doped to 2 x 10^{18} cm^{-3}; 1 μm n^+GaAs buffer (collector) layer, 2 x 10^{18} cm^{-3}; 1 μm n^-GaAs layer, low 10^{15} cm^{-3}; 168 ± 10 Å of undoped (AlGa)As; 1 μm n^+GaAs top (emitter) layer, 2 x 10^{18} cm^{-3}. The Al concentration varies from 37% (substrate side) to 32 ± 2%. They were processed into mesas of various diameters between 200 and 800 μm. The oscillatory structure for two of our devices is illustrated in Figure 2. Mesa A passes considerably less current than B for reverse bias less than 100 mV [3]. Hence the oscillatory structure is observed in A only at higher reverse biases. The difference between the J(V) characteristics of the

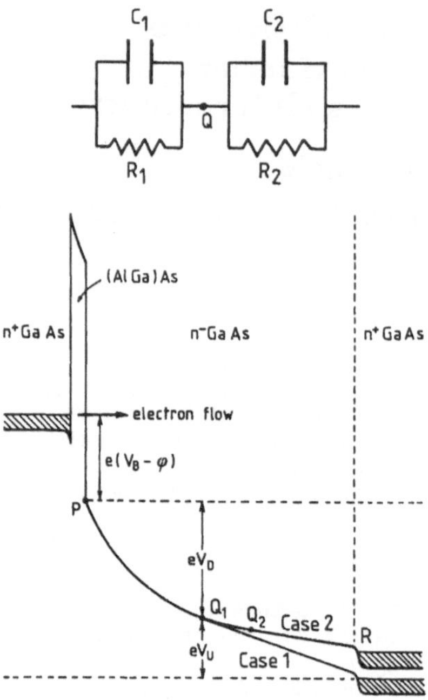

Fig. 1 Schematic diagram of device in reverse bias with equivalent circuit

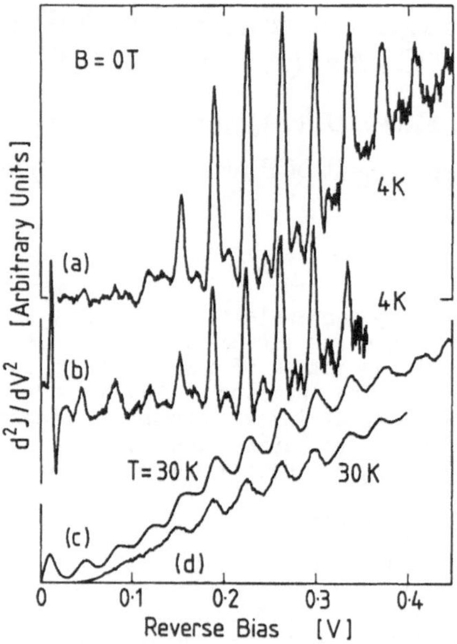

Fig. 2 d^2J/dV^2 vs -V: Curves a and d, mesa A; curves b and c, mesa B.

two mesas probably arises from differences in the charged defect density in the (AlGa)As barriers. C(V) measurements on our mesas reveal negative space charge in the (AlGa)As barrier. This results in increased depletion of the n^- GaAs layer even at zero bias. The effect of the negative space charge in (AlGa)As barriers has been discussed in detail by Hickmott et al [19]. The existence of this depletion region means that, at low reverse biases, electrons in the n^+ GaAs contact layer tunnel not only through the (AlGa)As barrier but also through a significant depletion layer thickness. The tunnel current at low reverse bias voltages is greatly attenuated by this additional barrier thickness. At higher reverse bias voltages (as shown in Figure 1), electrons tunnel only through the (AlGa)As layer thickness and at these voltages the J(V) of mesas A and B are identical. Derivative plots (dJ/dV or d^2J/dV^2) are necessary to fully reveal the oscillations. They are barely visible on the undifferentiated J(V) curves (typically $\Delta J/J \simeq 3 \times 10^{-3}$ at V = -0.35 V). Their amplitude remains fairly constant up to 10 K but falls rapidly at higher temperatures [4]. By increasing the modulation voltage the structure can be observed up to 50 K.

Our model for the origin of the oscillatory structure [5-8] can be understood by reference to Figure 1 in which the voltage applied to the device, $V = V_B + V_D + V_U$ is dropped across three distinct regions: the (AlGa)As barrier (V_B), the high-field depleted region of the n^- layer adjacent to the barrier (V_D) and the undepleted n^- region adjacent to the n^+ substrate (V_U). A correct description of the potential across the device is important since it determines the J(V) and C(V) curves and the oscillatory LO phonon-related structure. In our model, the potential distribution is determined self-consistently by Poisson's equation and current continuity. We show in section 3 that the voltage distribution in Figure 1 is maintained in a magnetic field, ie magnetic freeze-out does not occur throughout the entire length of the n^- GaAs layer as was originally supposed by other workers [1,2]. We stress this point as some models for the origin of the oscillatory structure [9,12,14] erroneously assume magnetic freeze-out throughout the n^- layer.

The kinetic energy gained by electrons in the barrier and depletion layer is lost principally by emission of an integral number of LO phonons, since acoustic-phonon emission is negligible. The energy distribution of the electrons entering the undepleted region will have peaks at kinetic energies $\varepsilon(n) = e(V_B + V_D) - n\hbar\omega_L$, where $n = 0,1,2, \ldots n_{max}$. Assuming some inelastic interaction between the incident electrons and the cold electrons in the neutral undepleted layer, the impedance R_U and corresponding potential drop V_U across the neutral region will be periodically modulated as the applied voltage changes. We emphasise that the current through the device is controlled by V_B and the oscillatory structure is due simply to the modulation of V_U. At low temperatures (< 10 K) most of the electrons in the undepleted region are bound in shallow hydrogenic donor states (binding energy $E_D = 5.5$ meV). We have previously suggested impact ionisation of neutral donors by incident hot electrons as a mechanism for modulating the resistance R_U [5,6]. We can reliably estimate the change in conduction electron density due to this mechanism from studies of the inelastic scattering of electrons by neutral hydrogen atoms [20-22]. Above the appropriate threshold energy E_t, the scattering cross-section for $1s \rightarrow 2p$ excitation ($E_t = 3E_D/4$) and ionisation ($E_t = E_D$) are both $\simeq \pi a_0^2$, where a_0 is the Bohr radius. Excitation to 2s, 3p etc states have smaller cross-sections. However, once a shallow donor is excited to 2s, 2p etc states it has a high probability of ionising either by field emission or thermal excitation. This effect is well known and used to advantage in photoconductivity experiments. Hence both excitation and ionisation processes contribute to the change in conduction electron density. Taking a total cross-section $\sigma_t \sim 3\pi a_0^2$ and equating the rate of impact ionisation $N_D\sigma_t F$ where F is the incident electron flux, to the recombination rate n/τ_r, where n is the conduction electron density and τ_r the recombination time, gives $n = N_D\sigma_t F\tau_r$. At a reverse bias of 0.3 V the current density $J \sim 10^3$ A m^{-2} and $F = J/e \sim 0.6 \times 10^{22}$ m^{-2}. From time-resolved photoconductivity experiments, it is known that the rate at which an excited donor relaxes back to the ground state is rather slow due to the phonon "bottleneck" between n = 2 and n = 1 states, and typically $\tau_r \sim 10^{-8}$ s for $N_D = 2 \times 10^{15}$ cm^{-3} [23]. This gives $n \sim 10^{14}$ cm^{-3} so that the hot-electron flux can maintain an appreciable ionisation of the shallow donors in the undepleted region. We assume that a significant fraction of the hot-electron flux is associated with the lowest peak in the energy distribution ($n = n_{max}$). When, with increasing applied bias, this peak passes the threshold energy for impact ionisation, there will be a significant change in conduction electron density and hence resistance R_U of the neutral region.

An additional mechanism for ionisation of the shallow donors is through interaction with optical phonons as suggested by Leburton [12]. An excess LO phonon density N_p arises in the undepleted layer from those hot electrons which emerge from the depletion region with energies $\varepsilon(n) > \hbar\omega_L$. The number of hot electrons per unit volume is $n_h = F/v$, where v is a mean velocity $\sim 4 \times 10^5$ ms^{-1} for electrons of energy $\hbar\omega_L$. Equating the phonon emission rate n_h/τ_{em} to the loss rate $N_p(\tau_i^{-1} + \tau_d^{-1})$ gives $N_p = n_h\tau_{em}^{-1}/(\tau_i^{-1} + \tau_d^{-1})$. Here τ_{em}^{-1} is the emission rate of LO phonons for one electron, τ_i^{-1} is the rate of ionisation of donors and τ_d is the phonon decay time. An estimate based on the known LO-phonon-electron interaction gives $\tau_i \sim 10^{-11}$ s for $N_D = 2 \times 10^{15}$ cm^{-3}, and taking $\tau_d \sim 10^{-11}$ s and $\tau_{em} \sim 0.5 \times 10^{-12}$ s [24] gives $N_p \sim 10^{11}$ cm^{-3}. Hence in equilibrium the conduction electron density is given by $n = N_p\tau_r/\tau_i \sim 10^{14}$ cm^{-3}.

There is considerable uncertainty in this estimate but nevertheless this indicates that the phonon ionisation mechanism may be comparable to the impact ionisation mechanism. Modulation of the conduction electron density follows when with increasing bias, the peak in the electron energy distribution passes the threshold for LO phonon emission which alters N_p and hence n.

It is straightforward to test whether this model gives a reasonable value for the observed amplitude $\Delta I/I$ of the oscillations. At T = 4.2 K, B = 0 and V = -0.35 V, we observe that for a 200 μm diameter mesa, $\partial I/\partial V = 2 \times 10^{-4}\ \Omega^{-1}$ and $\Delta I/I = 3 \times 10^{-3}$. The required resistance change ΔR_U of the undepleted layer is given by $\Delta R_U = (\Delta I/I)/(\partial I/\partial V) = 15\ \Omega$ (for derivation, see ref 6). In comparison, the equilibrium value of R_U for 0.5 μm of undepleted n^- GaAs at 4.2 K is $R_U \sim 300\ \Omega$. Therefore the relative change $\Delta R_U/R_U$ required by our model is not unreasonable considering the ionised donor densities estimated above.

We can test whether impact- or LO phonon-ionisation is the dominant mechanism giving rise to the oscillatory structure in J(V) by examining the amplitude of the LO phonon-related peaks in d^2J/dV^2 as a function of magnetic field. If LO phonon ionisation were dominant, one would expect a series of hot electron magnetophonon resonances (MPR) in the oscillatory amplitude at fields given by $\hbar\omega_L = N\hbar\omega_c + E_D(B)$ where $\omega_c = eB/m^*$ and N = 1,2,3 ... [25]. On the other hand, if impact ionisation were dominant, a peak in the amplitude of the oscillations should occur when $\hbar\omega_c = E_D(B)$. At this magneto-impurity resonance (MIR) field [26] the ionisation energy corresponds to the joint density of states maximum for hot electrons falling from the p = 1 to p = 0 Landau level. Hence the free carrier density n should increase resonantly at $\gamma_B = \hbar\omega_c/2E_D(0) =$ 0.77, corresponding to B_r = 5.2T in GaAs. As shown in Figure 5, we observe no MPR in plots of the oscillatory peak amplitude versus B. Instead a maximum occurs around B = 5T, in excellent agreement with the MIR field B_r corresponding to impact ionisation. In addition, as shown in Figure 2 (inset) the widths δV of the peaks in the LO structure narrow remarkably at around 5T, due to the effect of the singularity in the Landau level density of states. The absence of MPR also argues against Ihm's model [16].

We also note that the absence of oscillatory structure at B = 0 in the devices of Hickmott et al can be explained by their much lower current densities. The magnetic field (see Figure 3) sufficiently increases the resistance R_U to a value that it can be modulated over a wide enough range to give rise to the observable oscillatory structure in J(V). The effect of a magnetic field on the electrical properties of the devices is the subject of Section 3.

To summarise this section, we have shown that the modulation of R_U by ionisation of donors can explain the amplitude of the oscillatory structure in J(V) at 4.2 K and B = 0. The model also accounts for the temperature and magnetic field dependence of the amplitude and the extinction of the oscillations at large reverse biases ($\sim$1 V) when the depletion region extends across the entire n^- layer.

3. Magnetocapacitance and Magnetic Freeze-out

Hickmott et al made extensive use of C(V) measurements to analyse the distribution of space charge in their devices. At B = 0 they observed that the capacitance fell monotonically with increasing reverse bias voltage . This occurs as a result of the increasing width of the depletion region in the n^- layer. At high enough reverse bias voltages, the depletion layer fills the complete length t of the n^- GaAs layer giving a constant capacitance $C_{min} = \varepsilon_r\varepsilon_0/(b + t)$, per unit area, where b is the barrier thickness. However, at high fields (B > 4 T) the capacitance even at low reverse bias decreased

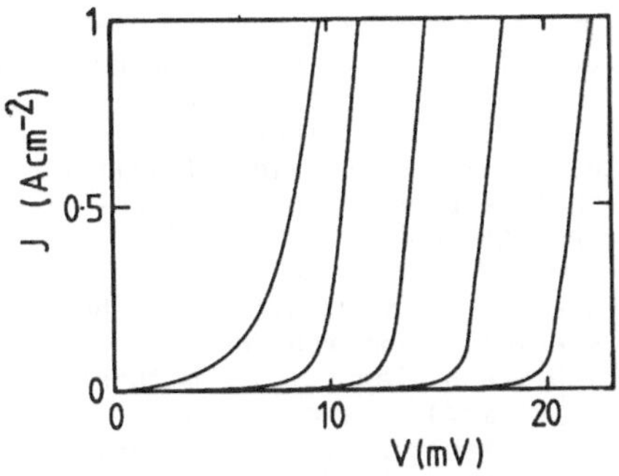

Fig. 3 J(V) Curves of 1 μm thick $n^+n^-n^+$ GaAs mesas (N_D-N_A = 2 x 10^{15} cm^{-3} in n^- layer) at 2K for various magnetic fields, left to right: 0,2.8,5.7,8.6,11.4T, B ⊥ J. These test structures lack the (AlGa)As barrier and are used for studying as a function of bias the magnetoresistance of the lightly doped n^- GaAs which forms the undepleted region of the tunnelling structures.

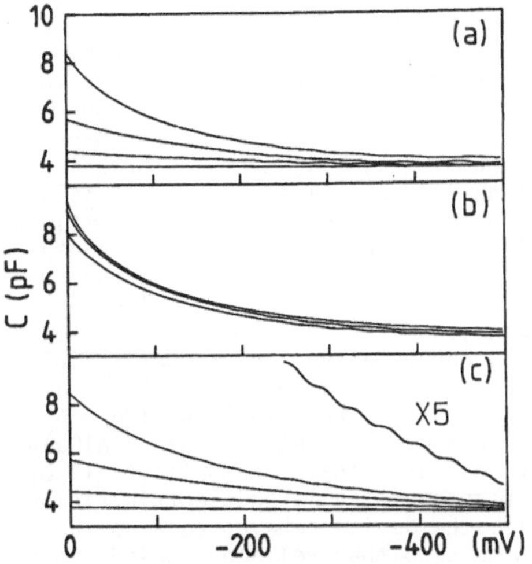

Fig. 4 (a) C(V) for 200 μm diam. mesa, 4.2K, top to bottom: B=0, 3, 5, 11.4T (b) As (a), but 17K, top to bottom: B=0,5,11.4T; (c) Simulation of data for T = 4.2K of Fig. 4a using equivalent circuit of Fig. 1. The top curve is expanded 5x to show structure in C(V). Top to bottom the simulations have $R_0(B)$ = 0.51, 1.8, 3.2 and 5.9KΩ.

towards this value C_{min} and at sufficiently high values of B, C(V) remained almost constant and close to C_{min} over a wide range of reverse bias. We have observed similar behaviour as shown in Figure 4(a) by the experimental values of C(V) measured at 1 MHz for a 200 μm diameter mesa in reverse bias with various values of magnetic field at 4.2 K. However, note that the magnetic field-induced fall in capacitance at 17 K (Figure 4(b)) is much smaller. For the remainder of this section we refer to the low and essentially voltage-independent values of C(V) measured at 4.2 K and high magnetic fields as the "anomalous" magnetocapacitance. The dependence of the reverse bais C(V) curves on B was interpreted by Hickmott et al. as arising from magnetic freeze-out of electrons onto shallow donors. This, it was claimed, neutralised all of the positive space charge in the depletion layer and led to a uniform electric field throughout the length of the n^- region. This interpretation has been followed by several workers who have subsequently attempted to explain the oscillatory structure [9,12,16]. We have previously pointed out that magnetic freeze-out under the high electric-field conditions of the experiment is inconsistent with the known electrical behaviour of n^- GaAs. Although magnetic freeze-out of electrons onto shallow donors is a well-established effect and indeed occurs in the low field part of the n^- layers of the single barrier heterostructures, it cannot occur throughout the length of the n^- layer in the presence of the high electric fields (> 1 kV/cm) corresponding to the reverse-bias voltages over which the anomalously low magnetocapacitance values are observed. At these electric fields, electrons are stripped from the donor states by field-emission or avalanche multiplication due to impact ionisation by hot carriers. Since the current through the device is low, the charge due to current carrying electrons in the depletion layer can be neglected and the space charge is given by (N_D - N_A)e.

Hickmott suggested several factors which may assist in sustaining magnetic freeze-out in the high electric fields [18]. One is the magnetic field-induced compression of the donor wavefunction and the associated increased binding energy. However, even if this process occurs to some extent for $\underline{B} \perp \underline{E}$, it does not explain the anomalous magnetocapacitance for $\underline{B} \parallel \underline{E}$ ($H_\perp$ in the notation of refs 1,2,18,19) since compression of the donor ground state wavefunction along $\underline{B}$ is relatively small at fields up to 10 T [27]. The other mechanisms proposed [18] to explain the anomalous magnetocapacitance behaviour (limitation of the current by the barrier and the long mean free path of the electrons in the n^- region) do not address the question of how the donors can remain neutral in the presence of electric fields large enough to cause field emission. This point is illustrated by the data in Figure 3 which show the J(V) curves at various magnetic fields for a specially prepared test structure: a simple $n^+n^-n^+$GaAs sandwich in which the n^- layer is 1 μm thick and doped to the same level ($N_D - N_A$ = 1-2 x 10^{15} cm^{-3}) as the n^- layer in both our and Hickmott et al's GaAs/(AlGa)As tunnelling structures. This experiment demonstrates that a magnetic field of ∿10 T cannot bind the donor in the presence of the large electric fields. It can be seen that even at fields of 11 T the device behaves as a voltage limiter due to field emission processes. (Note also that voltages considerably greater than $kT/e \simeq 0.3$ mV can be sustained since many inelastic processes occur down the 1 μm long conduction channel).

The apparently anomalous magnetocapacitance data can be understood by a re-appraisal of the way the capacitance is measured on this type of device. Both Hickmott and ourselves have measured C(V) using a commercial HP4274A or HP4275A multifrequency LCR meter. This type of meter analyses the complex impedance of any circuit or device as a single capacitance C* and a single resistor R* either in parallel or series. For the single barrier heterostructure, the parallel combination is usually more appropriate and is used in the measurements of Hickmott et al and ourselves. However, it must be borne in mind that the electronic behaviour of the device may be more complicated than a two-parameter parallel RC circuit. Hickmott noted that the resistance of the undepleted region must, in principle, be taken into account but he neglected its capacitance C_2. Using small-signal analysis, it can be shown [28] that the equivalent circuit appropriate to the single barrier heterostructure is the one shown in Figure 1. C_1 and R_1 represent the capacitance and dynamic conductance ($\partial I_r/\partial V$) of the barrier and depletion region (PQ). Here I_r is the current component which is in phase with the modulation voltage. C_1 is given by $\varepsilon_r\varepsilon_0/(b + s)$, where b is the barrier width and s is the depletion layer width. C_2 and R_2 are the corresponding values for the undepleted region (QR). It is straightforward to express R* and C* as measured by the LCR meter in the parallel circuit configuration as functions of R_1, C_1, R_2 and C_2 [28]. We can then compare our model with the observed values of R* and C* at any voltage, magnetic field or modulation frequency. Using this method, the apparently anomalous magnetocapacitance can be understood. Even at liquid helium temperatures the zero magnetic field value of R_2 is generally small compared to the impedance $1/\omega C_2$ at the measurement frequencies (0.1 - 2 MHz). Thus R_2 effectively short-circuits C_2. In addition, R_2 is small compared to R_1, the dynamic impedance of the barrier plus depletion layer so that, to a good approximation, $C^* \sim C_1$ and $R^* \sim R_1$ at zero magnetic field. This explains why the C(V) characteristics of the device at zero magnetic field can be modelled successfully in terms of the increase of the depletion layer width with increasing reverse bias voltage [18]. This approximation works even better at relatively high temperatures ($4 < T < 60$ K) when R_2 is small but when the temperature is still low enough for thermionic emission over the barrier to be negligible.

The crucial point apparently overlooked in refs [1,2,18] is that at $T \leq 4.2$ K the value of R_2 increases markedly with increasing B (see Figure 3). This results from the behaviour of lightly doped semiconductors at low temperatures in which the electrical conduction is dominated by hopping between impurities or by low mobility states close to the edge of the band. When a magnetic field is applied ($\underline{E} \perp \underline{B}$ or $\underline{E} \parallel \underline{B}$) the magnetoresistance can increase by several orders of magnitude [27]. When the resistance R_2 becomes sufficiently large, it is no longer appropriate to neglect C_2. In fact, at sufficiently high modulation frequencies ($\omega C_1 R_1 \sim 1$ and $\omega C_2 R_2 \sim 1$), $C^{*-1} \sim C_1^{-1} + C_2^{-1}$ the value corresponding to two capacitors in series. Thus C* falls significantly below C_1 even at low reverse bias. C* is then given approximately by its lowest value, $\varepsilon_r \varepsilon_0/(b + t)$. This description explains why the capacitance curves measured by Hickmott et al and by ourselves at liquid helium temperatures and high magnetic fields have low values which are essentially independent of the reverse bias voltage. It is incorrect to assume that because the capacitance at high magnetic fields has a low and constant value, all of the n^- GaAs layer is undergoing magnetic freeze-out. We stress that when the magnetic field is applied, a depletion region remains in existence. The measured capacitance decreases because the capacitance of the undepleted region must be taken into account at high magnetic fields. Hickmott's use of the van Gelder-Nicollian equation [29] to obtain Figure 5 of ref 18 assumes that the measured capacitance C* equals C_1 and ignores the impedance of the undepleted layer even in the presence of high magnetic fields. The three curves in Figure 4b taken at 17 K show clearly the effect of the lower value of R_2 where the temperature is high enough to maintain a significant free carrier concentration in the undepleted layer. In this case, the measured C* approximates more closely to C_1 since C_2 is shorted out by R_2. The equivalent circuit in Figure 1 also explains the frequency-dependent values of C* reported by Hickmott [18] and also observed by ourselves. For a simple two-parameter parallel R-C circuit, the measured value of C* should be independent of frequency. However, C* and R* are in fact frequency dependent [28]. This explains why at low modulation frequencies (100 kHz) we observe that the magnetic field has a much smaller effect on the measured capacitance C* than at 1 MHz since, at low frequencies, R_2 partially shorts out C_2 even when the magnetoresistance of the undepleted layer is large.

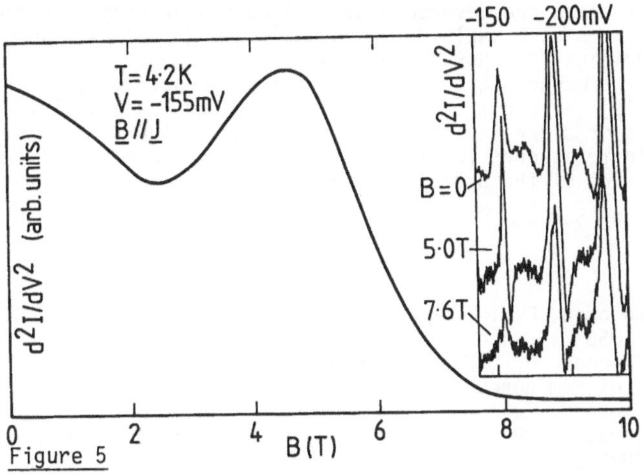

Figure 5

Variation in amplitude of the LO phonon peak at V = -155 mV in d^2I/dV^2 plot. Inset shows resonant narrowing of peak near the MIR field B = 5.2T compared to traces at 0 and 7.6T.

These qualitative ideas can be tested by numerically simulating the measured magnetocapacitance using the RC network of Figure 1. The results of such a simulation are shown in Figure 4(c). The value of C_1 is taken to be $\varepsilon_r\varepsilon_o/(b+s)$ where $s = [2\varepsilon_r\varepsilon_o|V-V_o|/(N_D - N_A)e]^{\frac{1}{2}}$ is the depletion layer width and V_o reflects the depletion at zero bias. Similarly $C_2 = \varepsilon_r\varepsilon_o/(t - s)$. The value of R_1 can be estimated from the measured dc reverse bias J(V) characteristics at temperatures ($\sim$ 20 K) for which the resistance of the undepleted region is low. The value of this impedance is considerably higher than $1/\omega C_1$ at the measurement frequency. Therefore, the exact choice of R_1 does not have a great effect on the simulated values of C*(V). For this reason and for simplicity, we take R_1 = 10 kΩ. The simulation of the magnetoresistance therefore involves essentially only one parameter R_2, the resistance of the undepleted region. Since the thickness of the undepleted region decreases with increasing reverse bias we represent R_2 by $R_2(V,B) = R_o(B)(1-s/t)$. The simulations shown in Figure 4(c) are obtained with values of $R_o(B)$ = 0.5 kΩ (B = 0 T), 1.8 kΩ (3 T), 3.2 kΩ (5 T) and 5.8 kΩ (11 T) chosen to fit the measured capacitance at V = 0. These values are consistent with the large magnetoresistance variation of lightly doped n^- GaAs test structure which we have measured at liquid helium temperatures [27]. The simulations are in fair agreement with the data for 4.2 K shown in Figure 4(a). The slightly greater slope of the observed C(V) at low bias probably arises from non-uniform doping near the barrier [29] and from the decrease of R_1 with increasing bias. We note that the ohmic behaviour of $R_o(B)$ is an approximation since at high values of V_U, $R_o(B)$ will voltage-limit as shown in Figure 3.

Our model also accounts for the oscillatory structure in the observed capacitance C*. This is due principally to the variation of R_2 rather than the much smaller variations of C_1 and C_2 arising from the modulation of the depletion layer width s. By writing $R_2(V) = R_o(1-s/t)(1 + a(V) \sin 2\pi V/\Delta V)$, we can simulate the weak oscillatory structure observed in the measured C* which arises from the variation of R_2. A value of $a(V) = 720\ V^2/R_o$ (with V in volts and R_o in Ω) provides a good simulation to the observed structure in both J(V) and C*(V) at 4.2 K, a result which lends strong support to our model. The weak oscillationsin C* induced by varying R_2 can be seen in the upper simulated curve in Figure 4(c), also shown in expanded form for clarity.

In summary, the model we proposed last year to describe the LO phonon-related oscillatory structure in the reverse bias J(V) curves of single barrier GaAs/(AlGa)As heterostructures also explains the anomalous magnetocapacitance and magnetic freeze-out originally reported in refs 1 and 18 and the oscillatory structure in the magnetocapacitance. Finally, we note that aside from other objections [7,15], a recent model [14] interpreting the oscillatory structure in J(V) in terms of a density of states variation in the n^+ substrate does not appear to be able to account for the oscillatory structure in C*(V) or the temperature dependence of the oscillatory amplitude.

This work is supported by SERC.

References

1. T.W. Hickmott, P.M. Solomon, F.F. Fang, F. Stern, R. Fischer and H. Morkoc Phys. Rev. Lett. 52, 2053 (1984).
2. T.W. Hickmott, P.M. Solomon, F.F. Fang, R. Fischer and H. Morkoc, Proc. Int. Conf. on Physics of Semiconductors, San Francisco (publ. Springer) pp.417-20 (1984).
3. L. Eaves, P.S.S. Guimaraes, B.R. Snell, D.C. Taylor, K.E. Singer, Phys. Rev. Lett. 53, 262 (1985).
4. P.S.S. Guimaraes, D.C. Taylor, B.R. Snell, L. Eaves, K.E. Singer, G. Hill, M.A. Pate, G.A. Toombs and F.W. Sheard, J. Phys. C: Solid State 18, L605 (1985).

5. D.C. Taylor, P.S.S. Guimaraes, B.R. Snell, L. Eaves, F.W. Sheard, G.A. Toombs and K.E. Singer, Physica 134B, 12 (1985).
6. L. Eaves, P.S.S. Guimaraes, F.W. Sheard, B.R. Snell, D.C. Taylor, G.A. Toombs and K.E. Singer, J. Phys. C: Solid State 18, L885 (1985).
7. L. Eaves, P.S.S. Guimaraes, B.R. Snell, F.W. Sheard, D.C. Taylor, G.A. Toombs J.C. Portal, L. Dmowski, K.E. Singer, G. Hill and M.A. Pate, Superlattices and Microstructures 2, 49 (1986).
8. D.C. Taylor, P.S.S. Guimaraes, B.R. Snell, F.W. Sheard, L. Eaves, G.A. Toombs J.C. Portal, L. Dmowski, K.E. Singer, G. Hill and M.A. Pate, Proc. Int. Conf. on Modulated Semiconductor Structures, Kyoto, to be publ. in Surface Science (1986). See also G.A. Toombs, F.W. Sheard and L. Eaves, Phonon Physics, publ. World Scientific, 561 (1985).
9. J.P. Leburton, Phys. Rev. B31 4080, (1985).
10. E.S. Hellman, J.S. Harris, C. Hanna and R.B. Laughlin, Physica 134B, 41 (1985).
11. J.R. Barker, ibid, p.22.
12. J.P. Leburton, ibid, p.32.
13. E.S. Hellman, J.S. Harris, Phys. Rev. B33, 8284 (1986).
14. J. Ihm, Phys. Rev. Lett. 55, 999 (1985).
15. C.B. Hanna and R.B. Laughlin, Phys. Rev. Lett. 56, 2547 (1986).
16. J. Ihm, Phys. Rev. Lett. 56, 2548 (1986).
17. T. Wang, J.P. Leburton, K. Hess and D. Bailey, Phys. Rev. B33, 2906 (1986).
18. T.W. Hickmott, Phys. Rev. B32, 6531 (1985).
19. T.W. Hickmott, P.M. Solomon, R. Fischer and H. Morkoc, J. Appl. Phys. 57, 2844 (1985).
20. R.L.F. Boyd and A. Boksenberg, Proc. Int. Conf. on Ionisation Phenomena in Gases, Uppsala, p.529 (publ. North Holland) (1959).
21. J.E. Golden and J.H. McGuire, Phys. Rev. Lett. 32, 1218 (1974).
22. J.W. McGowan, J.F. Williams and E.K. Curley, Phys. Rev. 180, 132 (1969).
23. J.M. Chamberlain, A.A. Reeder, L.M. Claessen, G.L.J.A. Rikken and P. Wyder Physica 134B, 426 (1985) and references therein.
24. J.A. Kash, J.C. Tsang and J.M. Huan, Phys. Rev. Lett. 54, 2151 (1985).
25. R.A. Stradling, L. Eaves, R.A. Hoult, A.L. Mears and R.A. Wood, Proc. Int. Conf. on Semiconductors, Boston (publ. USAEC/DTI) p.369 (1970).
26. L. Eaves and J.C. Portal, J.Phys.C: Solid State, 12, 2809 (1979).
27. See for example B.I. Shklovskii and A.L. Efros, Springer Series in Solid State Sciences 45, (1984), H. Kahlert, G. Landwehr, A. Schachetzski and H. Salow, Z. Phys. B24, 361 (1976), and D.C. Taylor et al to be published.
28. The derivation of the equivalent circuit in Figure 1(a) will be described in a paper by F.W. Sheard, G.A. Toombs and L. Eaves submitted to Semiconductor Science and Technology.

The circuit in Figure 1 can be represented by a single capacitor C* and impedance R* in parallel where

$$C^* = [(R_1 + R_2)(C_1R_1 + C_2R_2) - (C_1C_2)R_1R_2(1 - \omega^2C_1R_1C_2R_2)]/F^2,$$

$$R^{*-1} = [(R_1 + R_2) + \omega^2R_1R_2(C_1^2R_1 + C_2^2R_2)]/F^2,$$

$$F^2 = (R_1 + R_2)^2 + \omega^2(C_1 + C_2)^2R_1^2R_2^2 .$$

29. MOS Physics and Technology, E.H. Nicollian and J.R. Brews, Wiley p.385-90 (1982).

Aperiodic Quantum Magnetoresistance Oscillations in Submicron n^+GaAs Wires

R.P. Taylor[1], L. Eaves[1], P.C. Main[1], G.P. Whittington[1], S. Thoms[2], S.P. Beaumont[2], and C.D.W. Wilkinson[2]

[1]Department of Physics, University of Nottingham, Nottingham, NG7 2RD, U
[2]Department of Electronic and Electrical Enginerring, University of Glasgow, Glasgow, GL2 8QQ, UK

We have investigated the aperiodic oscillatory structure in the magnetoresistance of small n^+GaAs stripes of thickness 50 nm, length 10 μm and widths between 0.09 μm and 0.3 μm. The structure has been studied as a function of temperature (>1.5K) and magnetic field ($0 \leq B \leq 12T$). The rms amplitude of the oscillations is proportional to $T^{-\frac{1}{2}}$ in agreement with theoretical predictions. The angular dependence of peaks in the transverse magnetoresistance follows a $B = B_0/\cos(\theta+\delta)$ dependence where θ is the angle to the normal of the substrate and δ covers the range $-15^o < \delta < 15^o$ for different peaks.

Our understanding of the apparently random (aperiodic) yet reproducible structure in the magnetoresistance of narrow conducting channels (pinched-off Si MOSFETs, metal stripes) has increased considerably in recent years [eg refs 1-6]. This has been partly due to improvements in theoretical understanding [1,2] and to the detailed experimental work on narrow channel Si MOSFETs [3,4,5 The aperiodic conductance fluctuations ($\Delta G \sim e^2/h$) are of great fundamental interest. They appear to be a universal property of all conducting systems and arise from quantum interference and coherent scattering of single electrons. As semiconductor technology develops and the sizes of device structures and their interconnections become even smaller, the effect may also assume a technological importance. To date, most experimental work on the aperiodic structure in semiconductors has been restricted to n-type inversion layers of narrow channel Si MOSFETs. In this paper, we describe measurements on n^+GaAs stripes fabricated by electron beam lithography and dry etching [7] from 50 nm thick n^+ (mid 10^{18} cm^{-3}) GaAs layers grown on semi-insulating substrates by MBE A typical stripe is shown in the electron micrograph in Figure 1. We have investigated a range of stripe sizes with lengths between 2 and 10 μm and with widths between 0.09 μm and 0.3 μm.

The longitudinal ($\underline{J} \parallel \underline{B}$) and transverse ($\underline{J} \perp \underline{B}$) magnetoresistance of the wires were measured from 1.5K upwards in magnetic fields ($\underline{B}$) up to 11.5T. In the transverse geometry $\underline{B}$ could be rotated from parallel to the normal $\hat{n}$ of the substrate plane (T1 configuration) through to $\underline{B}$ in the plane of the substrate (T2 configuration). Figure 2 shows a series of transverse magnetoresistance plots taken at 4.2K for various angles θ, where $\cos\theta = \hat{n}.\underline{B}/|B|$; with $\underline{B} \perp \underline{J}$ for a sample with ℓ = 10 μm and w = 0.26 μm. Each curve is completely reproducible provided that the sample is maintained at liquid heliun temperature and properly screened. For example, at a fixed magnetic field configuration the aperiodic structure remains stable for a period of more than 24 hours. However, when the sample is heated to room temperature and then recooled, we observe qualitatively similar structure (i.e. similar amplitude and magnetic field interval, ΔB) but with differences in the detailed form. The curves are identical when B is reversed ($\underline{B}$ to $-\underline{B}$). Note that the aperiodic structure is superimposed on a negative magnetoresistance of comparable

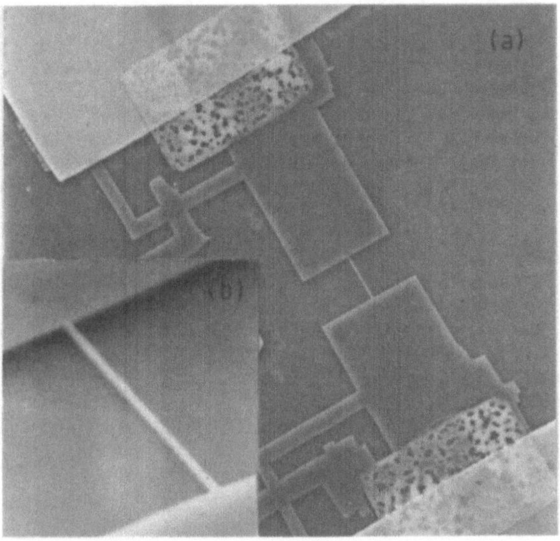

Fig. 1 (a) Bonding pad arrangement for a 0.26 x 10 μm^2 n^+GaAs stripe. The spongy areas are annealed Au/Ge/Ni ohmic contacts; the lighter areas are Ti/Au bonding pads. The wide n^+GaAs contact strip and the stripe itself are seen clearly as a plateau above the semi-insulating substrate layer. (b) Detail of the wire itself. Length = 10 μm

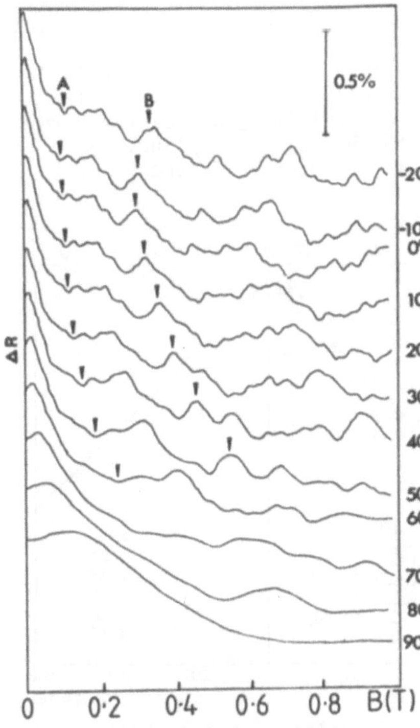

Fig. 2 Transverse magnetoresistance anisotropy of a 0.26 μm wide stripe for various θ. T = 4.2K. The 0.5% marker is relative to the resistance R_0 at B = 0, R_0 = 5.02 KΩ

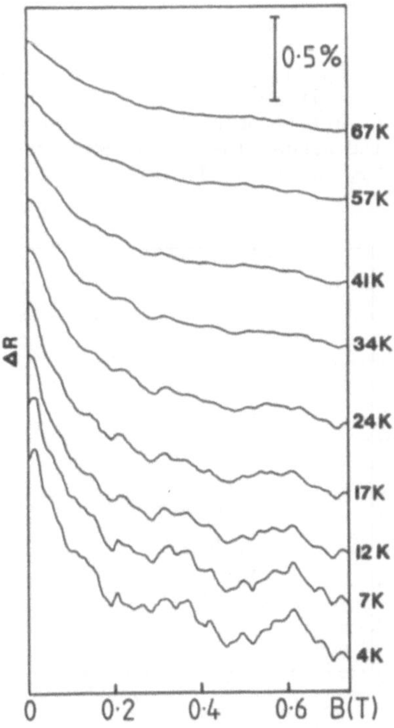

Fig. 3 Transverse magnetoresistance of a 0.26 μm wide stripe at various temperatures for θ = 0.

magnitude due to the presence of weak localisation. The slight positive magnetoresistance at low field may be due to spin-orbit scattering. As the temperature is increased the oscillatory structure and the negative magnetoresistance decrease in amplitude together, as can be seen in Figure 3. Remarkably, the structure remains visible up to 60K. Although the amplitude diminishes, the magnetic field corresponding to each feature remains constant. The rms amplitude $\overline{\Delta R}$ of all the structure shown in Figure 3 and magnetoresistance traces for similar wires is plotted as a function of temperature in Figure 4. The amplitude obeys the law $\Delta R/R_0 \propto T^{-\frac{1}{2}}$, consistent with the inelastic scattering length, $L_{in} \propto T^{-\frac{1}{2}}$ [5]. This behaviour has been confirmed by analysis of the negative magnetoresistance above 25K. Below that temperatur the oscillatory structure precludes such an analysis. At 25K $L_{in} \sim 0.2$ µm. There are several interesting features in the angular dependent magnetoresistance plots shown in Figure 2. First, each feature moves to higher field as θ is increased well away from 0. Several of the features, for example peaks A and B, can be followed over a wide range of θ. However, not all of the structure is symmetric around θ = 0° and no feature can be followed without ambiguity over more than 60°. Many of the features have a significant amplitud over only a restricted range of angle, not necessarily in the vicinity of θ = 0°. We can fit accurately the angular dependence of various peaks and troughs by the relation

$$B(\theta) = \frac{B_o}{\cos(\theta + \delta)}$$

where B_o and δ are adjustable parameters for a given feature. The results of such a fit are shown in Figure 5 for a number of peaks and troughs. It is important to note that the range of δ's required for fitting are distributed in the range -15° < δ < 15° around the normal to the epitaxial plane. A histogram of the distribution of δ's is shown in an inset in Figure 5.

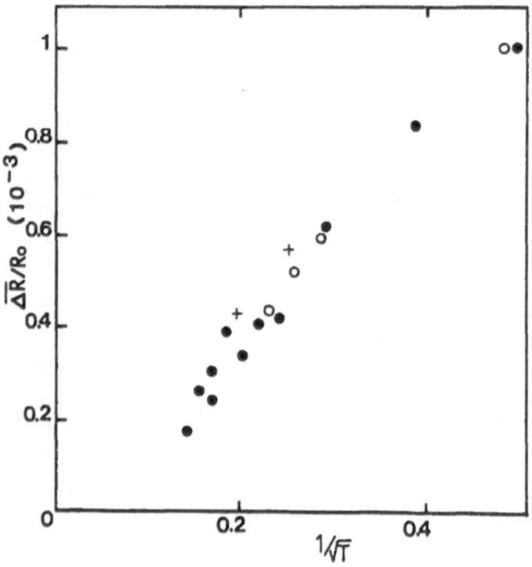

Fig. 4 The rms amplitude of oscillation ΔR, relative to the average resistance R_o, against $(T)^{-\frac{1}{2}}$ for various stripes with widths between 0.26 µm and 0.3 µm

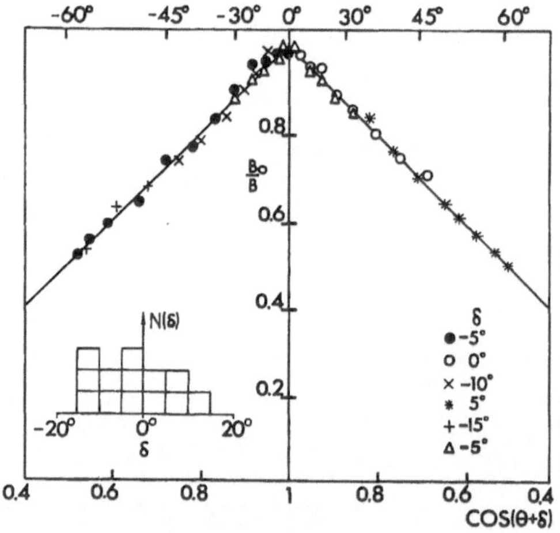

Fig. 5 Relative field position of various features in the transverse magnetoresistance plotted against cos(θ+δ). The inset histogram shows the distribution of parameter δ

At higher fields (1T < B < 11.5T) the aperiodic structure develops further in complexity and its amplitude increases. Above 2T, the rms amplitude is typically 0.3%. This work will be described in more detail elsewhere [8], but several interesting features are worth noting here. First, for $\theta = 0^o$, we observe a distinct Shubnikov-de Haas (SH) series in the transverse magnetoresistance for B > 5T indicating a Fermi energy ε_F = 136 meV and a volume electron density of 5.0 x 10^{18} cm^{-3}. Secondly, when B is tilted by even a few degrees the regular SH oscillations become lost in the aperiodic structure. At the lowest field (B ∿ 5T) where the SH oscillations can be resolved, the classical cyclotron orbital diameter at the Fermi energy is 0.15 μm. We note that this length is smaller than but comparable to the width of the wire. Also, at higher fields aperiodic structure is observed in the T2 configuration with B parallel to the plane of the substrate in contrast to the absence of structure for B < 1T shown in Figure 2. However, the typical field interval between adjacent peaks is much longer than in the T1 configuration. Structure with similar long field intervals is also observed with B parallel to the length of the wire.

The aperiodic structure has been described theoretically in terms of quantum interference of single-electron wavefunctions. The role of the magnetic field is to introduce phase differences between electronic trajectories. For macroscopic wires these effects average to zero due to the very small electron inelastic scattering length. The angular dependence at low magnetic fields clearly implies that these trajectories are approximately confined to a plane parallel to the substrate (see Kaplan and Hartstein [4]). However, the histogram in Figure 5 (inset) clearly shows a distribution away from this plane. It is unlikely that these finite values of δ correspond to trajectories confined to tilted planes; it is more likely that the trajectories have some three-dimensional meandering character shaped by the electronic potential. This type of trajectory is more likely in these GaAs wires than in the Si MOSFETs [4] since in the latter the electrons are confined by the insulating SiO_2 wall, whereas in our wires the confinement is by a softer depletion layer potential. This interpretation is consistent with our observation of structure in the T2 and longitudinal magnetoresistance.

This work is supported by SERC.

1. A.D. Stone, Phys. Rev. Lett. 54, 25 (1985).
2. P.A. Lee and A.D. Stone, Phys. Rev. Lett. 55, 15 (1985).
3. J.C. Licini, D.J. Bishop, M.A. Kastner and J. Melngailis, Phys. Rev. Lett. 55, 27 (1985).
4. S.B. Kaplan and A. Hartstein, Phys. Rev. Lett. 56, 22 (1986).
5. W.J. Skocpol, P.M. Mankiewicz, R.E. Howard, L.D. Jackel, D.M. Tennant and A.D. Stone, Phys. Rev. Lett. 56, 26 (1986).
6. C.P. Umbach, S. Washburn, R.B. Laibowitz and R.A. Webb, Phys. Rev. B30 4048 (1984).
7. S. Thoms, S.P. Beaumont and C.D.W. Wilkinson, Submitted to Conf. on Microcircuit Engineering, (1986), Zurich.
8. G.P. Whittington, P.C. Main, L. Eaves, R.P. Taylor, S. Thoms, S.P. Beaumont and C.D.W. Wilkinson, 2nd Int. Conf. on Superlattices, Microstructures and Microdevices, Gothenberg, Sweden, (1986).

Magnetic Crossover in Narrow 2D Electron Systems

A. Isihara[1], *K. Ebina*[1], *L. Smrčka*[2], *and H. Havlova*[2]

[1]State University of New York at Buffalo, Buffalo, NY 14260, USA
[2]Czechoslovak Academy of Sciences, 18040 Praha, Czechoslovakia

1. Introduction

Recently, there has been a surge of interest in narrow 2D electron systems in Si inversion layers or GaAs/GaAlAs quantum wells [1]. They show a variety of new phenomena, ranging from two dimensions to one. We are especially interested in the case with a perpendicular strong magnetic field because magnetic crossover in quantization can be expected. Since Landau levels correspond to infinite systems and no exact energy levels are known for finite width, the evaluation of the density of states and conductivity is very difficult.

We note that quantum transport in submicron devices is practically extremely important. However, when Si MOSFETs or GaAs quantum wells are closely packed in electronic devices, field leaking, heating, size and other problems occur. The narrow 2D systems enable a unique and systematic approach to quantum transport in submicron devices due to their variable width.

It is the purpose of the present article to report on the density of states and magnetoconductivity of these narrow systems. We shall discuss how the density of states and conductivity vary when the quantization conditions are changed. Especially, we shall report for the first time that size quantization causes oscillations in conductivity. When coupled with localization such oscillations are expected to result in strong quantum fluctuations. In what follows, we shall adopt two well defined models.

2. Hard-Wall Model

When a hard-wall model or "particle in a box" model is adopted, the system is characterized by a dimensionless parameter $\lambda = w/2a_o$, where w is the width and a_o is the magnetic length. When $\lambda >> 1$ and near $k = 0$, where $\hbar k$ is the momentum in the extended direction, the energy levels are approximately given by Landau levels. The dispersion curve in which the energy levels $E_n(k)$ are plotted against k, is flat near the origin and increases towards $k = w/2a_o^2$. The flat region shrinks as the energy becomes higher. If the level index n is increased, a new energy regime starts above the crossing point which is given by

$$E_{cr} = (\hbar^2/2m)(w/2a_o^2)^2 = (\lambda^2/2)\hbar\omega_c \quad (1)$$

where ω_c is the cyclotron frequency. Well above this point, the energy levels are approximately given by

$$E_n(k) \sim E_g(n+1)^2 + \hbar^2k^2/2m \quad (2)$$

where $E_g^* = (\pi\hbar/w)^2/2m$ is the ground state energy in this regime. The energy dispersion is parabolic, and the distance between the levels is not constant. Indeed, as n increases, the origin of the dispersion curve becomes higher in proportion to $(n+1)^2$ and the region where the energy expression prevails expands.

The density of states D(E) varies with λ. For $E_g^* = E_g/(\hbar\omega_c/2) \sim 0.1$, the distance between adjacent peaks is almost equal to $\hbar\omega_c$ up to $n \sim 10$, but D(E) does not vanish inbetween the peaks, unless E_g^* is extremely small.

Figure 1 illustrates the reduced density of states $D(E)/D_o$ as a function of the reduced energy $E/\hbar\omega_c$ for the case $E_g^* = 0.2$ or $w/a_o = \sqrt{5\pi}$. Here, D_o = eB/ch. Note that the distance between the peaks and the minima increase as the energy increases. The number of peaks becomes less for a given interval of energy if E_g^* is increased or λ is decreased. For large E_g^*, D(E) may be appropriately plotted against E/E_g instead of $E/\hbar\omega_c$. The gross feature of D(E) is analogous to that of Fig. 1. For a quasi 1D case, a similar graph has been reported by WHEELER et al [2]. On the other hand, if E_g^* is further reduced and λ is increased from those for Fig. 1, the peaks become more and more equidistant. For example, the experimental conditions of VON KLITZING et al [3] for the quantized Hall effect, i.e., w = 50 /μm/ and B = 18 /T/, correspond roughly to $E_g^* = 10^{-7}$. For such a small value of E_g^*, there are only few states at the skirt of D(E) when Landau levels are not broadened. In the other extreme, the number of Landau-type levels below E_{cr} is given by $(\lambda^2 + 1)/2$ so that there will be no such levels if $\lambda<1$.

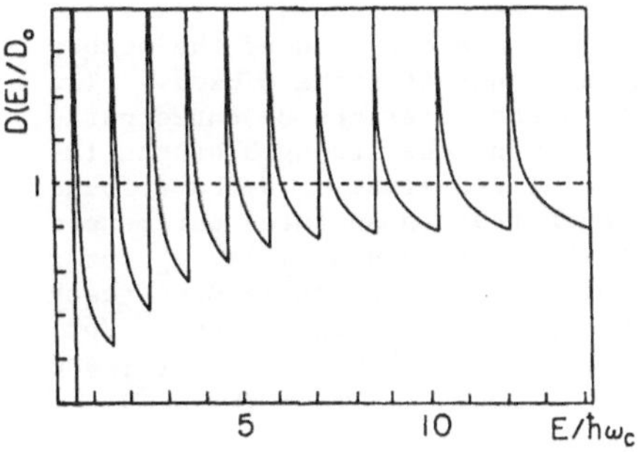

Fig. 1 Density of states for $E_g^* = 0.2$ and $w/a_o = \sqrt{5\pi}$.

3. Soft-Wall Model

For the above hard-wall case, we evaluated the energy levels numerically. Therefore, it is difficult to derive a conductivity formula. We have overcome this difficulty by adopting a soft-wall model with a parabolic confining potential given by

$$V(y) = m\Omega^2 y^2/2 \tag{3}$$

where Ω is the frequency representing the strength of the confinement. The width of the system in this case is energy dependent and is given by

$$w(E) = 2(2E/m)^{\frac{1}{2}}/\Omega \quad . \tag{4}$$

A great advantage of using the parabolic well model is that the energy levels are given by

$$E_n(k) = \hbar\tilde{\omega}(n+\tfrac{1}{2}) + \hbar^2k^2/(2\tilde{m}) \tag{5}$$

which are of the same form as the ordinary 3D case except that the modified cyclotron frequency and effective mass defined by

$$\tilde{\omega} = [\omega_c^2 + \Omega^2]^{\frac{1}{2}} \; ; \quad \tilde{m} = m(1 - \alpha^2)^{-1} \tag{6}$$

characterize the levels. Here, $\alpha = \omega_c/\tilde{\omega}$ is a crossover parameter. Also, a new effective magnetic length given by $\tilde{a}_o = (\hbar/m\tilde{\omega})^{\frac{1}{2}} = a_o\alpha^{\frac{1}{2}}$ plays a role. Thus, there is no essential difficulty in constructing the Green's function and evaluating the magnetoconductivity based on linear response theory. The Green's function, when averaged over impurity configurations, is characterized by a self-energy. The relation between the original and averaged Green's functions is given by a self-consistent equation.

When the conductivity is going to be studied as a function of electron energy or gate voltage, an elaborate determination of the self-energy is required because its imaginary part represents the relaxation time which is expected to depend on electron energy. If the energy is high, the electron will approach scattering centers closely so that the scattering rate will be high and the relaxation time will be small. If the energy is low, localization will become effective and the lifetime of the electron will decrease. These variations in electron scattering have actually been what experiments have shown. They are important in treating ordinary 2D systems, but even more so in narrow systems because the narrower the stronger localization.

We have employed pseudopotential theory in order to take into consideration of these variations. We have adopted a simple model in which only one localized state lies just below the conduction band threshold. In effect, we have employed the most important localized state and reduced the number of unknown parameters which would arize when many such states are used. The matrix element of the pseudopotential is then characterized by three parameters which depend on the impurity concentration, the strength of the impurity potential, and the location of the localized state. We have found that the relaxation time can be expressed such that these three parameters are determined by the energy E_σ [or E_τ] at which the conductivity σ_o [or relaxation time τ_o] in the absence of a magnetic field and confinement reaches a maximum, the maximum value of τ_o, and the impurity concentration. For their actual values, we have made use of the data of SKOCPOL et al [4] and of FANG et al [5] on the relaxation time. Our conductivity formula is

$$\sigma_{xx} = (1-\alpha^2)\sigma_o(1-\frac{\Delta D}{D_n}) + \frac{\alpha^2\sigma_o}{1+\tilde{\omega}^2\tau_o^2}[1+\frac{1+5\tilde{\omega}^2\tau_o^2}{2(1+\tilde{\omega}^2\tau_o^2)}\frac{\Delta D}{D_n}] \tag{7}$$

where $\Delta D/D_n$ is the ratio of the oscillating part to nonoscillating part of the density of states which can be determined from the Green's function following the work of ISIHARA and SMRČKA [6] and where

$$\tau_o = a\{E_\sigma E[E_\sigma E^2 + (E_\sigma^2 - 3E_\tau^2)E + E_\sigma E_\tau^2]^{-1}\}^2 \tag{8}$$

where a is inversely proportional to the impurity concentration.

Figure 2 illustrates our theoretical conductivity for narrow Si inversion layers at 2 /K/ and $m = 0.19m_o$. The parameters are chosen to reproduce the zero field conductivity σ_o [6]. Their values are $a = 5.2\text{x}10^{-10}$ /meV^{2}s/, $\varepsilon_\sigma = 17.5$ /meV/, and $\varepsilon_\tau = 8$ /meV/. The upper set of curves corresponds to the case with $\Omega = 1.2\text{x}10^{12}$ /s^{-1}/. The SdH oscillations are strong around E = 25 /meV/, particularly in the top curve for 8 /T/. The conductivity increases with gate voltage and approaches gradually an asymptotic value which depends on the magnetic field. As the zero field curve (second from

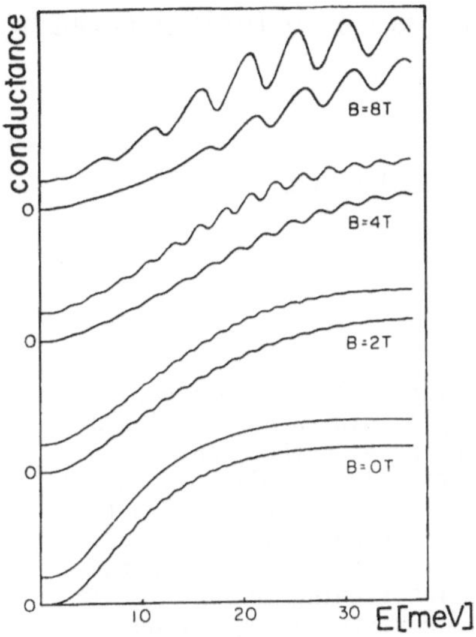

Fig. 2 Theoretical conductivity in narrow Si inversion layers at 2 /K/.
Upper set: $\Omega = 1.2 \times 10^{12}$ /s^{-1}/
Lower set: $\Omega = 1.8 \times 10^{12}$ /s^{-1}/

the bottom) shows, the effect of size quantization is still not visible at this value of Ω. The lower set of curves has been obtained for $\Omega = 1.8 \times 10^{12}$ /s^{-1}/, corresponding to a narrower width. The SdH oscillations are markedly suppressed now in both amplitude and period. Particularly interesting is the bottom curve which corresponds to zero magnetic field. This curve shows clearly that the conductivity oscillates due to size quantization. It is also interesting to observe that these quantum oscillations are strong at certain intervals of electron energy which depend on the magnetic field. These results agree extremely well with what Figs. 1(a) and 1(b) of SKOCPOL et al show.

Acknowledgement

This work was supported by the Office of Naval Research under Contract N000-14-K-84-0387.

References

1. A. Fowler, A. Hartstein and R. A. Webb: Phys. Rev. Lett. 48, 196 (1982), Physica 117/118B, 661 (1983).
2. R. G. Wheeler et al: Phys. Rev. Lett. 49, 1674 (1982), Surf. Sci. 142, 19 (1984).
3. K. von Klitzing, G. Dorda and M. Pepper: Phys. Rev. Lett. 45, 494 (1980).
4. W. J. Skocpol et al: Surf. Sci. 142, 14 (1984).
5. F. F. Fang, A. B. Fowler and A. Hartstein: Phys. Rev. B16, 4446 (1977).
6. A. Isihara and L. Smrčka: J. Phys. C: to be published.

Resonant 2D Magnetopolarons in Accumulation Layers on n-$Hg_{0.8}Cd_{0.2}Te$

J. Singleton, F. Nasir, and R.J. Nicholas

Clarendon Laboratory, Parks Road, Oxford OX13PU, USA

Cyclotron resonance (CR) measurements on accumulation-layer electrons at anodic oxide films n-(Hg,Cd)Te are reported. Up to six subband cyclotron resonances are observed, and discontinuities are seen in the subband effective masses, due to resonant polaron effects. The coupling is strongest for the higher (i = 3,4,5) subbands, where it is enhanced over that for the bulk, and occurs at the "HgTe-like" LO phonon frequency. The behaviour of the lowest, deeply bound (i = 0,1) subbands is more complex, with the coupling being weaker and the dominant interaction seeming to occur at the TO frequency for low values of N_s. The behaviour of the accumulation layer electrons is thus rather similar to the recently reported resonant 2D magnetopolarons observed in a variety of III-V heterostructures: in high carrier concentration (and hence heavily-screened) heterojunctions, the dominant interaction was at the TO phonon frequency, whereas in low carrier concentration quantum wells, the coupling occurred at the LO phonon frequency [1,2].

1. Introduction

There is a great deal of interest in the roles that confinement and screening play in the polaron effect in a quasi-two-dimensional electron gas (Q2DEG). In an infinitely thin, unscreened Q2DEG, theoretical models predict an enhancement of the electron-optic phonon coupling (e.g. [3]): the inclusion of screening and finite width quickly destroys this enhancement (e.g. [4]). Available experimental data seem system and sample-dependent, with enhanced polaronic effects reported in some cases (InSb [5]) but not in others (GaAs [6,7]). In this work we describe the observation of enhanced resonant 2D magnetopolarons in accumulation layers at anodic oxide films on n-$Hg_{0.8}Cd_{0.2}Te$, in contrast to earlier reports [8].

The growth of an anodic oxide film is frequently used to passivate the surfaces of $Hg_{0.8}Cd_{0.2}Te$ devices, and ionised deep traps in the oxide lead to the accumulation of $\sim 10^{12}$ cm^{-2} electrons at the surface of n-type material [9]. Such accumulation layers contain 5 or 6 populated 2D electric subbands, ranging from the deeply-bound i = 0 subband, to the i = 3, 4 and 5 subbands, which are bound by a few meV and penetrate several hundred Ångstroms into the bulk material [10]. These subbands have very well-defined occupancies, which are insensitive to variations in composition or doping level [9].

$Hg_{0.8}Cd_{0.2}Te$ exhibits two distinct reststrahlen bands due to "HgTe-like" phonons and higher energy "CdTe-like" phonons. The latter modes are much weaker, and Faraday geometry CR of bulk electrons shows resonant polaron coupling only at the "HgTe-like" LO phonon energy [11,12].

2. Experiment and Results

The samples used in this study are $\sim 2{\cdot}5$ mm × 1 mm × 10 μm monoliths of n-$Hg_{0.8}Cd_{0.2}Te$ and have been described in detail elsewhere [9]. Both front and back surfaces are anodically oxidised to produce two accumulation layers. Although the structures are ungated, N_s, the total surface carrier density can be varied at low temperatures using UV illumination [13]: in the experiments described below N_s was measured using the Shubnikov - de Haas effect [9].

The magnetotransmission of the samples was measured in perpendicular magnetic field at a number of wavelengths in the region 251 μm to 38 μm, and typical results are shown in figure 1: the CR appear as peaks in the absorption. The strong absorption at low fields is due to the bulk electrons whilst the varying number of resonances at higher fields are due to the surface electrons, which have larger effective masses. Strong polaron effects in the highest (i = 4,5) subbands are visible in the raw data: as $\hbar\omega_c$ approaches the "HgTe-like" reststrahlen region (73 μm $< \lambda <$ 83 μm), the large polaron contribution to the i = 4,5 subband effective masses progressively separates their CR from the bulk CR. This effect disappears above the reststrahlen band, as expected.

If the subband CR energies are plotted as a function of magnetic field (figure 2), resonant polaron coupling is observed as a displacement of the CR positions at energies above the "HgTe-like" phonons to lower magnetic fields than expected [5]. From inspection, the coupling strength appears to increase with subband index.

In order to compare the relative strengths of the bulk and surface polaronic effects, the subband effective masses from a sample with $N_s = 2{\cdot}0 \times 10^{12}$ cm^{-2} are normalised and plotted alongside normalised bulk data [11] in figure 3.

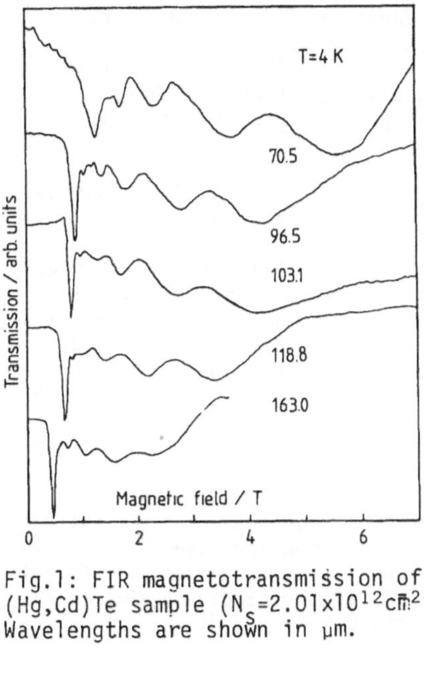

Fig.1: FIR magnetotransmission of (Hg,Cd)Te sample (N_s=2.01x$10^{12}$$cm^{-2}$). Wavelengths are shown in μm.

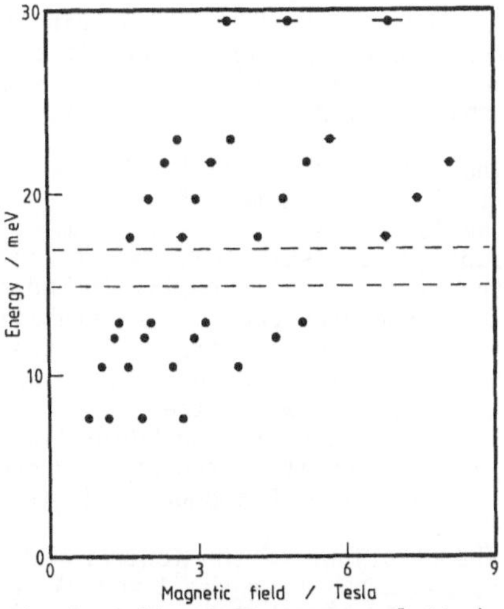

Fig.2: Subband CR energy plotted against magnetic field for a sample with N_s=2.5x$10^{12}$$cm^{-2}$.

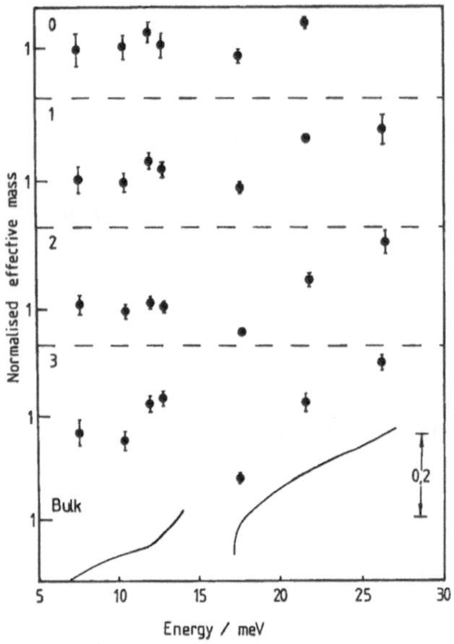

Fig.3: Normalised subband effective masses for i = 0 to 3 subbands of a sample with N_s=2.01x10^{12}cm^{-2}, compared with bulk data from /11/.

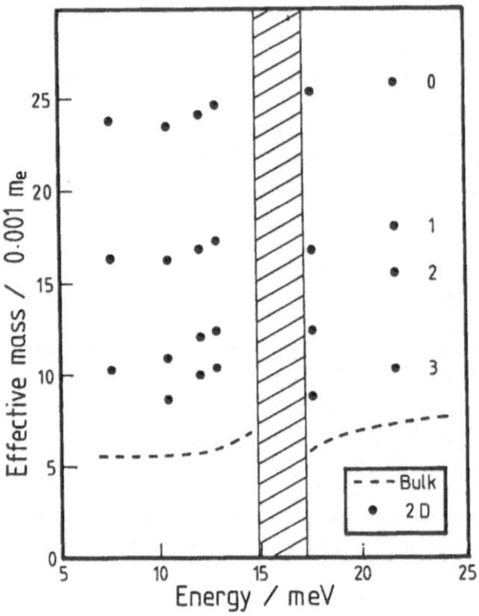

Fig.4: Subband effective masses for a sample with N_s=0.9x10^{12}cm^{-2}, compared with bulk data from/11/.

For the i = 0 and i = 1 subbands the polaronic effects are weaker than those in the bulk, being very small for i = 0. In the case of the i = 2 subband, the coupling is around the same strength as that in the bulk. Finally, in the case of the i = 3 subband, its ultraquantum limit is reached at a field corresponding to $\hbar\omega_c \simeq 12$ meV, so that the effective mass starts to increase due to band non-parabolicity at around this energy. Superimposed on this is the resonant polaron effect, which is enhanced over that in the bulk.

The behaviour of the i = 2 and i = 3 subbands is similar in all the samples studied: for example figure 4 shows the subband effective masses from a sample with $N_s = 0{\cdot}9 \times 10^{12}$ cm^{-2}. The populations of the i = 2 and i = 3 subbands are such that they reach their ultraquantum limits at $\hbar\omega_c \simeq 10$ meV and $\hbar\omega_c \simeq 7$ meV respectively and so the effective masses increase above these energies due to non-parabolicity: again, discontinuities in the i = 2 and i = 3 subband effective masses close to the "HgTe-like" reststrahlen band indicate resonant polaron coupling. In all of the samples studied, the phonon to which the i = 2 and i = 3 subbands are coupling appears to be closer in energy to the "HgTe-like" LO phonon than to the "HgTe-like" TO phonon: thus it is tentatively suggested that the i = 2 and i = 3 subbands couple to the LO phonon as do the bulk electrons [11,14].

The behaviour of the lower (i = 0,1) subbands is somewhat different. At high values of N_s (figures 2 and 3) very weak resonant coupling appears to occur at the "HgTe-like" LO phonon energy, and the interaction is stronger for the i = 1 subband than for the i = 0. In contrast, if the effective masses of the i = 0 and i = 1 subbands in the low N_s samples are examined as a function of energy (e.g. figure 4) a sharp increase is seen just below the "HgTe-like" TO phonon energy, behaviour more consistent with a resonant interaction at the TO phonon frequency. This increase is not due to band non parabolicity, as, for example, the ultraquantum limit of the i = 1 sub-band in figure 4 is reached at $\hbar\omega_c \simeq 30$ meV.

Recent experiments on a variety of (Ga,In)As-InP and (Ga,In)As-(Al,In)As heterostructures showed that weak resonant polaron effects occurred at the TO frequency in high carrier density heterojunctions, but that the dominant interaction was at the LO frequency in lower carrier density quantum wells [1,2].These results were interpreted as due to the difference in screening in the two cases. The subbands on n-(Hg,Cd)Te described above are perhaps behaving in the same way: the higher subbbands are analogous to the weakly screened electrons in the quantum wells, so that the resonant coupling occurs at the LO frequency. Likewise, the i = 0,1 subbands are strongly screened, as in the heterojunctions, so that at low N_s weak coupling at the TO frequency occurs, and further increases in N_s destroy this interaction. Attempts to explain this behaviour using static screening have been unsuccessful [14], and it is believed that a model incorporating full dynamical screening of the electron-phonon interaction [4] and the effects of a high magnetic field will be necessary in order to account for the observed results [15].

References

1. R.J. Nicholas et al.: Phys. Rev. Lett. 55 883 (1985)
2. L.C. Brunel et al.: Surface Science 170 542 (1986)
3. D.M. Larsen: Phys. Rev. B30 4595 (1984)
4. W. Xiaoguang, F.M. Peeters and J.T. Devreese: Phys. Stat. Sol. b133 229 (1986)
5. M. Horst, U.Merkt and J.P. Kotthaus: Phys. Rev. Lett. 50 754 (1983)
6. W. Seidenbusch et al.: Surface Science 142 375 (1984)
7. M. Horst et al.: Solid State Commun. 53 403 (1985)
8. J. Scholz et al.: Solid State Commun. 46 665 (1983)
9. J. Singleton et al.: J. Phys. C: Solid State Physics 19 35 (1986)
10. J. Singleton, F. Nasir and R.J. Nicholas: Solid State Commun. (in press)
11. M.A. Kinch and D.D. Buss: J. Phys. Chem. Solids 32 Supplément 1 461 (1971)
12. L. Swierkowski et al.: Solid State Commun. 27 1245 (1978)
13. F. Nasir, J. Singleton and R.J. Nicholas: to be published
14. J. Singleton, F. Nasir and R.J. Nicholas: Solid State Commun. 58 833 (1986)
15. F.M. Peeters: private communication.

Phonon Scattering in Multiquantum Well Structures

M. Singh

Department of Physics, The University of Western Ontario, London, Canada N6A 3K7

A theory of phonon scattering by carriers (electrons, holes) in a quasi two-dimensional gas has been developed for the multiquantum well structures (MQWS) in the presence of a magnetic field. We call it magneto-phonon scattering. The expressions for the Landau-level width (LLW) due to hole-phonon scattering and the cyclotron resonance line width (CRLW) due to electron-phonon scattering have been obtained. We found that the LLW and CRLW increases with the magnetic field and the temperature and decreases with the increase of the periodicity of the MQWS. It is found that the LLW of light holes is higher than the LLW of heavy holes. Finally the screened polar optical phonon contribution to the damping of the magneto-phonon oscillation in the MQWS is also discussed.

INTRODUCTION

The carrier-phonon interaction plays a very important role in the transport properties and the optical properties of quasi two-dimensional MQWS. Here carrier refers to electrons and holes. The electron-phonon scattering in zero magnetic field of a quasi two-dimensional electron gas in the presence of an infinitely deep quantum well (IDQW) has been calculated by several authors, /1/, Prasad and myself/1/ calculated the relaxation time using the Green's function approach for the IDQW. Recently myself and Chaubey/2,3/ have calculated the Landau level width due to electron-phonon scattering and the damping of the magneto-phonon oscillation (MPOD) due to electron-optical phonon scattering in the MQWS. In this paper we turn our attention to the calculation of the Landau-level width due to hole-phonon interactions and the cyclotron resonance line width due to electron-phonon interactions. We applied our theory to calculate the LLW and CRLW of a GaAs/GaAl superlattice. It is found that the LLW and CRLW increase with increasing temperature and magnetic field and decreases with an increase of the period of the MQWS. Theoretical results are in qualitative agreement with experimental results.

THEORY

The carrier-phonon interaction Hamiltonian of a two-dimensional carrier gas in MQWS subjected to a static magnetic field B applied perpendicular to the layers is given by

$$H_{ep} = \sum_{N,x'p,m} \sum_{N',x',p',m'} \sum_{\underline{Q},t} a^{+}_{N',x',p'} a_{N,x,p} (b^{+}_{-\underline{Q}} + b_{\underline{Q}}) \cdot \langle N',x',p',m' | V_{Qt} e^{i\underline{Q}\cdot\underline{r}} | N,x,p,m \rangle \qquad (1)$$

where N, p, m and x are the Landau quantum number, subband quantum number, carrier band quantum number and centre of the cyclotron orbit, respectively. a^{+}_{λ} and a_{λ} are the creation and annihilation operators, respectively, for the carrier in the Landau level $|\lambda\rangle = |N,x,p,m\rangle$, whereas b^{+}_{λ} and b_{λ} are the creation and annihilation operators, respectively, for a phonon wave vector

$\underline{Q}\cdot\underline{R}(r,z)$ is the position vector of a carrier in MQWS and $Q = (q_\perp, q_z)$, $q_\perp$ and q_z being the components of the phonon wave vector along the layer and perpendicular to the layer, respectively. t stands for phonon polarization. The values of V_{Qt} for electron-acoustic phonon and electron-optical phonon interactions are given in references /4/ and /5/, respectively. The value of $V_{Q,t}$ for hole-phonon interactions is written as

$$V_{Qt} = \left|\frac{3\hbar QE_d^2}{2\delta\Omega v_t}\right|^{\frac{1}{2}} \cdot \left[\left[\left(D_1+D_2 J_x^2\right)\hat{q}_x(e_t)_x+(J_xJ_y+J_yJ_x)\left[(\hat{e}_t)_x\hat{q}_y+(\hat{e}_t)_y\hat{q}_x\right)+C.P\right] \quad \ldots(2)$$

Here we consider the Bir and Pikus /6/ type of interaction Hamiltonian which includes the effects of internal strains. $D_1 = E_a/E_d$, $D_2 = \sqrt{3}(E_b/E_d)$, where E_a, E_b and E_d are the deformation potential. C.P. refers to terms obtained through cyclic permutation of indices and J_α is the αth component of the angular momentum $J = 3/2$. $\hat{q}$ is the unit vector along the direction of $\underline{Q}$ and $\hat{e}_t$ is the polarization vector in the (Q,t) mode. The self-energy $\Sigma_\lambda(E)$ of the retarded Green's function $G_\lambda(E)$ describing the effect of the scattering of a carrier in the Landau level $|\lambda\rangle$ is given by

$$\Sigma_\lambda = \Sigma_{\lambda'} \Sigma_{Q,t} |W_{Qt}^{\lambda\lambda'}|^2 \left[\frac{N_o(Q)}{E_\lambda - E_{\lambda'} - \hbar\omega_Q - \Sigma_{\lambda'}} + \frac{1 + N_o(Q)}{E_\lambda - E_{\lambda'} + \hbar\omega_a - \Sigma_{\lambda'}}\right]$$

$$|W_{Qt}^{\lambda\lambda'}|^2 = |\langle\lambda'|V_{Qt}e^{i\underline{Q}.\underline{r}}|\lambda\rangle|^2 = |F_{NN'}(q_\perp)|^2 |F_{pp'}(q_z)|^2 |C_{q\lambda}^{mm'}|^2 \quad \ldots(3)$$

where $N_o(Q) = [\exp(\hbar\omega_Q/k_BT)-1]^{-1}$ and E_λ is the energy of the $|\lambda\rangle$ state. Rewriting $\Sigma_\lambda = \Delta + i\Gamma_\lambda/2$ in equ. (3) and then calculating the imaginary part (which is the Landau level broadening LLB) of both sides of the same equation, we obtain the following equations for the LLB for the heavy hole ($m_j = \pm 3/2$), $(\Gamma_{N,p}^h)$ and light hole ($m_j = \pm\frac{1}{2}$), $(\Gamma_{N,p}^\ell)$

$$\Gamma_{N,p}^h = \Sigma_{Q,t}\Sigma_{N',p'}(2N_o(Q) + 1)\left[\Sigma_{m_j}^{\pm 3/2} \frac{F_{pp}^2 |C_{Qt}^{\lambda\lambda'}|^2 \Gamma_\lambda^h}{\left[E_\lambda^h - E_{\lambda'}^h\right]^2 + \left[\Gamma_\lambda^h\right]^2} + \Sigma_{m_j}^{\pm\frac{1}{2}} \frac{|C_{Qt}^{\lambda'\lambda}|^2 \Gamma_\lambda^\ell F_{pp}^2}{\left[E_\lambda^h - E_{\lambda'}^\ell\right]^2 + \left[\Gamma_\lambda^\ell\right]^2}\right] \quad \ldots(4a)$$

$$\Gamma_{N,p}^\ell = \Sigma_{Q,t}\Sigma_{N',p'}(2N_o(Q)+1)\left[\Sigma_{m_j=\pm 3/2}\left[\frac{F_{pp}^2 |C_{Qt}^{\lambda'\lambda}|^2 \Gamma_\lambda^h}{\left[E_\lambda^\ell - E_{\lambda'}^h\right]^2 + \left[\Gamma_\lambda^h\right]^2}\right] + \Sigma_{m_j=\pm\frac{1}{2}}\left[\frac{|C_{Qt}^{\lambda'\lambda}|^2 \Gamma_\lambda^\ell F_{pp}^2}{\left[E_\lambda^h - E_{\lambda'}^\ell\right]^2 + \left[\Gamma_\lambda^\ell\right]^2}\right]\right] \quad (4b)$$

The values of $J_{NN'}(q_\perp)$ and $F_{pp'}(q_z)$ are given in reference /5/, the value of $C_{Qt}^{\lambda'\lambda}$ will be published elsewhere /5/, it cannot be given here due to the limited space.

Here we have neglected the $\hbar\omega_{Qt}$ term with respect to $(E_\lambda - E_\lambda)$ in equ. (4). The expression for LLW of the electron-phonon interaction are given in ref. /4/. The above two equations are coupled in $\Gamma_{N,p}^h$ and $\Gamma_{N,p}^\ell$, which can hence be found by self-consistent calculation. Note that the LLW of light holes depends on the LLW of heavy holes and vice versa. This is due to the interaction between the two bands which plays a very important role in the transport and optical properties of the MQWS. The exact calculation of the LLW for the hole-phonon interaction is very difficult because of the complexity of the valence band structure. In the calculation of LLW the interaction between the valence bands appears through $C_{Qt}^{\lambda\lambda'}$ and the Landau level energy E_λ. In the present paper we will neglect the interaction between the valence bands and use the effective mass approximation. The result will still show the qualitative (but not quantitative) behaviour. To get quantitative results one should use a 4x4 Luttinger-type matrix Hamiltonian or $\vec{k}.\vec{p}$. Hamiltonian to calculate the Landau level. Work is in progress on this line.

Similarly one can also calculate CRLW for the MQWS. We present here the final expression in the elastic approximation when the interaction between the bands is neglected.

$$\gamma^2_{N,p} = \int \frac{d^3Q}{2\pi^3} |V_{Qt}|^2 [1+2N(Q)] [K_{NN}(q_\perp) + K_{N,N+1}(q_\perp)] |F_{pp}(q_z)|^2 \qquad \ldots(5)$$

The expression for K matrix is given by

$$K_{N,N'}(x) = \frac{N!}{N'!} e^{-x} x^{N'-N} L_N^{N'-N}(x) L_{N+1}^{N'-N-1}(x) \; ; \qquad N'>N$$

$$= \frac{N'!}{(N+1)!} x^{N'-N+1} e^{-x} L_{N'}^{N-N'}(x) L_{N'}^{N-N'+1}(x) \; ; \; N'\leq N$$

where $L_N^m(x)$ are the associate Laguerre polynomials. $x = q_\perp^2 \ell^2/2$, where ℓ is the Landau length ($= \sqrt{\hbar/eB}$; B is the magnetic field). A similar expression can also be obtained for the case of hole-phonon interactions.

RESULTS AND DISCUSSIONS

In this section we present the numerical results of LLW, CRLW and MPOD. In Table 1 we present the values of LLW for light and heavy holes. The parameters used in the calculations are b, the inter-quantum distance = 30Å; a, the width of the well = 30Å; W, the quantum well height = 0.05 eV, and N, the Landau quantum number = 0 in the extreme subband limit (i.e. p=1). It is clear from the table that the LLW increases as B increases and also that the LLW for light holes is higher than for heavy holes. We also found that the LLW increases with temperature and that it increases with decreasing periodicity d = a+b. These results are not shown here because of the limited space.

TABLE 1

The LLW for light and heavy holes at T = 10 K

B (Tesla)	LLW (meV)	
	heavy hole	light hole
5.0	.15	.38
9.0	.20	.43
13.0	.24	.50
17.0	.26	.56

In Table 2 we present the CRLW for GaAs/AlAs. The parameters used in the calculations are a = 30Å, b = 30Å, W = 0.12 eV and N = 0. It is clear from the table that the CRLW increases with an increase of magnetic field and temperature. We also found that it decreases with an increase of periodicity and Landau level number.

In Table 3 the MPOD due to electron-phonon interaction in GaAs/AlAs are presented /2/. We used the Thomas-Fermi type of screening in the present calculation. The parameters are B = 10 Tesla, T = 100 K, a = 50Å, b = 50 Å, W = 0.15 eV and N = 0. Here q_o is the Thomas-Fermi wave vector. One can see from the table that as q_o increases the MPOD decreases.

TABLE 2

The CRLW due to electron-phonon interactions in GaAs/AlAs

B (Tesla)	CRLW (meV)		
	T = 10 K	T = 40 K	T = 70 K
2.0	0.18	0.34	0.44
6.0	0.31	0.58	0.77
10.0	0.41	0.76	0.99
14.0	0.49	0.90	1.18
18.0	0.57	1.02	1.34
22.0	0.64	1.13	1.48

TABLE 3

$(q_o \ell)$	MPOD (meV)
0	1.23
0.3	0.98
0.6	0.89
0.9	0.83
1.2	0.78

ACKNOWLEDGEMENTS

The author is thankful to Prof. J. Nuttall and his students for computational help and to Dr. M.P. Chaubey for helpful discussions. The financial support from NSERC, Canada in the form of a Research Grant is acknowledged.

REFERENCES

1. B.K. Ridley, J. Phys. C15, 5899 (1982); L. Friedman, Phys. Rev. B32, 955 (1985); M. Prasad and M. Singh, Phys. Rev. B29, 4803 (1984).
2. M.P. Chaubey and M. Singh, Phys. Rev. B34, 3054 (1986).
3. M. Singh and M.P. Chaubey, Phys. Rev. B34, No. 5 (1986).
4. G.L. Bir and G.E. Pikus, Symmetry and strain-induced effects in semiconductors (Wiley, New York, 1974).
5. M. Singh, Phys. Rev. B (in press 1987).

Tunneling in Silicon Inversion Layers in High Magnetic Fields

U. Kunze

Institut für Elektrophysik, Technische Universität Braunschweig und Hochmagnetfeldanlage der Physikalischen Institute der Technischen Universität Braunschweig, D-3300 Braunschweig, Fed. Rep. of Germany

A review is given on recent tunneling studies of magnetic quantum effects in electron inversion layers on Si(001) surfaces. In perpendicular and tilted fields the magnetic quantization of surface and bulk states, and in parallel fields the diamagnetic rise of subband energies, and the effective mass change and spin splitting in the ground subband are discussed.

1. Introduction

In metal-oxide-semiconductor (MOS) systems an electron inversion layer is confined in a narrow potential well, which is formed by the oxide barrier and the semiconductor band bending. The band structure of this quasi two-dimensional (2D) surface layer consists of a series of subbands, whose minimum energies depend on the surface and depletion field, and on the effective mass for the motion perpendicular to the interface. Due to the anisotropic band structure of bulk Si, on Si(001) surfaces this results in two sets of subbands (E_n, $E_{n'}$). Figure 1 shows schematically the corresponding constant-energy contours /1/.

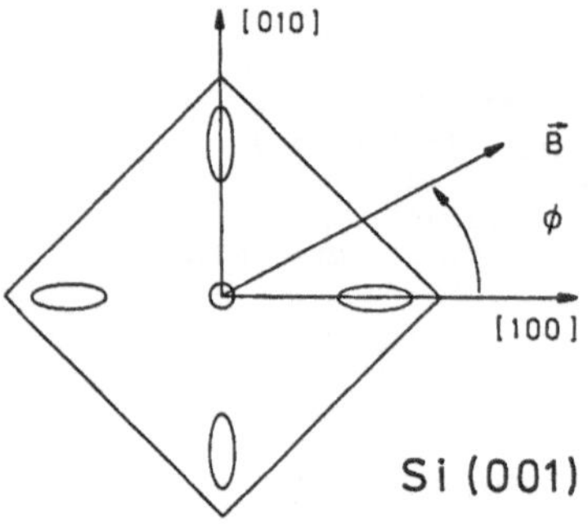

Fig. 1. Constant-energy contours inside the first Brillouin zone of a Si(001) electron space charge layer

Under high magnetic fields perpendicular to the surface each subband is quantized into discrete 2D Landau levels. A parallel magnetic field deforms the subband wave function, which results in a change of the $E(\vec{k})$ relation. When the electric field is strong enough so that the spread of the wave function perpendicular to the surface is less than the diameter of the cyclotron motion, this change can be expressed as a diamagnetic rise of the minimum energy and a displacement of the $E(\vec{k})$ parabola along the k-axis perpendicular to the magnetic field /1/, and as an increase of the effective mass in the same direction /2/. As the electric field decreases, e.g. for excited subbands in accumulation layers /3/, the hybrid electric-magnetic subbands change from electric into magnetic type /4/, until at vanishing electric field pure magnetic surface states are reached /5/ and the minimum energies converge into the bulk (3D) Landau levels.

The present paper gives a survey of recent results gained from tunneling in electron inversion layers on Si(001) surfaces under high magnetic fields. For the experiments the junctions have been mounted on a rotatable sample holder and cooled on T = 4.2 K. The cryostat has been placed into a 15.6 T Bitter Magnet. Magnetic field or bias voltage sweeps have been performed where the second derivative characteristics $d^2I/dV^2(V)$ of the junction have been recorded by means of the common modulation technique /6/.

2. Magnetic Quantization of Surface and Bulk States

Figure 2 shows the influence of a high perpendicular magnetic field on the second derivative characteristic $d^2I/dV^2(V)$ of an Mg-SiO_2-Si junction. In the absence of magnetic field the recording exhibits two dips at biases V_n, which reflect the energies of the lowest subbands E_0 and $E_{0'}$. From the position of the dip caused by E_0, V_0=48 mV, we obtain the areal density of electrons in the ground subband, $N_0 \cong D_0(E_F-E_0)$. Using the theoretical value for the density of states D_0 /1/, the result is $N_0(V_0)=7.6\times10^{12}$ cm^{-2}. At B=15.6 T the curve is imposed by short-periodic oscillations, which arise from the Landau level structure of the ground subband /7/. Evidently no indication is found either of spin and valley splitting or of Landau quantization of the 0' subband. This is in accordance with the scattering rate at zero magnetic field as determined from mobility measurements on similar samples in a field-effect geometry. Typical results are $\mu_{eff}\cong700$ cm^2/Vs at $N_s\cong8\times10^{12}$ cm^{-2} (Mg electrode) and $\mu_{eff}\cong1200$ cm^2/Vs at $N_s\cong2.5\times10^{12}$ cm^{-2} (Ti electrode), T = 4.2 K.

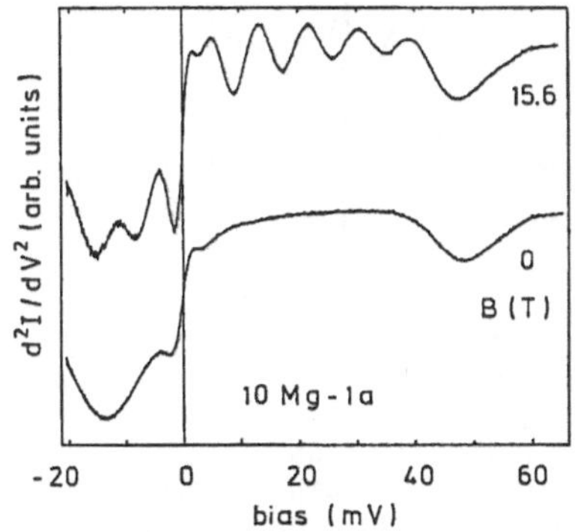

Fig. 2. d^2I/dV^2-vs-V characteristics of an Mg-SiO_2-Si junction in perpendicular magnetic fields B=15.6 T and B=0 at T=4.2 K. The substrate bias is V_s=-30 V.

Since the oscillations are nearly sinusoidal we assume, that the oscillatory component of the tunneling conductance, which is proportional to the density of states in the lowest subband /8/, may be described by /9/

$$\left(\frac{dI}{dV}\right)_{osc}=\left(\frac{2\pi^2kT}{\hbar\omega_c}\right)\mathrm{csch}\left(\frac{2\pi^2kT}{\hbar\omega_c}\right)\exp\left(-\frac{2\pi^2kT_D}{\hbar\omega_c}\right)\cos\left(2\pi\,\frac{E_F-E_0(V)-eV}{\hbar\omega_c}+\varphi\right) \qquad (1)$$

where $\hbar\omega_c$ is the Landau splitting and T_D is the Dingle temperature. According to (1) we have determined the cyclotron effective mass m_c from the period of the oscillations. The result is m_c/m_0=0.22±0.01, where the weak bias voltage dependence of $E_F-E_0(V)$ has been taken into account. This is clearly enhanced from the bulk value of $m/m_0=0.19$ as it has been found also in other investigations (e.g. /10-12/).

Another consequence of (1) is, that in a magnetic field sweep oscillations occur that are periodic in 1/B. Their period $\Delta(1/B)$ is proportional to the number of states N_k per unit area in the real space, which are enclosed by the constant energy contour $E=E_F-eV-E_0(V)$. The dependence of N_k on bias V respectively on energy E is displayed in Fig. 3. This graph is an alternative representation of the dispersion relation $k^2(E)$ /13/, because the ground subband is isotropic. It should be mentioned, that a constant density of states in the subband is established by the straight line up to energies of 65 meV above the band edge, which contradicts the assumption of nonparabolicity /12/. The density of states, which is obtained from the slope of the line, exactly corresponds to the cyclotron mass given above.

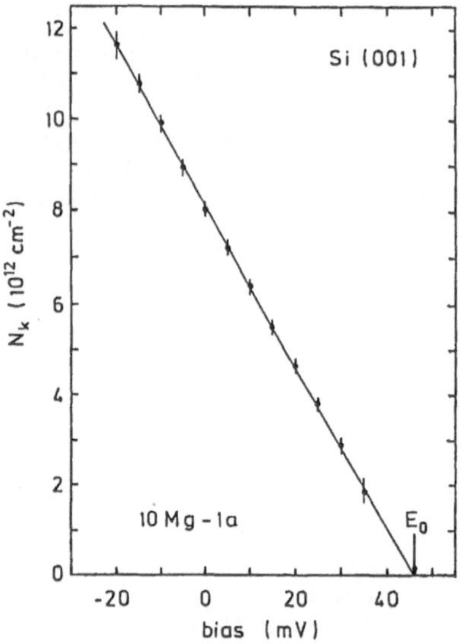

Fig. 3. Areal density of states in the ground subband up to an energy E that lies by -eV above the Fermi level (T=4.2K)

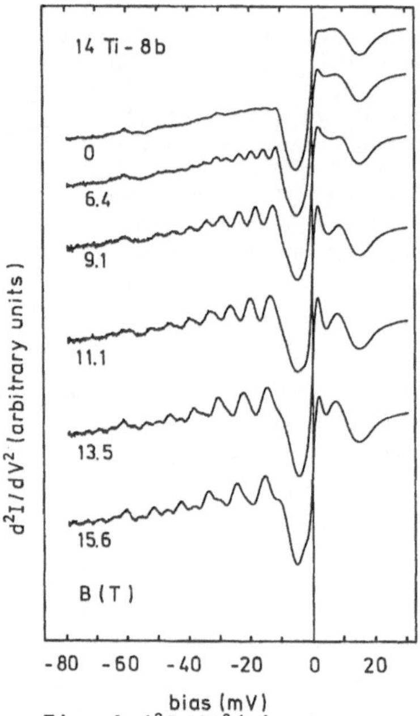

Fig. 4 $d^2I/dV^2(V)$ of an accumulation layer sample with Ti as counterelectrode under different perpendicular magnetic fields

The Dingle temperature T_D has been determined from the magnetic field dependence of the oscillation amplitude. At V=0 the corresponding scattering time is roughly in agreement with the mobility data, at $|V|>0$ the broadening increases due to additional relaxation processes as, e.g., phonon emission /6/

In measurements on accumulation layer samples a second period shows up in the reverse bias region (Fig. 4). By tilting the magnetic field from the normal into the [110] direction, we can exclude spin splitting or the Landau level structure of the 0' subband as origin. In contrast to spin splitting, which depends on the total magnetic field, and to 0' Landau splitting, which is determined only by the perpendicular component of B, the period is described by the cyclotron mass of bulk Landau levels arising from the four "primed" valleys. Compared with the other two valleys, these valleys are strongly preferred in tunneling due to their larger density of states and lower effective mass for the motion parallel and perpendicular to the interface, respectively /14/. However, a bulk orbital motion is only possible, because the states in accumulation layers are nearly extended at energies that lie a few tens of meV above the Fermi level, where the surface field is screened by the electron layer and the depletion field is vanishingly small.

3. Diamagnetic Effect in Parallel Magnetic Fields

Bulk Landau levels that are observed in accumulation layers under parallel magnetic fields are the minimum energies of the magnetic surface states. Since these surface states have a continuous spectrum of energies /5/ no additional structure occurs in the tunneling characteristics. When the depletion field increases the magnetic levels convert into electric type as illustrated in Fig. 5, where the subband energies change from the Landau ladder at $N_{depl} \cong 0$ into nearly electric quantum levels that are slightly increased by the diamagnetic effect. The diamagnetic rise of the minimum energy of the n-th hybrid subband is given by the first-order perturbational result /1/

$$E_n(B)-E_n(0) = \frac{e^2B^2}{2\,m_x}(\langle z^2\rangle-\langle z\rangle^2)_n \qquad (2)$$

where $\vec{B}=(0,B,0)$ and $(\langle z^2\rangle-\langle z\rangle^2)^{1/2}$ is the spread of the corresponding wave function perpendicular to the interface in the absence of magnetic field. We

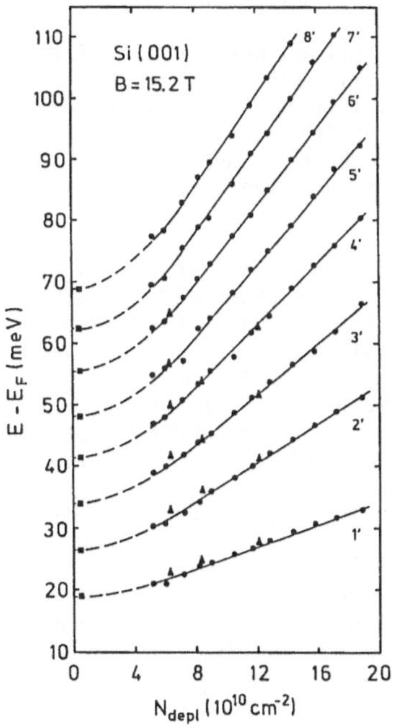

Fig. 5. Minimum energies of hybrid electric and magnetic subbands as a function of depletion charge density. The constant magnetic field is applied parallel to the [110] direction (T=4.2 K).

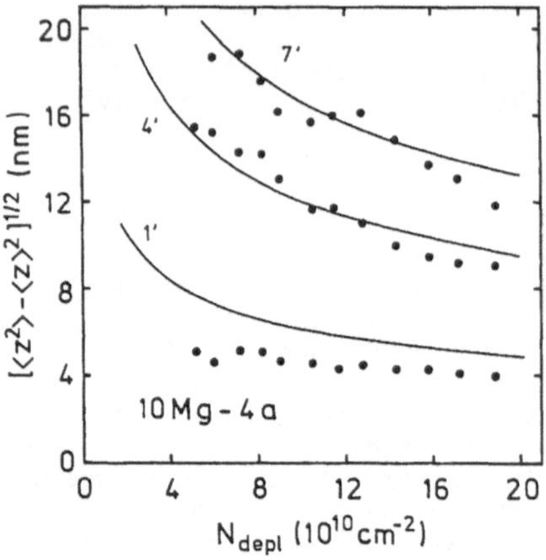

Fig. 6. Spread of the wave function of three different subband levels as a function of depletion charge density. Solid lines are calculated from the triangular potential model.

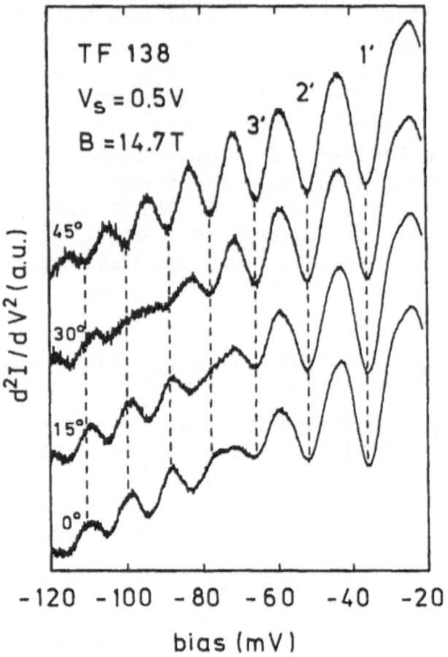

Fig. 7. $d^2I/dV^2(V)$ of a Ti-metallized junction under constant magnetic field, applied at various angles ϕ as given in Fig. 1. V_s denotes the substrate voltage (T=4.2 K)

use the proportionality of the energy shift to B^2 /15/ as an indicator, that (2) is applicable, i.e., that the subband energy is mainly determined by the electric field. Figure 6 shows the spread of some subband wave functions as a function of the depletion charge density. The experimental result is reasonably described by the solid lines calculated from the triangular-well model with a constant electric field given by the depletion field $F_{depl}=eN_{depl}/\varepsilon_{Si}$. Evidently the less extended 1' subband is more tightly bound than given by the depletion field alone, which reflects the influence of the surface field.

According to (2) the diamagnetic rise depends on the effective mass m_x perpendicular to the magnetic field direction. Hence the primed levels are only degenerate for B || [110] (see Fig. 1), whereas other field directions $\phi < 45°$ lift the degeneracy. This effect is demonstrated by the measurements shown in Fig. 7 /15/, where a rapid phase change occurs in the nearly sinusoidal oscillations in $d^2I/dV^2(V)$ due to the subband ladder, whenever the splitting between subband levels of equal index n exceeds the energy separation of adjacent levels of indices n and n+1 from different valley pairs.

At high depletion fields where the subband energies are raised far above the Fermi level, the spread of the wave functions $\delta_n=(\langle z^2\rangle-\langle z\rangle^2)_n^{1/2}$ shows a distinct deviation from the results of the triangular potential model. In Fig. 8 the ratio of the experimentally determined spread to that of the triangular approximation has been plotted as a function of the subband energies at B=0. Within the indicated error bars, the data points taken from various subbands at four different depletion fields obey roughly the same energy de-

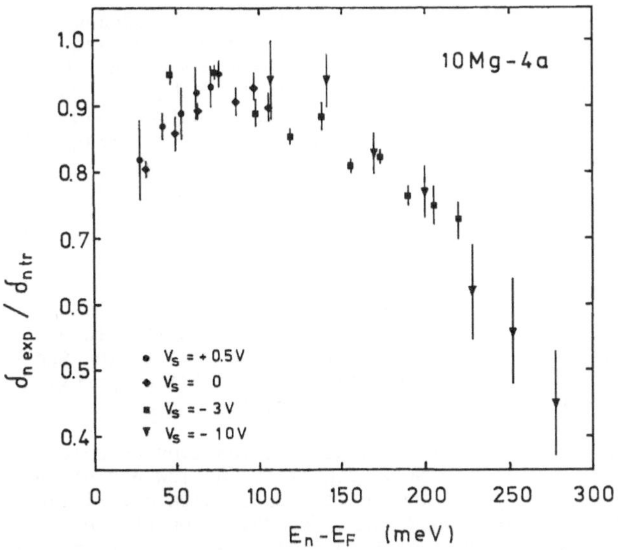

Fig. 8. Experimentally determined spread of the wave function of higher primed subbands, normalized by the result of the triangular well model, as a function of the energy position of the corresponding subband level

pendence, indicating a peculiarity of the surface potential well. The decrease at energies $E_n - E_F < 80$ meV can be attributed to the self-consistent potential of the surface electrons. The decrease at high energies, which is opposite to that expected from the curvature of the depletion potential, possibly reflects an increase of the effective mass m_x due to nonparabolicity or the penetration of the wave function into the barrier.

4. Effective Mass Change in Parallel Magnetic Fields

A careful measurement at high parallel fields reveals a small rise of the subband level E_0 proportional to B^2 as shown in Fig. 9. This decrease of $E_F - E_0$ cannot be explained by the diamagnetic shift of the ground subband /16/, although it is in the same order of magnitude. The reason is, that the dia-

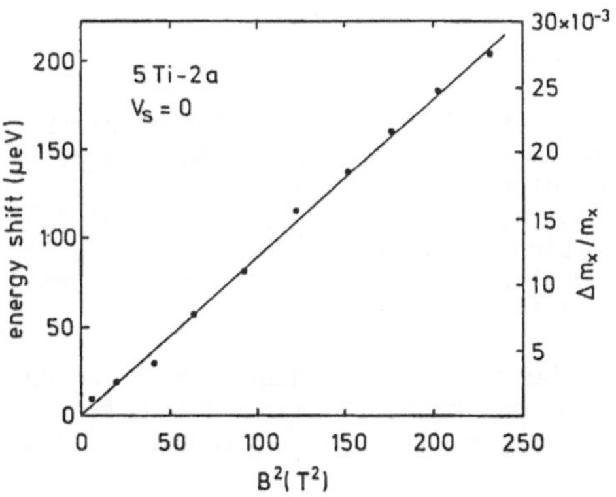

Fig. 9. Energy shift $\Delta(E_F - E_0)$ respectively effective mass change $\Delta m_x/m_x$ of the ground subband as a function of a parallel magnetic field

magnetic shift (2) is a magnetic field induced correction of the subband energy measured from the bottom of the potential well, whereas tunneling spectroscopy measures the energy relative to the Fermi level. Thus the change in $E_F-E_0 = N_0/D_0$ reflects an increase of the density of states D_0, because the electron density N_0 can be regarded as unaffected by the magnetic field. From the change in $D_0=2(m_x m_y)^{1/2}/(\pi\hbar^2)$ we get the relative change $\Delta m_x/m_x$ (Fig. 9), which can be expressed by the second-order perturbational result /2/

$$\frac{m_x(B)}{m_x(0)} = 1 + \frac{2e^2B^2}{m_x E_{10}} (\langle z^2\rangle - \langle z\rangle^2)_0 \tag{3}$$

where the subband separation $E_{10} = E_1 - E_0$ can be read from the tunneling characteristics. A second measurement directly gives the density-of-states increase of the ground subband: The contribution of this subband to the tunneling conductance is found to change proportional to B^2, yielding approximately the same values for $\Delta m_x/m_x$ as from the energy shift.

Figure 10 shows the spread of the ground subband wave function as obtained from (3) measured at different depletion charge densities. At low depletion charge density the data points in Fig. 10 are in accordance with a variational result /1/, which is indicated by the dashed line. The deviation of the experimental results from theory at high depletion fields may arise from a too slow decay of the variational wave function at large values of z, which varies as exp(-z) compared with $\exp(-z^{3/2})$ for the Airy function.

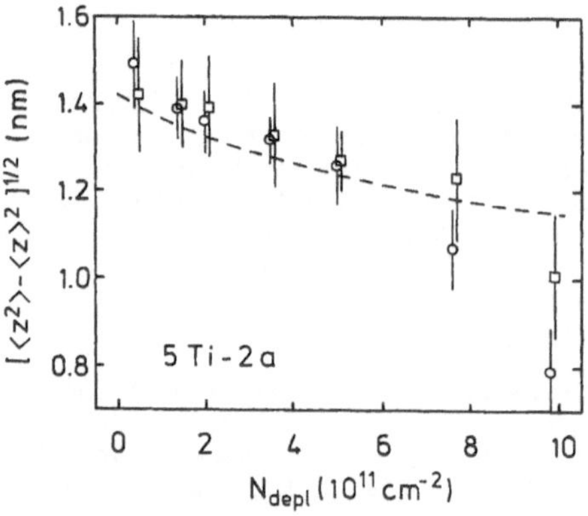

Fig. 10. Spread of the ground state wave function as a function of depletion charge density. Circles are obtained from the variation of $E_F-E_0(B)$ and squares are from the change of the contribution of the lowest subband to the tunneling conductance.

5. Spin Splitting in Parallel Magnetic Fields

A parallel magnetic field leads to a spin-Zeeman splitting of the electron states of $\Delta E=g^*\mu_B B$, where g^* is the effective g factor, μ_B the Bohr magneton and B the total magnetic induction. Theoretical investigations /17, 18/ predict an enhancement of the effective g factor from its bulk value $g^*\cong 2$, where at weak fields g^* is a smooth function of the electron density and at high perpendicular components of B g^* oscillates as a function of the Fermi energy position relative to the quantized levels.

A number of experimental studies have been performed under quantizing magnetic fields, where the effective g factor has been determined for different Fermi level positions, e.g. /19-21/. In the present work the magnetic field is applied parallel to the surface, thus it is the first attempt to measure the weak-field g factor.

The spin splitting of the ground subband splits the E_0 dip in $d^2I/dV^2(V)$ into two dips separated by $\Delta V=\Delta E/e$. Since here tunneling can be regarded as unsensitive to the electrons spin, the two dips are of the same shape as at zero magnetic field, except the slight increase of their amplitude due to the increase of the density of states. The superposition of the two dips result in one broadened dip of reduced amplitude. The magnitude of the spin splitting ΔV can be determined from the known lineshape and the reduction in amplitude, where the density-of-states effect has to be taken into account. Figure 11 shows the splitting obtained in this way from a Ti-metallized junction ($N_s \cong 2.4\times10^{12}cm^{-2}$) at different magnetic fields. The slope of the line corresponds to an effective g factor of $g^*=2.55\pm0.05$, which agrees with calculated values /18/.

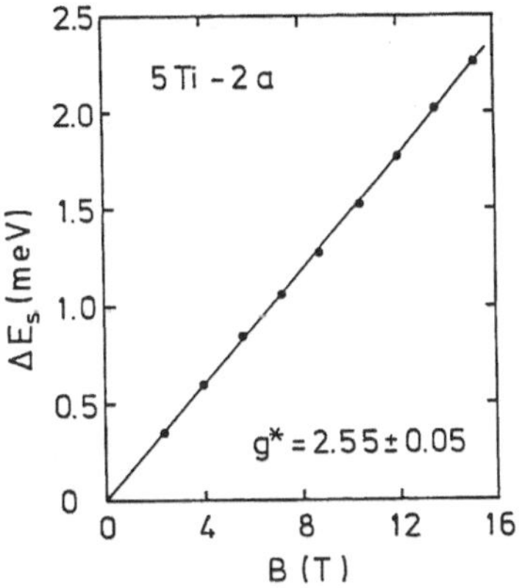

Fig. 11. Spin splitting of ground subband states as a function of parallel magnetic field. The solid line represents a fit to the experimental points

6. Concluding Remarks

This brief review demonstrates clearly, that tunneling spectroscopy has developed to a powerful method to study electric and magnetic quantum effects in Si inversion layers. However, considerable progress is still necessary in order to obtain tunnel junctions with high-mobility electron (or hole) surface layers, which are required for a deeper investigation of the energy spectrum of totally quantized space charge layers as, e.g., in the quantum Hall regime /22/.

Acknowledgement

The author is indebted to Prof. G. Lautz for valuable and stimulating discussions.

References

1. F. Stern, W.E. Howard: Phys. Rev. 163, 816 (1967)
2. F. Stern: Phys. Rev. Lett. 21, 1687 (1968)
3. T. Ando: J. Phys. Soc. Japan 39, 411 (1975)
4. U. Merkt: Phys. Rev. B 32, 6699 (1985)

5. R.E. Prange, T.W. Nee: Phys. Rev. 168, 779 (1968)
6. U. Kunze: J. Phys. C 17, 5677 (1984)
7. U. Kunze, G. Lautz: Surf. Sci. 142, 314 (1984)
8. D.J. BenDaniel, C.B. Duke: Phys. Rev. 160, 679 (1967)
9. L.M. Roth, P.N. Argyres: In Semiconductors and Semimetals, ed. by R.K. Willardson, A.C. Beer, Vol. 1 (Academic, New York 1966) p. 159
10. G. Abstreiter, J.P. Kotthaus, J.F. Koch: Phys. Rev. B 14, 2480 (1976)
11. A. Hartstein, F.F. Fang: Phys. Rev. B 18, 5502 (1978)
12. E. Batke, D. Heitmann: Solid State Commun. 47, 819 (1983)
13. G.M. Min'kov, V.V. Kruzhaef: Sov. Phys. Solid State 22, 959 (1980)
14. D.J. BenDaniel, C.B. Duke: Phys. Rev. 152, 683 (1966)
15. U. Kunze: Surf. Sci. 170, 353 (1986)
16. D.C. Tsui: Solid State Commun. 9, 1789 (1971)
17. T. Ando, Y. Uemura: J. Phys. Soc. Japan 37, 1044 (1974)
18. T.K. Lee, C.S. Ting, J.J. Quinn: Solid State Commun. 16, 1309 (1975)
19. F.F. Fang, P.J. Stiles: Phys. Rev. 174, 823 (1968)
20. H. Köhler, M. Roos: phys. stat. sol. (b) 95, 107 (1979)
21. Th. Englert, K. von Klitzing, R.J. Nicholas, G. Landwehr, G. Dorda, M. Pepper: phys. stat. sol. (b) 99, 237 (1980)
22. K. von Klitzing, Physica 126 B, 242 (1984)

Germanium Bicrystals in High Magnetic Fields

S. Uchida[1], *G. Reményi*[2], *and G. Landwehr*[2]

[1]Department of Applied Physics, University of Tokyo,
Hongo 7-3-1, Bunkyo-ku, Tokyo 113, Japan
[2]Physikalisches Institut, Universität Würzburg,
D-8700 Würzburg, Fed. Rep. of Germany

1. INTRODUCTION

An almost temperature-independent resistance was found more than twenty years ago in p-type space charge layers adjacent to the grain boundary of medium angle germanium bicrystals/1,2/. The bicrystals can be grown in a controlled fashion by pulling crystals in which the two halves are tilted by an angle Θ with respect to each other. This is schematically shown in Fig.1. In this configuration a grain boundary arises which is stabilized by a regularly spaced array of edge dislocations. In germanium bicrystals derived from (100) seed crystals the edge dislocations have dangling bonds which act as a line defect with accepter character. For tilt angles $\Theta > 8^\circ$ the space charge cylinders of adjacent dislocations overlap so much that narrow, rather uniform p-type layer exists with a hole concentration typically of the order of $5 \times 10^{12} cm^{-2}$. It can be expected that surface quantization is present in p-type inversion layers similar to silicon MOSFETs.

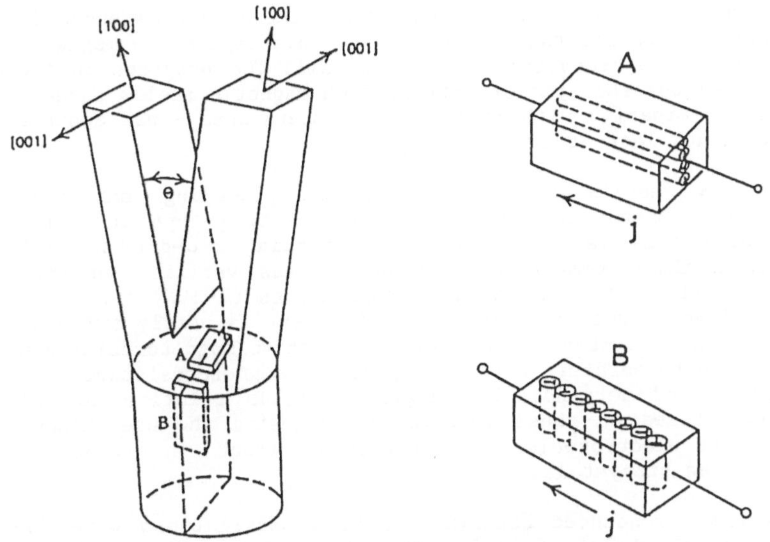

Fig.1: Schematic drawing of a germanium bicrystal, with two samples A and B of different orientation

It was shown already about 20 years ago by Landwehr and Handler/2/ that the magneto-transport properties of p-type inversion layers in Ge-bicrystals were anomalous. The conductivity was almost temperature independent below 20K as expected for a degenerate hole gas. However, the observed change in resistance in transverse and longitudinal magnetic-fields was by orders of magnitude larger at low temperatures than predicted by conventional theories. Eventually the experiments were terminated because no satisfactory explanation of the data was available in 1960's.

In the late 70's the resistivity of p-type space charge layers in Ge-bicrystals with tilt angles between 7° and 30° was systematically studiedby Vul and Zavaritskaya/3/. They observed an activated resistivity for specimens with $\Theta < 10°$ and an almost temperature independent one for larger tilt angles. The boundary between the two regimes is around a conductivity of about $5\times10^{-5}\Omega^{-1}$ not too far from the predicted minimum metallic conductivity for two-dimensional systems. They also found that there was a difference in the measured resistivity, depending on the current direction relative to the dislocations. In specimens (type B) in which the current was directed perpendicular to the dislocations the resistivity was larger than in specimens in which current and dislocations were parallel (type A). The anisotropy of the resistivity decreased with increasing tilt angle.

The situation changed in the 1980's when detailed theories for transport in two-dimensional disordered systems were developed. The scaling theory of Anderson localization/4/ predicted a logarithmic increase of resistance with decreasing temperature and a negative magnetoresistance /5/. Subsequently it was shown by Altshuler et al./6/ and by Fukuyama /7/ that the electron-electron interaction in the presence of impurity scattering can also cause a logT resistivity with a characteristic positive magneto-resistance /8/. Under these circumstances it seemed appropriate to resume the studies and to investigate the magneto-transport effects in Ge-bicrystals in more detail. The progress in the experimental techniques the availability of high magnetic fields exceeding 20 Tesla in combination with millikelvin temperatures - was another impetus to have a new look at the problem.

Our preliminary results were reported in the preceding conference (Grenoble, 1982). It was clearly demonstrated that p-type inversion layers in Ge-bicrystals are indeed an interesting two-dimensional system. A logarithmic temperature dependence was verified in both resistivity and Hall coefficient at low temperatures (T<10K) /9/. The slope of the logT-dependence of the Hall coefficient is nearly twice as large as that of the resistance. Whereas no negative magnetoresistance was observed, large temperature dependent positive magnetoresistance was found in parallel and perpendicular magnetic fields relative to the grain boundary. Based on these results we suggested the predominant contribution of the electron-electron interaction effects in p-type inversion layers of Ge-bicrystals.

In the meantime pronounced Shubnikov-de Haas oscillations were observed on particular samples (Θ=15°). This stimulated subband calculations by Bangert. It was expected that two-dimensional quantization due to the small thickness of the conducting layer of less than 100Å results in electrical subbands. The calculation showed that two subbands are occupied which can be attributed to light and heavy holes as in p-type bulk germanium /10/.

In the present paper emphasis is on the observations of the extraordinary transport properties in Ge-bicrystals. They are not in complete accordance with the current theories of two-dimensional systems. In particular unexpected results have been found in the millikelvin temperature range, part of which were reported briefly at the 2D-Conference at Oxford (1983)/11/.

2. QUANTUM OSCILLATIONS AND SUBBAND STRUCTURE

A series of bicrystals was grown at the Max-Planck Institut für Festkorperforschung in Stuttgart from (001) seed crystals which were tilted along a (100) axis. The bulk crystals were doped with Sb up to a doping level of $6 \times 10^{15}\,cm^{-3}$. Alloying small indium pills to the grain boundary area produces Ohmic contacts to the p-type space charge layer and rectifying contacts to the n-type bulk. The transport experiments were performed in the temperature range from 50mK to 4.2K with a superconducting coil generating fields up to 12.5 Tesla or between 1.5 and 4.2K in fields up to 23 Tesla generated with a polyhelix-Bitter type magnet at the MPI Hochfeld-Magnetlabor Grenoble. It was possible to rotate the samples relative to the magnetic field.

Typical magnetoresistance data are shown in Fig.2 for both A and B type samples with $\Theta=10^\circ$(indicated by the notation $10^\circ_{\parallel}$ and $10^\circ_{\perp}$, respectively). Except for the 15 specimens the magnetoresistance be-

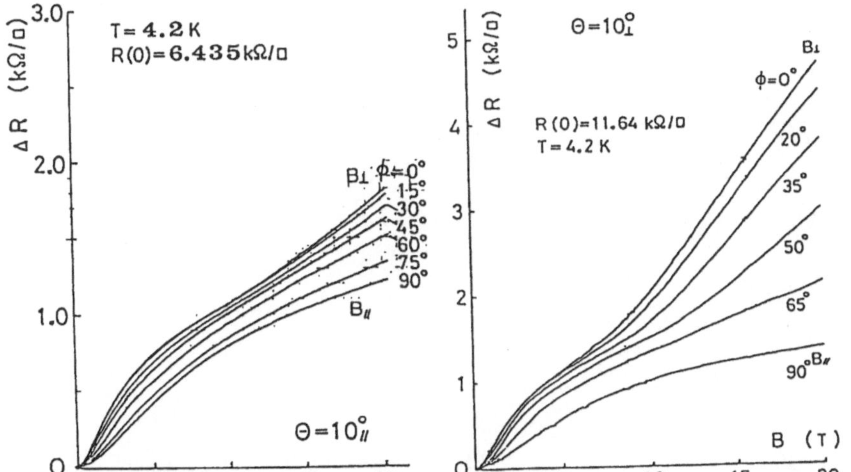

Fig.2: Magnetoresistance as a function of magnetic field with the angle between grain boundary normal and field as parameter for a sample of type A (left) and B (right)

haved-qualitatively although not quantitatively-alike for $\Theta>10°$. It is evident that there is a substantial magnetoresistance even when the field is oriented parallel to the grain boundary ($\phi=90°$). Also the positive magnetoresistance in the perpendicular fields ($\phi=0°$) is quite large in particular for B-type samples. In both field configurations the magnetoresistance is found to have a logarithmic functional form in a rather wide range of the magnetic fields and it increases in magnitude with decreasing temperature. This kind of behavior suggests that Zeeman effects in the weakly localized regime are important /8/.

The magnetoresistance data for the 15° samples differ from the rest. The concuctivity of this particular bicrystal is unusually high and very pronounced Shubnikov de Haas oscillations were observed beyond 4 Tesla for both $15°_{//}$ and $15°_{\perp}$ specimens as shown in Fig.3. In the figure the Hallresistance is also plotted and shows an oscillatory behavior. The period of the oscillation is almost identical for both types of samples and the dominant oscillations turn out to consist of a single period corresponding to the hole density of $3\times10^{12}\,cm^{-2}$. This density is substantially smaller than that derived from the low field Hall coefficient, $1\times10^{13}\,cm^{-2}$. By incorporating the subband calculation /10,12/, we can conclude that the holes reside in the ground heavy-hole subband as well as in the light-hole subband. This view is supported by the amplitude moduration of the SdH-oscillations, which can be explained by the presence of a second period. The calculations also predict a heavy-hole mass of $0.39m_0$. The light-hole mass is enhanced significantly over the bulk value $0.04m_0$ due to the existence of a strong electric field at the interface which causes considerable band mixing. The light-hole mass rises from $0.25m_0$ at a total hole density of $4\times10^{12}\,cm^{-2}$ to about $0.37m_0$ at $1\times10^{13}\,cm^{-2}$. The result is in reasonable agreement with the effective mass of $0.34m_0$ which was derived from the temperature dependence of the amplitude of the oscillations.

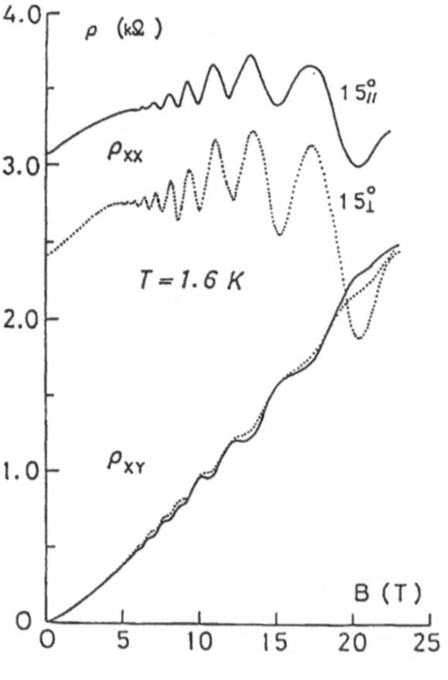

Fig.3: Resistivity (ρ_{xx}) and Hall resistance (ρ_{xy})at 1.6K as a function of transverse magnetic field for a sample of type A ($\Theta=15°_{//}$) and type B ($\Theta=15°_{\perp}$)

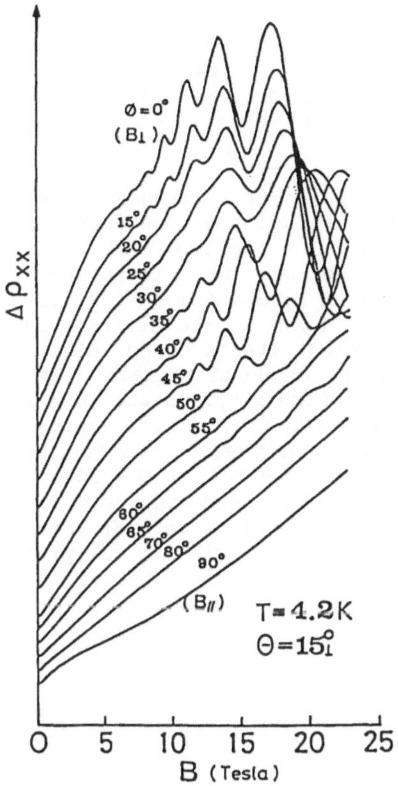

Fig.4 Increase of resistivity as a function of magnetic field with the angle ϕ between field and the normal to the grain as a parameter for a $15^{\circ}_{\perp}$ sample

A quite peculiar feature shows up in the angular dependence of the magnetoresistence as shown in Fig.4. If the angle between the magnetic field and the normal to the grain boundary is increased from 0°to 75°one observes an interference effect in the magnitude and the periodicity of the oscillations. For $\phi<30^{\circ}$ the parallel component of the magnetic depends does not affect the Landau quantization. The maxima and minima of the oscillations appear at constant values of the normal component of the field. However, for angles $\phi>30^{\circ}$the amplitude is significantly reduced and the periodicity of the oscillation is markedly disturbed. Normally this can be understood if one assumes that the landau level splitting depends only on the normal component of the magnetic field whereas the spin splitting depend on the total magnetic field. Thus, whenever the spin splitting is larger than the Landau level splitting, the periodicity and amplitude of oscillations are expected to be disturbed. If this is origin for the observed interference effect in the 15° samples, then one must presume quite a large g-factor for the holes responsible for the oscillations.

3. WEAK LOCALIZATION

During the last few years much attention has been paid to the investigation of the low temperature properties of two-dimensional electronic systems /13,14/. There has been a great deal of the theoretical progress in the understanding of the localization of electrons in disordered systems and experiments have verified most of the theoretical predictions, notably for the two-dimensional system in the weakly localized regime where a comparison between theory and experiment is possible.

We have investigated samples from Ge-bicrystals with tilt angles between 7° and 25° . A logarithmic increase of resistance with decreasing temperature was found for samples with $\Theta > 8°$. The low field Hall coefficient also increases logarithmically in the same temperature range. The magneto-resistance is positive and a negative magnetoresistance was never found except under special conditions in the millikelvin range /11/. All this indicates that localization effects do not play a significant role in the p-type inversion layers of Ge-bicrystals. The physical reason for thisshould be sought in the strong spin-orbit interaction in p-type germanium. It has turned out that the magnetotransport properties show many of the features which are predicted by the interaction theory. For a particular sample ($\Theta = 10°$) the investigation of the conductivity and magnetoresistance in magnetic fields up to 20 Tesla in a wide temperature range allows a complete determination of the interaction parameters which have been introduced by Fukuyama /7/.

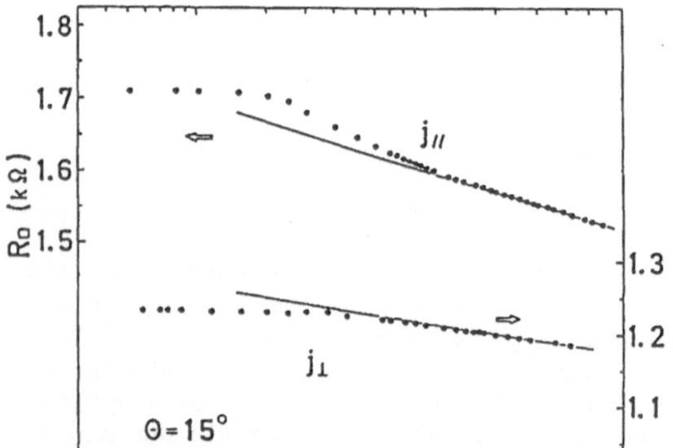

Fig.5: Resistivity as a function of temperature on a logarithmic scale for 15° samples with dislocations parallel (j//) and perpendicular ($j_\perp$) to the current

Here we show typical data on the bicrystals with $\Theta = 15°$ and discuss them in detail. They are the same samples for which the Shubnikov-de Haas oscillations were observed as shown in Figs.3 and 4. In Fig.5 we plot data of the zero-field resistivity over a wide temperature range from 50mK to 5K for two 15° samples which differ in the orientation of the dislocation lines relative to the current direction. In both samples the resistivity increases logarithmically with decreasing temperature down to about 1K, demonstrating effects characteristic for a disordered two-dimensional system. One can recognize in this figure that an anisotropy exists not only in the resistivity but also in the prefactor of its logT dependence. The magnitude of the prefactor is larger in the parallel specimen than that in the perpendicular specimen. This was also observed for other samples with different tilt angles.

When the resistivity measurements were extended to lower temperatures, it turned out that the logarithmic behavior did not continue in the mK range. The deviation from the logT dependence is remarkably different in the two types of samples. The resistivity levels off and saturates below 0.5K for the sample with perpendicular dislocations, whereas it increases more steeply for the sample with parallel dislocations. It was checked with great care that these effects are not due to heating nor miscalibration of the thermometer. External high frequency signals were effectively blocked off. The dissipated power due to the sample current was below 10^{-13} W and an increase of the power by about two orders of magnitude did not change the measured resistivity. A temperature independent resistivity in the mK range has also been observed in Si inversion layers but at much lower temperatures /15/. A very similar behavior in the mK range was also observed for 10° samples.

The magnetoresistance in fields perpendicular to the grain boundary has already been shown in Fig.3. With decreasing temperature the amplitude of the Shubnikov-de Haas oscillations increases and becomes saturated below 1K. From this the effective mass of holes was estimated /10/. On the other hand the magnetoresistance in parallel fields exhibits a very peculiar T-dependence. In the temperature range down to about 2K it scales rather well with B/T which indicates that it arises from spin Zeeman splitting /8/. Different from metallic thin films, orbital effects, including the effect of the spin-orbit interaction, are not expected in the present genuine two-dimensional system /16,17/.

Below 2K significant deviations from the B/T scaling show up which become more and more pronounced when the temperature is lowered further. Again deviations are observed in a quite different manner in the two types of samples. In the case of the type A, parallel sample a negative magneto-resistance shows up in the low temperature range, whereas the positive magnetoresistance seems to be enhanced with decreasing temperature in the case of the type B. A similar behavior has been found in samples with tilt angles 10° and 13°. Thus far it can be concluded that a negative magneto-resistance is observed only in the samples in which dislocations are oriented parallel to the current. These phenomena have never been observed in other two-dimensional systems, so their origin has probably to be sought in specific properties of the Ge-bicrystals as discussed in the following.

Linear Array of Edge Dislocations

Regularly spaced dislocations in the grain boundary should set up a superlattice potential which can interact with the free holes. A 'minigap' is expected to appear when some portion of the Fermi surface has contact with the mini-zone boundary. We suggested in a previous paper /18/ that open orbits could arise in high magnetic fields which might be responsible for the linear component of the magnetoresistance observed for the samples with perpendicular dislocations in transverse magnetic fields (see the data in Fig.2 for the 10° specimen). It is not clear, however, whether the anomalous low temperature properties such as the magnetoresistance in parallel fields are related to the presence of such a minigap or not.

As far as we know, there has been only one mechanism proposed which gives rise to a T-dependent negative magnetoresistance in magnetic fields parallel to two-dimensional layers. It was shown theoretically by Ohkawa et al. that the interplay of the spin scattering and weak localization in 2D systems leads to a 1/T divergence of the resistivity as well as to a negative magnetoresistance /19,20/. This implies that the ordinary logT divergence in the Kondo effect is enhanced by the normal impurity scattering. The results for the parallel specimens are in qualitative agreement with this theoretical prediction. It is unlikely that a substantial concentration of magnetic impurities is present at the grain boundary of our bicrystals, but one should keep in mind that the dangling bonds along the edge dislocations may have a high concentration and that a relatively small part of the dangling bonds becomes inactive due to the trapping of electrons from the bulk. We have not got any answer yet to the question why the 'Kondo' effect seems to be absent in the perpendicular specimens.

Two-Subband System

For the perpendicular samples the positive magnetoresistance increases with decreasing temperature in the mK range more than expected from the B/T- scaling. This is also observed for the parallel samples if one extracts the positive component from the data. It is possible to interpret this as an additional contribution from the holes in the second subband dominating at low temperatures. Actually a rather good quantitative fit with the experimental data shown in Fig.6 can be obtained as-

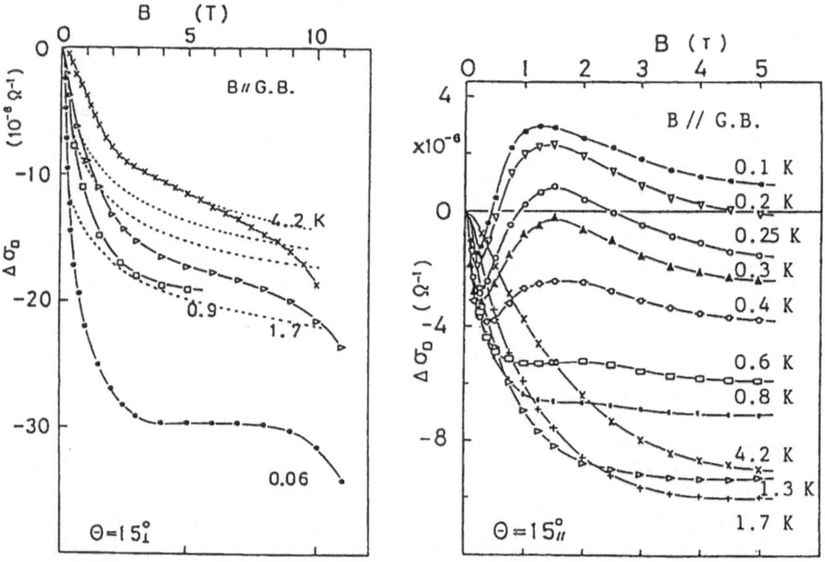

Fig.6: Change of conductivity as a function of magnetic field applied parallel to the grain boundary at various temperatures. The data are for samples with $\Theta = 15^\circ_\perp$(left) and $15^\circ_{/\!/}$ (right). The dashed curves indicate the contribution from one carrier with $g^* = 25$ and the solid curves are the guide for the eye

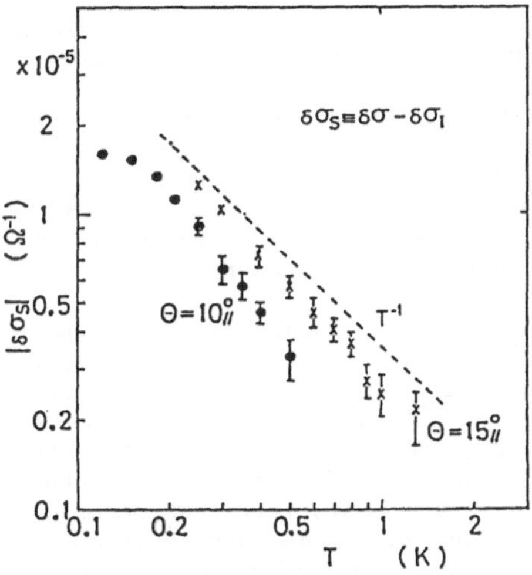

Fig.7: Plot of excess conductivity as a function of T on a log-log scale for two samples with 10°// and 15°// tilt angles. The values are estimated by subtracting the extrapolated conductivity from higher temperatures from the observed one

suming two types of carriers with very different g-factors, e.g., $g^* \simeq 25$ for one and $g^* \simeq 0.6$ for the other. The contribution from the carriers with smaller the g^* value becomes apparent only at low temperatures, because the Zeeman splitting is only significant when $g^* \mu_B B > k_B T$.

Here one should note that it is necessary to presume a large g-factor for one type of holes in order to interpret the positive etoresistance in parallel magnetic fields not only in this particular sample but also inothers of either conf figuration. This is compatible with the observed interference effect in the angular depend ence of the Shubnikov-de Haas oscillations.

So far the discussion has been based on a one-carrier or two-independent-carrier model. Scattering by impurities may lead to transitions between the heavy and light hole subbands. The effect of inter-subband scattering on the weak localization has briefly been investigated by Altshuler et al. /16/ and by Fukuyama /21/. Such phenomena as observed in Ge-bicrystals have not been predicted by either theory. A thorough investigation of the interband effects would be desirable.

Higher-Order Interaction

In the grain boundary region of bicrystals the interelectron potential is expected to be much stronger than at the interface of MOS structures where charges on a metal plate may produce a mirror image, making the interaction very short-ranged. Thus it is necessary to invoke higher-order contributions of the electron-electron interaction in the present case. Higher-order interaction effects have been studied by Fukuyama et al. /22/. They have shown that the interaction via diffusion channel is enhanced due to the higher-order effects whereas the interaction via Cooper channel is suppressed. This is in reasonable agreement with our experimental results. The estimated values of the interaction parameters for the diffusion channel are much larger than those for the Cooper channel /9,23/.

At about the same time, Finkelstein developed a scaling theory for a system of interacting electrons /24/. He arrived at the same conclusions as Fukuyama et al. The most important consequence of Finkelstein's theoryis the prediction that not only the resistance but also the interaction parameters, are scale dependent. The magnitude of the interaction parameter in the diffusion channel increases with decreasing temperature and as a result the resistance exhibits a leveling-off with decreasing temperature and even a fall-off at the lowest temperatures. An enhancement of the g-factor also results from the theory.

It is certain that these theories are still inadequate for a quantitative comparison with the present experimental results. For example, they would be inapplicable if there were two types of carriers present and the transitions between the two subbands were important. Nevertheless, Ge-bicrystals are one of the promising systems to test the predictions of recent interaction theories. Zharikov has reported that a maximum in the temperature dependence of the resistivity was observed in a bicrystal with $\Theta = 16°$/25/. Although our specimens have never showed a resistivity maximum, the saturation of the resistivity at not too low temperatures and the enhanced value of the g-factor seem to be in qualitative agreement with the theory.

4. CONCLUSION

The study of the magneto-transport properties of Ge-bicrystals has clearly demonstrated that p-type inversion layers adjacent to grain boundaries are a two-dimensional electronic system with disorder. The system is not simpleas a consequence of the complicated valence band structure of Ge and of the presence of a linear dislocation array at the grain boundaries. The band calculations have shown that under usual circumstances two subbands are occupied which can be attributed to light and heavy holes. The dislocation array should set up a superlattice potential which will make the system anisotropic. In spite of these complications the magneto-transport properties show many of the features predicted by the interaction theory in the weakly localized regime. In the mK range some of the properties are quite unusual and cannot be properly understood even in the framework of the most recently developed theory.

ACKNOWLEDGMENTS

Most of the experiments reported here were done at the Max-Planck-Institut fur Festkorperforschung, Hochfeldmagnetlabor Grenoble. The authors would like to thank Mr. A.Kohler for growing the bicrystals and Prof. H.Fukuyama for stimulating discussion. The support by Stiftung Volkswagenwerk is gratefuly acknowledged.

REFERENCES

1. See e.g., H.F.Matare: Defect Electronics in Semiconductors, Wiley Inter science (1971).
2. G.Landwehr and P.Handler: J.Phys.Chem.Solids 23, 891 (1962).
3. B.M.Vul and E.I.Zavaritskaya: Sov.Phys.JETP 49, 551 (1979).
4. E.Abrahams, P.W.Anderson, D.C.Licciardello and T.V.Ramakrishnan; Phys.Rev.Lett.42, 673 (1979).

5. S.Hikami, A.I.Larkin and Y.Nagaoka: Prog.Theor.Phys. 63, 707 (1980).
6. B.L.Altshuler, A.G.Aronov and P.A.Lee: Phys.Rev.Lett. 44, 1288 (1980).
7. H.Fukuyama: J.Phys.Soc.Jpn. 49, 644 (1980), ibid. 50, 3407 (1981).
8. A.Kawabata: J.Phys.Soc.Jpn. 50, 2461 (1981).
9. S.Uchida and G.Landwehr: In Application of High Magnetic Fields in Semi-conductor Physics, ed. by G.Landwehr, Springer Lecture Notes in Physics 177, 65 (Springer, Berlin, Heidelberg 1983).
10. S.Uchida, G.Landwehr and E.Bangert: Solid State Commun. 45, 869 (1983).
11. G.Remenyi, S.Uchida, G.Landwehr, A.Briggs and E.Bangert: Surf.Sci. 142,43 (1984).
12. G.Landwehr, E.Bangert and S.Uchida: Solid State Electronics 28,171(1985).
13. T.Ando, A.B.Fowler and F.Stern: Rev.Mod.Phys. 54, 437 (1982).
14. See e.g. Anderson Localization, ed. by Y.Nagaoka and H.Fukuyama, Springer Solid State Sciences 39 (1982).
15. e.g., M.J.Uren, R.A.Davies, M.Kaveh and M.Pepper, J.Phys.C14 5737 (1981).
16. S.Maekawa and H.Fukuyama: J.Phys.Soc.Jpn. 50, 2516 (1981).
17. B.L.Altshuler, A.G.Aronov, D.E.Khmelnitzkii and A.I.Larkin: Sov.Phys. JETP 54, 411 (1981).
18. G.Landwehr and S.Uchida: Surf.Sci. 170, 719 (1986).
19. F.J.Ohkawa, H.Fukuyama and K.Yoshida: J.Phys.Soc.Jpn. 52, 1701 (1983).
20. F.J.Ohkawa and H.Fukuyama: J.Phys.Soc.Jpn 53, 2640 (1984).
21. private communications
22. H.Fukuyama, Y.Isawa and H.Yasuhara: J.Phys.Soc.Jpn. 52, 16 (1983) and Y.Isawa and H.Fukuyama:J.Phys.Soc.Jpn. 53, 1415 (1984).
23. G.Landwehr and S.Uchida: In Localization and Metal Insulator Transitions, ed. by D.Adler and H.Fritzsche (Plenum, New York 1985) p. 379.
24. A.M.Finkelstein: Sov.Phys.JETP 57, 97 (1983) and Z.Phys.B56, 189 (1984).
25. O.V.Zharikov: Sov.Phys.JETP Lett. 38, 126 (1983).

Part V

Metal-Insulator Transition

The Metal-Insulator Transition in n-Type Silicon in High Magnetic Fields

T.G. Castner[1], *W.N. Shafarman*[1], *J.S. Brooks*[2], *K.P. Martin*[2], *and M.J. Naughton*[2]

[1]University of Rochester, Rochester, NY 14627, USA
[2]Boston University, Boston, MA 02215, USA

1. Introduction

The study of the metal-insulator (MI) transition in doped semiconductors in recent years has focused on the scaling behavior of the DC conductivity [1,2], the static dielectric 'constant' [3,4], and more recently the critical behavior of the spin susceptibility [5] and Mott variable range hopping (VRH) conduction [6]. Unlike the 2d inversion layer systems, where one can continuously vary the carrier density by changing a gate voltage, there is no simple way to continuously vary the impurity or carrier density for a bulk 3d system. For these systems experimentalists have employed uniaxial stress or magnetic fields to tune the critical density N_c and thereby vary $(N-N_c)$ and in many cases to successfully tune a sample completely through the transition. Uniaxial stress experiments have been particularly important and useful for the study of the n-type many-valley semiconductors Ge [7] and Si [2,4] because the stress removes the degeneracy of the conduction band minima, changes the screening, and thereby alters N_c [8]. PAALANEN et.al. [2,4] have utilized uniaxial stress to accurately determine the scaling behavior of the DC conductivity σ_{DC} and the 'static' dielectric response $\varepsilon'(N)$ for Si:P as $T\to 0K$. Since the classic YAFET, KEYES, and ADAMS [9] calculation demonstrating the donor wave function shrinkage in large static magnetic fields there have been many studies of InSb [10-13] demonstrating the tuning of metallic samples to insulating behavior with some threshold field H_c dependent on $(N-N_c)$. More recently there have been successful magnetic tuning studies on InP [14,15] and the interesting magnetic semiconductor $Gd_{3-x}V_xS_4$ [16]. The latter case, studied by VON MOLNAR et.al. [16], has shown $\sigma_{DC} \propto (H-H_c)$ as $T\to 0K$ with H_c a function of x and has been interpreted in terms of the magnetic polaron. There has also been substantial interest in $Hg_{1-x}Cd_xTe$ and other II-VI compounds, in addition to the enormous interest in 2d-inversion layers and the quantum Hall effect. However, this paper will be limited primarily to the high-field behavior of bulk 3d n-type Si and more specifically to the critical concentration range $0.8N_c<N<1.2N_c$ where the localization (correlation) length is much larger than the isolated Bohr radius (mean-free-path). The experimental results discussed herein give the first definitive, recognized evidence for the magnetic tuning of N_c and the localization length exponent ν with field H for bulk n-type Si.

The critical parameters determining the magnetic field regime, weak or strong, are the magnetic length $\lambda = (\hbar c/eH)^{1/2}$ and the donor Bohr radius a_D^*. For InSb $a^*\sim 600[\text{Å}]$ and it is easy to achieve the strong field limit $(\lambda \ll a^*)$ in modest magnetic fields since $\lambda=81[\text{Å}]$ at $H=10[T]$. It is also possible to achieve the strong field limit for InP and GaAs. For n-type Si, however, a_D^* is 16.7 and 15.4[Å] for P and As donors respectively and any hope of observing magnetic-tuning effects depends on studying samples very close to the MI transition, namely in a narrow range near N_c. The zero-field localization length is given by

$$\xi(N)=\xi_0|1-N/N_c|^{-\nu}=\xi_0\varepsilon^{-\nu}, \tag{1}$$

where the prefactor $\xi_0(N<N_c)$ is of the order of the donor Bohr radius a_D^*. The scaling theory of ABRAHAMS et.al. [17] yields $\sigma_{DC}\propto e^2/h\xi \propto (N/N_c-1)^{\nu}$ with $\nu \sim 1$ while the Si:P results [1,2] yield $\nu=0.51+0.05$ and Si:As results [18, 19] yield $\nu=0.6+0.1$. The Si:P dielectric 'constant' results [3,4] show $\varepsilon'(N)-\varepsilon_h=4\pi\chi'(N)\propto(1-N/N_c)^{-\zeta}$ and yield $\zeta=1.0$ demonstrating $\zeta=2\nu$. Until recently [20] there was no direct experimental evidence documenting the magnetic tuning of $\xi(N)$ in the critical regime $\varepsilon<<1$, although some magnetoresistance data [15] on some n-type semiconductors contained certain features similar to the Si:As results discussed below. The magnetic field dependence of $\xi(N,H)$ would be expected to arise from the field dependence of the parameters ξ_0, N_c, and ν, although it is necessary to recognize that a magnetic field introduces a new characteristic length into the problem — something that doesn't occur in the uniaxial stress case. For Si:As with $\xi_0 \sim a_D^*<<\lambda$ one expects from theory [21] only a very small change in ξ_0 with H. In the critical regime the important quantities varying with field are N_c and ν. The standard approach for obtaining the field dependence of $N_c(H)$ has been the use of a field-dependent Mott criterion [11-13], namely

$$N_c(H)a_{\perp}^2(H)a_{\parallel}(H)=N_c(0)a_D^*(a)^3 \sim (0.26)^3, \tag{2}$$

where $a_{\perp}^2(H)$ and $a_{\parallel}(H)$ have been given by YAFET et.al. [9]. This expression has been successful in explaining the magnetic tuning of $N_c(H)$ observed in InSb [10-13] and InP [14,15] studies, both of which involved reaching the high field limit ($\lambda<a_D^*$). In the weak field limit ($\lambda>a_D^*$) appropriate for Si:As, use of the MILLER-ABRAHAMS result [21] for $a_D^*(H)/a_D^*(0)$ yields $N_c(H)=N_c(0)[1+\eta H^2]$ with $\eta=6\times10^{-5}[T^{-2}]$ for Si with $a_{em}^*=20.2[\text{Å}]$. A smaller value of a_D^* for a specific donor will yield a smaller value of η since $\eta\propto a_D^{*3}$. Based on magnetoconductivity arguments in the critical regime, KHMEL'NITSKII and LARKIN [22] have found $N_c(H)=N_c(0)[1+(a^2/\lambda^2)^{\nu/2}]=N_c(0)[1+\text{const } H^{\nu/2}]$. SHAPIRO [23], in considering a magnetic phase diagram for $N_c(H)$, finds $N_c(H)$ first decreases with H based on the negative magnetoresistance result of KAWABATA [24]. The situation concerning the field dependence of $\nu(H)$ has received less consideration. HIKAMI [25], employing a nonlinear σ-model which neglects electron-electron interactions, has shown $\nu(0)=1$ (orthogonal symmetry — ordinary localization case [17]) and $\nu(H)=\frac{1}{2}$ (unitary symmetry). CASTELLANI et.al. [26] have considered the exponent ν for different universality classes and have also suggested, taking account of the Si:P experimental results [1, 2] $\nu(0)=0.5$, that $\nu(H)\rightarrow1$ in high fields. We are currently unaware of any theory that produces an actual field dependence of $\nu(H)$.

Three specific types of data will be considered, namely: 1) the magnetoresistivity ratio $\rho(N,H)/\rho(N,0)$ versus H as $T\rightarrow0K$ for barely metallic samples; 2) the low-temperature magnetocapacitance contribution of the donors for an insulating sample of the form $[\varepsilon'(N,H)-\varepsilon_h]/[\varepsilon'(N,0)-\varepsilon_h]$ versus H; and 3) the characterization of Mott VRH conduction at fixed magnetic fields for barely insulating samples, thus permitting a determination of the ratio $T_0(N,H)/T_0(N,0)$ where T_0 is the characteristic temperature in the Mott law $\sigma_{DC} = \sigma_0 \exp[(-T_0/T)^{1/4}]$. All three of the above results depend on the length ratio $\xi(N,H)/\xi(N,0)$ which is given by

$$\frac{\xi(N,H)}{\xi(N,0)} = \frac{\xi_0(H)}{\xi_0(0)}\frac{|1-N/N_c(H)|^{-\nu(H)}}{|1-N/N_c(0)|^{-\nu(0)}} . \tag{3}$$

Employing the weak-field result $N_c(H)=N_c(0)[1+\eta H^2]$ and $\varepsilon=|1-N/N_c(0)|$ one obtains for the ratio the result

$$\frac{\xi(N,H)}{\xi(N,0)} = \frac{\xi_0(H)}{\xi_0(0)}\left[\frac{1+\eta H^2}{1\pm\eta H^2/\varepsilon}\right]^{\nu(0)}\left[\frac{1+\eta H^2}{\varepsilon\pm\eta H^2}\right]^{\Delta\nu(H)}, \tag{4}$$

where $\nu(H)=\nu(0)+\Delta\nu(H)$ and the plus sign is for the insulator and the minus sign for the metal (assuming, of course, $\eta>0$). Neglecting the prefactor ratio, which is unimportant in the critical regime $\varepsilon<<1$, one observes the first term consists of a pure N_c-tuning contribution, while the second term consists of an exponent tuning contribution, but is only pure $\Delta\nu(H)$ tuning for $\eta=0$. The first term provides an antisymmetric change in the ratio decreasing for the insulator and increasing for the metal ($\eta H^2<\varepsilon$ required). The second term provides a symmetric increase (decrease) in the ratio for positive (negative) $\Delta\nu(H)$ for $\eta=0$. Note also that Eq. (4) yields the threshold field (for changing a metal to insulator) $H_c=(\varepsilon/\eta)^{\frac{1}{2}} \propto (N/N_c-1)^{\frac{1}{2}}$. This result is somewhat oversimplified, but is qualitatively correct. The experimental results for metallic and insulating samples will be analyzed in terms of the predictions of Eq. (4).

2. Experimental Considerations

The experimental procedures have already been discussed by SHAFARMAN et.al. [27], however several additional specific points should be made here. In Ref. 27 donor concentrations were determined from $\rho(RT)$ measurements employing a calibration curve obtained from $\rho(RT)$ data and neutron activation analysis by NEWMAN et.al. [28]. It was later determined there was a systematic discrepancy between our $\rho(RT)$ data and those at Cornell by comparing our $\rho(4.2K)/\rho(RT)$ ratios with those of NEWMAN and HOLCOMB [18,29]. This led to a 3 to 3.5% increase in our donor densities as discussed by SHAFARMAN and CASTNER [6]. The N_c value determined by NEWMAN and HOLCOMB [18] agrees with the values obtained from our results within experimental uncertainties. All Si:As samples were cut from the same Czochralski-grown ingot featuring an approximate linear As concentration gradient along the ingot axis. The acceptor concentration in this nominally uncompensated ingot has been stated to be $10^{14}/cm^3$ by the manufacturer (MACOM).

All magnetic field measurements shown are for the field transverse to the current (bar) axis. Longitudinal geometry ($\vec{H}||\vec{j}$) measurements gave virtually identical results in the low field range (H<4T), however a small anisotropy (approx. 10%) develops at high fields (H $\sim$ 19T). Preliminary Hall effect measurements by KOON [30] on similar samples (Van der Pauw geometry) show for $H||z$ that $\rho_{xy}/\rho_{xx} = \sigma_{xy}/\sigma_{xx}<<1$. A metallic sample ($N/N_c(0)\simeq1.13$) shows $\rho_{xy}/\rho_{xx}\simeq0.06$ at H=12[T]. At 1.1K, more dilute samples yield even smaller values of ρ_{xy}/ρ_{xx}. These results show that $\sigma_{xx}(H)\sim1/\rho_{xx}(H)$ even at the largest fields employed in these studies.

The magnetoresistance measurements of metallic samples and the magnetocapacitance measurements of an $N/N_c(0)=0.87$ insulating sample have been made in a dilution refrigerator in a Bitter magnet. Magnetic field sweeps, starting at about T=50[mK], to 19T have resulted in a 15 to 20% upward drift in temperature in the maximum field range. The Mott VRH conduction measurements on insulating samples have been made in a ^{3}He refrigerator employing either a field-stabilized Bitter magnet or a superconducting solenoid capable of 15T. The ^{3}He refrigerator provided the necessary temperature control to permit experimental determination of the logarithmic derivative [6] $d\ln\sigma_{DC}/d\ln(1/T)\simeq-(T^2/\sigma_{DC})(\Delta\sigma_{DC}/\Delta T)$, which was important in establishing whether the VRH conduction exponent m [$\ln\sigma_{DC} \propto -(T_0/T)^m$] varies with magnetic field.

3. Metallic Behavior in High Magnetic Fields Near N_c

The low-temperature magnetoresistance behavior of heavily doped n-type semiconductors has a history dating back more than several decades. It has been in the last six years or so that detailed theories [24,31] have been developed to explain all the features of the data including the widely observed negative magnetoresistance first observed by SASAKI et.al. [32] which was contrary to the classical theory of $\rho(H)/\rho(0)$. In 1963 ROTH et.al. [33] analyzed extensive data on metallic Ge:As samples and found a good fit to the relation

$$\Delta\rho(H)/\rho(0)=[\rho(H)-\rho(0)]/\rho(0)=aH^c+bH^2\,, \tag{5}$$

where the positive bH^2 term represents the normal magnetoresistance behavior and the first term represents the anomalous behavior. For sufficiently high As density ($N \gg N_C$) and sufficiently low temperatures, ROTH et.al. [33] found $c \simeq 0.5$ and $a(N)$ negative and decreasing in magnitude with increasing N. In recent years a large number of heavily doped semiconductors have shown low-field magnetoresistive behavior of the form

$$\Delta\rho(H)/\rho(0)=AH^{1/2}=(A_\ell+A_c)H^{1/2} \tag{6}$$

for the Zeeman energy $g\mu_BH \gg kT$ where A_ℓ ($A_\ell<0$) is the negative contribution first calculated by KAWABATA [24] and A_c represents the positive contribution resulting from electron-electron interactions and has been calculated by LEE and RAMAKRISHNAN [31]. Whether A_ℓ or A_c (i.e., localization effects or e-e interactions) dominate the $H^{1/2}$ behavior depends on the system (crystalline or amorphous) and how close the density N is to N_C. ROSENBAUM et.al. [34] first demonstrated for barely metallic Si:P samples that $A>0$ while as N increases A changes sign and is dominated by A_ℓ. Similar behavior has also been seen in Ge:Sb [35] and Si:As [36], a-$Si_{1-x}Au_x$ [37] and in III-V compounds like n-InSb [38] and InP [15]. The low-field magnetoresistive behavior has been reviewed by OOTUKA and KAWABATA [39].

In Fig. 1 the magnetoresistance of five metallic Si:As samples at $T \simeq 60$mK is shown in the lower field regime. Each sample shows $H^{1/2}$ behavior, however there is an expected deviation as $H \to 0$ because $g\mu_BH$ is no longer larger than kT. However, there is also a density-dependent upward deviation suggesting the appearance of the second term in Eq. (5). This deviation starts at smaller fields and is much larger for the $8.67 \times 10^{18}/cm^3$ sample which is closest to N_C. The deviation for the 9.91 and 10.4 samples is barely perceptible

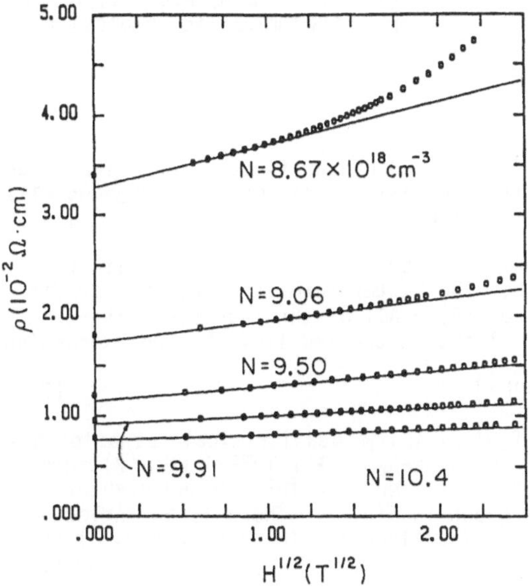

Fig. 1 $\rho(N,H)$ for five metallic samples vs. $H^{1/2}$ at $T \sim 60$ mK

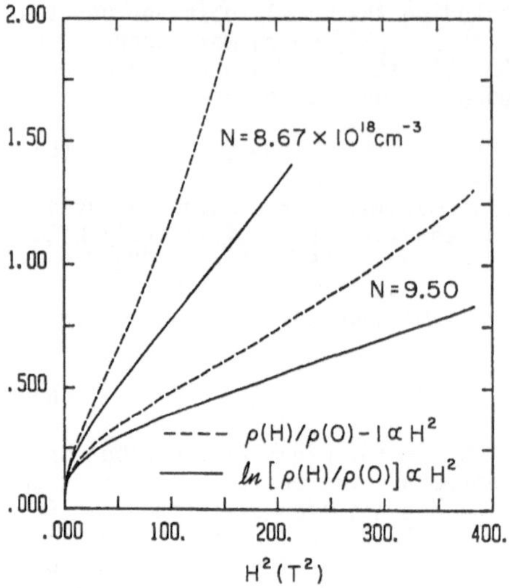

Fig. 2 Comparison of linear and exponential dependences of $\rho(H)/\rho(0)$ on H^2. The exponential fit is better at high fields.

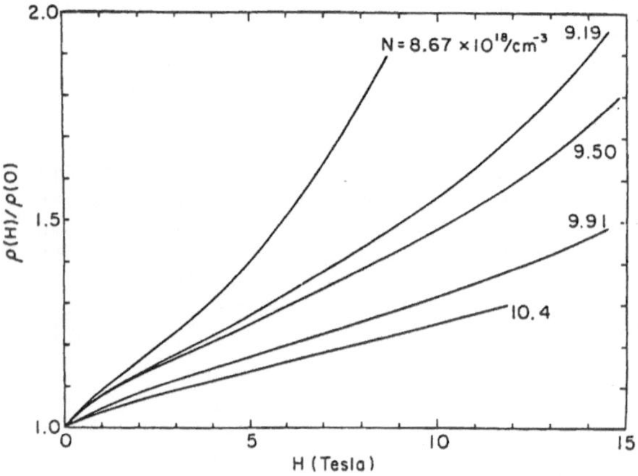

Fig. 3 $\rho(H)/\rho(0)$ vs. H for five metallic samples at $T\sim60$ mK

here and is better shown in Fig. 3 showing the higher field range. The values of $A=A_\ell+A_c$ decrease monotonically with increasing N as for Si:P [34] and will be quantitatively compared with theory elsewhere.

Fig. 2 shows the data for the 8.67 and 9.50 samples plotted as either $\Delta\rho(H)/\rho(0)$ or $\ln[\rho(H)/\rho(0)]$ versus H. The latter form is a much better fit than the usual H^2 term as given in Eq. (5). All of the magnetoresistance data for the metallic samples are well fit in the $T\to0$ limit by the expression

$$\rho(N,H)/\rho(N,0)=(1+A(N)H^{1/2})\exp[f(N)H^2], \quad (7)$$

where the experimental coefficient f(N) exhibits 'quasicritical' behavior becoming large, as illustrated in Fig. 5, as $N\to N_{c+}$. For $fH^2\ll1$ Eq. (7) agrees with the ROTH et.al. expression in Eq. (5). We note $fH^2=1.8$ and 0.48 at H=20T for the 8.67 and 9.50 samples shown in Fig. 2. There is an inflection field H (between decreasing and increasing slope of $\rho(H)/\rho(0)$ versus H) as shown in Fig. 3. From Eq. (7) one obtains $H_i=[A(N)/8f(N)]^{2/3}$. The substantial increase of H_i with N is dominated by the rapidly decreasing f(N) with increasing $N-N_C$.

Since $\sigma_{DC} \propto e^2/h\xi$ and $\sigma_{xy}/\sigma_{xx} \ll 1$ in the field range of these experiments, thus allowing $\rho_{xx}(H) \simeq 1/\sigma_{xx}(H)$, one employs $\rho(H)/\rho(0)=(1+AH^{1/2})\xi(N,H)/\xi(N,0)$ and Eq. (4) to derive the following expression valid in regime $\eta H^2 < \varepsilon \ll 1$

$$\ln\rho(H)/\rho(0) \simeq \ln(1+AH^{1/2})+[\nu(0)+\Delta\nu(H)]\eta H^2(1/\varepsilon+1)+\Delta\nu(H)\ln(1/\varepsilon). \qquad (8)$$

For $\Delta\nu(H) \ll \nu(0)$ the third term resulting from the exponent shift is negligible and the second term demonstrates that the f(N) in Eq. (7) is given by $f(N)=\nu(0)\eta(N/(N-N_C))$. The linear behavior for f(N) with $N/(N-N_C)$ is demonstrated in Fig. 4 and the slope $\nu(0)\eta=1.2\times10^{-4}[T^{-2}]$ yields $\eta=2\times10^{-4}[T^{-2}]$ for $\nu(0)=0.6$. This value of η is nearly a factor of ten larger than the value expected from Eq. (2) using $a_D^*(0)=15.45$[Å]. The point for the 8.67 sample ($1/\varepsilon>100$) below the linear part of the f vs. $N/(N-N_C)$ curve is not consistent with our expansion criterion $\eta H^2<\varepsilon$ in the high field limit and might represent a negative correction to f from the third term in Eq. (8) since $\Delta\nu(H)$ is negative as demonstrated for insulating samples. The insulating behavior, to be discussed next, shows $\Delta\nu(H) \propto a_1H^{1/2}+a_2H$. If $\Delta\nu(H)$ showed the same field dependence for metallic samples, then one would expect a downward deviation from the quadratic behavior of $\ln\rho(H)/\rho(0)$ in the high field regime. This is not observed leading to the conclusion that $\Delta\nu(H)$ is very small for the metallic Si:As samples and the principal effect of magnetic tuning is the tuning of N_C with H. As will be considered elsewhere, Eq. (7) can also describe well the data of more metallic n-type semiconductors where A<0.

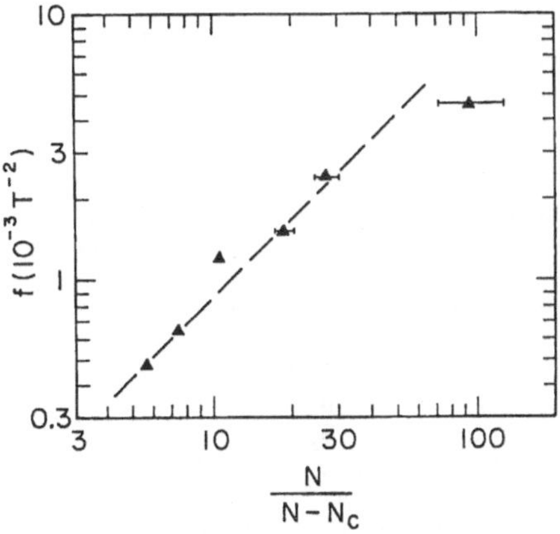

Fig. 4 f(N) vs. $N/(N-N_C)$ for metallic samples

4. Insulating Behavior in High Fields Near N_C

Earlier low-temperature magnetocapacitance measurements by NEW et.al. [40,41] in the intermediate concentration regime ($N<0.3\ N_C$) for Si:P and Si:As have shown a strongly temperature-dependent decrease in the donor electrical susceptibility $4\pi\chi'(N,T,H)=\varepsilon'(N,T,H)-\varepsilon_h$ that is proportional to $(H/T)^2$ until $g\mu_BH>10kT$. This behavior has been associated with different polarizabilities of different spin states of donor clusters. It has been difficult to obtain magnetocapacitance data in the critical regime at high fields because of the difficulty of achieving sufficiently low temperatures in fields above 10T. However, data has been obtained on an $0.87N_C$ Si:As sample at T∿60 mK in fields to 14T that definitely shows a decrease in the T→0 localization length with

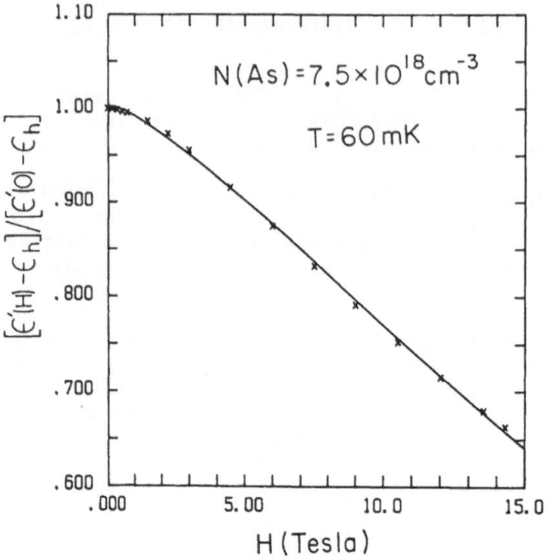

Fig. 5 $[\varepsilon'(H)-\varepsilon_h]/[\varepsilon'(0)-\varepsilon_h]$ vs. H for a 0.87 N_c sample

magnetic field. In Fig. 5 the ratio $[\varepsilon'(H)-\varepsilon_h)/(\varepsilon'(0)-\varepsilon_h]$ is shown versus H for T∿60 mK. At this temperature most of the thermally activated hopping has been removed and $\varepsilon'(0,T=60\text{ mK})-\varepsilon_h=55$, about 5% above the H=0 value of $\varepsilon'(0)-\varepsilon_h$ extrapolated to T=0. Initially $[\varepsilon'(H)-\varepsilon_h/\varepsilon'(0)-\varepsilon_h]$ decreases almost quadratically with H, but then flattens to a linear decrease with field. Neglecting the density-of-states and prefactor changes with magnetic field one finds the ratio $[\varepsilon'(H)-\varepsilon_h]/[\varepsilon'(0)-\varepsilon_h]$ equal to the square of Eq. (4). Although the N_c-tuning term in Eq. (4) yields a quadratic decrease in the ratio, the magnitude of the decrease is a factor of four too small for $\eta=2\times10^{-4}[T^{-2}]$. The $\nu(H)$ tuning term in the ratio for $\eta H^2 \ll \varepsilon \ll 1$ is approximately $(1/\varepsilon)^{-3\Delta\nu(H)}$. A relatively small shift of -0.009 for $\Delta\nu(H)$ can yield the observed magnitude at H=3.0[T]. At larger fields as ηH^2 starts to approach ε the field dependence changes to a slower linear dependence with both the N_c-tuning term and the $\Delta\nu(H)$-tuning term of comparable importance. At H=14.1 [T] a magnitude $\Delta\nu=-0.037$ can explain the data for $\eta=2\times10^{-4}[T^{-2}]$. These results demonstrate that relatively small exponent shifts have a substantial effect on the results because of the small value of ε. Because of the small values of $\Delta\nu(H)$ there is an obvious uncertainty as to how reliably the values of $\Delta\nu(H)$ are determined from these results. Nevertheless, one can say that positive values of $\Delta\nu(H)$ are ruled out and the present results are qualitatively consistent with the HIKAMI [25] theoretical prediction. Measurements closer to the transition with smaller values of ε will make the results more sensitive to $\Delta\nu(H)$, particularly in the field range where $\eta H^2<\varepsilon$, however such measurements will require temperatures much less than 60 mK to remove the thermally activated hopping.

An alternative approach for obtaining experimental data on the insulating side of N_c relating to the ratio $\xi(N,H)/\xi(N,0)$ in Eq. (4) is available from Mott VRH conduction since the Mott characteristic temperature $T_0 \propto (kN(E_F)\xi^3)^{-1}$. The critical behavior of T_0 in zero-field has been documented for insulating Si:As by SHAFARMAN and CASTNER [6], who made measurements in the temperature range 0.4K to 10K. The results for $(1-N/N_c)<0.07$ yield $T_0 \propto (1-N/N_c)^p$ with $2.3<p<2.9$. Neglecting $N(E_F)$ variations in this regime the $T_0 \propto \xi^{-3}$ relationship yields $0.77<\nu<0.97$. An explanation offered for this larger value of $\nu(0)$ results from the higher temperature range of these experiments, where the thermal energy is larger than the Coulomb gap width ($KT>\Delta$), the gap is filled in and $N(E)$ is slowly varying near E_F. We speculate the higher temperatures provide empty states below E_F and produce a situation analagous to compensation, which for Ge:Sb has been shown [42] to increase ν toward unity with increasing compensation.

There have been a number of theoretical and experimental studies of the magnetic-field dependence of VRH conduction. SHKLOVSKII [43] has found the magnetoresistance in the VRH conduction regime in the weak field limit to be given by $\ln\rho(H)/\rho(0)\propto H^2(T_0/T)^{3p}$ with $p=1/4$ for the noninteracting case and $p=1/2$ when $N(E)$ is determined by the Coulomb gap, presumably with $kT<<\Delta$, the gap width. In InSb TOKUMOTO et al. [44] have shown the Mott VRH [$\ln\sigma_{DC}\propto-(T_0/T)^m$] exponent m changes from 1/4 to 1/2 in modest magnetic fields, but nevertheless in the high field limit ($\lambda<a^*$) for InSb. SHKLOVSKII [45] has given a theoretical interpretation of this based on nonresonant subbarrier tunneling. The results on Si:As [20] in the critical regime $0.12<\varepsilon<0.01$ show different behavior than that observed in Ref. 44 and are not in agreement with the SHKLOVSKII [43] expressions. $\ln\sigma_{DC}$ versus $T^{-1/4}$ is shown in Fig. 6 for a $N/N_c(0)=0.94$ Si:As sample at fixed fields from 0 to 19T. The Mott exponent m remains 1/4, even at the highest field and this has been confirmed by logarithmic derivative plots. This is true for all insulating samples studied with $N/N_c(0)>0.9$. A $7.57\times10^{18}/cm^3$ sample shows $m\sim1/4$ at 4.3 and 6T, but then at 10.9T shows a slightly stronger temperature dependence at lower temperatures and finally at 19T shows behavior ($m\to1$) close to the SHKLOVSKII theoretical prediction [43]. Thus, the principal effect of the magnetic field is to increase the slope $[T_0(N,H)]^{1/4}$. There is also a change in the prefactor ratio $\sigma_0(H)/\sigma_0(0)$, but that will not be considered here. The $T_0(N,H)/T_0(N,0)$ ratio is shown in Fig. 7. For any one sample the field dependence takes the form $\ln T_0(H)/T_0(0)\propto H$ in the high field range and with perhaps an $H^{1/2}$ term dominating at smaller fields below 5T. At a particular field (best illustrated at 10.9T) the ratio $T_0(N,H)/T_0(N,0)$ increases rapidly as $N\to N_c(0)$ as a function of $1/\varepsilon$. Using $T_0(N,H)\propto[N(E_F,H)\xi(N,H)^3]^{-1}$, Eq.(4), and setting $N(E_F,H)/N(E_F,0)$ and $\xi_0(H)/\xi_0(0)$ equal to unity one obtains

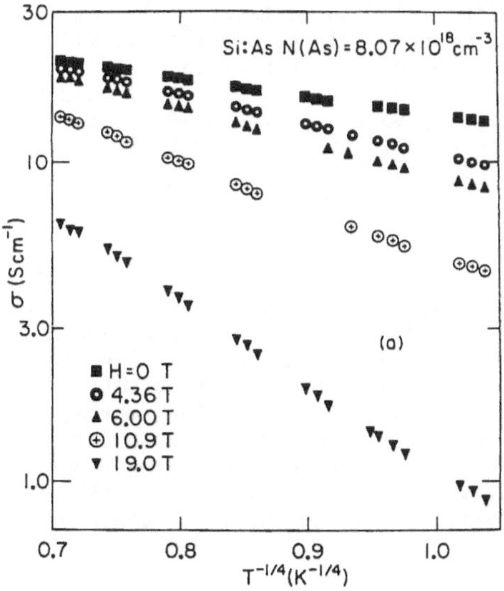

Fig. 6 $\ln\sigma$ vs. $T^{-1/4}$ for fixed fields for 0.94 N_c sample

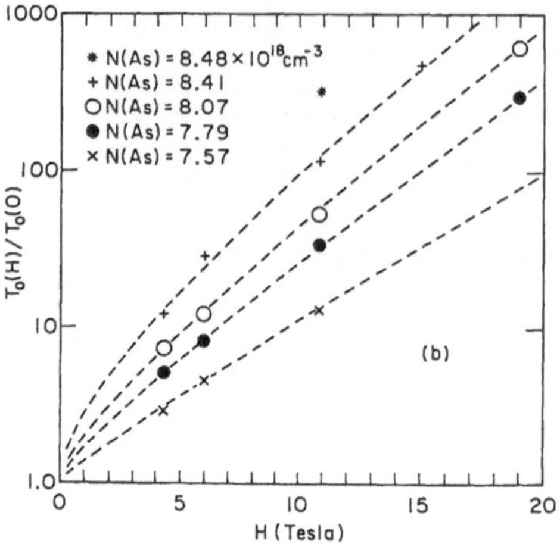

Fig. 7 $\ln T_0(H)/T_0(0)$ vs. H for insulating samples

$$\frac{T_0(N,H)}{T_0(N,0)} = \left[\frac{1+\eta H^2/\varepsilon}{1+\eta H^2}\right]^{3\nu(0)}\left[\frac{1+\eta H^2}{\varepsilon+\eta H^2}\right]^{-3\Delta\nu(H)} \tag{9}$$

The first term, the pure N_C-tuning case, leads to $\ln T_0(H)/T_0(0)\propto 3\nu(0)\eta H^2(1/\varepsilon-1)$ which is not only the wrong field dependence, but also more than a factor of ten too small for $\eta=2\times10^{-4}[T^{-2}]$ to explain the magnitude of $T_0(N,H)/T_0(N,0)$. The second term can qualitatively explain the results if $\Delta\nu(H)$ is negative with a field dependence of the form $-\Delta\nu(H)=(\nu(0)-\nu_f)g(H)$ where $g(H)=a_1H^{\frac{1}{2}}+a_2H$ with a_1 and a_2 as positive coefficients which may also be density-dependent. At fixed field $T_0(N,H)/T_0(N,0)\sim(1/\varepsilon)^{3[\nu(0)- f]g(H)}$ for $\eta H^2\ll\varepsilon\ll1$. As a comparison HIKAMI [25] finds $\nu(0)=1$ and $\nu_f=1/2$. At these temperatures it has been established [6] that $\nu(0)$ is closer to unity than the zero temperature result $\nu\sim0.6$ [18,19], thus leading to a much larger value of $\nu(0)-\nu_f$ than for the low-temperature magnetocapacitance data already discussed. We cannot obtain ν_f from the data but we note that with $\nu_f=1/2$ one obtains $g(H)>1$ at $H=19[T]$. The data in Fig. 7 when analyzed for $\eta=0$ yield $(\nu(0)-\nu_f)a_1$ varying with density between 0.04 and 0.15T while one obtains $(\nu(0)-\nu_f)a_2=.06[T^{-1}]$. The results change an insignificant amount for $\eta=2\times10^{-4}[T^{-2}]$. These values of $\Delta\nu(H)$ are a factor 25 larger than the $\Delta\nu(H)$ values at the same fields obtained at much lower temperatures from the magnetocapacitance results.

5. Discussion and Conclusions

The critical behavior of both metallic and insulating samples yields evidence that has been interpreted in terms of N_C-tuning for metallic samples and a combination N_C-tuning and $\Delta\nu(H)$-tuning for the insulating samples although the $\Delta\nu(H)$-tuning is dramatically larger for the higher temperature Mott VRH conduction characteristic temperature ratio $T_0(N,H)/T_0(N,H)$. The smaller values of $\Delta\nu(H)$ obtained from the low-temperature magnetocapacitance data are consistent with the $\Delta\nu(H)$ terms in the metallic case being unimportant. The negative values of $\Delta\nu(H)$ are qualitatively consistent with the HIKAMI [25] calculation, however no theory is available giving the field dependence of $\Delta\nu(H)$.

Although it is easy to show the field dependence of $N(E_F,H)/N(E_F,0)$ is negligible within a free-electron model for $g\mu_B H \ll E_F$, this may not be obvious for barely insulating samples in a disordered medium featuring localized states. From cluster theory (for large sizes, >30) there are a large number of states of different total spin distributed over an energy range W(N). For $g\mu_B H \lll W(N)$, as is clearly the case, the density-of-states change with field will be extremely small. Even though the single particle density $N(E_F)$ will go to zero in the Coulomb gap regime as $T\to 0$, LEE [46] has noted that in conduction cases $N(E_F)$ should be replaced by dN/dE_F which will vary slowly near E_F. In the dielectric response $N(E_F)$ should be replaced by dN/dE_F. For the Mott VRH conduction at higher temperatures the Coulomb gap is thermally filled in for $\varepsilon<0.1$ and one has dN/dE_F of order $N(E_F)$.

The least satisfactory part of the above analysis is the field-dependence of the Mott VRH conduction. SHKLOVSKII [43] notes that to be in the weak field regime one requires not only $\lambda \gg a^*$, but also $\lambda^2/a^* \gg N^{-1/3}$. The latter condition is violated when N is of order N_c, particularly if one replaces a^* by $\xi(N)$. In the VRH case SHKLOVSKII [43] shows the crossover between high and low field behavior is given by $H_c \sim (\hbar c/e)[N(E_F)kT]^{\frac{1}{4}}/a^{*5/4}$. With a^* replaced by $\xi(N)$ one can achieve the high field limit if $\nu(0)$ is closer to 1 than 1/2. The high field prediction [43] yields $\ln\rho(H)/\rho(0)=(T_0(H)/T)^{1/3}$ with a temperature dependence that would be extremely hard to distinguish from the Mott exponent $m \sim 1/4$. However, $T_0(H)$ has been predicted [43] to be given by $T_0(H) \sim 2(eH/\hbar c)/kN(E_F)a^*_H$ where a^*_H is Bohr radius along the field. This does not yield the extremely strong density-dependent field variation shown in Fig. 7, however a modified version may yield better agreement. The data in Fig. 6 seems more readily explainable by employing the high field limit predictions, but additional analysis is required.

6. Acknowledgements

We are grateful for technical assistance from L. Rubin, P. Tedrow, and B. Brandt of the Francis Bitter National Magnet Laboratory and also to D. Koon, M. Migliuolo, and V. Zarifis for assistance with experiments. This work was supported by the National Science Foundation through Grant No. DMR-8306106.

7. References

1. T.F. Rosenbaum, K. Andres, G.A. Thomas, R.N. Bhatt: Phys. Rev. Lett. 46, 568 (1981)
2. M.A. Paalanen, T.F. Rosenbaum, G.A. Thomas, R.N. Bhatt: Phys. Rev. Lett. 48, 1284 (1982)
3. G.A. Thomas, M. Capizzi, F. DeRosa: Phil. Mag. B42, 913 (1980)
4. M.A. Paalanen, T.F. Rosenbaum, G.A. Thomas, R.N. Bhatt: Phys. Rev. Lett. 51, 1896 (1983)
5. M.A. Paalanen, A.E. Ruckenstein, G.A. Thomas: Phys. Rev. Lett. 54, 1295 (1985)
6. W.N. Shafarman, T.G. Castner: Phys. Rev. B33, 3570 (1986)
7. M. Cuevas, H. Fritzsche: Phys. Rev. 139, A1628 and 137, A1847 (1965)
8. R.N. Bhatt: Phys. Rev. B26, 1082 (1982)
9. Y. Yafet, R.W. Keyes, E.N. Adams: J. Phys. Chem. Solids 1, 137 (1956)
10. R.W. Keyes, R.J. Sladek: J. Phys. Chem. Solids 1, 143 (1956)
11. J.L. Robert, A. Raymond, R.L. Aulombard, C. Bousquet: Phil. Mag. B42, 1003 (1980)
12. S. Ishida, E. Otsuka: J. Phys. Soc. Japan 43, 124 (1977)
13. R. Mansfield, M. Abdul-Gader, P. Rozonni: Sol. St. Elect. 28, 109 (1985)
14. G. Biskupski, H. Dubois, J.L. Wolkiewiez, A. Briggs, G. Remenyi: J. Phys. C: Solid State Phys. 17, L411 (1984)

15. A.P. Long, M. Pepper: Sol. St. Elect. 28, 61 (1985)
16. S. Von Molnar, A. Briggs, J. Flouquet, G. Remenyi: Phys. Rev. Lett. 51, 706 (1983)
17. E. Abrahams, P.W. Anderson, D.C. Licciardello, T.V. Ramakrishnan: Phys. Rev. Lett. 42, 673 (1979)
18. P.F. Newman, D.F. Holcomb: Phys. Rev. B28, 628 (1983)
19. W.N. Shafarman: Ph.D. thesis, Univ. of Rochester
20. W.N. Shafarman, T.G. Castner, J.S. Brooks, K.P. Martin, M.J. Naughton: Phys. Rev. Lett. 56, 980 (1986)
21. A. Miller, E. Abrahams: Phys. Rev. 120, 745 (1960)
22. D.E. Khmel'nitskii, A.I. Larkin: Sol. St. Commun. 39, 1069 (1981)
23. B. Shapiro: Phil. Mag. B50, 241 (1984)
24. A. Kawabata: Sol. St. Commun. 34, 431 (1980) and J. Phys. Soc. Japan 49, 628 (1980)
25. S. Hikami: Phys. Rev. B24, 2671 (1981)
26. C. Castellani, C. DiCastro, P.A. Lee, M. Ma: Phys. Rev. B30, 527 (1984)
27. W.N. Shafarman, T.G. Castner, J.P. Brooks, K.P. Martin, M.J. Naughton: Sol. St. Elect. 28, 93 (1985)
28. P.F. Newman, M.J. Hirsch, D.F. Holcomb: J. Appl. Phys. 58, 3779 (1985)
29. P.F. Newman: Ph.D. thesis, Cornell University
30. D. Koon: private communication
31. P.A. Lee, T.V. Ramakrishnan: Phys. Rev. B26, 4009 (1982)
32. W. Sasaki, Y. Kanai: J. Phys. Soc. Japan 11, 894 (1956)
33. H. Roth, W.D. Straub, W. Bernhard, J.E. Mulhern, Jr.: Phys. Rev. Lett. 11, 328 (1963)
34. T.F. Rosenbaum, R.F. Milligan, G.A. Thomas, P.A. Lee, T.V. Ramakrishnan, R.N. Bhatt: Phys. Rev. Lett. 47, 1758 (1981)
35. Y. Ootuka, S. Katsumoto, S. Kobayashi, W. Sasaki: Sol. St. Elect. 28, 101 (1985)
36. T.G. Castner, W.N. Shafarman: In Localization and the Metal-Insulator Transitions, ed. by H. Fritzsche, D. Adler (Plenum Publishing Co., London 1985) p. 9
37. N. Nishida, T. Furubayashi, M. Yamaguchi, K. Morigaki, H. Ishimoto: Sol. St. Elect. 28, 81 (1985)
38. S. Morita, N. Mikoshiba, Y. Koike, T. Fukase, S. Ishida, M. Kitigawa: Sol. St. Elect. 28, 113 (1985)
39. Y. Ootuka, A. Kawabata: Prog. of Theo. Phys. Suppl. 84, 249 (1985)
40. D. New, N.K. Lee, H.S. Tan, T.G. Castner: Phys. Rev. Lett. 48, 1208 (1982)
41. D. New, T.G. Castner, M.J. Naughton, J.S. Brooks: In Lecture Notes in Physics 177, ed. by G. Landwehr (Springer-Verlag, Berlin 1983) p. 475
42. G.A. Thomas, Y. Ootuka, S. Katsumoto, S. Kobayashi, W. Sasaki: Phys. Rev. B25, 4288 (1982)
43. B.I. Shklovskii, A.L. Efros: In Electronic Properties of Doped Semiconductors, Springer Ser. Solid-State Sci. Vol. 45 (Springer, Berlin 1984) Ch. 9
44. H. Tokomuto, R. Mansfield, M.J. Lea: Phil. Mag. B46, 93 (1982)
45. B.I. Shklovskii: Pis'ma Sh. Eksp. Fiz. 36, 43 (1982) [JETP Lett. 36, 51 (1982)]
46. P.A. Lee: Phys. Rev. B26, 5882 (1982)

Anderson Transition Induced by Strong Magnetic Fields

Y. Ono and T. Ohtsuki

Department of Physics, Faculty of Sciences, University of Tokyo, Hongo 7-3-1, Bunkyo-ku, Tokyo 113, Japan

Applying the self-consistent theory of the Anderson Localization to three-dimensional systems under quantizing magnetic fields, we discuss the metal-insulator transition induced by the magnetic field in the case where only the lowest Landau subband is occupied. The critical field and the critical behaviour of the diagonal conductivities are discussed for various zero-field values of the Fermi energy and the elastic relaxation time.

1. Introduction

It has been established that weak magnetic fields weaken the Anderson localization [1-4] by decreasing the quantum interference effect between the states with wave numbers k and -k. On the other hand, a classical argument leads to the conclusion that the transverse conductivities will vanish in the infinitely strong field limit because the cyclotron radius goes to zero in this limit. In the case of two-dimensional systems, it has been found that the weak magnetic fields never create the extended states [3,4] but that the extended states appear at the center of each Landau subband in the strong field limit [5-8]. The latter fact is closely related to the quantized Hall effect.

In this work we consider the field-induced metal-insulator transition in three-dimensional systems [9]. The field is assumed so strong that only the lowest Landau subband is occupied. In such a situation, the electronic motion becomes almost one-dimensional like along the field direction, and the effective Fermi wave length is increased as the field gets stronger. If the field is so strong that the Fermi wave length becomes of the same order as the mean free path due to the impurity scatterings, it is plausible to expect the Anderson localization of electrons. In the following, we calculated the diffusion coefficients by using the self-consistent treatment of the Anderson localization, which has provided successful results in two dimensions [3,7,8,10,11]. The original idea of the self-consistent theory was proposed by Götze [12] and it was reformulated in terms of the microscopic diagram method by Vollhardt and Wölfle (VW) [10]. The method of VW was later extended to the case without time-reversal symmetry by Yoshioka et al. [3,11] and to the case with strong magnetic fields by one of the present authors [7,8].

2. Formulation

We start with the Hamiltonian for three-dimensional non-interacting electrons in random impurity potentials and a magnetic field parallel to the z-axis. For the impurities we take those with a delta-function-type potential for simplicity. The field is assumed so strong that the Fermi level lies within the lowest Landau subband and that the spin degeneracy is also lifted. It is well known that the wave number and frequency dependences of the density relaxation function $\Phi(\mathbf{q},\omega)$, which has a simple relation with the density response function, yield essential information on the transport properties of the system. In the present case, $\Phi(\mathbf{q},\omega)$ takes the following form in the small wave number limit,

$$\Phi(\mathbf{q},\omega) = -N(E)/\{\omega + i[D_{\parallel}(\omega)\, q_z^2 + D_{\perp}(\omega)\, q_{\perp}^2]\}, \qquad (1)$$

where N(E) is the density of states at the Fermi energy E and the diffusion coefficients $D_{\parallel}$ and $D_{\perp}$ describe the electronic diffusive motions parallel and perpendicular to the magnetic field; $q_{\perp}^2 = q_x^2 + q_y^2$. The above form of $\Phi(\mathbf{q},\omega)$ can be derived through very general approximation-free arguments such as the number conservation [9]. It is also possible to decompose $\Phi(\mathbf{q},\omega)$ into a sum of the contributions of many diagrams. Then we obtain an exact relation between the diffusion coefficients and the irreducible vertex correction. It is, however, not possible at the moment to calculate the irreducible vertex correction exactly, and therefore we have to introduce an approximation. Götze and VW showed that the renormalization of the diffusion processes is essential in discussing the localization problem. VW succeeded to express this idea in terms of diagrams.

Here we follow their idea. In the present situation it is natural to consider $D_{\parallel}$ and $D_{\perp}$ separately. This separation can be made by assuming $q_{\perp} = 0$ or $q_z = 0$ for the external wave number. By putting $q_{\perp} = 0$, the calculation of $D_{\parallel}$ reduces to that in the one-dimensional system. Only difference from VW's treatment in the one-dimensional system is that we need not consider the contribution of the particle-particle channels which have no diffusion-pole-like singularity in the presence of the strong magnetic field. Thus $D_{\parallel}$ is approximately expressed in terms of particle-hole diffusion processes which depend on $D_{\parallel}$ and $D_{\perp}$. The perpendicular diffusion coefficient can be calculated by putting $q_z = 0$, and then the calculation becomes analogous to that in two-dimensional systems [7,8]. We use the same type of approximation as used in [7]. After lengthy calculations, we end up with an expression of $D_{\perp}$ in terms of the diffusion processes. Thus we obtain coupled self-consistent equations for $D_{\parallel}$ and $D_{\perp}$ which are schematically written as follows,

$$D_{\parallel} = f_{\parallel}(D_{\parallel}, D_{\perp}; \varepsilon_F, \tau_0, \omega_c), \qquad (2)$$

$$D_{\perp} = f_{\perp}(D_{\parallel}, D_{\perp}; \varepsilon_F, \tau_0, \omega_c), \qquad (3)$$

where ε_F, τ_0 and ω_c are the zero-field values of the Fermi energy and the elastic relaxation time and the cyclotron frequency; they characterize the carrier concentration, the purity of the system and the strength of the field, respectively. The real Fermi energy is determined uniquely by these three parameters. In order to avoid too much complication, we calculate the single particle Green functions within the self-consistent Born approximation. In writing down eqs. (2) and (3), the static limit has been taken.

The explicit forms of the self-consistent equations (2) and (3) are given in [9]. They are too complicated to be solved analytically. Some important conclusions, however, can be derived without explicitly solving them. Namely, if the transition from the metallic to the insulating phase occurs at a finite field, both of $D_{\parallel}$ and $D_{\perp}$ have to vanish at the same time, keeping the ratio $D_{\perp}/D_{\parallel}$ finite. Furthermore the critical exponent is expected to be unity.

3. Results and Discussion

The self-consistent equations (2) and (3) are solved numerically. Once we know the diffusion coefficients, it is straightforward to calculate the diagonal conductivities through the Einstein relation, $\sigma_\mu = e^2N(E)D_\mu$ ($\mu = /\!/$ or $\perp$). In Fig. 1, we have plotted the conductivities as functions of ε_F/ω_c for the case with $\varepsilon_F\tau_0 = 10$; this value of $\varepsilon_F\tau_0$ corresponds to a good metallic sample. The conductivities are scaled by the zero-field classical conductivity $\sigma_0 = ne^2\tau_0/m$ with n the electron concentration and m the effective mass of an electron. In the same figure, the conductivities calculated from the Kubo formula without the vertex corrections are also shown for comparison. Through similar calculations for various values of $\varepsilon_F\tau_0$, it has been concluded that both of $\sigma_{/\!/}$ and $\sigma_\perp$ vanish linearly at a critical field depending on ε_F and τ_0 and that the following universal relations are satisfied at the transition point,

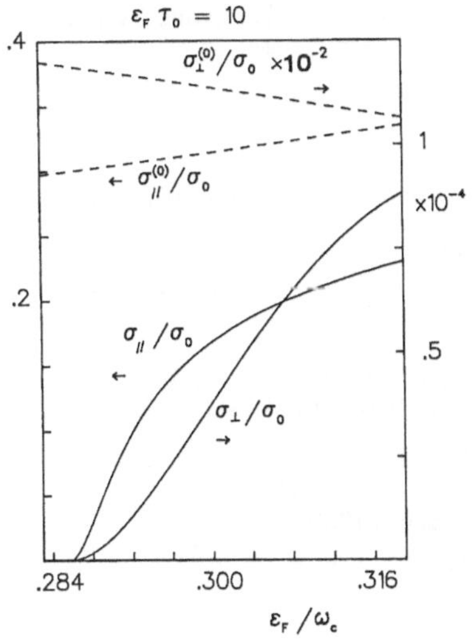

Fig. 1: The longitudinal and transverse conductivities near the critical field for $\varepsilon_F\tau_0 = 10$. For comparison those (with superscript (0)) obtained from the Kubo formula without vertex corrections are also shown. All the conductivities are scaled by σ_0 the classical one in vanishing field.

$$\left[\frac{\sigma_\perp / \sigma_\perp^{(0)}}{\sigma_{/\!/} / \sigma_{/\!/}^{(0)}}\right]_{crt} = 1.25 \times 10^{-3} \quad , \tag{4}$$

$$(\varepsilon_F\tau_0)^{1/4}(\varepsilon_F/\omega_c)_{crt} = 0.509 \quad . \tag{5}$$

Particularly the last relation implies that the system becomes an insulator in a sufficiently strong (but still finite) magnetic field however metallic it is in the absence of the field.

From eq. (4), we find that the anisotropy of the conductivities is much enhanced near the transition point compared to that obtained in the lowest approximation (i.e. without vertex corrections). Such a strong anisotropy may make it difficult to perform precise measurement of the transverse conductivity. Recently one of the present authors and Fukuyama [13] pointed out to determine the extremely anisotropic diffusion coefficients from the anisotropy of the ultrasonic attenuation, namely from the dependence of the attenuation coefficient on the angle between the directions of the sound propagation and the applied field. It will be useful to measure the conductivity and the ultrasonic attenuation coefficient in the same sample.

We have also compared the results of the present calculation with the field induced metal-nonmetal transition experimentally observed in n-InSb by Mansfield et al. [14]. The agreement is not so bad qualitatively though it may not be satisfactory quantitatively. In the experiment the critical exponent is found to be unity which agrees with the present result. If we choose the value of $\varepsilon_F\tau_0$ to fit the conductivity (σ_{zz}) at magnetic fields well below the critical one, the present theory predict a critical field smaller than the experimental one by a factor of about 1/2. Main reason of the discrepancy would be that the samples investigated by Mansfield et al. are not so metallic in the absence of the field; namely $\varepsilon_F\tau_0$ is not so large. Present theory stresses the diffusive motion of the electrons and therefore would be more appropriate for larger values of $\varepsilon_F\tau_0$. Furthermore the effect of magnetic freeze-out and electronic mutual interactions which are not included here may not be neglected in the real systems.

References

1. S. Hikami, A.I. Larkin and Y. Nagaoka: Prog. Theor. Phys. **63**, 707 (1980)
2. A. Kawabata: J. Phys. Soc. Jpn. **49**, 628 (1980)
3. Y. Ono, D. Yoshioka and H. Fukuyama: J. Phys. Soc. Jpn. **50**, 2143 (1981)
4. S. Hikami: Phys. Rev. B**24**,2671 (1981)
5. T. Ando: J. Phys. Soc. Jpn. **52**, 1740 (1983), **53**, 3101 (1984)
6. A. MacKinnon and B. Kramer: In Application of High Magnetic Fields in Semiconductor Physics, ed. by G. Landwehr (Springer, 1983)
7. Y. Ono: J. Phys. Soc. Jpn. **51**, 2055, 3544 (1982)
8. Y. Ono: Prog. Theor. Phys. Suppl. **84**, 138 (1985)
9. T. Ohtsuki and Y. Ono: J. Phys. Soc. Jpn. **55**, 2347 (1986)
10. D. Vollhardt and P. Wölfle: Phys. Rev. B**22**, 4666 (1980)
11. D. Yoshioka, Y. Ono and H. Fukuyama: J. Phys. Soc. Jpn. **50**, 3419 (1981) [Errata ibid. **51**, 340 (1982)]
12. W. Götze: J. Phys. C**12**, 1279 (1979)
13. Y. Ono and H. Fukuyama: J. Phys. Soc. Jpn. **54**, 4704 (1985)
14. R. Mansfield, M. Abdul-Gader and P. Fozooni: Solid State Electr. **28**,127 (1985)

Electron Transport in the Magnetically Induced M-I Transition in InSb

M. Abdul-Gader[2], *R. Mansfield*[3], *and P. Fozooni*[1]

[1]Department of Physics, Royal Holloway & Bedford New College, University of London, London, UK

[2]Department of Physics, University of Jordan, Amman, Jordan

[3]Department of Physics, University of Brunei, Brunei

Measurements of the components of the resistivity tensor of n-InSb in the region of the magnetic field-induced M-I transition are described. On the metal side of the transition the magnetic dependence of ρ_{xx}/σ_0 and ρ_{zz}/ρ_0 are in satisfactory agreement with the theory of Roth and Argyres providing the screening theory of Jog and Wallace is used. Measurements close to the transition which were reported in [1] showed that the critical magnetic field which induces the M-I transition is in good agreement with Mott's formula. On the insulator side the temperature dependence of ρ_{xx} and ρ_{zz} is of the form $\rho_0 \exp(T_0/T)^x$ with $x = 1/4$ for these heavily doped samples, in contrast to $x = 1/2$ previously observed for lightly doped samples. The possibility that the transition is due to a Wigner crystallisation is discussed but it is concluded that the transition is a Mott type followed on the insulator side by a magnetic freeze-out of the electrons.

1. Introduction

A magnetic field induces a metal-insulator transition in semiconductors with a small energy gap such as InSb and HgCdTe but it is difficult to determine the precise value of the magnetic field H_{MI} which induces the transition. Initial attempts on InSb used Arrhenius plots of the resistivity for various magnetic fields and H_{MI} was chosen as the field when activated behaviour was first observed. Following the lead of the Bell group who investigated the metal-insulator transition by conductivity measurements of a series of samples of varying doping levels and hence of varying donor separations. MANSFIELD, ABDUL-GADER and FOZOONI [1] studied the temperature variation of $\sigma(T)$ in InSb at closely spaced magnetic fields near the transition, which is equivalent to varying the size of donor centres. They found $\sigma(T)$ obeyed the relation used by the Bell group:

$$\sigma(T) = \sigma(0) + AT^{\frac{1}{2}} + BT \; . \tag{1}$$

Thus $\sigma(0)$ could be found and the field for which $\sigma(0) \to 0$ was determined and identified with H_{MI}. The critical field obtained this way was in good agreement with Mott's formula $n^{1/3}a_B \cong 0.26$ for the M-I transition using $a_B = (a_\perp^2 a_\parallel)^{1/3}$ where $a_\perp$ and $a_\parallel$ are either the Yafet, Keyes and Adams parameters defining the donor wave function or $a_\perp = (\hbar c/eH)^{\frac{1}{2}} = \lambda$ the magnetic length and $a_\parallel = a_B/2\ln(a_B/\lambda)$ both evaluated at H_{MI}. These detailed measurements were made on the resistivity in a longitudinal magnetic field ρ_{zz} since the conductivity can be calculated using the relation $\sigma_{zz} = \rho_{zz}^{-1}$. This is necessarily the case if measurements are made of ρ_{xx} since $\sigma_{xx} \neq \rho_{xx}^{-1}$. Finally ROSENBAUM et al. [2] estimated H_{MI} for HgCdTe from measurements of the Hall resistance $\rho_{xy} = -HR$ (R = Hall coefficient). They found a critical magnetic field where a rapid increase in ρ_{xy} with H occurred. This critical field increased linearly with temperature and the extrapolated field at $T \to 0$ was identified with H_{MI}. The fact that the critical field varied with temperature was attributed to the formation of a Wigner crystal since the melting temperature of such a crystal is a function of the field. The critical field determined from these measurements was found to be a factor of two greater than the expected value for the field to induce a Wigner transition. These results have been criticised by several groups who have suggested that they are not characteristic of the bulk material but are governed by a low resistivity surface layer. An anomalous feature of Fig.1 of ROSENBAUM et al [2] which is not discussed is that ρ_{xy} appears to be negative below H_{MI} which corresponds to a positive Hall coefficient. SHAYEGAN et al [3] have measured ρ_{xx}, ρ_{zz} and ρ_{xy} on both HgCdTe and InSb pointing out that both materials have very low electron effective masses and similar dielectric constants and hence one would expect the localization transition should be similar. Their measurements were confined to temperatures greater than 0.4 K and they confirmed that for both materials there was a critical magnetic field where a rapid increase in ρ_{xy} with H occurs and that this critical field varies linearly with temperature. The extrapolated field at T = 0 was also identified with H_{MI} and they concluded that the two systems behaved similarly. Recently [4] impurity cyclotron resonance (ICR) has been observed in HgCdTe where an optically induced transition of a donor bound electron occurs to an excited bound state. This energy separation is slightly greater than that of the related Landau levels which give rise to conduction-band cyclotron resonance (CCR). The ICR absorption is observed to increase and CCR to decrease as the temperature is reduced, showing freeze-out of the electron onto donor centres, and this is confirmed by altering the electron temperature by varying the power level of the far infra-red radiation. An anomalous feature of their results which does not fit any metal-insulator model is that ICR is observed at magnetic fields below that required for the M-I transition where the material should be metallic. It is suggested that even on the metallic side of the M-I transition the delocalized electrons are in donor-band states which are distinct from conduction band states. We suggest that an alternative explanation is that the value of H_{MI} which could not be determined on the low concentration sample and which had to be estimated by extrapolation of data on samples with a higher concentration might be too large. A check on our results on InSb using low concentration samples shows that there is a discrepancy between H_{MI} obtained from the rapid rise in ρ_{xy} and by other methods based on ρ_{xx} and ρ_{zz} and this discrepancy increases as the concentration decreases. For example, a rapid rise in ρ_{xy} in sample 1015 described in TOKUMOTO et al [5] does not occur until 30 kG whereas the M-I transition determined by the Mott formula and from measurements of the ρ_{xx} and ρ_{zz} is at 15 kG. This point is discussed in more detail later.

2. Results

The preliminary report in [1] only presented results on ρ_{zz} on two fairly heavily doped samples of InSb. Measurements taken at the same time on the transverse resistivity ρ_{xx} and Hall resistivity ρ_{xy} [6] are presented here, together with a more detailed discussion of the results on either side of the transition.

Figures 1 and 2 show field dependencies of ρ_{xx}, ρ_{zz} and ρ_{xy} and Figure 3 gives the temperature dependence of the Hall coefficient of sample 6715. The theoretical field dependencies also shown in Figs. 1 and 2 are of ρ_{xx} and ρ_{zz} in the metallic region, calculated using the theory of ROTH and ARGYRES [7] for a degenerate semiconductor in the extreme quantum limit in

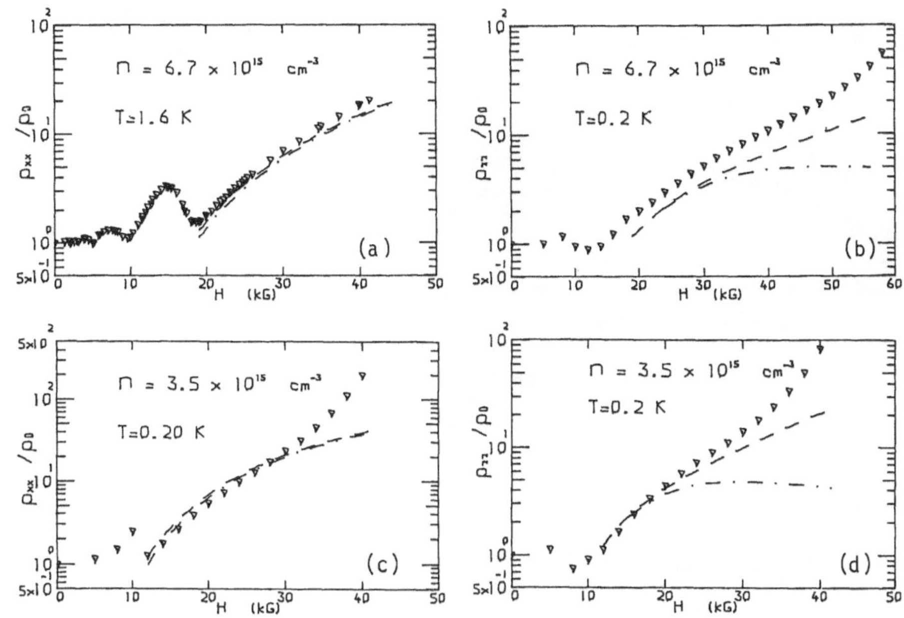

Fig.1. The ratio of the transverse ρ_{xx} and longitudinal ρ_{zz} resistivities in a magnetic field to the resistivity ρ_0 in zero magnetic field for (a) and (b) sample 6715 and for (c) and (d) sample 3515. Theoretical curves are shown using the theory of Roth and Argyres (dash-dot) and using the Jog-Wallace screening radius (dash).

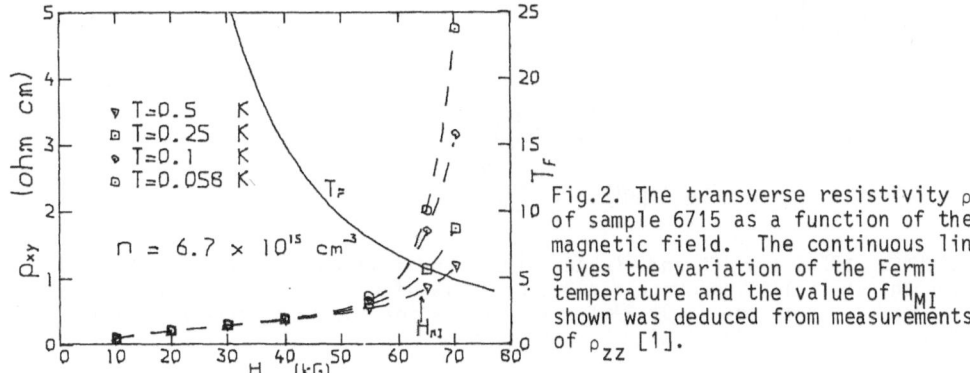

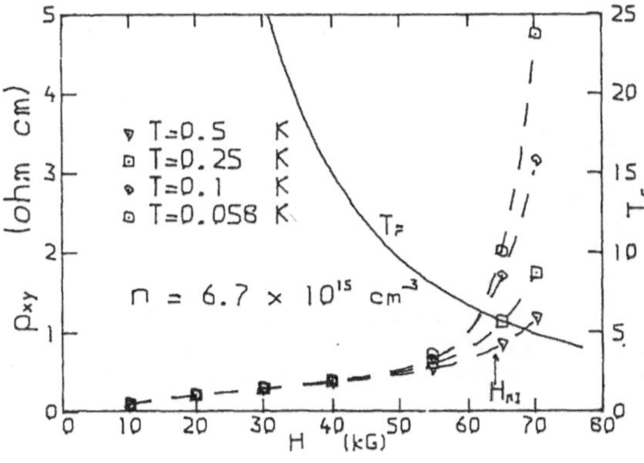

Fig.2. The transverse resistivity ρ_{xy} of sample 6715 as a function of the magnetic field. The continuous line gives the variation of the Fermi temperature and the value of H_{MI} shown was deduced from measurements of ρ_{zz} [1].

which there is long-range ionised impurity scattering. The appropriate expressions are

$$\frac{\rho_{zz}}{\rho_o} = \frac{1}{F(\zeta_o)}\,\frac{3\hbar\omega_c}{4\zeta_z+\varepsilon_s}\ , \tag{2}$$

$$\frac{\rho_{xx}}{\rho_o} = \frac{27}{128\,F(\zeta_o)}\left(\frac{\hbar\omega_c}{\zeta_o}\right)^3 \ln\left[\frac{(\hbar\omega_c)^2}{(4\zeta_z+\varepsilon_s)\varepsilon}\right]\ , \tag{3}$$

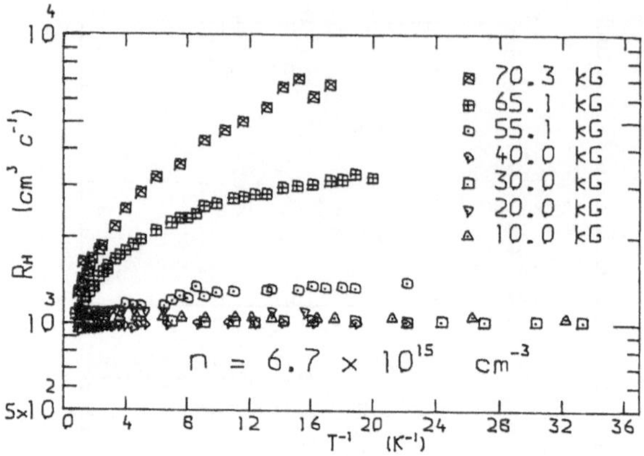

Fig.3. The Hall effect of sample 6715.

where

$$\rho_o = \left(\frac{4\pi Ze^2}{\kappa}\right) \frac{N_I\ F(\zeta_o)}{16\pi\hbar\zeta_o^{\ 3/2}} \left[\frac{\hbar^2}{2m^*}\right]^{\frac{1}{2}} \tag{4}$$

is the resistivity in zero magnetic field,

$$F(\varepsilon) = \ln\left[1+\frac{4\varepsilon}{\varepsilon_s}\right] - \left(1+\frac{\varepsilon_s}{4\varepsilon}\right)^{-1}, \tag{5}$$

$$\varepsilon_s = \frac{\hbar^2}{2m^*}\left[\frac{1}{R_{SC}^H}\right]^2,$$

$$\zeta_o = \frac{\hbar^2}{2m^*}(3\pi^2 n)^{2/3}, \qquad \zeta_z = \frac{16}{9}\ \frac{\zeta_o^{\ 3}}{(\hbar\omega_c)^2},$$

$$R_{SC}^H(\text{deg}) = \text{screening length} = \left[\frac{\zeta_z\kappa}{2\pi ne^2}\right]^{\frac{1}{2}}. \tag{6}$$

WALLACE (1974) and JOG and WALLACE (1975) [8] have shown that the screening length parallel to the magnetic field $R_{SC}^{\parallel}$ is different from the length perpendicular to the field $R_{SC}^{\perp}$. This gives the following screening radii for degenerate statistics:

$$R_{SC}^{\parallel} = \left[\frac{\kappa\zeta_z}{2\pi ne^2} + \frac{\hbar^2}{12m^*\ \zeta_z}\right]^{\frac{1}{2}}, \tag{7}$$

$$R_{SC}^{\perp} = \left[\frac{\kappa\zeta_z}{2\pi ne^2} + \frac{4\hbar^2}{m^*}\ \frac{\zeta_z}{(\hbar\omega_c)^2}\right]^{\frac{1}{2}}. \tag{8}$$

The predicted variation of resistivity with magnetic field is shown in Figs. 1 and 2 using the screening length given by ROTH and ARGYRES [7] and by WALLACE and JOG [8].

Considering that there are no adjustable parameters, the agreement between theory and experiment is satisfactory except when the field approaches the M-I transition. The effect of non-parabolicity of the effective mass [9] in the degenerate limit is to reduce the resistivity by an amount which is negligible at low fields but which becomes significant at the critical field. This effect increases the discrepancy between theory and experiment. In the degenerate extreme quantum limit the resistivity should be independent of temperature. Reference has already been made to a small temperature dependence of the conductivity in accordance to (1). The term $AT^{\frac{1}{2}}$ is due to Coulomb interactions which causes a temperature dependence of the density of states at the Fermi level and the term BT is due to localization effects. Both terms should be small correcting terms but as the magnetic field is increased to the critical value the temperature dependence is no longer small although (1) is still obeyed up to 1 K. The theoretical values of σ_{xx} and σ_{zz} should be compared to the conductivity extrapolated to zero temperature $\sigma(0)$ and since σ increases or ρ decreases with temperature the zero temperature resistivity would be higher than the experimental curves shown in the figures. This also widens the discrepancy between theory and experiment as the M-I transition is approached.

Figure 2 shows magnetic field dependence of the Hall resistivity ρ_{xy} of sample 6715 in the same way as in [2,3]. The magnetic field which induces the M-I transition in this sample deduced from the field for which $\sigma(0) \to 0$ is 64 kG [1] and one would obtain a similar value from the field when ρ_{xy} rises rapidly. However, Fig. 4 shows that this would not be the case for the purer sample 1015 described in [5]. The value of H_{MI} determined from ρ_{xx} and ρ_{zz} is 15 kG and this agrees with the Mott formula but ρ_{xy} does not start to increase rapidly with field until the field is 30 kG. The components of the resistivity tensor for sample 1015 are consistent with magnetic freeze-out of the electrons from the conduction band to donor centres during which the conductivities consist of the sum of a component due to electrons remaining in the conduction band σ_c and a component due to hopping conduction σ_I. At a fixed magnetic field and hence donor activation energy the Hall coefficient increases to a maximum as the temperature is decreased when the two components are roughly equal and then decreases as σ_I becomes important, probably to zero, since there is evidence that the Hall coefficient for hopping conduction is very small [10]. Similar behaviour occurs in Si and Ge with zero magnetic field and in these cases for small activation energy the variation in R with temperature is found to be small. When the temperature is constant and magnetic field is increased then R increases most sharply at the temperature corresponding to the maximum R

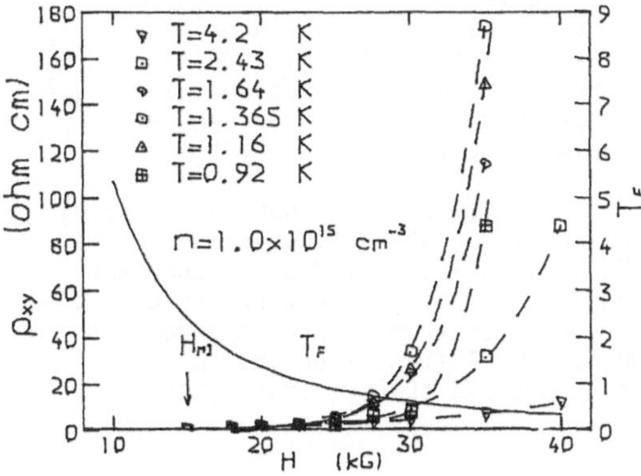

Fig.4. Transverse resistivity of ρ_{xy} of sample 1015 [5] as a function of the magnetic field. The continuous line gives the variation of the Fermi temperature. The value of H_{MI} is the field where ρ_{zz} starts to increase rapidly with field at 30 mK.

and for other temperatures the increase is less rapid. Also the increase only occurs for fields much larger than H_{MI}. For example at 22.5 kG the Hall constant of sample 1015 has a weak maximum at about 1.5 K and below 0.9 K the Hall angle becomes too small to measure. In contrast ρ_{xx} and ρ_{zz} increase by five order of magnitudes as the temperature is reduced to 0.03 K. This is a clear indication that at this field we are in the insulator region.

Consider next the nature of the M-I transition and the conduction process in the vicinity of the transition. The lightly doped InSb samples described in [5] were satisfactorily explained by a MOTT transition followed by magnetic freeze-out of the electrons onto donor centres as the field is increased. This is confirmed by impurity cyclotron measurements. It has been pointed out [3] that HgCdTe also has a low effective mass so the localization transition in this case should be similar, and as already mentioned ICR measurements [4] confirm that magnetic freeze-out occurs in lightly doped samples. Values of the H_{MI} determined in this investigation on InSb and in [3] on both InSb and HgCdTe on more heavily doped samples agree with the MOTT formula provided H_{MI} is identified with B_L in [3]. Measurements by de Vos et al [11] on magnetotransport in HgCdTe with $n = 2.4 \times 10^{14}\ cm^{-3}$ have also been satisfactorily interpreted in the same way. However, several papers have recently appeared claiming the magnetic field induces a Wigner crystallization [2,12]. The evidence in [2] has already been criticised because of its dependence on measurements of ρ_{xy}. We consider that a Wigner condensation is unlikely in any case with the type of samples used in these investigations. This is because the compensation ratio $K = N_A/N_D$ is small in the samples used. In [3] the compensation estimated from mobility measurements is described as relatively small (0.1 - 0.3) and since the mobilities quoted by the other workers were similar they must also be relatively lightly compensated. The theory however requires that the Wigner-Seitz cell around each lattice point should contain a uniform positive background charge and this could only be achieved if the number of electrons forming the lattice is small compared to the numbers of donors and compensating acceptors which provides the uniform jellium. With little compensation the background would be granular and a Wigner crystallisation unlikely. CARE and MARCH [13] quote a figure for K of 0.99 and MOTT [14] also emphasises that samples should be highly compensated. This would also make it unlikely that on the insulator side close to the transition there should be an electron-liquid flow in the Wigner condensation sense, and it is suggested that variable range-hopping occurs. The temperature dependence of the resistivity for $H > H_{MI}$ of samples 6715 and 3515 obey the relation $\rho = \rho_o \exp(T_0/T)^X$ with $X = 1/4$ [6] instead of $x = 1/2$ which was observed for the more lightly doped samples used in [5]. The value $x = 1/4$ is expected for Shklovskii wave functions in a large magnetic field [15,16] where the density of states at Fermi level is constant. The Hall effect increases as the temperature is reduced, see Fig.3, but there is no theory of the Hall effect in the variable range-hopping region in a high magnetic field with which to compare our results.

In conclusion, the resistivity of fairly heavy doped n-InSb on the metal side of the magnetically induced metal-insulator transition is in satisfactory agreement with the theory of ROTH and ARGYRES provided one uses the screening radius given by WALLACE and JOG. The agreement breaks down close to the transition where the resistivity becomes temperature dependent in accordance with (1). The transition is a MOTT transition and on the insulator side, variable range-hopping occurs and the resistivity obeys the $T^{-1/4}$ law.

ACKNOWLEDGMENTS

The authors are grateful to the SERC and Central Research Fund of the University of London for financial support. They are also grateful to A.K. Betts for technical assistance.

References

1. R. Mansfield, M. Abdul-Gader and P. Fozooni, Solid State Electronics, 28, 109 (1985).
2. T.F. Rosenbaum, S.B. Field, D.A. Nelson and P.B. Littlewood, Phys.Rev. Letters 54, 241 (1985).
3. M. Shayegan, H.D. Drew, D.A. Nelson and P.M. Tedrow, Phys.Rev.B 31, 6123 (1985); and M. Shayegan, V.I. Goldman, H.D. Drew, D.A. Nelson and P.M. Tedrow, Phys.Rev. 32,6952 (1985).
4. V.J. Goldman, H.D. Drew, M. Shayegan and D.A. Nelson, Phys.Rev. Letters 56, 968 (1986).
5. H. Tokumoto, R. Mansfield and M.J. Lea, Phil.Mag. B 93 (1982).
6. M. Abdul-Gader: Ph.D Thesis, London University 1984.
7. L.M. Roth and P.N. Argyres, Physics of III-V Compounds: Semiconductors and Semimetals, Ed: R.K. Willardson and A.C. Beer. Vol.1,159(Acad.NY,1966).
8. P.R. Wallace, J.Phys.C 7, 1136 and 4007 (1974), S.D. Jog and P.R. Wallace, J.Phys.C 8, 3608 (1975).
9. U.D. Phadke and S. Sharma, J.Phys.Chem.Sol. 36,1 (1975).
10. R.S. Klein, Phys.Rev.B 31, 2014 (1985).
11. G. de Vos, F. Herlach and H.W. Myron, J.Phys.C 19, 2509 (1986).
12. J.P. Stadler and G. Nimtz, Phys.Rev. Letters 56, 382 (1986).
13. C.M. Care and N.H. March, Adv.Phys. 24, 101 (1975).
14. N.F. Mott, The Metal-Non Metal Transition in Disordered Systems, Ed: L.R. Friedman and D.P. Tunstall, SUSSP Publication (1978).
15. B.I. Shklovskii, Sov.Phys. JETP Lett. 39, 51 (1982).
16. R. Mansfield and H. Tokumoto, Phil.Mag.B 48, 4 (1983).

Magnetic Field Induced Metal-Nonmetal Transition in GaAs-GaAlAs Heterostructures with a Spacer

A. Raymond, J.L. Robert, and C. Bousquet

Groupe d'Etude des Semiconducteurs (Associé au CRNS no. UA 357),
Université des Sciences et Techniques du Languedoc,
F-34060 Montpellier, France

The metal-nonmetal transition in GaAs-GaAlAs heterostructures has been investigated by transport experiments in the presence of a magnetic field and hydrostatic pressure. The binding energy of magnetodonors, composed of donor atoms in the doped layer of $Ga_{1-x}Al_xAs$ and electrons in GaAs separated from one another by a spacer, has been determined as a function of magnetic field for different surface densities controlled by the pressure. A simple model is presented which accounts qualitatively for the observed effects.

I - INTRODUCTION

In the last decade there have been a number of spectacular advances in constructing crystalline semiconductor heterostructures, which have had an important impact on both the fundamental physics and the potential for applications of semiconductor heterostructures and superlattices. Many of these advances are a direct consequence of the perfection of epitaxial crystal growth techniques, such as molecular beam epitaxy (MBE) or metal-organic chemical vapor-deposition (MOCVD).

The present work has been performed on modulation-doped GaAs-GaAlAs heterostructures grown by MBE and MOCVD [1]. In 1969 ESAKI and TSU [2] proposed separating the carriers from their parent impurities in order to reduce impurity scattering and in 1978, DINGLE et al. [3] successfully implemented such a concept in modulation-doped GaAs-GaAlAs superlattices, obtaining electron mobilities which far exceed the Brooks-Herring predictions. Soon after, STÖRMER et al. [4] reported the first observation of a two-dimensional (2D) electron gas at a semiconductor-semiconductor (GaAs-GaAlAs) interface. Later, in such modulation doped heterostructures, in order to achieve the separation between the 2D electrons and their parent donors, an undoped layer (spacer) immediately adjacent to the 2D channel was introduced. A typical modulation-doped structure used in the experiments is shown in Fig.1. The active layer which contains the 2D electron gas is an undoped GaAs epitaxial layer ~ 1μm thick grown on a semi-insulating GaAs substrate crystal. Electrons are introduced to the active layer from a $Ga_{1-x}Al_xAs$ layer, doped with Si donor atoms. Between the active GaAs layer and the doped $Ga_{1-x}Al_xAs$ layer an undoped $Ga_{1-x}Al_xAs$ layer constitutes the spacer.

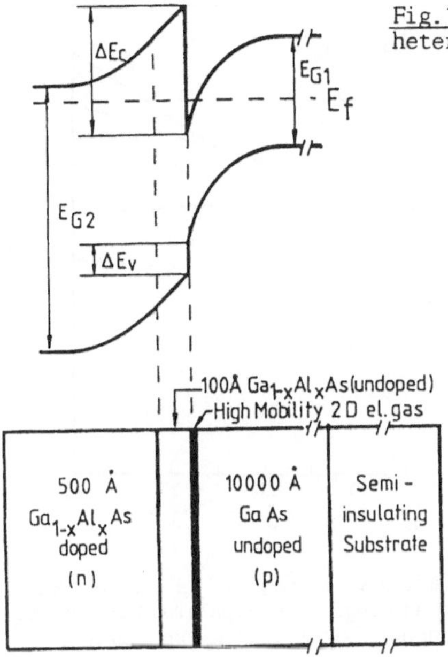

Fig.1. Modulation-doped n-type GaAs-$Ga_{1-x}Al_xAs$ heterostructure and its energy band structure

In this paper, we are interested in the localization of the 2D electrons in the presence of a high magnetic field and hydrostatic pressure in modulation-doped GaAs-GaAlAs heterostructures with a spacer. In the investigated samples the thickness of the spacer varies between 60 Å and 250 Å. As we will show, the observed localization effect cannot be ascribed to a Wigner condensation of a dilute 2D gas, but can be interpreted in terms of a metal-nonmetal transition induced by the magnetic field. As in 3D systems, the magnetic field induced metal-nonmetal transition is observed in the ultra quantum limit [5]. In order to reach the ultra quantum limit with available magnetic fields, it is necessary to deal with a 2D electron gas of a sufficiently low density. In our experiments such a condition is created by applying hydrostatic pressure.

In the first part we will explain the effect of hydrostatic pressure on Si-doped GaAs-GaAlAs heterostructures. In the second part we will give the experimental evidence of the magnetic field induced metal-nonmetal transition. In the third part a simple model will be presented which accounts qualitatively for the observed effects.

II - Effect of hydrostatic pressure on Si-doped GaAs-GaAlAs heterostructures

Hydrostatic pressure applied to a crystal leads to a change in the lattice constant. The relative variation is rather small (10^{-3}/kbar) but the effect on the electronic properties can be

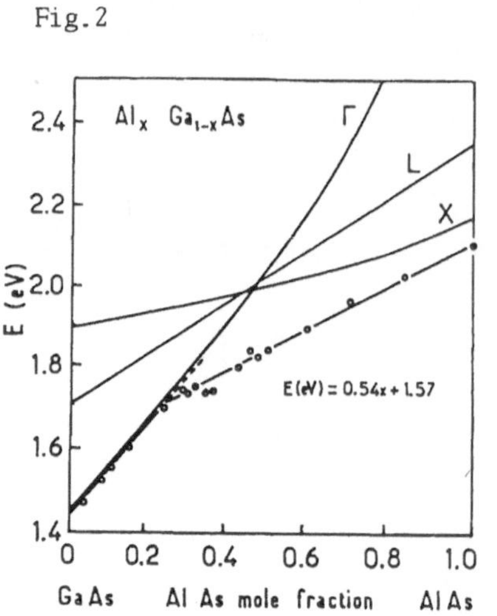

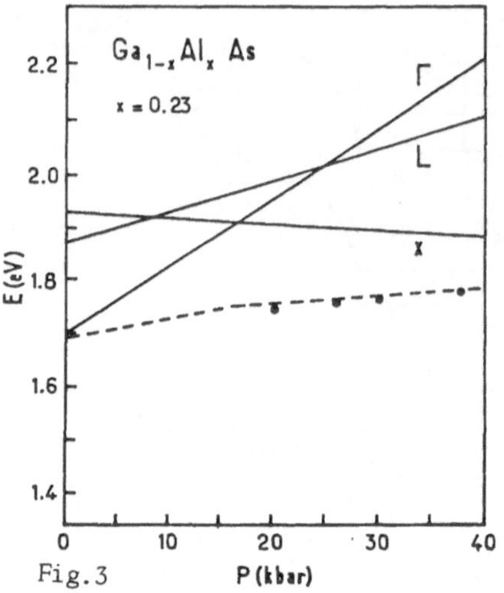

Fig.2. Variation of Γ, L and X band energy minima in $Ga_{1-x}Al_xAs$ as a function of AlAs content x [6]. The solid line through the experimental points represents the position of the Si dominant donor level in the band gap

Fig.3. Pressure dependence of the donor deep level and of the band structure of $Ga_{.77}Al_{.23}As$ [7]

important. Pressure induces a modification of the band structure. In the case of $Ga_{1-x}Al_xAs$ the effect is particularly interesting.

In Fig.2 we have reproduced the relative positions of the Γ, L and X minima as a function of x [6]. Also plotted are the measured values of the Si-donor activation energies of both deep and shallow levels with respect to the conduction band minima. The deep character of the Si donor states for x larger than 0.2 has been also observed by other authors [7,8].

Because of their deep character, the Si-donor states are very pressure dependent. As an example, Fig.3 shows the pressure dependences of the deep level donor states and of the band structure of $Ga_{1-x}Al_xAs$ for x = 0.23 [9] : when the hydrostatic pressure is increased, this deep-lying level, located in the energy gap of $Ga_{1-x}Al_xA_s$ shifts rapidly downward with respect to the conduction-band minimum. As a consequence, donor deionization takes place in GaAlAs, the energy diagram varies as shown on Fig.4 and the surface electron density n_s decreases. When the temperatures are lower than 100K the deep lying level is characterized by metastable occupation /10/. Because of lattice relaxation effects, the surface density n_s can be slightly modified for a given pressure, depending on the sample cooling speed. At sufficiently high pressures one can reduce n_s to values lower than 5×10^{10} cm^{-2}, even for highly doped samples.

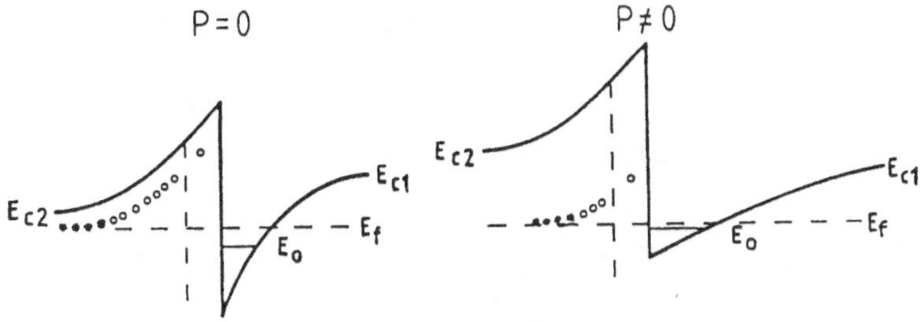

Fig.4. Schematic variation of the energy diagram of GaAs-$Ga_{1-x}Al_xAs$ heterostructures under hydrostatic pressure

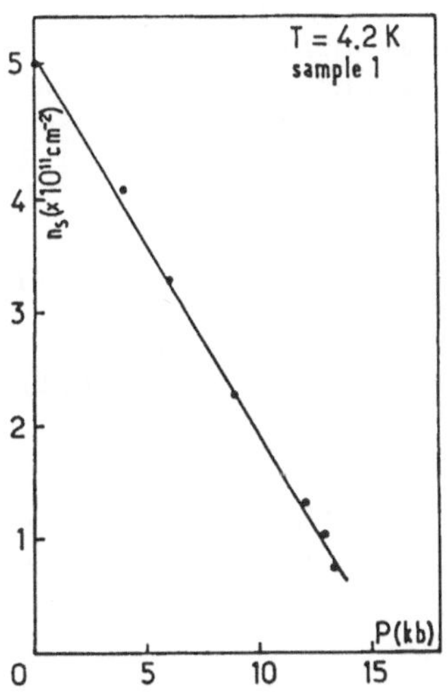

Fig.5. Typical experimental pressure dependence of the surface electron density n_s (sample 1)

This decrease of n_s with increasing pressure is shown in Fig. 5. The linear dependence has been observed at all the investigated temperatures (4.2K, 77K and 300K) and has been quantitatively explained /11/ assuming a pressure coefficient dE_d/dP = 11 meV/kbar and E_{do} = 60 meV for the Si donor level in $Ga_{0.7}Al_{0.3}As$.

III - Experimental procedure and experimental results

We have studied GaAs-GaAlAs heterostructures grown by MBE or MOCVD techniques with a spacer thickness d varying between 60 and 250 Å. The Hall coefficient and the transverse magnetoresistance were measured for two current and magnetic field directions in the temperature range 1.5 - 4.2K. Magnetic field intensities up to 19T were used. To stay within the Ohmic regime by keeping the electric field low enough to avoid impact ionization effects was a constant concern during the experiments.

In order to analyse the localization effect it is necessary to measure correctly the surface electron density n_s. In two-dimensional systems the Hall coefficient R_H can be written

$$R_H = \frac{1}{B} \frac{\sigma_{xy}}{\sigma_{xx}^2 + \sigma_{xy}^2} = \frac{1}{B} \rho_{xy} \qquad \text{(I)}$$

with, in the relaxation time approximation,

$$\sigma_{xx} = \frac{n_s e^2}{m^*} \left\langle \frac{\tau}{1 + \omega_c^2 \tau^2} \right\rangle,$$

$$\sigma_{xy} = - \frac{n_s e^2 \omega_c}{m^*} \left\langle \frac{\tau^2}{1 + \omega_c^2 \tau^2} \right\rangle ;$$

σ_{xx} and σ_{xy} are the conductivity tensor components and ρ_{xy} is the Hall resistivity equal to the Hall resistance in 2D systems. In the high magnetic field limit, i.e $\omega_c \tau >> 1$, one has $\sigma_{xy} = - n_s e/B$
On the other hand

$$\sigma_{xy} = \frac{- \rho_{xy}}{\rho_{xx}^2 + \rho_{xy}^2} = - \frac{R_H B}{\rho_{xx}^2 + R_H^2 B^2} \qquad \text{(II)}$$

which leads to $n_s = \dfrac{R_H B^2}{e(\rho_{xx}^2 + R_H^2 B^2)}$ (III)

where ρ_{xx} is the transverse magneto resistivity.

Thus, in high magnetic field conditions, the surface electron density n_s must be calculated by using equation (III) which

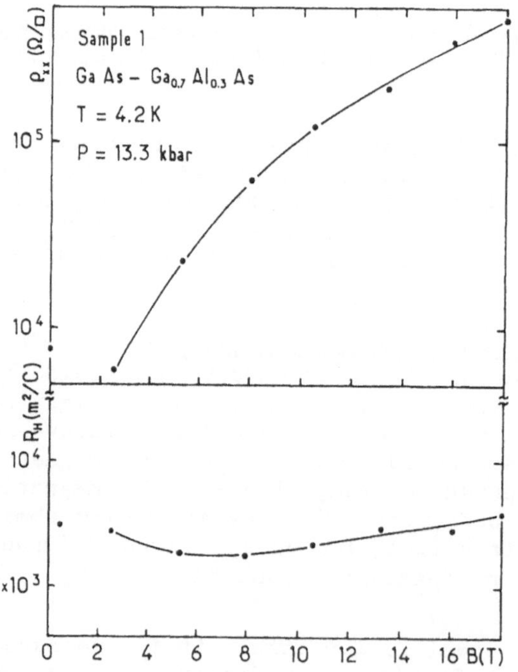

Fig.6. Magnetic field dependence of the Hall coefficient R_H and of the transverse magnetoresistivity ρ_{xx} for sample 1 at 4.2 K and under pressure of 13.3 kbar

TABLE 1

Sample characteristics: critical magnetic field B_c(T), surface electron densities n_s(cm^{-2}), surface electron densities for M-NMT n_{sc} (cm^{-2}) and mobility μ (cm^2/Vs) in GaAs-$Ga_{1-x}Al_xAs$ heterojunctions at different hydrostatic pressures. The last column gives the Mott-like criterion for the metal-nonmetal transition.

Sample	x	d (Å)	P=0 T=4.2K n_s 10^{10}	μ 10^4	P (kbar) T = 4.2 K p	n_s 10^{10}	μ 10^4	B_c	n_{sc} 10^{10}	$\sqrt{n_{sc}}\,L_c$
1	0.3	60	52	7.5	13.3	8.5	0.95	6	6	0.26
2	0.25	150	24	12.9	8.8	6.3	3	4.8	6.4	0.3
					8.8	5.7	2.36	4.2	5.6	0.29
					8.8	7.8	1.55	3.5	6	0.34
					8.8	5.1	1.84	3.3	4.5	0.3
3	0.27	250	20.8	5.2	5.9	12.5	2.0	10	11	0.27
4	0.3	250	35	41.9	13	6.5	6.82	8	6.5	0.23

in the magnetic freeze-out regime is never equivalent to the simple expression $n_s=1/R_He$.
(As we can see in Fig. 6 in this regime the condition $\rho_{xx}^2 \ll R_H^2 B^2$ is indeed never satisfied).

In Table 1, the sample characteristics at 4.2K (with and without pressure) are presented. The values of n_s and of the mobility are the Hall values, measured at low magnetic fields($B = 0.5T$).

Typical dependences of ρ_{xx} and R_H on magnetic field at 4.2K are shown in Fig. 6. Figure 7 represents the temperature and magnetic field dependence of n_s for sample 1 under 13.3 kbar. It is seen that above a critical field B_c the surface electron density n_s is thermally activated (magnetic freeze-out). This allows us to determine the critical value n_{sc} corresponding to the transition between metallic and nonmetallic types of conduction. Both quantities B_c and n_{sc} are reported in Table 1. A thermally activated density n_s can be described as $n_s = n_o \exp(-E_a/kT)$, where E_a is the activation energy. Figure 7 shows a

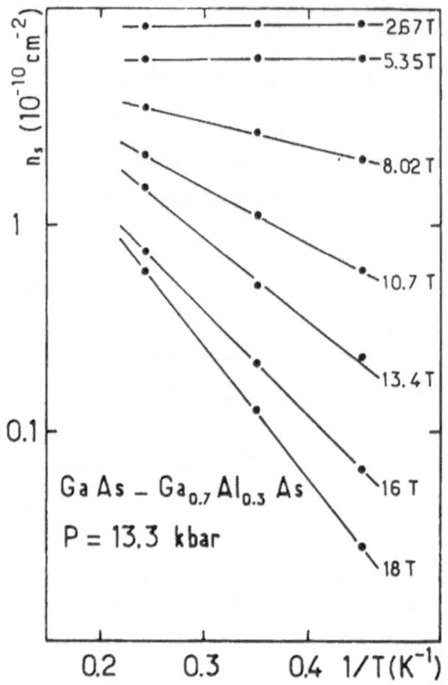

Fig.7. Temperature dependence of the surface density for different magnetic fields. Sample 1 under pressure of 13.3 kbars

linear dependence of ln (n_s) versus 1/T, then n_o is not temperature dependent and the activation energy E_a can be determined from the slopes of the curves.

Figure 8 represents the activation energies obtained for samples 1, 2, and 3 with various spacer thickneses (60, 150 and 250 Å). For samples 3 and 4, the activation energies were detectable at the higher magnetic fields only and remained very small.

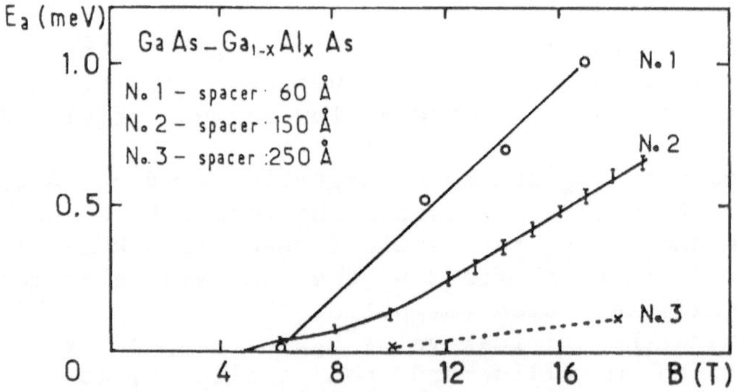

Fig.8. Activation energy E_a of magnetodonors (the inversion electron is separated from the donor atom by a spacer) for samples 1,2 and 3 versus magnetic field

For samples 1 and 2 the pressure was chosen such as to obtain approximately the same critical density n_{sc} and practically the same magnetic field B_c. Under these conditions, the overlap of the donor wave functions and the screening of the donor potentials by surface electrons are the same at the transition. Nevertheless Figure 8 shows that the magnetic field dependences of the activation energies are distinctly different for different samples : the activation energy decreases with increasing spacer thickness. This result suggests that the observed localization effect should not be ascribed to a Wigner condensation of a dilute 2D gas, which, in the case of similar electron densities, would lead to the same activation energy value.
On the other hand, if the observed decrease of n_s is due to a localization of electrons in GaAs on the donors in $Ga_{1-x}Al_xAs$, the E_a (B) variation would strongly depend on the distance between them, i.e. on the thickness of the spacer. This is exactly what is observed. This is further confirmed by the experiments of Mendez et al. /12/ with a larger spacer thickness of 520 Å and comparable surface electron density n_s, in which no temperature-activated process was observed even at high magnetic fields.

IV - A simple model of a magnetodonor

In order to be more precise we have proposed /1/ the following simple model : we consider a donor atom at a distance d from a GaAs-GaAlAs interface and for simplicity assume that the electrons in GaAs are perfectly 2D. If d is smaller than the radius λ of the donor wave function, a semiclassical picture of the situation may be used (cf. Fig. 9). Since the 2D electron gas may not penetrate into $Ga_{1-x}Al_xAs$, in the bound donor state its wave function is given approximately by a disc, resulting from the intersection of the sphere of radius λ with the interface plane. BASTARD has shown /13/, by variational

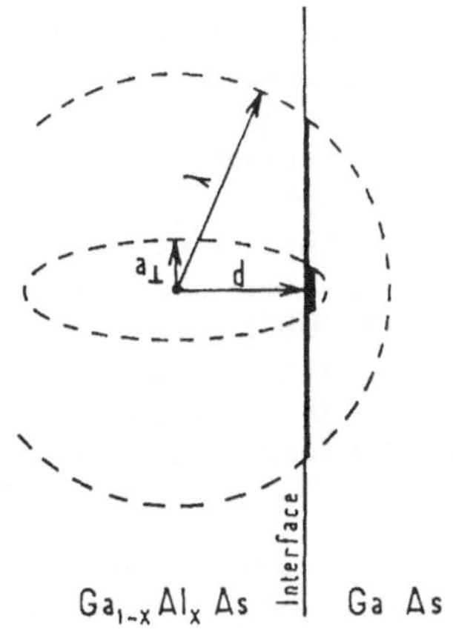

Fig.9. Semiclassical model for the surface electron in GaAs interacting with the Si donor atom in the doped layer of GaAlAs: d is the thickness of the spacer, λ the effective radius of the donor wave function at B = 0, and $a_\perp$ the transverse radius of the orbit of the 3D electron bound to a donor at B = 0. Thick lines indicate intersections of the donor wave functions with the interface for B = 0 and B $\neq$ 0 respectively

calculations, that the donor binding energy E_a in this case is a small fraction of the bulk effective Rydberg R_y* and its effective radius λ is much larger than the Bohr radius a^*_B (a^*_B = 100 Å for GaAs).

It is well known from bulk investigations that in the presence of a magnetic field the donor wave function has a cigar shape and that its dimensions decrease with increasing magnetic field /14/. It appears from Fig. 9 that the intersection of the donor wave function with the interface gives in this case a smaller disc, so that the electron is on average closer to the donor atom than without a magnetic field. As a result, the Coulomb binding energy is enhanced by the presence of the magnetic field. This results in a magnetic freeze-out which we do observe experimentally. It is also clear that under high field conditions the binding energy of such a magnetodonor should be much smaller than the one in bulk GaAs, since in the latter case the electron is on average closer to the donor atom. This is in fact what we observe : for B = 17 T we measure for sample 1 E_a = 1 meV, whereas for the bulk magnetodonor in GaAs at the same field one calculates E_d = 12.5 meV /14/. The above reasoning is qualitatively valid also for $\lambda < d$, although in this case one should not use the semiclassical picture.

A metal-nonmetal Mott-type transition is usually associated with an overlap of the impurity wave functions. As shown by YAFET et al./14/, at high magnetic fields, the 3D electron bound to a donor moves on orbit whose transverse radius $a_\perp$ is nearly equal to magnetic length $a_\perp \simeq L = (\hbar/eB)^{1/2}$. As follows from Fig. 9, one expects that the surface electron moves on orbit with the same radius. On the other hand, the average distance between surface electrons at the transition is $n_{sc}^{-1/2}$. Thus, the overlap condition for the Mott transition is approximately given by $2\ L_c \simeq n_{sc}^{-1/2}$, i.e., $n_{sc}^{1/2}\ L_c \simeq 0.5$. As it appears in Table 1, we find, in fact, that the product $n_{sc}^{1/2}(\hbar/eB_c)^{1/2}$ is close to that value. In Fig.10 we report the experimental activation energy for sample 2, measured at the same pressure P = 8.8 kbar, for which the surface density n_s has been varied by changing the sample cooling speed. In this case one deals with the same spacer thickness but we observe that, to a first approximation, and particularly for a magnetic fields much larger than the critical magnetic field, the activation energy increases with decreasing n_s. As this has been confirmed by the theoretical calculations of ZAWADZKI and KUBISA /15/ we believe that the screening of the donor potentials by surface electrons plays an important role in the metal-nonmetal transition : as n_s decreases, the screening becomes weaker and the magneto-donor binding energy increases.

These results show that in GaAs-GaAlAs heterostructures having not too wide a spacer, the Coulomb interaction between

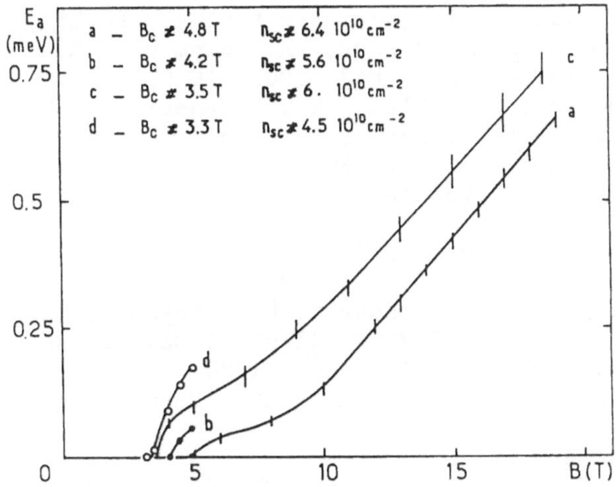

Fig.10. Magnetodonor activation energies E_a for sample 2, (under pressure of 8.8 kbar) versus magnetic field. The lines are drawn to guide the eye. Different E_a (B) dependences correspond to various n_s values, obtained by different speeds of cooling of the sample. (Cf. Table 1)

donor atoms in $Ga_{1-x}Al_xAs$ and inversion electrons in GaAs, is of essential importance, and it must not be neglected in the investigations of the quasi two-dimensional electron gas.

Acknowledgements

The authors would like to thank Drs. J.P. André, P.M. Frijlink (LEP), F. Alexandre and J.M. Masson (CNET) for providing the samples. The authors acknowledge SNCI-CNRS and Drs L. Konczewicz, E. Litwin-Staszewska, and R. Piotrskowski for their participation in some experiments supported by MRT and CNRS. The authors are grateful to Prof. W. Zawadzki and Dr. G. Bastard for many valuable discussions.

REFERENCES

1. J.L. Robert, A. Raymond, L. Konczewicz, C. Bousquet, W.Zawadzki, F. Alexandre, I.M. Masson, J.P. André, P.M. Frijlink: Phys. Rev. B33, 5935 (1986)
2. L. Esaki, R. Tsu: IBM Research Note RC 2418 (1969)
3. R. Dingle, H.L. Stormer, A.C. Gossard, W. Wiegmann: Appl. Phys. Lett. 33, 665 (1978)
4. H.L. Störmer, R. Dingle, A.C. Gossard, W. Wiegmann, M.D. Sturge: Solid State Commun. 29, 705 (1979)
5. A. Raymond: Proc. Int. Conf. "The Application of High Magnetic Fields in Semiconductor Physics". Grenoble 1982.

Lecture Notes in Physics, Vol. 177 (Springer-Verlag, Berlin, Heidelberg (1983) p.344; J.L. Robert, A. Raymond, R.L. Aulombard, C. Bousquet, Philos. Mag. B42, 1003 (1980)
6. N. Chand, T. Henderson, J. Klem, W.T. Masselink, R. Fischer, Y.C. Chang, H. Morkoc: Phys. Rev. B 30, 4481 (1984)
7. A.K. Saxena: J. Phys. C. 13, 4323 (1980)
8. H.J. Lee, L.Y. Juravel, J.C. Woolley: Phys. Rev. B21, 659 (1980)
9. A.K. Saxena: Appl. Phys. Lett. 36, 79 (1980)
10. R.J. Nelson: Appl. Phys. Lett. 31, 351 (1977)
11. J.M. Mercy, C. Bousquet, J.L. Robert, A. Raymond, G. Gregoris, J.Beerens, J.C. Portal, P.M. Frijlink, P.Delescluse, J. Chevrier, N.T. Linh: Surf. Sci. 142, 298 (1984)
12. E.E. Mendez, M. Heiblum, L.L. Chang, L. Esaki: Phys. Rev. B28, 4486 (1983)
13. G. Bastard (private communication).
14. Y. Yafet, R.W. Keyes, E.N. Adams: J. Phys. Chem. Solids, 1, 137 (1956)
15. W. Zawadzki, M. Kubisa: to be published.

Part VI

Semimagnetic Semiconductors, IV-VI Materials

High Field Investigations on IV–VI Semiconductors

H. Pascher[1] *and G. Bauer*[2]

[1]Physikalisches Institut der Universität Bayreuth, D-8580 Bayreuth, Fed. Rep. of Germany

[2]Institut für Physik, Montanuniversität Leoben, A-8700 Leoben, Austria

Introduction

In recent years the interest in IV-VI lead chalcogenides has grown since they are very useful substances for optoelectronic devices in the infrared region of the spectrum /1-3/. PbS has served as a very sensitive infrared detector for many years. From the pseudobinary alloys $Pb_{1-x}Sn_xTe$ and $Pb_{1-x}Sn_xSe$ diode lasers as well as detectors are manufactured in the range between 6 μm and 20 μm. Using $Pb_{1-x}Eu_xTe_{1-y}Se_y$, laser operation down to wavelengths of 3.8 μm has been achieved.

The semimagnetic alloy $Pb_{1-x}Mn_xTe$ shows an amount of spin splitting which is nearly independent of the applied magnetic field. This behaviour is very unusual for dilute magnetic semiconductors and has to be further investigated.

In the last few years, remarkable progress has been made in crystal growing technology. Either by hot wall epitaxy /4,5/ or by MBE, /2,3/ epitaxial films, quantum well structures, superlattices and nipi-crystals of high quality have been grown on various substrate materials. Carrier concentrations as low as $10^{16}/cm^3$ and mobilities up to 50 000 cm^2/Vs (77 K) can be obtained.

For the investigation of the electronic properties of such high quality material, magnetooptical methods are very well suited. Therefore in this paper we concentrate on the results of inter- and intraband absorption, luminescence and coherent Raman techniques in order to evaluate the bandstructure of various IV-VI compounds in high magnetic fields.

The outline of this paper is as follows:
Section I briefly reviews theoretical approaches to the description of the bands lying near the Fermi level. Section II describes the different magnetooptical methods with the example of PbTe. Section III shows the results for the other cubic compounds. Section IIIa deals with $Pb_{1-x}Mn_xTe$ and Section IV with heterostructures, superlattices and nipi crystals.

I. Band Model

The semiconductors considered here have direct gaps at the L-point of the Brillouin zone. Surfaces of constant energy are ellipsoids with their symmetry axis along the ⟨111⟩ direction. The angle between magnetic field and [111] axis we call Θ.

Near the Fermi level there are three valence and three conduction bands within an energy spread of less than 5 eV. Other energy levels are about 8 eV removed and can be neglected in band structure calculations valid for optical investigations.

The structure of the lowest conduction and highest valence band can be calculated by a $\vec{k}\cdot\vec{p}$ treatment as given by Mitchell and Wallis /6/. The interaction of the lowest conduction (L_{62}-) and highest valence (L_{61}-) level is treated exactly, the other four more distant levels are incorporated in k^2 approximation. The model has 11 parameters, which are listed in Table I.

Table I: Band parameters in Mitchell and Wallis notation

Two-band parameters	Far-band parameters
E_g: energy gap	m_t^+/m_o : contribution of the four remote bands to the transverse mass of the valence band
$2P_\perp^2/m_o$: interband matrix element	m_t^-/m_o : contribution of the four remote bands to the transverse mass of the conduction band
a_{cv} : $P_\perp/P_\parallel$ ($\perp$ and // are related to [111])	$a^+ = (m_l^+/m_t^+)^{1/2}$
	$a^- = (m_l^-/m_t^-)^{1/2}$
	g_t^+, g_l^+ : contribution of the four remote bands to the transverse and longitudinal g-factor, respectively, of the valence band
	g_t^-, g_l^- : contribution of the four remote bands to the transverse and longitudinal g-factor, respectively, of the conduction band

The Landau level energies are obtained by diagonalization of a 4x4 matrix for each Landau quantum number /7/. This model neglects the interaction of Landau levels with different quantum numbers /44/. This interaction is taken into account by Bangert /8/ and by Zawadzki /8a/.

Numerical calculations with the two models reviewed, the 4x4 model and the diagonalization of much bigger matrices, show very small differences. For example, with a set of band parameters valid for $Pb_{1-x}Sn_xTe$ with a Sn content $x = 0.135$, for a magnetic field B=4T, $\Theta=70.53^o$ and the conduction band level 4^+ we find the two results 103.494 meV and 103.551 meV , respectively. As can be seen, the difference is less than one in a thousand. Similar results are obtained when calculating transition energies for allowed optical transitions. For band parameters valid for PbTe the results are exactly the same for both models. In this paper therefore we always used the 4x4 matrix, which is much faster to calculate.

As noted above, nowadays most experiments are carried out using epitaxial films. With such samples problems arise with strain, caused either by a lattice mismatch or by differences in the thermal expansion coefficients between substrate and film /23/. A reasonable lattice match is obtained in the system PbTe/BaF_2 with growth direction [111] and we concentrate on this case. The lattice constant of bulk PbTe is larger than that of BaF_2 by about 3%. As a consequence, the cubic epitaxial PbTe layer contracts in the interfacial plane and has to expand in the growth direction. At low temperatures there are also differences in the thermal expansion coefficients, which lead to further dilatational strain in the PbTe film.

The strain lifts the fourfold degeneracy of the $\langle 111\rangle$ valleys. The energy shifts are as follows /8,9,20/:

$$\delta E_{cv}^{[111]} = (3D_d^{c,v} + U_u^{c,v})\varepsilon_d + 2D^{c,v}\varepsilon_s ,$$

$$\delta E_{cv}^{\langle\bar{1}11\rangle} = (3D_d^{c,v} + D_u^{c,v})\varepsilon_d - \frac{2}{3}D_u^{c,v}\varepsilon_s ,$$

where $\varepsilon_{s,d}$ are the shear and dilatational components of the strain tensor, respectively, and $D_{u,d}$ are the deformation potentials. It turns out that the shifts in the valence- and conduction bands are nearly symmetric. Usually no differences in the gaps are observed for the [111] and the ⟨$\bar{1}$11⟩ valleys within the experimental accuracy of about 0.5 meV - 1 meV. The energy shifts can only be observed in band filling effects.

II. Magnetooptical Experiments

Precise information on the electronic band parameters can be obtained from magnetooptical intraband transitions (between states in a single band) and interband transitions (involving states in the valence and conduction bands). Using laser radiation with fixed frequencies, the Landau levels of the sample can be tuned by an applied magnetic field and thus for different photon energies the dependence of the transition energies can be determined as a function of the magnetic field /10,11/.

Interband transitions can be observed in absorption, reflection or in luminescence. For bandstructure investigations absorption is preferable, since the wavelength of the luminescence light is strongly influenced by the Fermi level /41,42/.

Intraband transitions are observable either by cyclotron resonance or by Raman techniques. Since the selection rules are different, the two methods are complementary.

Zawadzki has calculated the matrix elements which determine the interband and intraband magnetooptical transitions. The results for $\vec{B}$//[111] are summarized in Table II. With other directions of the magnetic field in addition to the listed ones, weak transitions are possible. The $\Delta n=2$ interband transitions are also very weak compared to those with $\Delta n=0$.

Table II: Polarization selection rules

polarization	E_z		E_-		E_+	
transition	Δn	Δs	Δn	Δs	Δn	Δs
1-photon interband	0	0	0 +2	+1 -1	0 -2	-1 +1
2-photon interband	+1 -1	-1 +1	+1 +3	+1 -1	-1 -3	-1 +1
1-photon intraband	+1 -1	-1 +1	+1 0	0 +1	-1 0	0 -1

1. Interband Transitions

If the photon energy is chosen larger than the bandgap plus Fermi energy, interband transitions in a magnetic field can occur at $k_z=0$. Since there the combined density of states is maximal, in transmission and reflectivity experiments $k_z=0$ transitions are observed. The real part of the dielectric function has only small modulations with magnetic field, therefore transmission minima occur exactly at those magnetic fields where the energy difference of the levels involved equals the photon energy. The resonance positions are not influenced by changes in the reflectivity of the sample surfaces or by Fabry-Perot modes of thin epitaxial layers.

If the substrate is not transparent for light with wavelengths near the bandgap of the film, reflectivity can be measured as a function of the magnetic field. But the interband transitions cause only weak changes of the reflected intensity. Therefore magnetic field modulation techniques must be used. In such an experiment the first derivative of the reflectivity is recorded as a function of the magnetic field and thus the resonances are at the minima of these curves.

In principle all band parameters can be determined from interband transitions between different Landau levels in the Faraday and Voigt configurations measured as a function of the magnetic field. But the precision of the data can be improved if in addition intraband transition energies are included in the fitting procedure.

2. Intraband Absorption

IV - VI compounds, especially PbTe, have been the subject of numerous investigations of magnetooptical intraband transitions /7,8,19,21,22/. These materials have rather low transverse optic phonon frequencies (PbTe: ω_{TO} = 18 cm^{-1} at 4.2 K) and large LO-TO-phonon splittings (PbTe: ω_{LO} = 114 cm^{-1}). The relatively high free carrier concentrations ($n,p > 10^{16}/cm^3$) are the reason why the Fermi energy is at least of the order of few meV within either the conduction or valence band. Therefore the analysis of experimental far infrared transmittance or reflectance data is usually based on a line shape analysis using an appropriate magnetoplasma theory.

The dielectric response of a PbTe like semiconductor in a magnetic field has been calculated using the classical equation of motion method as well as a linear response theory.

Since we deal with thin films grown in a [111] growth direction Fig.1 illustrates the orientation of the applied magnetic field with respect to the four valleys at the L-point of the Brillouin zone in Faraday ($\vec{B}$// [111]) or Voigt geometry ($\vec{B}$//[$1\bar{1}0$], [$1\bar{1}2$]).

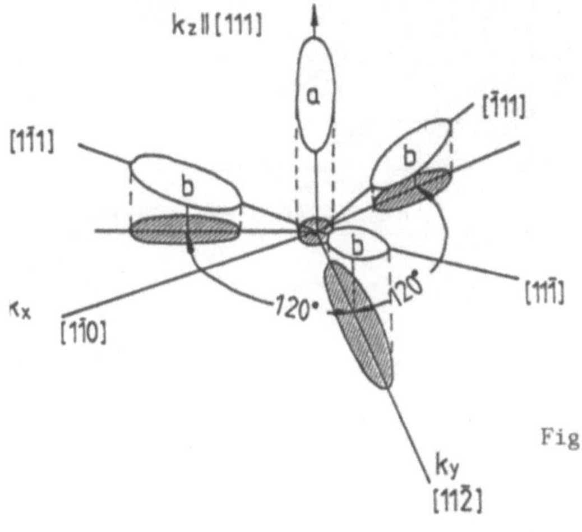

Fig.1: Three- and two-dimensional carrier pockets in the many-valley semiconductor PbTe

The propagation direction of radiation is parallel to the [111] direction. In the classical model, the free carriers of each L-valley are assumed to move according to a Drude equation:

$$\hat{m}\,\dot{\vec{v}} = q(\vec{E}+\vec{v}\times\vec{B}) - \omega_\tau \hat{m}\vec{v}$$

with a time dependence $\vec{v}(t) \sim \exp(-i\omega t)$; ω_τ denotes a damping parameter and $\hat{m}$ is the effective mass tensor. Since the current density $\vec{j} = n\,e\,\vec{v} = \hat{\sigma}\vec{E}$, the contributions of the high frequency conductivity $\hat{\sigma}^{(i)}$ due to the four valleys in the coordinate systems where $\hat{m}$ is diagonal are calculated. After the appropriate coordinate transformations the frequency-dependent total dielectric function is given by

$$\varepsilon = (\varepsilon_\infty + \chi_{ph}(\omega))\hat{1} + \left(\frac{1}{\varepsilon_\infty \omega}\right) \hat{\sigma}(\omega) \ ,$$

where $\varepsilon_\infty = 1+\chi_\infty$,

$\chi_{ph} = \Delta\chi\omega_{TO}^2/(\omega_{TO}^2-\omega^2-i\omega\Gamma)$,

Γ : phonon damping parameter,

$\Delta\chi$: oscillator strength,

χ_{fc} : free carrier contribution to suscpetibility.

In Faraday configuration, for the two circular polarizations ($\pm$) the free carrier contribution $\chi^{\pm}$ including damping was given by Burkhard et al. /7/.

In Voigt configuration ($\vec{B}//[1\bar{1}0]$, $\vec{k}//[111]$) the corresponding expressions for the ordinary mode ($\vec{E}//\vec{B}$), $n_o^2=\varepsilon_{xx}$, and for the extraordinary mode, $n_e^2=\varepsilon_{yy}-\varepsilon_{yz}\cdot\varepsilon_{zy}/\varepsilon_{zz}$ with $\vec{B}//[1\bar{1}0]$ and $\vec{E}\perp\vec{B}$, including explicitly free carrier damping as well, were given by Krost et al. /19/. For light propagation parallel to a [111] direction, using constant laser frequency and a varying magnetic field, in the ordinary mode just one resonance occurs at $\omega=\omega_{cb}= qB((2m_l+m_t)/(3m_t^2m_l))^{1/2}$ whereas in the extraordinary mode three resonances appear as shown in Fig.2a. One is the so-called hybrid resonance at a position close to $\omega = 1/2\,(\omega_{ca}^2+\omega_{cb}^2)^{1/2}$ where $\omega_{ca} = qB/\ m_t m_l$ (if the lattice contributions to the dielectric function were constant; for polar materials a

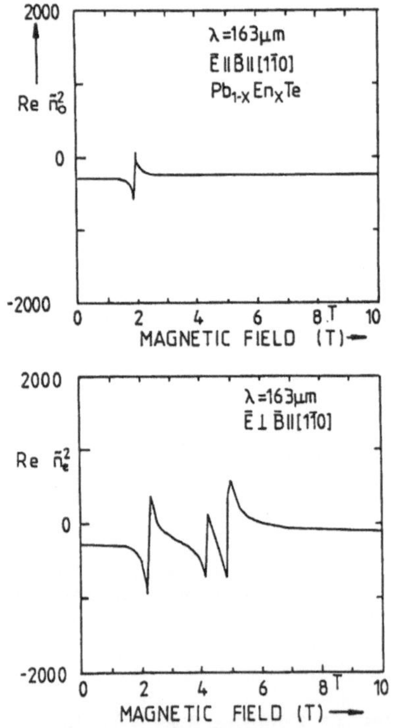

Fig.2a: Refractive indices of ordinary and extraordinary mode for PbEuTe. k//[111] ; parameters as /19/

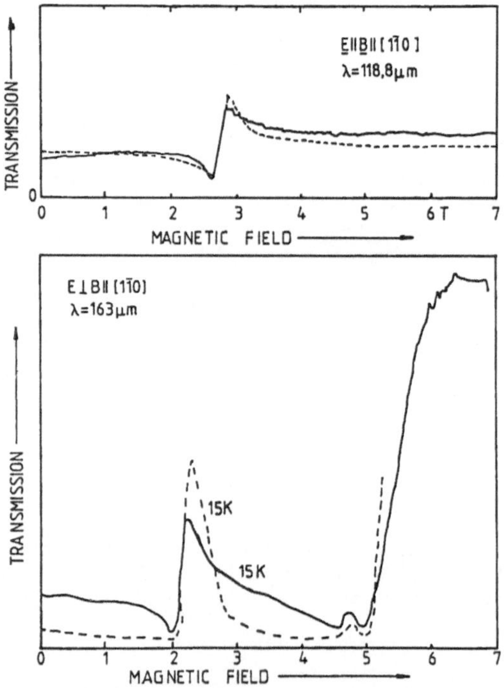

Fig.2b: Experimental data for ordinary and extraordinary modes together with calculated data (dashed) after /19/

shift towards ω_{cb} occurs). Then a resonance at $\omega \approx \omega_{ca}$ occurs and finally for higher magnetic fields a resonance due to a zero of ε_{zz} between ω_{ca} and the phonon frequency. These structures are clearly observable in the experimental data.

The laser frequency ω_L determines the shape of the reflectivity or transmission as a function of magnetic field. For $\omega_L < \omega_p^+ = (\omega_{LO}^2 + \omega_p^2)^{1/2}$ the resonances are associated with dielectric anomalies since the real part of the dielectric function is negative at B = 0. (ω_p denotes the plasma frequency.) For $\omega_L > \omega_p^+$, $Re\{\varepsilon\}>0$ and thus transmission minima correspond to resonances.

The classical model for the free carrier contribution to the dielectric function has several shortcomings: (i) the IV-VI compounds are narrow gap semiconductors and therefore strongly nonparabolic. The transitions cannot be characterized by a single cyclotron mass. The line shape of the spectra depends on the actual dependence of the Landau levels on the carrier momentum k_B along $\vec{B}$ and the position of the Fermi level. A linear response theory for the free carrier part of the dielectric function as formulated by Wallace /36/ should be used instead. The application to PbTe and $Pb_{1-x}Ge_xTe$ was described in Refs. /7,8/.

In the [111] oriented epitaxial film, the biaxial strain present due to the differences in the expansion coefficients between the IV-VI-film and the substrate lifts the degeneracy of the four carrier pockets and thus leads to unequal population /23/. For realistic strain splittings therefore the magnetic field dependence of the Fermi level has to be calculated in order to assign the transitions. Despite the fact that for B = 0 the [111] valley is oriented parallel to the field and is thus more populated, for sufficiently high fields it becomes depopulated and all carriers will be distributed among the three oblique valleys.

3. Coherent Raman Scattering

It is well known that semiconductors, especially those with narrow bandgaps have very large nonlinear optical susceptibilities. Since the cubic IV-VI compounds crystallize in the NaCl-structure, which is inversion symmetric, the third is the lowest nonvanishing order of nonlinear susceptibility. This one is responsible for coherent anti-Stokes Raman scattering, which is a well-known technique for observing Raman-allowed transitions. Two laser beams with frequencies ω_S and ω_L interact in the sample and generate radiation with frequencies $\omega_L \pm i(\omega_L - \omega_S)$, i being an integer. In the following we are concerned with $\omega_{AS} = 2\omega_L - \omega_S$. For the intensity I_{AS} the following equation holds:

$$I_{AS} = L^2 \frac{9}{16} \frac{\omega_{AS}^2}{c^4 n_{AS} n_L^2 n_S \varepsilon_o^2} |\chi|^2 I_L^2 I_S \left(\frac{\sin(\Delta\vec{k}\vec{x}_o L/2)}{\Delta\vec{k}\vec{x}_o L/2}\right)^2 ,$$

$$\Delta\vec{k} = 2\vec{k}_L - \vec{k}_S - \vec{k}_{AS} ;$$

$\vec{k}$ being the wave vectors, $\vec{x}_o$ a vector with length 1 parallel to $\vec{k}_{AS}$, L the sample length, n the linear refractive index, I the intensities and c the velocity of light. In thin epitaxial layers the phase factor can be approximated by 1. The susceptibility χ has nonresonant contributions caused by the nonparabolicity of the bands and is resonantly increased if $\omega_L - \omega_S$ is within the linewidth of a Raman transition. The interference of the real nonresonant and the complex resonant parts in general causes complicated line shapes of I_{AS} as a function of the magnetic field. These line shapes and the procedures to find the resonant magnetic fields are discussed in detail in /Ref. 12/.

Figure 3 shows as an example a recording of I_{AS} as a function of the magnetic field of a PbTe epitaxial film in Voigt configuration, measured with

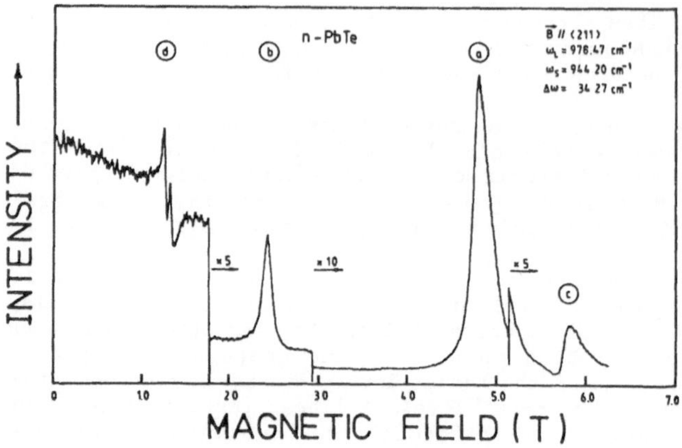

Fig.3: CARS intensity as a function of the magnetic field.

two simultaneously Q-switched CO_2 lasers. The resonances marked a, b and d are due to spin flip transitions ($\hbar(\omega_L - \omega_S) = g^*/u_B B$) in the valleys with $\Theta = 90^o$, 61.81^o and 19.47^o, respectively, resonance c is due to a combined spin flip ($\hbar(\omega_L - \omega_S) = \hbar\omega_c - g^*/u_B B$). As can be seen from this example, the coherent Raman technique allows the determination of effective g-values and masses as well.

III. a) Band Structure of Cubic Compounds

Table III shows the effective g-values and masses of different IV-VI semiconductors as directly measured by the methods described in the preceding section.

Table IV gives the band parameters of the materials, fitted from inter- and intraband data.

In PbTe the anisotropy $P_\perp^2/P_\parallel^2$ is different from the far-band anisotropies g_l^+/g_t^+, g_l^-/g_t^-, even in sign. In PbSe the quotient of the valence band masses differs most from $P_\perp^2/P_\parallel^2$. All these differences lead to nonspheroidal shapes

Table III: Effective g-values and masses

Parameter	PbTe	PbSe	$Pb_{0.865}Sn_{0.135}Te$	$Pb_{0.83}Sn_{0.17}Te$
m_l^{cb}/m_o	0.21	0.0685	0.175	0.14
m_t^{cb}/m_o	0.021	0.0370	0.0175	0.0135
m_l^{vb}/m_o	0.28	0.0655	0.17	0.16
m_t^{vb}/m_o	0.024	0.0360	0.017	0.014
g_l^{cb}	66.1	41.08	-	104.0
g_t^{cb}	16.7	32.5	-	31.8
g_l^{vb}	65.6	-	99.0	100.4
g_t^{vb}	13.5	30.6	28.0	31.8

Table IV: Bandparameters in Mitchell and Wallis notation

Parameter	PbTe	PbSe	$Pb_{0.865}Sn_{0.135}Te$	$Pb_{0.83}Sn_{0.17}Te$
E_g	189.7meV	146.3meV	113.0meV	94.6meV
$2P^2/m_o$	6.02eV	3.60eV	5.62eV	5.66eV
P /P	3.42	1.35	3.42	3.38
m_t^+/m_o	-0.102	-0.29	-0.095	-0.089
m_t^-/m_o	0.060	0.27	0.063	0.063
m_l^+/m_o	-0.92	-0.37	-0.63	-1.65
m_l^-/m_o	0.50	0.95	0.69	0.41
g_t^+	4.39	5.4	1.00	0.00
g_t^-	-1.39	-4.0	-2.33	-1.50
g_l^+	-2.61	7.3	-7.12	0.24
g_l^-	1.72	-8.1	6.46	-2.44

of the surfaces of constant energy. The most striking difference between the bandparameters of PbTe and PbSe are the contributions of the far-bands to g-values and masses. Whereas in PbTe the values g_l^-, g_t^-, g_l^+, g_t^+ are relatively small, in PbSe they reach 25%. On the other hand, in PbTe the far-band contributions to the masses are relatively large, but in PbSe small.
These differences between the two IV-VI compounds are caused by differences in the level ordering between the two materials. These different level orderings resulted from pseudopotential calculations by Bernick and Kleinmann /13/. Their pseudopotential parameters are quite correct for PbSe. With PbTe there are some discrepancies between their values and the correct g-factors /14/.

III.b) Semimagnetic IV-VI compounds

Semimagnetic semiconductors have recently attracted much interest due to their peculiar magnetic field and temperature-dependent properties. If Pb in PbTe or PbS is partially replaced by Mn^{2+} ions (half-filled $3d^5$ shell) or Eu^{2+} (half-filled $4f^7$ shell) one can expect some differences with respect to the semimagnetic II-VI compounds: The minimum gap at the L-point and the different host band structure will influence the exchange interaction between the localized Mn electrons and the mobile band electrons. Indeed, in PbMnTe and in PbMnS with Mn contents up to 4 %, a drastic effect, namely a splitting of the energy gap, is observed at vanishing external magnetic field /15,17/. On the other hand, susceptibility and magnetization measurements do not exhibit any evidence for ferromagnetic interaction but just the expected paramagnetic behaviour /18,37,38/.

From these data a weak antiferromagnetic exchange interaction between the Mn ions is deduced with exchange constants which are typically a factor of 4-5 smaller than for semimagnetic II-VI compounds. Magetooptical interband transitions on n- or p-type samples clearly reveal this splitting and indicate that, at a given temperature, exchange-induced corrections to the g-factors are negligible (for PbMnTe, x²0.03) /16/ or at least 4 times smaller than for semimagnetic II-VI compounds in the case of $Pb_{1-x}Mn_xS$ /17/. In the latter case the corrections decrease with increasing Mn content. The field dependence of the macroscopical magnetization which usually governs the exchange-induced corrections to the effective g-factor (within a mean field approximation) is not of importance.

The temperature dependence of the splitting of the gap does not follow the temperature dependence of the magnetization (or susceptibility). It decreases with increasing temperature much faster. Due to the different selection rules for magnetooptical interband transitions in the Faraday and Voigt geometries, measurements in these two configurations reveal unambiguously that more than 90% of the observed value of the splitting is in the valence band states. This is in agreement with magnetooptical intraband transitions in n-type samples, which do not show any indication for zero-field split conduction band states. Intraband experiments in Voigt geometry $\vec{E}//\vec{B}//[1\bar{1}0]$ reveal the presence of a spin-flip transition in n-PbMnTe in addition to the tilted orbit resonance. This transition is apparently induced by the presence of Mn ions and disappears for vanishing Mn content.

The semimagnetic lead compounds are typical examples of a breakdown of the usually applied mean field formalism treating the spin-spin exchange interaction. Recently von Ortenberg /35/ proposed considering the non-diagonal terms of this interaction in order to illustrate the possibility of a zero-field splitting.

IV. Heterostructures, Single and Multi-Quantum-Well Structures, and Superlattices

After the pioneering work of Schaber and Doezema /21/ on inversion and accumulation layers on bulk PbTe films, several groups have studied quasi two-dimensional carriers in single, multi-quantum-well structures and superlattices. PbTe/$Pb_{1-x}Sn_xTe$ MQWs and SLs were grown and studied by Fujiyasu /5/ and coworkers and Clemens et al. /4/. Partin /2/ has grown lattice-matched PbTe/$Pb_{1-x}Eu_xSe_yTe_{1-y}$ (x up to 0.1, y up to 0.096) single and multi-quantum-well structures using molecular beam epitaxy. Since mainly a [111] growth direction was used, according to Fig. 1, two systems of size quantized sublevels E_{1a}, E_{2a} ... as well as E_{1b}, E_{2b} associated with the a-valleys aligned along the [111] direction and the three b-valleys aligned oblique to it occur. The relevant m_z masses (z//[111]) responsible for the sublevel energy differences are $m_z = m_l$ (a-valley) and $m_z = 9m_t m_l/(m_t+8m_l)$ (b-valleys). Due to the large mass anisotropy ($K = m_l/m_t \approx 10$)the subband spacings of the two valley systems differ considerably. In addition, the biaxial strain present in the heterostructures usually causes different values of conduction and valence band discontinuities for the [111] and the three oblique $\langle 1\bar{1}1\rangle$ valleys /23/ (inset of Fig. 4).

For a single or multi-quantum well structure the two-dimensional Fermi surface consists of a circle and three ellipses with main axis m_t, m_t and m_t, $(m_t+8m_l)/9$, respectively, as shown in Figs.1,4 together with the appropriate orientations.

For high magnetic field investigations in the $B_\perp$ geometry ($\vec{B}\perp$ to the layers, i.e. parallel to the [111] growth direction) in this simple picture the motion in the xy-plane and in the z-direction are decoupled (for the carriers in the a-pocket). Thus the cyclotron mass is $m_c=m_t$ (the projected [111] ellipsoid, the circle in Fig. 4), whereas for the three oblique valleys $m_c= 1/3m_t(1+8m_l/m_t)^{1/2}=(m_x m_y)^{1/2}$.

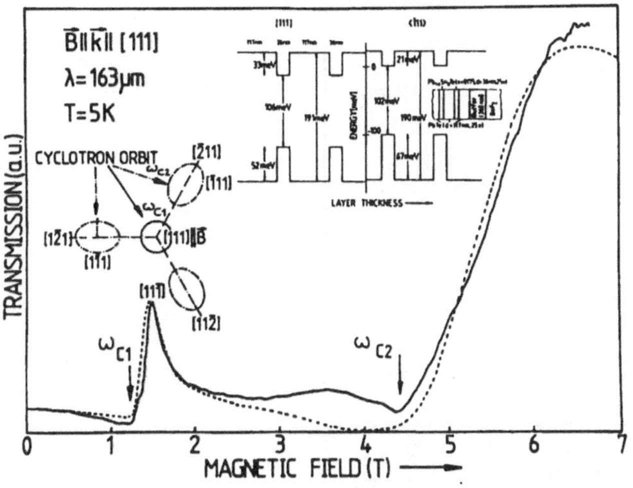

Fig.4: Magnetotransmission of a $PbTe/Pb_{1-x}Sn_xTe$ SL as indicated in the inset (d_1=36 nm, d_2=117 nm) 25 periods, on BaF_2 substrate.
The two cyclotron resonances correspond to the two possible cyclotron orbits as indicated. The two valley systems have unequal conduction and valence band offsets due to strain effects.

The oblique cyclotron mass is thus larger for quasi two-dimensional carriers than for the three dimensional cyclotron mass which is given by $1/m_c = 1/3(8/m_1m_t+1/m_t^2)^{1/2}$. This larger cyclotron mass in a MQW sample structure can be clearly seen from far infrared magentotransmission data as shown in Fig. 4, where ω_{C1} and ω_{C2} correspond to the resonance of the carriers in the a or b pockets.

For the model calculation of the magnetotransmission (see Fig. 4) the free carrier contribution to the total dielectric function has been modified and it is given by /39/

$$\chi^{\pm}_{fc} = -\frac{\omega_p^2}{4\omega}\cdot\frac{m_o}{m_t}\left(\frac{1}{\omega\pm\omega_{C1}+i\omega_\tau}\right) - \frac{\omega_p^2}{\omega}\,\frac{m_o\left[(\omega+i\omega_\tau)\frac{3}{8}\left[\frac{1}{m_t}+\frac{9}{m_t+8m_l}\right]\pm\omega_{C1}\,\frac{3}{8}\,\frac{18}{m_t+8m_l}\right]}{(\omega^2-\omega_{C2}^2-\omega^2+2i\omega\omega_\tau)}$$

Another interesting situation occurs for magnetic fields applied parallel to the layers ($B_{//}$) /40/. For sufficiently high fields and the center of the cyclotron orbit being situated in the center of the well, the cyclotron resonance frequency is close to its three dimensional value and given by the energetic distance of Landau states. For small or even vanishing magnetic fields the Landau states then collapse into the electrically quantized subbands. For an infinitely high potential barrier there would be just a 1:1 correspondence between the subband states and the Landau states (i.e. the n=0 Landau level emerges from the lowest electric subband, the n=1 Landau level from the next higher lying subband). However, for smaller magnetic fields the oscillator strength decreases for a transition with the electric field polarized in the plane of the layers since the transition becomes more like an intersubband resonance and not like a cyclotron resonance.

Whereas the cyclotron transition masses are quite different from the three- dimensional case, the coherent Raman scattering experiments on MQW structures yield electronic g-factors which are more or less identical to those of the material which forms the well. This is mainly due to the fact

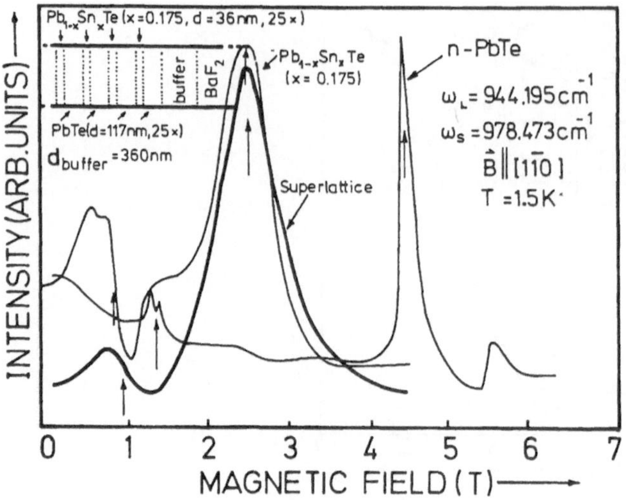

Fig.5a: CARS-intensity as a function of the magnetic field

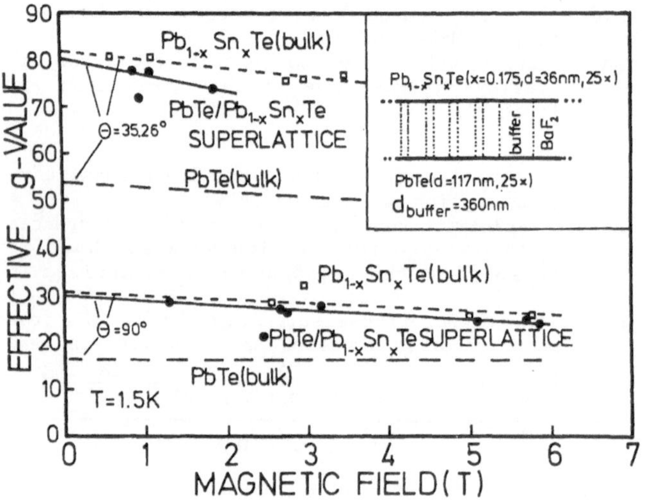

Fig.5b: Effective g-values of PbTe(--) PbSnTe (□) and of SL (○)

that although the energy of a Landau state depends strongly on the position of the center of a cyclotron orbit in real space, the two spin levels to the same Landau state have an equal dependence on this position. In Figs. 5a,b these data are shown for a PbTe/PbSnTe MQW sample and PbSnTe and PbTe films for comparison. Apparently the PbSnTe layers form the wells and the PbTe layers the barriers in our PbTe/$Pb_{1-x}Sn_xTe$ samples /29/.

The data presented are in agreement with a type I SL model, i.e. PbSnTe forms the wells and PbTe the barriers. (for x = 0.175, i.e. ΔE_c = 40 meV). A similar conclusion was obtained from luminescence measurements performed by Valenko et al. on PbTeSnTe MQWs. Murase and coworkers /26,27/ on the other hand have found a type I' SL model to apply for their PbTe/PbSnTe SLs with a conduction band offset of about 50 meV, but shifted in the other direction, i.e. the CB of PbTe lying higher than in PbSnTe.

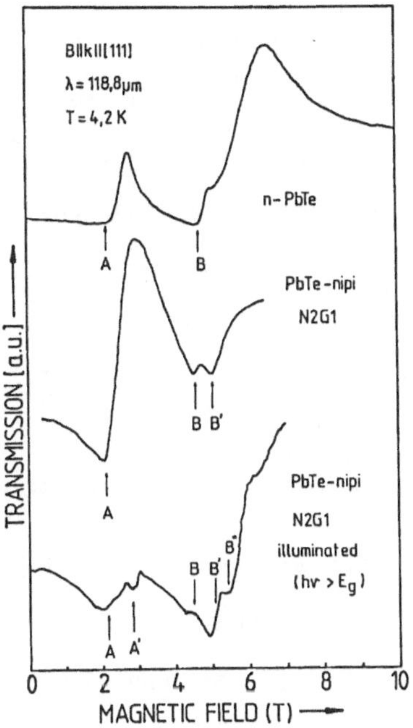

Fig.6: Magnetotransmission for n-PbTe. A,B: resonances in the a, b valley. Second curve: nipi-structure; A, B: electrons in the buffer, B': electrons in the n-layers. After illumination: additional hole resonances (A': $\Theta=0°$; B'': $\Theta=70.53°$)

Finally, in Fig.6 results of FIR mag-netotransmission on a PbTe doping superlattice are shown. The observed resonances differ from that of a bulk PbTe sample shown for comparison considerably. The nipi-samples are sensitive to band gap illumination and due to the drastic increase of the carrier lifetime both electrons and holes can be observed in the cyclotron resonance. These data are in good agreement with results of Hall and conductivity measurements performed under identical illumination conditions /41/. These many-valley doping superlattices offer interesting applications as extremely sensitive mid-infrared detectors.

REFERENCES

1. H. Preier, Appl. Phys. 20, 189 (1979).
2. D.L. Partin, C.M. Trush, Appl. Phys. Lett. 45, 193 (1984); J. Appl. Phys. 55, 678 (1984).
3. K.-H. Bachem, P. Norton, H. Preier, in Springer Series in Solid State Sci. Vol. 53, (Springer, Berlin, Heidelberg 1984) p. 147
4. H. Clemens, E.J. Fantner, G. Bauer, Rev. Sci. Instrum. 54, 685 (1983).
5. H. Kinoshita, H. Fujiyasu, J. Appl. Phys. 51, 5845 (1980); ibid. 52, 2869 (1981)
6. D.L. Mitchell, R.F. Wallis, Phys. Rev. 151, 581 (1966).
7. H. Burkhard, G. Bauer, W. Zawadzki, Phys. Rev. B19, 5149 (1979).
8. E. Bangert, G. Bauer, E.J. Fantner, H. Pascher, Phys. Rev. B31, 7958 (1985).
8a. W. Zawadzki, private communication.
9. L.G. Ferreira, Phys. Rev. 137A 1601 (1965).
10. G. Appold, Ph. D. Thesis, Würzburg
11. G. Appold, R. Grisar, G. Bauer, H. Burkhard, R. Ebert, H. Pascher, H.G. Häfele, Edinburgh (1978) Institute of Physics Conf. Series 43, 1101 (1978).

12. H. Pascher, Appl. Phys. B34, 107 (1982).
13. R.L. Bernick, L. Kleinmann, Solid State Commun. 8, 569 (1970).
14. H. Pascher, R. Grisar, G. Bauer, Physica 117+118B 398 (1983).
15. H. Pascher, E.J. Fantner, G. Bauer, W. Zawadzki, M. v. Ortenberg, Solid State Commun. 48, 461 (1983).
16. G. Elsinger, L. Palmetshofer, A. Lopez-Otero, 5th International Conference Ternary Compounds, il Nuovo Cimento, (in press).
17. G. Karczewski, M.v.Ortenberg, 17th Int. Conf. Phys. Semicond. San Francisco, (1984) p. 1435
18. M. Escorne, A. Mauger, J.L. Tholence, R. Tribulet, Phys. Rev., B29, 6305 (1984).
19. A. Krost, B. Harbecke, R. Faymonville, H. Schlegel, E.J. Fantner, K.E. Ambrosch, G. Bauer, J. Phys. C. 18, 2119 (1985).
20. J. Singleton, E. Kress, Rogers, A.V. Lewis, R.J. Nicholas, E.J. Fantner, G. Bauer, A. Lopez-Otero, J. Phys. C 19, 77 (1986).
21. H. Schaber, R. Doezema, Phys. Rev. B20, 5257 (1979).
22. S.W. McKnight, H.D. Drew, Phys. Rev. B21, 3447 (1980).
23. E.J. Fantner, G. Bauer, Springer Ser. Solid-State Sci., Vol. 53 (Springer Berlin, Heidelberg 1984) p. 207
24. G. Bauer, Surf. Sci. 168, 462 (1986).
25. H. Pascher, P. Pichler, G.Bauer, H. Clemens, E.J. Fantner, M. Kriechbaum, Surf. Sci. 170, 657 (1986).
26. M. Murase, S. Shimomura, S. Takaokab, A. Ishida, H. Fujiyasu, Superlattices and Microstructures 1, 177 (1985).
27. K. Shinohara, Y. Nishijiawa, H. Ebe, A. Ishida, H. Fujiyasu, Appl. Phys. Lett. 47, 1184 (1985).
28. D.L. Partin, Superlattices and Microstructures 1, 131, (1985).
29. H. Pascher, G. Bauer, H. Clemens, Solid State Commun. 55, 765 (1985).
30. M. Kriechbaum, K.E. Ambrosch, E.J. Fantner, H. Clemens, G. Bauer, Phys. Rev. B30, 3394 (1984).
31. M.V. Valenko, I.I. Zasavitskii, Sov. Phys.-JETP (1986), in press.
32. J. Heremans, D.L. Partin, P.D. Dresselhaus, M. Shayegan, H.D. Drew, Appl. Phys. Lett. 48, 928 (1986).
33. A. Golnik, and J. Spalek, J. Magn. and Magn. Mater. 54-57, 1207 (1986).
34. W.C. Galtsos, A.V. Nurmikko, D.L. Partin, Solid State Commun. (1986) in press.
35. M. von Ortenberg, Solid State Commun. 52, 111 (1984).
36. P.R. Wallace, phys. stat. sol. (b) 38, 715 (1970).
37. J.R. Anderson, M. Gorska, Solid State Commun. 52, 601 (1984).
38. G. Karczewski, M. von Ortenberg, Z. Wilamonski, W. Dobrowodski, J. Niedwodniczanska-Zawadzka, Solid State Commun. 55, 249 (1985).
39. P. Pichler, Ph.D. Thesis, Montanuniversität Leoben.
40. M. Kriechbaum, Springer Ser. Solid-State Sci., Vol. 67 (Springer Berlin, Heidelberg 1986), p.120
41. D.M. Gureev, I.I. Zasavitskii, B.N. Matsonashvili, P. Shatov, Sov. Phys.-Semicond. 12, 411 (1978).
42. D.M. Gureev, I.I. Zasavitskii, B.N. Matsonashvili, A.P. Shatov, Sov. Phys.-Semicond. 13, 1245 (1979).
43. T. Ichiguchi, S. Nishikawa, K. Murase, Solid State Commun. 34, 309 (1980).
44. M.S. Adler, C.R. Hewes, S.D. Senturia, Phys. Rev. B7, 5186 (1973).
45. P. Pichler, H. Clemens, G. Bauer, W. Jantsch, Surf. Sci., in press.

Recent Developments in Semimagnetic Semiconductors in Warsaw

M. Grynberg

Institute of Experimental Physics, University of Warsaw,
Hoza 69, 00-681 Warsaw, Poland

Semimagnetic semiconductors (SM-SC), or diluted magnetic semiconductors, are ternary or quaternary compounds made up of nonmagnetic and magnetic semiconductors.

SM-SC have been intensively studied in Warsaw for more than ten years. The first attempts were devoted to understanding the influence of the magnetic ion - delocalized electron (hole) exchange interaction on free electron (hole), in Td-structure materials. Due to delocalization of the carriers, the Mean Field Approximation (MFA) was the natural one to adopt. It has been shown that the exchange interaction not only splits the bands, but also introduces important band anisotropy. The substantial spin splitting is the origin of the many interesting effects characteristic of SM-SC, such as giant Faraday rotation, oscillatory behaviour of the Shubnikov-de Haas amplitude v.s. temperature, or "boiling-off" of the shallow acceptors in a magnetic field [1]. In studies of these effects, the influence of the localized spins of the magnetic ions on the carriers was taken into account. However, there exists a feedback, the influence of the carriers on the magnetic ions, which was neglected in that approach. The MFA is not sufficient for describing the influence of the localized electrons (holes) on the magnetic ions. The "feedback" effect was strongly related to magnetization fluctuation within a localized electron orbit, which is very important in a description of the Bound Magnetic Polaron (BMP) at finite temperature [2].

Although the magnetic properties of SM-SC have received considerable attention in the last decade, no coherent microscopic interpretation of these properties has as yet emerged.

1. MAGNETIC POLARONS, MAGNETIZATION FLUCTUATION

The first experimental evidence for the existence of the BMP in SM-SC has come from the work of Nawrocki et al. [3]. The term BMP is used here to describe an impurity electron (or hole) with the fluctuation of the spins of the magnetic ions within its orbit [2].

The influence of the BMP on the various magnetooptical phenomena is revealed only in small magnetic fields, in high fields the BMP disappears.

In the MFA, the spin splitting of the band states is proportional to the magnetization, as has been experimentally checked in the large magnetic field region. However, more precise experiments in low magnetic fields (low magnetization) show nonlinear behaviour of the Zeeman splitting of interband transitions v.s. magnetization [4]. This nonlinearity was explained using the magnetic polaron mechanism [5] and a significant influence of the magnetic fluctuations on the exciton in the region of its creation was shown.

In a recent analysis of far infrared hopping magnetoabsorption in n-type SM-SC, the BMP seems to be responsible for the experimentally observed magnetotransmission maximum in a low magnetic field [6]. Thermodynamic fluctuations of the magnetization lead to important modification of the shallow donor intraimpurity transitions [7], including the spin-flip transition. In compensated n-type crystals, thermodynamic fluctuations of magnetization cause a narrowing of the donor density of states band with an increasing magnetic field [8].

The zero-field splitting of the localized states in cubic crystals is now relatively well understood. The picture becomes more complicated for uniaxial crystals such as CdMnSe (wurtzite structure), where the upper valence band (split by the crystal field) is characterised by a strongly anisotropic exchange integral. This leads to anisotropy of the magnetic polaron bound to an acceptor [9].

For delocalization (band) states, the zero-field splitting, recently observed in different SM-SC crystals, is not yet theoretically described in a satisfactory manner [10], [11]. The first observation of delocalized states which show zero-field splitting comes from PbMnSe [12] (Figure 1).

The lead chalcogenide has a rather complicated band structure with anisotropic bands located at the L point of the Brillouin zone. The exchange interaction parameters for this material are still not well established. There is no satisfactory, even phenomenological, theory describing the ob-

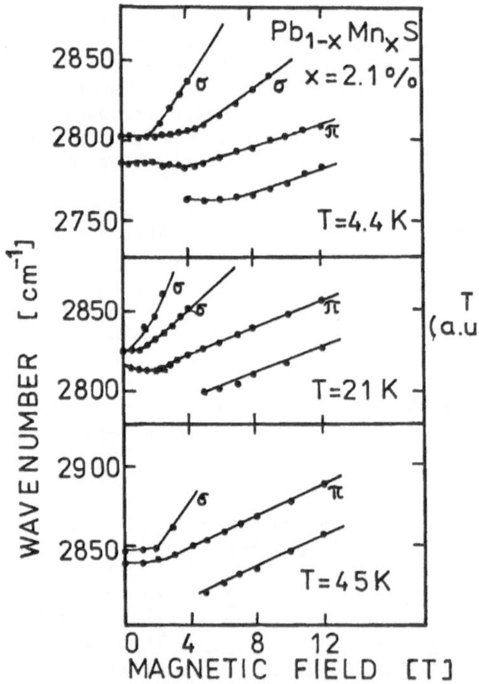

Fig.1. Emission energies vs. B of PbMnS diode [12]

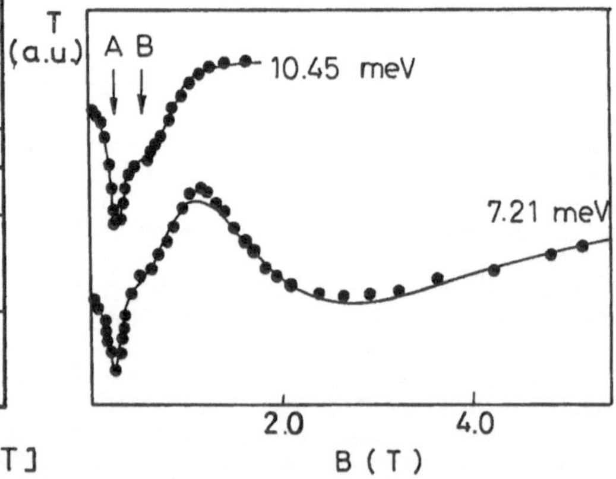

Fig.2. Combined resonances with free electron (A) and Mn electron (B) spin flip in HgMnTe. Voigt configuration E∥B [13]

served phenomenon. Recently, in n-HgMnTe two combined resonances have been observed in the Voigt (E∥B) configuration [13]. An explanation of the experimental data (Figure 2) is possible if formation of the Free Magnetic Polaron (FMP) is assumed. The proposed phenomenological model of the FMP assumes that the free electron wavefunction is delocalized and the electron interacts with all the paramagnetic ions, but the free electron adapts the spin direction to the magnetization fluctuations. The dimension of the volume in which the electron adapts its spin to the local magnetization was taken as a fitting parameter (a volume with N-270 Mn ions was obtained from a fit to the experimental data).

A combined resonance with the free electron spin flip (A) and the paramagnetic ion spin "flip" (B) was experimentally observed (Figure 2). This model leads to a finite spin splitting of the conduction band in the absence of a magnetic field. The splitting is proportional to the exchange integral and $\sim 1/\sqrt{N}$.

2. NEW SM-SC MATERIALS

Recently, A.Mycielski has undertaken an investigation of a new family of SM-SC based on HgSe and CdSe with Fe as a magnetic ion. In Figure 3 the band structure of the quaternary HgCdFeSe compounds is shown. Using various experimental techniques, the position of Fe^{+2} state v.s. the composition was established [14]. A new and interesting possibility offered by this system is that it allows us to study a SM-SC with a paramagnetic ion ground state degenerate with the conduction band ($N_{Cd} < 40\%$) or in the forbidden gap ($40\% < N_{Cd} < 70\%$).

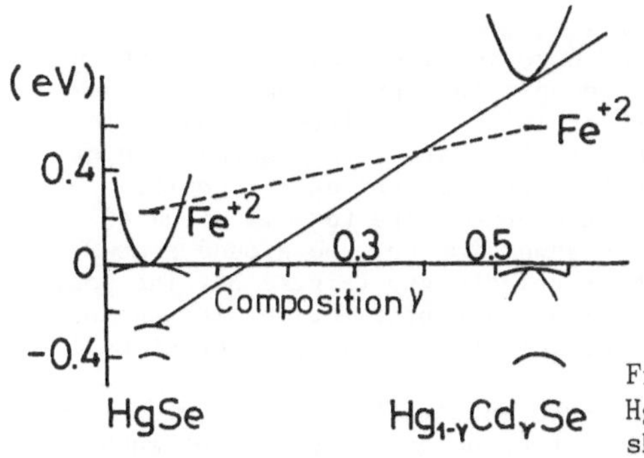

Fig.3. The schematic band structure of HgCdSe for 4K. The position of Fe^{+2} is shown [14]

A galvanomagnetic investigation of $Hg_{1-x}Fe_xSe$ has shown (Figure 4) that for Fe concentrations higher than $5x10^{18}$ cm^{-3}, the Fermi level is pinned to the Fe^{+2} resonance state. Surprisingly, the mobility of the free electrons in a crystal with Fe concentration of the order of 10^{20} cm^{-3} at liquid helium temperatures is abnormally high and the Dingle temperature is abnormally low. In order to explain this anomaly, J.Mycielski [15] has proposed a model in which the ionized donors (if the number of ionized resonance Fe^{+3} donors is much smaller than the number of neutral Fe^{+2} donors) due to the Coulomb interaction form a sublattice.

This model assumes a random distribution of donors, a long-life (narrow) resonant state, and a resonance state which is sufficiently well localized to avoid formation of an impurity band.

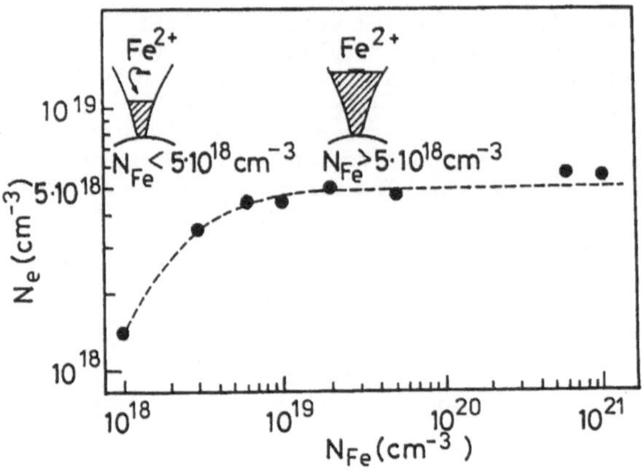

Fig.4. Experimental dependence of the free N_e electron concentration versus iron N_{Fe} concentration for HgFeSe (T = 4.2 K) [14]

The "crystallization" temperature of this sublattice was also calculated (∿ 4K). When the ionized resonant donors form such a sublattice, the potential of the ionized donor system is periodic, and thus the momentum transfer for a scattered electron must be equal to one of the reciprocal sublattice vectors. This excludes the low-momentum transfer scattering processes very important in Coulomb scattering. The calculated mobility and the Dingle temperature are in good agreement with the experimental data.

3. MAGNETIC PROPERTIES

Although the magnetic properties of SM-SC have received considerable attention in the last years, no coherent microscopic interpretation of them has as yet emerged. Actually, two "families" of SM-SC with different magnetic ions, the first with Mn, and the second with Fe, were investigated. One of the principal differences between the two systems is the energy position of the magnetic ion ground state. For Mn, the ground state (d^5) is situated about 3.5 eV below the top of the valence band; for Fe, the ground state (d^6) is in the conduction band or in the forbidden gap (Figure 3). The position of the ground state of the magnetic ion and hybridization with the band states may have a fundamental significance for the "ion-ion" interaction.

A systematic study of the high-temperature susceptibility of the Mn "family" SM-SC allow us to determine the dominant exchange integrals for each of the alloys [16]. The results are presented in Table 1. By comparing the results obtained with the cation-cation distances in the Mn "family" SM-SC (Figure 5), the authors come to the conclusion that superexchange is the dominant mechanism. However, recent results of spin-glass investigations of ZnMnSe and ZnMnTe [17] at very low temperatures and small Mn concentrations show the spin-glass transitions to be far below the percolation limit. From an analysis of the distance dependence of the Mn-Mn exchange interaction, the authors suggest that the contribution of the Bloembergen-Rowland exchange mechanism is not negligible. For the Fe "family", the exchange integrals obtained from the high-temperature susceptibility data [18] are also presented in Table 1. The large gap materials (ZnFeSe) exhibit a Van Vlack--type paramagnetism at low temperatures [19], an effect not observed in the small gap materials (HgFeSe, HgCdFeSe) [18]. As can be seen from Table 1, the strength of the antiferromagnetic exchange integrals diminishes with the increasing band gap. However, the situation is not that simple because, si-

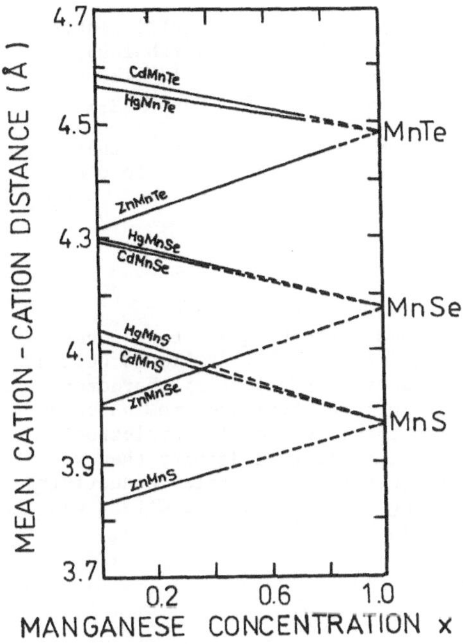

Fig.5. Cation-cation distance for Mn "family" SM-SC [16]

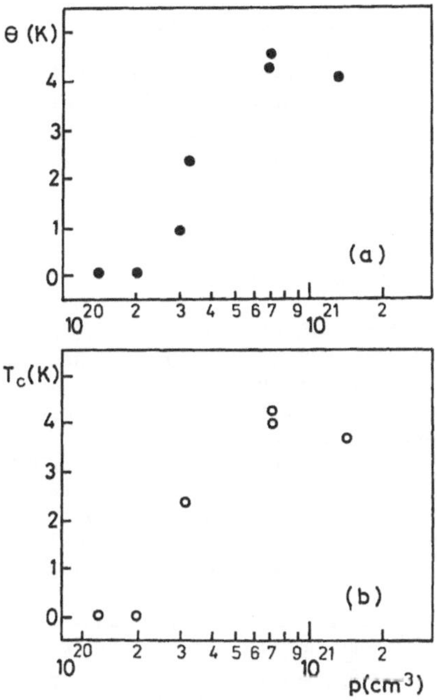

Fig.6. Carrier concentration dependence of (a) paramagnetic and (b) ferromagnetic Curie temperature of PbSnMnTe [20]

Table 1

	$2J_1/k_B$		$2J_1/k_B$
CdMnSe	21.2 ± 0.4	$Hg_{0.95}Fe_{0.05}Se$	29.0 ± 2
CdMnTe	13.8 ± 0.3	$Hg_{0.9}Fe_{0.1}Se$	31.0 ± 2
HgMnSe	21.8 ± 1.4	$Hg_{0.82}Cd_{0.11}Fe_{0.07}Se$	36.5 ± 2
HgMnTe	14.3 ± 0.5	$Hg_{0.83}Cd_{0.07}Fe_{0.10}Se$	28.5 ± 2
ZnMnTe	23.7 ± 0.5	$Hg_{0.54}Cd_{0.41}Fe_{0.05}Se$	24.2 ± 3
		$Cd_{0.95}Fe_{0.05}Se$	22.5 ± 3
		$Cd_{0.85}Fe_{0.15}Se$	22.5 ± 3

multaneously, the distance between the Fe^{+2} level and the top of the valence band also increases.

The authors [18] conclude that Bloembergen-Rowland exchange plays an important role in the Fe "family", but that the dominant contribution comes from superexchange. In p-type PbSnMnTe, for the first time in SM-SC the effect of free carriers on the magnetic properties was demonstrated [20]. A transition from a paramagnetic to a ferromagnetic phase at helium temperatures was observed. The ferromagnetic transition temperature increases with increasing free hole concentration (Figure 5). Of the known exchange mechanisms, only indirect exchange via carriers (RKKY) seems to be sufficiently long-ranged to account for the ferromagnetism of PbSnMnTe. Nevertheless, within the framework of the RKKY interaction alone, the threshold-like variation of the Curie temperature (Figure 6) is difficult to understand.

This very brief review of the magnetic properties of SM-SC shows that further experimental and theoretical investigations are necessary to clear up the problem.

4. 2DEG IN SM-SC

The first experiments with 2DEG in SM-SC dealt with grain boundaries in p-HgMnTe [21]. By means of the Shubnikov-de Hass effect, quasi-two-dimensionality of the electron gas was demonstrated (Figure 7). At liquid helium temperatures, the strongly temperature-dependent spin splitting characteristic of SM-SC, was observed. Recently, far-infrared cyclotron resonance of the space-charge layers in p-HgMnTe [22] was investigated. The cyclotron mass was found to increase more steeply with the electron density than was expected from $\overline{k \cdot p}$ calculations. The high-frequency surface hole conductivity was found to increase drastically with the magnetic field. This effect was not observed in a similar (HgCdTe) nonmagnetic crystal.

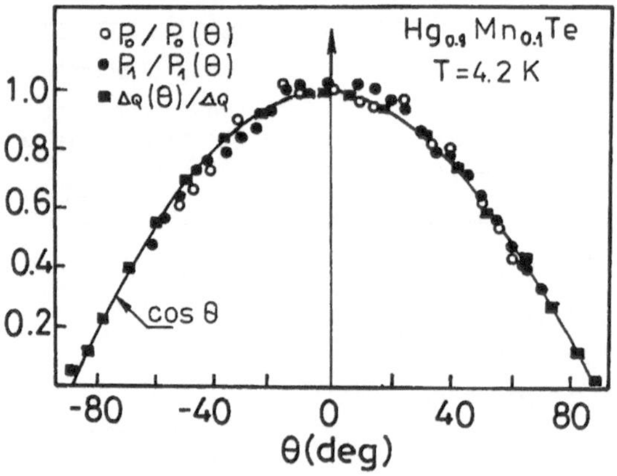

Fig.7. Relative periods of SdH oscillation vs. angle between magnetic field direction and the grain boundary plane [21]

5. GIANT SPIN SPLITTING AND LOCALIZATION

The greatly enhanced value of the electronic spin-splitting in SM-SC offers a possibility to study the influence of the spin splitting on conductivity in the Weakly Localized Regime (WLR). One of the most important mechanisms of Anderson Localization is the coherence between plane wave states with opposite wave vectors caused by scattering of the electrons by impurities.

If the coherence is destroyed by the magnetic field, one expects a negative magnetoresistance. However, the electron-electron interaction may change this picture and can lead even to positive magnetoresistance. The contribution to the e-e interaction is sensitive to the spin splitting [23], [24]. The recent experiments on n-CdMnSe and n-CdSe (Figure 8 and Figure 9) show that exchange-enhanced spin splitting leads to unusually large positive magnetoresistance. The authors conclude that the temperature and magnetic field dependence of the conductivity in CdMnSe at millikelvin temperatures cannot be explained by the existing theories.

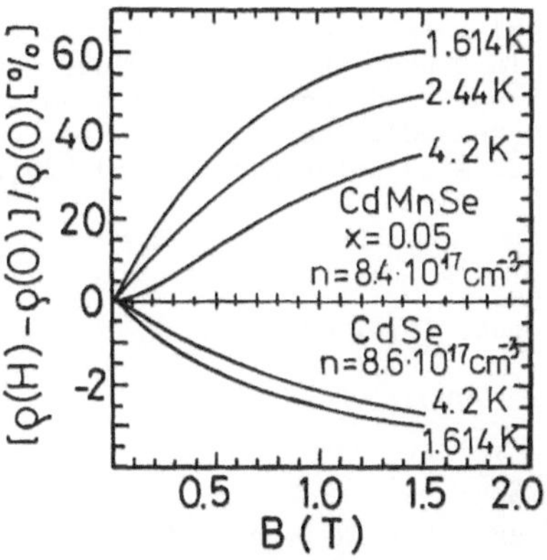

Fig.8. Magnetoresistance in CdMnSe and CdSe vs. B. Note the change of scale [23]

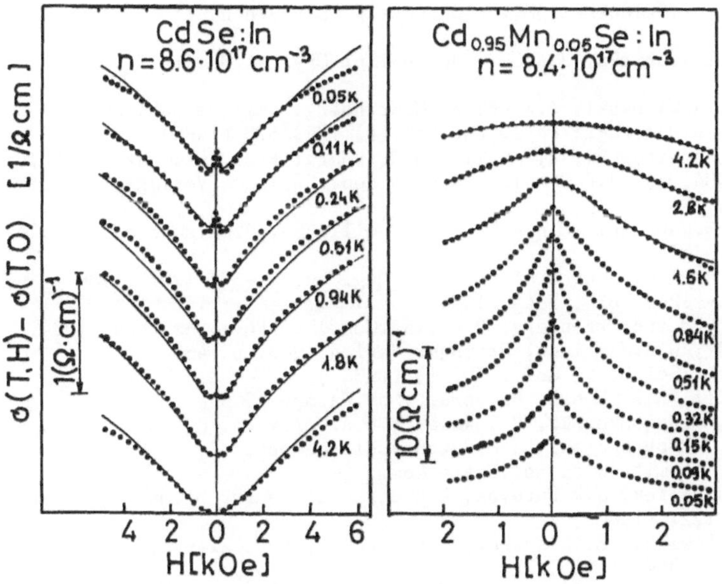

Fig.9. Measured (dots) and calculated (solid line) magnetoconductivity of CdS:In and CdMnSe:In [24]

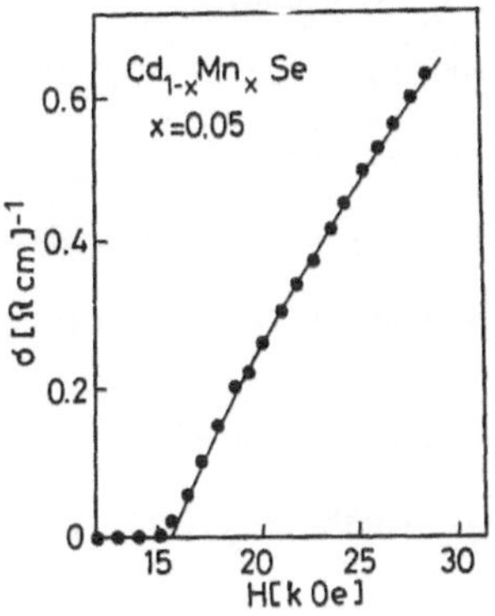

Fig.10. Extrapolated zero-temperature conductivity σ of n-CdMnSe as a function of the magnetic field [25]

In the same crystals (n-CdMnSe), a non metal-metal transition was observed [25] as the field was increased (Figure 10). This non-standard behaviour is characteristic of SM-SC materials (for normal semiconductors a metal-nonmetal transition occurs as the magnetic field is increased). The giant spin splitting causes the transition to occur only in one spin-polarized subband [25].

6. LITERATURE

1. For a review of fundamental properties see:
   R.R.Gałazka: Proc.Int.Conf. on Phys.Semicond. 1978, ed. B.L.H.Wilson Inst. of Phys.Conf.Proc. 43, 133 (1978);
   J.Gaj: J.Phys.Soc. Japan 49, Suppl.A, 787 (1980);
   M.Grynberg: Physica 117B, 461 (1983)
2. T.Dietl, J.Spałek: Phys.Rev.Lett. 48, 355 (1982);
   T.Dietl, J.Spałek: Phys.Rev. B28, 1548 (1983)
3. M.Nawrocki, R.Planel, G.Fishman, R.R.Gałazka: Phys.Rev.Lett. 46, 735 (1981)
4. A.Twardowski, P.Świderski, M.von Ortenberg, R.Pauthenet: Sol.State Comm. 50, 509 (1984)
5. A.Golnik, A.Twardowski, J.A.Gaj: Journal of Crystal Growth 72, 376 (1985)
6. J.Mycielski, A.Witowski, A.Wittlin, M.Grynberg: will be published
7. J.Mycielski, A.Witowski: Phys.Stat.Sol.(b) 134, 675 (1986)
8. J.Mycielski, A.Witowski, A.Wittlin, M.Grynberg: Proc.Int.Conf. on Phys. Semicond. 1984, ed. J.D.Chadi, W.A.Harrison, Springer Verlag 1985, p, 1415
9. D.Scalbert, M.Nawrocki, C.Benoit a' la Guillaume, J.Cernogora: Phys.Rev. B33, 4418 (1986)
10. A.Golnik, J.Spałek: Journ. of Mag. and Mag.Mat. 54-57, 1207 (1986)
11. J.Spałek: Phys.Rev. B30, 5345 (1984)
12. G.Karczewski, M.von Ortenberg: Proc.Int.Conf. on Phys.Semicond. 1984, ed. J.D.Chadi, W.A.Harrison, Springer Verlag 1985, p.1435
13. R.Stepniewski: Sol.State Comm. 58, 19 (1986);
    R.Stepniewski: Ph.D thesis, University of Warsaw 1986
14. A.Mycielski, P.Dzwonkowski, B.Kowalski, B.A.Orłowski, M.Dobrowolska, M. Arciszewska, W.Dobrowolski, J.M.Baranowski: accepted in J.Phys.C
15. J.Mycielski: submitted to Sol.State Comm.
16. A.Lewicki, J.Spałek, J.K.Furdyna, R.R.Gałazka: Journ. of Mag. and Mag. Mat. 54-57, 1221 (1986);
    J.Spałek, A.Lewicki, Z.Tarnawski, J.K.Furdyna, R.R.Gałazka, Z.Obuszko: Phys.Rev. B33, 3407 (1986)
17. A.Twardowski, C.J.M.Denissen, W.J.M.de Jonge, A.T.A.M.de Waele, M.Demianiuk, R.Triboulet: submitted to Sol.State Comm.
18. A.Lewicki, A.Mycielski, J.Spałek: Acta Phys.Polon. in press;
    A.Lewicki, J.Spałek, A.Mycielski: submitted to J.Phys.C
19. A.Twardowski, M.von Ortenberg, M.Demianiuk: J. of Cryst.Growth 72, 401 (1985)
20. T.Story, R.R.Gałazka, R.B.Frankel, P.A.Wolf: Phys.Rev.Lett. 56, 777 (1986)
21. G.Grabecki, T.Dietl, P.Sobkowicz, W.Zawadzki: Appl.Phys.Lett. 45, 1214 (1986)
22. M.Chmielowski, T.Dietl, F.Koch, P.Sobkowicz, J.Kossut: Acta Phys.Polon. in press
23. M.Sawicki, T.Dietl, J.Kossut: Acta Phys.Polon. A67, 399 (1985)
24. M.Sawicki, T.Dietl, J.Kossut, J.Igalson, T.Wojtowicz, W.Plesiewicz: Phys.Rev.Lett. 56, 508 (1986)
25. T.Wojtowicz, T.Dietl, M.Sawicki, W.Plesiewicz, J.Jaroszyński: will be published in Phys.Rev.Lett.

Recent High-Field Work on Diluted Magnetic Semiconductors

P.A. Wolff[1], D. Heiman[1], E.D. Isaacs[1], P. Becla[1], S. Foner[1], L.R. Ram-Mohan[2], D.H. Ridgely[3], K. Dwight[3], and A. Wold[3]

[1]Francis Bitter National Magnet Laboratory, MIT, Cambridge, MA 02139, USA

[2]Worcester Polytechnic Institute, Worcester, MA 01609, USA

[3]Brown University, Providence, RI 02912, USA

I. Introduction

Magnetic polarons are ferromagnetic spin clusters created by the exchange interaction of a carrier spin (electron or hole) with localized spins embedded in a semiconductor lattice. They were first studied in magnetic semiconductors [1]; more recently, there have been extensive investigations [2] of polaron behavior in diluted magnetic semiconductors (DMS), such as $Cd_{1-x}Mn_xTe$. DMS are favorable media for magnetic polaron studies because they have simple s-p bands, excellent optical properties, and can be grown in large single crystals. Two types of magnetic polarons have been identified in DMS - the bound magnetic polaron (BMP), whose carrier is localized by an impurity [3], and the free polaron (FP) consisting of a carrier trapped by its own, self-consistently-maintained, exchange potential [4].

For DMS with $x < 0.1$, a polaron theory [5] that assumes that unpaired Mn^{2+} spins (concentration $\bar{x} < x$) respond paramagnetically to the carrier exchange field gives good agreement with optical experiments [3,5,6]. The magnetic energies are then relatively small (10 - 20 meV), and polaron effects are only seen at low temperatures. To increase the polaron energy, and to exceed the threshold for FP formation [7], it is natural to extend these studies to larger x-values, where Mn^{2+} - Mn^{2+} interactions become important. For this purpose, we have developed a phenomenological theory of polarons that uses the measured high-field magnetization of the host crystal as an input to determine the Mn^{2+} response. This approach leads to a nonlinear Schrodinger equation for the polaron wave function, that we have solved in representative cases. The solutions determine BMP energies, shapes, and moments, as well as the conditions for FP formation. Large magnetic fields are needed in this problem since an external field, $B > 100$ T, is required to saturate a $Cd_{0.7}Mn_{0.3}Te$ sample. Pulsed field Faraday rotation experiments [8] were used to determine the magnetization of CdMnTe and CdMnSe crystals to 45 T.

Our calculations show that for large x, polaron formation is severely inhibited by the antiferromagnetic, nearest neighbor Mn^{2+} - Mn^{2+} interactions. The large BMP binding energies (≈ 0.3 ev at $x = 0.3$) implied by a paramagnetic model cannot be achieved in conventional, random DMS alloys. Pure FP also do not exist in such materials, though they may be stabilized in alloy-mediated, band gap fluctuations. These conclusions are consistent with recent time-resolved luminescence experiments [9].

To achieve large polaron energies in DMS, a host crystal with effective, free Mn^{2+} spin concentration, $\bar{x} \approx 0.2$, is required. Since next nearest neighbor Mn^{2+} - Mn^{2+} interactions are weak, this goal would be achieved if Mn^{2+} ions could be ordered on the cation sublattice, to eliminate nearest neighbor pairs. The paper concludes with a discussion of materials and structures that might provide such a desirable situation.

II. The Polaron Wave Equation

Magnetic polaron wave functions are usually approximated [5] by products of orbital and spin wave functions:

$$\psi(r,s;S_j) \simeq \phi(r)\chi(s;S_j), \tag{1}$$

where (r,s) are the coordinate and spin of the localized carrier, and S_j the spins of Mn^{2+} ions at sites R_j. The Schrodinger equation of the simplest magnetic polaron, the donor - BMP, takes the form:

$$\frac{-\hbar^2\nabla^2\phi}{2m^*} - \frac{e^2\phi}{\varepsilon r} - J\sum_j[\delta(r-R_j)\langle s\cdot S_j\rangle]\phi = E\phi, \tag{2}$$

where J is the electron - Mn^{2+} exchange constant. The thermal average over spin states, $\langle s\cdot S_j\rangle$, is evaluated from the spin Hamiltonian:

$$H_{spin} = -J\sum_j[|\phi(R_j)|^2(s\cdot S_j)] \equiv -\sum_j[K_j(s\cdot S_j)], \tag{3}$$

and is a functional of ϕ.

The correlation function, $\langle s\cdot S_j\rangle$, has been calculated [5] for the case of noninteracting Mn^{2+} spins (small x). At low temperatures the result is:

$$\langle s\cdot S_j\rangle \simeq 5/_4 B_{5/2}(\beta K_j)/_2, \tag{4}$$

where $B_{5/2}$ is the Brillouin function. In the continuum limit, the corresponding nonlinear Schrodinger equation takes the form:

$$\frac{-\hbar^2\nabla^2\phi}{2m^*} - \frac{e^2\phi}{\varepsilon r} - 5/_4 x(JN_0)B_{5/2}[\beta J|\phi(r)|^2/_2]\phi = E\phi. \tag{5}$$

This formula is equivalent, for small values of the Brillouin function argument, to one derived by SPALEK [10]. It also correctly describes Mn^{2+} spin saturation - that is crucial in FP formation - as $\beta J|\phi(r)|^2$ becomes large.

Equation (4) determines the alignment of the Mn^{2+} spin at site R_j in the local exchange field, $\mu g B_{exch}(R_j) = J|\phi(R_j)|^2/_2$. The corresponding factor $5/_2(xN_0)B_{5/2}$ in (5) is $(\mu g)^{-1}$ times the susceptibility of the noninteracting Mn^{2+} spin system. These observations suggest a generalization of (5) to the case of interacting Mn^{2+} spins. The proper recipe is to make the replacement:

$$5/_2 x N_0 B_{5/2}[\beta\mu g B_{exch}(r)] \rightarrow (\mu g)^{-1} M[B_{exch}(r)] \tag{6}$$

in (5), where M(B) is the <u>measured</u> magnetization of the crystal. This result can be derived, with reasonable approximations, from the Hamiltonian of the interacting carrier and Mn^{2+} spin systems.

To test this theory, we have numerically integrated (5) to determine the energy and form of $\phi(r)$ in the noninteracting Mn^{2+} spin case. The nonlinearity of the equation complicates the analysis since, for an arbitrary starting value, $\phi(0)$, the resulting wave function is usually not properly normalized. Thus, in effect, one must search a two-dimensional parameter space $[E,\phi(0)]$ to find a normalized eigenfunction of (5). In practice, this search was performed by an iterative procedure that converges rapidly. Figure 1 compares polaron wave functions at temperatures T = 1 K, 10 K, and 100 K (negligible polaron effect) for the interesting case x = 0.1, $m^* = 0.4\ m_0$, $(JN_0) = -0.88$ ev that provides a hydrogenic approximation to the acceptor - BMP in CdMnTe. The calculated energy, $E_{BMP} \simeq 80$ meV, is larger than the measured [6] value, $E_{BMP} \simeq 40$ meV, because Mn^{2+} spin pairing reduces the free spin concentration in the crystal to $\overline{x} = .035$.

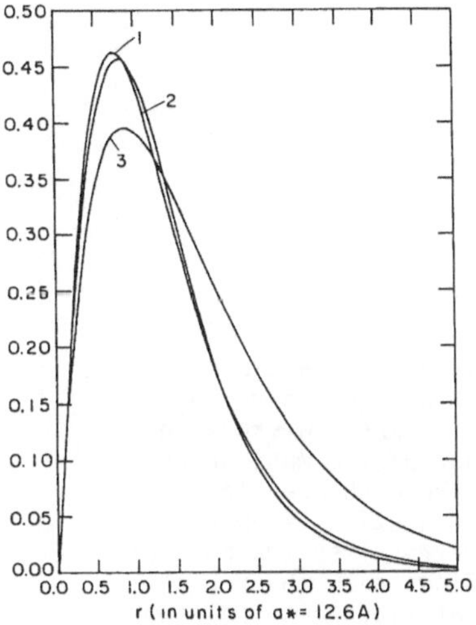

Fig. 1 BMP wave functions vs. radius (r/a_0^*) in p-$Cd_{0.9}Mn_{0.1}Te$. ① T = 1 K, ② T = 10 K, ③ T = 100 K

We have also used (5), without the Coulomb term, to calculate the properties of FP in p - CdMnTe (again for x = 0.1). The FP is weakly bound (E_{FP} = 21 meV) at 1.5 K, and becomes unstable above 2.5 K. These results imply that the threshold for FP formation in p - CdMnTe is $\overline{x} > 0.1$. Since $\overline{x} \leq 0.08$ in alloys with random Mn^{2+} distribution, pure FP are probably not stable in most DMS [9].

III. Phenomenological Calculations of BMP Properties

To calculate the properties of BMP in the large - x regime, we replace the free spin magnetization in (5) by the measured values for CdMnTe and CdMnSe crystals. Typical magnetization data are illustrated in Fig. 2. M(B) can be accurately decomposed into a Brillouin-like term, that saturates for B > 15 T, plus a linear term. The former results from alignment of small spin clusters whose number decreases rapidly above the percolation threshold (x = 0.17); the latter is reminiscent of the magnetization of an antiferromagnet.

Note that in the x = 0.5 sample, $\langle S_z \rangle$ only extrapolates to its maximum value at B = 240 T. The small slope of the M vs B curve drastically

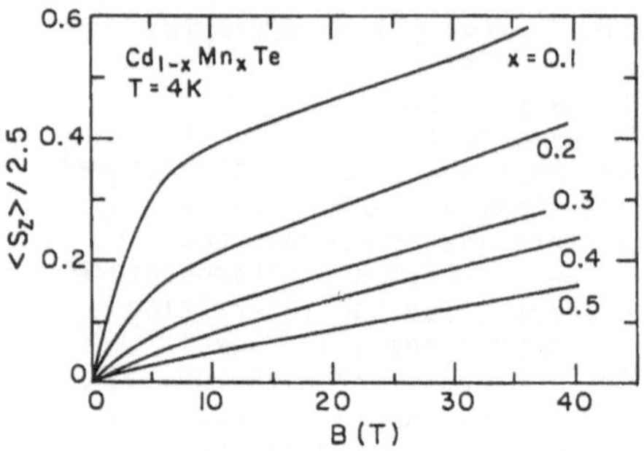

Fig. 2 Mn^{2+} spin alignment vs. field (in tesla) for $Cd_{1-x}Mn_xTe$ at 4 K

Table I
Calculated Acceptor - BMP Properties in $Cd_{1-x}Mn_xTe$

x	E_B [meV]	Radius [A]
0.1	21	11
0.2	19	11
0.3	13	11
0.4	7	12
0.5	4	12

reduces BMP energies. Table I lists the calculated radii and energies of acceptor - BMP in CdMnTe. Beyond x = 0.2, the polaron energy decreases with increasing x, whereas in the absence of Mn^{2+} - Mn^{2+} interactions a linear increase is predicted. For large x, most of the Mn^{2+} spins are "frozen out" of the polaron problem because the exchange field is not strong enough to break up their antiferromagnetic alignment. These conclusions are consistent with recent time-resolved luminescence experiments of ZAYHOWSKI [9]. For x = 0.2, he finds a 20 meV red shift of the luminescence with time after excitation, attributed to polaron formation, whereas in an x = 0.3 sample the shift is only 9 meV. Note, however, that the theory gives a lower value of the acceptor - BMP energy at x = 0.1 than that measured via dc luminescence experiments [6].

IV. Conclusions

Our calculations imply that Mn^{2+} - Mn^{2+} interactions severely inhibit magnetic polaron formation in DMS alloys with $x > 0.2$. Calculated and measured polaron energies are at least an order of magnitude smaller than those anticipated in a material of comparable x - value without Mn^{2+} - Mn^{2+} interactions. This disappointing conclusion suggests a search for DMS having a sizable concentration of essentially noninteracting Mn^{2+} spins. Since next nearest neighbor Mn^{2+} - Mn^{2+} interactions are quite small (in CdMnTe, $J_{NNN} \simeq -0.6$ K compared to $J_{NN} = -7.7$ K), such a material could be realized by ordering of the Mn^{2+} ions on the cation sublattice, to eliminate nearest neighbor Mn^{2+} pairs. In such a crystal the large, hole exchange field ($B_{exch} \simeq 30 - 100$ T) of the acceptor-BMP would overcome weak, next nearest neighbor Mn^{2+} - Mn^{2+} interactions to cause fairly complete polaron formation.

Rather surprisingly, there is a large class of tetrahedrally bonded, diamond-type semiconductors - the stannites - that satisfies these conditions

[11]. Examples include Cu_2FeSnS_4, Cu_2MnGeS_4, Cu_2MnSnS_4, $Cu_2MnSiSe_4$, etc. These crystals are effectively DMS with x = 0.25 and no nearest neighbor transition metal pairs (the nearest Mn^{2+} - Mn^{2+} distance in Cu_2MnGeS_4 is approximately equal to the next nearest neighbor spacing in CdMnTe). The stannites have small Néel temperatures ($T_N \approx 10$ K), expected of antiferromagnets with weak exchange interactions.

An ordered Mn^{2+} sublattice, similar to that found in stannites, might also be fabricated by MBE - growth of suitably alternated CdTe and MnTe layers. Such a structure would be produced in superlattices grown in the (120) direction, with three CdTe layers alternating with a single MnTe layer (x = 0.25 for the overall structure). As indicated in Fig. 3, which shows two layers of the CdMnTe cation lattice, the (120) superlattice has no Mn^{2+} ions in every other (010) plane, and square arrays of Mn^{2+} ions, at the next nearest neighbor distance, in the intervening ones. The structure has no nearest neighbor Mn^{2+} ions.

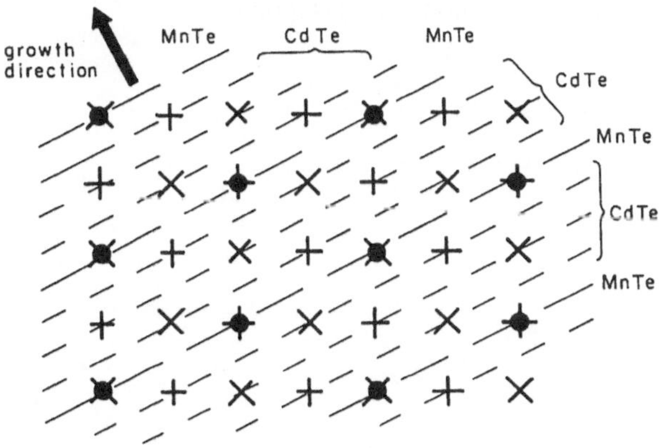

Fig. 3 Cation layers in ordered $Cd_{0.75}Mn_{0.25}Te$ grown by (120) MBE. (+ = top layer; x = next layer)

Interesting, partially ordered superlattices could also be produced by (100) MBE - growth of alternating CdTe and CdMnTe layers, or by (110) growth of two CdTe layers interspersed with a single CdMnTe layer. If one ignores small next nearest neighbor exchange interactions, the former becomes a random 2D antiferromagnet, and the latter a random 1D antiferromagnet.

Though speculative, these examples indicate a variety of ways of achieving fairly concentrated, "magnetically free" DMS. Such materials are of considerable inherent interest, and might be valuable in applications of DMS.

This research was supported by NSF Grant DMR-8504366.

Bibliography

1. See E. Nagaev, Physics of Magnetic Semiconductors (MIR Publishers, Moscow, 1983).
2. P. Wolff and J. Warnock, J. Appl. Phys. 55, 2300 (1984).

3. A. Golnik, J. Gaj, M. Nawrocki, R. Planel, and C. Benoit á la Guillaume, Proc. XV Intl. Conf. Phys. Semiconductors, Kyoto, 1980 (J. Phys. Soc. Japan, Suppl. A49, pg. 819); M. Nawrocki, R. Planel, G. Fishman and R. Galazka, ibid, pg. 823.
4. A. Golnik, J. Ginter and J. Gaj, J. Phys. C16, 6073 (1983).
5. T. Dietl and J. Spalek, Phys. Rev. Letters 48, 355 (1982); T. Dietl and J. Spalek, Phys. Rev. B28, 1548 (1983); D. Heiman, P. Wolff, and J. Warnock, Phys. Rev B27, 4848 (1983).
6. T. Nhung and R. Planel, Proc. XVI Intl. Conf. Phys. of Semiconductors, Montpellier, Physica 117B - 118B, 488 (1980).
7. T. Kasuya, A. Yanase, and T. Takeda, Sol. State Comm. 8, 1543 (1970).
8. D. Heiman, E.D. Isaacs, P. Becla, and S. Foner (to be published); D. Heiman, E.D. Isaacs, S. Foner, A. Wold, K. Dwight, and D. Ridgely (to be published).
9. J. Zayhowski, Ph.D. Thesis, MIT, 1985.
10. J. Spalek, Phys. Rev. B30, 5345 (1984).
11. W. Schäfer and R. Nitsche, Mat. Res. Bull. 9, 645 (1974).

Magnetic Field Dependence of the Acceptor Binding Energy in Open-Gap HgMnTe: Photo- and Magnetoconductivity

J. Wróbel[1], *T. Wojtowicz*[1], *A. Mycielski*[1], *A. Raymond*[2], *J.L. Robert*[2], *F. Kuchar*[3], *and R. Meisels*[3]

[1]Instytut Fizyki Polska Akademia Nauk, Al. Lotników 32/46, PL-02-668, Warszawa, Poland
[2]Groupe d'Etudes des Semiconducteurs, CNRS, LA 357, USTL, Place Eugene Bataillon, F-34060 Montpellier Cedex, France
[3]Ludwig Boltzmann Institut für Festkörperphysik, Kopernikusgasse 15, A-1060 Wien, Austria

Optical and transport experiments have been performed to study the binding energy E_a of acceptor levels in open-gap HgMnTe. Clear experimental evidence of a non-monotonic magnetic field dependence of E_a is given.

1. INTRODUCTION

The exchange interaction is responsible for a strong spin-splitting of band and impurity states in semimagnetic semiconductors. To facilitate a discussion the Landau level scheme of open gap HgMnTe (calculated in the framework of a modified Pidgeon-Brown model /1/ is shown on Fig. 1. The acceptor binding energy (or ionization energy) E_a is the energy difference between the ground acceptor level $|-3/2>$ and the uppermost valence band

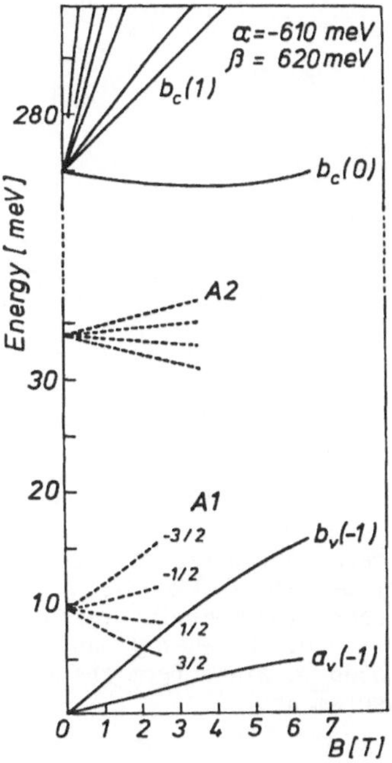

Fig. 1 The calculated energy levels of HgMnTe (Mn content 15 %) at 4.2 K. The split impurity levels are shown schematically (dashed lines): A1-shallow acceptor, A2-deep acceptor which reveals smaller exchange splitting as compared to A1 /6/. The values of the exchange parameters α and β have been obtained in the present work

sublevel $b_v(-1)$. It has been predicted theoretically /2,3/ that E_a of shallow acceptors first decreases in a magnetic field (since at low fields the impurity splitting is weaker than the band splitting) and then starts to increase.

The aim of the present paper was to extend previous investigations /4,5/ to the range of strong magnetic fields and to verify the above-mentioned prediction.

2. RESULTS AND DISCUSSION

P-type HgMnTe samples with Mn contents of 15, 11 and 10 % have been studied. Hall-effect measurements enabled us to determine the difference between acceptor and donor concentrations: we obtained $N_A - N_D = 0.6 \times 10^{16}$, 3×10^{16} and 4×10^{16} cm^{-3} for samples 1, 2 and 3, respectively.

Measurements of the near-infrared photoconductivity edges have been performed in magnetic fields up to 7 T for samples with the smaller acceptor concentration. Figure 2 shows typical photoconductivity spectra obtained for a sample with a thickness of d~250 μm. For B>0 additional structures in the photomagnetoconductivity edges are observed (for energies lower than the zero-field energy gap). The analysis of the results for samples with different thicknesses, together with available magneto-optical data /7/ leads us to the following interpretation: The structure marked B in the Faraday configuration is connected with $b_v(-1) \rightarrow b_c(0)$ transitions and the peak in the Voigt configuration is due to the $a_v(-1) \rightarrow b_c(0)$ photoionization. The peak A (strong in σ^- polarization, similar to B) may be caused by transitions

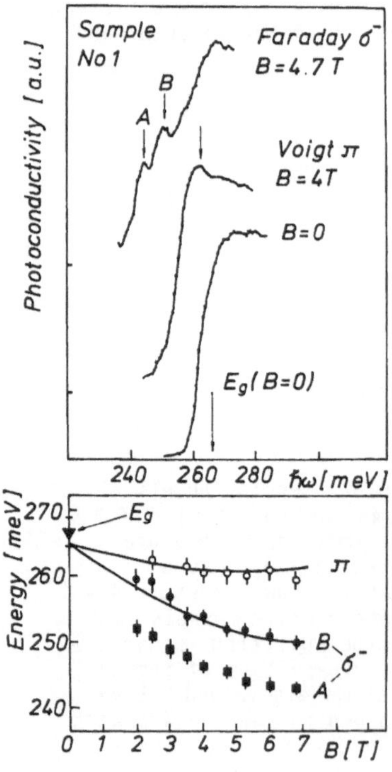

Fig. 2 Near-infrared photoconductivity edges for B=0 and B≠0 at 4.2 K. Below: Transition energies versus magnetic field. The energy gap at B=0 has been obtained from absorption measurements

from the impurity level $|-3/2\rangle$ to the conduction band sublevel $b_c(0)$. Such an assignment has enabled us to determine the binding energy in the magnetic field range 3 T - 7 T.

To determine E_a at lower fields, the far-infrared, low-temperature photoconductivity has been measured for sample No. 1, using a Fourier transform spectrometer. For B=0 at the energy ~ 10 meV a threshold has been observed, connected with the acceptor-valence band transitions. The spectra have been measured in magnetic fields up to 9 T, however the precise determination of $E_a(B)$ has not been possible due to the presence of a reststrahlen band in HgMnTe near an energy of 12 meV. Far-infrared measurements are much more suitable for studies of the deep acceptor in HgMnTe with E_a~30 meV /6/.

The determination of $E_a(B)$ (starting at B=0) was performed through transport experiments. We have measured the transverse magnetoresistance and the Hall coefficient in static magnetic fields up to 7 T for sample 1 and up to 18T for samples 2 and 3. In the extrinsic conductivity range, in addition to a previously observed negative magnetoresistance /5/ (due to the magnetic "boil-off" effect) we have found an increase of resistivity in fields higher than 6T (see Fig. 3) caused by magnetic "freeze-out" of carriers. The experimental ionization energy has been determined from log σ_{xx} vs 1/T plots at a constant magnetic field (sample 2, 3) and at a constant magnetization (sample 1).

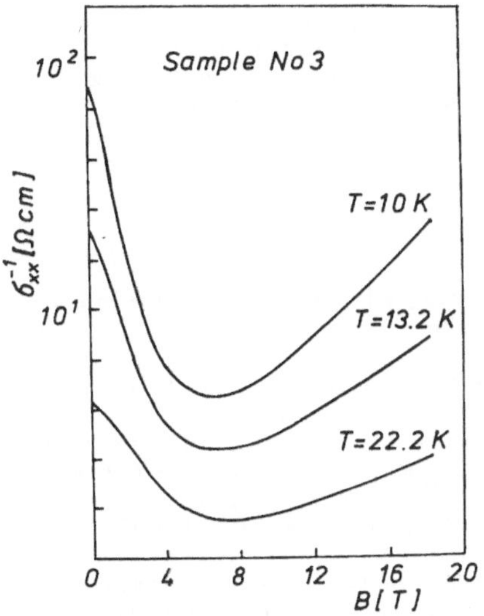

Fig. 3 Inverse of σ_{xx} components of the conductivity tensor versus magnetic field in the temperature range of extrinsic conduction

The optical and transport experiments give consistent results which reveal the following characteristic features: (see Fig. 4)

1. The dependence of E_a on B is non-monotonic for shallow acceptors. The decrease for small fields is well described by the theory but the minimum value of $E_a(B)$ is smaller than predicted and occurs at higher fields than expected.
2. The quantum-limit (QL) approach (i.e. the construction of the $|-3/2\rangle$ level wave function only from the uppermost Landau valence subband /8/)

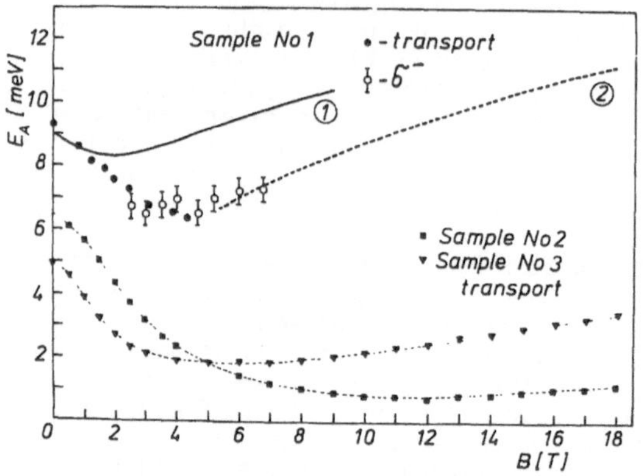

Fig. 4 Binding energy versus magnetic field in the temperature range 4.2-15 K for samples 1, 2 and 3, respectively. Lines correspond to the theoretical predictions: 1-spherical model with the addition of an exchange term /3,12/, 2-quantum limit approach. The calculations have been performed with a static dielectric constant ε_0=17. The high field E_a for sample 2 is not a pure ε_1, but rather mixed with ε_1 and ε_3

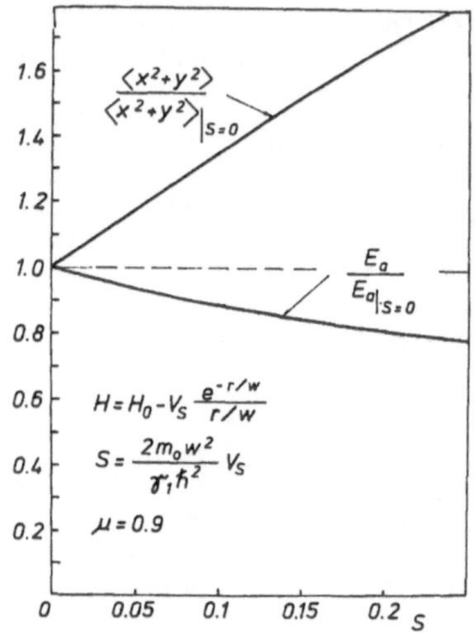

Fig. 5 Binding energy and average value of (x^2+y^2) as a function of the Yukakawa-type repulsive potential strength /11/ (w=0.03*effective Bohr radius)

describes well the binding energy for samples with small acceptor concentration. For samples 2 and 3 E_a is smaller.

The diamagnetic term in the effective mass Hamiltonian (proportional to B and dependent on the localization of the wave function) "shifts" in magnetic fields acceptor levels towards the valence band /9/. This term is however quenched in zincblende type semiconductors, the bigger the parameter μ , the stronger the quenching /10/ ($\mu = 2\bar{\gamma}/\gamma_1$, where $\bar{\gamma}$, γ_1 are the Luttinger parameters, μ is within the range 0-1). For open-gap HgMnTe $\mu \sim 0.9$.

The theoretical calculations /11/, performed in the low-field limit, show that a short-range repulsive potential added to the effective mass Hamiltonian almost does not change the spin splitting of a shallow acceptor. Moreover, this type of "central-cell" correction leads only to a slight decrease of E_a but to a relatively strong delocalization of the impurity wave function (for B=0), see Fig. 5. So we suggest that a short-range addition to the Coulomb potential delocalizes the acceptor ground state and thus is responsible for the stronger decrease of E_a in a magnetic field.

A theory able to describe our experimental results would have to include also many-body interactions. The role of these effects is very large since E_a (at B=0) decreases fast with increasing impurity concentration. This change of E_a is in qualitative agreement with a theory which takes into account the influence of the polarizability and the neutral acceptor concentration on the reduction of the zero-field ground state energy /12/. For B>0 the radius of an acceptor wave function increases, thus the polarizability of the neutral acceptor is bigger than it is at B=0 and the dielectric constant becomes magnetic field-dependent. This may be the second reason why E_a decreases more strongly in magnetic fields than the theory for isolated impurities predicts.

The authors wish to thank Prof. R.R. Gałązka and Prof. M. Grynberg for discussions and valuable remarks and Dr. E. Litwin-Staszewska for the help in the high magnetic field measurements.

REFERENCES

1. See, for example, R.R. Gałązka, J. Kossut, Lecture Notes in Physics 133, 245 (1980)
2. J. Mycielski, C. Rigaux, J. Physique 44, 1041 (1983)
3. T.R. Gawron, J. Phys. C 19, 21 (1986)
4. A. Mycielski, J. Mycielski, J. Phys. Soc. Japan Suppl. A 49, 807 (1980)
5. T.Wojtowicz, A. Mycielski, Acta Phys. Polonica A67, 363 (1985)
6. J. Wróbel, F. Kuchar, R. Meisels, Acta Phys. Polonica A67, 369 (1985)
7. G. Bastard, C. Rigaux, Y. Guldner, A. Mycielski, J.F. Furdyna, D.P. Mullin, Phys. Rev. B 24, 1961 (1981)
8. T.R. Gawron, J. Mycielski, Phys. Stat. Sol. (b) 125, 341 (1984)
9. P. Janiszewski, private information
10. D. Bimberg, A. Baldereschi, Proc. 14th Int. Conf. Semic. Edinburgh, 1978, B.L.H. Wilson ed, 403
11. J. Wróbel, to be published
12. R. Buszko, Acta Phys. Polonica A69, 1025 (1986).

Spin Modulated Energy State by Shubnikov-de Haas Analysis in $Hg_{1-x-y}Cd_xMn_yTe$ at High Magnetic Fields

S. Takeyama[1], J. Kossut[2], and S. Narita[3]

[1]Institute for Solid State Physics, University of Tokyo, 7-22-1 Roppongi, Minato-ku, Tokyo 106, Japan

[2]Institute of Physics, Polish Academy of Sciences, Al. Lotników 32/46, PL-02-668 Warsaw, Poland

[3]Department of Material Physics, Faculty of Engineering Sciences, Osaka University, Toyonaka, Osaka 560, Japan

1. Introduction

The dilute magnetic(or semimagnetic) semiconductor(DMS) alloys have presented lots of interesting phenomena not only concerning Landau spin subband energy states influenced by the exchange interaction between band electrons and the localized magnetic moment on Mn^{2+} ions, but also related to the magnetic properties themselves/1/. In order to study the relation between the magnetic properties and energy states in the host material more accurately and systematically, the present quaternary DMS; $Hg_{1-x-y}Cd_xMn_yTe$ is the most convenient system, since it has the advantage that the energy gap of the host semiconductor is controlled by the Cd composition, while leaving the magnetic properties determined only by the Mn composition/2/. Our system of present study is regarded as a "dilute Mn spin doped $Hg_{1-x}Cd_xTe$ alloy", which covers the energy gap range, from negative (ie. $E_g<0$)to open gap (ie. $E_g>0$)sides.

In the Shubnikov-de Haas(SdH), the most prominent spin modification is expected to be realized at low magnetic fields and low temperatures, where the paramagnetic behavior of the Mn moment changes drastically, obeying the Brillouin function $B_s(z)$. At higher magnetic fields, the modification occurs slowly for a wide range of temperatures, since $B_s(z)$ is a function of $z=g^*\mu_B B/k_B T$.

In DMS's it is known that the Landau levels N=0 or 1 are quite sensitive to the spin modification, and the low number Landau energy states in SdH oscillations are observed at higher magnetic fields. However, the last part of the SdH oscillations are typically quite broad, and this broadness obscures the exact positions of the Landau spin energy states, particularly in the case when two peaks appear close-by. Hence it is necessary to make a direct calculation of the magnetoresistance taking into account the energy states realized in narrow gap DMS's. We have developed the formula for the transverse magnetoresistance applicable to our system and analyzed our SdH data.

2. Spin-modulated SdH oscillations

Highly homogeneous,large size single crystals of $Hg_{1-x-y}Cd_xMn_yTe$ were grown by the modified Bridgman method for x<0.2, and for x>0.2 by the modified two-phase mixture method/3/. Carrier concentrations are controlled by In doping from 3×10^{16} to 4×10^{17} cm^{-3}. The prepared samples used for the measurements have alloy compositions y around 0.014 and x ranging from 0.01 to 0.23. Both the transverse ρ_{xx} and longitudinal ρ_{zz} magnetoresistance are measured by sweeping a magnetic field up to 6.7 T and changing the temperature from 1.5 K to 30 K.

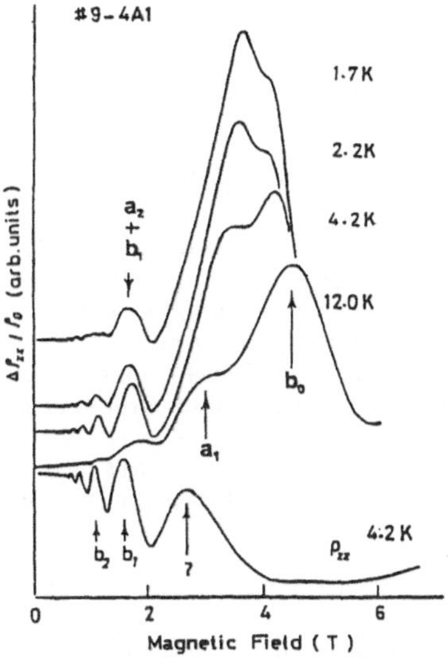

Fig.1 Typical temperature shift of SdH peaks of ρ_{xx}, and comparison with ρ_{zz} for $E_g<0$.

Figure 1 demonstrates typical transverse SdH oscillations ρ_{xx} for different temperatures with magnetic field up to 7 T for the negative energy gap sample 9-4A1 with E_g=-0.256 eV at 7 K and carrier concentration $n=3.5\times10^{16}$ cm^{-3}. In this paper we limit our attention only to the higher magnetic field region. As the temperature is decreased from 12 K down to 1.7 K, two componentsof the split peak observed around around B=4 T, interchange their intensity and become narrower, also the splitting between the peaks is reduced. At the bottom trace of Fig.1 the longitudinal SdH ρ_{zz} trace at 4.2 K is presented for comparison. A noticeable phase shift between ρ_{xx} and ρ_{zz} is observed as in the case of $Hg_{1-x}Cd_xTe$ /4/. Figure 2 displays the typical case of an open-gap alloy 5-6-2 with E_g=0.181 eV at 7 K. The peak at the highest field shows a small shift to lower magnetic field and the split peaks around B=3 T do not present a drastic temperature change of their position. The longitudinal ρ_{zz} is demonstrated in the bottom trace in Fig.2 for T=4.2 K. The phase shift is almost negligible for this case.

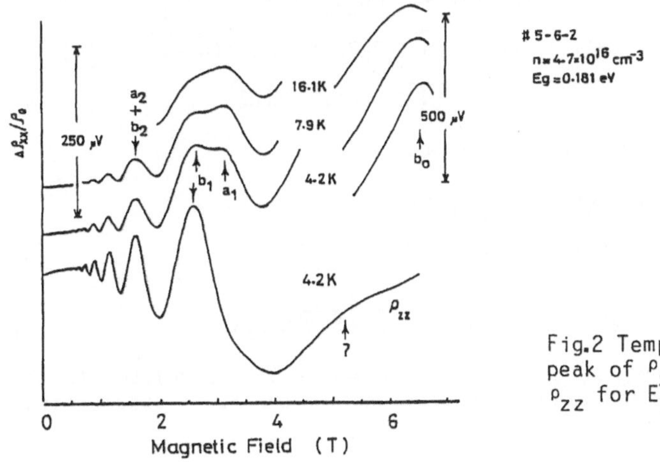

Fig.2 Temperature change of SdH peak of ρ_{xx} and comparison with ρ_{zz} for $E_g>0$.

3. Theory of the magnetoresistance ρ_{xx}

Starting with the well-known Kubo formula /5/, the transverse conductivity is expressed as,

$$\sigma_{xx} = \int \cdot dE \left(- \frac{\partial f_0}{\partial E} \right) \cdot \sum_{N=0} \cdot \sum_{N'=0} \cdot \sum_{\sigma=\pm} \cdot \sum_{\sigma'=\pm} A_{N,\sigma} \cdot A_{N',\sigma'} W^{\sigma,\sigma'}_{N,N'} \quad , \qquad (1)$$

where, N and σ denote the Landau number and the spin states a or b-set, respectively. $A_{N,\sigma}$ in the above formula represents the spectral function which takes into account the non-parabolicity by the quasi-parabolic approximation developed in eq.(17) of ref./6/. The energy of each Landau spin level; $E_{N,\sigma}$ is calculated from the modified Pidgeon-Brown model/7/, which includes the exchange interaction Hamiltonian between band electrons and the Mn magnetic moment.

The transition matrix element $W^{\sigma,\sigma'}_{N,N'}$ in eq.(1) can be calculated, assuming a short-range impurity scattering potential, for $E_g>0$ with wide energy gap to a simple form as,

$$^{s}W^{\sigma,\sigma'}_{N,N'} = \delta_{\sigma\sigma'} \cdot V_0^2 \; \frac{\ell^2}{2\pi} \, (N + N' + 1) \, , \qquad (2)$$

where, ($l=\sqrt{\hbar c/eB}$=magnetic length), the potential V_0 is taken with the value 1.3×10^{-22} $eV\cdot cm^3$/8/. For $E_g<0$, one has to take the spin-orbit interaction as well as the strong non-parabolic effect into consideration. The transition matrix element can be calculated from the electron wavefunction in the presence of an external magnetic field taken from those derived by Kacman and Zawadzki/9/. Since the electron wavefunction is not a pure spin function but a mixture of spin-up and down functions, the spin-flip transition becomes an allowed term. Hence, for spin-conserved transitions,

$$\begin{aligned} ^{c}W^{a,a}_{N,N'} &= |\langle \Psi_{N,k_y,k_z,a} | V | \Psi_{N',k'_y,k'_z,a} \rangle|^2 \\ &= \frac{l^2}{2\pi} [V_s^2 |A^*_a A'_a|^2 (N+N'+1) + V_p^2 \{ (N+N'+3) I^{a\,2}_1 + (N+N'+1) I^{a\,2}_2 + (N+N'-1) |C^*_a C'_a|^2 \\ &\quad -2(\sqrt{(N+1)\cdot(N'+1)} \, I^a_3 + \sqrt{(N+2)\cdot(N'+2)} \cdot |H^*_a H'_a| I^a_2 \} + 2V_s \cdot V_p |A^*_a A'_a| \cdot \{ (N+N'+1) I^a_2 \\ &\quad - \sqrt{(N-1)\cdot(N'-1)} \cdot |C^*_a C'_a| - \sqrt{(N+2)\cdot(N'+2)} \, I^a_1 \}] \end{aligned} \qquad (3)$$

and

$$\begin{aligned} ^{c}W^{b,b}_{N,N'} &= \frac{l^2}{2\pi} [V_s^2 |A^*_b A'_b|^2 (N+N'+1) + V_p^2 \{ (N+N'+3) \, |B^*_b B'_b|^2 + (N+N'+1) I^{b\,2}_2 \\ &\quad + (N+N'-1) I^{b\,2}_4 - 2(\sqrt{(N+1)\cdot(N'+1)} \; |C^*_b C'_b| + \sqrt{NN'} \, |B^*_b B'_b| + \sqrt{(N-1)\cdot(N'-1)} \, |H^*_b H'_b|) I^b_2 \\ &\quad + 2V_s V_p \{ (N+N'+1) I^b_2 - \sqrt{(N-1)\cdot(N'-1)} \; I^b_4 - \sqrt{(N+2)\cdot(N'+2)} \, |B^*_b B'_b| \; \}] \end{aligned} \qquad (4)$$

in addition, for the spin-flip transitions,

$$\begin{aligned} ^{c}W^{a,b}_{N,N'} &= |\langle \Psi_{N,k_y,k_z,a} | V | \Psi_{N',k'_y,k'_z,b} \rangle|^2 \\ &= \frac{l^2}{2\pi} V_p^2 \{ (N+N'+2) K_1^2 + (N+N') K_2^2 - 2\sqrt{(N+1)\cdot N'} \; K_1 K_2 \; \} \end{aligned} \qquad (5)$$

$$^{c}W^{b,a}_{N,N'} = \frac{l^2}{2\pi} V_p^2 \{ (N+N') K_3^2 + (N+N'+2) K_4^2 - 2\sqrt{(N'+1)\cdot N} \; K_3 K_4 \} \, . \qquad (6)$$

In the above expressions, I^j_i and K_i (i=1,⋯,4, j=a or b) are related to the coefficients in Kacman and Zawadzki's wavefunction/9/. It is assumed that the potential matrix element $V_s=\langle S|V|S\rangle$ and $V_p=\langle X|V|X\rangle$ take the same value as V_0. The collision broadening is roughly estimated to be the value $\Gamma=3.6\times10^{-3}$ eV as to fit roughly the broadening of SdH data.

4. Discussion

The fitting procedure used in eq.(1-6) is applied to the observed SdH data. The value of E_g and the momentum matrix element P are used from the value determined from the SdH data analysis for the low magnetic field region/2/. The higher band parameters γ_1, γ_2, γ_3, κ, and F are taken from the data for HgTe/10/. Then the value of the exchange parameters $\alpha=\langle S|J|S\rangle$ and $\beta=\langle X|J|X\rangle$ are fitted to the SdH data for different temperatures, using least square fitting method. The final results are demonstrated for sample 9-9B1 in Fig.3 for two different temperatures. The change of the peak position as well as the peak height with decreasing temperatures are well represented by the calculation. From the fitting for all the measured samples, the values $\alpha=-0.2$ eV and $\beta=0.73$ eV were obtained.

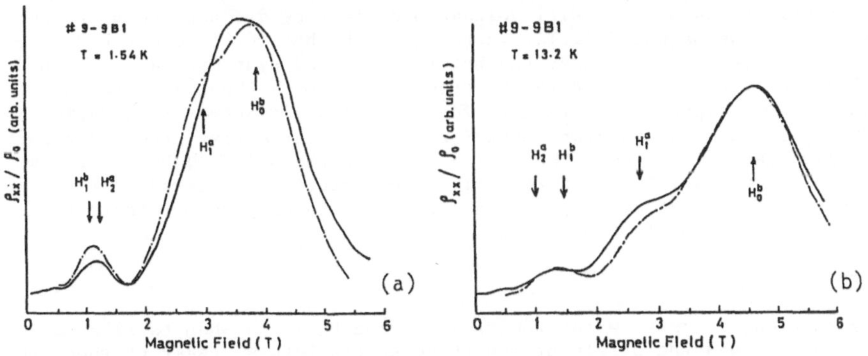

Fig.3 Calculated ρ_{xx}(broken line) and experimental SdH traces(solid line) (a) for T=1.54K and (b) for T=13.2K.

References

1. R.L.Aggarwal, S.N.Jasperson, P.Becla, R.R.Galazka: Phys. Rev. B, **32**, 5132(1985)
2. S.Takeyama, S.Narita: J. Phys. Soc. Jpn. **55**, 274 (1986)
3. S.Takeyama, S.Narita: Jpn. J. Appl. Phys. **24**, 1270 (1985)
4. K.Suizu, S.Narita: Phys. Lett. **73**, 353 (1973)
5. R.Kubo, S.Miyake, N.Hashizume: Solid State Phys. **17**, 269, ed by F.Seitz and D.Turnbull (Academic Press, New York,1965)
6. J.Kossut, J.Hajdu: Solid State Commun. **27**, 1401 (1978)
7. R.R.Galazka, J.Kossut: Lecture Notes in Phys.**133**, 245(Springer Verlag, 1980)
8. J.J.Dubowski: Phys. Stat. Solidi.(**b**) **85**, 663 (1978)
9. P.Kacman, W.Zawadzki: Phys. Stat. Solidi. (**b**) **47**, 629 (1971)
10. S.H.Groves, R.N.Brown, C.R.Pidgeon: Phys. Rev. **161**, 779 (1967)

Phase Differences Between Quantum Oscillations of the Magnetoresistance and the Hall Effect in $Hg_{1-x}Mn_xTe$ and $Hg_{1-x}Cd_xTe$

R.G. Mani[1], *J.R. Anderson*[1], *and W.B. Johnson*[2]

[1]Department of Physics and Astronomy, University of Maryland, College Park, MD 20742, USA

[2]Laboratory for Physical Sciences, 4928 College Avenue, College Park, MD 20740, USA

1 Introduction

During the course of investigation of magnetotransport in narrow-gap semiconductors, we noticed phase differences between Shubnikov-de Haas (ShdH) oscillations and Hall voltage oscillations. That is, the peaks in the Shubnikov-de Haas oscillations and the Hall voltage oscillations did not occur at the same values of the magnetic field. Although such phase differences had been observed previously [1-10], they had not been studied systematically nor had there been any attempt to compare the observations with theory. Here we present some of our experimental results for two ternary systems, $Hg_{1-x}Cd_xTe$ and $Hg_{1-x}Mn_xTe$, in which oscillatory behavior is observed easily. From a comparison of these data with recent theoretical work on the ShdH and Hall oscillations [11-14], we have found it possible to correlate the phase differences with semiconductor scattering parameters and thus to support our interpretation of previous experimental observations.

2 Theory

In this section, we examine the theories for the Hall and ShdH oscillations in order to understand the origin of the phase shifts between peaks of these oscillatory quantities.

It can be shown from the work of KULEEV and NOVOKSHONOV [11] that the density of states at the Fermi energy, ζ, is

$$\left.\frac{dn}{dE}\right|_H = \left.\frac{dn}{dE}\right|_0 \{1 + \frac{1}{2} (\frac{\hbar\omega}{\zeta})^{1/2} \delta^{-1/2}\} , \quad (1$$

where

$$\delta^{-1/2} \sim \sum_{\nu=1}^{\infty} \frac{(-1)^{\nu}}{\nu^{1/2}} \cos (\frac{2\pi\zeta}{\hbar\omega} \nu - \frac{\pi}{4}) . \quad (2$$

Here ω is the cyclotron frequency and $\left.\frac{dn}{dE}\right|_0$ is the density of states in the absence of a magnetic field. Clearly (1) exhibits oscillatory behavior in the presence of a magnetic field.

The problem of interest is to determine the effect of the magnetic field on the galvanomagnetic properties of a degenerate semiconductor. For magnetotransport measurements in the transverse configuration ($E \perp B$), the diagonal conductivity σ_{xx} vanishes as a consequence of $\vec{E} \times \vec{B}$ drift of the carriers unless scattering is included. If scattering is included, the diagonal conductivity σ_{xx} reflects the oscillatory variation of the density of states at the Fermi surface with magnetic field and the peaks in the conductivity correspond roughly to peaks in the density of states [5].

For a delta function scattering potential, ZYRYANOV and KULEYEV [13] have shown that the diagonal element of the conductivity tensor, σ_{xx}, is given by

$$\sigma_{xx} = \frac{|e|c}{H} \frac{\gamma(\zeta)I(\zeta)}{\hbar\omega}, \text{ where} \tag{3}$$

$$\gamma(\zeta) = n_s v_o^2 2\pi \sum_{N'} [\zeta-(N'+1/2)\hbar\omega]^{-1/2} = \gamma_0(\zeta)[1 + \frac{1}{2} (\frac{\hbar\omega}{\zeta})^{1/2} \delta^{-1/2}] , \tag{4}$$

γ_0 is the zero-field scattering parameter, and

$$I(\zeta) = \frac{1}{V} \sum_{N} \frac{\hbar\omega(N+1/2)}{[\zeta - (N+1/2)\hbar\omega]^{1/2}} = n [1 + \frac{3}{4} (\frac{\hbar\omega}{\zeta})^{1/2} \delta^{-1/2}] . \tag{5}$$

Note that $\delta^{-1/2}$ is an oscillatory function of the magnetic field [see (2)] and n is the electron density.

In the presence of a magnetic field, the transport coefficients can be expanded in a power series in terms of $(\omega\tau)^{-1}$, where τ is the scattering lifetime. The diagonal component σ_{xx} involves only the odd powers in the parameter $(\omega\tau)^{-1}$ while the off-diagonal component σ_{xy} involves only the even powers beginning with the zeroth power [12]. Thus, σ_{xy} does not exhibit oscillations to first order in scattering, at least, for a single band.

The experimentally measured quantities are the Hall voltage and the magnetoresistance, which are proportional to the Hall constant R_H, and the resistivity, ρ, respectively. In terms of the components of the conductivity tensor, one can write

$$\rho = \frac{\sigma_{xx}}{\sigma_{xx}^2 + \sigma_{xy}^2} \text{ and } R_H H = \frac{\sigma_{xy}}{\sigma_{xx}^2 + \sigma_{xy}^2} . \tag{6}$$

For one-band conduction in the limit $\mu H >> 1$,

$$\rho \simeq \frac{\sigma_{xx}}{\sigma_{xy}^2} \text{ and } R_H H \simeq \frac{1}{\sigma_{xy}} . \tag{7}$$

Thus, to first order in the scattering, oscillatory peaks in the resistance correspond to maxima in the conductivity while the Hall voltage does not exhibit oscillations.

Experimental observations of weak Hall oscillations in GaSb [1] and InSb [6] have suggested, however, that higher order terms in the scattering must be included. GUSEVA and ZYRYANOV [12] have shown that if terms up to the second order in the scattering are retained, then an additional scattering term, σ_{xy}^{sc}, has to be included in the off-diagonal conductivity component such that

$$\sigma_{xy} = \sigma_{xy}^1 + \sigma_{xy}^{sc} , \tag{8}$$

where σ_{xy}^1 is the zeroth order, non-oscillating term and

$$\sigma_{xy}^{sc} = \frac{ec}{H} (\frac{\gamma(\zeta)}{\hbar\omega})^2 I(\zeta) . \tag{9}$$

For a δ-function scattering potential, to second order in scattering,

$$\sigma_{xy}^{sc} = \frac{|e|cn}{H} (\frac{\gamma_0(\zeta)}{\hbar\omega}) [1 + (\frac{7}{4}) (\frac{\hbar\omega}{\zeta})^{1/2} \delta^{-1/2} + (\frac{\hbar\omega}{\zeta}) \delta^{-1} + \frac{3}{16} (\frac{\hbar\omega}{\zeta})^{3/2} \delta^{-3/2}] \tag{10}$$

$$\text{and} \quad \sigma_{xx} = \frac{|e|c\,n}{H} \left(\frac{\gamma_0(\zeta)}{\hbar\omega} \right) \left[1 + \frac{5}{4} \left(\frac{\hbar\omega}{\zeta}\right)^{1/2} \delta^{-1/2} + \frac{3}{8} \left(\frac{\hbar\omega}{\zeta}\right) \delta^{-1} \right] . \tag{11}$$

Taking into account the oscillating terms in the numerator and denominator of (6) for the Hall coefficient and the resistivity, we find for a delta function potential that

$$\frac{\Delta R_H}{R_H} = (\omega\tau_o)^{-2} \left\{ \frac{3}{4} \left(\frac{\hbar\omega}{\zeta}\right)^{1/2} \delta^{-1/2} + \frac{21}{16} \left(\frac{\hbar\omega}{\zeta}\right) \delta^{-1} + \frac{3}{4}\left(\frac{\hbar\omega}{\zeta}\right)^{3/2} \delta^{-3/2} + \frac{9}{64} \left(\frac{\hbar\omega}{\zeta}\right)^2 \delta^{-2} \right\} \tag{12}$$

$$\text{and} \quad \frac{\Delta\rho}{\rho} = \frac{5}{4} \left(\frac{\hbar\omega}{\zeta}\right)^{1/2} \delta^{-1/2} + \frac{3}{8} \left(\frac{\hbar\omega}{\zeta}\right) \delta^{-1} . \tag{13}$$

It can be shown [11] that if Landau level broadening is included, then for small values of z, where z represents the ratio of Landau level splitting to Landau level broadening (Γ) divided by the Landau level number number,

$$z = \left(\frac{\hbar\omega}{\zeta}\right)^{1/2} \delta_m^{-1/2} \sim \frac{\hbar\omega}{(\zeta\Gamma)^{1/2}} . \tag{14}$$

Here δ_m is the amplitude of the oscillating term, ρ and $R_H H$ oscillate as $\delta^{-1/2}$, and there is no phase shift. Conversely, for large values of z, $R_H H$ oscillates as δ^{-2} while ρ oscillates as δ^{-1} so that the phase shift is approximately 90°. As the parameter z varies with the magnetic field the phase shift also should increase with the field.

KULEEV and NOVOKSHONOV [11] have also considered the case where the scattering potential is long range, i.e. $\Delta = 2R_o^2/\ell^2 >> 1$, where R_o = range of potential and ℓ = magnetic length ($\ell = (\hbar/eH)^{1/2}$). Their results for the long-range Gaussian potential were independent of Δ, and a phase shift of $3\pi/4$ between the resistance and Hall oscillations was predicted.

3 Experiment

Oscillations of the Hall voltage and the transverse magneto-resistance were studied in a series of annealed $Hg_{1-x}Mn_xTe$, $0.05 \leq x \leq .145$, and $Hg_{0.8}Cd_{0.2}Te$ samples. The measurements were carried out in a superconducting magnet to 50 kOe and in an iron-core magnet to 28 kOe. The samples were mounted in the standard Hall bar configuration so that the Hall voltage and the transverse magnetoresistance could be measured simultaneously. The oscillations were observed both with and without field modulation. Measurements were made for "normal" and "reverse" fields and also for both current directions and then combined in the standard way.

4 Data and Discussion

Comparison of results for both $Hg_{1-x}Mn_xTe$ and $Hg_{1-x}Cd_xTe$ indicates that the phase relations are insensitive to the exchange interaction within experimenta error. In Fig. 1, we show the second derivatives of the Hall voltage and the magnetoresistance as a function of the applied magnetic field in the $Hg_{0.8}Cd_{0.2}Te$ sample W5 [15] for temperatures of 8 K and 16 K. The Hall and resistance oscillations have the same periodicity in reciprocal field and the frequency of oscillation, F, is temperature independent, F ≅ 10 kOe, up to T = 25 K. The g-factor, as measured from the spin-splitting of the oscillation peaks, is also temperature independent up to 25 K. The amplitudes of the resistance oscillations are about 2-4 times larger than the corresponding Hall peaks. In this sample, there is a 90° phase shift between the Hall and the resistance peaks for all observable oscillations. The carrier concentration ob-

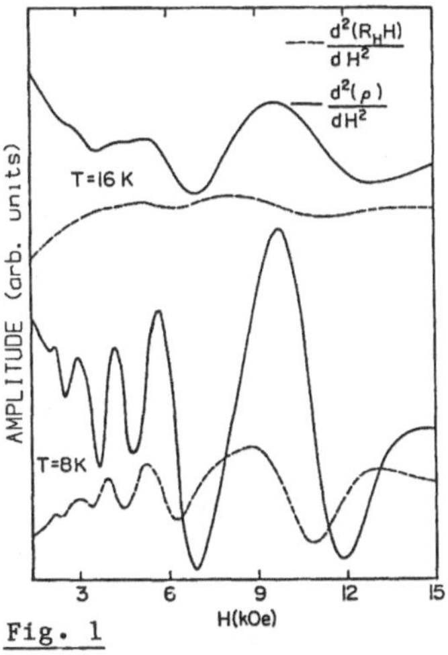

Fig. 1

Hall and resistance oscillations showing the temperature-independent 90° phase shift at T = 16°K and T = 8°K in the $Hg_{1-x}Cd_xTe$ sample W5.

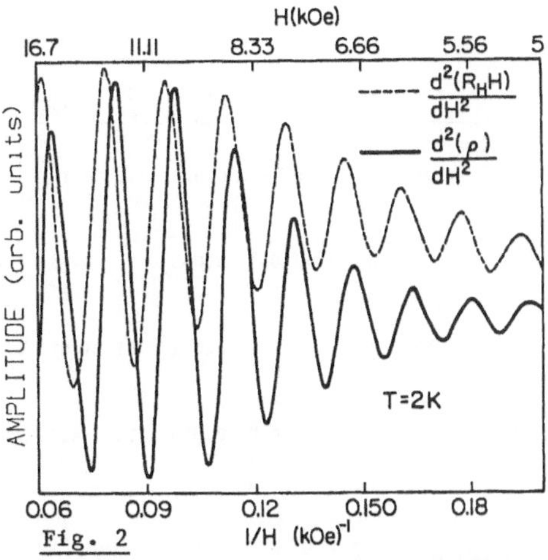

Fig. 2

ShdH and Hall voltage oscillations for $Hg_{1-x}Cd_xTe$ sample RM4 at 2 K. The phase difference is ~ 55°.

tained from the ShdH frequency agreed with the measured concentration, n, within experimental error. The Hall mobility μ was constant for T < 25 K while the phase shift remained fixed at 90°. A comparison of the temperature dependence of the mobility with the results of DUBOWSKI et al.[16] indicates that for T < 25 K, charged center scattering is responsible for limiting the electron mobility. A 90° phase shift between the Hall and the resistance peaks was also observed in the undoped, high mobility $Hg_{1-x}Mn_xTe$ (x ~ 0.06) sample K10AS1b, but, other HMT samples showed smaller phase shifts (see Table I). In Fig. 2, we plot the Hall and ShdH oscillatory patterns in the undoped $Hg_{1-x}Mn_xTe$ (x ~ 0.075) sample RM4 versus the inverse magnetic field. Here the average phase shift is approximately 55°. The figure suggests that the phase shift increases with magnetic field, but more careful measurements have to be made to verify this result. For an In doped $Hg_{1-x}Mn_xTe$ (x = .145) sample XLY25D3 [17] with low mobility (μ ~ 14000 cm^2/v-sec) the resistance maxima coincided either with the Hall maxima or the Hall minima depending upon the direction of the magnetic field. Thus, the phase difference was zero degrees, modulo π, in this sample.

In Table I, we summarize a few relevant parameters for four representative samples. The carrier concentrations shown in the table were obtained from the ShdH frequency, the electron effective mass ratio was obtained from a fit of the temperature dependence of the ShdH amplitudes, the Hall mobility was converted to a mobility temperature using

$$T_\mu = e\hbar/(2\pi k_B m^* \mu) \quad , \tag{15}$$

and the Fermi energy was calculated with the Kane model.

In order to apply the results of KULEEV and NOVOKSHONOV [11], it is necessary to estimate the range of the scattering potential in our samples. The range of the potential, R_0, is difficult to estimate when charged impurity

8 Table 1

Sample	x	E_0 (meV)	n cm^{-3}	m^*/m	ζ (meV)	T_μ (K)	R_0 cm	λ_f cm	Δ_0	Δ_1	Δ_2	$\frac{\hbar\omega}{\sqrt{k_B T_\mu \zeta}}$	$\Delta\phi$
W5*	0.20	65	1.5×10^{15}	0.015	7.5	0.14	5.4×10^{-6}	3.5×10^{-6}	.8	4.7 (5kOe)	9 (10kOe)	13	90°
K10AS1b**	0.06	-60	1.1×10^{15}	0.003	6	0.76	6×10^{-6}	7.8×10^{-6}	1.2	1.1 (1kOe)	3.4 (3kOe)	5	~90°
RM4**	0.075	0	9×10^{16}	0.012	85	1.01	1.4×10^{-6}	1.2×10^{-6}	3	0.3 (5kOe)	1.5 (25kOe)	3	55°
XLY25D3**	0.145	280	2.6×10^{16}	0.033	11	4.5	2×10^{-6}	1.9×10^{-6}	2.2	1.3 (10kOe)	2.5 (20kOe)	1.5	0°

* $Hg_{1-x}Cd_xTe$

** $Hg_{1-x}Mn_xTe$

scattering dominates because screening effects have to be taken into consideration. We expect, however, R_0 to be small because it is known that the carriers cannot be frozen out on the residual donors even at the lowest temperatures [18]. Thus, we have estimated R_0 from the charge center density assuming the charged impurity potentials do not overlap. The spatial extent of the electronic wavefunction is given by the smaller of the two quantities, the de Broglie wavelength (λ_F), and the magnetic length. In Table I, we include for comparison the ratio Δ_0, where

$$\Delta_0 = 2\,[R_0/(\lambda_F)\,]^2 \quad , \tag{16}$$

as well as the ratios determined by the magnetic length, Δ_1 and Δ_2, which correspond to the values of the range parameters at the low-field and high-field limits of observation of ShdH oscillations, respectively. Since $1 < \Delta < 10$ for all the samples, inter-Landau level scattering plays a significant role. Thus, the expressions for the Hall and resistance oscillations that take into account short-range potentials appear applicable to our measurements.

In the limit of short-range scattering, the Hall and resistance oscillations are specified by (12) and (13), respectively. Phase shifts occur between the two quantities when higher order terms in z become important in the two expressions, that is, when the Landau level number is small and the Landau level spacing is significantly greater than the broadening. Thus, the phase shift at any given field depends upon the size of z which is related through (14) to the width of the Landau level (Γ), usually defined in terms of the Dingle temperature, $\Gamma = \pi k_B T_D$. Since we were unable to accurately measure the Dingle temperature due to an insufficient number of oscillations, we have estimated the size of the parameter z by replacing the Dingle temperature with the Hall mobility temperature, i.e., $\Gamma \propto k_B T_\mu$. Therefore, in Table I we use for z the the parameter $\bar{z} = \hbar\omega/\sqrt{k_B T_\mu \zeta}$.

From Table I, we see that the measured phase shift correlates well with the magnitude of the parameter $\bar{z}$. The $Hg_{1-x}Cd_xTe$ sample W5 shows the largest phase shift ($\Delta\phi = 90°$) and has a correspondingly large value for $\bar{z}$, $\bar{z} \sim 13$. Conversely, the $Hg_{1-x}Mn_xTe$ sample XLY25D3 exhibits no phase shift and also has the smallest value for $\bar{z}$, $\bar{z} \sim 1$. Thus, our experimental results are in qualitative agreement with the theoretical predictions [11] for short-range potentials. In addition, a preliminary indication of the phase shift increasing with field in sample RM4, where the phase shift is neither 0° nor 90°, could be due to the field dependence of the parameter $\bar{z}$, i.e. $\bar{z} \propto H$, which controls the contribution of higher order terms in (12) and (13).

5 Acknowledgements

The $Hg_{1-x}Mn_xTe$ crystals were grown by Mr. Eric Larsen and Mr. Ken Hankin. This research was supported in part by the U.S. Army Research Office and the U.S. Department of Defense Advanced Research Projects Agency (DARPA) under Grant No. DAA629-85-K-0052.

6 Literature

1. T. O. Yep and W. M. Becker, Phys. Rev. 156, 939 (1967).
2. A. H. Kahn and H. P. R. Frederikse, in Solid State Physics, ed. F. Seitz and D. Turnbull (Academic, New York, 1959), vol. 9, p. 257.
3. A. I. Ponomarev and I. M. Tsidilkovskii, Sov. Phys. Semicond. 3, 773 (1969).
4. H. P. R. Frederikse and W. R. Hosler, Phys. Rev. 110, 880 (1958).
5. S. S. Shalyt and A. L. Efros, Sov. Phys. Sol. St. 4, 903 (1962).
6. G. A. Antcliffe and R. A. Stradling, Phys. Lett. 20, 119 (1966).
7. L. M. Bliek and G. Landwehr, Phys. Stat. Sol. 31, 115 (1969).
8. A. I. Ponomarev, G. A. Potapov, and I. M. Tsidilkovskii, Sov. Phys. Semicond. 11, 24 (1977).
9. R. A. Stradling and G. A. Antcliffe, J. Phys. Soc. Jpn. 21, 374 (1966).
10. V. I. Ivanov-Omskii, N. N. Konstantinova, R. V. Parfen'ev, V. V. Sologub and I. G. Tagiev, Sov. Phys. Semicond. 7, 496 (1973).
11. I. G. Kuleev and S. G. Novokshonov, Sov. J. Low Temp. Phys. 2, 64 (1976).
12. G. I. Guseva and P. S. Zyryanov, Phys. Stat. Sol. 25, 775 (1968).
13. P. S. Zyryanov and I. G. Kuleyev, Fiz. Metal. Metalloved. 28, 16 (1969).
14. I. G. Kuleev, S. G. Novokshonov and S. P. Perevalov, Sov. J. Low Temp. Phys. 3, 147 (1977).
15. The $Hg_{1-x}Cd_xTe$ samples were grown at the North China Research Institute of Optico-Electronics and kindly provided for us by Dr. Wu Yang Xian.
16. J. J. Dubowski, T. Dietl, W. Szymanska and R. R. Galazka, J. Phys. Chem. Solids 42, 351 (1981).
17. The $Hg_{1-x}Mn_xTe$ sample XLY25D3 was grown at the Purdue University Materials Research Facility and kindly provided by Dr. R. Holm.
18. B. L. Gelmont and M. I. Dyakonov, Sov. Phys. JETP 35, 377 (1972).

Magnetic Field Driven Insulator-to-Metal Transition in Semimagnetic Semiconductors

T. Wojtowicz, M. Sawicki, T. Dietl, W. Plesiewicz, and J. Jaroszyński

Institute of Physics, Polish Academy of Sciences, Al. Lotników 32/46, PL-02-668 Warszawa, Poland

Results of millikelvin magnetoresistance measurements of anisotropic $p-Hg_{0.915}Mn_{0.085}Te$ and isotropic $n-Cd_{0.95}Mn_{0.05}Se$ are presented and interpreted in terms of Anderson-Mott transition in a spin polarized band.

1. INTRODUCTION

It has recently been realized that impure magnetic systems may undergo an insulator-to-metal transition as the magnetic field increases [1,2]. The transition occurs in classically weak magnetic fields, $L_H > \ell$, where $L_H = (c\hbar/eH)^{1/2}$ is the magnetic length and ℓ microscopic mean free path for elastic collisions. In such fields, the influence of the Landau quantization on the density of states can be neglected, and the experimental results may thus provide a meaningful test of the current theories [3,4] of the Anderson-Mott transition [5].

In this paper, we present results of millikelvin magnetoresistance measurements on narrow gap ($E_g \cong 50meV$) $p-Hg_{0.915}Mn_{0.085}Te$ and wide-gap ($E_g \cong 1.9eV$) $n-Cd_{0.95}Mn_{0.05}Se$. This paper contains some of the results of a larger study performed by us, the results of which have been partially reported in an earlier work [2].

2. EXPERIMENT AND SAMPLES

The crystals of HgMnTe and CdMnSe were grown by the Bridgman method. The samples were cut from carefully selected single-crystalline grains. Six leads were indium-soldered to bar-shaped samples of typical dimensions $8\times1\times0.4mm^3$. Just before measurements, the samples were etched in a 10% solution of bromine in methanol. The above precautions were found to be especially important in the case of narrow gap HgMnTe, where highly conducting n-type inversion layers can be formed at grain boundaries [6] or at the oxidized surface [7]. The four probe magnetoresistance measurements were carried out in a dilution refrigerator down to 30mK. In the case of p-HgMnTe, the measurements were performed for the current either perpendicular or parallel to the magnetic field, whereas in the case of n-CdMnSe, the tranverse configuration has been used, since the magnetoresistance of n-CdMnSe had been established [8] to be isotropic.

We now present the experimental results for two $p-Hg_{0.915}Mn_{0.085}Te$ samples cut from an ingot doped with compensating In impurities to the level of $7\cdot10^{15}cm^{-3}$. One of the samples was annealed for one week in saturated mercury vapour at $320°C$. The net acceptor concentrations, $p \equiv N_A - N_D$, estimated from the plateau values of the Hall coefficient between $50-70$K, were found to be $9\cdot10^{16}cm^{-3}$ and $2\cdot10^{17}cm^{-3}$ for the annealed and as-grown samples, respectively. The origin of the acceptors is unknown, but their sensitivity to annealing is usually taken to indicate that mercury vacancies are involved. Our $n - Cd_{0.95}Mn_{0.05}Se$ sample had a net In-donor concentration, $n \equiv N_D - N_A$, of $4 \cdot 10^{17} cm^{-3}$, as deduced from Hall data at 300K.

3. RESULTS AND DISCUSSION

The experimental results which show that an insulator-to-metal transition takes place for p-HgMnTe as the magnetic field is increased, are collected in Fig.1. We see that below a critical field H_c, the resistivity ρ strongly (exponentially) increases with decreasing temperature T (Fig.1a), and thus becomes infinite as $T \to 0$. Above H_c, the conductivity $\sigma \equiv 1/\rho$ can be described by $\sigma(T, H) = \sigma(0, H) + A(H)T^{1/2}$, as shown in Fig.1b. Figure 1c presents the data obtained at 28mK, which demonstrate that the transition is continuous and that the low-temperature value of σ increases essentially linearly with $H - H_c$, for $H > H_c$. It is interesting to note that very similar dependances have been observed in a different

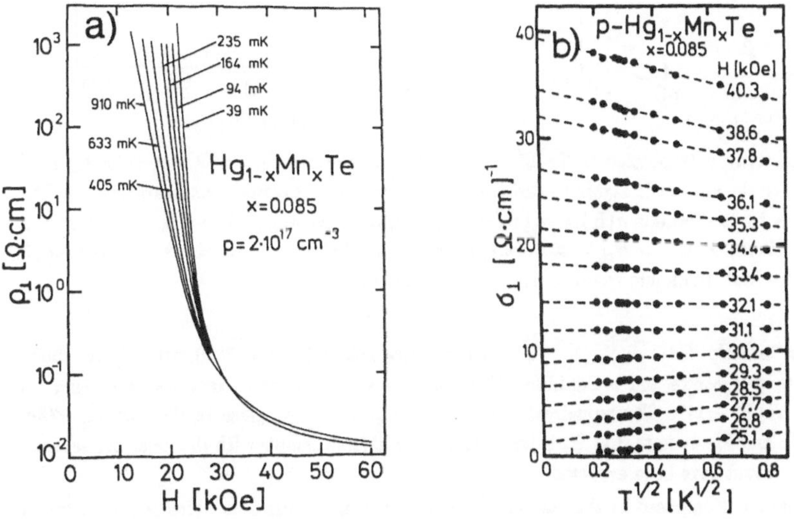

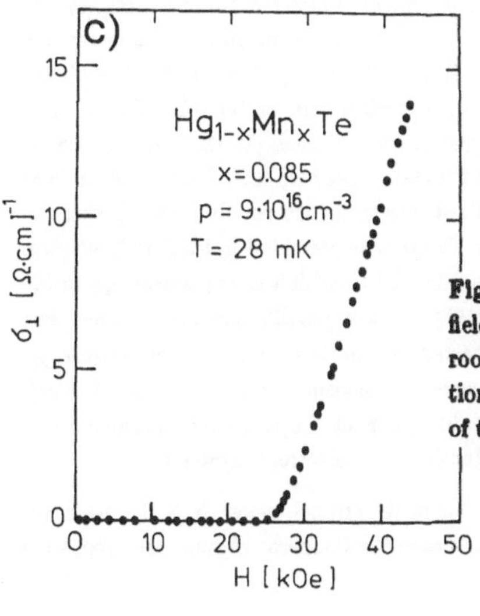

Fig.1 (a) Resistivity at selected temperatures vs. magnetic field and (b) conductivity in various magnetic fields vs. square root of temperature for HgMnTe with net acceptor concentration $p = 2 \cdot 10^{17} cm^{-3}$. (c) Conductivity at 28mK as a function of the magnetic field for $p = 9 \cdot 10^{16} cm^{-3}$.

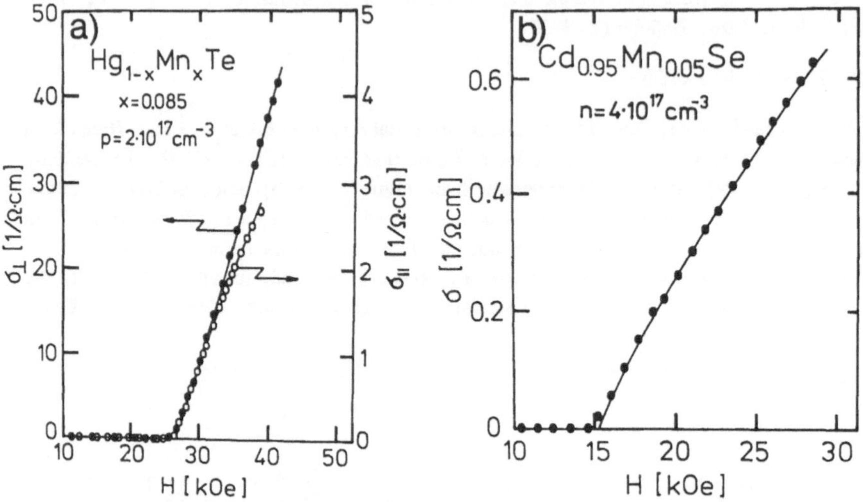

Fig.2 (a) Conductivity extrapolated to T=0K as a function of the magnetic field perpendicular ($\sigma_\perp$) and parallel ($\sigma_\parallel$) to the current in HgMnTe and (b) in CdMnSe in the transverse configuration (σ). The solid lines are fits by the formula $\sigma(H) = \sigma_c(H/H_c - 1)^\nu$, which give $\sigma_{c\perp} = 81.7 \pm 2$, $\sigma_{c\parallel} = 5.85 \pm 2$, $\sigma_c = 0.7 \pm 0.1$ in $(\Omega cm)^{-1}$; $H_{c\perp} = H_{c\parallel} = 26.6 \pm 0.1$, $H_c = 15.3 \pm 0.3 kOe$; $\nu_\perp = 1.08\pm^{0.15}_{0.05}$, $\nu_\parallel = 0.94\pm^{0.15}_{0.05}$, $\nu = 0.85\pm^{0.3}_{0.1}$. The Mott minimum conductivity is about $7(\Omega cm)^{-1}$.

magnetic system, namely, $Gd_{3-x}v_xS_4$ [1]. The conductivities extrapolated to T=0, $\sigma(0, H)$, are shown in Fig.2 for both p-HgMnTe and n-CdMnSe. The solid lines in that graph represent fits using the formula $\sigma = \sigma_c(H/H_c - 1)^\nu$. The numerical values of σ_c, H_c, and ν are given in the caption. When comparing Fig.1c and 2a, we note that H_c is larger in the p-HgMnTe sample with the lower net acceptor concentration, as could have been expected.

The principal question posed by the data concerns the physical mechanisms which cause the transition. It appears [2] that there are two main effects in magnetic systems which can produce insulator-to-metal transition in high magnetic fields. The first of them is destruction of magnetic polarons by the magnetic field [1,8,9]. The second effect is the giant exchange spin-splitting $\hbar\omega_s$ of the electronic states in magnetic systems [10]. The splitting causes a redistribution of the carriers between the spin-subbands and pushes the Fermi energy E_F into the region of states with higher conductances [11,12,2]. In the case of p-HgMnTe, the redistribution is connected with a change of the transverse hole mass from heavy to light [10,13]. This latter effect has two important implications. First, the transition should be much sharper, and hence σ_c much larger in p-HgMnTe than in n-CdMnSe. The results presented in Fig.2 fully confirm this expectation. Secondly, the conductivity of p-HgMnTe is expected to exhibit strong anisotropy in the magnetic field. Figure 2a shows that this is indeed the case, since in the metallic phase $\sigma_\perp$ is about one order of magnitude larger than $\sigma_\parallel$. A remarkable aspect of the data is that, in spite of the large anisotropy, the transition occurs at the same magnetic field for the transverse and longitudinal configurations (Fig.2a). These findings [2] have provided the first experimental confirmation of the theoretical predictions of WÖLFLE et al. [4] concerning the critical behavior of conductivity in anisotropic systems.

We now turn to the field and temperature dependence of σ in the critical region, $H \geq H_c$. Here, the coherence of backscattering and the electron-electron interactions in the Cooperon channel are suppressed by the magnetic field [3]. Furthermore, because of the large spin-splitting ($\hbar\omega_s \geq 0.5E_F$), the majority of carriers reside in one spin subband. The remaining carriers are most probably localized and their role seems to be limited to generation of an additional static random potential [2]. Recently, FINKELSTEIN [14] has derived the renormalization group equations for the transition occuring in a totally spin polarized electron gas. These equations, quoted in Ref.2 lead to $\nu = 1$, in agreement with our data. Moreover, they give the dependence of σ on T which is compatible with our experimental results [2].

REFERENCES

1. S. von Molnar, A. Briggs, J. Flouquet, G. Remenyi: *Phys. Rev. Lett.* **51**, 706 (1983); S. Washburn, R.A. Webb, S. von Molnar, F. Holtzberg, J. Flouquet, G. Remenyi: *Phys. Rev.* **B30**, 6224 (1984); S. von Molnar, J. Flouquet, F. Holtzberg, G. Remenyi: *Solid State Electron.* **28**, 127 (1985)
2. T. Wojtowicz, T. Dietl, M. Sawicki, W. Plesiewicz, J. Jaroszyński: *Phys. Rev. Lett.* **56**, 2419 (1986)
3. A.M. Finkelstein: *Zh. Eksp. Teor. Fiz.* **86**, 367 (1984) [*Sov. Phys. JETP* **59**, 212 (1984)]; C. Castellani, C. Di Castro, P.A. Lee, M. Ma: *Phys. Rev.* **B30**, 527 (1984)
4. P. Wölfle, R.N. Bhatt: *Phys. Rev.* **B30**, 3542 (1984); R.N. Bhatt, P. Wölfle, T.V. Ramakrishnan: *Phys. Rev.* **B32**, 569 (1985)
5. For a review on recent experiments, see, e.g., G.A. Thomas: *Philos. Mag.* **B52**, 479 (1985); for a review on the theory, see, e.g., P.A. Lee, T.V. Ramakrishnan: *Rev. Mod. Phys.* **57**, 287 (1985)
6. G. Grabecki, T. Dietl, P. Sobkowicz, J. Kossut, W. Zawadzki: *Appl. Phys. Lett.* **45**, 1214 (1984)
7. W. Scott, R.J. Hager: *J. Appl. Phys.* **42**, 803 (1971); S. Narita, T. Kuroda: *Nuovo Cimento* **39B**, 834 (1977)
8. T. Dietl, J. Antoszewski, L. Świerkowski: *Physica* **117-118 B+C**, 476 (1983); M. Sawicki, T. Dietl, J. Kossut, J. Igalson, W. Plesiewicz: *Phys. Rev. Lett.* **56**, 508 (1986)
9. M. Sawicki, T. Wojtowicz, T. Dietl, J. Jaroszyński, W. Plesiewicz, J. Igalson: in *Proc. 18th Int. Conf. on Physic of Semiconductors*, Stockholm 1986, to be published
10. see, e.g., J. Mycielski: in *Proc. 6th Int. Conf. on Ternary and Multinary Compounds*, ed. by B.R. Pamplin et al. (Pergamon, Oxford, 1985); see also, R.R. Gałązka: in Ref.9, and M. Grynberg, P.A. Wolff, and M. von Ortenberg: This volume
11. H. Fukuyama, K. Yoshida: *Physica* **105B+C**, 132 (1981)
12. Y. Shapira, D.H. Ridgley, K. Dwight, A. Wold, K.P. Martin, J.S. Brooks: *J. Appl. Phys.* **57**, 3210 (1985); Y. Shapira, D.H. Ridgley, K. Dwight, A. Wold, K.P. Martin, J.S. Brooks, P.A. Lee: *Solid State Commun.* **54**, 593 (1985)
13. J.A. Gaj, J. Ginter, R.R. Gałązka: *phys. stat. sol.(b)* **89**, 655 (1978)
14. A.M. Finkelstein, private communication

Magnetospectroscopy of Semimagnetic $Zn_{1-x}Mn_xSe$

M. von Ortenberg[1], *W. Erhardt*[1], *A. Twardowski*[2], *and M. Demianiuk*[3]

[1]Physikalisches Institut der Universität Würzburg,
Röntgenring 8, D-8700 Würzburg, Fed. Rep. of Germany
[2]Institute of Experimental Physics, University of Warsaw,
Hoza 69, PL-00-681 Warsaw, Poland
[3]Institute of Technical Physics, Military Technical Academy,
PL-00-908 Warsaw, Poland

1. Introduction

Semimagnetic semiconductors belong to an interesting kind of new material exhibiting the combined effects of canonical semiconductors and magnetism. Due to the exchange interaction of the statistically distributed, localized spins of the paramagnetic component, here. Mn++, the spin properties of the quasi-free charge carriers are essentially modified. Considerable effort has been made to investigate the nature of this exchange interaction, both experimentally and theoretically /1,2/. Whereas most of the experiments were performed on different narrow-gap semiconductors, the results on wide-gap materials are not so numerous. Especially concerning the problem of a "zero-field spin-splitting", which has been observed in various narrow-gap semiconductors /3-5/, no corresponding data were available so far for wide-gap materials. Therefore the principal objective of the present investigations was the study of such effects by suitable experimental techniques in a typical wide-gap semimagnetic semiconductor as Zn(1-x)Mn(x)Se for $0 < x < 0.05$.

2. Experimental

Pure Zn(1-x)Mn(x)Se is a highly resistive material because of the lack of quasi-free charge carriers, so that an artificial population of the conduction band becomes necessary. Due to the pronounced exciton effects /6/ one might expect, that the life time of optically, by interband absorption induced charge carriers is suffienctly large for the study of subsequent FIR-absorption. In our experiment, however, no indication of optically induced population changes was detected. As alternative methods we applied both surface-charge modulation in a MIS-configuration and a Zn-diffusion-doping technique as applied successfully by ref. /7/. The present results are based on samples treated by the latter method, which reduced the room-temperature resistivity of the samples to the order of k-Ohm*cm. As experimental techniques we applied DC-magneto transport and submillimeter-magneto spectroscopy, both as function of temperature, and magnetization measurements. We should emphasize, that very often only the combined results of DC- and high-frequency investigations yield a more comprehensive understanding of the microscopic mechanism involved.

2.1. DC-Transport

The temperature dependence of the resistivity, as plotted in Fig. 1 for a 1%-Mn sample, shows clearly that the transport mechanism at low temperature is dominated by MOTT's law /8/:

$$R = R_0 \exp[(T_0/T)0.25]. \qquad (1)$$

Fig. 1

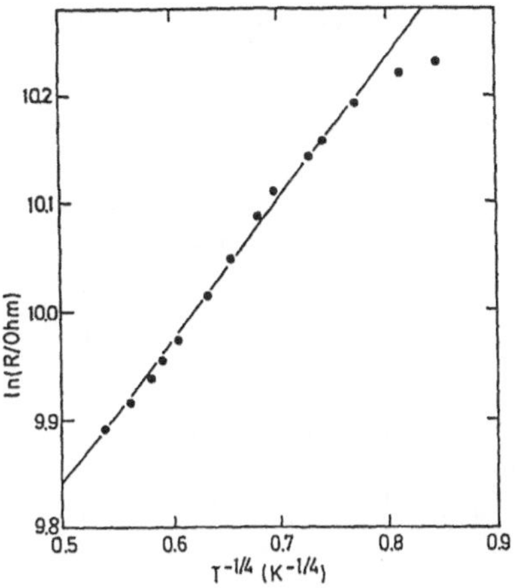

Fig. 2

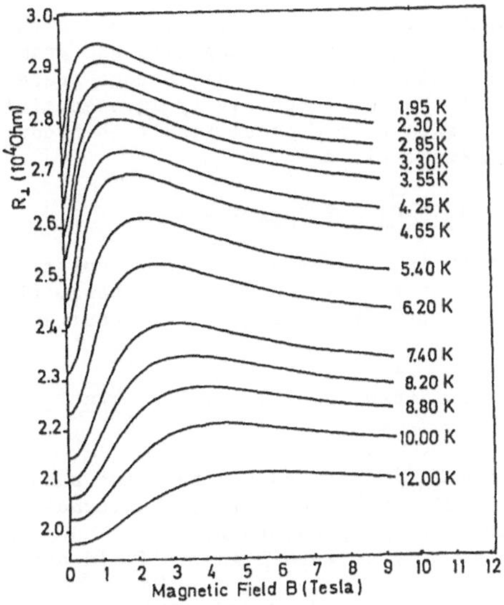

Fig. 1 The temperature dependence of the resistivity shows clearly an $\exp[(T_0/T)0.25]$-law

Fig. 2 The magneto-resistance for different temperatures

The very small value of the parameter T_0 = 3 K indicates, however, that the application of MOTT's Variable Range Hopping for the interpretation of the data might be questionable. As a matter of fact the unscreened BOHR radius in ZnSe is of the order of a = 33 Å, using $\varepsilon = 9.2*\varepsilon_0$ and $m* = 0.147*m$ /9/. MOTT's theory would result in a density of states at the FERMI level much to high:

$$N(E_F) = \frac{18.1}{kT_0*a^3} \quad 2*10^{21}\ (\text{meV*ccm})^{-1}\ . \tag{2}$$

That means, that in the following we will use equ. (1) essentially as a semi-empirical dependence, but avoid the usual microscopic picture.

All samples exhibit a pronounced structure in the transverse magnetoresistance as shown representatively in Fig. 2 for the same sample. Corresponding curves were reported for Cd(1-x)Mn(x)Se by DIETL et al. /10/ for T = 1.5 K, showing a similar negative differential magneto-resistance, which was qualitatively explained as an effect of the bound magnetic polaron. The interesting feature of our results, however, is the strong temperature shift of the maximum in the magneto resistance indicating the influence of the s-d exchange interaction on the hopping process. Note that a similar, however much weaker dependence of the magnetoresistance has been found in Ge and Si /11./ For these results the theoretical understanding was given by MOVAGHAR and SCHWEITZER /12,13/ in the theory of spin-flip hopping and further modified by OSAKA /14/. According to this theory the hopping conductivity is produced by "normal" and "anomalous" transfers in that sense, that the "anomalous" hop involves a simultaneous spin-flip of the electron, whereas the spin of the "normal" transfer is conserved. Due to occupation an "anomalous" nearest-neighbour hop is more probable than a "normal" second-nearest-neighbour transfer. This means that the net conductivity is essentially determined by spin-flip hopping. The spin-flip process itself is produced by the transverse part $\vec{B}_\perp$ of the internal field $\vec{B}_i$:

$$\vec{B}_i = \vec{B}_d + \vec{B}_{hyp} + \vec{B}_{ex}, \tag{3}$$

which is a superposition of dipolar, hyperfine and exchange fields. Whereas for Si and Ge the exchange field is negligible /12/, in our case the latter is dominant. The resultant magnetoresistance is given by /14/:

$$\Delta R/R(B=0) = c(T)*\left\{\frac{B_\perp^2}{B_h^2}\left[1 - \frac{B_h^2}{B^2}\ln(1 + \frac{B^2}{B_h^2})\right] - b^2\ln(1 + \frac{B^2}{B_h^2})\right\}. \tag{4}$$

Here T_0 is related to the value used in equ. (1) but not necessarily exactly the same. B is the external magnetic field and b is related to the average g-shift /12/. The "hopping field" B_h is related to the hopping frequency at zero magnetic field and given by /13/:

$$B_h = B_h^0 * \exp[-(T_0/T)0.25] \tag{5}$$

and reflecting the temperature dependence. It should be noted that in equ. (4) only the average of the square of the transverse internal field enters, which depends, of course, via the exchange interaction also on the external magnetic field. At this point we modify the MOVAGHAR and SCHWEITZER theory and assume, that the $\vec{B}_\perp$ is directly related to the exchange coupling matrix element α and has the following form:

$$\overline{B_\perp^2} = B_{\perp 0}^2*(S(S+1) - \overline{\langle S_z\rangle^2})/S(S+1), \quad S = 5/2, \quad B_{\perp 0} = x*\alpha/(2*/u_B). \tag{6}$$

Here x is the chemical composition parameter. This means that the transverse component of the internal field $\vec{B}_\perp$ is essentially determined by the magnetization of the sample, which is shown in Fig. 3 and exhibits the usual GAJ-like behaviour /15/.

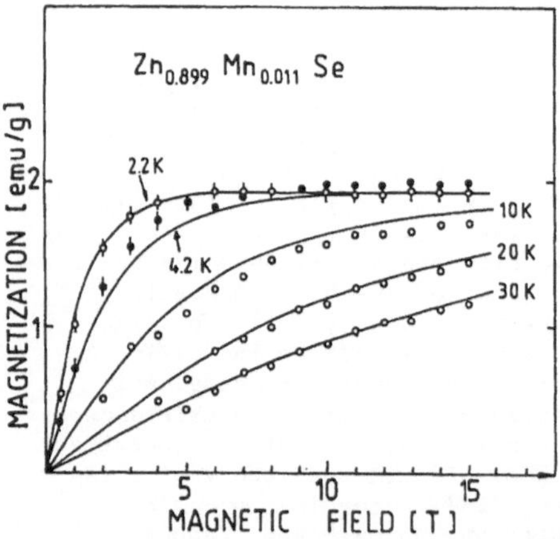

Fig. 3 The magnetization for different values of the temperature

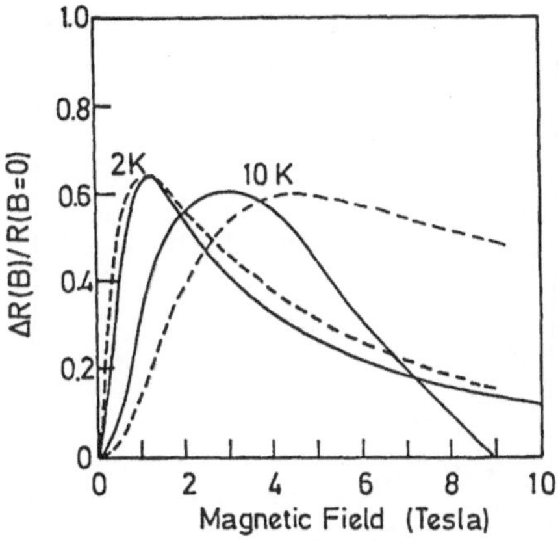

Fig. 4 The simulated magneto-resistance (solid curves) in comparison with the experimental data (broken curves)

In Fig. 4 we have plotted the simulation using equ. (5) and equ. (6) by solid curves in comparison with the experimental data as represented by dotted curves. The fit values are:

$$B_h^0 = 13.2 \text{ Tesla}, \; b = 0.95 \qquad (B_{\perp 0} = 0.01 * \alpha/(2\mu_0) = 27.7 \text{ Tesla}). \quad (7)$$

Where we have used for T_0 = 3K and α = 0.32 eV as derived in the next paragraph.

We like to point out that the temperature shift of the maximumn in the magneto resistance to higher fields with increasing temperatures is a direct consequence of the growing value of $\vec{B}_\perp$. For a temperature independent $\vec{B}_\perp$ as used by KUIVALAINEN et al. /11/ for Si, the maximum is shifted to lower magnetic fields for higher temperatures. The comparison of experimental and theoretical results in Fig. 4 demonstrates clearly the importance of the spin-flip components of the exchange interaction for the transport process, which are solely responsible for the shift to higher magnetic fields with temperature. For higher temperatures the normal hopping transfer without spin-flip becomes nonnegligible, so that the agreement is less satisfactory.

2.2. FIR Magneto Spectroscopy

Whereas DC-transport properties give essential information on the scattering process involved in the transfer, high-frequency investigations reveal in addition the resonance structure of the dielectric function. In our experiment the undoped samples did not show any structure in the FIR-laser-magneto-transmission spectra, after diffusion doping the spectra in Fig. 5 and Fig. 6 were recorded, for the two different radiation energies of 5.4 meV and 1.4 meV, respectively. Astonishingly the dominant background modulation corresponds to the structure of the DC-magneto transport and is essentially independent of the radiation frequency as can be seen in Fig. 7. For the interpretation of the pronounced transmission minimum at zero magnetic field we do not adopt the interpretation of DOBROWOLSKA et al. given for Cd(1-x)Mn(x)Se as 1s->2p transition /17/. As a matter of fact the photo-conductivity signal, as compared in Fig. 8 with the DC-magnetoresistance and FIR-transmission curves, can be quantitatively explained by a thermo-modulation of the DC-magnetoresistance and

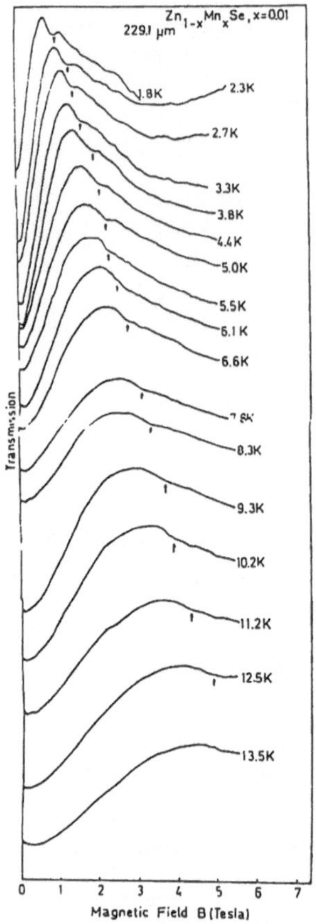

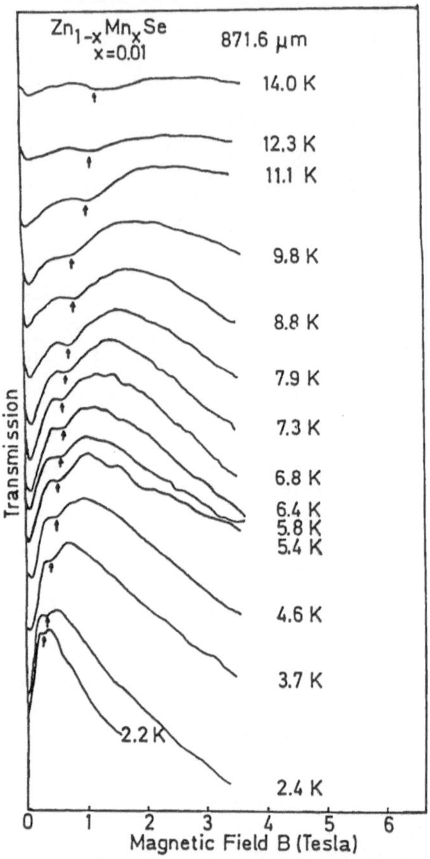

Fig. 5 and Fig. 6 The magneto-transmission spectra for the wavelengths of $\lambda = 229\ \mu m$ and $\lambda = 871\ \mu m$ respectively. The parameters indicate the temperature.

shows for all radiation energies up to 14 meV no indication of any impurity excitation. This means that at least for this range of binding energies no resonant donor states are present. The background modulation of the transmission spectra can be explained by the dielectric response due to nonresonant, high-frequency hopping-transfer including spin-flip according to MOVAGHAR and SCHWEITZER /12/. If we assume that the magnetic field dependence of the conductivity is not essentially changed in the FIR, we obtain the typical pseudo-resonance structure at zero magnetic field.

Superimposed on the broad "hopping modulation" of the transmission spectra we observe a very weak, but strongly temperature dependent resonance. With increasing temperature from 2 K to 13 K the resonance field is increased to more than 500%. Similar spectra have been observed by DOBROWOLSKA et al. /16, 17/ in Cd(1-x)Mn(x)Se and have been explained as electric-dipole induced spin-flip of a donor-bound electron. In contrast to the data on Cd(1-x)Mn(x)Se our resonance positions scale and extrapolate to zero energy for zero magnetic field as shown in Fig. 9. This means

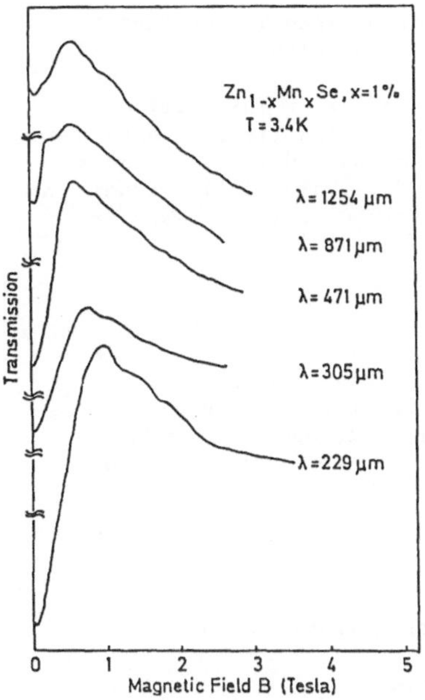

Fig. 7 The FIR-transmission is essentially independent of the radiation frequency

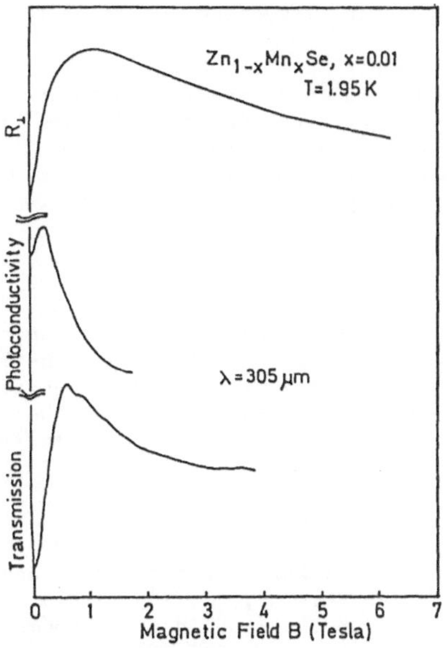

Fig. 8 The DC-magneto resistance, photoconductivity, and FIR-transmission as function of B

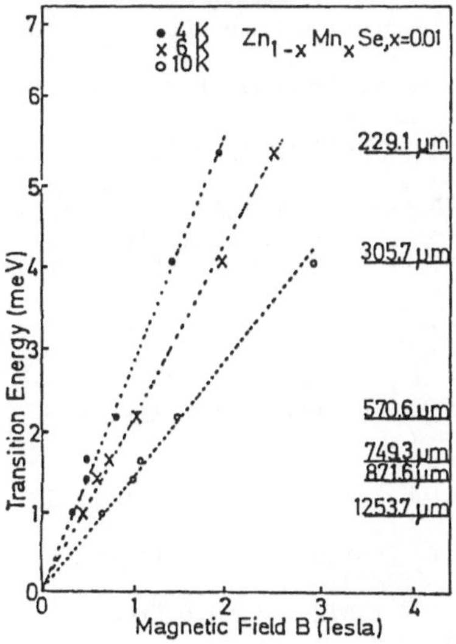

Fig. 9 The resonance position scales with frequency for all T

that no zero-field spin-splitting of a free or bound magnetic polaron is present. This result is quite interesting, because our DC-data have underlined the importance of the spin-flip part. We also like to point out, that no indication of a cyclotron resonance is observed in our spectra. Evidently the mobility of the hopping process in our samples is so low, that the cyclotron motion is nearly completely suppressed in contrast to the spin precession, where no transfer in space is involved.

Applying the KOSSUT-GALAZKA /2/ theory of the exchange interaction we are able to derive from both the spin-resonance and magnetization data the exchange coupling matrix element of the conduction band directly without interference of the valence band to α =0.32 eV. This value is in rather good agreement with the value α = 0.26 eV as obtained from the giant exciton splitting by TWARDOWSKI et al. /15/.

3. Summary

The DC- and FIR-magneto spectra of Zn-doped Zn(1-x)Mn(x)Se are essentially determined by spin-flip processes. Whereas the transport process is dominated by a non-resonant space transfer accompanied by simultaneous spin-flip, for the electric-dipole excited spin-resonance no hopping process is involved thus yielding a sharp resonance structure.

4. Literature

1. see references in: N.B. Brandt and V.V. Moshchalkov: Advances in Physics, 33, 193 (1984)
2. J. Kossut: Solid State Commun. 27, 1237 (1978)
3. H. Pascher, E.J. Fantner, G. Bauer, W. Zawadzki, M. von Ortenberg: Solid State Commun. 48, 461 (1983)
4. G. Karczewski and L. Kowalczyk: Solid State Commun: 48, 461 (1983)
5. G. Karczewski and M. von Ortenberg: Proc of the "17th Intnl. Conf. on the Physics of Semicond.", San Francisco, p. 1435 (1984)
6. A. Twardowski, T. Dietl, M. Demianiuk: Solid State Commun. 48, 845 (1983)
7. J. Gautron, C. Raisin, and P. Lemasson: J. Phys. D: Appl. Phys. 15, 153 (1982)
8. see in: H. Böttger an V.V. Bryksin: Hopping Conduction in Solids, Weinheim; Deerfield Beach, Fl.: VCH (1985)
9. H.W. Hölscher, A. Nöthe and Ch. Uihlein: Physica 117B&118B, 395 (1983)
10. T. Dietl: Springer Series in Solid State Sciences 24 "Physics in High Magnetic Fields", ed. S. Chikazumi and N. Miura, Springer-Verlag Berlin Heidelberg New York, p. 344 (1981)
11. P. Kuivalainen, J. Heleskivi, M. Leppihalme, U. Gyllenberg-Gästrin, and H. Isotalo: Phys. Rev. B26, 2041 (1982)
12. B. Movaghar and L. Schweitzer: phys. stat. sol. b80, 491 (1977)
13. B. Movaghar and L. Schweitzer: J. Phys. C: Solid State Phys. 11, 125 (1978)
14. Y Osaka: J. Phys. Soc. Japan 47, 729 (1979)
15. A. Twardowski, M. von Ortenberg, M. Demianiuk, and R. Pauthenet: Solid State Commun. 849 (1984)
16. M. Dobrowolska, H.D. Drew, J. K. Furdyna, T. Ichiguchi, A. Witowski, and P.A. Wolff: Phys. Rev. Letters 49, 845 (1982)
17. M. Dobrowolska, A. Witowski, J.K. Furdyna, T. Ichiguchi, H.D.Drew, and P.A. Wolff: Phys. Rev. B29, 6652 (1984)

Part VII

Magneto-Optics in 3D Systems

Nonlinear Spectroscopy of Narrow Gap Semiconductors in High Magnetic Fields *

D.G. Seiler[1] *and C.L. Littler*[2]

[1]Center for Applied Quantum Electronics, Department of Physics, North Texas State University, Denton, TX 76201, USA
[2]Central Research Laboratories, Texas Instruments, Dallas, TX 75265, USA

1. INTRODUCTION AND BACKGROUND

The realm of linear optical spectroscopy (including magneto-optical studies) has proven to be one of the most scientifically interesting and yet very important for characterizing semiconductors. Semiconductor properties such as energy band structure, excitons, impurities, carrier lifetimes, symmetries, absorption coefficients, etc., can be investigated and determined. A new frontier now exists for studying and characterizing semiconductors using the nonlinear optical interaction of laser light with matter. Bloembergen has concluded that "the field of linear spectroscopy is more than a century old and still flourishing. It may therefore be predicted with some confidence that nonlinear spectroscopy, which can augment linear spectroscopy in significant and non-trivial ways, will have a long and useful life."[1] In this paper, we concentrate on two-photon magnetospectroscopy (TPMS) and emphasize the following topics: (1) a review of past work on semiconductors and (2) a demonstration of the utility of TPMS with specific new results on InSb.

Narrow gap semiconductors (NGS) have long been recognized for their special characteristics that give rise to not only interesting physical effects but also useful technological applications. They can be defined as solids having an energy gap less than 0.5 eV.[2] Thus, it is clear that NGS are found among a broad spectrum of elements, compounds, and alloys. Our primary focus here will be on the narrow gap semiconductor, InSb, although the principles and techniques are equally applicable to the study of other NGS along with wider band gap semiconductors and even superlattices.

Two-photon absorption (TPA) processes in semiconductors have been the subject of extensive theoretical and experimental work for more than two decades. The transition probability rate per unit volume $W^{(2)}$ for a direct electronic transition from an initial state v to a final state c accompanied by the simultaneous absorption of two similar photons of frequency ω is proportional to $\delta(E_c-E_v-2\hbar\omega)$. The TPA coefficient K_2, the quantity usually measured in an experiment is defined by $K_2=2\hbar\omega W^{(2)}(k)/(1-R)^2I^2$, where I is the incident intensity and R the sample reflectivity. The calculation of reliable numerical values for K_2 is conceptually easy, but difficult in practice since it requires knowledge of the interaction Hamiltonian matrix elements among all the eigenstates of the crystal and summations over all the energy bands. In the past, many approximations and simplifying assumptions regarding the energy bands, momentum matrix elements, and intermediate states were made to reduce the complexity of the calculations. Theoretical calculations now give satisfactory results for values of K_2 and reasonable agreement with the data.[3-5]

The general experimental TPA setup involves two light sources of energy $\hbar\omega_1$ and $\hbar\omega_2$, each of which passes easily through the sample. When both beams are simultaneously present (both in the same space and time) and when $\hbar\omega_1+\hbar\omega_2$ equals an allowed transition energy, the sample simultaneously absorbs one quantum from each beam. Both light polarization and magnetic field direction control the TPA selection rules.[6,7] The beam configurations can be parallel, antiparallel, perpendicular, or inclined to one another. The light sources themselves can take various combinations such as: (1) both can be lasers, (2) one beam can be from a laser and the other from any number of sources--flash lamp and monochromator, stimulated Raman scattering, optical parametric oscillator. . . . Each configuration and combination of light sources may be particularly advantageous for different experimental conditions.

There are a number of experimental methods which have been used extensively to investigate the absorption of two photons: (1) transmission, (2) luminescence, and (3) photoconductivity. Transmission is the most direct method, but it is not as sensitive as the others. Photoconductivity measures the creation of free carriers induced by the absorption of light and is extremely sensitive, as shown by the following example. Consider a sample with a dark carrier concentration $n_0=10^{14}$ cm^{-3}. With modern electronic detection techniques, values of $\Delta n/n_0 \leq 0.1\%$ can easily be seen, giving a sensitivity of $\Delta n=10^{11}$ cm^{-3}. In this paper we show that the photo-Hall effect is also a valuable tool for nonlinear studies. It provides a sensitive but more direct measure of the number of photoexcited carriers.

2. PRESENT STATE OF TWO-PHOTON MAGNETOSPECTROSCOPY (TPMS)

Table 1 shows the results of studies on various semiconductors which demonstrate that TPMS is a powerful tool for extracting quantitative information on band and exciton parameters. In many cases, unique and detailed information can be obtained.

3. RESULTS OF TRENDS AND NEW DEVELOPMENTS

3.1 Eigenstate Analysis

Several theoretical papers have pointed out that two-photon spectroscopy in an external magnetic field allows in principle the direct determination of both the conduction and valence band effective masses.[38-41] Unfortunately, direct confirmation of this simple, but basic idea has been difficult to obtain in NGS until now. This was due to the complex nature of the bands and the difficulty in interpreting the TPMS spectra. High-resolution two-photon spectra[27,28] on InSb obtained with cw lasers now makes possible the determination of both the initial and the final eigenstates of the two-photon transitions. We now develop an analytical model for the two-photon transition energies using simple, but adequate expressions for the light-hole valence and conduction band Landau level energies. Analysis of the data results in values for the energy gap and the effective masses and g-factors of the electrons and light holes of InSb.

With the energy zero at the top of the valence band, the Landau level energies for the conduction $E(N_c,S_c)$ and light-hole valence $E(N_v,S_v)$ bands can be written simply as[42]

Table 1. Two-photon magneto-absorption in semiconductors

Material (Year) [Ref.] Photon Combination	Method Used and Important Results
CdS (1977) [8-10] dye & CO_2 lasers	Absorption. Created spin oriented 2P(A) excitons. g-values. Fast modulator for CO_2 laser beam.
(1982,83) [11,12] dye and Raman-shifted YAG	Photoconductivity. Zeeman splittings and diamagnetic shifts of 2P and 3P A and B free excitons. m*'s, E_g's, g's, R*'s.
CuCl (1983) [13] dye & YAG laser	Absorption. g's. g versus k. Polariton, longitudinal exciton, and paraexciton studied.
Cu_2O (1981,82) [14-16] dye & Raman-shifted YAG	Absorption. Splittings of exciton lines. Polarization dependence. g's.
GaAs (1985) [17] Raman-shifted dye	Photoconductivity. TPMA spectra obtained. Results described by Pidgeon-Brown band model.
InSb (1966,69,71) [18-20] CO_2 laser	Photoconductivity. Theory, experimental and theoretical TPMA studies.
(1971) [21,22] CO_2 laser	Absorption. Absorption of light by free holes created by TPMA. Perturbation theory.
(1973) [23] CO_2 laser	Photoconductivity, Absorption and saturation observed. Stimulated recombination at 5.3 μm.
(1976) [6]	Theory. Matrix elements and selection rules.
(1976) [25] twice CO_2 laser photon energy	Absorption. Observed TPMA spectra at 5 μm's and high fields with frequency doubling of CO_2 laser.
(1981) [26] two CO_2 lasers	Photoconductivity. 1st TPMA experiments in solids using only cw lasers.
(1982) [7]	Theory. Selection rules, polarization dependence.
(1982-84) [27-29] CO_2 laser	Photoconductivity, photovoltaic. Transition assignments. Intensity, polarization dependence. Band parameters extracted. Anisotropy. Lifetime resonances.
(1984,85) [30,31] CO_2 laser	Photo-Hall. 1st use of photo-Hall effect to determine number of TPA produced carriers. E_g versus T_L.
(1984) [32,33] CO_2 laser	Photoconductivity. Photo-Hall, Relationship of TPMA to shallow donor levels. Nonlinear coupled rate eqnts. used to describe results.
(1985) [34] CO_2 laser	Luminescence. Intensity and polarization dependence. Recombination with heavy holes.
ZnO (1970) [35] flash lamp & YAG	Absorption. Observed narrow 2P exciton lines compared to 1S. m*'s and g's.
ZnSe (1983, 85) [36,37] dye & Raman-shifted YAG	Absorption. Studied P excitons. Determined m*'s, g's, R*'s.

$$E(N_c,S_c) = \hbar eB(N_c+1/2)/m_c^* + S_c\beta g_c^* B+E_g \text{ and}$$
$$E(N_v,S_v) = -\hbar eB(N_v+1/2)/m_v^* - S_v\beta g_v^* B, \qquad (1)$$

where $N_c(N_v)$ is the Landau level number of the conduction (valence) band and $S_c(S_v)$ is the spin quantum number ($\pm 1/2$) for the conduction (valence) band. We note that S=-1/2 corresponds to an "a spin state" and S=+1/2 to a "b spin state." The effective masses m* and effective g-factors g* (treated here as magnitudes) depend upon the band parameters of the semiconductor. The Bohr magneton $\beta=e\hbar/2m_0=5.77 \times 10^{-3}$ meV/kG=5.77 $\times 10^{-2}$ meV/T.

A TPA transition occurs when $2\hbar\omega = E(N_c,S_c) - E(N_v,S_v)$ or

$$2\hbar\omega = E_g + \hbar eB[(N_c+1/2)/m_c^* + (N_v+1/2)/m_v^*] + \beta B[S_c g_c^* + S_v g_v^*]. \qquad (2)$$

The selection rules[6,7] in the spherical approximation governing transitions between initial and final states for the different light polarizations are for σ_L: $\Delta N=2, \Delta S=0$; σ_R: $\Delta N=-2, \Delta S=0$; σ: $\Delta N=0, \Delta S=0$; π: $\Delta N=0, \Delta S=0$. For σ and π transitions $N_c=N_v=N(\Delta N=0)$ and $S_c=S_v=S$ ($\Delta S=0$). Thus

$$2\hbar\omega = E_g + \hbar eB\,(N+1/2)/\mu + S\beta g^* B, \qquad (3)$$

where the reduced mass μ is given by $\mu^{-1} = m_c^{*-1} + m_v^{*-1}$ and the combined effective g-factor by $g^*=g_c^*+g_v^*$. In a parabolic band model where $1/\mu$ and g^* do not depend upon the magnetic field, a plot of $2\hbar\omega$ versus B gives straight lines with the intercept=E_g and the slope=$\hbar e\,(N+1/2)(1/\mu) + S\beta g^*$. This explains the almost linear behavior of the data and illustrates how E_g can be easily determined from a "fan chart" plot. The actual nonparabolicity of the bands is important to take into account and can easily be done with more complex band models. For our purpose of developing a simple analytical model to show the basic physics, however, the parabolic model is sufficient.

Figure 1 shows such a "fan chart" plot of the data for the σ or π transitions that were observed at liquid helium temperatures.[27,28] The

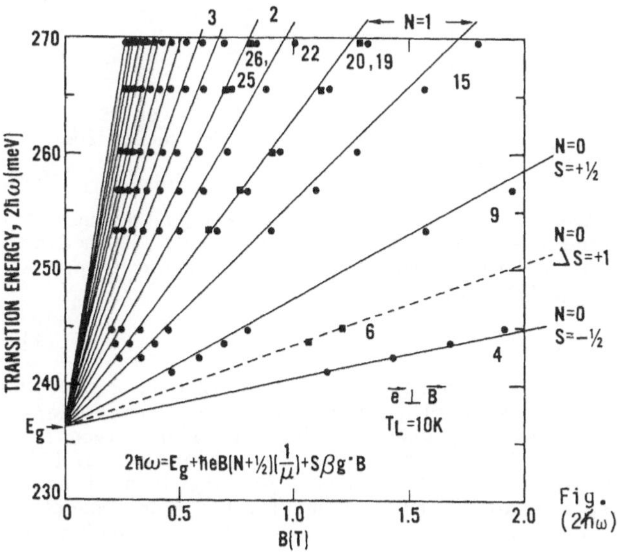

Fig. 1. TPA transition energy ($2\hbar\omega$) versus B

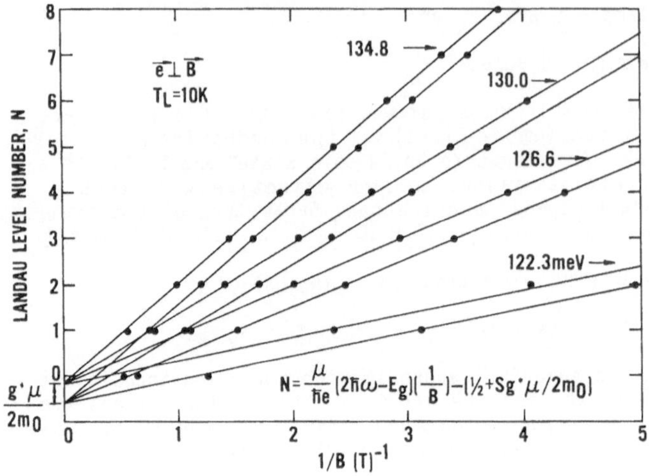

Fig. 2. Landau level number (N) versus 1/B

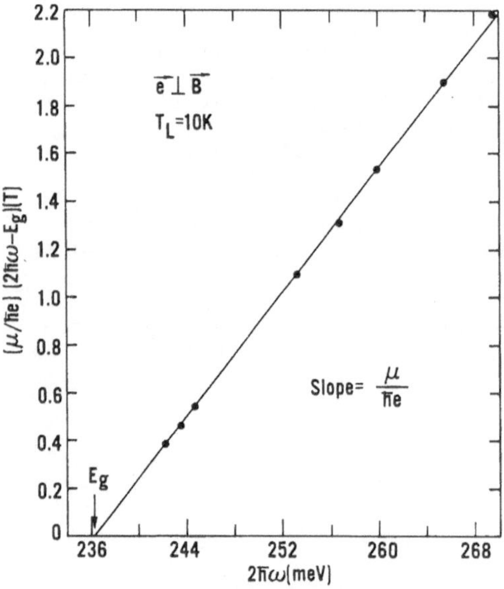

Fig. 3. Slope (of N versus 1/B plot in Fig. 2) versus $2\hbar\omega$

solid lines represent calculated results using Eq. (3) and the values of E_g=236.3 meV, μ=0.0075 m_o and g*=120 (to be discussed later). The major deviations of the data from the lines comes from not including nonparabolicity. Rearrangement of Eq. (3) gives

$$N = (\mu/\hbar e)(2\hbar\omega-E_g)\ (1/B) - [1/2+Sg^*\mu/2m_o] \qquad (4)$$

Thus a plot of N versus 1/B for a constant $\hbar\omega$ as shown in Fig. 2 should give two straight lines with the same slope of $[\mu/\hbar e(2\hbar\omega-E_g)]$ and intercepts (with S=±1/2) of $-[1/2+(Sg^*\mu'/2)]$, where $\mu'=\mu/m_o$. Data are given for four different values of $\hbar\omega$ and they confirm the predictions of Eq. (4). Next, a plot of this slope (of N versus 1/B) versus $2\hbar\omega$ should be linear with an intercept on the $2\hbar\omega$ axis of E_g and a slope of μ/he. Consequently, Fig. 3 shows the straight line behavior with a slope giving $\mu' \approx 0.0075$ or

$1/\mu' \approx 133$ and a value of E_g=236.3 meV as the intercept. Knowing the value of μ', the difference in "y-axis" intercepts in Fig. 1 is equal to $g^*\mu/2m_o$ from which $g^* \approx 120$ is found.

The determination of m_c^* and m_v^* individually can be done by making use of the ΔN=+2 and the ΔN=0 transitions. For example, for transition #20, $b^+(0) \rightarrow b^c(2)$ and for #9, $b^+(0) \rightarrow b^c(0)$.[26,27] The energy difference between these two transitions at a constant magnetic field of $B_{20}=B_9=B$ is $2\hbar\omega_{20}-2\hbar\omega_9=2\hbar eB/m_c^*$, which only involves m_c^* as a parameter. At B=1T, this splitting is =15.3 meV and thus m_c^*=0.0151 m_o from which m_v^*=0.0150 m_o. Likewise using transitions #19 and #26 at 0.5T gives m_c^*=0.0149 m_o and m_v^*=0.0152 m_o. If one uses transition #25 then $2\hbar\omega_{25}-2\hbar\omega_{20}=2\hbar eB/m_v^*$ from whence m_v^*=0.0156 and m_c^*=0.0145 m_o. These mass values are in reasonable agreement with each other and with values expected from previous results of both intra- and interband experiments.

In order to determine g_c^* and g_v^* separately a spin flip transition must be used. These transitions are weak, but are fortunately observable. Transition #6 is $a^+(0) \rightarrow b^c(0)$ and can be used along with #4 $a^+(0) \rightarrow a^c(0)$ to give at constant B $2\hbar\omega_6-2\hbar\omega_4=\beta Bg_c^*$=2.7 meV at 1T, from which g_c^*=47 and g_v^*=73. Alternatively transitions #9 and #6 give $2\hbar\omega_9-2\hbar\omega_6=\beta Bg_v^*$=3.8 meV at 1T, resulting in g_v^*=66 and g_c^*=54. Thus, g_c^*=50 and g_v^*=70 are reasonable estimations.

In summary, this section shows how to analyze complex TPMS data in InSb with a simple analytical model. The two-photon transitions shown in Fig. 1 are adequately explained by this simple model, which gives good physical insight to the nature of two-photon spectroscopy. From the σ or π transitions (ΔN=0), values for E_g, μ, and g^* can be obtained. Additional use of $\sigma_L(\Delta N=+2)$ or $\sigma_R(\Delta N=-2)$ transitions then allows m_c^* and m_v^* to be determined. Finally observation of spin-flip transitions allows g_c^* and g_v^* to be calculated. Reasonable values of these quantities have been determined for InSb:[43] E_g=236.3 meV, μ=0.0075, g^*=120, m_c^*=0.0148 m_o, m_v^*=0.0153 m_o, g_c^*=50, g_v^*=70.

3.2 Two-Photon Photo-Hall Magnetospectroscopy

We now discuss the photo-Hall effect and its advantages for studying TPA. We will show that, by judicious choices of temperature and magnetic field, photo-Hall techniques can be used to understand the role of impurity absorption (IA) in the observed signals (often composed of TPA and IA) as well as the photo-excited carrier dynamics. Although not mentioned here, the two-photon photo-Hall method has been recently used to acquire TPMS data and determine the temperature dependence of the energy gap for InSb.[31]

Figure 4 shows[33] the Hall coefficient versus magnetic field observed in the dark (R_H^D) and for two CO_2 laser wavelengths: (1) λ=10.81 μm where no TPA is possible and (2) λ=9.66 μm where TPA does occur. The increase in R_H^D with increasing magnetic field indicates the presence of magnetic freeze-out effects that are well known. We first note that for λ=10.81 μm, there is a decrease in R_H, indicating an increase in the number of conduction band electrons due to IA from shallow and deep impurity or defect levels present in the sample. Also shown is resonant behavior in the field dependence of R_H for λ=9.66 μm occurring at magnetic field values where TPA resonances have previously been identified.[27] Since the electron mobility is much greater than the hole mobility in n-InSb, a one-band Hall coefficient model is used to calculate the variation of the number of photo-

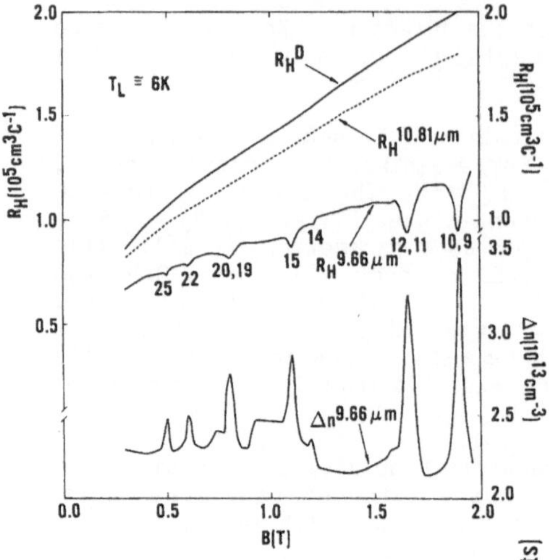

Fig. 4. Magnetic field dependence of the Hall coefficient in the dark R_H^D, at λ=10.81 μm, and 9.66 μm; lower half is Δn for 9.66 μm

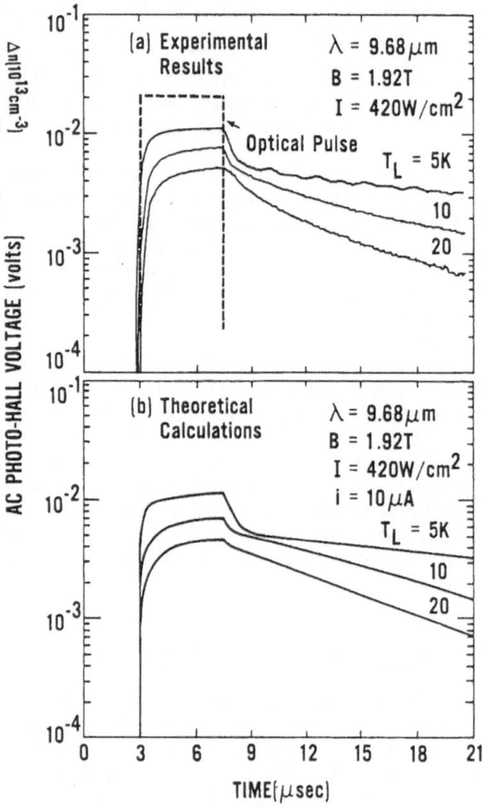

Fig. 5. AC photo-Hall voltage versus time for TPA transition 10, 9

excited electrons $\Delta n = n - n_0$ (n_0 is the concentration of electrons in the dark) shown in the lower half of Fig. 4. Resonant increases in n are seen and the nonresonant background is due largely to IA.[30] The direct determination of Δn is valuable since the dependence of the induced free carrier concentration on external parameters such as intensity and lattice temperature allows direct comparison with the predictions of rate equations. This in turn leads to a better understanding of kinetics of two photo-carrier generation and recombination when impurities or defects are present.

An example of time-resolved photo-Hall measurements is shown in Fig. 5. The waveforms shown in the upper half of Fig. 5 were obtained at B=1.92 T, or at the peak of the TPA resonance labeled 10, 9 in Fig. 4. The theoretical synthesis of the observed time-resolved response is shown in the lower half of Fig. 5 and was obtained using a rate equation approach which describes the photo-excited carrier dynamics of multiple absorption mechanisms. The relevant rate equations for this situation are given by

$$\begin{aligned} dn/dt &= gp' + (A_p+A_T)\, m + A_q\, q - r_s n(N_d-m) - r_q n(N_q-q) - r_b np, \\ dm/dt &= -(A_p+A_T)\, m + r_s n(N_d-m), \\ dq/dt &= -A_q q + r_q n(N_q-q), \text{ and} \\ dp/dt &= gp' - r_b np, \end{aligned} \tag{5}$$

where $A_p m$ and $A_q q$ are the one-photon generation rates out of shallow and deep impurity levels, respectively, $A_T m$ is the rate of thermal generation of carriers out of the shallow impurity level, r_s, r_q, and r_b are the coefficients associated with carrier recombination to the shallow level, deep level and valence band respectively, N_d and N_q are the concentrations of the shallow and deep impurities, respectively, gp' is the rate of generation of two-photon produced carriers, p is the number of photo-excited holes generated by TPA, and $p'=p_o-p$, where p_o is the number of optically coupled states. Equation (5) constitutes a set of four nonlinear, coupled rate equations which can be solved numerically for n. In the absence of TPA, the system reduces to three levels involving the population of electrons in the conduction band (n) and the shallow (m) and deep (q) levels. Solutions to the three level system for the values $K_s=5.5 \times 10^{-3}$ cm^{-1}, $K_q=2.5 \times 10^{-4}$ cm^{-1}, $r_s=2.7 \times 10^{-8}$ cm^3/sec, $r_q=2.5 \times 10^{-9}$ cm^3/sec, and $N_q=2.8 \times 10^{12}$ cm^{-3} can be found in a recent paper.[44] K_s and K_q are the IA coefficients which determine A_p and A_q. Thus, for our discussion, the theoretical calculations shown in the lower half of Fig. 5 were obtained by adjusting only the TPA coefficient K_2 ($g=K_2I^2(1-R)^2/2\hbar\omega p_o$) and the bimolecular rate r_b. The time response of the photo-Hall voltage shown in the lower half of Fig. 5 was synthesized by solving Eq. (5) for n(t), allowing optical generation to continue for the length of the optical pulse, then shutting off photo-excited carrier generation by setting all absorption coefficients equal to zero for times after the optical pulse had passed. The photo-Hall voltage was then calculated from n(t) and compared with experimental results, as shown in Fig. 5. From the analysis, K_2 and r_b were found to be 16 cm/MW, and 1×10^{-9}, 4×10^{-9}, and 1.5×10^{-8} cm^3/sec for 5, 10, and 20 K, respectively. The value of K_2 agrees with values generally accepted for InSb,[5,21,45,46] and r_b agrees with values obtained for bimolecular recombination of two-photon generated carriers in InSb at low temperatures.[27,47]

In addition to explaining the time-resolved results, solutions to Eq. (5) were used to describe the intensity dependence of n shown in Fig. 6. Saturation effects are clearly seen at the highest intensities for 5 and 10 K. From the theoretical analysis, it was seen that p_o, was largely responsible for this behavior and the values 7×10^{13} cm^{-3}, 1×10^{14} cm^{-3}, and 2×10^{14} cm^{-3} for 5, 10, and 20 K were obtained. It can be seen from both Figs. 5 and 6 that the major features of the photo-Hall response are explained well by the rate equations given in Eq. (5). In particular, all regions of the time response of the photo-Hall voltage are reproduced faithfully by the predictions of the rate equations, including both the time response of the photo-Hall voltage during photo-excited carrier generation and the two-stage photo-Hall decay. Similar two-stage decays of TPA signals have been seen previously in the photoconductive response of n-InSb;[47,48] the first stage being identified with heating of electrons by the laser pulse and the second stage due either to bimolecular recombination,[48] or the trapping of holes.[49] However, the photo-Hall

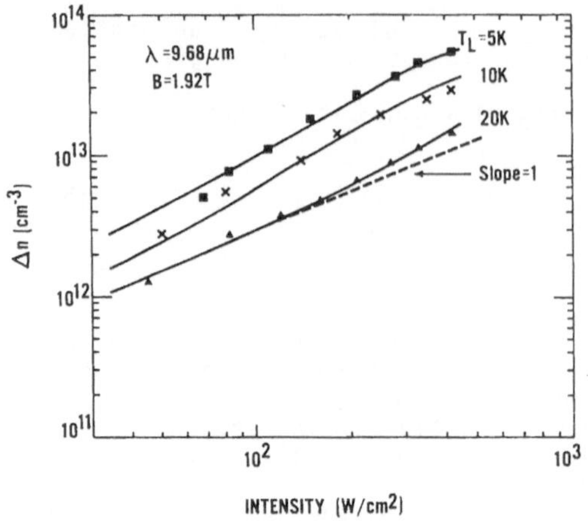

Fig. 6. Intensity dependence of Δn. The solid lines represent calculations from Eq. (5) and the symbols represent experimental results

effect clearly provides a more definitive identification of the two decay times. The fast decay seen here results from recombination of photo-excited carriers to the unoccupied shallow donor levels, while the slower decay is due largely to bimolecular recombination, with a small contribution from the recombination to deep impurity levels.[44]

From Fig. 6, another interesting feature is evident. Note that at all temperatures, both the theory and the results predict an initial linear dependence of Δn on the laser intensity. However, as the intensity is increased and contributions from TPA and impurity absorption become roughly equal, the slope increases to ≈1.5. Finally, at the largest intensities, the value of Δn saturates due to the depletion of the states optically coupled by TPA. The photo-excited carrier dynamics in the presence of competing absorption mechanisms shown in Fig. 6 can be more clearly understood by examining the solutions of Eq. (5) for Δn, m, q, and p. This is illustrated in Fig. 7. Figure 7 shows the calculated intensity dependence of Δn, m, q, and p for the same conditions as shown in Fig. 6. The populations of the shallow donor level and the valence band are of most interest to our discussion since both of these levels make significant contributions to the total value of Δn measured. At the low intensities the population of the shallow donor level, m, is seen to decrease, corresponding to photo-excited carrier generation from the shallow level by one-photon absorption. At the higher intensities, however, the photo-excited carrier dynamics become more complicated. The generation of TPA carriers and their subsequent recombination via shallow or deep levels influence the populations of these levels. The extent of this influence is strongly controlled by the rates of decay of the photo-excited carriers into each level. For intensities greater than 200 W/cm^2, the population of the shallow level not only stops decreasing, but actually increases with increasing I. This behavior is due jointly to the large two-photon generation rate and the fast recombination to the shallow level as compared to the slower bimolecular and deep impurity level recombination. This particular behavior has been shown to give rise to resonant carrier lifetimes in n-InSb.[32,33]

In summary, we have shown that the photo-Hall effect provides new insight into the photo-excited carrier dynamics associated with TPA, particularly when impurities are present. Figure 8 illustrates this quite well; we are able to see both the region where IA dominates over TPA, and where the reverse is true. In addition, saturation effects are clearly

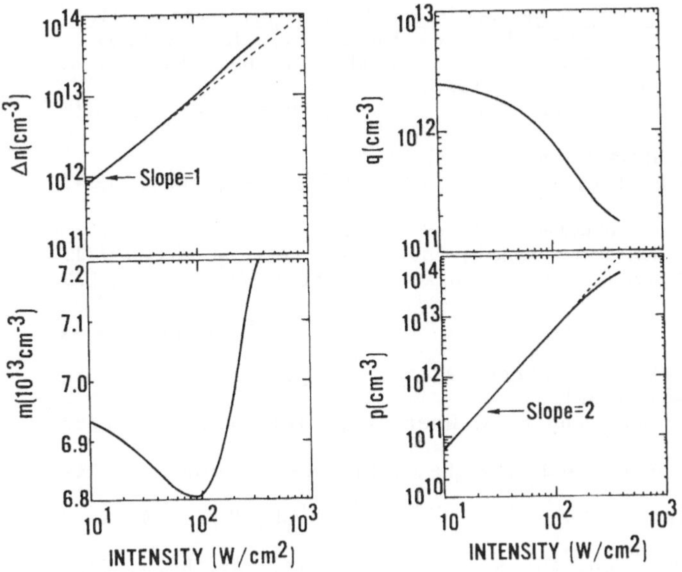

Fig. 7. Intensity dependence of Δn, m, q, and p for $\lambda = 9.68$ m

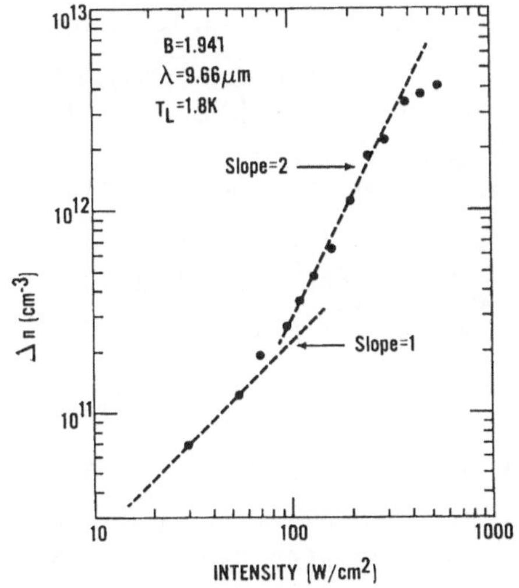

Fig. 8. Observed intensity dependence of Δn for TPA transition 10, 9 at B=1.94 T and $\lambda = 9.66\ \mu m$

seen at the highest intensities. Finally, the sensitivity of photo-Hall techniques is demonstrated in Fig. 8, allowing the detection of $\approx 7 \times 10^{10}$ photo-excited carriers/cm^3.

4. REFERENCES

*Work supported in part by NSF Grant ECS-8310625, LTV Aerospace and Defense Co., and a Faculty Research Grant from NTSU.

1. N. Bloembergen, in Proc. Int. School of Phys. Enrico Fermi on Nonlinear Spectroscopy (Academic Press, New York, 1977), p. 1.

2. L. Sosnowski, in Proc. Narrow Gap Semiconductors, Physics and Applications, Nimes, 1979 (Springer-Verlag, New York, 1980), V. 133, p. 1.
3. C. R. Pidgeon, B. S. Wherrett, A. M. Johnston, J. Dempsey, and A. Miller, Phys. Rev. Lett. 42, 1785 (1979).
4. A. M. Johnston, C. R. Pidgeon, and J. Dempsey, Phys. Rev. B 22, 825 (1980).
5. M. H. Weiler, Solid State Commun. 39, 937 (1981).
6. W. Zawadzki and J. Wlasak, J. Phys. C: Solid State Physics 9, L663 (1976).
7. M. H. Weiler, Solid State Commun. 44, 287 (1982).
8. V. T. Nguyen, T. C. Damen, E. Gornik, Appl. Phys. Lett. 30, 33 (1977).
9. T. C. Damen, V. T. Nguyen, and E. Gornik, Solid State Commun. 24, 179 (1977).
10. V. T. Nguyen, T. C. Damen, E. Gornik, and C. K. N. Patel, Appl. Phys. Lett. 31, 603 (1977).
11. D. G. Seiler, D. Heiman, R. Feigenblatt, R. L. Aggarwal, and B. Lax, Phys. Rev. B 25, 7666 (1982).
12. D. G. Seiler, D. Heiman, and B. S. Wherrett, Phys. Rev. B 27, 2355 (1983).
13. D. Frohlich, H. Holscher, and A. Nothe, Solid State Commun. 48, 217 (1983).
14. D. Frohlich, R. Kenklies, Ch. Uihlein, and C. Schwab, Phys. Rev. Lett. 43, 1260 (1979).
15. Ch. Uihlein, D. Frohlich, and R. Kenklies, Phys. Rev. B 23, 2731 (1981).
16. D. Frohlich and R. Kenklies, Phys. Stat. Sol. (b) 111, 247 (1982).
17. D. G. Seiler, C. L. Littler, and D. Heiman, J. Appl. Phys. 57, 2191 (1985).
18. K. J. Button, B. Lax, M. H. Weiler, and M. Reine, Phys. Rev. Lett. 17, 1005 (1966).
19. M. H. Weiler, R. W. Bierig, and B. Lax, Phys. Rev. 184, 709 (1969).
20. M. H. Weiler, R. W. Bierig, and B. Lax, Proc. 3rd Int. Conf. on Photoconductivity (Pergamon Press, NY, 1971), p. 145.
21. V.-T. Nguyen and A. R. Strnad, Optics Commun. 3, 35 (1971).
22. V.-T. Nguyen, A. R. Strnad, and Y. Yafet, Phys. Rev. Lett. 26, 1170 (1971).
23. S. K. Manlief and E. D. Palik, Solid State Commun. 12, 1071 (1973).
24. G. Favrot, R. L. Aggarwal, and B. Lax, in Proc. 13th Int. Conf. Phys. of Semiconductors (Tipografia Marves, Rome, 1976), p. 1035.
25. D. G. Seiler, M. W. Goodwin, and M. H. Weiler, Phys. Rev. B 23, 6806 (1981).
26. D. G. Seiler, M. W. Goodwin, and M. H. Weiler, in Proc. 4th Int. Conf. on Phys. of Narrow Gap Semicond., Linz, Austria, 1981 (Springer-Verlag, NY, 1982), p. 192.
27. M. W. Goodwin, D. G. Seiler, and M. H. Weiler, Phys. Rev. B 25, 6300 (1982).
28. D. G. Seiler, M. W. Goodwin, S. W. McClure, and L. A. Veilleux, in Application of High Magnetic Fields in Semiconductor Physics (Springer-Verlag, NY, 1983), p. 297.
29. D. G. Seiler and S. W. McClure, Solid State Commun. 47, 17 (1983).
30. C. L. Littler, D. G. Seiler, and S. W. McClure, Solid State Commun. 50, 565 (1984).
31. C. L. Littler and D. G. Seiler, Appl. Phys. Lett. 46, 986 (1985).
32. S. W. McClure, D. G. Seiler, and C. L. Littler, J. Appl. Phys. 56, 1655 (1984).
33. D. G. Seiler, C. L. Littler, and S. W. McClure, Optics Commun. 50, 359 (1984).
34. M. A. Alekseev, M. S. Bresler, O. B. Gusev, I. A. Merkulov, and A. O. Stepanov, Sov. Phys. Semicond. 19, 443 (1985).

35. R. Dinges, D. Frohlich, B. Staginnus, and W. Staude, Phys. Rev. Lett. 25, 922 (1970).
36. H. W. Holscher, A. Nothe, and Ch. Uihlein, Physica 117B, 395 (1983).
37. H. W. Holscher, A. Nothe, and Ch. Uihlein, Phys. Rev. B 31, 2379 (1985).
38. F. Bassani and R. Girlanda, Optics Commun. 1, 359 (1970).
39. F. Bassani, in Proceedings Int. School of Phys. "Enrico Fermi" Course LII (Academic Press, NY, 1972), E. Burstein, editor, p. 592.
40. F. Bassani and A. Baldereschi, Surface Science 37, 304 (1973).
41. R. Girlanda, Nuovo Cimento 39B, 593 (1977).
42. This is a reasonable approximation for the light hole Landau levels except for the two lowest which are "heavy-hole like" because of the strong coupling to the heavy mass states. These states have a quantum number N_v=-1 and are not describable in this simple model.
43. For comparison purposes, see Table I of C. L. Littler, D. G. Seiler, R. Kaplan, R. J. Wagner, Phys. Rev. B 27, 7473 (1983).
44. C. L. Littler and D. G. Seiler, J. Appl. Phys. 60, 261 (1986).
45. J. M. Doviak, A. F. Gibson, M. F. Kimmit, and A. C. Walker, J. Phys. C: Solid State Phys. 6, 593 (1973).
46. A. Miller, A. Johnson, J. Dempsey, S. D. Smith, C. R. Pidgeon, and G. D. Holah, J. Phys. C: Solid State Phys. 12, 4839 (1979).
47. H. J. Fossum and B. Ancker-Johnson, Phys. Rev. B 8, 2850 (1973).
48. E. K. Guseinov, R. I. Ibraginov, V. G. Koratin, D. N. Nasledov, and Yu. G. Popov, Sov. Phys.-Semicond. 5, 1549 (1972).
49. R. E. Slusher, W. Giriat, and S. R. J. Brueck, Phys. Rev. 183, 758 (1969).

Energy Relaxation of Hot Carriers in GaAs in a Strong Magnetic Field Studied with Picosecond Photoluminescence

R.W.J. Hollering[2], *T.T.J.M. Berendschot*[1], *H.J.A. Bluyssen*[1], *H.A.J.M. Reinen*[1], *and P. Wyder*[3]

[1]Research Institute for Materials, University of Nijmegen, Toernooiveld, NL-6525 ED Nijmegen, The Netherlands
[2]Philips Reserach Laboratories, NL-5600 JA Eindhoven, The Netherlands
[3]Max-Planck-Institut für Festkörperforschung, Hochfeld-Magnetlabor, 166 X, F-38042 Grenoble Cedex, France

Energy relaxation by optical phonon emission of photoexcited carriers in bulk GaAs has been studied previously by cw and timeresolved picosecond spectroscopy [1]. In the picosecond experiments substantial deviations were observed between calculated and experimentally determined energy relaxation rates, which were ascribed to screening of the carrier-phonon interaction by the high carrier density [2,3] and generation of a nonequilibrium optical phonon distribution [4].

In this paper we present time-energy resolved photoluminescence measurements on GaAs in the presence of strong magnetic fields up to $B = 20$ T. Analysis of the measurements yields the time evolution of the temperature $T_{e,h}(t)$ and the density $n_{e,h}(t)$ of the carriers (e is electron and h is hole) and of the Landau level linewidth $\Gamma(t)$. The results show that the energy relaxation rate is reduced with increasing value of the magnetic field, and that the carrier cooling is adequately described by the kinetics of the coupled carrier-phonon system using a magnetic field-dependent phonon generation rate term.

The experiments were carried out on bulk GaAs which is grown on an n-type GaAs substrate, and consists of a 0.25 µm thick GaAs layer, confined between 0.1 µm thick $Al_{.6}Ga_{.4}As$ barrier layers. These confining layers are transparent to the excitation wavelength and avoid surface recombination, carrier diffusion to the substrate and reduce the gradient in carrier density perpendicular to the surface. The sample was cooled down to a temperature of 1.5 K in a bath cryostat which was mounted in the hybrid magnet of the High Field Magnet Laboratory of the University of Nijmegen. Optical excitation was achieved with picosecond light pulses (2 ps, $h\nu = 2.05$ eV) which generate instantaneously "mono-energetic" electron and hole distributions in respectively the conduction- and valence bands. Within a picosecond the photoexcited carriers thermalize by carrier-carrier interaction to a temperature $T_{e,h}$, which differs substantially from the lattice temperature. For $T_{e,h} \geq 50$ K the excess energy of the carrier gas reduces mainly by emission of optical phonons. Timeresolved detection of the luminescence radiation, which is due to electron-hole recombination during the relaxation process, is performed with the use of an upconversion technique [5,6].

Figure 1 shows time-energy resolved photoluminescence spectra at different times after excitation for $B = 16$ T. The slopes on the high-energy side of the spectra reflect the decreasing carrier temperature due to phonon emission. The low-energy side gives the reduced bandgap which shifts with time to higher energies due to the decreasing carrier density. At 35 ps after excitation both $T_{e,h}$ and τ, the carrier scattering time, have reached values such that $kT_{e,h} < \hbar\omega_c$, the Landau level splitting, and $\omega_c\tau > 1$, and Landau level structure is observed. Due to both cooling and recombination of the electrons and holes, occupation of the higher Landau levels decreases which is clearly shown by the range of the luminescence spectra and results in a population of only the lowest Landau level at 750 ps after excitation.

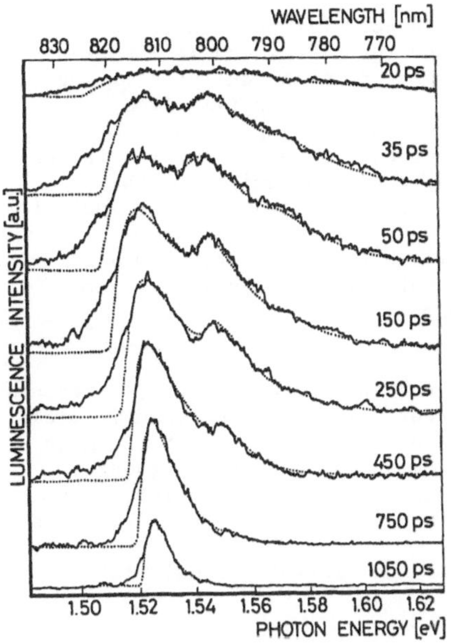

Figure 1 Measured and calculated timeresolved luminescence spectra in the presence of a magnetic field of B = 16 T.

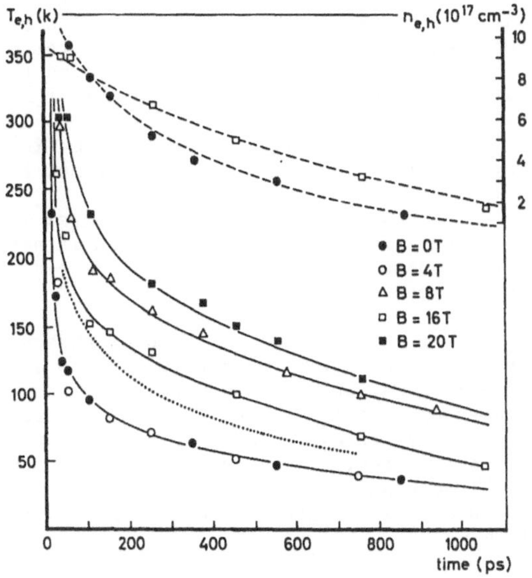

Figure 2 Evolution of temperature and density of the carriers in GaAs for B=0,4, 8,16 and 20 T. The solid lines show cooling curves calculated with a model containing the magnetic field dependent kinetics of the coupled carrier-phonon generation and decay rates. The dashed lines are drawn as a guide to the eye.

In order to obtain the time evolution of the temperature $T_{e,h}(t)$ and the density $n_{e,h}(t)$ of the carriers and the Landau level linewidth $\Gamma(t)$, the spectra from Fig. 1 are fitted with an expression for free carrier band to band recombination [7] given by: $I(\hbar\omega) \sim \int g_c g_v f_c f_v dE$. The Fermi distribution functions $f_{c,v}$ for electrons and holes contain the quasi-Fermi levels which are determined from the relation $n_{e,h} = \int g_{c,v} f_{c,v} dE$. The density of states (dos) functions for the three-dimensional carriers in the presence of a magnetic field are given by [8]:

$$g^{3D}_{c,v}(E) = \frac{1}{2\pi l^2} \left(\frac{m^\star}{\pi\hbar^2}\right)^{\frac{1}{2}} \sum_{N=0}^{\infty} \left[\frac{(E-E_N) + ((E-E_N)^2 + \Gamma^2)^{\frac{1}{2}}}{(E-E_N)^2 + \Gamma^2}\right]^{\frac{1}{2}}, \qquad (1)$$

where $l = (\hbar/eB)^{\frac{1}{2}}$ is the magnetic length, $m^\star$ the effective carrier mass, $E_N = (N+\frac{1}{2})\hbar\omega_c$ with N the Landau level index and ω_c the cyclotron frequency. The Landau levels have a Lorentzian broadening, as presented by $\Gamma = \hbar/2\tau$, introduced by DINGLE [9] where $1/\tau$ presents the carrier scattering rate. However to obtain a fit which gives the correct energy splitting between the peaks in the luminescence spectra for $m_e = 0.068m_0$ and $m_h = 0.5m_0$, we had to use a dos-function in which the $N = 0$ term was replaced by a more complex function and the Landau level linewidth is given by $\Gamma = \hbar/2\tau(\omega_c\tau)^{1/3}$ [10,11]. For large Landau level broadening this dos-function approaches that of zero field. Reabsorption effects and spin splitting can be neglected [5].

The dotted lines in Fig. 1 show a reasonable good fit between the calculated and experimental spectra at the high-energy side which is important for the determination of the carrier temperature. From the analysis of the spectra at B = 0,4,16 and 20 T, we find that $n_{e,h}$ varies in time from $9\times10^{17}cm^{-3}$ to $1\times10^{17}cm^{-3}$, as shown in the upper part of Fig. 2. Due to different experimental conditions at B = 8 T the carrier density was a factor of three higher during the entire time interval. For comparison of the energy relaxation rates this difference is taken into account in the analysis below.

Values obtained for the Landau level linewidth $\Gamma(t)$ vary from 7 meV at 35 ps to 2 meV at 1000 ps which are in agreement with the absolute value of a few tenths of a picosecond and the time dependence of the carrier-phonon scattering time.

In Fig. 2 the time evolution of the carrier temperature $T_{e,h}(t)$ as obtained from the analysis is presented for the different values of the magnetic field. It should be noted that apparently the time evolution of $T_{e,h}$ at 8 T is much slower than at 16 T. However it follows from the analysis that for equal carrier densities, $T_{e,h}(t)$ at 8 T is faster than at 16 T, as is indicated by the dotted curve. Thus it is found that the energy relaxation rate decreases with increasing magnetic field.

To analyze $T_{e,h}(t)$ the energy relaxation of hot carriers is calculated from [4]:

$$\frac{dE}{dt} = -\sum_q \sum_i^{LO,TO} \hbar\omega_q \left(\frac{dN_q}{dt}\right)_{c,i} , \qquad (2)$$

where summation runs over the different carrier phonon interaction modes (LO,TO) and over the wavevectors q of the phonons with energy $\hbar\omega_q$ which are involved in the energy relaxation process.

The terms $(dN_q/dt)_{c,i}$ depend on both emission and absorption rates for LO and TO phonons by the carriers and on the number N_q of the phonons which follows from

$$\frac{dN_q}{dt} = \sum_i^{LO,TO} \left(\frac{dN_q}{dt}\right)_{c,i} - \frac{N_q - N(T_L)}{\tau} . \qquad (3)$$

In this expression the first term on the right-hand describes generation of LO and TO phonons, the second their nonelectronic decay. Further τ is a wavevector-independent phonon lifetime of the order of 30 ps [12], while T_L is the lattice temperature. $N(T_L)$ is given by $(\exp(E_O/kT_L)-1)^{-1}$, where E_O is the optical phonon energy.

In the presence of a magnetic field the phonon generation term $\left(\frac{dN_q}{dt}\right)_{c,i}$ is described by an expression given by BAUER et al. [13] containing degenerate Fermi statistics and phonon occupation numbers N_q:

$$\sum_q \left(\frac{dN_q}{dt}\right)_{c,i} = \frac{m^\star N}{(2\pi\hbar)^3 2l^2 n_{e,h}} \sum_{n,m} \int dq_z^2 \frac{1}{q_z^2} \times \int dq_\perp C^2(q)$$

$$\times [N(T_{e,h}) - N_q] [f(E,T_{e,h}) - f(E+E_O,T_{e,h})] \times |M_{nm}(\xi)|^2 . \qquad (4)$$

Here $m^\star$ represents the carrier mass, N the number of carriers, N_q the number of phonons with wave vector q, $N(T_{e,h}) = (\exp(E_O/kT_{e,h})-1)^{-1}$, $C^2(q)$ is the carrier-phonon interaction matrix element [11], $f(E,T_{e,h})$ are the Fermi functions with E the carrier energy [13]. Summation runs over the Landau level transition matrix elements $M_{nm}(\xi)$ with $\xi = l^2 q_\perp^2/2$ and l is the magnetic length. Since the spherical symmetry is broken the integral over the phonon momentum phase space is separated in directions normal ($q_\perp$) and parallel (q_z) to the magnetic field. It should be noted that no Landau level broadening is taken into account. Using $(dE/dt) = (dE/dT_{e,h})\,(dT_{e,h}/dt)$ where $(dE/dT_{e,h})$ is the specific heat of the carrier gas, the temporal evolution of both the carrier temperature and phonon occupation numbers is obtained by parallel integration of equations (2) and (3) with use of equation (4). To fit the experimental data the initial temperature T_0 and a multiplication factor C in front of dE/dt are used as the only adjustable parameters. The solid lines represent calculated cooling curves for $T_0 = 800$ K. Since for B = 4 and 8 T a density of states function with unbroadened Landau levels is questionable, the zero-field model is used [4], and values for C = 0.5 at B = 4 T and C = 0.1 at B = 8 T are respectively obtained. For B = 16 T and B = 20 T the carrier cooling is adequately described by using the magnetic field-dependent carrier phonon interaction rates where a value of C = 0.2 was obtained. The values of $C \neq 1$ may be due to neglection of phonon generation during the pulse [4] and/or dynamical screening of the carrier-phonon interaction due to the high carrier density [15].

In summary (1) our measurements show a decreasing hot carrier energy relaxation rate with increasing magnetic field strength and (2) the carrier cooling is adequately described by the kinetics of the coupled carrier-optical phonon system using a magnetic field-dependent phonon generation term.

Acknowledgement - We thank A.F. van Etteger for his assistance with the laser equipment. Part of this work has been supported by the "Stichting voor Fundamenteel Onderzoek der Materie" (FOM) with financial support of the "Nederlandse Organisatie voor Zuiver Wetenschappelijk Onderzoek" (ZWO).

1. For a review see J. Shah: J. Phys. (Paris) 42C-7, 445 (1981)
2. W. Graudzus and E.O. Goebel: Physica (Utrecht) 117B+118B, 555 (1978)
3. E.J. Yoffa: Phys. Rev. B21, 2145 (1980)
4. W. Poetz and P. Kocevar: Phys. Rev. B28, 7040 (1983)
5. R.W.J. Hollering, T.T.J.M. Berendschot, H.J.A. Bluyssen, P. Wyder, M. Leys, J. Wolter: Solid State Commun. 57, 527 (1986)
6. R.W.J. Hollering, T.T.J.M. Berendschot, H.J.A. Bluyssen, P. Wyder, M. Leys, J. Wolter: Physica (Amsterdam) 134B+C, 422 (1985)
7. S. Tanaka, H. Kobayashi, H. Saito and S. Shionoya: J. Phys. Soc. Jpn. 49, 1051 (1980)
8. For a discussion on Landau level broadening see: L.M. Roth and P.N. Argyres: Semiconductor and Semimetals, Vol. 1 (Academic Press Inc., London 1966)
9. R. Dingle: Proc. R. Soc. London A 211, 517 (1962)
10. R. Smetsers and H.J.A. Bluyssen: to be published in Semicond. Science and Technology
11. R. Kubo, N. Hashitsume and S.J. Miyake: Solid State Phys. 17, 269 (1966)
12. A. Mooradian, G.B. Wright: Solid State Commun. 4, 431 (1966)
13. G. Bauer, H. Kahlert and P. Kocevar: Phys. Rev. B11, 968 (1975)
14. M. Pugnet, J. Collet and A. Cornet: Solid State Commun. 38, 531 (1981)
15. C.H. Yang and S.A. Lyon: Physica (Amsterdam) 134B+C, 309 (1985)

Population Inversion and Anomalous Voigt Effect Due to J x H Force in InSb

T. Morimoto and M. Chiba

Institute of Atomic Energy, Kyoto University, Uji, Kyoto 611, Japan

It is shown that a large phase-shift is expected for light propagating in semiconductors from the notable influence of electromagnetic forces in the Voigt configuration ($k \perp H$).

Recently we have found a new phenomenon of valence to conduction bands breakthrough due to the electromagnetic excitation by a J x H force in InSb at 80 - 300 K, in the transverse configuration of current density J and magnetic field H, which leads to a highly inverted population resulting in lasing[1]. The aim of this paper is to point out another important effect of such electromagnetic excitation on the Voigt ellipticity expected for the light propagating in the Voigt configuration ($k \perp H$), k being the wave vector of light, with emphasis on the useful application in controlling the phase shift by the current flow.

Let us first consider the Voigt effect[2], in the usual case of J = 0, for electron-hole plasmas comprising of n_1 electrons per cm^3 and n_2 holes per cm^3. Suppose that $k//\hat{y}$, $H//\hat{z}$, and assume $\omega\tau \gg 1$, where ω and τ are the angular frequency of the incident light and the relaxation times of electrons and holes, respectively. For the propagating light with frequency ω, which satisfies $E_g > \hbar\omega > \hbar\omega_p$, E_g and ω_p being, respectively, the energy gap and the plasma frequency, we can derive an expression for the phase shift, δ, per unit path, between the ordinary and extraordinary modes having, respectively, the wave vectors k_o and k_e as follows[3];

$$\delta = k_o - k_e = \frac{\sqrt{\varepsilon_\infty}}{2c\,\omega^3}(\omega_{p1}^2\,\omega_{c1}^2 + \omega_{p2}^2\,\omega_{c2}^2), \tag{1}$$

where ε_∞ is the high-frequency dielectric constant, ω_{pi} is the plasma angular frequency, ω_{ci} is the cyclotron angular frequency, and the suffix i = 1 denotes electrons and i = 2 denotes holes. This is the general expression of the Voigt ellipticity for the 2-band model of electron and holes. For compensated plasmas, putting $n_1 = n_2 = n_o$ in eq.(1), we have

$$\delta = \frac{2\pi n_o e^4 H^2}{\sqrt{\varepsilon_\infty}\, c^3\,\omega^3}\left[\left(\frac{1}{m_1}\right)^3 + \left(\frac{1}{m_2}\right)^3\right], \tag{2}$$

where n_o is the electron concentration in thermal equilibrium, e the electronic charge, c the light velocity, m_1 the effective mass of electrons, and m_2 that of holes. Equation (1) indicates that the Voigt effect is additive with respect to electrons and holes, nevertheless the Faraday rotation vanishes when $n_1 = n_2$.

Next consider the effect of the J x H force on the Voigt ellipticity when $J \neq 0$. In this case, the angular frequency of light, ω, sensed by the drifting electrons and holes with drift velocity $v_d (//\hat{y})$, should be Doppler shifted to $\omega' = \omega \pm kv_d$. However, this shift can be neglected, since $\omega \gg kv_d$ when $\omega \gg \omega_p$[4]. Thus, the most important effect of the J x H force emerges as the increase in the carrier number due to the excitation, especially through the interband breakthrough[1], as long as the parabolic band model is valid. Getting the analytical expression for the effective carrier concentration n* is difficult when the breakthrough takes place, but the value could be evaluated from the voltage-current characteristics. Therefore, in this case we can merely replace the electron concentration n_o by the effective concentration n* in eq.(2), and then we have the expression for the Voigt phase-shift per unit path[3];

$$\delta = \frac{2\pi n^* e^4 H^2}{\sqrt{\varepsilon_\infty}\, c^3 \omega^3} \left[\left(\frac{1}{m_1}\right)^3 + \left(\frac{1}{m_2}\right)^3 \right]. \tag{3}$$

Now, let us consider the population inversion caused by the electromagnetic excitation at high magnetic fields. Figure 1 shows the voltage-current characteristics of n-InSb at 80 K for given transverse magnetic fields. The dimensions of the sample were 4 mm long, 0.55 mm wide, and 0.2 mm thick. The carrier concentration n_o and the Hall mobility μ at 77 K were 1×10^{15} cm^{-3} and 3×10^5 cm^2/Vs, respectively. As can be seen in the figure, a negative resistance arises under the application of high magnetic fields. After some computations, it has been confirmed that the negative resistance is caused by the valence to conduction bands breakthrough, which takes place when the condition

$$(J_c H/n^* c) l^*(H) \simeq E_g \tag{4}$$

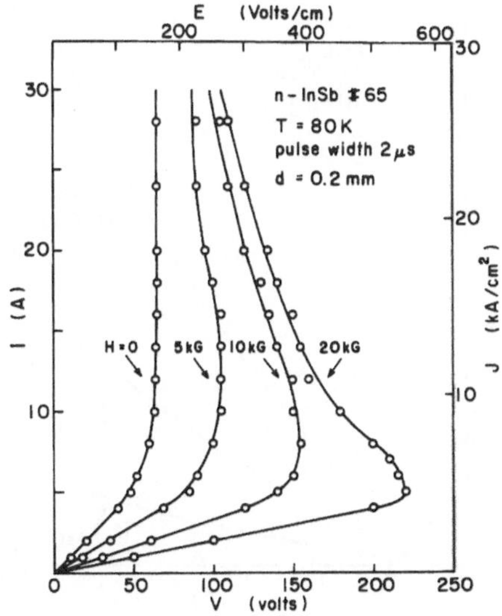

Fig. 1. V - I characteristics at 80 K. J ⊥ H.

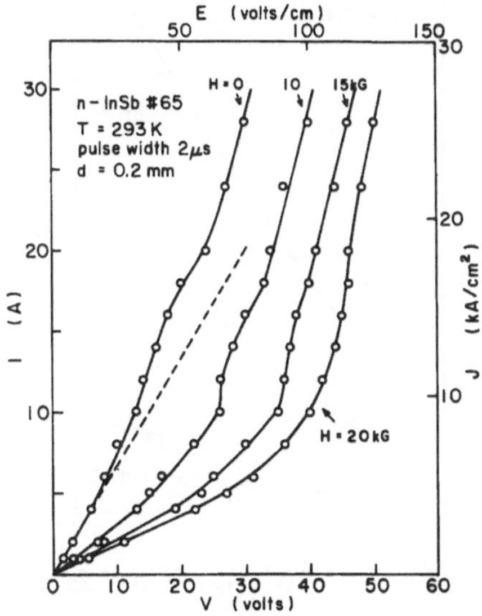

Fig. 2. V - I characteristics at 293 K. J ⊥ H.

is satisfied at high currents and high magnetic fields[1], indicating the important action of the J x H force in compensated plasmas under high magnetic fields. Here, J_c is the critical current density at which dE/dJ = 0, n* is the effective carrier concentration, l*(H) is an effective ambipolar diffusion length, being interpreted as the distance which the excess carriers excited by the J x H force can move along the lines of force (y-direction) until they are annihilated by the recombination. In this case, the values of l*(H) are reasonably estimated to be 1.3 μm, 0.8 μm and 0.5 μm, respectively, for magnetic fields of 5, 10 and 20 kG.

By application of high magnetic fields, H, along the z-direction, the effective (induced) electric field, E*, equivalent to the J x H force (along the y-direction), becomes stronger than the longitudinal electric field, E, along the x-direction ($J \parallel \hat{x}$). In fact, for the case of H = 20 kG and J_c = 4.5 kA/cm^2, the value of E_c* defined by $E_c = (J_c H/n^*ec)$ becomes as high as 4.9 kV/cm. Thus, we can see that such a high value of E* could easily cause the interband breakthrough in narrow-gap semiconductors, such as InSb, which leads to a highly inverted population resulting in lasing for light of wavelength, $\lambda_o \simeq ch/E_g$, h being Planck's constant[1]. The value of the effective carrier concentration, n*, for the case of H = 20 kG and J = 30 kA/cm^2, is estimated to be 1.8×10^{16} cm^{-3} at 80 K from the steady state V-I characteristics of Fig. 1, although this value of n* is never the peak value at the excitation but the considerably smaller mean value during the excitation and the recombination relaxation[1].

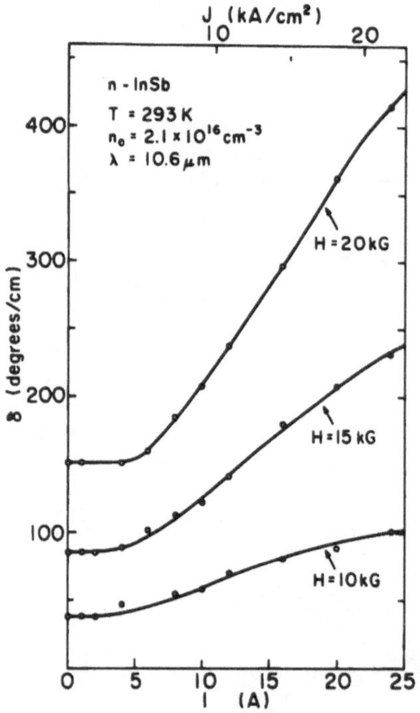

Fig. 3. Current dependence of the Voigt phase-shift per unit path, δ, at 293 K. k//J x H.

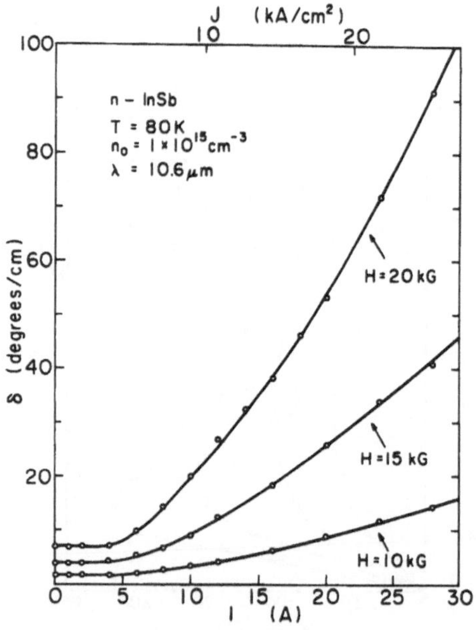

Fig. 4. Current dependence of the Voigt phase-shift per unit path, δ, at 80 K. k//J x H.

At room temperature, a similar interband breakthrough has been observed, as shown in Fig.2, at high currents and high magnetic fields, indicating a notable increase in the effective carrier concentration, n*, and hence a remarkable increase of the Voigt phase-shift at high currents. In Figs.3 and 4, the Voigt phase-shift per unit path, δ, calculated from eq.(3) by using the value of n* estimated from the V-I characteristics at 293 and 80 K, are shown for incident light of wavelength 10.6 μm. The calculated value at 80 K for J = 0, of course, coincides with the experimental value by TEITLER et al.[5], if the difference in the values of carrier number and wavelength is taken into account. The relatively large value of δ at room temperature comes from the large value of n* owing to the large value of the intrinsic carrier concentration. Anyway, it is interesting to see that the value of δ increases rapidly with increasing current in the nonlinear region of the V-I curve even at room temperature. This outstanding result must have useful applications in controlling the phase shift and hence modulating the intensity of the incident light by the current flow in combination with a suitable polarizer.

In the case of $E_{F1} - E_{F2} > \hbar\omega > E_g$, where E_{F1} and E_{F2} are the quasi Fermi levels for electrons and holes, respectively, the incident light is to be amplified by means of the stimulated emission, besides the expected phase shift. In this case, therefore, the quantum mechanical treatment is needed in a strict sense. Nevertheless, the classical expression (3), using effective carrier concentration n*, seems to be still valid, at least formally, when it is taken into account that the polarization of photons has to be conserved at the stimulated emission. In the simple estimate, the value of δ for this amplifying frequency-range ($\lambda \sim 5.3$ μm for InSb) becomes ~ 1/8 of the value estimated for 10.6 μm light because of the λ^3 dependence. However, the value might be much larger than that, when the multiple reflection of light by the surfaces has to be taken into account for the lasing mode. The evaluation of the details is a problem of the future.

The authors would like to thank S. Ueda for technical assistance and Dr. S. Akai for helpful advice, and Prof. G. Landwehr for comments on the stylistic improvement of the manuscript.

1. T.Morimoto and M.Chiba:Bull.Inst.Atomic Energy, Kyoto Univ. 70, 43(1986)
2. See, for example, C.R.Pidgeon: In Handbook on semiconductors, Vol.2 ed. T.S.Moss (North-Holland, Amsterdam, 1980) ch.5
3. T.Morimoto: Bull. Inst. Atomic Energy, Kyoto Univ. 68, 37 (1985)
4. For the effect of the Doppler shift due to the drift motion of electrons and holes in the Faraday configuration when $\omega < \omega_p$, see, J.Bok and P.Nozieres: J. Phys. Chem. Solids, 24, 7809 (1963)
5. S.Teitler, E.D.Palik, and R.F.Wallis: Phys. Rev. 123, 1631 (1961)

Interband Magneto-Optics in Narrow-Gap Semiconductors

M. Singh[1] *and P.R. Wallace*[2]

[1]Department of Physics, University of Western Ontario, London, Canada N6A 3K7
[2]Department of Physics, McGill University, Montreal, Canada

We develop the theory of magneto-optical transitions between any of the valence bands and the conduction band for semiconductors described by the Kane model or its anisotropic generalization, the Bodnar model. We find the selection rules $\Delta n = \pm 1$, $\Delta s = 0$ and $\Delta n = \pm 1$, $\Delta s = \pm 1$ when the magnetic field makes an arbitrary angle with respect to the c-axis of the crystal.

INTRODUCTION

In earlier papers /1, 2/ we developed a detailed theory of the intraband magneto-optics of semiconductors described by the Bodnar model in both the Faraday and Voigt configurations. Any results obtained with the Bodnar model /3/ are readily specialized to the Kane model, simply by putting the anisotropy parameters equal to zero. In the present paper we turn our attention to the range of frequencies which excite interband transition from either the two light hole bands or the heavy hole band to the conduction band. Because of the wide range of application of our results we will not be able to present numerical results covering all cases of interest. We shall therefore give analytic formulae, reserving numerical calculation to the chosen example of cadmium arsenide. A partial list of materials covered by our theory, divided according to categories, is as follows:

Bodnar-type
- positive gap (Cd_3P_2, $Cd_3As_{2-x}P_x$, $Cd_{3-x}Zn_xAs_2$, ...)
- inverted gap (Cd_3As_2, $Cd_3As_{2-x}P_x$, $Cd_{3-x}Zn_xAs_2$, ...)

(For the alloys indicated, the sign of the gap depends on relative concentration; for a particular composition, the gap becomes zero.)

Kane type
- positive gap (InSb, $Hg_{1-x}Cd_xTe$, $Hg_{1-x}Cd_xSe$, ...)
- inverted gap (HgTe, HgSe, $Hg_{1-x}Cd_xTe$, $Hg_{1-x}Cd_xSe$, ...)

THEORY

In frequency dependence the dielectric function in a sufficiently strong magnetic field is

$$\epsilon_{\lambda\mu} = \epsilon^L_{\lambda\mu} + 4\Pi i \ \sigma_{\lambda\mu}/\omega, \qquad [1]$$

where $\epsilon^L_{\lambda\mu}$ is the phonon contribution and $\sigma_{\lambda\mu}$ is the magneto-conductivity tensor. We shall in general express this tensor in the polarization representation with respect to the magnetic field axis, if the z-axis is along the magnetic field B and the x- and y-axes are arbitrarily chosen perpendicular to it. Ignoring level broadening, the formula for the conductivity tensor is

$$\sigma_{\lambda\mu} \sim \frac{i}{\omega} \sum_{\alpha,\alpha'} \frac{f(E_{\alpha'}) - f(E_{\alpha})}{E_{\alpha'} - E_{\alpha} - \omega - i\eta} \langle\alpha|V_{\lambda}|\alpha'\rangle\langle\alpha'|V_{\mu}|\alpha\rangle. \qquad [2]$$

The states $|\alpha\rangle$ which we have calculated have eight components in the Bodnar model, each component being expressible in terms of Landau functions $|n,k_y,k_B,s\rangle$, where n is the Landau quantum number, k_B the wave vector along the magnetic field, k_y a degeneracy parameter characterizing the origin of the wave function and s is the spin quantum number. The components of the velocity operator are an 8x8 matrix in the Bodnar model. The components of σ transverse to the field are given by λ,μ = (+,−,3). The components of the vector $\underline{V}$ are $V_+ = (V_x+iV_y)/\sqrt{2}, (V_x-iV_y)/\sqrt{2}$ and $\sigma_{+-} = -\sigma_{-+}$.

From ϵ, the transmission, reflection or absorption of the wave may be calculated by classical optics. In fact, when the sample is thin ($a\omega/c \ll 1$), and near an absorption peak ($\epsilon_r \gg 1$) the absorption coefficient is

$$A \cong a\omega|\epsilon_i|/2c, \qquad [3]$$

where ϵ_i is the imaginary part of the dielectric function.

The sums in [2] are over eight states (conduction band and three valence bands, all spin-split. In each, there is a sum over k_y, k_B, and the Landau quantum number n. The sum over k_y may be carried out, giving merely the numerical factor $1/4\Pi^2\ell^2$. The sum over k_B may be replaced by an integral for each term. This gives a real part which is a principal part integral and an imaginary part which is the integral of $i\delta(E_{\alpha'}-E_{\alpha}-\omega)$. The integral over k_B may then be replaced by an integral over the final state energy $E_{\alpha'}$ for fixed initial state $|\alpha\rangle$. In making this change of variable the quantity $dk_B/dE_{\alpha'}$ appears in the integrand. But the equation for the energy levels is in the form

$$K_B^2 = \Phi_\sigma(E); \; dk_B/dE = (d\Phi_\sigma/dE)/2\sqrt{\Phi_\sigma}. \qquad [4]$$

This factor is proportional to the density of states. The zeros of Φ_σ give the energies of the extrema of the bands; hence, as is well known from the quantized Landau states, the density of states has an inverse square root singularity at the band extrema. Thus, the most important transitions will clearly be those between states near the band extrema, and the absorption will be greatest when ω corresponds to transitions between states adjacent to these band extrema.

We shall therefore confine our attention to the calculation of matrix elements between such states. In fact, we have already seen in /1/ that intraband matrix elements vary only very slowly with k_B around these energies. The same is found to be true for the interband case.

The matrix elements are calculated for the right circular polarized light (RCP) and the left circular polarized light (LCP), respectively. We found the following eight selection rules and their corresponding intensities (≈matrix elements square). The expressions for the intensities are not given here because of limited space but will be presented elsewhere /4/. We denote $|\alpha\rangle \equiv |n_v, s\rangle$ for the valence band and $|\alpha\rangle \equiv |n_c,s\rangle$ for the conduction band, $\Delta n = (n_c-n_v)$ and $s = \uparrow,\downarrow$.

Selection rules:

(1) $\Delta n=+1$, $\Delta s = 0$, i.e. $|n_v,\uparrow\rangle \rightarrow |n_c,\uparrow\rangle$, $n_v=n_c-1$

(ii) $\Delta n=-1$, $\Delta s = 0$, i.e. $|n_v,\uparrow\rangle \rightarrow |n_c,\uparrow\rangle$, $n_v=n_c+1$

(iii) $\Delta n=+1$, $\Delta s = 0$, i.e. $|n_v,\downarrow\rangle \rightarrow |n_c,\downarrow\rangle$, $n_v=n_c-1$

(iv) $\Delta n=-1$, $\Delta s = 0$, i.e. $|n_v,\downarrow\rangle \rightarrow |n_c,\downarrow\rangle$, $n_v=n_c+1$

(v) $\Delta n=+1$, $\Delta s =+1$, i.e. $|n_v,\downarrow\rangle \rightarrow |n_c,\uparrow\rangle$, $n_v=n_c-1$

(vi) $\Delta n=-1$, $\Delta s =+1$, i.e. $|n_v,\downarrow\rangle \rightarrow |n_c,\uparrow\rangle$, $n_v=n_c+1$

(vii) $\Delta n=+1$, $\Delta s =-1$, i.e. $|n_v,\uparrow\rangle \rightarrow |n_c,\downarrow\rangle$, $n_v=n_c-1$

(viii) $\Delta n=-1$, $\Delta s =-1$, i.e. $|n_v,\uparrow\rangle \rightarrow |n_c,\downarrow\rangle$, $n_v=n_c+1$

The expressions of the intensities for RCP and LCP are different.

But when $\underline{B}||\underline{C}$ axis we have only the following selection rules:

(a) RCP

(i) $\Delta n=1$, $\Delta s = 0$, i.e. $|n_v,\uparrow\rangle \rightarrow |n_c,\uparrow\rangle$, $n_v=n_c-1$

(ii) $\Delta n=1$, $\Delta s = 0$, i.e. $|n_v,\downarrow\rangle \rightarrow |n_c,\downarrow\rangle$, $n_v=n_c-1$

(b) LCP

(i) $\Delta n=-1$, $\Delta s = 0$, i.e. $|n_v,\uparrow\rangle \rightarrow |n_c,\uparrow\rangle$, $n_v=n_c+1$

(ii) $\Delta n=-1$, $\Delta s = 0$, i.e. $|n_v,\downarrow\rangle \rightarrow |n_c,\downarrow\rangle$, $n_v=n_c+1$

The expressions of the intensities of the $\underline{B}||\underline{C}$ case are given in the appendix. It is interesting to note that for the Kane type of semiconductors only the selection rules $\underline{B}||\underline{C}$ are present.

RESULTS AND DISCUSSION

We calculated the matrix elements square for the eight selection rules for the case of Cd_3As_2. In the calculation the Bodnar parameters are used and their values are given in the appendix. We took B=2T and the final conduction Landau level n = 2; B forms an angle of $\Theta = 60^o$ with the c-axis. The results are presented in Table 1 in arbitrary units. These results are very sensitive to anisotropic parameters $P_\perp$, $P_{||}$ and δ. In other words, if the values of the anisotropic parameters are large the values of the matrix elements corresponding to the selection rules which are not present in Kane type semiconductors are also large. Hence the probability of observing these selection rules in experiments is greater.

In table 1 in case of RCP, selection rule (i) corresponds to the cyclotron resonance transition. The selection rules (iv), (v) and (viii) are about 1% to 10% of the cyclotron resonance transition. Of course it depends on the angle between the B and C axis. These selection rules are absent in Kane type semiconductors. A similar argument can be made for the LCP case.

TABLE 1

The square of the matrix elements when B makes an angle of $\theta = 60^0$ with the c-axis of Cd_3As_2. The final conduction band Landau level n_c=2 and B =2T are considered in the calculation. The values are presented in arbitrary units.

Selection rules	RCP	LCP
(i)	1.3×10^4	295.00
(ii)	1.0	57.0
(iii)	299.0	9.0
(iv)	596.0	2.1×10^4
(v)	467.0	44.0
(vi)	18.0	806.0
(vii)	12.0	537.0
(viii)	700.0	66.0

ACKOWLEDGEMENTS

The authors are thankful to NSERC, Canada, for financial support in the form of a Research Grant.

APPENDIX

We present here the square of the matrix elements denoted by I in the case of $\underline{B}||\underline{C}$:

(a) RCP

for selection rule (i), $I = n\ A^2\ (E_{n_c}\uparrow, E_{n_v}\uparrow)$
for selection rule (ii), $I = n\ A^2\ (E_{n_v}\downarrow, E_{n_c}\downarrow)$

(b) LCP

for selection rule (i), $I = (n+1)A^2(E_{n_v}\uparrow, E_{n_c}\uparrow)$
for selection rule (ii), $I = (n+1)A^2(E_{n_c}\downarrow, E_{n_v}\downarrow)$

where $A(X,Y) = \beta\ (x)\ [\lambda_1(y)-\sqrt{2}\xi_1(y)] + \beta(y)[\sqrt{3}\ \nu_2(x)]$;

$$\beta(E) = N_c(E);\ \lambda_1(E) = N_cP_\perp(E+\Delta+\delta)/\sqrt{3}D_n\ell;$$

$$\nu_2(E) = P_\perp N_c(E)/E\ell;\ \xi_1(E) = -\sqrt{2/3}\ P_\perp N_c(E)(E+\delta)/D_n(E)\ell.$$

Here ℓ is the Landau length; $D_n(E)$ and $N_c(E)$ are defined in reference /1/. The values of the Bodnar parameters are given in reference /1/.

REFERENCES

1. M. Singh, P.R. Wallace: J. Phys. C16, 3877 (1983)
2. M. Singh, P.R. Wallace, J. Leotin: J. Phys. C17, 1385 (1984)
3. J. Bodnar: Proc. Conf. Narrow-Gap Semiconductors (Warsaw, 1977), p. 311.
4. M. Singh, P.R. Wallace: J. Phys. C. [in press 1987].

Far Infrared Thermomodulation in $Hg_{0.8}Cd_{0.2}Te$

M. van der Burgt, P. Janssen, L. van Bockstal, and F. Herlach

Laboratorium voor Lage Temperaturen en Hoge-Veldenfysika,
Katholieke Universiteit Leuven, Celestijnenlaan 200 D,
B-3030 Leuven, Belgium

1. Introduction

The narrow gap semiconductor $Hg_{1-x}Cd_xTe$ has been extensively studied for its peculiar magnetotransport properties [1 - 3]. Magnetotransmission in the far infrared was investigated by GOLDMAN et al [4]. It was our intention to study the magnetophotoconductivity in this material in order to observe conduction electron cyclotron resonance (CCR) and possibly impurity cyclotron resonance (ICR). With the laser chopped at a frequency of 20 Hz, a fairly strong signal was observed which contained the expected resonances. However, it was then discovered that this signal persisted when the sample current was turned off, and further investigation revealed that practically no photoconductive response could be detected in these samples in the low-field region ($B < 1$ T). We have come to the conclusion that the observed signal is due to a modulated thermoelectric voltage resulting from inhomogeneous heating of the sample by the chopped laser radiation.

2. Experimental Technique

The samples were placed in an 8 Tesla superconducting magnet, and cooled by liquid helium. Temperatures down to 1.5 K were achieved by reducing the helium vapor pressure.

An optically pumped FIR laser was used to obtain laser lines with wavelengths from 103 μm to 699 μm, with HCOOH, CH_3OH, CH_3OD and CD_3OD as laser gases. The laser radiation was chopped at about 20 Hz and guided through a system of mirrors and lightpipes to the sample. To avoid photovoltaic effects the electrical contacts on the samples were screened from the FIR radiation by a tantalum screen or a conducting silverpaint. However, this did not have a pronounced effect on the signal; this is not surprising since the FIR radiation is rather diffuse.

For photoconductive measurements a DC current from 10 μA to 100 μA was passed through the sample. The thermomodulation voltage was measured in the direction parallel to the magnetic field. After preamplification this signal was detected by a phase-sensitive detector (PSD), and then recorded by an analog x-t recorder.

In order to separate a possible photoconductive effect from thermomodulation and magnetoresistance signals a system with two phase sensitive detectors was employed. An AC current of 10 kHz and 100 μA was passed through the sample. The sample signal was then detected by a first PSD with a 10 kHz square wave as reference input. After cutting off the low-frequency magnetoresistance signal by a highpass filter, this was passed to a second PSD synchronised with the chopper frequency. The signals obtained in this way are purely photoconductive.

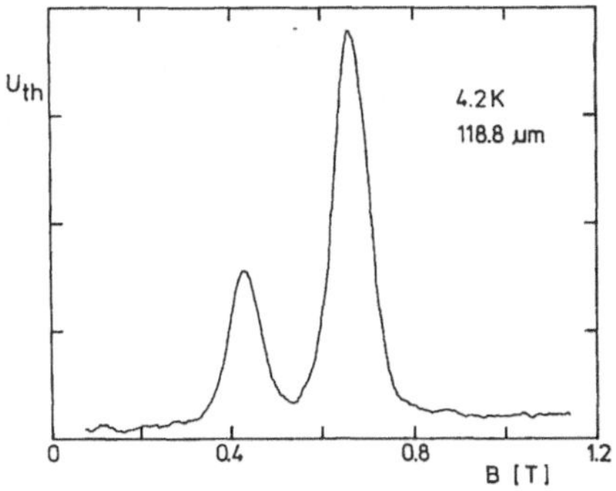

Fig. 1 Thermomodulation voltage versus magnetic field for sample H5 at 4.2 K and with the laser wavelength 118.8 μm (2523.5 GHz, 10.44 meV)

3. Experimental Results and Discussion

Samples of n-type $Hg_{0.8}Cd_{0.2}Te$ were provided by Prof. Nimtz from the University of Cologne, with the following characteristics : carrier density $n = 3.4\cdot10^{14}$ cm^{-3} and mobility $\mu_e = 4.6\cdot10^5$ cm^2/Vs.

A typical result is shown in Fig. 1 : the thermomodulation voltage of sample H5 versus magnetic field shows two separate peaks, with the laser wavelength 118.8 μm (2523.5 GHz, 10.44 meV). The amplitude is about a few microvolts. These double resonances also appeared at other laser lines (103.1 μm, 232.9 μm, 699.4 μm) and with other samples. The signal amplitude was strongly dependent on the chopper frequency. It was reduced by a factor 6 when changing the chopper frequency from 12.5 Hz to 25 Hz. From this we conclude that the observed voltage across the sample is not photovoltaic; following a suggestion by M. von Ortenberg we interpret this as a thermomodulation voltage. Due to imperfections in mirrors and lightpipes, and the diffuse character of FIR radiation the surface of the sample is irradiated inhomogeneously by the chopped laser. This results in an inhomogeneous heating of the sample. The electrical contacts on the sample are made with indium; therefore the sample can be seen as a $In/Hg_{0.8}Cd_{0.2}Te/In$ thermocouple, and the inhomogeneous heating results in a thermoelectric power. When the sample is in resonance the absorption increases and thus the temperature will increase.

Because we measured in the Voigt configuration our results are not directly comparable with the transmission experiments in Faraday configuration of GOLDMAN [4]. We can identify the resonance at about 0.66 T in Fig. 1 as a conduction electron cyclotron resonance. This is in good agreement with the values found by GOLDMAN [4] at the same photon energies. The resonance at 0.43 T can be interpreted as a resonance between impurity levels. However, it can not be completely excluded that the two observed resonances are due to two different laser lines excited by the same pump line : e.g. 170.5 μm appearing together with 118.8 μm.

Figure 2 gives the CCR fields (B_{CCR}) versus laser frequency for sample H3. The parabolic behaviour of B_{CCR} is due to the magnetic field dependence of the effective mass. Calculated values of the effective mass are given in table 1.

While the conduction cyclotron resonance has been consistently observed in our experiments, the behaviour of the second line which we have tentatively identified with ICR has been somewhat erratic. Before drawing strong con-

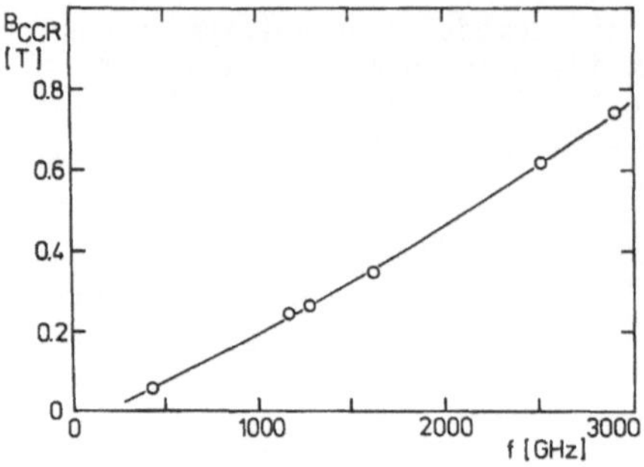

Fig. 2 The conduction cyclotron resonance field (B_{CCR}) versus laser frequency for sample H3; the full line represents a parabolic fit

Table 1 Calculated values of the efective mass of sample H3 for different resonance fields (B_{CCR}) and laser wavelengths

λ [μm]	B_{CCR} [T]	m^*/m_0 [10^{-3}]
103.1	0.74	7.2
118.8	0.62	7.0
184.0	0.35	6.3
232.9	0.27	6.3
255.0	0.24	6.4

clusion regarding this line, further experiments are needed, in particular at wavelengths below 100 μm. A laser for this range is now under construction.

4. Acknowledgment

We should like to express our gratitude to Professor G. Nimtz and J.P. Stadler at the University of Cologne for providing the samples and for getting us started with the experiments.

5. Literature

1. J.P. Stadler, G. Nimtz, G. Remeyeni: Solid State Commun. 57, 459 (1986)
2. S.B. Field, D.H. Reich, B.S. Shivaram, T.F. Rosenbaum, D.A. Nelson, P.B. Littlewood: Phys. Rev. B33, 5082 (1986)
3. G. De Vos,F. Herlach, H.W. Myron: J. Phys. C19, 2509 (1986)
4. V.J. Goldman, H.D. Drew, M. Shayegan, D.A. Nelson: Phys. Rev. Lett. 56, 968 (1986)

Part VIII

Magneto-Transport in 3D Systems, Bandstructure

Theoretical Aspects of Wigner Condensation

R.R. Gerhardts

Max-Planck-Institut für Festkörperforschung,
D-7000 Stuttgart 80, Fed. Rep. of Germany

This paper will review theoretical work on magnetic-field-induced Wigner condensation of three-dimensional electron systems in semiconductors and discuss some closely related work on two-dimensional systems.

1. INTRODUCTION

Recently the concept of a possible magnetic-field-induced Wigner condensation (WC) of both three- and two-dimensional electron systems in semiconductors has gained a renewed interest. Magneto-transport anomalies of the three-dimensional narrow gap semiconductor $Hg_{1-x}Cd_xTe$ (with $x \approx 0.2$) have been detected and interpreted as a possible WC by Nimtz et al. /1/ already in 1979. Subsequent investigations /2/ of several aspects of the phenomenon seemed to confirm this interpretation, but only the recent work of Rosenbaum et al. /3/, who reported such an anomaly at lower values of the temperature and magnetic field and propagated this result as the first unambiguous demonstration of a three-dimensional Wigner-crystal, has apparently drawn broader attention to this interesting subject. The question remains, however, if the experimental evidence of a sudden change of the electrical transport properties can decisively prove the realization of the phase transition of the conduction electrons driven by their mutual Coulomb interaction which had been predicted by Wigner /4/ half a century ago. One has to be aware of the existence of competing mechanisms such as carrier localization due to magnetic freeze-out, but also of the mere qualitative nature of the existing theory of WC in three dimensions which at present is not able to predict unambiguously a phase diagram to compare with the experimental findings. Experimental aspects of the WC will be discussed in the contribution by G. Nimtz. In the present paper I will try to summarize the state of the theory.

For a better understanding of some microscopic results it will be useful to consider also the two-dimensional (2D) case. There is a large amount of theoretical work on the magnetic-field-induced Wigner crystal in two dimensions, and it is presently discussed in the context of the fractional quantum Hall effect as a possible alternative to the liquid-like state proposed by Laughlin /5/. Since a fair review of this work far exceeds the scope of the present article, I will consider only selected 2D investigations which are closely related to the 3D analogue.

2. CALCULATIONS OF THE PHASE DIAGRAM

In contrast to the experiments /1-3/, which investigate transport properties, the theory of the three-dimensional WC has considered so far only the nature and stability of thermodynamic equilibrium states. Moreover, the theoretical considerations are based on the convenient jellium model, i.e. the positive charges, which in real (HgCd)Te samples are more or less localized at donors or near Te vacancies, are smeared out to a homogeneous neutralizing background.

Calculating the ground state energy of the degenerate electron gas in the absence of external fields, Wigner discovered that (within the jellium model) the homogeneous ground state becomes unstable for small values of the electron density n and a crystalline state becomes more favorable for the electron system /4/. It is now generally believed that the presence of a strong homogeneous magnetic field B enhances the tendency of the electron system to undergo such a phase transition. There is, however, up to now no reliable calculation of the transition temperature as a function of B and n, even for the simple jellium model. On the contrary, as I will discuss in the following, different plausible approaches yield qualitatively different results. The difficulty is related to the fact that for given values of temperature T and electron density n the nature of the electron system may change from degenerate at low magnetic field to non-degenerate at high magnetic field.

2.1 A Simple Estimate

A simple-minded approach to Wigner condensation starts with a comparison of the mean kinetic energy $\langle E_{kin}\rangle$ and the mean potential energy $\langle V\rangle$ per electron. If the ratio

$$\Gamma = \langle V\rangle/\langle E_{kin}\rangle \tag{1}$$

exceeds a critical value Γ_c, it becomes energetically favorable to minimize the Coulomb energy $\langle V\rangle = e^2/r$, where $r = (4\pi n/3)^{-1/3}$ is the mean distance between electrons. This is achieved by maximizing the distance between every two electrons, i.e. by arranging them on a lattice. For a degenerate electron gas in the absence of a magnetic field one has $\langle E_{kin}\rangle \sim E_F \sim r^{-2}$, where $E_F = \hbar^2(3\pi^2 n)^{2/3}/2m$ is the Fermi energy. Thus, with decreasing density the crystalline phase becomes more favorable. For free electrons, the ground state becomes unstable /4/ for $r \gtrsim 7$ a, with $a = \hbar^2/m_0e^2$ the Bohr radius, which means $n \lesssim 5\cdot 10^{21}$ cm^{-3} and leads to $\Gamma_c \approx 8$. To adapt this to the semiconductor system, we have to replace the free electron mass m_0 by the effective mass m and a by $a^* = a\,\varepsilon\, m_0/m$, (for $Hg_{0.8}Cd_{0.2}Te$: $m/m_0 = 4.9\cdot 10^{-3}$, $\varepsilon = 17$, $a^* = 1.8\cdot 10^{-5}$ cm). The transition is then expected only for $n \lesssim 10^{12}$ cm^{-3}.

For the narrow gap semiconductor $Hg_{0.8}Cd_{0.2}Te$, the magnetic quantum limit (MQL), in which only the lowest Landau level and one spin direction is occupied, is reached for $B \gtrsim 0.2$ T. Then the degeneracy of the quasi-one-dimensional energy spectrum of the motion parallel to the magnetic field is proportional to B, so that the Fermi energy varies as B^{-2},

$$E_F = \hbar^2(2\pi^2\ell^2 n)^2/2m, \quad \ell = (\hbar c/eB)^{\frac{1}{2}}, \tag{2}$$

with ℓ the magnetic length, and the condition (1) for $\Gamma_c \approx 8$ and for realistic values of the density, $n \gtrsim 10^{14}$ cm^{-3}, is satisfied for $B \gtrsim 0.6$ T. It will be assumed in the following that the spacing of the spin-split Landau levels $[\sim \frac{1}{2}\hbar\omega = Ry^*(a^*/\ell)^2$, with $Ry^* = e^2/(2\varepsilon a^*) = 0.22$ meV the effective Rydberg energy$]$ is larger than E_F, than the mean Coulomb energy per electron $\langle V\rangle = 2Ry^*(a^*/r)$, and than the thermal energy kT, so that the restriction on a single Landau and spin state in the MQL is justified.

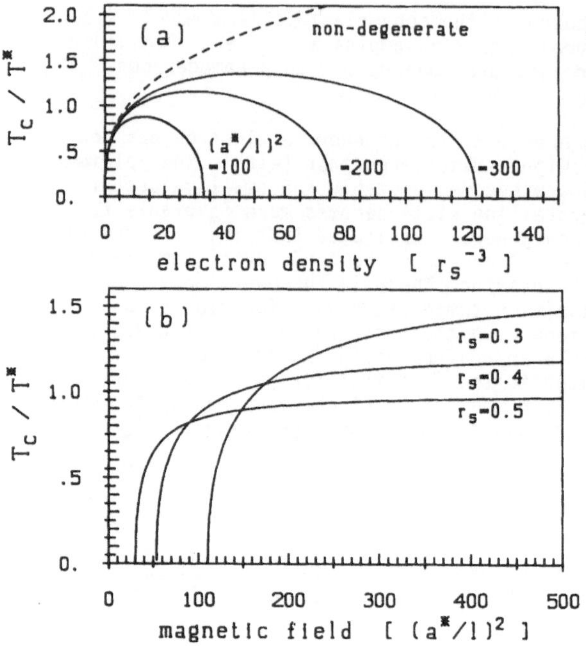

Fig. 1: Phase diagram calculated from (1) to (5) for $\Gamma = 8$: (a) in the T-n-plane for several values of B, (b) in the T-B-plane for three values of n. Effective atomic units are used with $r_s = (4\pi n a^{*3}/3)^{-1/3}$. The non-degenerate limit $T_c/T^* = 4/\Gamma r_s$ is indicated in (a). The high density limits in (a) and the threshold fields in (b) are determined by $(a^*/\ell)^2 = (3\pi^2\Gamma/8)^{1/2}\, r_s^{-5/2}$, which holds in the degenerate limit.

To get an impression of the possible phase diagram, one may use (1) with a fixed value of Γ_c and evaluate $\langle E_{kin}\rangle$ as a function of temperature T. The resulting transition curves, T_c vs. n for several values of B, are shown in Fig. 1a, obtained from

$$n\langle E_{kin}\rangle = \int_{-\infty}^{\infty} dE\; E D(E) f(E) \;, \qquad (3)$$

$$n = \int_{-\infty}^{\infty} dE\; D(E) f(E) \quad , \; f(E) = [\exp \beta(E-\mu)+1]^{-1} \qquad (4)$$

with the quasi-onedimensional density of states

$$D(E) = \hbar^{-1}(2\pi\ell)^{-2}\,(2m/E)^{\frac{1}{2}}\,\Theta(E) \qquad (5)$$

appropriate for the MQL. As expected from (2), for larger values of B the condensed phase extends to higher values of the electron density n. For small values of n, the curves are determined by the fact that the electron system becomes non-degenerate. In the non-degenerate limit the Fermi-Dirac-integrals in (3), (4) can be evaluated analytically to yield $\langle E_{kin}\rangle = \frac{1}{2}\, kT$. This classical expression is correct in the limit $B\to\infty$ and yields with (1) $T_c \sim n^{1/3}$, as indicated by the broken line in Fig. 1a. In Fig. 1b the corresponding transition curves in the T-B-plane are shown for different values of n. The most important features can be obtained from the asymptotic behavior of (3) and (4): There is a threshold value of B proportional to $n^{5/6}$ for the onset of the WC. Near the threshold the electron system is degenerate. For a fixed value of n, the increase of T_c with increasing B becomes weaker as the system enters the non-degenerate regime and for $B\to\infty$ T_c approaches the "classical" limit $\sim n^{1/3}$.

2.2 Ground State Calculations

Similar results have been obtained by Kleppmann and Elliott /6/ (KE) from ground state calculations. Assuming trial wave functions $\psi(\vec{r}-\vec{R})$ of the form

$$\psi(\vec{r}) \sim \exp[-\tfrac{1}{4}(x^2+y^2)/\ell^2-\delta^2 z^2] \tag{6}$$

centered about the lattice points $\vec{R}=(X,Y,Z)$ of a regular lattice (hexagonal or tetragonal), KE minimized the cohesive energy per particle with respect to the spread $1/\delta$ of the wavepacket in the direction of the magnetic field (z-direction) and to the lattice constant. They found a Wigner lattice consisting of elongated electron distributions, with $2\ell\delta < 1$ and only a negligible overlap of adjacent distributions, slightly more favorable the hexagonal "rod state" with translational symmetry in B-direction, which had been predicted by Kaplan and Glasser /7/. In contrast to the rod state, exchange contributions are negligible in the ground state obtained by KE.

A melting temperature T_m is estimated from the cohesive energy $\Delta\varepsilon$ using the phenomenological rule $kT_m/\Delta\varepsilon \sim 1/8$ satisfied by inert-gas crystals /6/. The result is similar to that of Fig. 1b, although the slope of the T_m vs B curve is smaller than that of the corresponding T_c vs. B curve and remains finite at the threshold. Within this approach the result is interpreted as follows. With increasing magnetic field the spatial extent of the localized wavefunctions decreases. This reduces the overlap integrals and favors the formation of a Wigner crystal. With increasing density, B must exceed increasing threshold values to produce a sufficiently strong localization. With increasing B the wavefunctions shrink, and in the limit $B\to\infty$ the cohesive energy approaches the Madelung energy of a classical point crystal, which increases with increasing density. This sounds plausible, and so does the phenomenological melting criterion, since the thermal energy needed to melt the crystal should be roughly proportional to the cohesive energy.

Nevertheless, the results of KE remain questionable for the following reasons. First, it is not evident that the rule $kT_m/\Delta\varepsilon \sim 1/8$ read off from inert-gas data may be applied to the Wigner crystal in a strong magnetic field, since, besides the spatial anisotropy, the long-range Coulomb interaction and the Fermi statistics of its constituent electrons distinguish it from inert-gas crystals. Second, near the melting transition one expects a considerable overlap of electronic wavefunctions localized in adjacent lattice cells. Then exchange effects become important which play no role in the KE ground state. Third, the zero point motion is not treated correctly in the KE approach. In summary, the approach of KE considers only the classical aspects and their estimate for the melting curve is not reliable for low temperatures near the threshold field, where the electrons form a degenerate Fermi system.

Subsequent work improved the situation somewhat. Kuramoto /8/ performed a variational calculation of the ground state allowing for overlapping wavefunctions in adjacent lattice cells and paying due attention to the exchange effect. He found that the ground state changes gradually with increasing density from an anisotropic Wigner crystal at low density over a "chain phase", which is the "rod state" with a periodic modulation in B-direction, to a CDW state for which the exchange interaction becomes important along every direction in space. Kuramoto /9/ also calculated the long-wavelength excitations of a Wigner crystal in a strong magnetic field and found transverse modes which exist only if the shear modulus is finite and which should be observable in ultrasound absorption. He emphasized that, besides direct observation of the crystal structure by neutron diffraction /6,8/, observation of these shear modes could provide clear evidence for the WC. A simplified vibrational spectrum (not including the effect of finite shear stress) has been used to discuss the zero-point motion and the stability of the Wigner crystal /10/.

2.3 Finite Temperature Calculations

A direct approach to the transition temperature is to investigate the thermodynamic stability of the homogeneous electron gas state. Fukuyama /11/ investigated the static density response function $P(\vec{q})$ within the Hartree-Fock approximation (HFA), in which the vertex of the density-density correlation function is approximated by a RPA term and a ladder term (the corresponding exchange term). He found an instability with respect to the formation of a charge density wave (CDW) with wavevector $\vec{q}$, which occurs if a condition of the form

$$P_0(q_z)[u(q_\perp) - v(\vec{q})\exp(-\tfrac{1}{2}\ell^2 q_\perp^2)] = 1 \qquad (7)$$

can be satisfied, where $q_\perp = (q_x^2 + q_y^2)^{\frac{1}{2}}$.

$$P_0(2q) = -\frac{1}{(2\pi\ell)^2} \int_{-\infty}^{\infty} dp \; \frac{f(\varepsilon_{p+q}) - f(\varepsilon_{p-q})}{\varepsilon_{p+q} - \varepsilon_{p-q}} > 0 \qquad (8)$$

is the bare quasi-onedimensional polarization function appropriate to the MQL, $v(\vec{q})$ is the Fourier transform of the Coulomb potential, and $u(q_\perp)$ is an effective potential describing the exchange effect. Since $P_0(2k_F)$, where $k_F = 2\pi^2\ell^2 n$, diverges logarithmically in the zero-temperature limit and since the maxium value of the square bracket in (7) is positive, the condition (7) can be satisfied at a finite temperature. T_c is the highest temperature at which (7) can be satisfied for any possible choice of $\vec{q}$. In the degenerate case this implies $q_z = 2k_F$ and (8) can be evaluated to yield /12/

$$T_c = \gamma\, T_F \exp[-\pi(T_F/T^*)^{\frac{1}{2}}/\phi_0] \quad , \quad (T_c \ll T_F) \qquad (9)$$

where $T_F = E_F/k$ is the Fermi temperature, $T^* = Ry^*/k$ the Rydberg temperature ($T^* = 2.7$ K for $Hg_{0.8}Cd_{0.2}Te$), and the maximum value of the square bracket in (7), which occurs at $q_\perp \approx 1.5/\ell$, is written as $\phi_0\; 4\pi\ell^2 e^2/\varepsilon$. If one linearizes the energy spectrum $\varepsilon_p = \hbar^2 p^2/2m$ near the Fermi energy, as is usual in discussions of the Peierls instability of onedimensional conductors, one obtains $\gamma = 1.14$ and (9) is Fukuyama's formula /11/, if not, one obtains $\gamma = 4.56$ /12/.

Recently, Heinonen and Al-Jishi /13/ emphasized that in strongly polar semiconductors the electron-phonon interaction may lead to a Peierls instability with a permanent distortion of the electron-phonon system. Investigating the renormalized phonon propagator, they found that the singular polarization function (8) leads to a soft-phonon instability. Evaluating (8) in the degenerate limit, they obtain Fukuyama's formula (9), however, with a different ϕ_0, now describing an effective electron-phonon interaction potential. Heinonen and Al-Jishi claim that their transition temperature can be larger by several orders of magnitude than the one obtained by Fukuyama. This may be true for very small T_c, $T_c \ll T_F$, where (9) is dominated by the exponent. With increasing B the exponent becomes smaller, T_c increases towards $T_F \sim (n/B)^2$, see (2), runs through a maximum, and finally (9) yields $T_c \approx \gamma\, T_F$ independent of the interaction mechanism described by ϕ_0. It should be noticed, however, that near the maximum, and beyond, the system is no longer degenerate and (9) does not hold. Most of the experiments on (HgCd)Te have been performed in this non-degenerate regime /1/.

An alternative approach within the HFA which introduces explicitly CDW's and can be continued into the condensed phase has also been discussed previously /12/. The criterion for the possible onset of a unidirectional CDW perturbation of the homogeneous state is again (7). Numerical evaluation of (8) yields T_c vs. B curves as shown in Fig. 2. In the non-degenerate limit, (8) can be evaluated analytically to yield

$$T_c = (4/\pi)(T_F T^*)^{\frac{1}{2}} \phi_0 \quad , \quad (T_c \gg T_F) \quad . \tag{10}$$

In view of the preceeding sections the most unexpected result of the calculations within the HFA is that in the strong-magnetic-field limit T_c decreases with increasing B instead of showing the plausible saturation behavior. Indeed, there are indications that this decrease is an insuffiency of the approximation /12/. The crosses on the curves of Fig. 2 separate the small-B degenerate regime from the large-B non-degenerate regime. Whereas the optimum $\vec{q}$ in the former has $q_z = 2k_F$, it has $q_z = 0$ in the latter. In the non-degenerate regime one finds that a triangular-symmetric superposition of three CDW's with wavevectors satisfying $\vec{q}_1+\vec{q}_2+\vec{q}_3 = 0$ is more stable than a unidirectional CDW and leads to a first order transition at a higher temperature $T_1 > T_c$. This situation is similar to the 2D case, where it has been shown /14/ that inclusion of higher harmonics leads to higher transition temperatures, approaching the classical limit for $B \to \infty$. At the same time, the optimum $q_\perp$-value, which for a single unidirectional CDW is determined by the value of B, becomes n-dependent and approaches the value expected for the classical Wigner crystal.

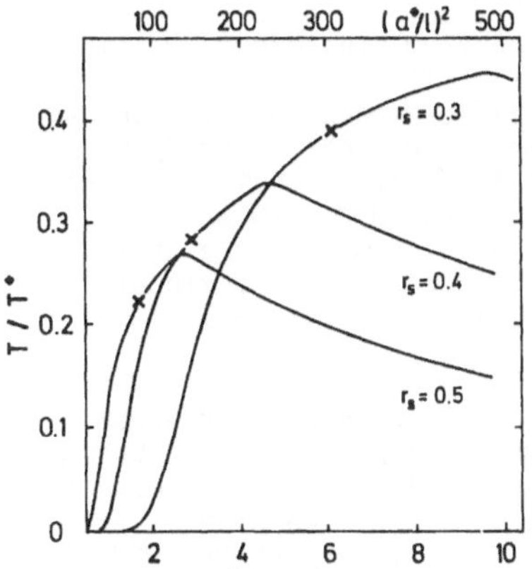

Fig. 2: Transition temperatures calculated from (7) with a screened Coulomb potential $v(\vec{q}) = 4\pi e^2/[(q^2+\kappa^2)\varepsilon]$, where $(\kappa a^*)^2 = (12/\pi)^{2/3} r_s^{-1}$ /12/. The lower scale refers to $Hg_{0.8}Cd_{0.2}Te$, for which $T^* = 2.7$ K and the density parameters $r_s = 0.5$, 0.4, and 0.3 correspond to n= 3.1, 6.0, and $14.0 \cdot 10^{14} cm^{-3}$, respectively. Without screening, $\kappa=0$, similar curves with higher T_c values result: an enhanced maximum (factor ≈ 2) occurs at lower B (factor ≈ 0.7)

In the degenerate case, on the other hand, it is not possible to satisfy $\vec{q}_1+\vec{q}_2+\vec{q}_3 = 0$ for optimum $\vec{q}$-vectors ($q_{iz} = \pm 2k_F$) and a superposition of CDW's is not more favorable than a single one: we obtain a second order transition at T_c. This indicates that the HFA approach is suitable in the degenerate case, where the phase transition is driven by the exchange scattering at the Fermi energy of the quasi-onedimensional spectrum, but not in the non-degenerate case, where spatial correlations are more important than the two-particle exchange.

3. COMPARISON WITH THE 2D CASE

Since the different approaches discussed in the preceeding section apparently do not lead to a very coherent picture of the WC in three dimensions, it is instructive to consider the 2D analogue.

The situation is clear for the electron crystal on a liquid-He surface /15/. The melting curve is well described by (1) with $\langle V \rangle = e^2(\pi n_{2D})^{\frac{1}{2}}$, $\langle E_{kin} \rangle = kT$, and $\Gamma = 131 \pm 7$. The Γ-value is much larger than one expects from simple theories. It has been understood, however, within the theory of dislocation-mediated melting and with a sophisticated treatment of the temperature-dependence of the shear modulus /16/.

For 2D electron systems in semiconductors the situation is by far not so clear, in spite of a tremendous amount of literature on this subject. The fractional quantum Hall effect clearly indicates the occurrence of new collective states. However, whether these are liquid-like, translational-invariant or some kind of a Wigner crystal is at present under dispute /5,17/.

In view of the 3D case it is interesting to note that within the HFA the instability of the homogeneous state under CDW formation occurs at a temperature T_c satisfying $kT_c \approx 0.557\, \nu e^2/\varepsilon \ell \sim n/B^{\frac{1}{2}}$ for $\nu = 2\pi \ell^2 n \to 0$ /18/. Again this T_c decreases with increasing B instead of approaching the classical limit $T_c^{cl} \sim n^{\frac{1}{2}}$ for $n/B \to 0$. To obtain this limit, one has to consider a superposition of many harmonics of a fundamental CDW /14/. Higher harmonics are also needed to see that the cohesive energy per particle in the CDW state vanishes as $n^{\frac{1}{2}}$ in the low-density limit /19/. This shows that the naive result $T_c \sim n/B^{\frac{1}{2}}$ obtained in the HFA is not correct for large B- values, where the correct result approaches the classical limit.

A transition temperature can also be estimated from Lindemann's melting criterion if one calculates the thermal fluctuations from the collective phonon modes of the Wigner crystal /20/. From simple-minded calculations of a phonon spectrum, e.g. the Einstein or Debye model, one obtains a T_c which for intermediate values of n and B, $0.1 \lesssim \nu \lesssim 0.5$, is of the same order of magnitude as the result of the HFA, which corresponds to $\Gamma \sim 10$ in (1) /14,20/. Calculations of the shear modulus /21/ and estimates of a critical temperature for topological melting due to the dissociation of dislocations lead to much lower T_c-values /20,21/ consistent with the classical value $\Gamma \sim 130$.

These results indicate the possibility of a complicated phase diagram /20/ with a Wigner-crystal ground state which melts at a low temperature ($\Gamma \sim 130$) into a "liquid-crystal-like" highly correlated state, whereas the transition to the usual electron gas state occurs at the much higher T_c ($\Gamma \sim 10$) found in the HFA. Then the CDW state considered in the HFA would correspond to phase with short- or intermediate-range order /14,18/, whereas the Wigner-crystal is characterized by its finite shear modulus.

Finally, the effect of a fluctuating external potential due to the background charges in a real system should be mentioned. It will distort the Wigner crystal locally, but not destroy it /22,23/. On the contrary, pinning to the background potential may stabilize the crystal and eliminate the long-wavelength excitations which hinder true long-range order in two dimensions /23/.

The observation of the quantum Hall effect probably rules out a pinned Wigner crystal, but may be consistent with a sliding CDW state.

4. CONCLUDING REMARKS

If we accept that also in the 3D case a sufficiently sophisticated ansatz for the CDW state will within the HFA lead to a non-decreasing function $T_c(B)$ which saturates for $B\to\infty$ approaching the classical limit, then the different approaches discussed in Sect. 2 yield qualitatively the same phase diagram. Only the correct behavior of the T_c vs. B curve in the degenerate limit for small values of B and T_c remains unclear. The experimental results by Rosenbaum et al. /3/ seem to agree best with the $T_c(B)$-curves estimated by KE /6/, which neglect the effects of quantum statistics and, therefore, from a theoretical point of view have the poorest justification in this low-temperature limit.

As in the 2D case, one may speculate on several condensed phases, e.g. a viscous-liquid-like or weakly pinned sliding-CDW phase separating the high-temperature electron gas from a low-temperature pinned Wigner crystal /12/. This picture is consistent with the experimental results. Most results of the Nimtz group would refer to the CDW phase /1,2/. Indication for the solid-liquid transition comes from recent results of Field et al. /3/, which they interpret as depinning of a Wigner crystal, and from information about the electronic specific heat, extracted by Nimtz and Stadler /24/ from hot carrier experiments. One should, however, be aware of the fact that this appealing picture of a WC in (HgCd)Te is mainly based on intuitive arguments. A reliable equilibrium theory, and a forteriori a transport theory of the condensed phase is still missing.

On the other hand, I see no serious argument /25/ excluding a possible WC in pure (HgCd)Te samples with a high concentration of conduction electrons ($n > 10^{14}$ cm^{-3}). It is known /26/ that most of these electrons come from Te-vacancies, which in contrast to shallow donors do produce resonance states high in the conduction band but not near the conduction band edge, so that a single-particle localization due to magnetic freeze-out is unlikely. Observation of donor states in samples with very low carrier density /27/ or of single-particle localization in heavily doped and compensated samples /28/ does not rule out the possibility that under more favored conditions the conduction electrons undergo a phase transition driven by their mutual Coulomb interactions. A quantitative distinction between WC and single-particle localization from transport data /29/ may, however, be very difficult.

REFERENCES

1. G. Nimtz, B. Schlicht, E. Tyssen, R. Dornhaus, L.D. Haas, Solid State Commun. 32, 669 (1979); G. Nimtz, B. Schlicht, H. Lehmann, E. Tyssen, Appl. Phys. Lett. 35, 640 (1979)
2. B. Schlicht, G. Nimtz: In Physics of Narrow Gap Semiconductors, Vol. 152 of Springer Lecture Notes in Physics, ed. E. Gornik, H. Heinrich, L. Palmetshofer (Springer, Berlin, 1982), p. 383; J.P. Stadler, G. Nimtz, B. Schlicht, G. Remenyi, Solid State Commun. 52, 67 (1984); J. Gebhardt, G. Nimtz, B. Schlicht, J.P. Stadler, Phys. Rev. B32, 5449 (1985); Phys. Rev. Lett. 55, 443 (1985)
3. T.F. Rosenbaum, S.B. Field, D.A. Nelson, P.B. Littlewood, Phys. Rev. Lett. 54, 241 (1985); 55, 444 (1985); S.B. Field, D.H. Reich, B.S. Shivaram, T.F. Rosenbaum, D.A. Nelson, P.B. Littlewood, Phys. Rev. B33, 5082 (1986)
4. E.P. Wigner, Phys. Rev. 46, 1002 (1934); Trans. Faraday Soc. 34, 678 (1938)

5. S. Kivelson, C. Kallin, D.P. Arovas, J.R. Schrieffer, Phys. Rev. Lett. 56, 873 (1986); S.T. Chui, T.M. Hakim, K.B. Ma, Phys. Rev. B33, 7110 (1986)
6. W.G. Kleppmann, R.J. Elliott, J. Phys. C 8, 2729 (1975)
7. J.I. Kaplan, M.L. Glasser, Phys. Rev. Lett. 28, 1077 (1972)
8. Y. Kuramoto, J. Phys. Soc. Japan 44, 1572 (1978)
9. Y. Kuramoto, J. Phys. C 12, 2033 (1979)
10. N.A. Usov, F.R. Ulinich, Solid State Commun. 30, 783 (1979)
11. H. Fukuyama, Solid State Commun. 26, 783 (1978)
12. R.R. Gerhardts, Solid State Commun. 36, 397 (1980)
13. O. Heinonen, R.A. Al-Jishi, Phys. Rev. B33, 5461 (1986)
14. R.R. Gerhardts, Phys. Rev. B24, 1339 (1981); 4068 (1981)
15. C.C. Grimes, G. Adams, Phys. Rev. Lett. 42, 795 (1979)
16. R.H. Morf, Phys. Rev. Lett. 43, 931 (1979)
17. A.H. MacDonald, S.M. Girvin, Phys. Rev. B33, 4009 (1986)
18. H. Fukuyama, P.M. Platzman, P.W. Anderson, Phys. Rev. B19, 5211 (1979)
19. D. Yoshioka, P.A. Lee, Phys. Rev. B27, 4986 (1983)
20. Yu.E. Lozovik, D.R. Musin, V.I. Yudson, Sov. Phys. Solid State 21, 1132 (1979); Yu.E. Lozovik, S.M. Apenko, A.V. Klyucknik, Solid State Commun. 36, 485 (1980)
21. K. Maki, X. Zotos, Phys. Rev. B28, 4349 (1983)
22. H. Aoki, J. Phys. C 12, 633, (1979)
23. A.G. Eguiluz, A.A. Maradudin, R.J. Elliott, Phys. Rev. B27, 4933 (1983)
24. J.P. Stadler, G. Nimtz, Phys. Rev. Lett. 56, 382 (1986)
25. An argument given by H. Schulz and H. Keiter [J. Low Temp. Phys. 11, 181 (1973)] that fluctuations prohibit a magnetic-field-induced phase transition within the jellium model has been disproved by P. Schlottmann and R.R. Gerhardts [Z. Physik B34, 363 (1979]
26. G. Nimtz, B. Schlicht: In Festkörperprobleme XX, ed. J. Treusch (Vieweg, Braunschweig, 1980), p. 360
27. V.J. Goldman, H.D. Drew, M. Shayegan, D.A. Nelson, Phys. Rev. Lett. 56, 968 (1986)
28. Yu.G. Arapov, A.B. Davydov, M.L. Zvereva, V.I. Stafeev, I.M. Tsidil'kovskii, Sov. Phys. Semicond. 17, 885 (1983)
29. M. Shayegan, H.D. Drew, D.A. Nelson, P.M. Tedrow, Phys. Rev. B31, 6123 (1985).

On the Magnetic Field Induced Electron Condensation in n-HgCdTe

G. Nimtz

Zweites Physikalisches Institut, Universität zu Köln,
Zülpicher Straße 77, D-5000 Köln 41, Fed. Rep. of Germany

1. Introduction

The narrow gap semiconductor n-$Hg_{0.8}Cd_{0.2}Te$ represents the very material to study the magneto-transport in the extreme magnetic quantum limit. In consequence of the unusually light conduction band electrons ($m^* = 0.005\ m_0$) the magnetic quantum limit, where all the carriers are confined to the lowest Landau level, is passed at magnetic fields of only 0.5 T in samples with carrier densities between 10^{14} and 10^{15} electrons cm^{-3} /1/. Another favourable property of this material is, that without adding chemical impurities it can be made n-type by small stoichiometry deviation. A Te vacancy acts as a donor, but, in contrast to usual donors, it has an energy level being resonant with the conduction band, placed some eV above the conduction band edge. As a result of this special property thermal and magnetic freeze-out of free electrons should not take place. If other donors are negligible, the stoichiometric defects make n-type $Hg_{0.8}Cd_{0.2}Te$ a dilute metal rather than a semiconductor.

Some years ago we discovered a magneto-transport anomaly in this special alloy, when measuring the longitudinal (ρ_l) and transverse (ρ_t) magneto-resistivity /2/. At 4.2 K both $\rho_l(B)$ and $\rho_t(B)$ are approximately proportional to power laws of the type $\rho \propto B^n$ (see Fig. 1), a behaviour expected from one-electron model calculations for the extreme magnetic quantum limit /2/. However, after cooling the sample down to e. g. 1.5 K the resistivity, especially $_l$, increases much stronger with magnetic field /2/. Another interesting feature at 1.5 K is that the pronounced anisotropy of the resistivity observed at 4.2 K almost disappears with increasing field since $\rho_l(B)$ approaches $\rho_t(B)$ from below. (For the sample of Fig. 1, the resistivities have a tendency to grow more slowly above magnetic fields of 7 T at 1.5 K. This effect is due to a low resistivity n_s^+-film on the surface of $Hg_{1-x}Cd_xTe$ crystals, as was elaborated by STADLER et al. /3,4/ recently.)

Because the measured Hall-coefficient is not much dependent on magnetic field at both 4.2 and 1.5 K, suggesting a constant carrier concentration, the magneto-transport anomaly can be characterized by the following special

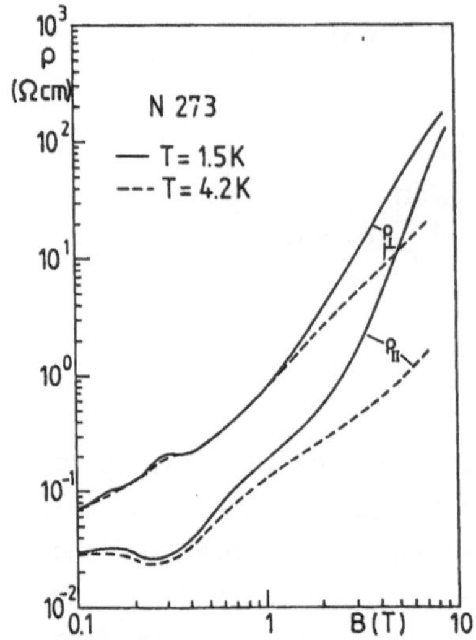

Fig. 1 Transverse and longitudinal magneto-resistivity for two temperatures. The last SdH oscillation is seen near 0.4 T.

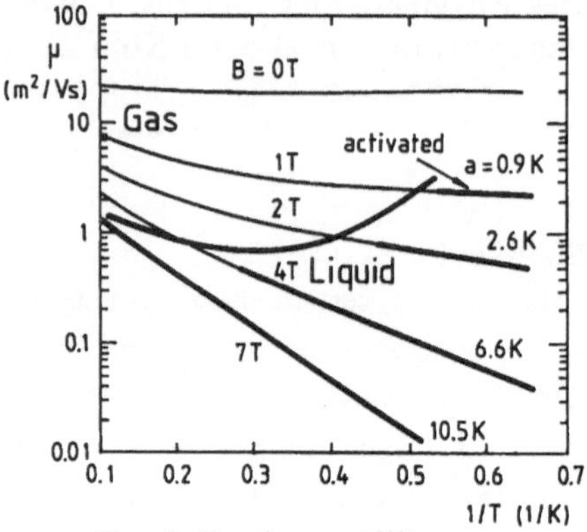

Fig. 2 Carrier mobility vs reciprocal temperature. The transition non-activated (gas) to activated mobility (liquid) is indicated by a solid line.

features:

(a) The mobility is thermally activated and can be described by a relation

$$\mu(T) = \mu_0 \exp(-\ a(B)/T), \tag{1}$$

with μ_0 = 150 000 cm^2/Vs and a(B) being the magnetic field dependent activation temperature (measured in K). This behaviour is demonstrated in Fig. 2: Below a critical temperature, which does depend on the magnetic field /5/, the mobility becomes activated.

(b) The current-voltage characteristic is strongly non-ohmic as shown in Fig. 3. From the plotted curve it seems there are two electric field regimes with different derivatives dI/dU. It was shown by SCHLICHT /6/ that the strongly nonlinear characteristic is a hot carrier effect on the current density j in consequence of the activated mobility:

$$j = n\,e\,\mu\,E = n\,e\,\mu_0\,E\,\exp(-a(B)/T_e(E)), \tag{2}$$

where T_e is the electron temperature in the presence of the electric field E.

All three unusual transport properties made us conclude that a magnetic field-induced electron condensation takes place in n-$Hg_{0.8}Cd_{0.2}Te$ /2,7/. The condensation may be based on the electron-electron interaction alone /8,9/ or may even be enhanced by an electron-phonon interaction /2,10,11/. The transition from the electron gas to an electron condensate is expected to

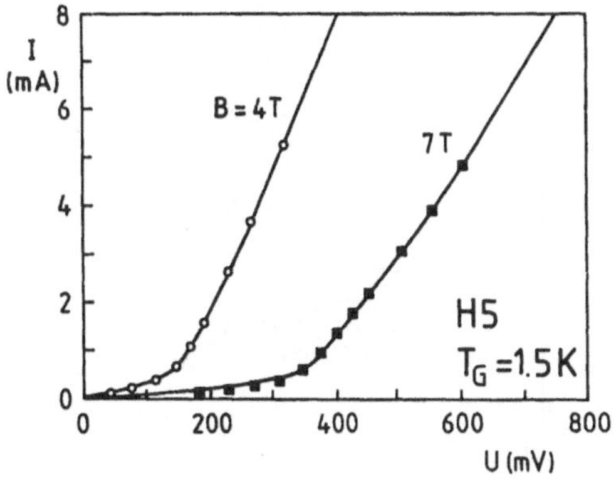

Fig. 3 Current-voltage characteristics, parameter is the magnetic field.

proceed via the evolution of charge density waves (CDW), which eventually may result in a solid at still lower temperatures and increasing magnetic fields /9/. The CDW state corresponds to an electron liquid and the solid ground state to a Wigner crystal or a Wigner glass. As a result of inherent defects in real semiconductors only a short-range order is expected in the electron condensate. A critical report on theoretical studies of a magnetic field-induced electron (Wigner) condensation is given by R. R. Gerhardts in these Conference Proceedings.

2. Recent Studies

Recently a number of different investigations have supported our first interpretation of the magnetic field induced transport anomaly. Among them are the following studies: an analysis of the high field Hall effect /6,12/, the photo-conductivity in the range of activated mobility /13/, the specific heat of electrons /14/, and the depinning of the electron condensate /15/ (including an estimate of tens of micrometers for the disorder-induced correlation length in the electron condensate) at temperatures as low as 10 mK and in magnetic fields higher than 6 T. It is important to point out that the latter study was carried out with an InSb-like composition of the $Hg_{1-x}Cd_xTe$ alloy, namely $x = 0.24$ with $m^* = 0.013\ m_0$ at 4.2 K.

3. Hall Effect in the Extreme Magnetic Quantum Limit

The interpretation of the Hall effect in the extreme magnetic quantum limit has been subject of various discussions, see e. g. /12,16/. I am going to present some details of the dispute. Typical experimental data of the Hall coefficient

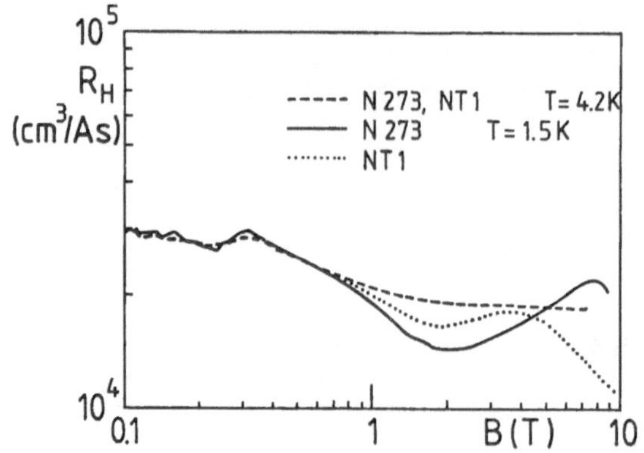

Fig. 4 Hall coefficient vs magnetic field for two samples. (NT1: $n(77K) = 3.5 \cdot 10^{14} cm^{-3}$, $\mu(77K) = 2.5 \cdot 10^5 cm^2/Vs$; N273: $3.0 \cdot 10^{14} cm^{-3}$, $2.8 \cdot 10^5 cm^2/Vs$)

$$R_H = \rho_H B \tag{3}$$

are displayed in Fig. 4 as a function of B where ρ_H is the Hall resistivity and B the magnetic field. Results of two samples obtained at T = 4.2 K and 1.5 K are shown. Only minor differences between the R_H values of the two samples were found at 4.2 K, therefore only one curve is plotted for the higher temperature. Above the last Shubnikov-de Haas (SdH) oscillation (≈ 0.4 T), R_H drops by a factor of 1.4 at T = 4.2 K and approaches a constant value at fields above 2 T. Relative changes in R_H of up to 30 % are observed when cooling down the samples to 1.5 K. The high field behaviour is somewhat different for the two samples in spite of the fact that their transport data were nearly identical at 4.2 K. The oscillatory behaviour beyond 1 T could indicate the presence of surface quantum oscillations, caused by the n^+-surface layer /3,4/.

From the measured values of ρ_l, ρ_t, and ρ_H the components of the conductivity tensor can be calculated as follows:

$$\sigma_{xx} = \rho_t / (\rho_t^2 + \rho_H^2), \tag{4}$$

$$\sigma_{xy} = \rho_H / (\rho_t^2 + \rho_H^2). \tag{5}$$

According to the analysis of the experimental data of ρ_H and ρ_t follows that the ratio $\sigma_{xx}/|\sigma_{xy}|$ is approximately 0.2 in the SdH regime, but increases with magnetic field to a value of 10 at 9.6 T and 1.5 K /12/. The latter result $\sigma_{xx}/|\sigma_{xy}| \gg 1$ is equivalent to the condition $\mu \cdot B \ll 1$.

Following the suggestions by MANSFIELD (see /12/) one has to consider two contributions to the ratio of the conductivity tensor components provided that surface effects can be neglected:

$$\sigma_{xx} / |\sigma_{xy}| = 1 / (\mu B) + \sigma_{xy,i} / |\sigma_{xy,c}|, \tag{6}$$

the first term being the free carrier contribution and the second arising if hopping conduction has to be taken into account (the subscripts i and c refer to impurity and conduction band, respectively). DE VOS et al. /16/ have performed measurements of ρ_t, ρ_l and ρ_H on $Hg_{0.8}Cd_{0.2}Te$ samples at magnetic fields up to 15 T. The experimental results agree essentially with those previously obtained by NIMTZ and SCHLICHT /12/. In both cases the Hall coefficient is only weakly dependent on the magnetic field. DE VOS et al. interpret their data with magnetic freeze-out. They come to this conclusion after having deduced an activated behaviour of the Hall-conductivity σ_{xy} from their data. However, their assumptions that the electron mobility is constant and the concentration n is given by

$$n = \sigma_{xy} \cdot B/e \tag{7}$$

is not well founded.

In the analysis of our data we have used the relation for the free carrier density

$$n = (R_H \cdot e)^{-1} = (\rho_H \cdot B \cdot e)^{-1}, \tag{8}$$

which is valid for a one-band situation in high magnetic fields and approximately correct under the condition $\rho_{xx}^2 >> \rho_{xy}^2$. A one-band approximation seems appropriate, because there is no evidence for the existence of an impurity band. The experimental data yield a constant carrier density, which is independent of temperature and magnetic field. A field and temperature independent concentration was also deduced from photoconductivity experiments which determine the carrier density independent of the Hall effect /13/. I have to point out, that this result is only found in HgCdTe samples with an n-type conductivity dominated by Te vacancies. Chemical impurities, however, cause freeze-out as we are used to from e. g. InSb.

Recently SHAYEGAN et al. /18/ claimed to have observed carrier freeze-out in some n-type HgCdTe samples at fields and temperatures, at which we claimed to have observed the activation of the carrier mobility, but not a freeze-out of carriers. They measured an increase of the Hall coefficient with magnetic field similarly to the data (II) presented in Fig. 5. The data in question were obtained with samples having a ratio of length to width = 2 /19/, which seems insufficient /20/. As a consequence of non-homogeneous current distribution in too short samples, both longitudinal and transverse resistivities do contribute to the measured Hall resistivity and yield an incorrect result. The influence of the sample geometry is demonstrated in Fig. 5. Sample (I) has a ratio of length to width of 6 and does not show a significant increase of R_H with magnetic field. The short sample displays a strong increase with B and gives the impression that a carrier freeze-out happens /21/. The claimed observation of carrier freeze-out by SHAYEGAN et al. /18/ might well be the result of an improper sample geometry.

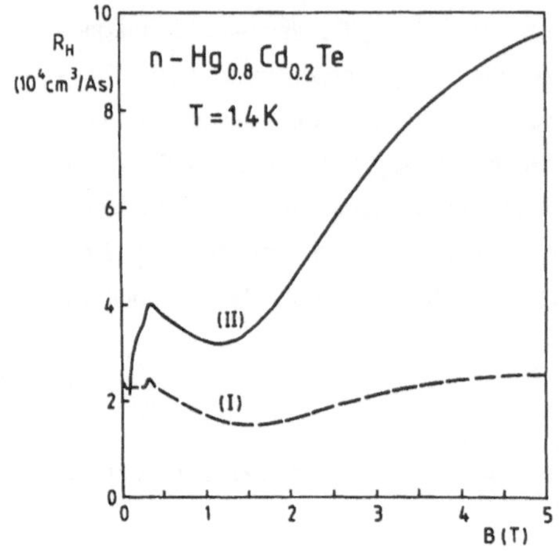

Fig. 5 Hall coefficient vs magnetic field. The data were obtained from a long and a short sample of the same crystal ($n(77K) = 3\cdot10^{14}$ cm^{-3}). (I) are data of the long sample with size 12 x 2 mm^2, (II) the short sample with size 4 x 2 mm^2 shows an increasing $R_H(B)$ above 2 T. Both curves were recorded at the same temperature.

Besides the sample geometry there is another important problem involved in the measurement of the Hall effect. On the surface of $Hg_{0.8}Cd_{0.2}Te$ crystals there is always an n_s^+-layer present, which appears to be rather independent of the sample preparation /3,22/. The characteristics of the film are about $n_s^+ \approx 10^{12}$ electrons/cm^2, layer thickness ≈ 30 nm, low temperature mobility 10^3 - 10^4 cm^2/Vs. Such a layer, with a resistivity, which does not much depend on temperature and magnetic field /3,5/ may not only influence the activated longitudinal resistivity of the bulk but also the Hall effect /23/. An estimate of the influence of the surface layer on the measured Hall resistivity indeed shows, that a significant freeze-out of bulk carriers, as claimed in Refs. /16,18/ could not become evident at temperatures below 1 K. In this case the Hall conductivity of the n_s^+-layer may dominate that of the bulk. Thus in order to determine the carrier density reliable experiments independent of the Hall effect are compulsory. Such experiments are the photoconductivity and the energy relaxation time which are described in detail in Refs. /13,14/ respectively.

4. On the Activation Energy

As mentioned in the introduction, one of the anomalous properties of an electron condensate is the thermally activated charge transport. This behaviour is analogous to the transport of atoms or molecules in a viscous liquid. The activation energy was found to be a function of the magnetic field (see Fig. 2). Recently the activation energy was determined in fields up to 20 T by STADLER et al. /5/. Data of different samples are shown in Fig. 6. As seen by inspection of Fig. 2, the activation of the electron mobility dis-

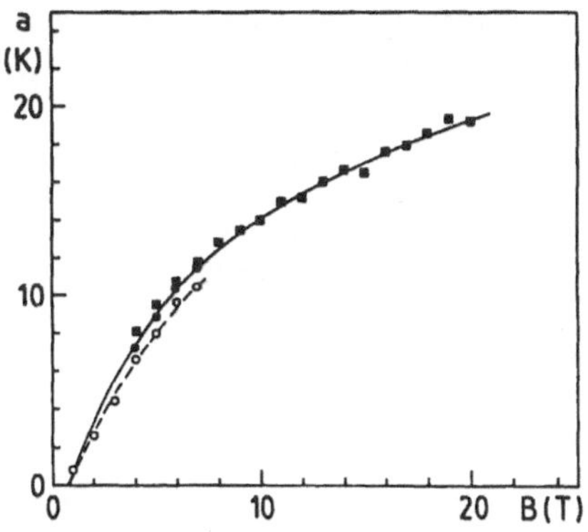

Fig. 6 Activation energy (in K) vs magnetic field B. The different symbols represent data of different samples.

appears smoothly as the temperature of the sample approaches the value of the activation temperature. We take the activation temperature as an estimate for the transition temperature from the viscous electron liquid to the electron gas state. As far as we know, theoretical calculations of this "temperature of evaporation" of an electron condensate in a magnetic field have not been made yet. For the time being we may compare the measured activation energy with data available from two model calculations: (i) the "melting temperature" of the Wigner lattice from ground state calculations by KLEPPMANN and ELLIOT /8/ and (ii) the critical temperature for the formation of a charge density wave state calculated within a Hartree-Fock approximation by GERHARDTS /9/. The qualitative agreement between theory and experiment is excellent, but, the absolute value of the measured activation temperature is about one order of magnitude higher than the results of both calculations. The discrepancy may be based on an additional electron-phonon interaction in the condensation process /2,10,11/. In the calculations published so far only the electron-electron interaction was considered for the electron-electron correlation /8,9/. Model calculations published by HEINONEN and AL-JISHIN taking into consideration both mechanisms are not representative for the extreme magnetic quantum limit /11/.

The observed thermally activated mobility of condensed electrons can be understood analogous to particle transport in a viscous liquid on the basis of the Eyring-Debye model. In a liquid some atoms or molecules are pinned for example at a wall, whereas in our dilute metal some electrons are expected to be pinned at lattice defects. In order to move a particle positioned next to a pinned one, the particle-particle interaction has to be overcome. For instance, in the case of water one hydrogen bond has to be broken in order to move a molecule to the next equilibrium position. This process is illustrated in Fig. 7. In the case of an electron liquid the activation temperature corresponds to the electron-electron interaction potential U.

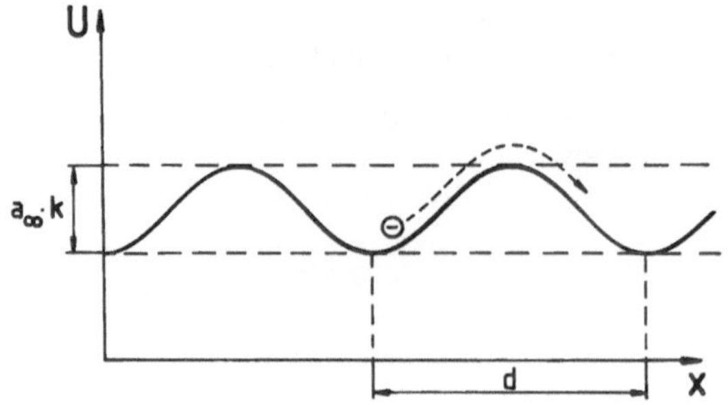

Fig. 7 Sketch of the electron potential vs interelectron distance. $a_\infty k$ and d are the electron-electron interaction energy for $B \rightarrow \infty$ and d the interelectron equilibrium distance, respectively.

The magnetic field dependence a(B) with the tendency of a saturation in high fields (Fig. 6) is predicted by theory. This can be made plausible by realizing that the energy competing with the electron-electron energy, namely, the Fermi energy, decreases with B^{-2}. That means with increasing magnetic field the observed activation energy approaches the electron-electron interaction potential which in turn is almost independent of B and is in charge of the condensation process.

Conclusion

Narrow-gap n-type HgCdTe shows a remarkable magneto-transport behaviour at low temperatures in high magnetic fields. The magnetoresistance and Hall-data obtained by several groups have been explained either by magnetic freeze-out or by magnetic field-induced electron condensation. In this review the arguments in favour of the latter model have been brought forward.

Acknowledgements

I would like to thank J. Gebhardt, P. Marquardt and J. Schilz very much for their help in the preparation of the manuscript.

References

1. R. Dornhaus, G. Nimtz, and B. Schlicht, Narrow Gap Semiconductors, Springer Tracts in Modern Physics Vol. 98 (1983)
2. G. Nimtz, B. Schlicht, E. Tyssen, R. Dornhaus, and L. D. Haas, Solid State Commun. 32, 669 (1979)
   G. Nimtz and B. Schlicht, Festkörperprobleme 20, 369 (1980)
3. J. P. Stadler, G. Nimtz, H. Maier and J. Ziegler, J. Phys. D 18, 2277 (1985)
4. G. Nimtz, J. Gebhardt, B. Schlicht, and J. P. Stadler, Phys. Rev. Letters 55, 443 (1985)
5. J. P. Stadler, G. Nimtz, and G. Remenyi, Solid State Commun. 57, 459 (1986)
6. B. Schlicht, Dissertation Mathematisch-Naturwissenschaftliche Fakultät, Universität zu Köln (1983)

7. G. Nimtz and B. Schlicht, J. Phys. Soc. Japan 49 Suppl. A, 313 (1980)
8. W. G. Kleppmann and R. J. Elliot, J. Phys. C8, 2729 (1975)
9. R. R. Gerhardts, Solid State Commun. 36, 397 (1980)
10. H. Fröhlich and C. Torreaux, Proc. Phys. Soc., London 86, 233 (1965)
11. O. Heinonen and R. A. Al-Jishin, Phys. Rev. B 33, 5461 (1986)
12. B. Schlicht and G. Nimtz, Physics of Narrow Gap Semiconductors in: "Lecture Notes in Physics 152", 383 (1982)
13. J. Gebhardt, G. Nimtz, B. Schlicht, and J. P. Stadler, Phys. Rev. B 32, 5449, (1985)
J. Gebhardt and G. Nimtz, Solid State Commun. 56, 131 (1985)
14. G. Nimtz and J. P. Stadler, Physica B 134, 359 (1985) and J. P. Stadler and G. Nimtz, Phys.Rev.Letters 56, 382 (1986)
15. S. B. Field, D. H. Reich, B. S. Shivaram, T. F. Rosenbaum, D. A. Nelson, and P. B. Littlewood, Phys. Rev. B 33, 5082 (1986)
16. G. de Vos, F. Herlach, and H. W. Myron, J. Phys. C 19, 2509 (1986)
17. O. Beckman, E. Hanamura, and L. I. Neuringer, Phys. Rev. Letters 18, 773 (1967)
18. M. Shayegan, V. J. Goldman, H. D. Drew, Proceedings of 18th Int. Conf. Phys. Semicond. Stockholm (1986)
V. J. Goldman, M. Shayegan, H. D. Drew, Phys. Rev. Lett. 57, 1056 (1986)
19. H. D. Drew, private communication, September 3rd, 1986
20. E. H. Putley, The Hall Effect and Related Phenomena, Butterworth London (1960) p. 42 - 62
21. G. Nimtz, J. Gebhardt and J. P. Stadler, to be published
22. J. B. Mullin and, A. Royle, J. Phys. D 17, L69 (1984);
L. F. Lou and W. H. Frey, J. Appl. Phys. 56, 2253 (1984)
23. G. Landwehr, private communication

High Field Magnetoresistance of Organic Conductors and Disordered Metals

J.C. Ousset, J.P. Ulmet, J. Léotin, and S. Askénazy

Laboratoire de Physique des Solides,
Service CNRS des Champs Magnétiques Intenses, I.N.S.A.,
Av. de Rangueil, F-31077 Toulouse-Cedex, France

A high magnetic field is a powerful tool for exploring the physical properties of materials as different as low-dimensional organic conductors and amorphous metallic alloys. These materials, whose conductivity is comparable to that of degenerate semiconductors, exhibit easily measurable magnetoresistances in the quasi-static pulsed field (up to 35 T) of the high-magnetic-field group at Toulouse.

In the case of organic conductors such as $(TMTSF)_2ClO_4$ and $(TMTSF)_2PF_6$, belonging to the family of the so-called "Bechgaard salts", the high field magnetoresistance exhibits quantum Shubnikov-de-Haas-like oscillations and rapid slope changes. These features are related to the change of state from metal to spin-density-wave state and to the appearance of closed orbits on the Fermi surface.

Concerning the amorphous metallic alloys, we have clearly observed the effect of Anderson localization on their transport properties. We have discussed the accuracy of the theoretical models and measured a large magnetoresistance due to incipient strong localization in the system $V_{1-x}Si_x$ near the metal-semiconductor transition.

1. Organic conductors

The organic salts from the TMTSF family exhibit not only a well-known superconducting behaviour [1] but also many remarkable properties at low temperature under a magnetic field, due to a phase transition from the metallic state to a spin-density-wave (SDW) one [2]. Two of these compounds have been intensively studied during the last four years: $(TMTSF)_2ClO_4$, a superconductor at atmospheric pressure and $(TMTSF)_2PF_6$, a superconductor under a pressure of some kilobars. Many surprising results in the transport at very low temperature and under a magnetic field [3, 4] have been explained satisfactorily by the theory of field-induced transitions and quantized nesting developed by Heritier et al [5]. At higher temperatures and fields the pockets of electrons and holes resulting from a longitudinal nesting of the Fermi surface (FS) could account for the Shubnikov-de Haas (SDH) like oscillations observed on the magnetoresistance.

1.1 Magnetoresistance of $(TMTSF)_2ClO_4$

The thin needles of the compound are slowly cooled down to 40 K to avoid cracks. From 40 K to 4.2 K the cooling rate does not exceed 0.3 K/minute in order to allow the anion ordering at 24 K and thus produce the so-called "relaxed state". The transverse magnetoresistance for B//c* (c* direction perpendicular to the crystal axes a and b) has been investigated between 2 and 12 K.

Below 5 K many oscillations appear, the temperature dependence showing that two series "a" and "b" having the same frequency are in fact present. A good periodicity in 1/B makes it possible to define a fundamental field B_F.

At 4.2 K the two series have comparable amplitudes at high field but at lower field the "b" series tends to vanish (Fig. 1). The rapid change of slope of the magnetoresistance at B_c is related to the metal-to-SDW transition.

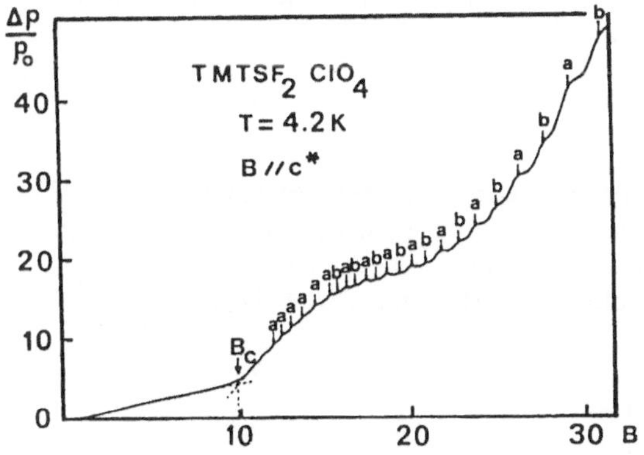

Fig. 1 Transverse magnetoresistance of $(TMTSF)_2ClO_4$ for B//c* and T = 4.2 K.

Between 4.2 K and 4.8 K, the "b" series disappears very quickly. At temperatures higher than 5 K only the "a" series remains visible (Fig. 2). No rapid change of the slope of the monotonic magnetoresistance is seen anymore.

From the field positions of the oscillations, B_N, it is possible to calculate the fundamental field B_F and the area S of the FS cross-section perpendicular to B :

$$1/B_N = 1/B_F \times (N - N_o) \quad (1)$$

$$S = \frac{2\pi e}{\hbar} B_F \quad (2)$$

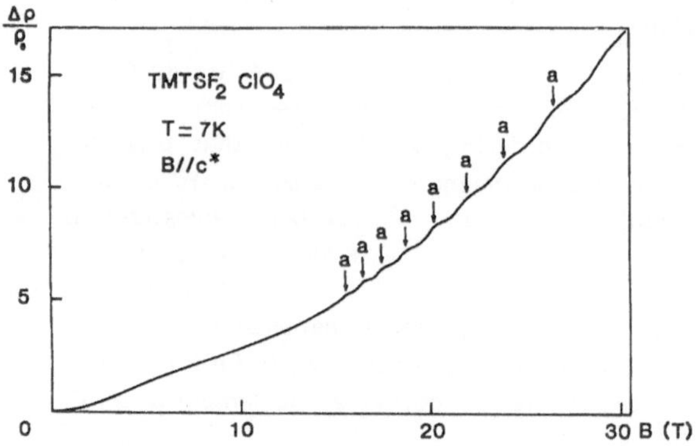

Fig. 2 Transverse magnetoresistance of $(TMTSF)_2ClO_4$ for B//c* and T = 7 K.

The area S is found to be about 3.4 % of the Brillouin zone with $B_F \simeq 260$ T [6]. A longitudinal nesting vector $Q_L = (2\,k_F,0,0)$ could give equal pockets of electrons and holes in good agreement with the observation of two series having the same frequency.

1.2 Magnetoresistance of $(TMTSF)_2PF_6$

The crystals are also slowly cooled but the centrosymmetric anions present no ordering. A set of magnetoresistances is displayed in Fig. 3. At high field, B > 18 T, SDH-like oscillations become visible. Only one series is observed whose fundamental field is close to 230 T for B//c* [7].

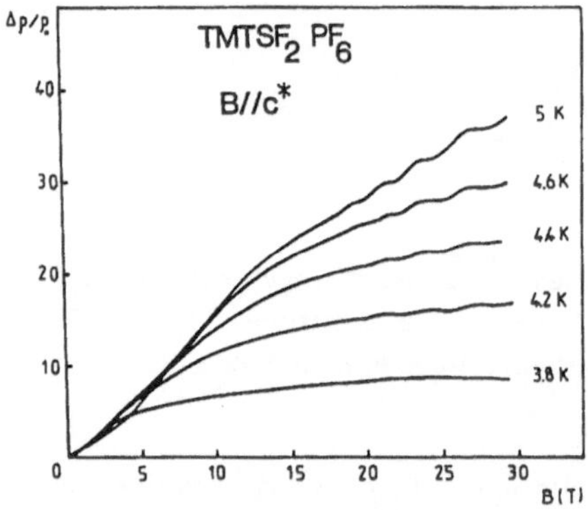

Fig. 3 Set of magnetoresistances of $(TMTSF)_2PF_6$ for B//c*

The corresponding area of orbits represents 3 % of the Brillouin zone, indicating the probability of a longitudinal nesting as in TMTSF perchlorate.

The amplitudes of the peaks are damped when the temperature is increased, the oscillations remaining visible up to around 10 K. When the temperature is lowered, a rapid vanishing of the oscillations occurs around 4 K together with a saturation of the magnetoresistance at high field. Both phenomena could be explained by a transition towards an open FS. It is interesting to point out that Takahashi, from NMR experiments on this compound, has observed some sort of transition at exactly the same temperature [8].

2. Localization effects in amorphous alloys

Weak localization effects in a disordered system were predicted by Anderson [9] in 1958. They have been widely studied during the last years in two dimensional systems [10]. More recently they were observed in three dimensional systems [11] and the amorphous metallic alloys [12 - 15] constitute an appropriate tool for these studies. Simultaneously theoretical models have been developed and they allow the calculation of the magnetoresistance arising from weak localization [16, 17] and electron-electron interaction [17, 18].

The magnetoresistance due to weak localization has been calculated by various authors [10 - 12] and is given by the expression :

$$\frac{\Delta\rho}{\rho} = \rho \frac{e^2}{2\pi^2\hbar} \left(\frac{eH}{\hbar}\right)^{\frac{1}{2}} \left[\frac{1}{2} f_3\left(\frac{H}{H_i}\right) - \frac{3}{2} f_3\left(\frac{H}{H_{so}}\right)\right] \quad (3)$$

where

$$H_i = \frac{\hbar}{4\,eD} \tau_i^{-1} \text{ and } H_{so} = \frac{\hbar}{4\,eD} \left(\tau_i^{-1} + 2\,\tau_{so}^{-1}\right) = H_i + H'_{so} \;; \quad (4)$$

τ_i and τ_{so} are the electronic relaxation times for inelastic and spin-orbit scattering.

In the weak localized regime, the contributions to the magnetoresistance arising from electron-electron interactions can be computed if one knows the values of the diffusion constant D and of the screening parameter for coulombic interactions F. So we have determined F and D in a free electron model using resistivity measurements.

Under high magnetic fields, in a wide range of temperature, we have measured the magnetoresistance of amorphous alloys $V_{1-x}Si_x$ prepared by co-evaporation in the Laboratoire de Physique des Solides de Nancy. These materials for $0.5 < x < 0.86$ are not superconductors.

We present in Fig. 4 the magnetoresistance after removing the interelectronic contributions for the alloy $V_{0.48}Si_{0.52}$ and also the best fit obtained from the theoretical models of weak localization (eq. 3). We must use a prefactor $\alpha = 1.5$. The value of this term would be equal to one in

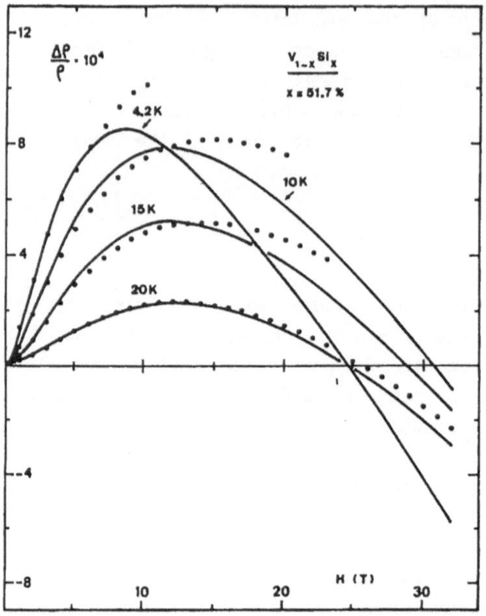

Fig. 4 Magnetoresistance due to weak localization for $V_{0.48}Si_{0.52}$ (full lines) and best fit obtained from equation 3 (points)

the case of free electrons, but in V-Si alloys the increase of α is probably due to the d-character of conduction electrons.

The fit becomes better when the ratio H/T decreases. It is excellent at high temperature. This good agreement at high field (for T = 20 K makes it possible to determine with reasonable precision the parameter H'_{so} and then the spin-orbit scattering frequency $\tau_{so}^{-1} = 1.43\ 10^{12} s^{-1}$. Indeed the function $f_3(H/H_{so})$ has important values only in high magnetic field if the spin-orbit coupling is strong.

In Table 1, we give the values of the frequencies of inelastic and spin-orbit scattering for three silicon concentrations and for various temperatures.

For x = 0.52 we have found a temperature dependence of τ_i^{-1} like T^2 which is characteristic of a predominant inelastic scattering by phonons. At low temperature (T < 8 K) a deviation appears which has been observed by other authors [19].

The spin-orbit scattering frequency decreases with vanadium concentration. This result is consistent because the spin-orbit coupling is stronger on the vanadium sites.

Table 1 Parameters determined from the best fit of the weak localization contribution to the magnetoresistance for our $V_{1-x}Si_x$ alloys.

X		0.52	0.60	0.75
$\tau_i^{-1}\ s^{-1}$	4.2 K	$0.98\ 10^{11}$	$0.77\ 10^{11}$	$0.28\ 10^{11}$
	10 K	$1.96\ 10^{11}$		
	15 K	$3.83\ 10^{11}$		$1.26\ 10^{11}$
	20 K	$7.60\ 10^{11}$		$3.21\ 10^{11}$
	25 K		$6.78\ 10^{11}$	
$\tau_{so}^{-1}\ s^{-1}$		$1.43\ 10^{12}$	$1.02\ 10^{12}$	$6.74\ 10^{11}$
α		1.5	1.5	1.6

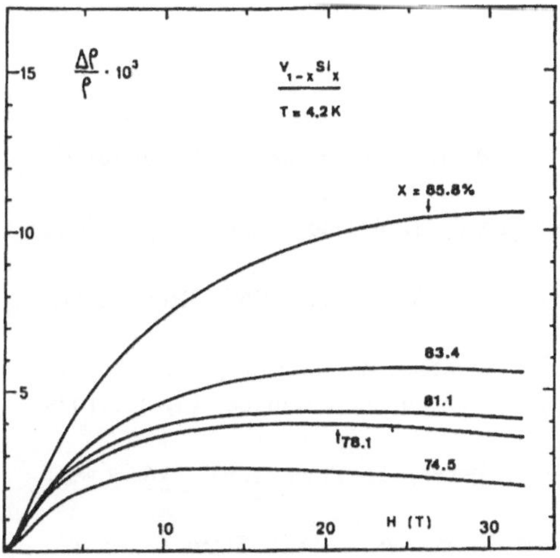

Fig. 5 Magnetoresistance of $V_{1-x}Si_x$ amorphous alloys with high silicon concentrations

For silicon concentrations higher than 75 % the magnetoresistance of the vanadium-silicon alloys increases, as it can be seen from Fig. 5.

We think that the large magnetoresistance observed for $x = 0.858$ probably arises from incipient strong localization and from the enhancement of the electron-electron interactions near the metal-semiconductor transition. Indeed we expected this transition for $x \simeq 0.84$. However by resistivity measurements we could not observe this one because the system presents two phases when $x > 0.80$, one of them being metallic. We are going to study vanadium silicon alloys with $x > 0.80$ prepared by sputtering in order to estimate the influence of the elaboration process on the structural properties of the amorphous alloys $V_{1-x}Si_x$. At present no theoretical model allows one to take into account the contribution of strong localization to the magnetoresistance.

Conclusion

The experiments we have presented show clearly the interest of high magnetic fields in the study of transport phenomena. For instance, in the case of $(TMTSF)_2PF_6$, no oscillation is visible under 18 T and in the range 18 - 32 T only four oscillations appear which is the minimum necessary to calculate accurately the fundamental field.

Concerning amorphous metallic alloys, when a strong spin-orbit coupling is present, the frequency of spin-orbit scattering can be determined with a good accuracy only in a high magnetic field, through the study of Anderson localization effects on the magnetoresistance.

References

1. K. Bechgaard, K. Carneiro, M. Olsen, F.B. Rasmussen and C.S. Jacobsen, Phys. Rev. Lett., 46, 852 (1981)
2. T. Takahashi, D. Jerome and K. Bechgaard, J. Physique Lett., 43, L 656 (1982)
3. J.F. Kwak, J.E. Shirber, R.L. Greene and E.M. Engler, Phys. Rev. Lett., 46, 1296 (1981)
4. K. Kajimura, H. Tokumoto, M. Tokumoto, K. Murata, T. Ukachi, H. Anzai, T. Ishiguro and G. Saito, J. Physique, Colloque 44, C3-1059 (1983)
5. M. Heritier, G. Montambaux and P. Lederer, J. Phys. Lett., 46, L-831 (1985)
6. J.P. Ulmet, A. Khmou, P. Auban and L. Bachere, Solid State Commun., 58, 753 (1986)
7. J.P. Ulmet, P. Auban, A. Khmou and S. Askénazy, J. Phys. Lett., 46, L-535 (1985)
8. T. Takahashi, Y. Maniwa, H. Kawamura and G. Saito, to be published in Proceedings of the Yamada Conf. (Japan, May 1986).
9. P.W. Anderson, Phys. Rev., 109, 1492 (1958)
10. G. Bergmann, Phys. Rev. Lett., 43, 1357 (1979)
11. R.C. Dynes, T.H. Geballe, G.H. Hull and J.P. Garno, Phys. Rev. B27, 1588 (1983)
12. J.B. Bieri, A. Fert, G. Creuzet, J.C. Ousset, Sol. Stat. Commun., 49, 849 (1984)

13. R.H. Cochrane, J.O. Strom-Olsen, Phys. Rev., B29, 1080 (1984)
14. M.A. Howson and D. Greig, J. Phys. F, 13, L155 (1983)
15. J.C. Ousset, H. Rakoto, J.M. Broto and S. Askenazy, Sol. St. Commun., 56, 29 (1985)
16. A. Kawabata, Sol. St. Commun., 34, 431 (1980)
17. B.L. Alt'Shuler, A.G. Aronov, A.I. Larkin and D.E. Khmel'Nitski, Sov. Phys. JETP, 54, 411 (1981)
18. P.A. Lee and T.V. Ramakrishnan, Phys. Rev., B26, 4009 (1982)
19. J.B. Bieri, A. Fert, G. Creuzet, A. Schuhl, to be published.

The Magnetophonon Effect in Narrow Gap Semiconductors

K. Takita

Institute of Materials Science, University of Tsukuba, Ibaraki, 305 Japan

New types of magnetophonon phenomena observed in narrow gap semiconductors under high electric field are presented: (1) magnetophonon resonance recombination with emission of two TA-phonons in LPE-HgTe and LPE-$Hg_{1-x}Mn_xTe$; and (2) magnetophonon resonance trapping of electrons with emission of multiple LO-phonons in $Hg_{1-x}Cd_xTe$. It is deduced that the TA-phonon density of states of HgTe has two peaks at 2.6 and 3.1meV at low temperatures. Peaks of spin-flip transitions are well resolved and the exchange integral between the conduction electron and the localized moment of Mn is deduced in $Hg_{1-x}Mn_xTe$; β=0.62eV. Furthermore, an impurity state is observed at 2meV above the top of the valence band. The trap level is deduced and the lifetime limiting recombination processes can be determined in $Hg_{1-x}Cd_xTe$, depending on the crystals and the temperatures.

1. Introduction

As is well known, most of the narrow gap semiconductors have electrons with very small effective mass m^*. Accordingly such electrons are very sensitive to magnetic fields and the high field limit can be attained easily. For example, in $Hg_{1-x}Cd_xTe$ with an energy gap Eg of 100meV, the Landau-level-spacing between the first and the second level in the conduction band reaches about 50meV at 10T and the high field condition of $\omega_c\tau \gg 1$ is fulfilled above 1T in most cases. (ω_c is the cyclotron frequency and τ is the scattering time of the carrier). Thus the narrow gap semiconductors have the potential of providing a wide variety of interesting subjects in the research on the application of high magnetic fields in semiconductor physics [1,2].

As far as experiments on the magnetophonon effect are concerned, however, the magnetic fields of 10T are too high for the observation of resonance transitions associated with single LO-phonons, because the LO-phonon energy in $Hg_{1-x}Cd_xTe$ is about 17meV. The possibility of magnetophonon resonance under ohmic conditions is limited to magnetic fields below 1.5T for $Hg_{1-x}Cd_xTe$ with Eg=100meV [3].

On the other hand, so-called zero-gap semiconductors such as HgTe have a very small energy gap induced by the magnetic field. The energy gap is only about 6~7meV at 10T in HgTe because of the quantum-effects first investigated theoretically by Luttinger [4,5]. Thus a field of 10T is too small to observe a magnetophonon transition between the conduction and the valence bands with emission of LO-phonons in these semiconductors. Such a transition in HgTe has been reported around 30T by Gel'mont et al. [6].

We have found some new types of magnetophonon phenomena in the magnetic field range 1~10T where the high field condition is fulfilled, but conventional magnetophonon resonance can not be expected [2,7~10]. In this paper, we give a review of some of these phenomena and present some new results, which were successfully observed under high electric fields using a second derivative method. They are (1) magnetophonon resonance recombination with emission of two TA-phonons in HgTe and $Hg_{1-x}Mn_xTe$ which were grown by LPE method, and (2) magnetophonon resonance trapping of electron with emission of multiple LO-phonons in $Hg_{1-x}Cd_xTe$.

2. Magnetophonon Resonance Recombination with Emission of Two TA-Phonons

2.1 HgTe

Magnetophonon resonance recombination of heated electrons and holes with emission of two TA-phonons was first reported in HgTe [8]. The observation was made for the first time by using thin layers of high-quality crystals grown by the liquid phase epitaxy (LPE) on CdTe. It has become possible to measure the resistance under heated electron conditions without using pulse technique because of the use of thin samples immersed in liquid helium. The energy of the TA-phonon at low temperatures has been precisely determined by the experiment.

The second derivatives of the longitudinal magnetoresistance under heated electron conditions were investigated for several different samples where the current direction was along different crystallographic orientations. Fig.1 shows a typical example of an oscillation and the proposed

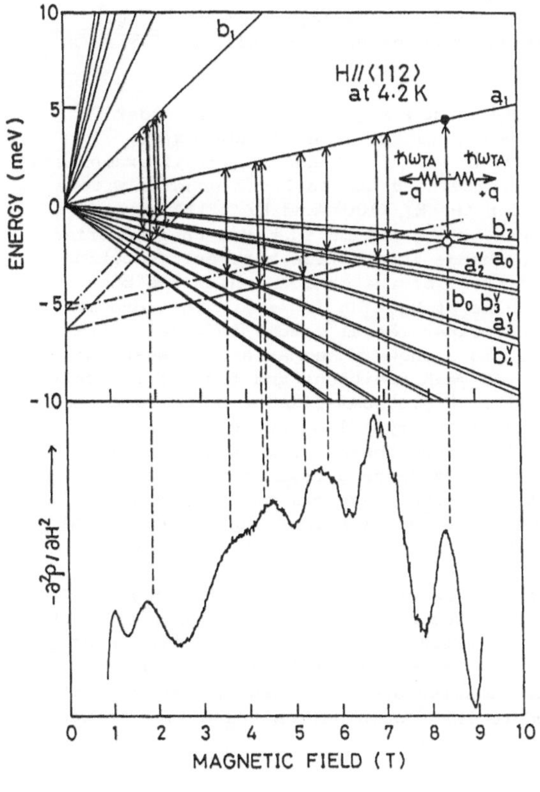

Fig.1
Landau-level scheme of HgTe and possible magnetophonon resonance recombination process accompanied by two TA-phonon emission for $2\hbar\omega_{TA}$=5.2meV and =6.2meV including the schematic drawing of the process (upper part). The lower part is an example of the observed resonance.

interpretation. The recombination process of heated electrons and holes accompanied by two TA-phonon emission is schematically illustrated in Fig.1. An electron in a Landau-level of the conduction band with energy E_{Lc} and a hole in a Landau-level of the valence band with energy E_{Lv} recombine emitting two TA-phonons which satisfy the energy and momentum conservation laws. The resonance condition is expressed as follows;

$$2\hbar\omega_{TA} = E_{Lc} - E_{Lv} , \qquad (1)$$

where $\hbar\omega_{TA}$ is the energy of the TA-phonons concerned.

According to the interpretation, the near zone edge TA-phonon energy at low temperatures was obtained by comparison of the experimental data with a calculation. The result indicates that the TA-phonon density of states has two peaks at 2.6 and 3.1meV, respectively. The calculated peak positions based on these phonon energies are indicated in the Landau-level scheme of HgTe for H//<112> case in Fig.1. The agreement between the observed oscillations and the calculation is excellent.

2.2 $Hg_{1-x}Mn_xTe$

2.2.1 Determination of the Exchange Parameter β and Observation of Spin-Flip Transitions

A magnetophonon resonance study similar to the method described in the preceding section was applied to $Hg_{1-x}Mn_xTe$ using the crystals grown by LPE from a Hg-solution. Band structure parameters could be precisely determined based on the previously determined TA-phonon energy in HgTe, because the electronic state is affected strongly by the alloying of Mn to HgTe while the phonon structure will not be affected substantially as far as x is not too large.

The second derivatives of the longitudinal magnetoresistance under heated electron conditions were investigated for several samples of $Hg_{1-x}Mn_xTe$ with a composition x of about 1%. Fig.2 shows typical examples of second derivative curves and the calculated Landau-level schemes with assigned recombination transitions for the H//<100> and H//<111> cases, respectively. The calculation of the energy scheme is based on a modified Pidgeon-Brown model including the exchange interaction between conduction electrons and electrons from the 3d shell of the Mn ions which has been proposed for the magnetic energy scheme of $Hg_{1-x}Mn_xTe$ [11]. The agreement between the calculated values and the experimental result is excellent, as can be seen in Fig.2. Furthermore, Fig.3 shows a comparison between the temperature dependence of the observed peak positions and that of the calculated ones for the case of H//<111>. Fig.4 summarizes the observed peak positions of samples with $x\simeq1\%$ for various crystallographic directions. The calculated peak positions are also shown there. From the comparison between experiment and calculation, the exchange parameter β between the conduction electrons and the localized spins is determined as β=0.62eV. The excellent agreement between the experimental points and the calculated ones means that the value of β has been obtained with high accuracy.

Furthermore it strongly supports the result, previously obtained in HgTe, that the TA-phonon density-of-states has two peaks at 2.6 and 3.1meV. Four peaks observed in the case of H//<111> are divided into a couple of paired peaks as can be seen in Fig.2. The paired peaks are the manifestation of the fact that the TA-phonon density-of-states has two maxima. The two peaks around 4~5T are assigned to the recombination transition between

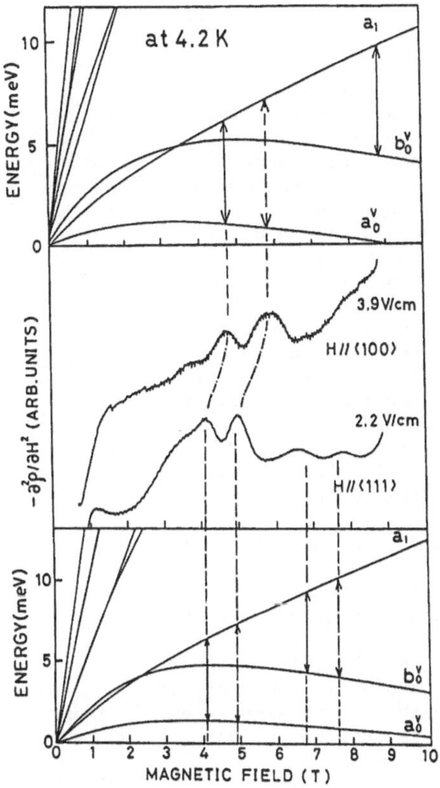

Fig.2
Typical examples of second derivative curves of $Hg_{1-x}Mn_xTe$ with x≃1% and the calculated Landau-level schemes including possible transitions with emission of two TA-phonons for H//<100> (upper) and H//<111> (lower), respectively.

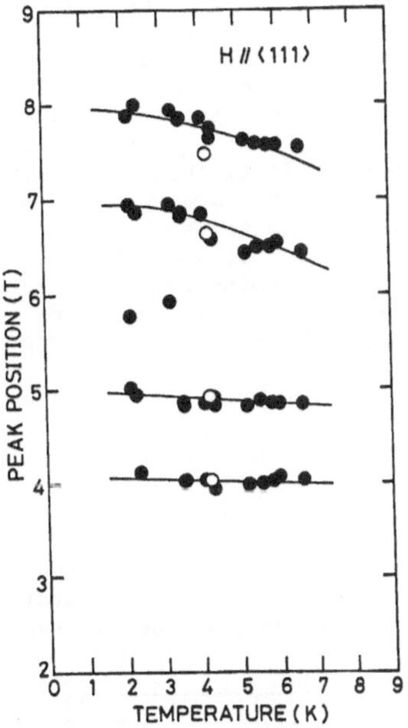

Fig.3
Temperature dependence of the observed peak positions for H//<111> in the sample with x≃1%. The solid lines show a calculated temperature dependence based on the modified Pidgeon-Brown model.

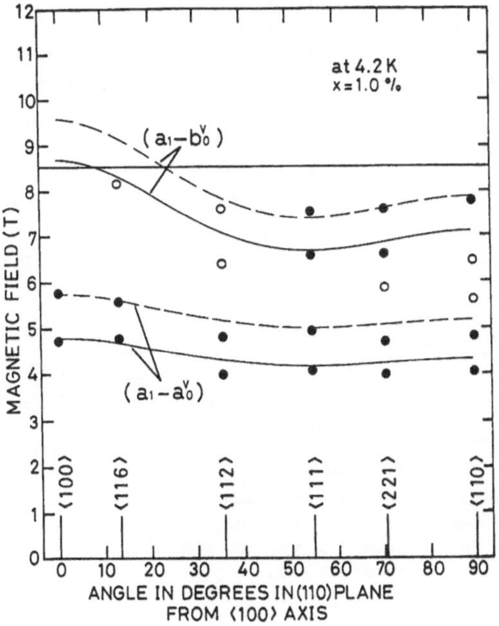

Fig.4
Observed peak positions for various directions of magnetic field H are compared with calculated peak positions for x=1%. Closed circles indicate clear peaks in the experiment and open circles are very small peaks.

a_0^v-valence band and a_1-conduction band while two small peaks around 7T are between b_0^v-valence band and a_1-conduction band. This means that the latter process is the spin-flip transition caused by acoustic phonons. This interpretation is strongly supported by the fact that the former peaks at lower fields are larger than the latter peaks at higher fields. It must be emphasized that this experiment shows well-resolved peaks due to spin-flip transitions in contrast to the spin-conserving transitions. The results provide a clear evidence for spin-flip transitions in narrow gap semiconductors, which have been discussed extensively by Zawadzki [12] and Zawadzki et al. [13]. A peculiar feature of the Landau-levels in $Hg_{1-x}Mn_xTe$, i.e. a large energy difference between b_0^v and a_0^v caused by the Mn ions, makes it possible to observe these well-resolved peaks due to spin-flip transitions —differing from the case of HgTe. Because of the large energy difference between b_0^v and a_0^v, the resonance peak positions of the spin-flip transitions shift to the high field side relative to the spin-conserving peaks. According to Zawadzki, the probability of spin-flip transitions is finite for the scattering by short-range potentials in InSb- and HgTe-type semiconductors, because the Landau-levels are not the pure spin states due to the large spin-orbit interaction in these semiconductors. The spin-flip transitions observed in the present experiment are of a new type in the sense that they are caused by two TA-phonon scattering.

2.2.2 Magnetophonon Resonance Recombination via an Impurity State

According to some previous papers [14~16], it was suggested that an acceptor-like impurity state may exist around 1 to 3meV above the top of the valence band in HgTe. In Fig.5, the Landau-level schemes near the band edge are shown for HgTe, $Hg_{1-x}Mn_xTe$ with x=0.5% and x=1%, respectively, which were calculated based on the parameters used in the preceding sections. The impurity state at about 2meV above the valence band is indicated by broken lines in the magnetic field-induced energy gap. As can be

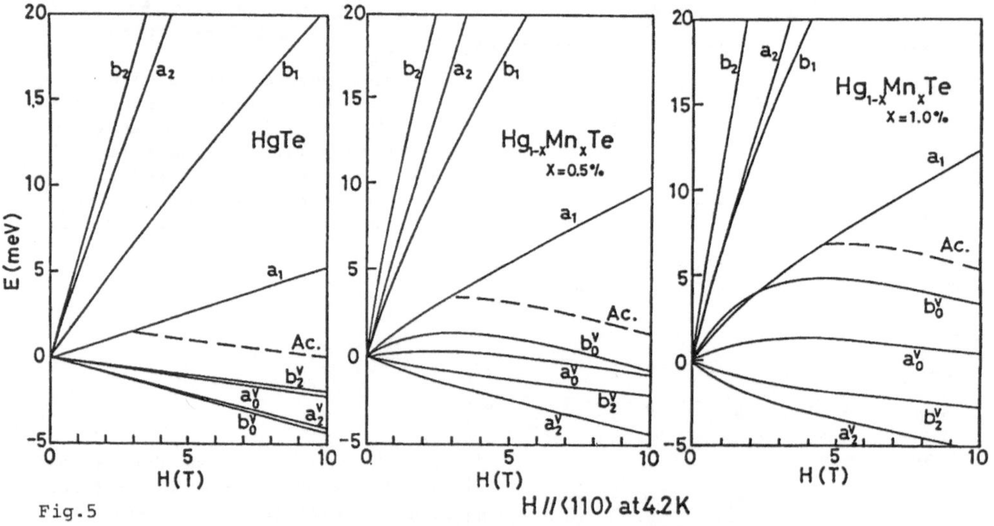

Fig.5
A comparison of the Landau-level schemes near the band edge for $Hg_{1-x}Mn_xTe$ with x=0, 0.5% and 1%, respectively, in the case of H//<110> at 4.2K. An impurity state 2meV above the top of the valence band is indicated by broken line denoted as Ac..

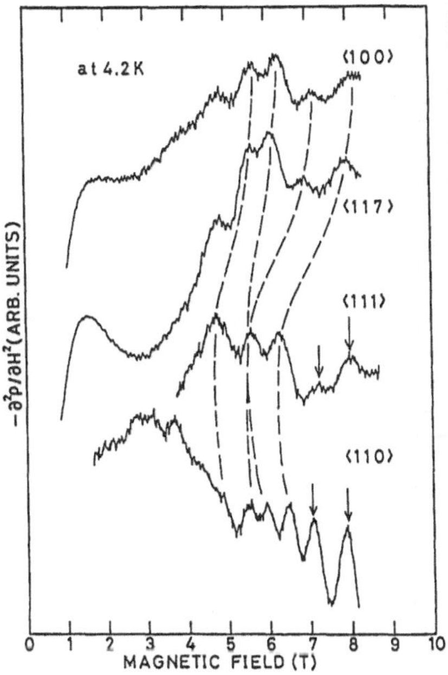

Fig.6
Typical examples of the magnetophonon resonance recombination oscillation of $Hg_{1-x}Mn_xTe$ with $x \simeq 0.5\%$ for four kinds of samples with various crystallographic orientation. Peaks assigned to the transitions between Landau levels are indicated by the broken lines connecting the peaks. A couple of peaks which are not assigned to the transition between Landau levels are indicated by arrows.

seen in the figure, it is expected that magnetophonon resonance transitions associated with the impurity state can be observed by the experiment of two TA-phonon resonance in magnetic fields up to 10T, if we use a sample of $Hg_{1-x}Mn_xTe$ with x=0.5%. For this sample, the magnetic field-induced energy gap is larger than those for the other two cases mentioned above.

Fig.6 shows typical examples of the magnetophonon resonance recombination oscillations of $Hg_{1-x}Mn_xTe$ with $x \simeq 0.5\%$ for various crystallographic directions. On the higher field side among the observed peaks, a couple of maxima were observed around 7~8T in the cases of H//<111> and H//<110>, which can not be assigned to a transition between Landau-levels. These peaks are interpreted to be associated with the impurity state existing in the magnetic field-induced energy gap.

Figure 7 shows the comparison of the observed oscillations with the Landau-level scheme and the impurity level scheme. According to the observed peak positions assigned to the impurity transition, the impurity levels are determined to be located 2.2meV and 1.8meV above the valence band for H//<111> and H//<110>, respectively. In these calculations, we used values for the parameters of a modified Pidgeon-Brown model [11] as follows:

$$P=8.3\times10^{-8}\,\mathrm{eV\cdot cm},\quad \Delta=1.08\,\mathrm{eV},\quad E_g=-250\,\mathrm{meV},\quad \beta=0.62\,\mathrm{eV},$$

$$\gamma_1=3.0,\quad \gamma_2=-0.5,\quad \gamma_3=1.0\quad \text{and}\quad \varkappa=-1.3.$$

β is as determined in the preceding section and the set of γ-parameters is that of Groves et al. [5] but modified in γ_3 to express the observed anisotropy of the band.

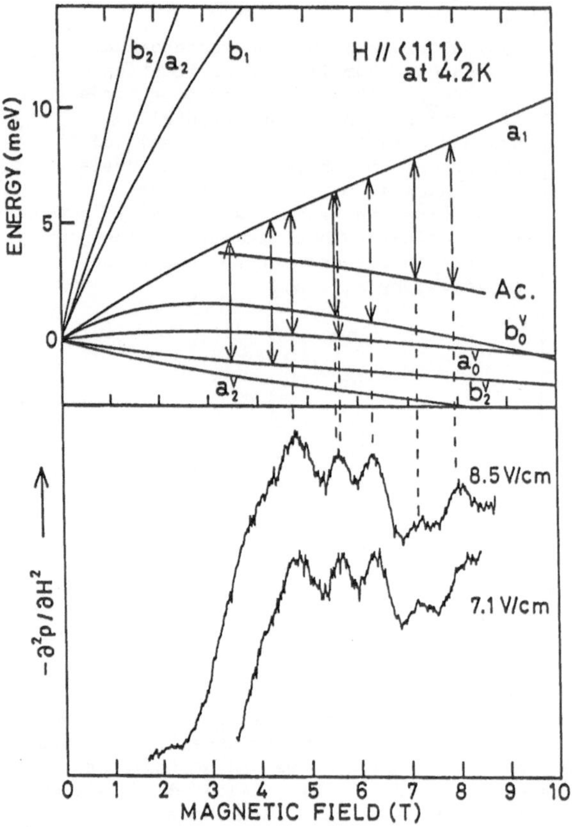

Fig.7
A comparison of the observed oscillations with the calculated Landau-level and impurity level scheme, for the sample with $x \simeq 0.5\%$.

3. Magnetophonon Resonance Trapping of Electrons with Emission of Multiple LO-phonons

A new type of magnetophonon effect of electron trapping with multi-LO-phonon emission was observed in n-$Hg_{1-x}Cd_xTe$ ($x \simeq 0.2$) together with oscillations due to Auger recombination. The observation was successfully performed in a wide temperature range by the combination of a pulse technique and a second derivative method.

Figure 8 shows typical examples of the second derivative curves of a typical sample measured in electric fields up to about 30V/cm. The carrier concentration n and the mobility μ_H of this sample are $n=2\times10^{15}cm^{-3}$ and $\mu_H=2\times10^{4}cm^{2}/Vsec$ at 4.2K, respectively. Two large broad peaks were observed below 15K as indicated by arrows and the peak positions shifted to the lower field side with increasing temperature as shown in the figure. On the contrary, four sharp peaks were clearly observed above 20K (as indicated by arrows) and the peak positions shifted to the higher field side

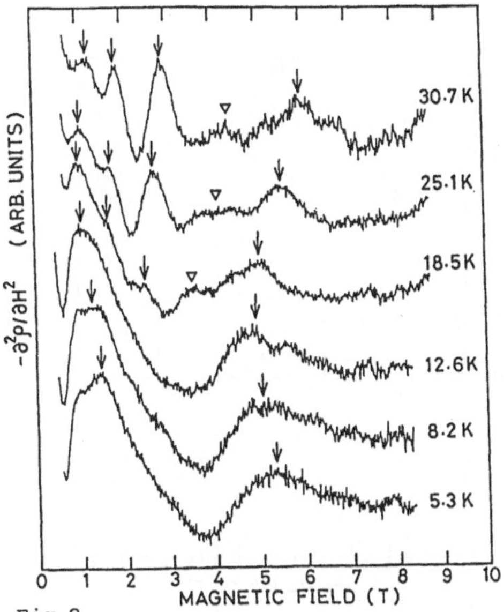

Fig.8
Typical examples of the second derivative curves of the longitudinal magnetoresistance in $Hg_{1-x}Cd_xTe$ ($x\simeq0.2$) under pulsed high electric fields. Results for a typical sample with $n=2\times10^{15}cm^{-3}$ and $\mu_H=2\times10^4cm^2/Vsec$ at 4.2K are shown for various temperatures.

with the increase of temperature. It was found that the temperature dependence of all the observed peaks above 20K coincides with a calculation based on the interpretation that the magnetoresistance peaks are due to the decrease of excess carriers through onset of Auger recombination at the resonance magnetic field. Such an effect of magneto-Auger-recombination oscillations at 4.2K and a preliminary result obtained at higher temperatures has been reported in our previous paper [1].

In contrast to the higher temperature side, only two broad peaks were observed below about 20K and the temperature shift of them is opposite to the higher temperature case. This behavior was found to be consistent with an interpretation that the peaks are due to a decrease of excess carriers at a certain field through onset of the magnetophonon resonance trapping with emission of multi-LO-phonons associated with Shockley-Read recombination. The resonance condition for this process may be expressed as,

$$n\hbar\omega_{LO} = Eg(H) - E_I \qquad (2)$$

where $n\hbar\omega_{LO}$ is the LO-phonon energy associated with n-phonon emission, Eg(H) is the magnetic field-dependent energy gap, and E_I is the energy of the trap center measured from the top of the valence band and assumed to be independent from the magnetic field. Calculated temperature dependences of the peak positions of this process agree well with the observed peak positions below 15K. Resonance transitions with emission of two- and three-phonons are observed in this case. By comparison between the calculation and the experiment, E_I=31meV was deduced for this sample, which is consistent with previously reported values of a trap center in $Hg_{1-x}Cd_xTe$ with $x\simeq0.2$, measured by DLTS [17].

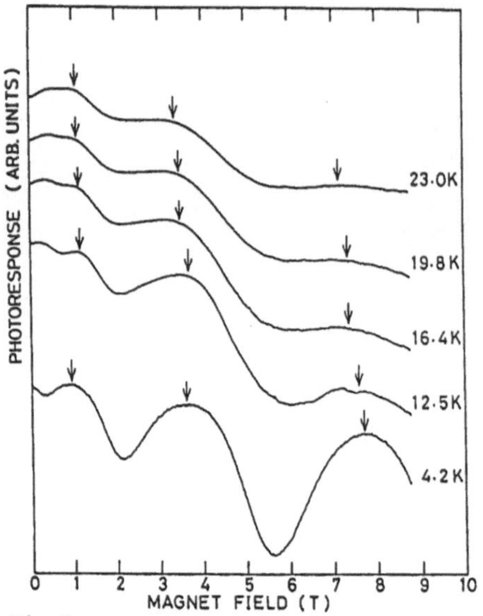

Fig.9
An example of the magnetic field dependence of photo-response. Result for the sample with smaller band gap and with lower mobility than that of Fig.8.

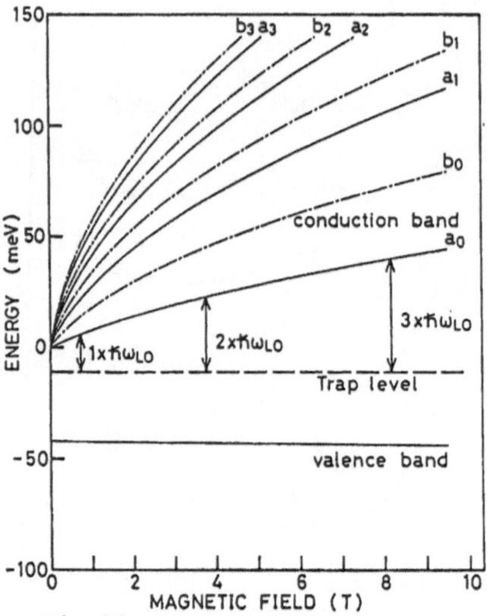

Fig.10
Schematic illustration of the magnetophonon resonance trapping of electrons with emission of LO-phonons, indicating the calculated resonance positions by arrows.

The proposed resonance trapping process was observed also in photoconductivity measurements in magnetic fields. A typical example of the photo-response under constant current conditions is shown in Fig.9, which was obtained for a 1.15μm-light from He-Ne laser. As this sample has a smaller band gap, three resonance peaks corresponding to n=1,2 and 3 in Eq.(2) could be observed. The proposed interpretation of magnetophonon resonance trapping with emission of LO-phonons is schematically illustrated in Fig.10.

A result for a high mobility sample with $n=2\times10^{15}cm^{-3}$ and $\mu_H=3\times10^5cm^2/Vsec$ (4.2K) indicated that the magneto-Auger-recombination oscillation is dominant in the magnetoresistance even below 10K, while the photoconductivity result showed a behavior typical of the trapping process. In the sample with lower mobility, the temperature range where the magnetophonon resonance trapping was observed, became wider. Thus, in the moderate quality samples, recombination through a deep trap center is dominant at low temperatures while Auger recombination is dominant at higher temperatures.

Acknowledgment
This work was performed in collaboration with prof. K. Masuda of the University of Tsukuba. The experiment was performed with the help of T. Ippōshi, T. Uchino and T. Gochou. This work was supported in part by a Grant-in-Aid for Scientific Research of the Ministry of Education, Science and Culture.

References

[1] K. Takita, T. Ippōshi, A. Suzuki and K. Masuda, Solid State Commun., 56, 599 (1985).
[2] K. Takita, T. Ippōshi and K. Masuda, to be published in the proceedings of 18th Int. Conf. Physics of Semiconductors (Stockholm).
[3] K. Takita, A. Suzuki and K. Masuda, Solid State Commun., 58, 209 (1986).
[4] J. M. Luttinger, Phys. Rev., 102, 1030 (1956).
[5] S. H. Groves, R. N. Brown and C. R. Pidgeon, Phys. Rev., 161, 779 (1967).
[6] B. L. Gel'mont, V. I. Ivanov-Omskii, N. N. Konstantinova, D. V. Mashovets, R. V. Parfen'ev and I. N.Yassievich, Sov. Phys.-JETP, 44, 823 (1976).
[7] K. Takita, T. Ippōshi, H. Otake and K. Masuda, Proc. 17th Int. Conf. Physics of Semiconductors, (1984), p.1185, (Edited by J. D. Chadi and W. A. Harrison), Springer-Verlag, New York, (1985).
[8] K. Takita, T. Ippōshi and K. Masuda, Solid State Commun., 52, 1021 (1984).
[9] K. Takita, T. Uchino, T. Ippōshi and K. Masuda, Solid State Commun., 56, 603 (1985).
[10] K. Takita and K. Masuda, Solid State Commun., 56, 283 (1985).
[11] M. Jaczynski, J. Kossut and R. R. Galazka, phys. stat. sol. (b), 88, 73 (1978).
[12] W. Zawadzki, Proc. 3rd Int. Conf. Physics of Narrow-Gap Semiconductors Warsaw, (1977), p.281, (Edited by J. Rauluskiewiez, M. Gorska and E. Kaczmarek), PWN-Polish Scientific, Warsaw and North Holland, Amsterdam, (1978).
[13] W. Zawadzki, A. Mauger, S. Otmezguine and C. Verie, Phys. Rev., B15, 1035 (1977).
[14] G. Bastard, Y. Guldner, C. Rigaux, N'guyen Hy Hau, J. P. Vieren, M. Menant and A. Mycielski, Phys. Lett., 46A (1973).
[15] W. Knap, I. Roschger, W. Szuszkiewicz, H. Krenn, A. M. Witowski and M. Grynberg, Proc. 17th Int. Conf. Physics of Semiconductors, (1984), p.659, (Edited by J. D. Chadi and W. A. Harrison), Springer-Verlag, New York, (1985).
[16] M. Dobrowolska and W. Dobrowolski, J. Phys. C, 14, 5689 (1981).
[17] C. E. Jones, V. Nair, J. Lindquist and D. L. Polla, J. Vac. Sci. Technol., 21, 187 (1982).

Hopping Conduction in n-InP in a Magnetic Field

S. Abboudy[1], *R. Mansfield*[2], *and P. Fozooni*[1]

[1]Department of Physics, Royal Holloway & Bedford New College, University of London, London, UK
[2]University of Brunei, Brunei

Measurements of the resistivity of n-InP in magnetic fields up to 70 kG and for temperatures down to 60 mK on three samples with a range of doping levels on the insulator side of the metal-insulator transition are described. Results in the nearest neighbour hopping region for the weak field range confirm previously reported measurements and the anisotropy of the magnetoresistance agrees closely with the expected $(H/4H_c)^2$ dependence ($H_c = N_D^{1/3}\,c\hbar/ae$). The field dependence in the intermediate range $2H_c < H < 6H_c$ is discussed. In the variable range-hopping region the resistivity of two of the samples agrees closely with a $\rho_0 \exp(T_0/T)^{1/4}$ dependence. The magneto resistance $\rho(H)/\rho(0)$ in the weak-field range varies as $T^{-3/4}$ and H^2 as expected for variable range hopping and a constant density of states at the Fermi level.

SHKLOVSKII and EFROS [1] have reviewed up to 1983 the measurements on InP in the hopping region and have shown that for nearest neighbour hopping reasonable agreement is obtained between theory and experiment for the resistivity in zero field and for the weak magnetic field region [2,3,4]. The predicted high magnetic field behaviour has also been observed although at fields below the high field limit [4]. In the variable range-hopping region however fewer measurements have been made and there is controversy over the temperature variation of the resistivity in zero magnetic field [5,6]. Measurements of the effect of a magnetic field on the resistivity in this region appear to be confined to one relatively heavily doped sample close to the metal-insulator transition [7].

This is a preliminary report of an investigation of the components of the resistivity tensor of n-InP in magnetic fields up to 70 kG and for temperatures down to 60 mK. Three samples were prepared from a single slice of InP and two of them were irradiated by thermal neutrons to produce a range of doping levels on the insulator side of the metal-insulator transition. The electron concentrations ($N_D - N_A$) of the samples at 77 K were 4.8, 5.9, 7.8 x 10^{15} cm^{-3} which are close to the lighter doped sample described in [4]. Our results on the temperatures and magnetic field variation of the resistivity in the nearest neighbour region agree with [4] so we will confine ourselves therefore to the anisotropy of the resistivity in the nearest neighbour hopping region and to the temperature and magnetic field dependence in the variable range hopping region. Figure 1 shows the typical family of resistivity versus T^{-1} curves for various fields for one of our samples. Each curve is divided into an ε_1 region where freeze-out of the carriers occurs, an ε_3 region where nearest neighbour hopping occurs and a variable range hopping region in the lowest temperature range. The activation energy ε_3 for hopping conduction is very much less than the theoretical value of $0.99\ e^2 N^{1/3} \kappa^{-1}$ (N = electron concentration, κ = the dielectric constant) and this is attributed, as with other semiconductors, to

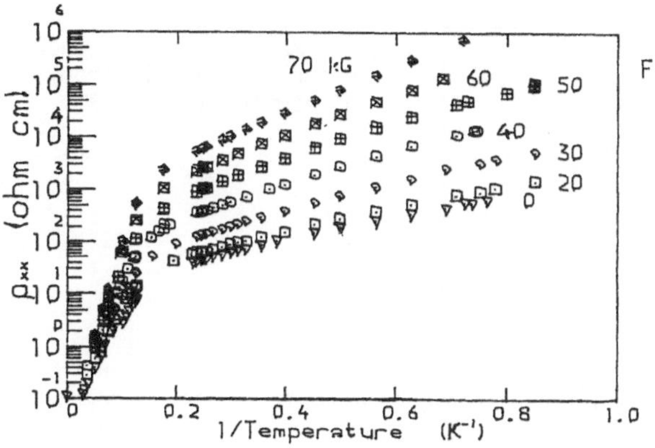

Fig.1. Resistivity of sample II as a function of T^{-1}.

correlation effects. The increase in ε_3 with field is due to the field reducing the correlation effects. In the low magnetic field region $H < 2H_C$ ($H_C = N_D^{1/3}\,c\hbar/ae$ where N_D = donor concentration and a = donor radius) the magnetoresistivity $\rho_3(H)/\rho_3(0)$ is in agreement with the theoretical expression

$$\frac{\rho_3(H)}{\rho_3(0)} = \exp\left[\frac{tae^2H^2}{N\,c^2\,\hbar^2}\right], \tag{1}$$

where t = 0.036.
This confirms measurements described in [4]. The high field region, when the magnetoresistance is expected to vary as $H^{\frac{1}{2}}$, should start when $H > 6H_c$ which for our samples is greater than 90 kG and is higher than the fields available to us. However, an $H^{\frac{1}{2}}$ dependence appears to be obeyed at field much less than this limit, as also observed in [4], but only over a restricted range of fields, and alternative variations could not be ruled out. IOSELEVICH [8] has provided a treatment covering the range of fields between $2H_C$ and $6H_C$. A universal curve between the reduced percolation parameter ξ^* which is determined from $\ln(\rho_H/\rho_0)$ and the reduced field H^* is provided. These parameters are both decreased when allowance is made for the effect of the magnetic field on the donor activation energy by the factor $(E(0)/E(H))^{\frac{1}{2}}$ where E(0) and E(H) are the activation energies in zero field and in a field H. Identifying these energies with the resistivity activation energy ε_3 results in a correction which is far too large, but this method of finding E_D has recently been criticized [9]. If, however, we use the Yafet, Keyes and Adams theoretical ratio E(H)/E(0) appropriate to the magnetic field then good agreement with the Ioselevich curve is obtained for high magnetic fields.

Figure 2 shows the magnetic field dependence of the anisotropy of samples II and III. The anisotropy is plotted in accordance with the theoretical relation

$$\frac{\rho_{xx} - \rho_{zz}}{\rho_{xx}} = \left[\frac{H}{4H_c}\right]^2 \tag{2}$$

given in [1] for the weak field case. An H^2 behaviour is observed but the right-hand side has to be multiplied by a factor 0.71 and 0.55 for samples II and III respectively.

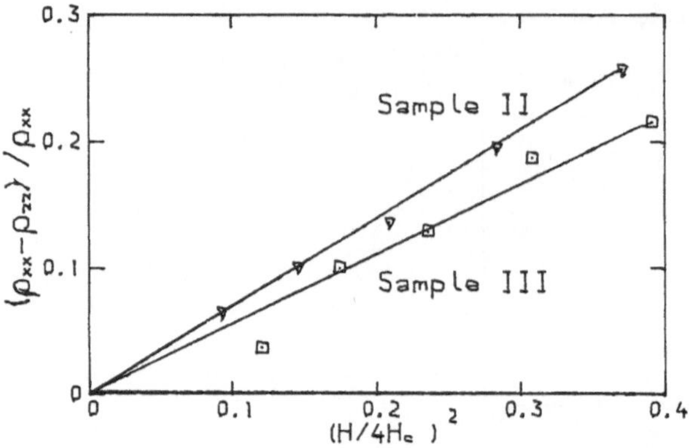

Fig.2. Magnetic field dependence of the anisotropy of samples II and III.

The controversy over the temperature variation of the resistivity in the variable range-hopping region with no field present is whether the dependence is $T^{-1/4}$ [5] or $T^{-1/2}$ [6]. The first dependence was derived by Mott for the case of a constant density of states at the Fermi energy and the second by Shklovskii and Efros for when there is an energy gap at the Fermi level. The two variations are shown for each sample in Figs. 3 and 4. Excellent agreement with a $T^{-1/4}$ law is found for samples I and II which confirms [5] but over a much wider temperature range and consequent

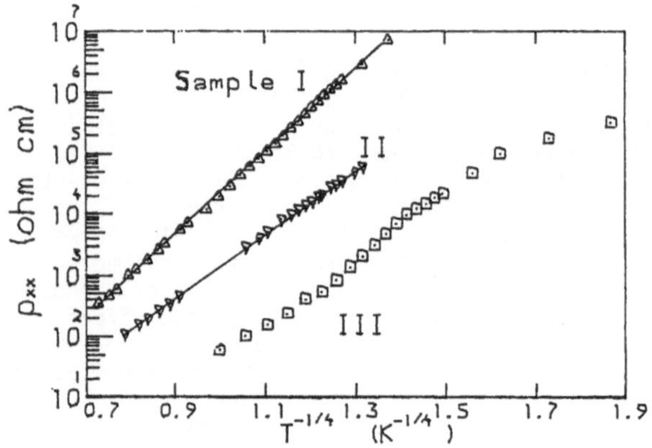

Fig.3. Resistivity of samples I, II and III as a function of $T^{-1/4}$ (no magnetic field).

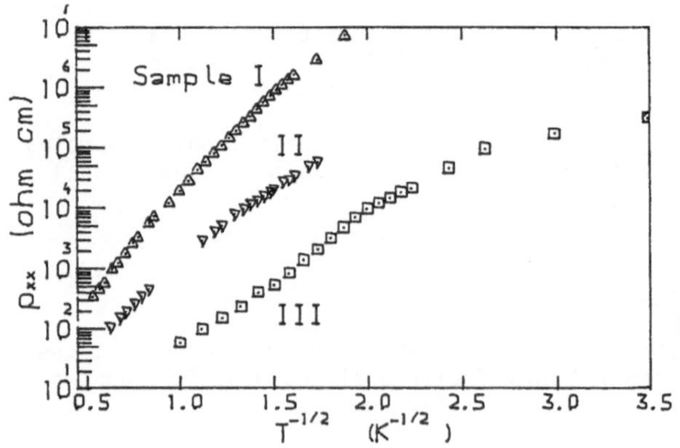

Fig.4. Resistivity of samples I, II and III as a function of $T^{-1/2}$ (no magnetic field).

variation in ρ. Both the samples in [5] and samples I and II have similar concentrations of electrons. Neither law can account for the variation of resistivity of samples III over the whole temperature range. A $T^{-\frac{1}{2}}$ law is obeyed down to 0.2 K but the variation becomes less rapid below this temperature. Results on the magnetoresistances which are discussed below indicate that this sample should follow a $T^{-1/4}$ law.

According to Shklovskii and Efros the temperature and magnetic field dependence of the resistivity for variable range hopping in the weak magnetic field limit is given by

$$\ln\left[\frac{\rho(H)}{\rho(0)}\right] = t^1 \frac{a^4}{\lambda^4}\left[\frac{T_0}{T}\right]^{3/4} \quad (3)$$

for a constant density of states at the Fermi level, $t^1 = 0.0025$ and

$$\ln\left[\frac{\rho(H)}{\rho(0)}\right] = t^2 \frac{a^4}{\lambda^4}\left[\frac{T_0}{T}\right]^{3/2} \quad (4)$$

for a density of states $g = g_0\varepsilon^2$ where ε is energy measured from the Fermi level, g_0 is a constant and $t^2 = 0.0015$. Figure 5 shows that the temperature variation of $\ln(\rho(H)/\rho(0))$ of samples II and III in a field of 30kG is in fair agreement with a $T^{-3/4}$ variation, whereas Sample I is anomalous in that it does not depend on either $T^{-3/4}$ or $T^{-3/2}$. Figure 6 shows that the magnetic field dependence is also in agreement with the H^2 dependence predicted by both (3) and (4). The value of t^1 determined,

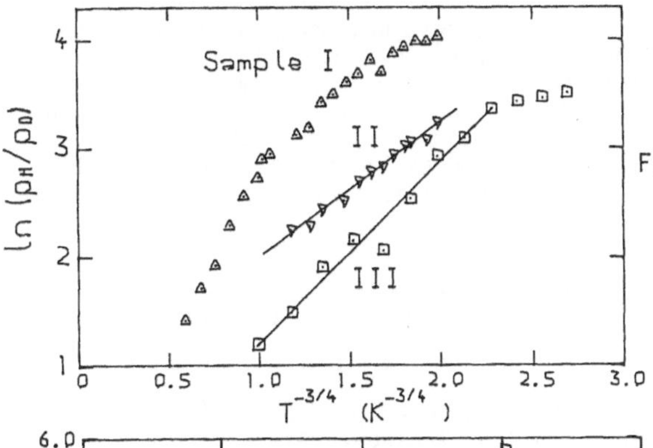

Fig.5. Resistivity of samples I, II and III in a magnetic field of 30 kG as a function of $T^{-3/4}$. Samples I and III were longitudinal and II transverse to the magnetic field.

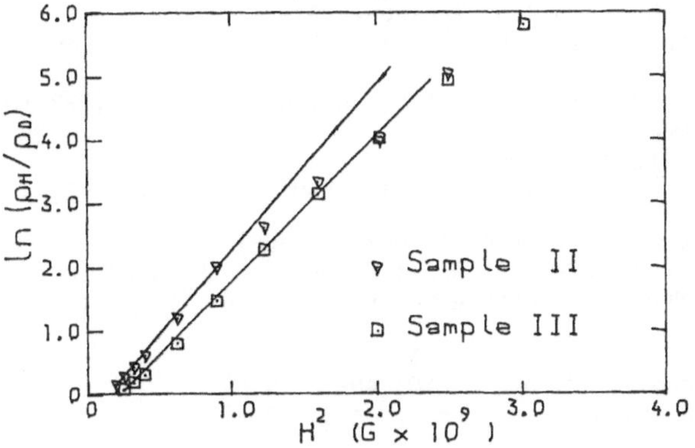

Fig.6. Magneto resistance in the variable range hopping region. Variation of the resistivity with H^2.

assuming value of T_0 obtained from the zero magnetic field $T^{-1/4}$ plots, are 0.0065 and 0.010 for samples II and III. These are substantially greater than 0.0025 given in [1]. It is possible that T_0 might increase with magnetic field, as is observed in InSb [10]. This would give a smaller value of t^1. Since there is some uncertainty about the zero-field dependence of sample III the value of t^2 obtained using (4) and T_0 derived from a $T^{-1/2}$ plot was calculated. A value of 0.11 was obtained which differs by two orders of magnitude from the theoretical value of 0.0015 and is evidence for a constant density of states at the Fermi energy.

Table 1 summarises the results. H_c has been derived assuming an estimated compensation ratio of 0.35.

Table 1

Sample	$(N_D - NA)cm^{-3}$	H_c(kG)	T_0(K)	t	t^1
I	4.8×10^{15}	15.3	5.8×10^4	.045	
II	5.9×10^{15}	16.4	2.4	.033	.0065
III	7.8×10^{15}	18.0	3.6	.031	.010
Theory				.036	.0025

Good agreement between experiment and theory is obtained for the dependence of the anisotropy on temperature in the nearest neighbour hopping region. The temperature dependence of the resistivity in zero field in the variable range hopping region is $T^{-1/4}$ for the two lighter doped samples, but the more heavily doped sample does not obey either a $T^{-1/4}$ or $T^{-1/2}$ variation over the whole temperature range. The temperature and magnetic field dependence of the magneto resistance of samples II and III are in agreement with theory but the behaviour of sample I is anomalous.

ACKNOWLEDGMENTS

We are grateful for the skilful technical assistance of Mr. A.K. Betts and to the SERC and Central Research Fund of the University of London for financial assistance.

1. B.I. Shklovskii and A.L. Efros, Electronic Properties of Doped Semiconductors, Ed: M. Cardona (Berlin: Springer 1984).
2. O.V. Emel'yanenko, K.G. Masugutov, D.N. Nasledov, I.N. Timchenko, Sov. Phys. - Semicond. 9, 330 (1975).
3. G. Biskupski, H. Dubois, O. Laborde, X. Zotos, Phil.Mag. B42, 19 (1980).
4. G. Biskupski and H. Dubois, (See Biskupski, G., 1982, Thesis Lille)
5. M. Benzaquen, K. Mazuruk, D. Walsh and M.A. di Forte-Poisson, J.Phys.C, 18, L1007 (1985).
6. D.M. Finlayson and P.J. Mason, J.Phys.C, 19, L299 (1986).
7. G. Biskupski, H. Dubois and O. Laborde, Application of high magnetic fields in semiconductor physics: Lecture notes in physics (Berlin: Springer 1983).
8. A.S. Ioselevich, Sov.Phys. - Semicond. 15, 1378 (1981).
9. T.H.H. Vuong and R.J. Nicholas, J.Phys.C, 18, 4021 (1985).
10. H. Tokumoto, R. Mansfield and M.J. Lea, Phil.Mag.B, 93 (1982).

Conduction Electrons in Bulk GaAs at High Magnetic Fields

W. Zawadki and P. Pfeffer

Institute of Physics, Polish Academy of Sciences, PL-02-668 Warsaw, Poland

Five-level k·p model for electrons in GaAs in the presence of a magnetic field is developed. Inversion asymmetry of the material is consistently taken into account. It is shown that inclusion of two higher conduction bands provides an adequate description of nonparabolicity and nonsphericity of the Γ_6 conduction band. Polaron effects are also considered since their magnitude in GaAs is comparable to nonparabolic effects resulting from the interband k·p interaction.

1. Introduction

It has been commonly assumed that the conduction band in GaAs is spherical and parabolic, or, when this assumption proved manifestly insufficient, a three-level nonparabolic k·P model has been used. However, as observed by Hermann and Weisbuch [1], the three-level model does not describe correctly the observed values of the effective mass m_o^* and the spin-splitting factor g_o^* at the conduction band-edge. The reason for this inadequacy is that the three-level model is suitable for small-gap semiconductors, in which the k·P interaction between the conduction level and the nearest valence levels dominates. In GaAs this holds only very approximately, since the fundamental gap is 1.5 eV, while the energy interval to higher conduction levels is 3 eV. Rössler [2] has extended the k·p model to five levels in order to describe ε(k) dependence in absence of external fields.

Ogg [3] and Braun and Rössler [4] have used expansions in powers of magnetic field B in order to calculate magnetic energy levels in the presence of an external magnetic field.

2. Five-level Model

The initial eigenvalue problem for an electron in a periodic potential in the presence of an external magnetic field $\vec{B}$, written in the Luttinger-Kohn representation, is [5]

$$\sum_{l}[(\frac{1}{2m_o}P^2+E_l-E)\delta_{l'l}+\vec{\kappa}_{l'l}\cdot\vec{P}+\mu_B\vec{B}\cdot\vec{\sigma}_{l'l}+H^{so}_{l'l}]f_l = 0 \; ; \qquad (1)$$

f_l are envelope functions, m_o is the free-electron mass, $\vec{P}=\vec{p}+e\vec{A}$ is the kinetic momentum, E_l are band-edge energies, $\vec{\kappa}_{l'l}$ and $\vec{\sigma}_{l'l}$ are interband matrix elements of $\vec{p}/m_o$ and of the Pauli spin operators $\vec{\sigma}$, respectively. Indices l and l' run over all bands.

If the Luttinger-Kohn basis is chosen to diagonalize exactly the spin-orbit interaction, the interband spin-orbit terms $H^{so}_{l'l}$ vanish. However, we choose the usual basis functions (cf. [5]), which diagonalize the spin-orbit interaction within $(\Gamma_8,\Gamma_7)_v$ and $(\Gamma_8,\Gamma_7)_c$ sets. Then an interband spin-orbit term between the above sets remains (cf. Pollak et al. [6]), which we call $\bar{\Delta}$.

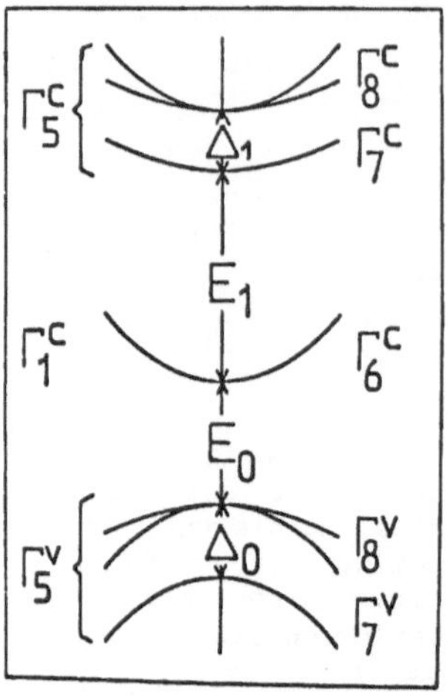

Fig.1. Five-level model for the conduction band of GaAs near the Γ point of the Brillouin zone

We truncate the infinite set of coupled differential equations (1) into a finite set of 14 equations considering a five-level model at k = 0 (cf. Fig.1). The distant level contributions to the conduction band are included later in a parabolic approximation. The truncated set involves four different matrix elements: $\langle S|p_x|X_5^v\rangle = P_o$; $\langle S|p_x|X_5^c\rangle = P_1$; $\langle X_5^v|p_y|Z_5^c\rangle = -\langle X_5^c|p_y|Z_c^v\rangle = Q$; and $\bar{\Delta}$. Together with distant-level contributions C and C' to m_o^* and g_o^*, respectively (cf. [1]), they constitute six parameters of the model, since the energy gaps are known from optical experiments (cf. Table 1).

Table 1. Energy intervals in GaAs used in the five-level k·p model (after [7,8]).

E_o = 1.519 eV	Δ_o = 0.341 eV
E_1 = 2.969 eV	Δ_1 = 0.171 eV

Due to appearance of the matrix element Q the resulting conduction Γ_6^c band is nonspherical. In consequence, solutions of the truncated set (1) are not given by simple harmonic oscillator functions, as in the case of the three-level model [5]. Following a procedure of Evtuhov [9], we look for envelope functions f_1 in the form of sums of harmonic oscillator functions. The differential eigenvalue problem can be then represented in the form of an infinite number matrix, in which different Landau states are coupled by matrix elements involving Q. We are interested in magnetic energies for $k_z = 0$. Diagonalising the matrix for a given Landau state n± it is enough to include the nearest states coupled to the state in question by Q elements. This corresponds to the second order perturbation theory in Q which is a very good approximation, since Q affects the conduction Γ_6^c band only indirectly.

Corresponding truncated matrices for $\vec{B}||(001)$ are 21×21, for $\vec{B}||(011)$: 35×35, and for $\vec{B}||(111)$: 28×28.

The adjustable parameters have been determined in the following way. Taking tentative values of C and C', and using the value of $\bar{\Delta} = -0.053$ eV [10] we have calculated P_o^2 and P_1^2 from the known values of m_o^* and g_o^*. Then the parameter Q has been determined by fitting a measured anisotropy of the spin doublet splitting of cyclotron resonance [11] (cf. also Fig.7). It should be mentioned that various measured quantities are still not too well established since they depend on the quality of samples and, consequently the adjusted band parameters should not be considered as final.

It is tempting to describe the calculated nonparabolic dispersion relation $E(\vec{k})$ by an effective two-level formula

$$\frac{\hbar^2 k^2}{2m_o^*} = E(1 + \frac{E}{E_g^*}) , \qquad (2)$$

in which E_g^* is an "effective" energy interval between the two levels, characterising band's nonparabolicity (stronger nonparabolicity corresponds to smaller E_g^*). It turned out to be possible to describe an average (over three principal $\vec{k}$ directions) E(k) relation for not too high k values by Eq. (1) in which the effective is given in Fig.2. This approximation is reasonable for the orbital electron properties but not for its spin properties.

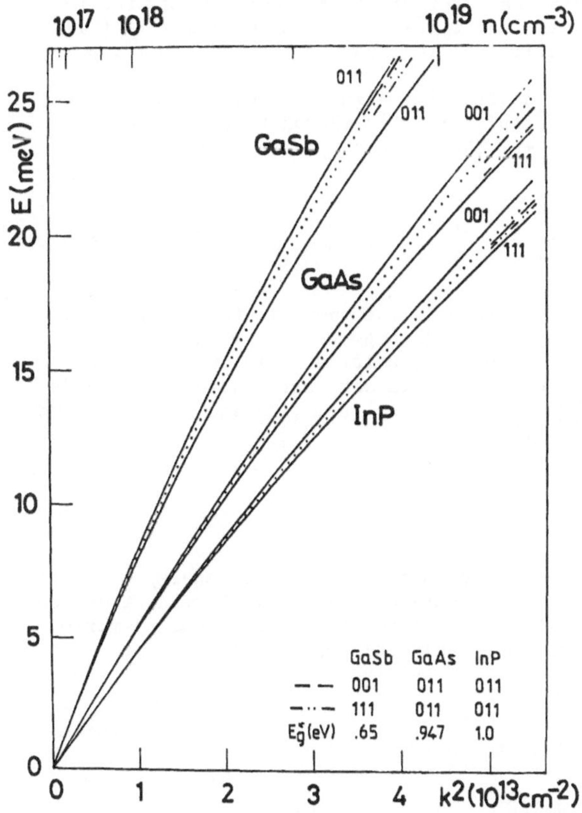

Fig.2. Dispersion relations $E(\vec{k})$ for GaSb, GaAs and InP calculated for different directions of the wavevector. The band parameters used for GaAs: P_0^2 = 28.66 eV, P_1^2 = 1.67 eV, Q^2 = 14.88 eV, $\bar{\Delta}$ = -0.053 eV, C = -3, C' = 0.01, m_o^* = 0.0659, g_o^* = -0.44. Free electron concentrations n corresponding to k = k_F are indicated on the upper abscissa. (After Pfeffer and Zawadzki [12])

Finally, it should be mentioned that an approximation of Q = 0 (which results in a spherical conduction band and numerically reduces to solutions of 7×7 matrix for each spin direction) is not good. It gives results half-way between the five-level and three-level k·p models.

3. Polaron Effects

Now we turn to the polaron effects in GaAs, which, as mentioned before, are comparable in this material to the effects resulting from interband k·p coupling. We concentrate on resonant aspect of magneto-polarons. Experimentally resonant polarons in GaAs have been recently investigated in cyclotron resonance by Sigg et al.[13] and in interband magneto-absorption by Hornung [16].

We are interested in the energy of the upper Landau state n perturbed by the polar interaction with the Landau state n=0, accompanied by a virtual or real longitudinal optic phonon emission at low temperatures. In order to describe both lower and upper polaron branches we need the Green function formalism, because the Wigner-Brillouin perturbation theory is insufficient. For resonant magneto-polarons this scheme was first used by Nakayama [17]m We extend it by introducing an additional residual scattering, always present in a crystal.

We consider the initial electron state $|n\rangle = |n, k_x = 0, k_z = 0, s = +\rangle$, which decays by emission of a longitudinal optic phonon to the state $|0\rangle = |0, k_x' = -q_x, k_z' = -q_z, s = +\rangle$. The resonant part of selfenergy of the state $|n\rangle$ is

$$\Sigma_n(E) = \sum_q \frac{|\langle 0|\ H_F\ |n\rangle|^2}{E+i\Gamma_o-E_o^+-\hbar\omega_L} - i\Gamma_o \ . \tag{3}$$

The phenomenological parameter Γ_o represents the above-mentioned residual scattering. It eliminates the nonphysical divergence at the energy $E = E_o^+ + \hbar\omega_L$ and gives a physically observed polaron behaviour: at magnetic fields below the resonance the lower branch is dominant, whereas at fields above the resonance the upper branch takes over (cf. [18] and [19]. Next we calculate the spectral function $A_n(E)$:

$$A_n(E) = \frac{1}{\pi} \frac{\Gamma_n}{(E-E_n^+-\Delta_n)^2+\Gamma_n^2} \ , \tag{4}$$

where Δ_n and Γ_n are the real and the imaginary parts of selfenergy, respectively: $\Sigma_n = \Delta_n - i\Gamma_n$. In pure crystals, where magneto-optical transitions occur at low k_z values, maxima of the spectral function correspond to maxima of the density of states, i.e. to observable polaron energies.
There is

$$|\langle 0|\ H_F|n\rangle|^2 = C^2 \frac{x^n}{n!} \exp(-x) \ , \tag{5}$$

where

$x = q_\perp^2 \cdot L^2/2$, with $q_\perp = (q_x^2+q_y^2)^{1/2}$ and $L = (\hbar/eB)^{1/2}$. Further

$$C^2 = \frac{2\pi \cdot e^2 \cdot \hbar\omega_L}{V} \left(\frac{1}{\kappa_\infty} - \frac{1}{\kappa_o}\right) \frac{1}{q^2} \ . \tag{6}$$

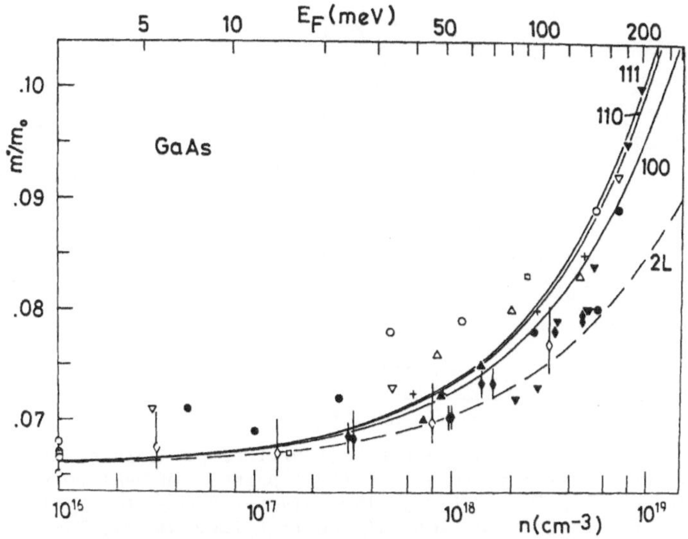

Fig.3. Cyclotron effective masses in GaAs calculated for three magnetic field directions using the five-level k·p model. Calculations for Q=0 and for the two-level (2L) model are also indicated, as well as experimental data of various authors (After Pfeffer and Zawadzki [12])

Fig. 4

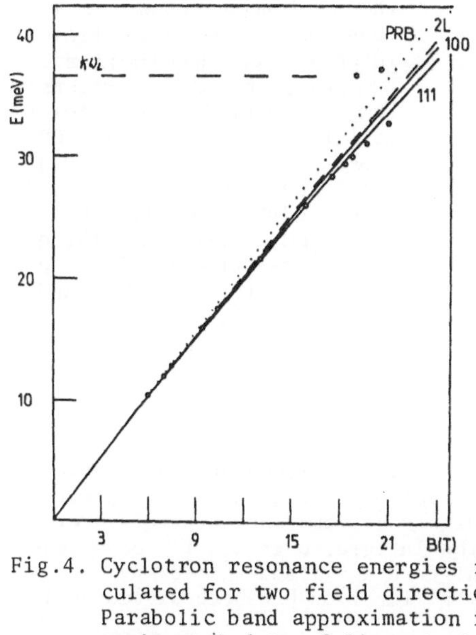

Fig. 5

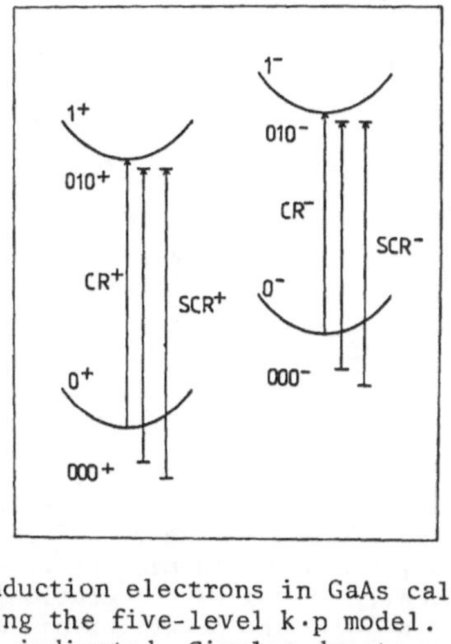

Fig.4. Cyclotron resonance energies for conduction electrons in GaAs calculated for two field directions using the five-level k·p model. Parabolic band approximation is also indicated. Circles denote experimental data of Sigg et al. [10] for $\vec{B}||(001)$. (After Pfeffer and Zawadzki [12])

Fig.5. Optical transitions of the cyclotron resonance and of the donor-shifted cyclotron resonance (for two donor species). Spin-up and spin-down transitions do not have the same energy due to band's nonparabolicity. (After Zawadzki et al. [14])

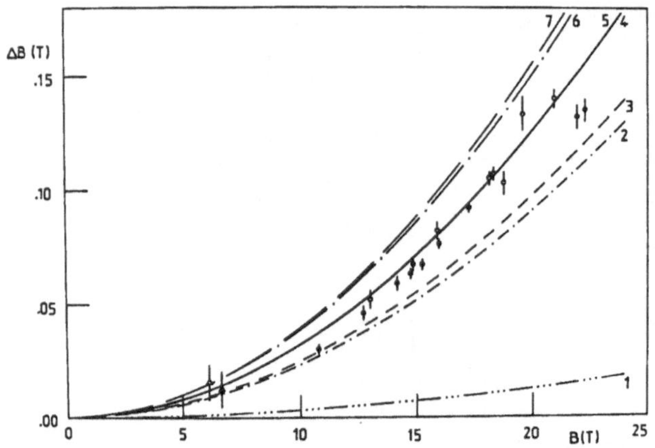

Fig.6. Spin doublet splitting of the cyclotron resonance (empty points) and of the donor-shifted cyclotron resonance (full points) in GaAs versus magnetic field intensity for $\vec{B}||(001)$. The curves 2-7 show theoretical calculations for various versions of the five-level model. The two-level theoretical results (1) are also indicated (After Zawadzki et al. [14])

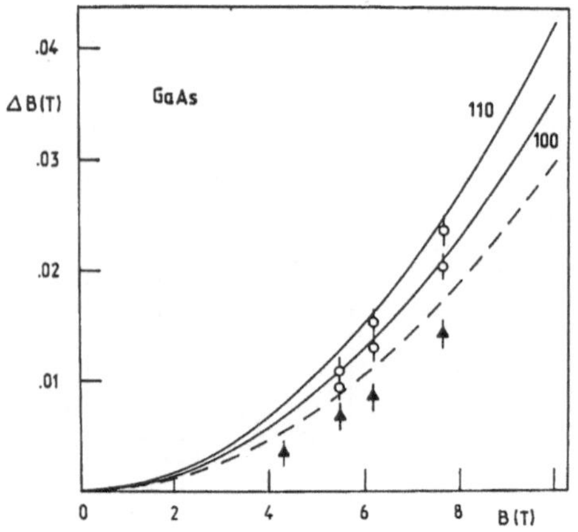

Fig.7. Spin doublet splittings of the cyclotron resonance in GaAs versus magnetic field intensity for $\vec{B}||(001)$ and $\vec{B}||(011)$ field directions (empty points) and field difference of average CR peaks for the above field directions (full points) after Golubev et al. [15]. Solid lines are calculated using the five-level $k \cdot p$ model. (After Pfeffer and Zawadzki [12])

Here V is the crystal volume, κ_∞ and κ_0 are the high-frequency and static dielectric constants, respectively. The summation over q indicated in Eq. (3) is performed in the cylindrical coordinate system. The integrals over q_z and ϕ can be performed analytically, the integral over $q_\perp$ is to be done numerically. Detailed calculations for the case $n = 1$ and $\Gamma_0 = 0$ may be found in [17], the behaviour of the spectral function has been discussed in [18].

We take the following values of the low-temperature parameters for GaAs: $\hbar\omega_L = 36.2$ meV and $\hbar\omega_T = 33.2$ meV (from Raman measurements [20], $\kappa_0 = 12.79$ [21] (a similar value of $\kappa_0 = 12.56$ has been determined in [22]), and $\kappa_\infty = 10.77$ (calculated, using the Lyddane-Sachs-Teller relation).

As to the unperturbed electron energies, appearing in Eqs. (3) and (4), we consider separately two cases. For $2 \leqslant n \leqslant 11$ the polaron resonances oc-

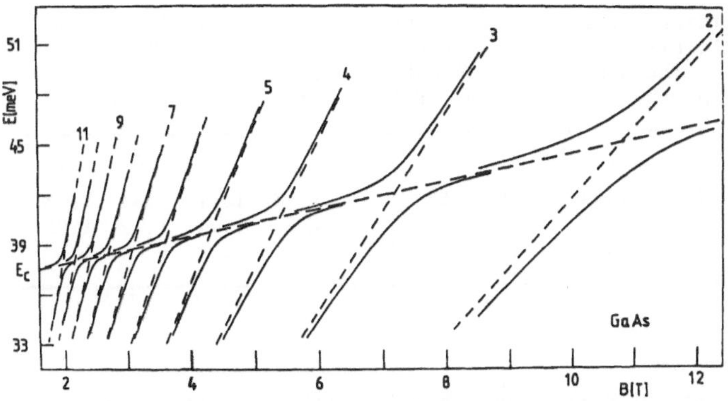

Fig.8. Calculated energies of resonant magneto-polarons in the conduction band of GaAs for the Landau states $2 \leq n \leq 11$ versus magnetic field, using the polar constant $\alpha=0.073$. The dashed lines indicate unperturbed magnetic energies. (After Pfeffer and Zawadzki [23]).

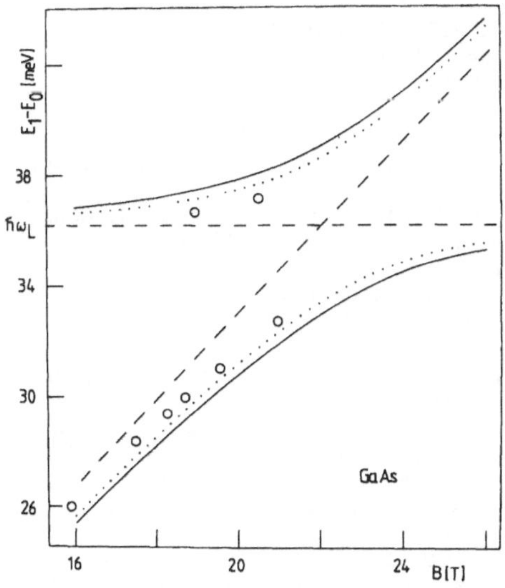

Fig.9. Calculated cyclotron energy $E_1^+ - E_0^+$ of a resonant magneto-polaron in GaAs versus magnetic field for B//(001). Solid lines - theory for dielectric parameters corresponding to the polar constant $\alpha=0.073$, pointed lines - theory for $\alpha=0.058$, circles - experimental data of Sigg et al [13]. Dashed lines indicate unperturbed magnetic energies, calculated according to the five-level k·p model. (After Pfeffer and Zawadzki [23])

cur at not too high magnetic fields and in the vicinity of resonance a parabolic-band approximation to the energies is acceptable. However, in order to describe the real field intensities at which the resonances are observed [16], we use the nonparabolic five-level k·p model. The calculated energies can be described by the parabolic approximation with slightly modified values of the effective mass and the Lande factor g^*. Thus we used the standard form: $E_n^+ = (\hbar eB/m^*)(n+1/2) + g^* \cdot \mu_B B/2$ with $m^* = 0.06905\ m_0$ and $g^* = -0.218$. For the case of n = 1 (cf. Fig.9) band's nonparabolicity becomes important also in the vicinity of the resonance and we have calculated E_0^+ and E_1^+ using directly the five-level k·p model for B//(001), with band parameters quoted in Ref. [14].

Table 2. Polaron splittings in GaAs at resonant magnetic fields for Landau states $1 \leq n \leq 11$. First entry - theoretical values for $\alpha = 0.073$, second entry - the same for $\alpha = 0.058$, third entry - experimental values after Hornung [16]. (After Pfeffer and Zawadzki [23])

n	1	2	3	4	5	6	7	8	9	10	11
th(0.073)	5.95	3.86	2.94	2.44	2.08	1.85	1.66	1.56	1.41	1.31	1.23
th(0.058)	5.13	3.30	2.52	2.08	1.79	1.58	1.42	1.30	1.21	1.12	1.05
exper.				2.16		1.46	1.31		1.13		1.1

Acknowledgments

We thank Prof. M.Cardona, Dr H.Sigg, Dr J.A.A.J.Perenboom, Miss Margaret Hopkins and Dr R.Nicholas for making available their work to us prior to publication.

References

1. C. Hermann and C. Weisbuch, Phys.Rev. B15, 823 (1977).
2. U. Rössler, Solid State Commun. 49, 943 (1984).
3. N.R. Ogg, Proc. Phys. Soc. 89, 431 (1966).
4. M. Braun and U. Rössler, J. Phys. C: Sol. St. Phys. 18, 3365 (1985).
5. W. Zawadzki, in "Narrow-gap Semiconductors. Physics and Applications" (Ed. W. Zawadzki), Lecture Notes in Physics, Vol. 133, Springer Verlag, (1979), p.85.
6. F.H. Pollak, C.W. Higginbotham, and M. Cardona, Proc. 11, Intern. Conf. Phys. Semicond., J. Phys. Soc. Japan 21 (Supplement), 20 (1966).
7. D.D. Sell, Phys. Rev. B6, 3750 (1972).
8. D.E. Aspnes and A.A. Studna, Phys. Rev. B7, 4605 (1973).
9. V. Evtuchov, Phys. Rev. 125, 1869 (1962).
10. M. Cardona, to be published.
11. H. Sigg, J.A.A.J. Perenboom, P. Pfeffer and W. Zawadzki, to be published.
12. P. Pfeffer and W. Zawadzki, to be published.
13. H. Sigg, H.J.A. Bluyssen, and P. Wyder, Sol. St. Commun. 48, 897 (1983).
14. W. Zawadzki, P. Pfeffer and H. Sigg, Solid State Commun. 53, 777 (1985).
15. V.G. Golubev et al. Pisma Zh. Eksp. Teor. Fiz. 40, 143 (1984).
16. T. Hornung "Bandkantenspektron von GaAs in Magnetfeld", PhD Thesis, University of Dortmund 1984, unpublished.
17. M. Nakayama, J. Phys. Soc. Japan, 27, 636 (1969).
18. L. Swierkowski and W. Zawadzki, J. Phys. Soc. Japan, Suppl. A., 49, 767 (1980).
19. J.P. Vigneron, R. Evrard and E. Kartheuser, Phys. Rev. B18, 6930 (1978).
20. U. Rohrer, "Raman-Spektroskopie an N-Dotiertem und Semiisolirendem GaAs", PhD Thesis, University of Munich (1982).
21. T. Lu, G.H. Glover and K.S. Champlin, Appl. Phys. Lett. 13, 404 (1968).
22. G.E. Stillman, D.M. Larsen, C.M. Wolfe and R.C. Brandt, Solid State Commun. 9, 2245.
23. P. Pfeffer and W. Zawadzki, Solid State Commun. 57, 897 (1986).

How Strong Are Resonant Polaron Effects in 2D and 3D Semiconductor Systems?

F. Malcher, G. Lommer, and U. Rössler

Institut für Theoretische Physik, Universität Regensburg, Postfach 397, D-8400 Regensburg, Fed. Rep. of Germany

Deviations of the cyclotron resonance energy from $\hbar\omega_B = \hbar eB/m^*c$, where m^* is the bulk electron mass, are calculated for bulk GaAs and AlGaAs/GaAs heterostructures and compared with previous theoretical and experimental results. We find that the resonant magneto-polaron effect becomes significant at higher magnetic fields in 2D than in 3D systems.

1. INTRODUCTION

The resonant magneto-polaron effect has been observed first in magneto-optic experiments on InSb /1/ at magnetic fields for which the electron cyclotron energy $\hbar\omega_c^*$ is in the region of the longitudinal-optical phonon energy $\hbar\omega_{LO}$. Due to the polar electron-phonon coupling the energy of the zero-phonon Landau level, $|0_{ph},N\rangle$, gets pinned for increasing magnetic field at the energy of the one-phonon Landau level, $|1_{ph},N'\rangle$ for $N'<N$. This pinning shifts the cyclotron resonance to a lower energy (corresponding to a larger cyclotron mass) than expected for an equidistant Landau ladder. Unfortunately, the nonparabolicity of energy bands causes a similar shift and experimental cyclotron data always contain the combined effect of electron-phonon coupling and nonparabolicity. Therefore, a theoretical interpretation of these data requires a correct treatment of either effect.

More recently a large amount of experimental data have become available for bulk-GaAs /2-6/ and for AlGaAs/GaAs heterostructures /5,7,8/ and have been accompanied by theoretical efforts with the aim of treating simultaneously nonparabolicity and resonant polaron aspects /2,5,9/. These theories make use of approximations in describing the nonparabolicity, the electron-phonon coupling, and the subband wave function in a heterostructure, but achieve agreement with experimental data by assuming e.g. too large values of the Fröhlich coupling constant /2/ or by using the electric field of the heterostructure as fitting parameter /5,9/, and thus eventually compensate possible errors of different origin. It is the purpose of this contribution to present quantitative results for nonparabolicity effects in Landau level energies of bulk GaAs and AlGaAs/GaAs heterostructures, which allow to ascribe the discrepancies between experimental data and these calculations to the magneto-polaron effect.

2. LANDAU LEVELS IN 2D AND 3D SEMICONDUCTORS

The aspect of nonparabolicity in the energy band dispersion and Landau level systems of semiconductors has been of current interest since KANE's early work on InSb /10/. It is known that a quantitative understanding of magneto-optic spectra in this narrow-gap material is possible only within a 8×8 model, which correctly includes not only the dominant coupling between conduction (Γ_{6c}) and valence band ($\Gamma_{8v}+\Gamma_{7v}$) but also higher band contributions /11,12/. In GaAs, for which the energetic separation of the lowest conduction band Γ_{6c} from the valence band $\Gamma_{8v}+\Gamma_{7v}$ is comparable with that from the higher conduction bands $\Gamma_{7c}+\Gamma_{8c}$, a 14×14 model turns out to be more appropriate than the 8×8 model /13,14/.

While these models consider multiband matrix-Hamiltonians with matrixelements bilinear in the wavevector components k_α, an alternative way to describe the conduction band dispersion and Landau levels is to diagonalize a single-band Hamiltonian with matrixelements containing higher order terms in k_α. Such a Hamiltonian, which without magnetic field is 2×2 because of spin, can be obtained by higher order perturbation theory from the extended 14×14 Kane-Hamiltonian /15/. The advantage of this model is its straightforward applicability to the problem of electron states in heterostructures /16/. If terms of sufficiently high order in k_α are considered, this 2×2 model should in principle yield the same conduction band Landau energies as the 8×8 model with remote band contributions /12/ or the 14×14 model /14/. Results for bulk GaAs and AlGaAs/GaAs heterostructures show, that a calculation including only up to fourth order terms in k_α overestimates the nonparabolicity by about 30%. If, however, the $\underline{k}\cdot\underline{p}$ coupling between Γ_{6c} and Γ_{8v} is included in all orders of k by using for this contribution the well-known expression

$$E(k) = - E_g/2 + \{(E_g/2)^2 + \frac{2}{3} P^2k^2\}^{1/2} , \qquad (1)$$

where $P=\hbar\langle S|p_x|x\rangle/m$ is Kane's matrixelement, than for bulk GaAs the Landau energies of the 8×8 and 2×2 models are almost identical and are in quantitative agreement also with the 14×14 model /14/. These results demonstrate the power of the 2×2 model, which can be considered as a state of the art nonparabolicity model.

In /2/ nonparabolicity has been considered in a k^4 approximation with contributions from Γ_{8v} and Γ_{7v} only, which for the energy difference of subsequent Landau levels yields

$$\Delta E_N = E_{N+1} - E_N = \hbar\omega_B - 2A(N+1)(\hbar\omega_B)^2 , \qquad (2)$$

where $\hbar\omega_B$ is the cyclotron energy calculated with the bulk mass. While for the parameters given in /2/ $2A=1.087eV^{-1}$, the k^4 version of our 2×2 model yields $2A=1.36eV^{-1}$, i.e. in /2/ the nonparabolicity coefficient is underestimated by 25%. It turns out, however, that this fact is accidentally compensated by the overestimated nonparabolicity of the k^4 approximation.

The nonparabolicity expressions used in /5,8/ turn out to overestimate the deviation of the Landau level spacing $\hbar\omega_c^*$ from $h\omega_B$ by about 10 to 15%.

3. COMPARISON WITH EXPERIMENTAL DATA

In Fig. 1 we compare experimental results for the deviation of the observed cyclotron resonance energies from $\hbar\omega_B$ in bulk GaAs /2,5/ with calculated data from our 2×2 model. For $\hbar\omega_B$ we used the bulk electron mass $m^*=0.066m_0$, which is to be considered as polaron mass. The parameters in our calculations are those of /13/ except for P=10.511eVÅ, P'=4.563eVÅ, C=-2.0, C'=-0.02 in order to be consistent with the bulk mass of $0.066m_0$. It should be mentioned that using a set of unrenormalized parameters with a bare electron mass of $0.0653m_0$ does not change the calculated results of Fig. 1. A close agreement

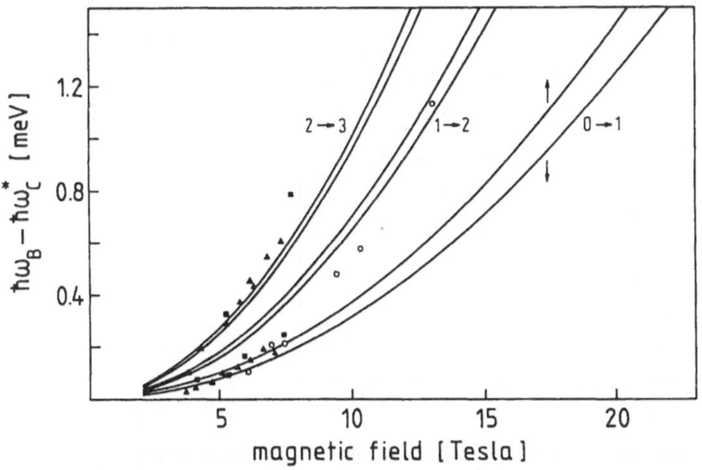

Fig. 1: Deviation of the cyclotron resonance energy $\hbar\omega_c{}^*$ from $\hbar eB/m^*c$ for $m^*=0.066m_0$ versus magnetic field. Experimental data from /2/ (full symbols, for 0→1 and 1→2 transitions) and from /5/ (open symbols for 0→1 transition) in comparison with calculated data (full lines, showing spin-splitting).

between theory and experiment for the 0→1 transition is observed in Fig. 1 up to 7 Tesla. Deviations for higher magnetic fields can be ascribed to the resonant magneto-polaron effect, which for the 1→2 transition is expected to dominate at about 8 Tesla. Unpublished data of NICHOLAS and HOPKINS /6/ for the 0→1 transition would coincide in Fig. 1 with our calculated results for the 1→2 transition; these data seem to be not consistent with the data of /2/ and /5/.

For AlGaAs/GaAs heterostructures we repeated our calculations reported in /16/ but included the higher order contributions of the Γ_{6c}-Γ_{8c} coupling (see (1)). This change turns out to lower the deviations from $\hbar\omega_B$ as shown in Fig. 2, where our previous (broken lines) and present results (solid lines) are compared with experimental data. As the transition between a certain pair of Landau levels can be observed only in a limited range of magnetic fields (indicated by the ends of the broken lines in Fig. 2) the observed cyclotron resonance energies at lower fields correspond to higher transitions which as in bulk GaAs are already significantly shifted due to the resonant magneto-polaron effect. For the 0→1 transition (observed above 8 Tesla) the relative agreement between theory and experiment persists up to about 15 Tesla above which the resonant polaron effect becomes important. The different energy scales in Fig. 1 and 2 demonstrate that nonparabolicity is much stronger in 2D than 3D systems. This is expected because cyclotron resonance in a degenerate 2D electron gas is determined by electron states well above the conduction band minimum. Experimental data from different sources /5/ and /8/, show remarkable scatter (Fig. 2).

4. CONCLUSIONS

Deviations of cyclotron resonance energies in 2D and 3D semiconductor systems are caused by band nonparabolicity and resonant magneto-polaron effects. We have separated these effects by comparing calculated cyclotron resonance en-

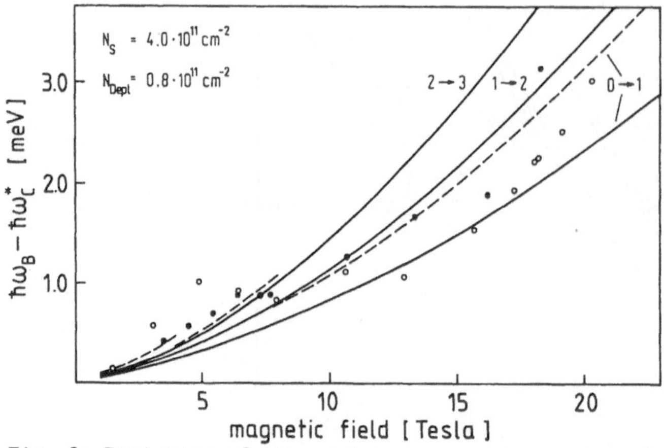

Fig. 2: Deviation of the cyclotron resonance energy $\hbar\omega_c{}^*$ from $\hbar eB/m^*c$ from $m^*=0.066m_0$ versus magnetic field. Experimental data from /5/ (open symbols) and from /8/ (full symbols) in comparison with calculated results in the k^4-approximation (broken lines) and including all higher orders according to (1) (full lines). The ends of the broken lines at about 4 Tesla and 8 Tesla mark the magnetic fields above which the 1→2 and 0→1 transitions can be observed, respectively.

ergies with experimental data and find that nonparabolicity is stronger in 2D than in 3D systems and that resonant polaron effects become significant at higher magnetic fields in heterostructures than in bulk material.

1. E.J. Johnson, D.M. Larsen: Phys. Rev. Lett. 16, 655 (1966)
2. G. Lindemann, R. Lassnig, W. Seidenbusch, E. Gornik: Phys. Rev. B28, 4693 (1983)
3. V.G. Golubev, V.I. Ivanov-Omskii, I.G. Minervin, A.V. Osutin, D.G. Polyakov: Sov. Phys. JETP 61, 1214 (1985)
4. T. Hornung, Thesis, Universität Dortmund 1984, unpublished
5. H. Sigg, P. Wyder, J.A.A. Perenboom: Phys. Rev. B31, 5253 (1985)
6. R. Nicholas, M. Hopkins, private communication
7. G. Lindemann, W. Seidenbusch, R. Lassnig, J. Edlinger, E. Gornik: Physica 117B&118B, 649 (1983)
8. M. Horst, U. Merkt, W. Zawadzki, J.C. Maan, K. Ploog: Solid State Commun. 53, 403 (1985)
9. W. Zawadzki: Solid State Commun. 56, 43 (1985)
10. E.O. Kane: J. Phys. Chem. Solids 1, 249 (1957)
11. C.R. Pidgeon, R.N. Brown: Phys. Rev. 146, 575 (1966)
12. H.R. Trebin, U. Rössler, R. Ranvaud: Phys. Rev. B20, 686 (1979)
13. U. Rössler: Solid State Commun. 49, 943 (1984)
14. W. Zawadzki, P. Pfeffer, H. Sigg: Solid State Commun. 53, 777 (1985)
15. M. Braun, U. Rössler: J. Phys. C18, 3365 (1985)
16. G. Lommer, F. Malcher, U. Rössler: Superlat. and Microstr. (in press)

Part IX

Very High Field Work

The State of the Art of Generating Pulsed Magnetic Fields

F. Herlach[1] *and N. Miura*[2]

[1]Laboratorium voor Lage Temperaturen on Hoge-Veldenfysika, Katholieke Universiteit Leuven, Celestijnenlaan 200 D, B-3030 Leuven, Belgium

[2]Institute for Solid State Physics, University of Tokyo, 7-22-1 Roppongi, Minato-ku, Tokyo 106, Japan

This is a general overview and an update of the previous review by MIURA and HERLACH [1] which contains more detail, in particular examples of many experiments and a complete list of references.

1. INTRODUCTION

The technology of water-cooled laboratory electromagnets has now reached a peak of its development. About 15 T can be generated in a 5 cm bore with 5 MW and ∿ 20 T with 10 MW [2]. In a hybrid magnet, about 10 T are added from a large external superconducting coil; ultimately this is aimed at generating d.c. fields of the order 35 T which may be considered as an uppermost limit for the foreseeable future. Much higher fields can be obtained by means of pulsed magnets; this development was initiated by P. Kapitza in the twenties and is still far from conclusion. The performance of water-cooled coils is mainly limited by the available power and by the cooling system which must remove the total power in the form of heat, typically by a stream of water of the order of 100 litres per second. In a pulsed magnet, Joule heat is absorbed by the heat capacity of the coil and the power supply is an energy storage device such as a capacitor bank, a flywheel or an inductor. In exceptional cases, pulsed power of the order 10 MW can be obtained directly from the mains, with power conditioning by thyristors and passive filters [3]. The most common type of power supply is a capacitor bank with a typical energy of 100 kJ and peak voltage in the range 3-10 kV for fields < 100 T and up to 40 kV for fields > 100 T. Exceptionally, capacitor banks up to 1 MJ are used for obtaining a longer pulse duration. The design of a suitable pulsed power supply is well within the state of the art of modern electrical engineering. In particular, the technology of high power semiconductor switching devices is now in rapid development and provides excellent devices for power conditioning.

The most severe problem in generating magnetic fields larger than 20-30 T is given by the magnetic stress. This amounts to 1 GPa (gigapascal) at 50 T, 4 GPa at 100 T and 100 GPa = 1 Mbar at 500 T = 5 megagauss. The energy density expressed in kJ/cm^3 is equal to the pressure in GPa. Given the fact that the strongest construction materials such as maraging steel and fiber composites have an ultimate tensile strength of the order of 2-3 GPa, it is evident that magnetic fields of more than 100 T cannot be contained in a stable mechanical structure. On basis of this criterion, pulsed magnetic fields can be divided loosely into three categories: "nondestructive" fields below ∿ 60 T with a relatively long pulse duration (10 ms - 1 s) determined by the energy stored in the power supply which must be matched by the heat capacity of the coil, "megagauss" fields > 100 T where destruction is inevitable and the pulse duration (< 5 μs) is limited by the propagation of shock waves through the confining structure, and finally an intermediate region between 60 T and 100 T where attempts are made to design nondestructive coils with partial inertial confinement and the pulse duration is of the order of 0.1 ms.

2. NONDESTRUCTIVE PULSED FIELDS

The mechanics of a nondestructive coil is similar to that of a vessel containing high pressure. However, by contrast to hydrostatic pressure which is applied only at the inner wall of the vessel, the magnetic force is distributed throughout the volume of the coil. The magnetic stress is proportional to the product of the current density which is constant in a wire-wound coil, and the magnetic field which decreases almost linearly from the inner to the outer radius. Although the magnetic stress is a decreasing function of the radius, the outer layers have a tendency for greater expansion because the total force exerted on a layer of the coil is proportional to the radius. This results easily in a negative radial stress whereby the outer layers of the coil exert a pull on the inner layers. Even a modest negative radial stress has the effect of strongly increasing the tangential stress at the inner radius [4]. A wire-wound coil consists of alternate layers of insulating and conducting material. This results in a complicated, inhomogeneous mechanical system. Therefore it is not feasible to make exact predictions of the mechanical behaviour under extreme magnetic stress which will drive the inner layers into the limit of plastic deformation [5]. The relative softness of the insulating material attenuates the transmission of radial stress. Therefore, a reasonable estimate of the peak field that can be supported by a wire-wound coil is given by the equation for the peak stress in a coil with free-standing windings:

$$B_{max} = \sqrt{2\mu_0\sigma}\,\sqrt{2}\,(1 - 1/\alpha), \tag{1}$$

where α is the ratio of the outer and the inner radius, σ the ultimate tensile strength and $\mu_0 = 4\pi\, 10^{-7}$ Vs/Am.

The electrical resistance of the coil increases during the field pulse due to ohmic heating. This has a strong influence on the pulse shape and on the peak field, in particular for pulses of long duration. Precooling of the coil to liquid nitrogen temperature not only increases the amount of heat that can be absorbed by the coil without damage to the insulation, it is essential in decreasing the resistance to allow the flow of high current for a given supply voltage. In practice, most coils are wound from copper wire with rectangular cross section to obtain a good filling factor and to avoid slipping of the wires against each other. For additional containment, the coils are potted in epoxy and tightly enclosed in a high-strength cylinder. The highest fields up to 68 T have been obtained with wire that incorporates a large number of fine niobium strands in a copper matrix [6, 7]. This type of wire was originally developed for the manufacture of superconducting coils; in pulsed field coils it is not used in its superconducting state but only for its mechanical strength. The record field of 68 T with a pulse duration of almost 10 ms was obtained by FONER [7] with a wire specifically designed for high strength.

A wire-wound coil connected to a capacitor bank of 100-200 kJ is an elegant and compact laboratory instrument to generate 40 T comfortably in a 20 mm bore with a pulse duration (half period) of 10 ms. Most modern capacitor banks for this application are switched by high-power thyristors (typical performance data are 4 kV, 25 kA single shot per thyristor with the possibility of series and parallel connection), some with trigger generators using fibre optics. At voltage reversal, the discharge is crowbarred by means of diodes with damping resistors. Thyristor switching is very smooth and reliable; the diode crowbar provides for a modest extension of the pulse duration and protects the capacitors from voltage reversal which would shorten their life expectation.

Nondestructive pulsed fields are presently available for research at several laboratories [1]. Many of these accommodate guest experiments on bases of mutual collaboration, in particular those at M.I.T., Osaka, Tokyo, Toulouse and Leuven. It is not really feasible to run a pulsed field laboratory as a general user's facility in the style of the big national d.c. magnet laboratories. Experimentation with pulsed fields requires much experience and special experimental techniques which are developed by the in-house research staff.

The Osaka laboratory is different from most others in that this facility has been set up with the goal of generating 100 T nondestructively. The coils consist of two or three concentric solid helices machined from maraging steel and designed for optimal sharing of the total magnetic force. The low inductance of these coils results in a short pulse duration of the order 0.1 ms. This makes it feasible to use a highly resistive material such as maraging steel without precooling (which would not be efficient anyhow because the residual resistance of this material is large). These coils can be heated to much higher temperatures than it is admissible for copper coils, provided that temperature-resistant insulating materials are used. So far, fields up to 60 T have been used in many experiments. Recent reports indicate that 80 T have been obtained with a triple helix. However, even if these coils could be made strong enough to contain a megagauss field, the temperature would become excessive above 80 T [8].

The Amsterdam magnet laboratory [3] is the prototype for a facility that uses pulsed power directly from the mains. While a capacitor discharge results in a damped sine wave, the pulse shaping by controlled rectification of three phase current provides more flexibility. In any case, the field must be rapidly increased to its peak value such that the peak field is not decreased by the resistance of the heated coil. The decrease of the field can be programmed in a number of steps. The coil is cooled by liquid neon. With the given supply voltage, this results in a higher current because of the smaller resistance, but the lowered heat conductivity results in cooling times of the order of three hours. A typical cooling time for a liquid nitrogen-cooled coil of average dimensions is 20 minutes. An alternative method to generate a field pulse with a flat top of a few milliseconds duration is the discharge of a series of capacitors and inductors that form a transmission line [10].

Since the publication of the previous review [1], two more pulsed field laboratories have come to our attention. At the University of Poznan (Poland), a coil has been designed which is called hybrid as it consists of three helices made from different materials, each connected to a separate capacitor bank [11]. The innermost coil with a 6 mm bore and 12 mm o.d. is connected to a 22.5 kJ, 3 kV bank, the intermediate coil is made from copper-beryllium with a 14 mm bore and 24 mm o.d. connected to 135 kJ at 6 kV and the outer helix is made of copper with a 26 mm bore and 50 mm o.d. and connected to 270 kJ at 6 kV. The peak field is 92 T with a rise time of 0.3 ms, a flattened top of 0.2 ms duration and a decay time of $\sim$ 1 ms. At the Technical University of Vienna, preliminary experiments have been carried out with a 25 kJ, 2.5 kV capacitor bank and coils made from copper wire with a reinforcement of fibre composites between the layers and on the outside. Peak fields were 26 T in liquid nitrogen and 28 T in liquid helium [12]. It is now planned to build a large pulsed field facility of the Amsterdam type.

3. MEGAGAUSS FIELDS

Megagauss fields are characterized by the high-energy density which becomes much larger than the energy density of chemical binding (as an example, take the heat of combustion of fossil fuels which is of the order of 40 kJ/cm^3). The effects of the high energy density come into play at the interface between the magnetic field and the conducting wall that confines it. As the field increases, electromagnetic energy flows into the conductor at a rate given by the Poynting vector

$$|\vec{E} \times \vec{H}| = v_f \mu_0 H^2, \qquad (2)$$

where $v_f = E/B$ is a speed that can be interpreted as describing the flow of magnetic flux in the direction perpendicular to the field lines. The form of this equation suggests that approximately one half of the energy is converted into Joule heat while the other half remains in the form of magnetic energy density $\mu_0 H^2/2$; this is true under the condition that the skin depth is small compared to the thickness of the conductor in the direction of the energy flow. The skin depth can be estimated by comparison to an exponential field rise which results in a field profile

$$B(x,t) = B_0 e^{\nu t} e^{-x/a} \quad \text{with the skin depth} \quad a = \sqrt{\rho/(\mu_0 \nu)}. \qquad (3)$$

It turns out that in coils of practical size the skin depth is indeed smaller than the conductor thickness. As a consequence, the Joule heating becomes independent of the resistivity and depends only on the specific heat per volume (which is of the order 3 J/(cm^{3}K) for most metals) and the square of the magnetic field. For example, the melting point of copper is reached at 110 T and the boiling point at 150 T. Beginning at the surface, the conductor will thus melt and vaporize at such a rate that it literally explodes. This explosion proceeds into the material at a speed of the order km/s which is related to the rate of energy input and the heat of vaporization [13]. This sets a limit to the duration of the field pulse but there is another effect which forces the pulse duration to be even shorter: the volume compression of the conductor material by the magnetic stress. Due to the rapid increase of the field already dictated by the Joule heating, the compression proceeds into the conductor material as a shock wave, and the wall recedes from the magnetic field at the speed of the medium (the "particle speed") behind the shock wave. The relations between particle speed, shock speed and pressure are well known as they represent the equation of state of the material [1]. It is evident that this effect provides a natural limit for the pulse duration of the order of a few microseconds or less.

In practice, two methods are in use to generate megagauss fields for experimental applications: the direct discharge of electromagnetic energy into a small single turn coil and magnetic flux compression by the rapid implosion of a conducting shell. As primary energy sources, capacitor banks and high explosives have been used for both methods. Although high explosives are much more reliable and easier to use than most people would believe, the capacitor discharge is of course the preferred method for laboratory use. With single turn coils and a 40 kV, 100 kJ capacitor bank, 150 T have been reproducibly generated in 10 mm diameter with a rise time of 2 μs and 250 T in 4 mm diameter [14]. In a single turn coil, the magnetic field protects the sample against destruction by the exploding conductor, while in an implosion system the sample is always destroyed by impact shortly after peak field. With explosive-driven flux compression, fields up to 1000 T have been generated for experimental applications [15].

In a flux compression experiment, the inner wall of the imploding conductor is decelerated by shock compression i.e. the implosion speed is diminished by the particle speed in the shock wave. Thus, the achievable peak field is directly related to the implosion speed - approximately 3 km/s for 500 T and 5 km/s for 1000 T [1, 16]. This is close to the upper limit that can be achieved in cylindrical implosions driven by massive charges of high explosive. This limitation is due to the finite speed of sound in the detonation products which sets a limit to the energy transfer from the explosive to the metal "liner" (the imploding cylinder is a lining to the explosive charge). Electromagnetic acceleration does not suffer from this limitation as magnetic energy can propagate at the speed of light. Many practical problems remain to be solved until this technique will be ready to match and eventually surpass the explosive method. Even now, electromagnetic flux compression is a useful tool for laboratory experiments with megagauss fields. This has been demonstrated at the Institute for Solid State Physics (The University of Tokyo) where a large experimental facility has been established for this purpose [17].

There has been much speculation about different methods to generate magnetic fields up to 10^4 T, mostly involving plasma implosions or high power lasers. So far, none of these has yielded substantial results. Recently, one technique has emerged which may hold some promise for development into a practical device. If a high power CO_2 laser is focused on a metal plate, a puff of plasma is emitted which contains very hot electrons. Very high current with an extremely short rise time of the order nanosecond can thus be induced in an external circuit connected to a parallel plate capacitor consisting of this metal plate and a second plate at a small distance (< 1 mm), with a hole for the laser beam in second plate. Multimegagauss fields could easily be obtained by connecting a small single turn coil to this capacitor, because the electrical energy is generated on the spot with a rise time at least 1000 times shorter than in a conventional fast capacitor discharge or in an implosion device. The problem is in the limited energy presently available from pulsed high power lasers. Unless this can be increased by at least one or more orders of magnitude, the highest fields would only be available in extremely small volumes with linear dimensions of the order 0.1 mm.

4. EXPERIMENTS

It may appear difficult to do experiments in megagauss fields, and it certainly is! The useful volume is not only small, it is directly surrounded by the violent phenomena due to the interaction of the megagauss field with the confining conductor. The rapid variation of the magnetic field is accompanied by electric fields of the order 10^6 V/m which interfere with electrical measurements and cause eddy currents resulting in sample heating. However, it has been demonstrated that suitable experimental techniques can be developed and that optical experiments in particular are well suited [17]. With megagauss fields, resonances related to electrons are shifted into the range of 10 μm or smaller which is covered by the CO_2 and CO lasers. These experiments have been extended even into the temperature range of liquid helium. For most solid state experiments, it is indeed the combination of high fields and low temperatures which brings out the interesting effects. One important aspect of experimental techniques for use in pulsed fields is the miniaturization of the experimental setup. The techniques developed by the electronics industry have barely been used for this purpose; their application, together with other innovative methods, will greatly enhance the potential of experiments with pulsed fields.

There is one problem that must be given careful consideration in all pulsed field experiments. This is the heating of conducting samples by eddy currents. If the skin depth is large compared to the sample thickness, the Joule heating can be estimated by assuming a current distribution which is a linear function of the distance from the center of the sample where the eddy current density is zero. The spatial average of Joule heating power per volume V is given by

$$\frac{1}{V}\frac{d\overline{W}}{dt} = \frac{d^2}{12\rho}\left(\frac{dB}{dt}\right)^2 , \tag{4}$$

where d is the sample thickness perpendicular to the field and ρ the resistivity. If the skin depth becomes small compared to the sample thickness, the heating is concentrated at the surface as it has been discussed for the coils, and in addition the sample is subjected to mechanical stress resulting from the magnetic stress (Maxwell tensor) which is converted into mechanical stress within the skin depth.

The single turn coil is most attractive for experimental applications because of its simplicity and ease of operation, the survival of the sample, the relatively large field volume, and the sinusoidal waveform which permits the taking of data for both increasing and decreasing field. However, this technique is presently limited at 300 T and optimistic estimates for the foreseeable future would not dare to go beyond 500 T. For experiments with higher fields, an implosion technique must be used. On the whole this is more difficult but for some experiments the different waveform is advantageous. With the single turn coil, the derivative of the waveform jumps to its highest value at the beginning of the field pulse, and peak field occurs already about two microseconds after this sharp voltage spike. As eddy current heating is proportional to $(dB/dt)^2$ this can cause problems with samples of high conductivity. Electrical measurements may be strongly disturbed by transients induced in the recording equipment by the initial voltage spike. These specific problems are almost absent in an implosion experiment where the derivative dB/dt increases progressively from zero in an interval of the order of 10 μs or longer, which is long enough for the initial transients to be effectively attenuated. Examples of the waveforms are shown in Figs. 1 and 2 [19].

For most experiments, a precise measurement of the magnetic field is required. This is an easy matter in a pulsed magnetic field: the voltage induced in a simple pick-up coil is exactly proportional to the derivative of the magnetic field. This can be integrated either electronically or numerically, or by a combination of both methods. This field measurement can be precisely calibrated by comparison to a resonance with a well-known g-factor such as the electron spin resonance in DPPH [20]. This has a g-factor of 2.0037 and a linewidth of 30 mT at liquid helium temperature. The electron spin resonance of ruby was pursued into the megagauss range and the

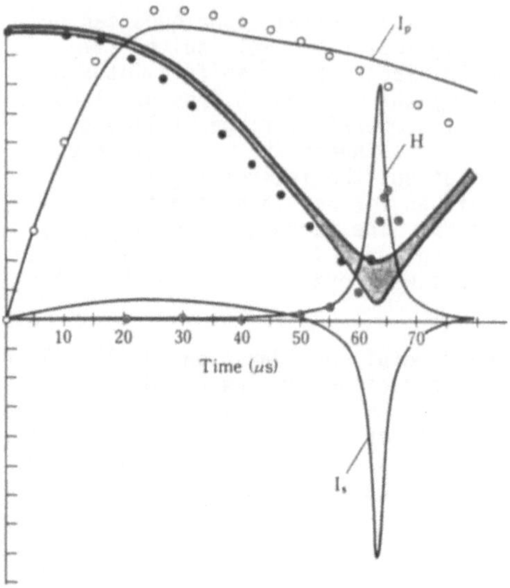

Fig. 1 Waveforms from an electromagnetic flux compression experiment. The solid lines are from a computer simulation.
Maxima: (experimental / calculated)

Experimental points:
○ primary current (3.5 / 3.28 MA)
● inner radius of the liner
◉ compressed field (320 / 590 T)
capacitance 5.01 mF
voltage 33 kV
bank inductance 53.9 nH
bank resistance 1.23 mΩ
initial field 2.78 T
driving coil radius 80 mm
length 60 mm
liner radius 75 mm
length 70 mm
thickness 1.5 mm

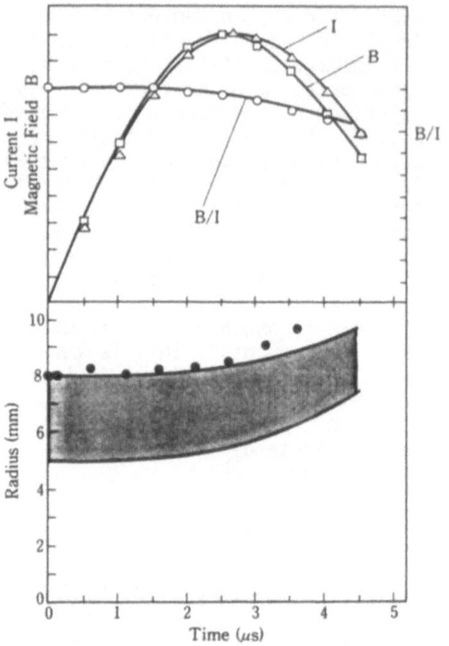

Fig. 2 Waveforms from a single turn coil. The solid lines are from a computer simulation.
Maxima: (experimental / calculated)
Experimental points:
□ magnetic field B (152 / 158 T)
△ current I (2.07 / 2.17 MA)
○ B/I ratio (80 / 75 T/MA)
● outer radius of the coil
capacitance 131 μF
voltage 40 kV
bank inductance 23 nH
bank resistance 2.96 mΩ
coil inner radius 5 mm
wall thickness 3 mm
axial length 10 mm

g-factor was found to be constant and thus well suited for calibration purposes in this range [21].

Pulsed fields are still not much in general use for experiments in physics. Most researchers have felt that there were enough experiments of interest to be done in the conveniently available d.c. fields. As this is now reaching a point of saturation and at the same time the pulsed field technology is coming out of its infancy, a gradual shift of research interests to pulsed fields can be expected. Eventually, this will pave the way to general experimentation with megagauss fields.

REFERENCES

1. N. Miura, F. Herlach: In [2], chapter 6
2. F. Herlach, ed.: "Strong and Ultrastrong Magnetic Fields and Their Applications", Topics Appl. Phys. Vol. 57 (Springer, Berlin, Heidelberg 1985)
3. R. Gersdorf, F.A. Muller, L.W. Roeland: Rev. Sci. Instrum. 36,1100 (1965)
4. J. Witters, F. Herlach: J. Phys. D16, 255 (1983)
5. F. Herlach, G. De Vos, J. Witters: Proc. 8th Int. Conf. on Magnet Technology 1983, J. Physique 45, supplément C1-915 (1984)
6. V.I. Ozhogin, K.G. Gurtovoj, A.S. Lagutin: In "High Field Magnetism", M. Date, ed. (North-Holland Amsterdam 1983), p. 267
7. S. Foner, E. Bobrov: to be published in [8]
8. Proc. 4th International Conference on Megagauss Magnetic Field Generation and Related Topics, 14-17 July 1986, Santa Fe, N. Mex. (to be published by Plenum Press, New York)
9. E.S. Bobrov, J.E.C. Williams, P.J. Raboin: to be published in [8]
10. N. Miura, T. Goto, K. Nakao, S. Takeyama, T. Sakakibara, F. Herlach: J. Magnetism & Magn. Mater. 54-57, 1409 (1986)
11. M. Surma: Journal de Physique supplement au No. 1, Vol. 45, p. C1-45 (1984); P. Czarnecki, M. Surma: Rev. Sci. Instrum. 54, 1202 (1983)
12. S. von Gründorfer, H. Fillunger, R. Grössinger: ELIN-Zeitschrift, Heft 1/2, p. 44 (1985)
13. A.R. Bryant: In "Proc. Int. Conf. on Megagauss Magnetic Field Generation by Explosives and Related Experiments", Frascati, Sept. 21-23, 1965, H. Knoepfel and F. Herlach, eds. (EURATOM Brussels 1966)
14. K. Nakao, F. Herlach, T. Goto, S. Takeyama, T. Sakakibara, N. Miura: J. Phys. E18, 1018 (1985)
15. A.I. Pavlovskii, N.P. Kolokolchikov, O.M. Tatsenko, A.I. Bykov, M.I. Dolotenko, A.A. Karpikov: In "Megagauss Physics and Technology", Proc. 2nd Int. Conf. on Megagauss Magnetic Field Generation and Related Topics, Washington, D.C., May 30 - June 1, 1979, ed. P. Turchi (Plenum, New York 1980) p. 627
16. F. Herlach: to be published in [8]
17. N. Miura: In this volume
18. H. Daido, F. Miki, K. Mima, M. Fujita, K. Sawai, H. Fujita, Y. Kitagawa, S. Nakai, C. Yamanaka: Phys. Rev. Lett. 56, 846 (1986)
19. N. Miura, K. Nakao: to be published in [8]
20. F. Herlach, P. de Groot, P. Janssen, G. De Vos, J. Witters: In "High Field Magnetism", M. Date, ed. (North Holland, Amsterdam 1983) p.229
21. G. Kido, N. Miura: Appl. Phys. Lett. 41, 569 (1982)

Note: A collection of papers describing most of the active magnet laboratories is contained in "High Field Magnetism", M. Date, ed. (North-Holland, Amsterdam 1983)

Recent Investigations on Semiconductors in the 100 Tesla Range

N. Miura

Institute for Solid State Physics, University of Tokyo, 7-22-1 Roppongi, Minato-ku, Tokyo 106, Japan

A review is presented of the recent investigations of semiconductor physics in very high magnetic fields up to a few megagauss. It includes magneto-optical spectroscopy of excitons in AlGaAs-AlAs alloy quantum wells, anthracene and BiI_3, cyclotron resonance in n- and p-type GaAs-AlGaAs superlattices and n-PbTe, and a magnetic field-induced SDW transition in an organic superconductor $(TMTSF)_2ClO_4$.

1. INTRODUCTION

Recent progress of high magnetic field technology has greatly extended the available field range for solid state research. At present, steady fields up to 30T can be produced by hybrid magnets, and pulsed fields in the megagauss range (B>100T) can be routinely used for solid state experiments/1/. Megagauss fields are in a special field range which can be reached only by destructive methods. Despite various experimental difficulties, a variety of solid state experiments can be performed by properly designing the experimental apparatus /2/. Application of megagauss fields in semiconductor physics is very useful as well as in other areas of solid state physics, because they give rise to an enormous effect on the electronic state in solids. For example, the energy of cyclotron motion $\hbar\omega_c$ becomes 11.6 meV or 134K at 100T for free electrons. In usual semiconductors, the effective mass m^* is smaller than the free electron mass m_0, so that this energy becomes even larger by a factor of m_0/m^*. With such a large cyclotron energy, the effect of magnetic fields may exceed the effects of various interactions or excitation energies in semiconductors. Thus new phenomena such as various non-linearities or phase transitions are expected to occur.

At ISSP of the University of Tokyo, pulsed high magnetic fields up to several megagauss are generated by various techniques. Fig.1 shows schematic illustrations of three magnet systems presently employed at ISSP and typical waveforms of the magnetic field produced by each system. First, let us make a brief survey of the characteristics of each field. Fig.1(a) illustrates the coil configuration of the electromagnetic flux compression, and (b) shows the profile of the primary current and the magnetic field. This method provides really high fields of more than 300T, although it is very destructive in nature /1-3/. The rise time of the field is rather long at the beginning including that of the seed field, but the field rise slope gradually increases and becomes steepest near the highest field. This type of field profile is sometimes convenient for the measurement in conducting substances, because the sample magnetoresistance is increased with a slow rise time before it is exposed to a large dB/dt, thus avoiding excess Joule heating /4/. Fig.1(c) shows the coil for the single-turn coil technique and (d) shows the waveforms of the current I and field B for the standard coil with an inner diameter of 10 mm/5/. The waveforms are almost sinusoidal and

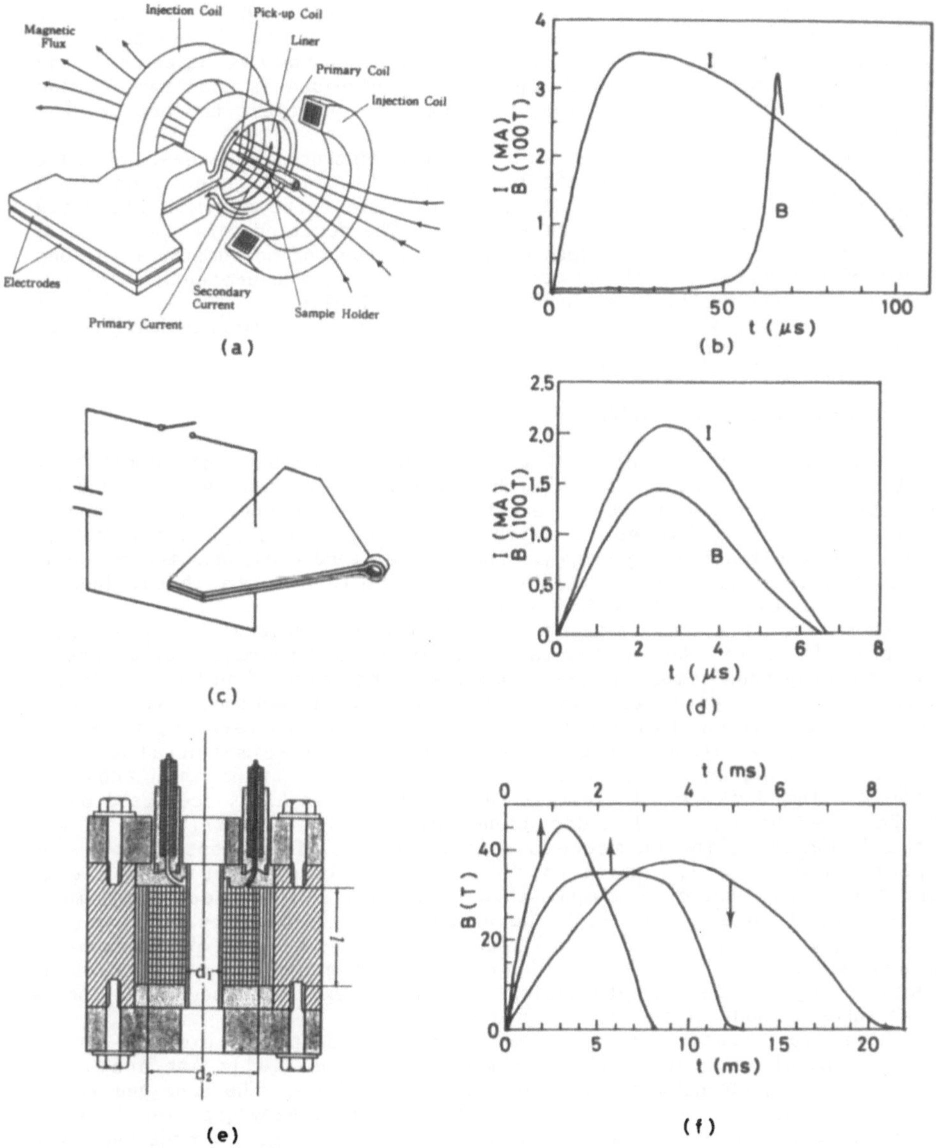

Fig.1. Various magnet systems currently used at ISSP, and the profiles of the magnetic field produced by each system : (a)(b) Electro magnetic flux compression, (c)(d) Single turn coil system, (e)(f) Non-destructive long pulse magnet for submegagauss fields. The primary current I is shown in (b). The current in the single turn coil I is shown in (d). In (f), various waveforms are shown including the flat-top pulse field produced by a PFN type condenser bank /1/.

the field change becomes slowest near the maximum field, which is convenient for increasing the accuracy of the measurement. Moreover, we can carry out two measurements at the same field in one pulse on the rising and falling field slopes. The greatest advantage of this technique is that samples and probes are not destroyed every time, so that we can repeat the measurement on the same sample.

Fig.1(e) shows a cross-section of the non-destructive wire-wound pulse magnet, and (f) shows the waveforms of magnetic fields generated using different sizes of coils and condenser bank /1,6/. The maximum field can reach 40-45T depending on the duration time. The duration time (half cycle) is as long as 20ms for the largest coil. These types of non-destructive long pulse fields are very useful for accurate preparatory measurements which are required for investigations in higher megagauss fields.

In this paper, we review recent investigations of semiconductor physics by means of these high fields.

2. MAGNETO-OPTICS OF EXCITONS

Magneto-optical spectra of excitons have long been of interest in relation to a fundamental problem of hydrogen atoms in high magnetic fields. In particular, for Wannier-type excitons, very high magnetic fields enable a condition to be achieved in which the energy of the cyclotron motion exceeds the exciton binding energy. For Frenkel-type excitons, the magnetic field effect is usually small, but considerable effects should become observable in the high magnetic field range.

In semiconductor multi-quantum wells, excitons have a two-dimensional character. It is well known that the two-dimensional hydrogen atom has a binding energy four times larger than the three-dimensional case. As the width L_z of the quantum well is reduced, the two-dimensional character is enhanced, so that the binding energy is increased. In large L_z limit, on the other hand, excitons in quantum wells tend to three-dimensional excitons with the binding energy of the bulk excitons. The transition from the three- to two-dimensional excitons should occur when L_z is decreased below the Bohr radius a_B. Such a dependence of the dimensionality on L_z was actually observed in the magneto-optical spectra of excitons in GaAs-AlAs superlattices /7/. The quantitative L_z dependence of the binding energy cannot be explained by a simple theory which assumes a heavy hole mass neglecting the heavy hole-light hole coupling. In reality, the valence bands in quantum wells have complicated dispersions because of the coupling between the heavy and light hole bands /8,9/. Therefore, analysis of the binding energy or the magneto-optical spectra of excitons requires a theory which takes account of the coupling properly.

Here, we discuss first the magneto-optical spectra of excitons in $Al_x Ga_{1-x}As$-AlAs alloy quantum wells. The alloy quantum wells are promising material for semiconductor lasers in the visible range. The band gap at the Γ point increases as x is increased. The indirect band gaps for X and L minima also increase with increasing x, but the direct gap exceeds the X and L gaps at x = 0.43 and x = 0.48, respectively. For larger x, the alloy becomes an indirect semiconductor. At zero field, the exciton absorption peaks were descernible up to x = 0.51/10,11/. In high magnetic fields, oscillatory magneto-absorption was observed as well as the shift of the ground state exciton levels. The binding energy of heavy hole excitons deduced from the magneto-optical spectra increases with increasing x because of the effective mass increase /10/. For the same reason, the diamagnetic shift of the ground state decreases with increasing x. It was found that these x dependences cannot be explained quantitatively by the theoretical model of excitons in quantum wells, if we assume that the x dependence of

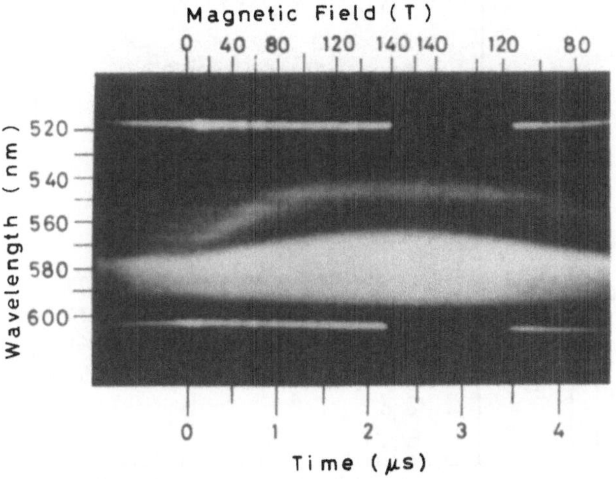

Fig.2. Streak picture of the magneto-absorption spectra of excitons in $Al_{0.51}Ga_{0.49}As$-AlAs alloy quantum wells. T = 41 K. L_z = 90 A.

the reduced mass is the same as in the bulk /10/. This result implies that the reduced mass in quantum wells is not simple because of the valence band degeneracy /8/.

In low magnetic fields, the ground state exciton wave function should be similar to that of a hydrogen atom, with an exponential function, and the diamagnetic shift is quadratic with magnetic field due to the H^2 term in the Hamiltonian. In the high magnetic field limit, the wave function tends to a Gaussian function. Then the diamagnetic shift becomes Landau level-like, as the Coulomb term becomes less important in comparison to the magnetic term. Thus the shift becomes a linear function of field. Such a cross-over of the diamagnetic shift from H^2- to H- dependence should occur near the field where the cyclotron motion energy becomes comparable with the binding energy. The diamagnetic shift of the heavy hole exciton ground state was measured in megagauss fields by using the single turn coil system and a streak spectrometer which comprises a monochromator and an image converter camera /11/. A typical example of the streak pictures of the magneto-optical spectra for a sample with x = 0.51 is shown in Fig.2. The ground state line of heavy hole excitons as well as the first excited state are clearly seen to shift to the high energy side as the field is increased. Such streak photographs are analyzed by a micro-photo-densitometer using a CCD camera which enables the two-dimensional data to be digitized and fed into a computer. An example of the micro-photo-densitometer traces is shown in Fig. 3 for a sample with x = 0.51. In such figures, we can obtain the diamagnetic shift at various magnetic fields.

In all the measured samples, the diamagnetic shift was found to be a linear function of magnetic field in the high field range. Fig.4 shows the diamagnetic shift of the heavy hole exciton as a function of magnetic field for various samples with different x but nearly the same value of well width $L_z \simeq$ 90A. It can be seen that in the low field range, the diamagnetic shift is almost proportional to H^2, but it gradually tends to show a H linear dependence in the high field range. In the low field region, the magnitude of the H^2 diamagnetic shift varies from sample to sample appreciably depending on x via the dependence on the reduced mass of

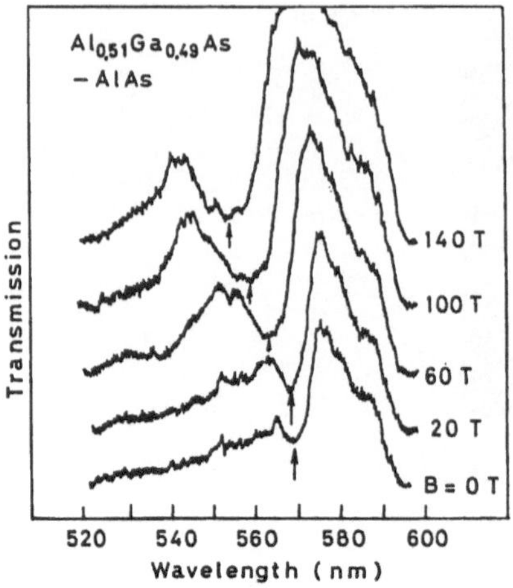

Fig.3. Micro-photo-densitometer traces of the streak spectra for a sample with x = 0.51.

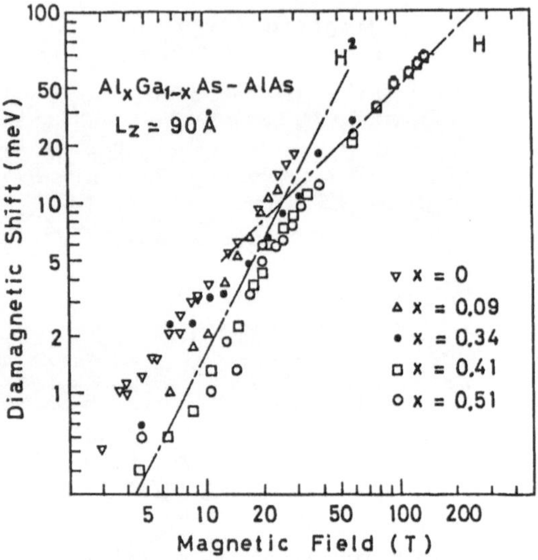

Fig.4. Diamagnetic shift of heavy hole excitons v.s. magnetic field for various samples with different x. $L_z \simeq 90$A.

excitons. In the H linear region, however, the dependence becomes smaller because of the smaller dependence of the shift on the reduced mass μ in the linear region than in the quadratic region. The overall field dependence of the diamagnetic shift should be analyzed taking account of the complexity of the magnetic energy levels in the valence band.

In an x = 0.51 sample, a considerable broadening or increase of the absorption peak occurred in high magnetic fields. It is not clear at this moment, whether this is due to the effect of the indirect gap.

In contrast to Wannier-excitons, Frenkel-type excitons which are generally observed in molecular crystals are localized within each molecule, and shows a very small effect of the magnetic field. However, sufficiently high magnetic fields cause significant effects. Such an example is seen in magneto-optical spectra of excitons in anthracene. Anthracene is a molecular crystal which has two molecules in a unit cell. A dipole-dipole interaction causes a so-called Davydov splitting of the exciton band for each k-vector into two bands with different polarizations. For two different incident light polarizations E // a and E // b, exciton absorption peaks are observed at different energies. Takeyama et al. made the first observation of the magneto-optic effect of the excions in anthracene in megagauss fields /12/. Fig.5 shows the streak spectra for E // a. In the middle of the spectra, the 0-0 exciton line is distinctly observed. It shows a small shift to the low energy side in high fields. It was found that the 0-0 line for E // b shows a shift in the opposite direction (high energy side), resulting in the decrease of the Davydov splitting as the field is increased. Although the origin has not been clarified at present, the significant effect of magnetic field on the Davydov splitting is an interesting problem to be solved in relation to the magnetic field effect on organic molecules. In addition, the spectra in Fig.5 shows a brighter background in the high field region, suggesting field-enhanced luminescence.

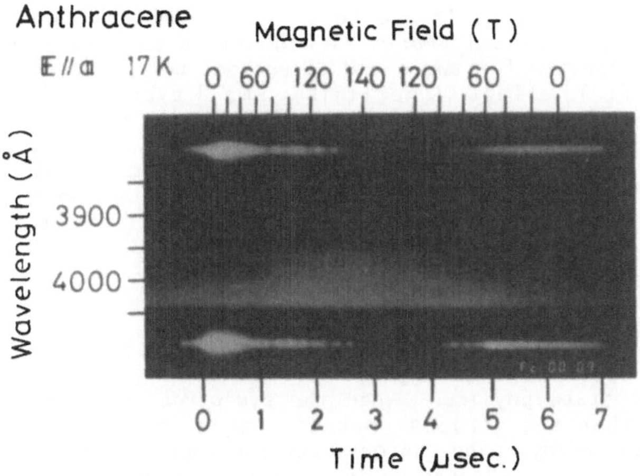

Fig.5. Streak spectra of magneto-absorption in anthracene. E // a. T = 17 K

In some layered semiconductors, excitons have Wannier-type character, but their wave function is localized in a small radius. In BiI_3, very sharp absorption lines called Q, R, S and T are observed from higher energy in the vicinity of indirect exciton absorption edge. Since these lines form an inverse hydrogenic series, Gross et al. proposed a bielectron model as their origin /13/. However, more detailed study showed that stacking faults in the crystal are responsible for the lines /14/. These stacking fault excitons have very small radius for the internal motion of electrons and holes, due to the cationic nature of the band edge excitons and the perturbation by the stacking faults. Komatsu et al. performed a magneto-

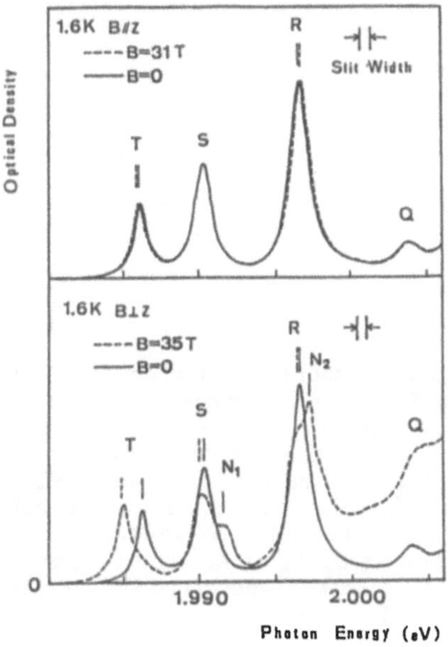

Fig.6. Absorption spectra of stacking fault excitons in BiI_3 in the presence and absence of magnetic field. (a) B // z (b) B ⊥ z.

optical measurement of the stacking fault excitons in BiI_3 /15/. Fig.6 shows a typical example of experimental traces of high field spectra as compared with the zero field spectra for two field directions B // z and B z. For B // z, the magnetic field effect is negligibly small even at 31T. For B⊥z, magnetic fields give rise to a considerable change in the spectra, including the shift of the lines and the appearance of new lines N_1 and N_2. The energy shifts of the lines show quadratic dependences on magnetic field. This behavior can be explained consistently with the cationic exciton model combined with the effect of the stacking fault. Measurements in megagauss fields are in progress and preliminary results have been obtained for the line shift for B // z.

3. CYCLOTRON RESONANCE

Infrared cyclotron resonance in megagauss fields is a powerful experimental means for solid state physics. When the cyclotron resonance energy $\hbar\omega_c$ becomes larger than the various characteristic energies in semiconductors, e.g. LO phonon energy, band gap or a certain energy barrier, new aspects of the magnetic energy levels can be investigated. Examples are polaron cyclotron resonance in CdS and CdSe, cyclotron resonance in a highly non-parabolic band in InSb etc., cyclotron resonance involving the magnetic break-through effect in GaP and so forth /16/.

Landau levels in GaAs-AlGaAs superlattices are of interest in terms of the magnetic field effect on quasi-two dimensional systems. Fig.7 shows experimental traces of the infrared cyclotron resonance in megagauss fields for n- and p-type GaAs-AlGaAs superlattices and n-type PbTe. For n - GaAs - AlGaAs superlattices, collaborative work with Kido and Sakaki is in progress. In modulation doped samples, the effective mass was investigated

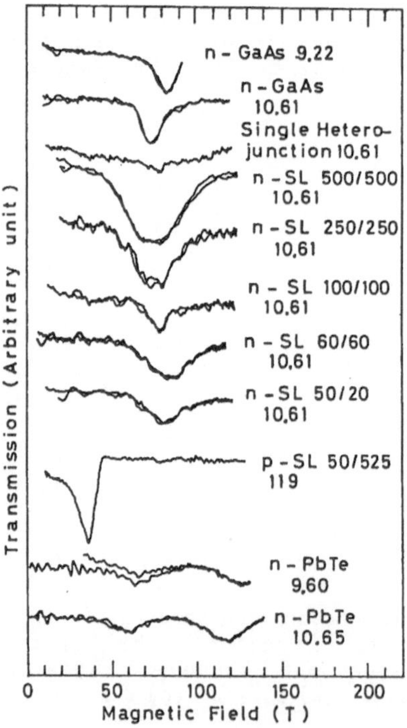

Fig.7. Experimental recordings of the infrared cyclotron resonance in megagauss fields. The wavelength of the measurement is indicated in μm for each curve. For GaAs-AlGaAs superlattices (SL), the well width L_z and the barreir width L_B are shown as L_z/L_B in Å. The temperature is room temperature except for p-SL (T = 13K).

for different quantum well widths L_z. A CO_2 laser was used to obtain radiation in the wavelength range between 9.2 and 10.9 μm. As L_z is reduced, the effective mass increases because of the non-parabolicity, being characteristic of the two-dimensional nature of the subband.

In p-type samples, the cyclotron resonance is more complicated because of the complex Landau level structure in the valence band. A collaborative work with NTT group has been done on the quantum cyclotron resonance in p-GaAs-AlGaAs superlattices in the submegagauss range /9/. With fixed frequencies in the far-infrared range, many absorption peaks were observed in different field positions, corresponding to the transitions between different combinations of the initial and final Landau states. The magnetic field dependence of the observed absorption lines showed a complicated behavior including bending or repulsion. The overall features of these these data were successfully explained by a recent calculation of Ando /9,17/. It was concluded that the 60 - 40 % rule of the band discontinuity is more favorable than the 85 - 15 % rule to explain the experimental results. In Fig.7 is shown an experimental trace up to the megagauss range for a sample with L_z = 50 Å and $p = 1.6 \times 10^{11}$ cm^{-2}. An H_2O laser was used as a radiation source for a wavelength of 119 μm ($\hbar\omega$ = 10.4 meV). A resonance peak was observed at B = 36T and no other peaks were seen in higher fields. The observed peak corresponds to the main transition 1a → 2a, as shown in Fig.8. This transition between the lowest Landau levels reflects the dispersion of the lowest valence sub-band formed in the quantum well. At low fields (low hole energy), the resonant photon energy increases almost linearly with increasing field. In the field range about 10 - 30T, the $\hbar\omega$ vs. B curve bends downwards indicating the increase of the mass due to the large coupling between heavy and light hole sub-bands. In higher fields than 30T, the mass starts to decrease again, as the energy range gets out of the strong coupling regime. Such a field dependence of the mass was confirmed by the high field experiment /17/.

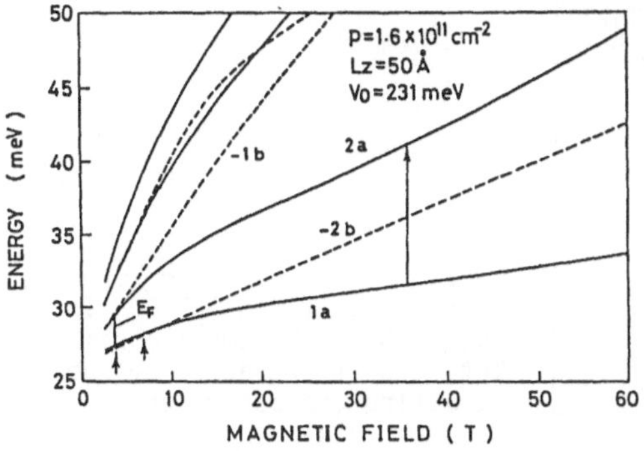

Fig.8. Landau levels in p-type GaAs-AlGaAs quantum wells and the cyclotron resonance transition at = 119 μm. The lines were calculated for $p = 1.6 \times 10^{11}$ cm^{-2}, L_z = 50A.

Lead salt IV-VI compounds such as $Pb_{1-x}Sn_xTe$ or $Pb_{1-x}Ge_xTe$ show a ferroelectric phase transition involving the crystal transformation. Collaborative work with Bauer has been started to investigate the high field effect on the phase transition. In Fig.7 is shown a preliminary result of the cyclotron resonance in n-PbTe at room temperature. The measurement was performed on an epitaxial film grown on BaF_2 substrates with the normal along <111> axis. Two resonance peaks were observed corresponding to the two kinds of valleys. A large non-parabolicity effect was observed because of the large photon energy more than one half of the band gap. Measurements at lower temperatures are in progress.

4. MAGNETIC FIELD-INDUCED-SDW TRANSITION IN ORGANIC SUPERCONCUCTOR

Another interesting area of the high magnetic field semiconductor physics is the field-induced electronic phase transition. The modification of the electron wave function by very high magnetic fields may give rise to various electronic phase transitions through the change of the interaction between electrons. Examples are the Wigner crystallization /18/, the excitonic phase /19,20/, and the charge density wave (CDW) state /4,21/ in high magnetic fields.

Recently, a series of quasi-one dimensional organic substances, such as $(TMTSF)_2X$, or BEDT-TTF salts have attracted much attention, because of their superconductivity. Among others, $(TMTSF)_2X$ salts are known to exhibit not only superconductivity but also the transition into a spin density wave (SDW) state at low temperatures, depending on pressure. In $(TMTSF)_2ClO_4$, a transition to the superconducting state takes place at T = 1.2K at ambient pressure if the sample is cooled slowly through the anion ordering temperature 24K to suppress the SDW transition (R-state). In the presence of a magnetic field along the c^*-axis, it undergoes a second orderphase transition to the field-induced SDW state at a certain threshold field B_{th}.

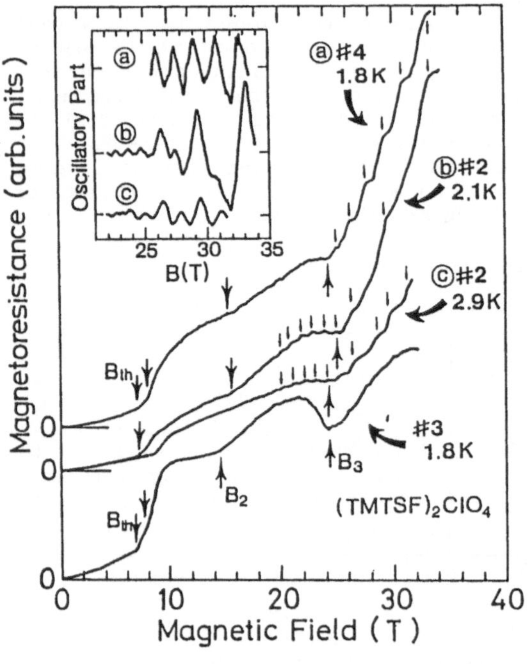

Fig.9. Transverse magnetoresistance in $(TMTSF)_2ClO_4$. The inset shows the small oscillatory part extracted by substracting the background.

This field-induced SDW state consists of many sub-phases, and above B_{th}, the first order transitions take place between the sub-phases successively, giving rise to a series of successive sharp structures in the magnetoresistance /22/. The origin of the field-induced SDW is the nesting of the Fermi surface which has an open orbit in the b direction. When a magnetic field is applied perpendicular to the open orbit (c^* direction), electrons near the Fermi level are characterized by a periodic sinusoidal orbit along the open orbit with a wave number $G = eBb/\hbar c$. The wave number G introduces a new translational symmetry to the system and plays a role of the

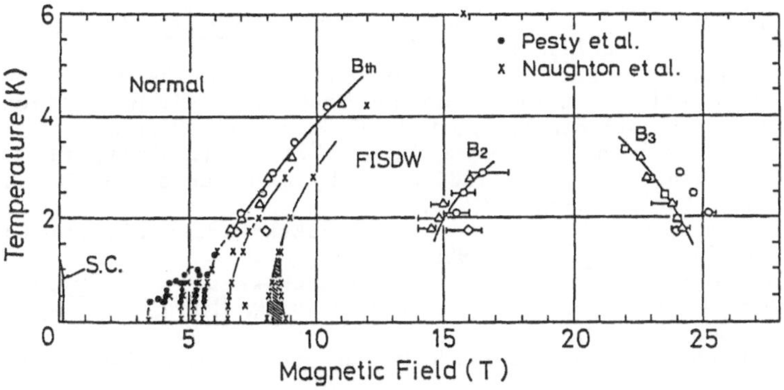

Fig.10. Phase diagram of $(TMTSF)_2ClO_4$. Open symbols represent the newly observed phase transition in the present work. Other points stand for previous data of specific heat and magnetization.

reciprocal vector in the k_x direction. The corresponding eigen energy is a function of k_x only and the dispersion is one-dimensional. Therefore, the system is unstable against an infinitesimal periodic potential with the wave vector connecting two points on the Fermi surface.

Osada et al. investigated the magneto-resistance of $(TMTSF)_2ClO_4$ in submegagauss fields extending the previous measurements to higher fields /23/. Examples of the observed magneto-resistance curves are shown in Fig.9. In addition to the structure at B_{th}, large structures are observed at B_2 and B_3 in higher fields. Since these fields B_2, B_3 depend on temperature, they are considered to correspond to phase transitions. Plotting the positions of B_2 and B_3 on the field-temperature plane, a new phase diagram is constructed as shown in Fig.10. The origin of the new phase transitions B_2 and B_3 is still an open question to be solved.

In addition to the large structures, Shubnikov-de Haas type small oscillations were observed as shown enlarged in the inset of Fig.9. The amplitude of the small oscillations showed a drastic change at B_3.

The small oscillations should not be the usual Shubnikov-de Haas effect because the Fermi level is always located in the gap between Landau levels of the pocket in the field-induced SDW state. As for the small oscillations, there has been much controversy. It is probably due to the oscillatory behavior of the matrix element involved in the carrier scattering, which arises from the oscillatory character of the wavefunction /24/. It would be of interest to extend the measurement to the megagauss range, since new kinds of transitions may occur, although the experiments would be difficult.

ACKNOWLEDGMENT

The author is indebted to his coworkers and collaborators who contributed to the works discussed in this paper, T. Goto, K. Nakao, S. Takeyama, T. Sakakibara, F. Herlach, T. Ando, Y. Iwasa, T. Osada, G. Saito, Y, Nagamune, G. Kido, H. Sakaki, H. Okamoto, S. Tarucha, T. Komatsu, Y. Kaifu, M. Kobayashi, A. Matsui, K. Mizuno and G. Bauer.

REFERENCES

1. N. Miura, T. Goto, K. Nakao, S. Takeyama, T. Sakakibara and F. Herlach: J. Mag. Mag. Mater. 54-57 1409 (1985).

2. N. Miura and F. Herlach : In Strong and Ultrastrong Magnetic Fields ed. by F. Herlach (Springer-Verlag, 1985) P.247.
3. T. Goto, N. Miura, K. Nakao, S. Takeyama and T. Sakakibara : To be published in Proc. 4th Int. Conf. Megagauss Magnetic Field Generation and Related Topics. (Santa Fe, 1986) ed. by C. M. Fowler (Plenum).
4. N. Miura, T. Osada and T. Goto : Proc. 17th Int. Conf. Phys. Semiconductors (San Francisco, 1984), ed. by J. D. Chadi and W. A. Harrison (Springer-Verlag, 1985). P.973.
5. K. Nakao, F. Herlach, T. Goto, S. Takeyama, T. Sakakibara and N. Miura : J. Phys. E. Sci. Instrum. 18 1018 (1985).
6. N. Miura, G. Kido, H. Miyajima, K. Nakao and S. Chikazumi : In Physics in High Magnetic Fields, ed. by S. Chikazumi and N, Miura, (Springer-Verlag, 1981) P.64.
7. S. Tarucha, H. Okamoto, Y. Iwasa and N. Miura : Solid State Commun. 52 815 (1984).
8. T. Ando : J. Phys. Soc. Jpn. 54 1528 (1985).
9. Y. Iwasa, N. Miura, S. Tarucha, H. Okamoto and T. Ando : Surface Science 170 587 (1986).
10. S. Tarucha, H. Iwamura, T. Saku, H. Okamoto, Y. Iwasa and N. Miura : To be published in Surface Science.
11. N. Miura, S. Takeyama and Y. Iwasa : To be published in Proc. 18th Int. Conf. Phys. Semiconductors (Stockholm, 1986).
12. S. Takeyama, M. Kobayashi, A. Matsui, K. Mizuno and N. Miura : In this Proceedings.
13. E. F. Gross, V. I. Perel and R. I. Shekhmametev : JETP Lett. 13, 229 and 357 (1971).
14. Y. Kaifu, T. Komatsu and T. Aikami : Nuovo Cimento 38 449 (1977).
15. T. Komatsu, Y. Kaifu, S. Takeyama and N. Miura : Submitted to Phys. Rev. Lett.
16. N. Miura : In Infrared and Millimeter Waves vol. 12. ed. by K. J. Button (Academic Press, 1984) P.73.
17. Y. Iwasa, N. Miura, S. Takeyama and T. Ando : In this Proceedings.
18. T. F. Rosenbaum, S. B. Field, D. A. Nelson and P. B. Littlewood : Phys. Rev. Lett. 54 241 (1985).
19. N. B. Brandt and S. M. Chudinov : J. Low. Temp. Phys. 8 339 (1972).
20. N. Miura, K. Hiruma, G. Kido and S. Chikazumi : Phys. Rev. Lett. 49 1339 (1982).
21. Y. Iye, L. E. McNeil, G. Dresselhaus, G. S. Boebinger and P. M. Berglund : In Proc 17th Int. Conf. Phys. Semiconductors, ed. by J. D. Chadi and W. A. Harrison (Springer-Verlag, 1985) P.981.
22. P. M. Chaikin, M. Y. Choi, J. F. Kwak, J. S. Brooks, K. P. Morton, M. J. Naughton, E. M. Engler and R. L. Greene : Phys. Rev. Lett. 51 2333 (1983).
23. T. Osada, N. Miura and G. Saito : To be published in Solid State Commun.
24. K. Yamaji : To be published in J. Phys. Soc. Jpn.

Davydov Splitting of Excitons in Anthracene Single Crystals in Megagauss Fields

S. Takeyama[1], M. Kobayashi[2], A. Matsui[3], K. Mizuno[3], and N. Miura[1]

[1]Institute for Solid State Physics, University of Tokyo, 7-22-1 Roppongi, Minato-ku, Tokyo 106, Japan

[2]Department of Material Physics, Faculty of Engineering Sciences, Osaka University, Toyonaka, Osaka 560, Japan

[3]Department of Physics, Konan University, Okamoto, Kobe 658, Japan

1. Introduction

Frenkel-type excitons, which are generally observed in molecular solids, have small spacial extension and are highly localized within each molecule. Consequently magnetooptical studies of Frenkel excitons are extremely difficult, since very large magnetic fields are required to achieve a significant perturbation of the exciton wavefunction. Anthracene ($C_{14}H_{10}$) is one of best studied molecular crystals. The crystal has a monoclinic structure with C_{2h}^5 space group symmetry and contains two molecules in each unit cell; each molecule has an anisotropic optical transition dipole moment. Recent developments in the sample preparation has enabled sufficiently pure and thin single crystals to be grown. This has enabled a study of various interesting phenomena, such as free, self trapped and quasi-free exciton dynamics and their co-existence, particularly by photoluminescence studies in a high pressure cell.

We report the first magnetoabsorption measurements of the lowest singlet excitons in high purity anthracene single crystals. Davydov splitting /1/ originates from a dipole-dipole intermolecular Coulomb interaction, and is observed on the 0-0 exciton peak around 3950 Å (25300 cm^{-1}). The splitting amounts to about 30 Å (200 cm^{-1}) (at room temperature) if a-axis (E//a) or b-axis (E//b) polarized light is incident perpendicular to the ab-plane (ie. the (001) plane). The application of an extremely high magnetic field (ie. H>100 T) is expected to cause a shrink of the exciton radius or influence the intermolecular interactions, contrary to the effect of the pressure.

2. Experiment

Magnetic fields up to 150 T, with a pulse duration of about 5 μs, are generated by a single turn coil system from the discharge of a 100 kJ, 40 kV fast condenser bank. This method destroys the field generating coils, but the sample and cryostat inside the coil are left intact, since the coil explodes outwards. Optical transmission of the sample was observed in both (E//a) and (E//b) polarizations by a streak spectrometer system with an image converter camera. The experimental arrangement and the detailed measuring techniques are described in ref./2/. Sufficient transmitted light for a relatively long optical path is obtained for monocrystalline flakes of anthracene of 1000 Å thickness by 2 mm diameter. Samples were set in a small liquid He flow cryostat made of thin phenolic resin pipes, which are strong enough to survive the shock induced during a pulse field shot (Fig.1). The transmitted streak spectra were taken by a polaroid camera and compared with the magnetic field signal recorded by a transient

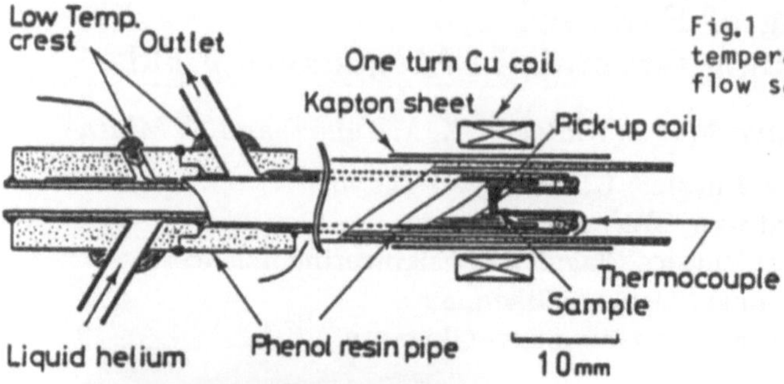

Fig.1 Phenolic resin low temperature liquid helium flow sample holder.

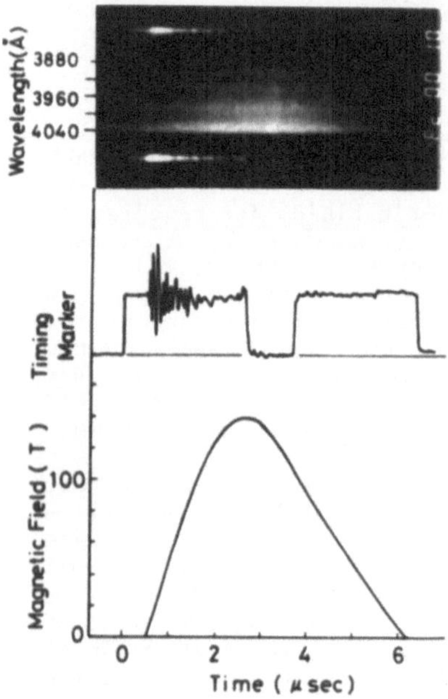

Fig.2 Streak spectra photograph of the magneto-absorption for(E//b) at T=37 K and pulse magnetic field up to 140 T.

recorder using a double pulse photo diode marker recorded to both the polaroid film and the transient recorder (Fig.2).

3. Results

The streak photograph was treated by a microphoto-densitometer and is presented in Fig.3(a) and (b) for (E//b) and (E//a) polarizations, respectively. Figure 3(a) displays wavelength aginst transmission ((E//b) polarization) for sevral different magnetic field strengths at constant temperature(T=37 K). Three absorption peaks (marked by arrows) can be clearly observed on this diagram. The longest wavelength absorption at 4007 Å is believed to be due to a quasi-free exciton, while the peaks at 3961 Å and 3929 Å are known as the $0\text{-}0_b$ and $0\text{-}1_b$ exciton bands, respectively, where the notation 0-0,0-1 describes the vibronic mode of the exciton, and the subscript indicates the polarization direction. The position of the $0\text{-}0_b$ band at 3961 Å corresponds well with the value 3966 Å obtained by Brodin et al./3/. As the magnetic fields increases, the $0\text{-}0_b$

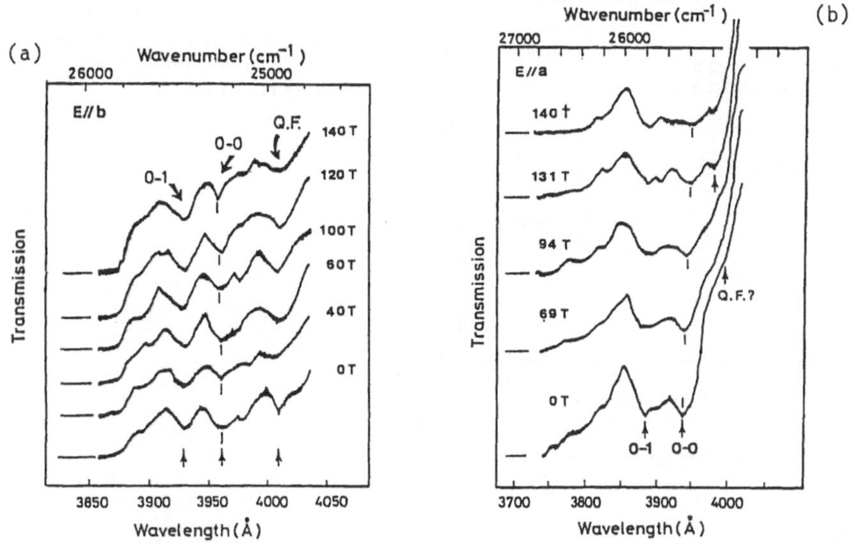

Fig.3 Microphoto-densitometer traces of the streak spectra at different magnetic fields (a) for (E//b), T=37 K and (b) for (E//a), T=57 K.

peak moves to shorter wavelength. At 139 T the $0\text{-}0_b$ peak has moved 3 Å to shorter wavelength with respect to zero field and the relative strength of absorption to the dark level becomes weaker.

The transmission spectra for (E//a) polarization at T= 57 K are illustrated in Fig.3(b). The $0\text{-}0_a$ absorption band at 3932 Å agrees well with the value obtained by Brodin et al. The $0\text{-}1_a$ exciton absorption peak occurs at 3877 Å. In contrast to the $0\text{-}0_b$ exciton peak, the $0\text{-}0_a$ exciton peak moves to longer wavelengths, and grows in intensity with increasing field. At 140 T, the shift of the $0\text{-}0_a$ exciton peak amounts to 12 Å with respect to zero field. It is noticed that a shoulder (marked Q.F.) appears at 3989 Å in the absence of magnetic field. This shoulder moves to shorter wavelengths, and evolves into a new distinctive peak at 3979 Å as the field increases up to 130 T. At higher fields (i.e. 140 T), the peak overlaps with the $0\text{-}0_a$ band . At a lower temperature (i.e. T=17 K) for (E//a) polarization, no shoulder can be observed even at the highest available magnetic field. It is tentatively suggested that this absorption peak is due to a quasi-free exciton.

4. Discussion

The shift of the 0-0 exciton band position against the applied magnetic field is plotted for each polarization (E//a) and (E//b),respectively, in Fig.4. In the case of (E//b) polarization , two different samples show the same rate of shift towards shorter wavelengths with increasing field. However, the (E//a) polarization spectra measured at 57 K shows a faster increase of the peak position, as a function of wavelength, with increasing magnetic field as compared to the measurements of 17 K. This enhancement of the shift at T=57 K may be caused by the quasi-free exciton peak at 3989 Å (B=0) in close vicinity of the $0\text{-}0_a$ peak, influencing the peak position. Hence the value of the shift at T=13 K is more reliable. In

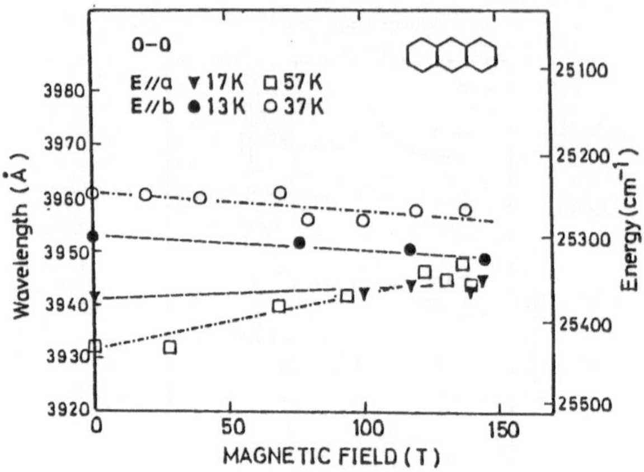

Fig.4 Shift of the 0-0 exciton bands position as a function of magnetic field.

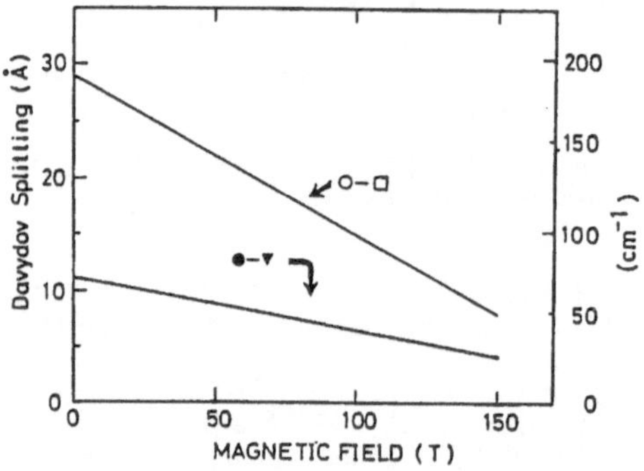

Fig.5 Change of the 0-0 exciton band Davydov splitting as a function of magnetic field.

Fig.5 the change of the Davydov splitting Δ with field is illustrated, where the value Δ is determined from the difference of the peak position between $0\text{-}0_b$ and $0\text{-}0_a$, and dependence of Δ on magnetic field is approximately equal to $\Delta(H)=0.05$ 0.14 Å/Tesla.

Initially, one can explain the decrease of the Davydov splitting with increasing magnetic field by a shrinkage of the exciton radius. This results in an effective increase of the exciton localization within each molecule. Thus inter-molecular interaction would be decreased, reducing the Davydov splitting. However, the splitting is reduced by more than 50 % at 150 T, much more than expected. Å problem arises whether this field strength is sufficient to reduce the spacial extension of the wave-function of such an incompressible Frenkel exciton (the exciton mass for (E//a) is estimated to be almost 10 m_0 /4/). Pressure acts to decrease the dipole-dipole distance, and results in an increase of the Davydov splitting. Å pressure study of the Davydov splitting by Otto et al. showed the splitting increase was much larger than expected for purely dipolar interaction /5/. If the shrinkage of the exciton wavefunction is responsible for our results, it also suggests a breakdown of the usual multipole expansion as mentioned by Otto et al. Other possible reasons are the magnetic field induced orientation of the polarization moment /6/ or the change of the electron distribution in the lowest singlet excited state. Further experimental and theoretical work on the magnetic induced effect on the Davydov splitting is required to confirm the results.

Acknowledgement

We are greatly indebted to Mr.Y.Nagamune and Prof.T.Goto of ISSP for their kind assistance during the measurements.

References

1. A.S.Davydov: Theory of Molecular Excitons (McGraw-Hill Publ. Co.,New York 1962)
2. G. Kido, N. Miura, H. Katayama, S.Chikazumi: J. Phys.: Sci. Instrum., 14, 349 (1981)
3. M.S.Brodin, S.V. Marisova: Optics and Spectroscopy, 10, 242 (1961)
4. D. M. Burland: Phys. Rev. Lett. 30, 833 (1974)
5. A.Otto, R. Keller, A. Rahman: Chem. Phys. Lett. 49, 145 (1977)
6. A. Yamagish, E. Nagao, M.Date: J. Phys. Soc. Jpn. 53, 928 (1984)

Index of Contributors